CONCRETE CONSTRUCTION HANDBOOK

OTHER McGRAW-HILL HANDBOOKS OF INTEREST

CONCRETE CONSTRUCTION HANDBOOK

JOSEPH J. WADDELL, *Editor-in-Chief*

Consulting Engineer
Construction Materials and Methods
Riverside, California

Second Edition

McGraw-Hill Book Company

New York   *St. Louis*   *San Francisco*   *Düsseldorf*   *Johannesburg*
Kuala Lumpur   *London*   *Mexico*   *Montreal*   *New Delhi*
Panama   *Paris*   *São Paulo*   *Singapore*   *Sydney*
Tokyo   *Toronto*

Library of Congress Cataloging in Publication Data

Waddell, Joseph J
  Concrete construction handbook.

  Includes bibliographies.
  1. Concrete construction. I. Title.
TA681.W23   1974   624'.1834   73-17385
ISBN 0-07-067654-2

CONCRETE CONSTRUCTION HANDBOOK

1234567890 KPKP 7654

*The editors for this book were Harold B. Crawford and
Ross J. Kepler and its production was supervised
by Stephen J. Boldish.   It was set in Lino 21
by The Maple Press Company. .*

Printed and bound by The Kingsport Press.

CONTRIBUTORS

Robert F. Adams, Supervisor, Concrete Laboratory, California Department of Water Resources, Sacramento, Calif. *Placing Concrete*

Joseph F. Artuso, Assistant to the Vice President, Pittsburgh Testing Laboratory, Pittsburgh, Pa. *Proportioning Mixes and Testing*

William A. Cordon, Professor of Civil Engineering, Utah State University, Logan, Utah. *Repair of Concrete*

David R. Harney, General Manager, Concrete Form Consulting Service, Torrance, Calif. *Formwork and Shoring*

John K. Hunt, President, Hunt Control Systems, Inc., Urbana Illinois (formerly Chief Engineer, Johnson Operations, Road Division, The Koehring Company, Champaign, Illinois). *Batching, Mixing, and Transporting*

John H. Ittel, Research Manager, Lupton-Olin, Cornwell Heights, Pa. *Steel for Reinforcement and Prestressing*

Robert Leroy Jones, President, Labrado Forms, Irvine, Calif. *Esthetics in Concrete*

F. E. Legg, Jr., Associate Professor of Construction Materials, University of Michigan, Ann Arbor, Mich. *Aggregates*

Charles J. Pankow, President, Charles Pankow, Inc., Altadena, Calif. *On-site Precasting and Tilt-up; Slipform Construction of Buildings*

Charles F. Peck, Jr., Professor, Newark College of Engineering, Newark, N.J. *Steel for Reinforcing and Prestressing* (with John H. Ittel)

Perry H. Petersen, Director of Engineering, Master Builders, Division of Martin-Marietta Corporation, Cleveland, Ohio. *Special Concrete and Techniques*

Robert E. Philleo, Chief, Concrete Branch, Civil Works, Office of Chief of Engineers, Washington, D.C. *Cracking and Surface Blemishes; Cooling and Grouting of Concrete in Place*

Raymond J. Schutz, Vice President, Research and Development, Sika Chemical Corporation, Lyndhurst, N.J. *Admixtures, Curing Mediums, and Other Materials*

Joseph J. Waddell, Consulting Engineer, Riverside, Calif. *Cement; Lift-slab Construction; Specialized Practices; Prestressed Concrete; Precast Concrete; Finishing and Curing*

George W. Washa, Professor of Engineering Mechanics, University of Wisconsin, Madison, Wisc. *Properties of Concrete*

PREFACE TO THE SECOND EDITION

Since publication of the first edition of the handbook in 1968, the technology of concrete construction has continued its ongoing evolution, in some areas almost dramatically. Construction is basically a stable industry, yet in many ways it is a dynamic industry. It is the need to keep abreast of the expansion and development of construction practices that made this second edition necessary.

Population and industrial growth, urban expansion, environmental limitations, and ecological controls all are indications of the external forces that catalyze, as it were, changes in concrete construction practices. An example is the impending shortage of natural aggregate sources in our urban areas, a shortage brought on in part by the expansion of residential and commercial developments into areas containing potential aggregate deposits. Such a shortage will force the industry to develop alternate sources, possibly synthetic aggregates of some kind. It is changes of this nature that quicken the constant evolution of construction, changes that this and future editions of the handbook will report.

The continued objective of this book is to assist the practicing construction man in his use of portland cement and concrete. Extensive revisions have been made in the sections on building construction systems, cement handling, pumping concrete, mixers, and esthetics. All standards and specifications have been reviewed and brought up-to-date with their latest revisions. As pointed out in the first edition, the user should check the ASTM or other standard to make sure that he is using the latest revision.

A book of this nature represents the work of many people—not only the present authors, but builders, engineers, contractors, researchers, and many others engaged in the concrete construction industry who have, through the years, contributed so much to the knowledge of concrete. To these we are extremely appreciative, and we have tried to give credit whenever possible.

The contributing authors are well-known experts in their respective fields, and I am grateful to them for the many hours of work that went into their contributions. Most of the original authors participated in the revision for the second edition, although four have been lost through resignation, death, or retirement. There are no new authors.

JOSEPH J. WADDELL, P.E.
Editor-in-Chief

CONTENTS

Section 7. Finishing and Curing

Section 8. Special Concrete and Techniques

Section 9. Advanced Building-construction Systems

Section 10. Specialized Practices

Section 11. Precast and Prestressed Concrete

Section 12. Cracking and Surface Blemishes

Section 13. Cooling and Grouting of Concrete in Place

Section 14. Repair of Concrete

Index follows Chapter 48.

CONCRETE CONSTRUCTION HANDBOOK

Section 1

MATERIALS FOR CONCRETE

Chapter 1

CEMENT

JOSEPH J. WADDELL

Portland cement is a hydraulic cement; that is, when mixed in the proper proportions with water, it will harden under water (as well as in the air). It derives its name from the similarity of its appearance, in the hardened state, to the Portland stone of England that was commonly being quarried and used in building construction during the early part of the nineteenth century, at the time the first cement of this type was made.

PRODUCTION

1 Manufacture

Raw Materials. The basic ingredients for portland cement consist of (1) lime-rich materials, such as limestone, seashells, marl, and chalk, that provide the calcareous components; (2) clay, shale, slate, or sand to provide the silica and alumina; and (3) iron ore, mill scale, or similar material to provide the iron, or ferriferous, components. The number of raw materials required at any one plant depends upon the composition of these materials and the types of cement being produced. A typical blend of raw materials might be proportioned so that the clinker, after burning, would consist of approximately 62% CaO, 22% SiO_2, 5% Al_2O_3, and 3% Fe_2O_3, with lesser proportions of other minor constituents. To accomplish the proper blend, the raw materials are continually sampled and analyzed, and adjustments are made to the proportions while they are being blended.

Processing. After excavation, the first processing operation is crushing, as shown in Fig. 1. Quarry stone is passed through the primary crusher, usually a gyratory crusher that accepts rock fragments as large as 5 ft in diameter, crushing them to about 6 in. in diameter. The crusher product is then passed to the secondary crusher, a hammer mill that reduces it to about ⅜ in. in size. At this point, blending of the several ingredients is accomplished, and the blend is conveyed to raw storage piles. Samples, mentioned above, are obtained at this point and immediately analyzed. In a modern plant this sampling and testing are the source of data fed into a digital computer controlling composition of stored and blended raw feed. A stacker may be used for blending the material in long storage piles from which it is removed by a reclaiming machine. In some plants, the material is blended and stored in piles by means of belt conveyors or cranes, and is reclaimed by the same type of equipment.

At this point in the process, there is a temporary divergence in the methods, depending on whether a dry or a wet process is being used. United States production is divided about equally between the two processes. Figure 2 is an aerial view of a dry-process plant.

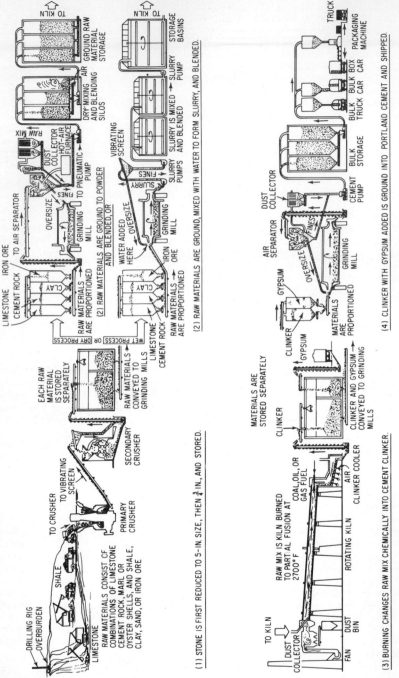

DRILLING RIG
OVERBURDEN
SHALE
LIMESTONE
RAW MATERIALS CONSIST OF
COMBINATIONS OF LIMESTONE
CEMENT ROCK, MARL OR
OYSTER SHELLS, AND SHALE,
CLAY, SAND, OR IRON ORE

TO CRUSHER
PRIMARY CRUSHER
TO VIBRATING SCREEN
SECONDARY CRUSHER

EACH RAW
MATERIAL
IS STORED
SEPARATELY

RAW MATERIALS
CONVEYED TO
GRINDING MILLS

(1) STONE IS FIRST REDUCED TO 5-IN. SIZE, THEN $\frac{3}{4}$ IN., AND STORED.

WET PROCESS OR DRY PROCESS

LIMESTONE   IRON ORE
CEMENT ROCK
CLAY
RAW MATERIALS
ARE PROPORTIONED

OVERSIZE
GRINDING MILL
FINES
TO AIR SEPARATOR
TO PNEUMATIC PUMP
DUST COLLECTOR
RAW MIX
HOT-AIR FURNACE

DRY-MIXING
AND BLENDING
SILOS

AIR

GROUND RAW
MATERIAL
STORAGE

TO KILN

(2) RAW MATERIALS ARE GROUND TO POWDER AND BLENDED, OR

LIMESTONE
CEMENT ROCK
CLAY
IRON ORE
RAW MATERIALS
ARE PROPORTIONED

WATER ADDED HERE
OVERSIZE
IRON ORE
GRINDING MILL
SLURRY
FINES
VIBRATING SCREEN
SLURRY PUMPS

SLURRY IS MIXED
AND BLENDED

SLURRY PUMP
TO KILN
STORAGE BASINS

(2) RAW MATERIALS ARE GROUND, MIXED WITH WATER TO FORM SLURRY, AND BLENDED.

TO KILN
DUST COLLECTOR
DUST BIN
FAN

RAW MIX IS KILN BURNED
TO PARTIAL FUSION AT
2700° F

COAL, OIL, OR
GAS FUEL
ROTATING KILN
AIR
CLINKER COOLER

MATERIALS ARE
STORED SEPARATELY

CLINKER

CLINKER AND GYPSUM
CONVEYED TO GRINDING
MILLS

(3) BURNING CHANGES RAW MIX CHEMICALLY INTO CEMENT CLINKER.

GYPSUM
CLINKER
GYPSUM
MATERIALS
ARE PROPORTIONED

AIR
SEPARATOR
OVERSIZE
GRINDING MILL
FINES
DUST COLLECTOR
CEMENT PUMP

BULK STORAGE

BULK
TRUCK
BULK
CAR
BOX
CAR
BOX
TRUCK
PACKAGING
MACHINE
TRUCK

(4) CLINKER WITH GYPSUM ADDED IS GROUND INTO PORTLAND CEMENT AND SHIPPED.

FIG. 1.   Steps in the manufacture of portland cement.   (*Courtesy of Portland Cement Association.*)

Dry Process. In the dry process, the material is now removed from the blending piles and is stored in bins. Drying, if necessary, is accomplished in rotary driers, similar in design and operation to horizontal kilns, except that the material is heated only enough to drive off the free moisture. From the storage bins, the material is delivered to the raw grinding department where it is reduced in size to about 80 to 90% passing the 200-mesh screen. Grinding is accomplished in mills with a capacity of over 100 tons per hr per mill, usually operating in closed circuit with an air separator.

Dry-process mills may be tube mills, ball mills, or compartment mills. Sometimes the feed goes first through a ball mill, then through a tube mill for final grinding. A ball mill is a horizontal cylinder rotating at a speed of 15 to 20 rpm containing a charge of steel balls 4 to 6 in. in diameter. As the cylinder revolves, the steel balls continuously cascade with the feed material, pulverizing the latter. A tube mill is similar to a ball mill except that it is more slender, contains smaller

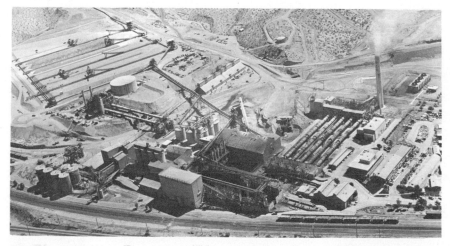

Fig. 2. Aerial view of a dry-process cement plant. Storage piles for raw kiln feed are in background. (*Courtesy of Riverside Division, American Cement Corp.*)

balls (between 1 and 2 in. in diameter), and accomplishes finer grinding. A compartment mill consists of two or three compartments separated by perforated-steel-plate bulkheads. Different-sized balls are in each compartment, graduated from large ones in the feed end to small ones in the last compartment, and the mill accomplishes the same grinding as the ball- and tube-mill combination. In any of these mills, a high-velocity air stream passes through the interior (hence the name "air-swept"), picking up the ground material and carrying it to the air separator that separates the fines from the coarse. Fines are conveyed to storage for kiln feed, and coarse oversize is returned to the mills for further grinding.

Wet Process. In the wet process, the raw feed is transferred from raw storage piles to the grinding mills, which are substantially the same as the ball, tube, or compartment mills used for dry grinding. Water is introduced into the mill along with the feed. Mills are usually operated in closed circuit with some type of classifying equipment, such as hydroseparators, screws, or rake classifiers, that separate the fines from the coarse. The fine slurry is then pumped to the thickeners, which remove much of the water, then into blending tanks or storage tanks, while the coarse material is returned to the mill for additional grinding. Storage tanks or basins are kept agitated to prevent settlement or segregation of the slurry before it is fed to the kilns.

Burning. Dry kiln feed or slurry is drawn from storage and fed into a rotary kiln that gradually heats the material to about 2700°F. At this temperature the material starts to melt or fuse into small lumps of clinker about the size of walnuts, and the lime, alumina, silica, and iron oxide are changed into new compounds. The material is in the kiln for 3 or 4 hr, during which time it is constantly cascaded or tumbled as it slowly moves through the length of the kiln[1].

The kiln, which may be 16 ft in diameter and 600 ft long, is inclined from ½ to ¾ in. per ft to facilitate the movement of the clinkering material through it. Burners at the discharge end may use natural gas, oil, or coal for fuel. The kiln is lined with a refractory lining to protect the steel shell and provide insulation against the loss of heat. Speed of rotation is about 1 rpm. Wet-process kilns require greater length than dry-process ones of similar capacity, and usually have more elaborate heat exchangers to increase the amount of slurry exposed to the combustion gases. Newer installations provide centralized process control by means

FIG. 3. Central process control for a modern cement plant. (*Courtesy of Riverside Division, American Cement Corp.*)

of a digital computer that keeps a continuous record of all kiln instrumentation from which operational changes can be computed so as to provide maximum efficiency at all times. Closed-circuit television enables the operator to keep the whole firing operation constantly under view. Figure 3 shows a modern control room.

Most mills are equipped with preheaters to heat the kiln feed with kiln stack gases prior to entering the kiln, thus improving fuel efficiency.

Upon leaving the kiln, the clinker is passed through a clinker cooler, then to storage or to finish grinding. Clinker may be stored either outdoors or indoors, depending on climatic conditions.

Finish Grinding. After cooling, the clinker goes to the final grinding mills, which are similar to the raw grinding mills previously described, the coarse material being recirculated for additional grinding. During finish grinding, from 3 to 5% of gypsum is interground with the cement to control the setting time and enhance strength. Sometimes other additions are interground at this time to facilitate grinding or to provide certain desirable properties in the cement.

From the finish mill the cement is conveyed to storage silos or stockhouse, where it is ready for shipment. Conveying of finish cement is usually accomplished by suspension in air in pipes. Shipment may be in bulk or in 94-lb paper sacks. Special hauling units for both rail and truck transport of bulk cement are available.

COMPOSITION OF CEMENT

2 Chemical Composition

Chemical Analysis. Portland cement cannot be shown by a chemical formula, as it is a complex mixture of several compounds, as shown in Table 1. However,

Table 1. Analysis of Type II Cement

Chemical analysis, %		Physical properties	
SiO_2	21.6	Fineness:	
Al_2O_3	4.6	Blaine, sq cm/g	3,410
Fe_2O_3	2.8	Wagner, sq cm/g	1,920
CaO	62.8	Minus 325-mesh, %	95.4
MgO	3.2	Compressive strength, psi:	
SO_3	2.1	1-day cube	1,300
Ignition loss	1.5	3-day cube	2,550
Insoluble residue	0.1	7-day cube	3,700
Free CaO	1.2	28-day cube	5,650
Na_2O	0.41	Setting time, hr and min:	
K_2O	0.24	Initial	3:25
Total alkalies	0.57	Final	5:55
		Air content, %	10.4
Compound composition, %		Autoclave expansion, %	0.25
		Specific gravity	3.15
C_3S	50		
C_2S	24		
C_3A	7		
C_4AF	8		

there are four compounds, computed from the oxide analysis of the cement, that for all practical purposes may be considered as comprising the cement. Minor constituents play their part also, as described below. The main compounds, with their commonly accepted abbreviations, are

$$\text{Tricalcium silicate, } 3CaO \cdot SiO_2 = C_3S$$
$$\text{Dicalcium silicate, } 2CaO \cdot SiO_2 = C_2S$$
$$\text{Tricalcium aluminate, } 3CaO \cdot Al_2O_3 = C_3A$$
$$\text{Tetracalcium aluminoferrite, } 4CaO \cdot Al_2O_3 \cdot Fe_2O_3 = C_4AF$$

These compounds, or "phases" as they are called, are not true compounds in the chemical sense, but the computed proportions of these compounds reveal valuable information concerning the cement. Strength-developing characteristics of the cement depend on the C_3S and C_2S, which comprise about 75% of the cement. The C_3S hardens rapidly and therefore has a major influence on setting time and early strength; hence a high proportion of C_3S results in high early strength and high heat of hydration. C_2S on the other hand hydrates more slowly and contributes to strength gain after about a week. C_3A contributes to high early strength and high heat but results in undesirable properties of the concrete, such as poor sulfate resistance and volume change. The use of iron oxide in the kiln feed contributes to lower C_3A but leads to the formation of C_4AF, a product that is little more than a filler that should be kept to a minimum, although it lowers clinkering temperature.

Compound composition is determined from the chemical analysis as follows:

$$\text{Tricalcium silicate} = (4.071 \cdot \% \text{ CaO}) - (7.600 \cdot \% \text{ SiO}_2)$$
$$- (6.718 \cdot \% \text{ Al}_2\text{O}_3) - (1.430 \cdot \% \text{ Fe}_2\text{O}_3)$$
$$- (2.852 \cdot \% \text{ SO}_3)$$
$$\text{Dicalcium silicate} = (2.867 \cdot \% \text{ SiO}_2) - (0.7544 \cdot \% \text{ C}_3\text{S})$$
$$\text{Tricalcium aluminate} = (2.650 \cdot \% \text{ Al}_2\text{O}_3) - (1.692 \cdot \% \text{ Fe}_2\text{O}_3)$$
$$\text{Tetracalcium aluminoferrite} = 3.043 \cdot \% \text{ Fe}_2\text{O}_3$$

Some producers use a "net" or "modified" C_2S, which is computed by subtracting 4.07 times the percentage of free CaO from the "gross" C_3S as computed in the above equation.

A small amount of gypsum, up to 5% $CaSO_4 \cdot 2H_2O$, is added to the clinker while it is being ground, to control the setting time of the cement by its action in retarding the hydration of tricalcium aluminate. Excessive amounts of free lime, CaO, result from underburning of the clinker in the kiln and may cause ultimate expansion and disintegration of the concrete. Free CaO should not exceed 2%. Magnesium oxide, because it results in unsoundness of concrete, especially when moist, is limited by specifications to not more than 5%.

Prehydration and carbonation of the cement result from moisture absorption and exposure to the air, and are revealed by high ignition loss, which should not exceed 3% (2.5% for Type IV). Water and carbon dioxide are the principal materials driven off when the cement sample is heated to about 1800°F.

The alkalies, Na_2O and K_2O, are important minor constituents, as they can cause very rapid expansive deterioration of concrete when certain types of siliceous aggregates are used. For this reason, "low-alkali" cement is specified in areas where these aggregates are present. Low-alkali cement contains not more than 0.60% total alkalies, computed as the percentage of Na_2O, plus 0.658 times the percentage of K_2O.

False set, also called premature stiffening, gum set, or rubber set, is manifested by a stiffening or loss of consistency of concrete shortly after mixing. It disappears after prolonged mixing or remixing. It results from dehydration of the gypsum in the cement, probably caused by too high a temperature in the finish grinding mill which changes the gypsum, $CaSO_4 \cdot 2H_2O$, into the anhydride, $CaSO_4$, or the hemihydrate. False set itself is harmless, but the early stiffening of the mix may cause trouble on the job because of a temporary decrease in workability of the concrete.

Heat of Hydration. The reaction of cement and water is exothermic; that is, heat is generated in the reaction, the average amount being about 120 cal per g during complete hydration of the cement. In normal construction, structural members are of such size that dissipation of the heat is no problem. As a matter of fact, by insulating the forms, this heat is used to advantage when concreting during below-freezing weather (see Winter Concreting in Chap. 30). However, dams and other massive structures require that means be taken to remove the heat by proper design and construction of the structure, by circulating cold water in embedded pipe coils, etc. Another method of controlling the heat evolution is to modify the compound composition of the cement by reducing the percentages of the high-heat compounds, C_3A and C_3S, and by slightly coarser grinding, the resulting cement being a Type IV cement.

Czernin[2] gives the following values for the total quantity of heat evolved for complete hydration of the cement:

cal/g

Tricalcium silicate	120
Dicalcium silicate	62
Tricalcium aluminate	207
Tetracalcium aluminoferrite	100

He also gives maximum and minimum values of the heat of hydration at various times, the exact amounts depending on the composition of the cement:

Curing time, days	Heat of hydration, cal/g	
	Min	Max
3	40	90
7	45	100
28	60	110
90	70	115

Figure 4 shows the temperature-rise characteristics of several types of cement. If we assume that the heat generated by Type I cement during the first 7 days of hydration is 100%, then

Type II will generate about 85%
Type III will generate as much as 150%
Type IV will generate about 50%
Type V will generate 60 to 75%

of the heat generated by Type I cement.

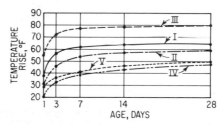

FIG. 4. Temperature rise of concrete. Tests of mass concrete with 4½-in.-maximum-size aggregate, containing 376 lb of cement per cu yd in 17 × 17-in. cylinders, sealed and cured in adiabatic calorimeter rooms. (*U.S. Bureau of Reclamation,* "*Concrete Manual.*")

3 Physical Properties of Cement

Fineness. Cement particles, because of their small size, cannot be separated on sieves, and other methods of measuring particle size are necessary. The specific-surface method is commonly used, in which the particles of cement are considered to be spheres. The specific surface is the summation of the surface area, in square centimeters, of the particles in 1 g of cement.

The Wagner turbidimeter is based on the principle that turbidity is a measure of the surface area of a sample of cement. Turbidity is determined at intervals by measuring with a photoelectric cell, the light passing through a suspension of cement in kerosene. Most manufacturers and agencies now use the Blaine air-permeability method for determining fineness. This method depends upon the observation that the rate of air flow through a prepared bed of cement in the apparatus is determined by the size and number of pores, which is a function of the particle size.

Fineness affects the strength-gaining properties, especially at ages up to about 7 days. For this reason, Type III cement is ground finer than the other types.

ASTM specifications place a minimum on fineness, but most cements exceed this minimum by a considerable margin.

Setting Time. As soon as water joins the cement, a reaction commences, resulting in, among other things, the formation of a gel that causes the paste to stiffen. This stiffening, however, does not affect workability until initial set takes place. The initial set is an arbitrary period of time, measured from the time water and cement are combined until a specified pat of neat cement paste will just support, without appreciable indentation, a steel needle $\frac{1}{12}$ in. in diameter weighing $\frac{1}{4}$ lb. Final set is similarly determined with a $\frac{1}{24}$-in. needle weighing 1 lb. Apparatus for this determination is known as the Gillmore apparatus. Another apparatus, called the Vicat needle, is based on the length of time required for a weighted needle to penetrate 25 mm into a neat cement paste in a small ring mold.

The rate of setting is not necessarily synonymous with the rate of hardening. A rapid-setting cement can be produced by varying the composition of the cement, if, for example, only a small amount of gypsum is introduced during the final grinding; or it can be caused by using certain admixtures such as sodium carbonate. On the other hand, a Type III cement may have nearly the same setting characteristics as a Type I cement from the same mill, but its early-strength development is more rapid, once the cement has set.

Strength. Formerly, the tensile test was the standard test for cement strength, in which small mortar briquettes are tested in tension. Present trend leans toward the use of 2-in. mortar cubes for compressive-strength tests. Mortar for the test consists of 1 part cement, 2.75 parts graded standard sand by weight, with sufficient water to produce a flow beween 100 and 115. Flow is the increase in average base diameter of a sample subjected to 25 drops of a standard-flow table. Comparative strength developments for the five types of cement are shown in Table 2, assuming that the strength of Type I cement is 100% at each age shown.

Table 2. Relative Compressive-strength Values*

Type of cement	Age of specimens			
	1 day	7 days	28 days	3 months
I	100%	100%	100%	100%
II	75	85	90	100
III	190	120	110	100
IV	55	55	75	100
V	65	75	85	100

Portland Cement Assoc. Bull. ST 92.

The rate of hardening of cement depends upon the chemical and physical properties of the cement and the curing conditions, i.e., temperature and moisture conditions during curing. The water-cement ratio affects the actual value of the ultimate strength, mainly by virtue of the effect of water on the porosity of the cement paste, a high water-cement ratio producing a paste of relatively high porosity.

Soundness. The effect of an excess of free lime or magnesia in cement is ultimate expansion and disintegration of concrete made with that cement. Potential expansion and soundness are determined by testing $1 \times 1 \times 11\frac{1}{4}$-in. neat cement specimens in an autoclave at a steam pressure of 295 psi for 3 hr. ASTM specifications place an upper limit of 0.80% on the increase in length of the specimen before and after autoclaving. Specifications also limit the amount of free lime and magnesia.

Sticky Cement. A relatively infrequent problem sometimes encountered by bulk-cement users is a reluctance on the part of the cement to flow. This stickiness, as it is called, has no effect on the concrete-making properties of the cement. Usually it is found that the trouble is moisture, static electricity, improper design of cement-handling facilities, or holding the cement too long without disturbing or moving it.

A stickiness or pack set occurs in bulk rail cars or trucks and results from the vibration of the vehicle during the journey to the unloading site. A similar condition arises when cement is permitted to rest in a silo or bin for a long period of time, resulting from compaction or compression of the cement. No hydration of the cement has occurred, but the condition interferes with transfer of the cement. Usually the use of air jets fluffs the cement enough to cause it to flow. On rare occasions, it has been found that the use of certain grinding aids in the finish mill improves the flow characteristics of the cement. Modern air-handling equipment, proper bin vibrators, and correctly designed silos and hoppers experience the least trouble. Round bins and conical bottoms are better than square bins with flat or pyramidal bottoms.

A "warehouse set" occurs when cement is held in storage too long, either in sacks or, rarely, in bulk. It results from surface hydration of the cement by absorbed moisture, and carbonation. The obvious solution is to keep the cement in storage for as short a period as possible. "First in, first out" should be the procedure here.

TYPES OF CEMENT

There are five types of standard portland cement in common use. Typical compound composition of these cements is shown in Table 3. In addition, there

Table 3. Typical Compound Composition of Cements

Type of cement	C_3S	C_2S	C_3A	C_4AF	MgO	SO_3	Ignition loss	Free CaO
Type I.....	49	26	11	8	3.0	2.2	1.3	1.0
Type II....	46	30	6	12	2.1	2.1	1.5	1.2
Type III...	55	14	10	7	2.1	2.8	1.5	1.6
Type IV...	30	47	5	13	2.1	2.1	1.4	0.8
Type V....	41	36	4	10	2.8	1.9	1.3	0.8

Notes:
1. Type I and Type II cements as manufactured today frequently meet the chemical and physical requirements of both specifications.
2. These are typical analyses to show approximate relative values for comparison only. Actual values from any one mill may vary somewhat from these.

are several special portland cements for which there are ASTM specifications, and numerous other cements designed to meet specific conditions of usage or exposure. Some cements are available only in certain areas. Some local cements of very limited usage are not included in the following list.

4   Standard Portland Cements

Type I, ASTM: C 150. Common or ordinary cement. Usually supplied unless another type is specified. Frequently meets Type II specification also.

Type II, ASTM: C 150. Modified. C_3A maximum limit is 8%, providing moderate resistance to sulfate attack. Moderate heat of hydration is usually specified. Heat of hydration is less than for Type I and is liberated at a slower rate.

Type III, ASTM: C 150. High-early-strength. Obtained by finer grinding and larger proportions of C_3S and C_3A. Concrete made with Type III cement has a 3-day compressive strength about equal to that made with Type I at 7 days, and 7-day strength equal to Type I at 28 days. Ultimate strength is about the same.

Type IV, ASTM: C 150. Low-heat. Proportions of C_2S and C_4AF are relatively high, and percentages of C_3A and C_3S are low. Heat of hydration is less than for the other types and develops more slowly. Strength development is slower. Used in massive dams, piers, footings, etc. Available only on special order. Requires more curing than Type I, usually specified to be 21 days.

Type V, ASTM: C 150. Sulfate-resistant. Very low C_3A content, not over 5%. Used in concrete exposed to soil alkali, sulfate groundwater, or seawater. Heat generation is slightly in excess of Type IV. Normally available from a few mills, otherwise requires special order.

5   Special Portland Cements

Types IA, IIA, and IIIA, ASTM: C 175. Air-entraining. Same as Types I, II, and III, except that an air-entraining agent is interground during manufacturing.

Types IS and ISA, ASTM: C 205. Portland blast-furnace slag cement, plain and air-entraining, respectively. Produced by intergrinding blast-furnace slag in an amount between 25 and 65% of the total cement, with C 150 cement clinker. Fairly low in heat evolution. Hardens more slowly than standard portland cement, but final strength is about the same. Reacts favorably under steam curing.

Suffix MS indicates moderate sulfate resistance.

Suffix MH indicates moderate heat of hydration.

Suffix MH-MS indicates both qualities.

Types IP and IPA, ASTM: C 340. Portland-pozzolan cement, plain and air-entraining, respectively. Produced by intergrinding pozzolan, in an amount between 15 and 40% of the total cement, with C 150 clinker. Some pozzolans provide protection against alkali-aggregate reaction, and attack by sulfates, seawater, or low-pH water. Possesses low shrinkage rate and low heat liberation. See Pozzolans in Chap. 4.

Types N and NA, ASTM: C 10. Natural cement, plain and air-entraining, respectively. Made by calcining natural clayey limestone at a temperature just high enough to drive off the carbonic acid gas. Usually high in MgO and free lime. Gradually disappearing from the market.

Masonry, ASTM: C 91. Type I is for masonry where high strength is not required. Type II is for high-strength masonry. Usually contains a fine-ground limestone as a filler and as a workability agent.

White, ASTM: C 150. White cement meets all the requirements of standard Type I cement. Special materials low in iron and manganese are specially processed to give a pure white cement. White cement can be used exactly the same as C 150 gray cement.

Oil Well API Standard 10A (Several Classes). Designed to meet the special high pressure and temperature conditions encountered in oil-well grouting. Produces a low-viscosity slow-setting slurry that remains as liquid as possible to ease pumping pressure in deep oil wells. Most oil-well cements contain a retarder or a friction-reducing additive. Low in C_3A. Rather coarse grind.

Plastic and Gun Plastic. Plastic cement is produced by intergrinding a plasticizing agent with portland-cement clinker that meets the requirements of ASTM specifications for Type I or Type II cement. The Uniform Building Code permits plasticizing agents to be added to portland cement Type I or II in the manufacturing process provided they are not in excess of 12% of the total volume. Plastic cement made in this manner must meet the requirements of ASTM: C 150 except with respect to the limitations on insoluble residue, air entrainment, and additions subsequent to calcination. Developed primarily for portland-cement plaster and stucco, its superior plasticity and workability also lend themselves very well for use in masonry mortar mixes. Because of the relatively high amount of air entrained by plastic cement, it is not recommended for general concrete construction.

Gun plastic cement was developed for application of plaster by guns or pumps and is designed to overcome segregation during pumping. It is manufactured by blending a small amount of fine asbestos fibers into the plastic cement described above.

Block. Block cement is designed to produce blocks of uniform light color and early strength and to facilitate block production through improved workability and performance. Similar to Type III in fineness and early-strength development.

Pipe. Specially made for use in the manufacture of centrifugally spun pipe. Its coarse grind facilitates movement of the mixing water out of the concrete or mortar during the spinning process, resulting in an improved and denser pipe comparatively free from blisters and other defects.

Waterproof. Portland cement containing water-repellent additions such as stearates, oleates, and tallow. Some degree of water repellency is imparted to concrete and mortar containing this cement, but alkalies liberated during hydration of the cement react with the waterproofing agent and tend to lessen its effectiveness.

Expansive. Expansive cement is a cement which, when mixed with water, forms a paste that increases significantly in volume during and after setting and hardening. It has been used to inhibit shrinkage of concrete, usually with regular portland cement, thus minimizing cracking. It has also been used as a "self-stressing" cement in prestressed concrete work. Not available from all cement manufacturers.

Expansive cement Type K is a mixture of portland cement compounds, anhydrous calcium sulfoaluminate ($4CaO;3AlSO_3;SO_3$), calcium sulfate ($CaSO_4$), and lime (CaO). The anhydrous calcium sulfoaluminate is a component of a separately burned clinker that is interground with portland clinker or blended with portland cement; or alternatively, it may be formed simultaneously with the portland clinker compounds during the burning process.

Expansive cement Type M is either a mixture of portland cement, calcium aluminate cement, and calcium sulfate, or an interground product made with portland cement clinker, calcium aluminate clinker, and calcium sulfate.

Expansive cement Type S is a portland cement containing a large C_3A content and modified by an excess of calcium sulfate above the usual amount found in other portland cements.

The concept of expansive cements is not new. Their history goes back as far as 1890 when a French researcher named Candlot reported that the interaction of certain cement compounds could cause expansion in concrete. The initial major development of today's expansive concrete started in France about 25 years ago, and since 1953 intensive studies on expansive cements and concrete have been made in the Soviet Union. Our present expansive cements in the United States began with the development of Type K at the University of California. The first commercial production of this type of cement was made in January, 1963 on the West Coast.

Shrinkage compensating cement is designed to do one thing: to neutralize or compensate for the drying shrinkage forces in concrete. In a structural member or grade slab containing reinforcing steel the shrinkage compensating cement causes the concrete to expand slightly over several days inducing a mild stress in the reinforcing steel. This stress in turn results in the concrete being in compression. Thus, when normal drying shrinkage takes place the shrinkage is accompanied by a relief of the compression and not a buildup of tensile stresses which otherwise could result in drying shrinkage cracking.

Self-stressing concrete is an expansive cement concrete in which expansion, if restrained, induces compressive stresses of a high enough magnitude to result in significant compression in the concrete after drying shrinkage has occurred.

Regulated-Set Cement. This cement was recently developed in the laboratories of the Portland Cement Association and is now being produced in limited amounts by several manufacturers. It is a hydraulic cement whose setting time can be controlled from approximately 1 to 2 min to 30 min or so with corresponding rapid strength development. It is a modified portland cement that can be manufactured in the same kiln used for manufacture of standard portland cement.

This modified portland cement incorporates set-control and early-strength development components.

Mortars and concretes made with regulated-set cement set more rapidly and develop a greater very early strength than comparable mixes made with standard portland cements. Except for these differences, which are of significance during the first few hours in particular, the physical characteristics of the resulting concrete in tests made to date are similar to comparable mixes made with standard portland cement.

Curing procedures are the same as those applicable to standard portland cement concretes.

With respect to the freshly mixed mortars or concretes, handling, placing, consolidating, and finishing operations must be completed within the handling time available. Revibration or reworking after initial hardening is not feasible. Another factor to be considered is the significantly greater heat evolution due to the very early and rapid hydration which may be advantageous or disadvantageous depending upon circumstances.

With respect to hardened mortar or concrete, there appear to be no different design considerations from those pertaining to standard portland cement concrete. As a structural material, the observed performance to date is essentially identical to that of comparable conventional concretes.

Ferrari. This is the European counterpart of United States Type V. It contains no C_3A and possesses low heat of hydration and low shrinkage.

6   Other Cements

Types S and SA, ASTM: C 358. Slag cement, plain and air-entraining, respectively, consists of at least 60% pulverized water-quenched blast-furnace slag and hydrated lime. British supersulfated cement is a slag-calcium sulfate cement. It is somewhat sulfate-resistant.

Aluminous. Also called high-alumina, cement fondu, or lumnite, it is made from limestone and bauxite, and thus contains calcium aluminates instead of calcium silicates. It gives high early strength and has the property of being refractory. It is low in lime (25 to 40%) and high in Al_2O_3 (about 40%).

Magnesite. Also called magnesium oxychloride cement, or Sorel cement, it is made by combining magnesium oxide with a saturated solution of magnesium chloride. It provides good strength and has elastic properties. It is easily polished, hence is used in terrazzo, but is easily damaged on exposure to water. Water resistance is provided by incorporating a small amount of copper in the mix. The copper furnishes a germicidal value that discourages fungus and vermin.

Keene's. A nonhydraulic hard-finish gypsum plaster.

7   Handling Cement

Conveying. In-plant moving of cement is accomplished in pneumatic trough conveyors, pipes, screw conveyors, and bucket elevators. Bucket elevators, however, are rarely used today.

The pneumatic trough conveyor, also known as an airfloat, air trough, or "Airslide," is a rectangular trough divided longitudinally by a permeable membrane (Fig. 5). This membrane, constituting a false bottom for the material conveying chamber, permits the continuous passage of air from the air chamber below into the material chamber. Cement, being fluidized by the low-pressure air passing through the membrane, flows by gravity down the slightly inclined conveyor, literally floating on a cushion of air. Cement is conveyed silently and without dust. There are no moving parts other than the air compressor, feeders, and gates. Length is limited by the amount of headroom available; systems a mile in length have been reported. Two types are available: open and closed. The open type is

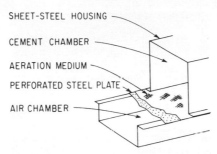

SHEET-STEEL HOUSING

CEMENT CHAMBER

AERATION MEDIUM

PERFORATED STEEL PLATE

AIR CHAMBER

FIG. 5. The pneumatic trough conveyor consists of a longitudinal air chamber on the bottom from which low-pressure air passes through the aeration medium into the material chamber above. The permeable aeration medium may be of fabric, ceramic, or plastic.

installed in the bottom of a bin, truck, or rail car to discharge cement from the container; the closed type is designed to transport cement about the plant. Sizes and capacities vary from manufacturer to manufacturer and are approximately as shown in Table 4.

Table 4. Pneumatic Trough Conveyors

Size (width), in.	Compartment height, in.		Capacity, cu ft per hr	Weight per ft, lb
	Upper	Lower		
6	4	2–3	1,000	9–10
8	6	3	2,000	11–13
10	6	3	3,000	13–16
12	8	3	4,000–6,000	15–19
14	10	3	6,000–8,000	17–28
16	10	3	8,000–10,000	25–33
20	11–14	3	14,000	40
24	12–18	4	18,000–21,000	58–75

Note: Above values are approximate. Consult manufacturer for exact data for any certain installation.

For conveying cement through a pipe, the pressure vessel, which acts as a pneumatic pump, is sometimes used. Cement is fed by gravity into the vessel (Figs. 6 and 7), the vessel is then closed, and air under pressure is introduced through an aeration system. The resulting air-cement mixture is then forced into the conveying pipe line. Delivery of cement to the pipeline is intermittent, a condition that is normally not objectionable. If constant flow is necessary, however, by installing a pair of vessels on one conveying line, a more constant delivery is possible, as one vessel can be filling while the other is discharging. The vessel can be installed in a pit to receive cement from rail cars or trucks, or it can be installed on the plant floor beneath a bin or silo. Vessels are available in sizes from 20 to 320 cu ft.

Another type of pump that utilizes a pipeline for delivery consists of a feed hopper that gravity-feeds the cement into a short screw in an enclosed barrel housing. Cement enters the pump through a rotary valve and is advanced into the barrel by the impeller screw. As the cement advances into the barrel, it

FIG. 6. Cement is fed by gravity into the pneumatic vessel; the vessel is closed and pressurized, forcing the air-cement mixture into the conveying pipeline. (*Special Products Division, Halliburton Services.*)

is compacted by the decreasing pitch of the impeller screw; then it enters the fluidizing chamber, where it is fluidized by low-pressure air from the blower, and finally enters the transport line. There being only one moving part (the screw), maintenance is relatively simple. A pump of this type can be installed in a shallow trench beneath the rails to receive cement from hopper-bottom rail cars, or can be similarly installed in a roadway, if delivery of cement is by hopper-bottom trucks. In these installations, a canvas boot, fastened to a flange on the pump housing, can be attached to a flange on the bottom of the car or truck (Fig. 8). This type of pump can be installed at floor level to receive cement from bins or silos. Capacity of a single pump can be in excess of 40 tons per hour, the actual rate depending on distance and elevation pumped as well as the general geometry of the layout.

A screw conveyor consists of a spiral screw rotating in an enclosed trough. Rotation of the screw moves the cement along the trough. A screw can operate horizontally, or can move cement up a slight incline. Feed and discharge of cement are accomplished through openings in the cover and in the bottom respectively.

It is common practice to arrange cement handling equipment in a sort of closed circuit: a single pump, for example, in a concrete batching plant pumping from rail cars into a silo, batching bin, or transport truck; from silo to truck or batching bin; or from truck into the silo. Figure 9 is a typical schematic installation diagram. In many cases, the cement handling plant is made up of a combination of components from different systems.

Regardless of the type of pneumatic cement handling system, it is imperative that the air be free of moisture, oil, and other contaminants.

Transporting. By far the greatest proportion of cement is shipped in bulk in rail cars and trucks. Two types of carriers are used. Bottom-dump cars and trucks have sloping bottoms, the cement flowing by gravity through the gate in the bottom of the carrier, thence through a boot into the receiving hopper of the conveying unit. The other type, commonly called an "air truck" or "air car," can be pressurized to force the cement out through a pipeline. A pneumatic trough or screw feeder in the bottom of the vehicle moves the cement to the

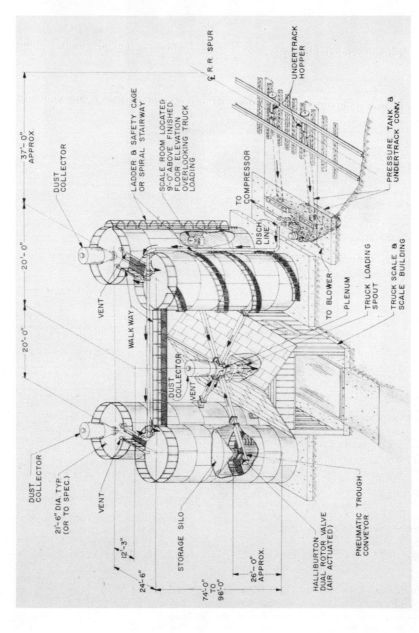

FIG. 7. A Halliburton pressure conveying system installation for a typical satellite distribution terminal receiving bulk cement by rail and delivering to bulk trucks. (*Special Products Division, Halliburton Services.*)

Fig. 8. Under-track installation of Fuller-Kompact screw pump. The 5-in. cement delivery hose leads off to the right. (*Fuller Company/GATX.*)

aeration area, where the air stream picks up the cement and carries it into the pipeline.

Large quantities of cement can be transported in barges and ships, most of which are of the self-unloading type.

When cement is being distributed at some distance from the manufacturing plant, it has been found economical to ship by rail in bulk to a distribution terminal. At the terminal, the cement is stored in silos from which it is loaded into trucks for delivery to customers in the immediate area. Under favorable conditions, cement can be shipped 300 miles or more from the source.

Cement in 94-lb paper sacks can be transported either by rail in box cars or on flatbed trucks. In most cases, the sacks are stacked on wooden pallets to facilitate loading and unloading with fork-lifts. A pallet usually contains 30 or 35 sacks.

Storage. Bulk cement is stored in steel or concrete silos and bins. Concrete silos are suitable for permanent locations, while steel silos are used in either permanent installations or, because they can be dismantled and moved, in temporary plants, such as those for large construction jobs. Cement, because it has less tendency to hang up in a conical silo bottom, moves more efficiently out of a round silo with a conical bottom than it does out of a square bin or silo with a pyramidal bottom. Occasionally the cement will not flow because it has "arched" and formed a plug a short distance above the outlet. Movement of the cement can be facilitated by admitting air through aerators inserted in the wall of the silo cone near the bottom.

When cement is put into a silo that is nearly full, it is not unusual for a core to form in the center of the stored cement through which the new cement flows by gravity directly to the outlet at the bottom of the silo, bypassing the old cement on the periphery of the silo. For this reason, it is a good idea to draw down the silo every 3 or 4 months to remove all cement stored therein.

Temporary storage of cement can be accomplished in a portable horizontal silo (sometimes called a "guppy") that can be transported empty over the highway as a semi-trailer. When the silo is positioned at the site, outriggers are lowered for support. One type is a completely self-contained unit with a cubage of over

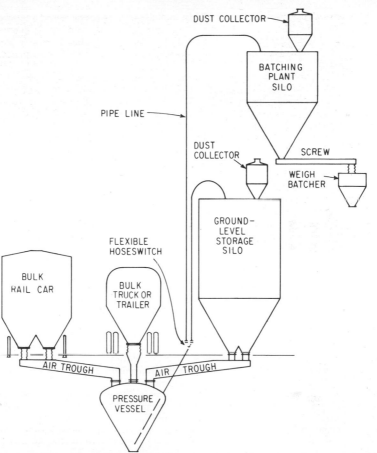

DUST COLLECTOR

BATCHING PLANT SILO

PIPE LINE

DUST COLLECTOR

SCREW

WEIGH BATCHER

GROUND-LEVEL STORAGE SILO

FLEXIBLE HOSESWITCH

BULK RAIL CAR

BULK TRUCK OR TRAILER

AIR TROUGH

AIR TROUGH

PRESSURE VESSEL

Fig. 9. Arrangement of cement conveying and storage components in a batching plant receiving cement by bulk in rail cars or trucks.

Fig. 10. Bulk cement is delivered by rail in either bottom-dump or pneumatic rail cars and transferred to the horizontal storage silos, from which it is pumped into bulk trucks for delivery. Transfer can also be made direct from the pneumatic car to the truck or trailer. (*Butler Manufacturing Company.*)

3,900 cu ft, empty weight of 22,000 lb including a gasoline engine power source, air blower, and the necessary valves and air controls. Cement is moved by air through a piping system between pneumatic rail cars, storage, gravity loading-out bin, batching plant, and pneumatic trucks. Figure 10 includes several of the units just discussed.

Sacked cement can be kept for several months if stored properly. Obviously, storage sheds should be well constructed and weather-tight. The maximum height of piles is usually not over 14 or 15 sacks high. An allowance of about 1 foot should be made between the sacks and exterior walls of the building, but the sacks should be placed closely together to reduce circulation of air. Sacks on pallets are easily stacked with a fork-lift, and the pallets protect the bottom layer of sacks from any moisture that might come through the floor.

Cement temporarily stored at the jobsite should be on a raised platform where it will be out of the way, and the pile should be covered all around with a waterproof cover, such as heavy canvas or polyethylene sheeting.

A "warehouse set" sometimes occurs in sacked cement that has been stored for some time. This is an apparent hardening of the cement and in no way impairs its quality. Warehouse set can usually be broken up by rolling the sack around on the floor or otherwise moving the cement. If the lumps can be readily broken up in the hand, the cement is satisfactory.

8 Air-pollution Control

With the increased emphasis on environmental contamination during the last few years as a result of serious deterioration of the atmosphere, manufacturers and users of portland cement have been faced with the necessity of preventing the loss of gases and particulate matter from their operations. Laws have been enacted. Air-pollution control districts have been established to police industries and require compliance with the laws. General policy has been to give no blanket approvals; instead, each installation has been judged on its own merits.

Cyclone separators are employed in a number of installations. A cyclone separator works on the same principle as a cream separator, using centrifugal force to throw solid particles out of a swirling stream of gas inside a cone-shaped vessel. Initial cost, power consumption, and maintenance cost are all relatively low. Cyclones are sometimes used as a primary control, followed by a bag house.

Bag houses may be used singly or in conjunction with primary cleaning with cyclones. A bag house, which works on the same principle as the common household vacuum cleaner, consists of an enclosure containing a number of long, cylindrical bags woven from cotton, glass fiber, or similar material. Gas, introduced under pressure into the bags, leaves its contaminants on the bags as it passes through to the atmosphere. Periodically the gas flow is stopped for a few moments while the bags are shaken and the dust gathered in hoppers at the bottom of the house.

Electrical precipitators employ oppositely charged high-voltage electrodes to collect the dust. Single-stage heavy-duty precipitators, which may use a voltage as high as 100,000 volts, efficiently collect particles over a wide range of sizes, but require fairly constant values of temperature and concentration of dust. In the two-stage precipitator, the gas is first passed through an ionizing field before passing into the collection field. Voltage is in the range of 10,000 to 13,000 volts.

REFERENCES

1. Peray, K., and J. J. Waddell: "The Rotary Cement Kiln," Chemical Publishing Company, Inc., New York, 1972.
2. Czernin, Wolfgang: "Cement Chemistry and Physics for Civil Engineers," Chemical Publishing Company, Inc., New York, 1962.

Chapter 2

AGGREGATES

F. E. LEGG, JR.

Approximately three-fourths of the volume of conventional concrete is occupied by aggregates consisting of such materials as sand, gravel, crushed rock, or air-cooled blast-furnace slag. It is inevitable that a constituent occupying such a large percentage of the mass should contribute important properties to both the plastic and hardened product. Additionally, in order to develop special lightweight, thermal-insulating, or radiation-shielding characteristics, aggregates manufactured specifically to develop these properties in concrete are often employed. Characteristics of such special aggregates and concretes are given in Sec. 8.

PURPOSE OF AGGREGATES

1 Aggregates in Fresh, Plastic Concrete

When concrete is freshly mixed, the aggregates really are suspended in the cement–water–air bubble paste. Behavior of this suspension (i.e., the fresh, plastic concrete), for instance, ease of placement without segregation causing rock pockets or sand streaks, is importantly influenced by selection of the amount, type, and size gradation of the aggregate. Depending upon the nature of the aggregates employed, a fairly precise balance between the amount of fine- and coarse-sized fractions may have to be maintained to achieve the desired mobility, plasticity, and freedom from segregation, all lumped under the general term "workability." Selection of mixture proportions is aimed to achieve optimum behavior of the fresh concrete consistent with developing desired properties of the hardened product. This selection is covered in Chap. 11.

2 Aggregates in Hardened Concrete

The aggregates contribute many qualities to the hardened concrete. The strength-giving binding material holding concrete together results from the chemical union of the mixing water and cement and is, of course, the basic ingredient. This hardened cement–water–air bubble paste would, by itself, be a very unsatisfactory building material, not to speak of its high cost. Indicative of the cost is the observation that if such paste were used alone at water contents of average concrete, it would contain from about 19 to 26 cwt of cement per cubic yard. The paste, subsequent to initial hardening, unless restrained by contained aggregates, undergoes an intolerable amount of shrinkage upon drying. The exposed portions of such pastes dry out first, and differential shrinkage between the outside and inside portions often results in cracking. The presence of aggregates provides an enormous contact area for intimate bond between the paste and aggregate surfaces. Rigidity

2–1

of the aggregates greatly restrains volume change of the whole mass. Figure 1 shows the amount of shrinkage of concrete relative to paste as the amount of aggregate in concrete is increased. At the usual aggregate content of about 75%, by absolute volume, shrinkage of concrete is only one-tenth that of paste. Thus, the aggregate is not only cost-conserving but is really essential.

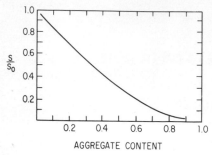

FIG. 1. Ratio of shrinkage of concrete S to shrinkage of paste S_0 as a function of aggregate content.[1]

In some cases, cement grouts containing little or no aggregate are employed successfully for underground work where severe drying is not expected. Expanding agents are also used to compensate for shrinkage where drying is anticipated such as in grouting tendons in posttensioned prestressed members (see Chap. 41).

The conclusion becomes inescapable that aggregates are not simply fillers used to dilute the expensive water-cement paste and thus make a cheaper product. Economics are important, but significant improvements in the workability of the fresh concrete are contributed by proper choice of aggregates. Such choice influences highly important properties of the hardened concrete as well, such as volume stability, unit weight, resistance to destructive environment, strength, thermal properties, and pavement slipperiness. The latter factors are discussed more fully later in this chapter.

DEFINITIONS

The great variety of granular material incorporated in concrete makes formulation of an entirely satisfactory definition of "aggregate" very difficult. For instance, some finely divided pozzolans such as fly ash may simultaneously act as a supplement to the fine aggregate and also contribute cementitious properties binding the mass together. Similarly, finely divided air bubbles characteristic of those in air-entrained concrete may contribute workability to fresh concrete akin to that achieved by adding very fine sand. Extremely lightweight cellular concrete is made with the "aggregate" constituent consisting partly or entirely of air bubbles. Such special concretes and those made containing manufactured lightweight aggregates are treated in Chap. 31. A definition recently adopted in the field of road and paving materials is given in ASTM Designation: D 8-70: "Aggregate—a granular material of mineral composition such as sand, gravel, shell, slag, or crushed stone, used with a cementing medium to form mortars or concrete, or alone as in base courses, railroad ballasts, etc." The designations "natural aggregate" or "mineral aggregate" are frequently applied to those derived from rock fragments used in their natural state except for washing, screening, beneficiation, or crushing.

For the purpose of this chapter where conventional aggregates are emphasized, the definitions given by the American Society for Testing and Materials will be used as found in Standard Definitions of Terms Relating to Concrete Aggregates, ASTM Designation: C 125.[2]

3 Coarse Aggregate

(1) Aggregate predominantly retained on the No. 4 (4.75-mm) sieve; or (2) that portion of an aggregate retained on the No. 4 (4.75-mm) sieve. (Note)

4 Fine Aggregate

(1) Aggregate passing the ⅜-in. sieve and almost entirely passing the No. 4 (4.75-mm) sieve and predominantly retained on the No. 200 (75 μm) sieve; or (2) that portion of an aggregate passing the No. 4 (4.75-mm) sieve and retained on the No. 200 (75 μm) sieve. (Note)

5   Gravel

(1) Granular material predominantly retained on the No. 4 (4.75-mm) sieve and resulting from natural disintegration and abrasion of rock or processing of weakly bound conglomerate; or (2) that portion of an aggregate retained on the No. 4 (4.75-mm) sieve and resulting from natural disintegration and abrasion of rock or processing of weakly bound conglomerate. (Note)

6   Sand

(1) Granular material passing the ⅜-in. sieve and almost entirely passing the No. 4 (4.75-mm) sieve and predominantly retained on the No. 200 (75 μm) sieve, and resulting from natural disintegration and abrasion of rock or processing of completely friable sandstone; or (2) that portion of an aggregate passing the No. 4 (4.75-mm) sieve and predominantly retained on the No. 200 (75 μm) sieve, and resulting from natural disintegration and abrasion of rock or processing of completely friable sandstone. (Note)

Note. The definitions are alternatives to be applied under differing circumstances. Definition (1) is applied to an entire aggregate either in a natural condition or after processing. Definition (2) is applied to a portion of an aggregate. Requirements for properties and grading should be stated in specifications. Fine aggregate produced by crushing rock, gravel, or slag is known as "manufactured" sand.

7   Crushed Stone

The product resulting from the artificial crushing of rocks, boulders, or large cobblestones, substantially all faces of which have resulted from the crushing operation.

8   Air-cooled Blast-furnace Slag

The material resulting from solidification of molten blast-furnace slag under atmospheric conditions. Subsequent cooling may be accelerated by application of water to the solidified surface.

9   Crushed Gravel

The product resulting from the artificial crushing of gravel with substantially all fragments having at least one face resulting from fracture.

SOURCES OF AGGREGATES

As indicated by the previous definitions, conventional concrete aggregates here considered, and relative to their source, may be divided into three major classifications:
1. Sand, gravel, and crushed gravel
2. Crushed stone
3. Air-cooled blast-furnace slag
Concrete aggregates of the first two classifications are derived from natural deposits in the earth's crust. Strictly speaking, the same is true for blast-furnace slag. However, the ore, coke, and limestone used in charging the furnace undergo a man-made processing which renders the slag an entirely different product from its raw constituents.

10   Sources of Sand, Gravel, and Crushed Gravel

Geologic history dictates the character and extent of aggregates derived directly from the earth of which man can make use as a concrete ingredient. Considering first sand and gravel, both are the result of being eroded from parent rock, and they usually have then been washed and sorted over the ages by action of flowing water. Wind-deposited sand dunes and loess areas are not generally useful to the concrete technologist. The origin of cobbles and boulders from which crushed gravel or crushed stone may be made is similar, but because of either their inherent greater resistance to breakdown or their less rigorous environment, they have not been reduced to such small sizes as the sands and gravels.

Natural deposits of sand–gravel–cobble materials of interest to the construction engineer may be broadly classified into those of glacial and nonglacial origin.

Glacial Origin. Much of the northern part of the United States and all of Canada was covered at one time or another with sheets of glacial ice, in some places reaching great thicknesses. Over the past million years, the climate apparently changed at least four times so as to alter the extent of these ice coverings. Prolonged, and sometimes torrential, volumes of water flowed during the recessions of these ice fronts, either on top, underneath in fissures or crevasses in the ice, or at the edges. In addition, movement of the ice created enormous grinding action on the underlying rock. The eroded material was deposited in snakelike ridges by the flowing water into formations called *eskers* or conical mounds called *kames*. At the outfall, the water spread the deposits out into *outwash plains*. Additionally, rainfall must have occurred to augment the meltwater to spread out the eroded material occasionally on *terraces* even at higher elevations. Figure 2 gives areas where glacial action occurred. However, granular deposits of value to the concrete engineer are not necessarily found throughout the glaciated region.

FIG. 2. Map of geographical areas where glaciation occurred.

Nonglaciated Deposits. Water-deposited supplies of sand and gravel occur in many regions south of the glaciated areas. These deposits were derived from either ancient rivers no longer existing or from present streams and are then known as *river deposits*. Sandy-soil mixtures may be deposited far from the stream source. Nearer the mountains from which the material originated, the more gravelly material may be laid down as *alluvial fans, bars, terraces,* etc.

11   Sources of Crushed Stone

Crushed stone is derived primarily from quarry operations. Bedrock occurs in many areas in the country either as small outcroppings, or with a sufficiently thin soil overburden that it can be economically quarried, or as mountains. "Granite rocks, used for the production of mineral aggregates, are to be found in widely scattered sections of the country. Trap rock, as well as other igneous material, is quarried extensively in the Connecticut Valley, New Jersey, several sections of the Piedmont in the East, the Columbia Plateau, and certain sections of the western ranges. Certain metamorphic rocks, such as quartzites, are found in South Dakota, Wisconsin, and Minnesota; other desirable metamorphic materials are found in the Piedmont sections in the East and in many of the mountainous regions of the country. Limestone is the most important sedimentary rock used for aggregate production. It occurs extensively throughout the Middle West, Kentucky, Tennessee, the Ozark Plateau, and widely scattered areas in the Great Plains and some small areas of the Far West."[3]

12 Air-cooled Blast-furnace Slag

Blast-furnace slag is a by-product of blast-furnace operation and therefore occurs only in the areas where pig iron is produced. For the year 1969, approximately 65% of the slag marketed was produced in Indiana, Illinois, Michigan, Ohio, and Pennsylvania. Substantial quantities were also produced in Alabama, California, Colorado, Kentucky, Louisiana, Maryland, Minnesota, New York, Texas, Utah, and West Virginia.

Because of the increasing practice of iron-ore concentration prior to entering the blast furnace, production of blast-furnace slag has increased slowly in recent years despite greater pig-iron production. As little as 335 lb of slag has been produced per ton of pig iron by carefully selecting the furnace charge.

In its production the molten slag is dumped into cooling pits immediately after withdrawing from the blast furnace. Stratification of the layers in the pit results from pouring hot molten slag over partially cooled slag, causing cracking so that upon further cooling it can be readily picked up with power shovels. Limited water quenching is also employed to aid fragmentation. After cooling (to 200°F or lower) the slag is transported to a plant for crushing and screening to sizes appropriate for concrete use.

Granulated slag ("popcorn slag") used for lightweight block or as a constituent of portland blast-furnace-slag cement is chilled much more rapidly than air-cooled slag by complete immersion in water. Expanded slag used in lightweight concrete and masonry units is chilled by aid of controlled amounts of water or steam, yielding a product of intermediate density and strength between air-cooled and granulated slag.

Most blast-furnace slags will fall within the following range of chemical analyses:

	%		%
Silica (SiO_2).................	33–42	Sulfur (S)*...................	1–3
Alumina (Al_2O_3)	10–16	Iron oxide (FeO).............	0.3–2
Lime (CaO)	36–45	Manganese oxide (MnO)	0.2–1.5
Magnesia (MgO)	3–12		

* In form of calcium sulfide or calcium sulfate.

Reference 4 provides a wealth of additional information on air-cooled blast-furnace slag.

AGGREGATES PRODUCED

Cement-production data provide a useful index for the estimation of annual consumption of concrete aggregates. Cement produced in the United States has grown from negligible at the beginning of the century to more than 70 million tons annually in 1970. Detailed production figures are available in the annual "Minerals Yearbook," vol. 1, prepared by the staff of the U.S. Bureau of Mines, which can be procured from the Superintendent of Documents, Washington, D.C. 20402 (price $6.00). Data are also provided in the above volume for sand and gravel, crushed stone, and blast-furnace slag. Aggregate-production data exclusively for use in concrete, however, are furnished only in the case of slag. As a consequence, it is necessary to make assumptions, based upon reported cement shipments, to predict the quantity of concrete aggregates consumed annually.

Table 1 shows the 1970 estimated tonnage of concrete aggregates used in each state, based upon the assumption that 95% of the cement shipped was destined for use in concrete containing 1.5 tons of aggregate per cu yd of concrete with a cement content averaging 500 lb per cu yd. It is observed that California far leads the states with 48 million tons, Texas is second with 31 million, and in descending order, Florida, New York, Ohio, Illinois, and Pennsylvania. The

Table 1. Estimated Tonnage of Concrete Aggregates Used in 1970*

Destination	Portland cement, thousands of tons	Estimated tonnage of concrete aggregates,§ thousands
Alabama......................	999	5,690
Alaska.......................	‡	‡
Arizona......................	1,060	6,040
Arkansas.....................	615	3,510
Northern California.............	2,853	16,260
Southern California.............	5,699	32,480
Colorado.....................	1,041	5,930
Connecticut†..................	834	4,750
Delaware†....................	162	920
District of Columbia†..........	182	1,040
Florida......................	3,603	20,540
Georgia......................	1,904	10,850
Hawaii.......................	460	2,620
Idaho........................	508	2,900
Illinois.......................	3,308	18,860
Indiana......................	1,604	9,140
Iowa.........................	1,596	9,100
Kansas.......................	964	5,490
Kentucky.....................	974	5,550
Louisiana.....................	1,902	10,840
Maine........................	206	1,170
Maryland.....................	1,449	8,260
Massachusetts†................	1,353	7,710
Michigan.....................	2,758	15,720
Minnesota....................	1,578	8,990
Mississippi...................	814	4,640
Missouri......................	1,747	9,960
Montana......................	319	1,820
Nebraska.....................	834	4,750
Nevada.......................	301	1,710
New Hampshire†...............	167	950
New Jersey†...................	2,067	11,780
New Mexico...................	429	2,450
New York.....................	3,359	19,150
North Carolina................	1,523	8,680
North Dakota†.................	289	1,650
Ohio.........................	3,229	18,400
Oklahoma.....................	1,236	7,040
Oregon.......................	644	3,670
Eastern Pennsylvania...........	2,095	11,940
Western Pennsylvania...........	1,150	6,550
Rhode Island†.................	188	1,070
South Carolina................	783	4,460
South Dakota.................	242	1,380
Tennessee....................	1,376	7,840
Texas........................	5,413	30,850
Utah.........................	419	2,390
Vermont†.....................	108	620
Virginia......................	1,772	10,100
Washington...................	1,136	6,480
West Virginia.................	469	2,670
Wisconsin....................	1,530	8,720
Wyoming.....................	186	1,060
Total United States...........	71,432	407,160

* Cement data from U.S. Bureau of Mines.
† Not a cement producer.
‡ Withheld to avoid disclosing manufacturers' confidential data.
§ Estimated on basis that 95 % of cement is used in concrete containing 1.5 tons of aggregate per cu yd with a cement factor of 500 lb per cu yd.

remaining states are each estimated to have used less than 18 million tons of concrete aggregates in 1970.

Prices of concrete aggregates display a slowly rising trend over the years. In 1972, per ton costs of typical aggregates were quite variable in the major metropolitan areas; for instance, 1½-in. gravel was highest in Pittsburgh at $5.30 and lowest in Montreal at $1.80; sand varied from a high of $6.25 in Atlanta to $1.05 in Detroit; 1½-in. crushed stone was highest in Minneapolis at $6.00 and lowest in Toronto at $1.50, and 1½-in. crushed slag (where available) varied from a high of $5.70 in Los Angeles to a low of $2.00 in Montreal.[5] Based upon a roughly estimated average price of $3.00 per ton, and since concrete production definitely increased between 1970 and 1972, the aforementioned seven states leading in cement consumption would each have a corresponding 1972 concrete-aggregate market estimated to exceed $50 million annually.

ROCK TYPES AND CHARACTERISTICS

Each rock of interest to the concrete technologist, whether in the form of a sand grain, pebble, cobble, or crushed ledge rock fragment, generally consists of

Table 2. General Classification of Rock*

Class	Type	Family
Igneous	Intrusive (coarse-grained)	Granite† Syenite† Diorite† Gabbro Peridotite Pyroxenite Hornblendite
	Extrusive (fine-grained)	Obsidian Pumice Tuff Rhyolite†,‡ Trachyte†,‡ Andesite†,‡ Basalt† Diabase
Sedimentary	Calcareous	Limestone Dolomite
	Siliceous	Shale Sandstone Chert Conglomerate§ Breccia§
Metamorphic	Foliated	Gneiss Schist Amphibolite Slate
	Nonfoliated	Quartzite Marble Serpentinite

* From Ref. 6.

† Frequently occurs as a porphyritic rock.

‡ Included in general term "felsite" when constituent minerals cannot be determined quantitatively.

§ May also be composed partially or entirely of calcareous materials.

an identifiable assembly of "minerals." Minerals thus constitute the elementary building blocks from which rocks are made and each has a definite chemical formula, specific crystalline structure, and characteristic molecular structure. Thus the mineral quartz (silicon dioxide) predominates as a constituent of such rocks as sandstone and quartzite. On the other hand, the silicon dioxide may be chemically combined with aluminum, potassium, sodium, calcium, or magnesium compounds in such a manner as to furnish a series of very complicated minerals demanding the assistance of a professional to make positive identification.

13   Rock Classification

Table 2 gives a general classification of rocks relative to their origin: igneous, sedimentary, or metamorphic. The *igneous* rocks are formed from molten material;

Table 3. Mineral Composition of Rocks[a]

Name of rock	No. of samples tested	Essential mineral composition, %[b]											
		Quartz	Orthoclase, microcline	Plagioclase	Augite	Hornblende	Mica	Calcite	Chlorite	Kaolin	Epidote	Iron ore	Remainder
Igneous rocks:													
Granite	165	30	45	(8)	...	...	6	...	....	(6)	...	...	5
Biotite granite	51	27	41	9	...	...	11	...	....	(7)	...	...	5
Hornblende granite	20	23	34	12	...	13	4	...	....	(10)	...	...	4
Augite syenite	23	(4)	52	7	8	...	4	...	(3)	(11)	(3)	(4)	4
Diorite	75	8	7	30	...	27	(4)	...	(3)	(8)	(5)	(3)	5
Gabbro	50	...	...	44	28	9	...	...	(3)	(6)	...	...	10
Rhyolite	43	32	45	(3)	...	...	(5)	...	(4)	(3)	...	(4)	4
Trachyte	6	(3)	42	...	...	6	...	(3)	(3)	(14)	(8)	(7)	5c
Andesite	67	...	...	48	14	3	...	...	(6)	....	(3)	(8)	6c
Basalt	70	...	...	36	35	...	...	...	....	....	...	(3)	5c
Altered basalt	196	...	...	32	31	...	...	...	(9)	(4)	...	(4)	8c
Diabase	29	...	...	44	46	...	...	...	....	....	...	(4)	6
Altered diabase	231	...	...	35	26	...	...	...	(15)	(9)	...	(4)	11
Sedimentary rocks:													
Limestone	875	(6)	...	...	...	...	...	83d	....	....	...	...	3
Dolomite	331	(5)	...	...	...	...	...	11d	....	....	...	...	2
Sandstone	109	79	(5)	...	...	...	...	...	....	(4)	...	(9)	3
Feldspathic sandstone	191	35	26	...	...	...	...	(3)	(3)	(22)	...	(4)	7
Calcareous sandstone	53	46	(3)	...	...	...	...	42	....	....	...	(3)	6
Chert	62	93	...	...	...	...	...	...	...	....	...	...	7e
Metamorphic rocks:													
Granite gneiss	169	34	35	(4)	...	...	20	...	....	....	...	...	7
Hornblende gneiss	18	10	16	15	(3)	45	(4)	...	....	....	...	...	7
Mica schist	59	36	14	(1)	...	...	40	...	....	....	...	...	9
Chlorite schist	23	11	...	10	...	(5)	...	...	39	...	28	(4)	3
Hornblende schist	68	10	(3)	12	...	61	...	...	....	....	(7)	...	7
Amphibolite	22	(3)	...	8	...	70	...	...	....	....	12	...	7
Slate	71	29	(4)	...	...	...	55	...	....	....	...	(5)	7
Quartzite	61	84	(3)	...	...	...	(4)	...	....	....	...	...	9
Feldspathic quartzite	22	46	27	...	...	...	(7)	...	(3)	(10)	...	...	7
Pyroxene quartzite	11	29	19	15	24	...	...	...	....	....	...	(5)	8f
Marble	61	(3)	...	...	...	...	...	96	....	....	...	...	1

[a] From Ref. 6.
[b] Values shown in parentheses indicate minerals other than those essential for the classification of the rock.
[c] Includes 10 to 20 % rock glass.
[d] Limestone contains 8 % of the mineral dolomite; the rock dolomite contains 82 % of this mineral.
[e] Includes 3 % opal.
[f] Includes 3 % garnet.

if cooling was slow, the mineral grains are relatively large and the rocks are classified as *intrusive,* and if cooling was rapid, the grains are small and the fine-grained rocks are classified as *extrusive.* Two rocks both of igneous origin, such as granite and rhyolite, may have identical mineral composition, but their names and engineering properties may be quite different depending upon grain size (see Table 3). As implied by the name, the *sedimentary* rocks are formed from sediments deposited in water or by wind. Limestones and dolomites so frequently used as concrete aggregates are of sedimentary origin and contain a predominance of the mineral calcite (calcium carbonate) in the case of *limestone*

Table 4. Average Values for Physical Properties of the Principal Types of Rocks*

Type of rock	Bulk specific gravity	Absorption,† %	Loss by abrasion, %	
			Deval‡	Los Angeles§
Igneous:				
Granite...............	2.65	0.3	4.3	38
Syenite...............	2.74	0.4	4.1	24
Diorite...............	2.92	0.3	3.1	
Gabbro...............	2.96	0.3	3.0	18
Peridotite............	3.31	0.3	4.1	
Felsite...............	2.66	0.8	3.8	18
Basalt...............	2.86	0.5	3.1	14
Diabase..............	2.96	0.3	2.6	18
Sedimentary:				
Limestone............	2.66	0.9	5.7	26
Dolomite.............	2.70	1.1	5.5	25
Shale.................	1.8–2.5			
Sandstone............	2.54	1.8	7.0	38
Chert................	2.50	1.6	8.5	26
Conglomerate.........	2.68	1.2	10.0	
Breccia..............	2.57	1.8	6.4	
Metamorphic:				
Gneiss...............	2.74	0.3	5.9	45
Schist................	2.85	0.4	5.5	38
Amphibolite...........	3.02	0.4	3.9	35
Slate.................	2.74	0.5	4.7	20
Quartzite.............	2.69	0.3	3.3	28
Marble...............	2.63	0.2	6.3	47
Serpentine............	2.62	0.9	6.3	19

* From Ref. 6.   † After immersion in water at atmospheric temperature and pressure.
‡ AASHO Method T 3.   § AASHO Method T 96.

or the mineral dolomite (54% calcium carbonate, 46% magnesium carbonate) in the case of the rock *dolomite.* As in the case of so many rock types, calcite and dolomite may be intermixed in the same rock, leading to the designation dolomitic limestone or calcitic dolomite. The sedimentary rocks containing silica, particularly shale and chert, deserve careful consideration before using as concrete aggregate (see Art. 27).

Prolonged heat and/or pressure over geologic times may eventually alter one rock type into another type giving rise to the *metamorphic* class. This alteration may be beneficial, such as transforming sandstone into the harder and tougher quartzite. The *foliated* metamorphic rocks possess a more or less parallel layered structure of mineral grains such as does schist or slate; the *nonfoliated* rocks, for example, quartzite and marble, are more random in structure.

Tables 4 and 5 give some physical and engineering properties of the principal types of rocks, taken from Ref. 6.* This 17-page pamphlet is highly recommended as a concise source of mineralogical information for the engineer.

Table 5. Summary of Engineering Properties of Rocks*

Type of rock	Mechanical strength	Durability	Chemical stability	Surface characteristics	Presence of undesirable impurities	Crushed shape
Igneous:						
Granite, syenite, diorite.........	Good	Good	Good	Good	Possible	Good
Felsite.............	Good	Good	Questionable	Fair	Possible	Fair
Basalt, diabase, gabbro	Good	Good	Good	Good	Seldom	Fair
Peridotite..........	Good	Fair	Questionable	Good	Possible	Good
Sedimentary:						
Limestone, dolomite..	Good	Fair	Good	Good	Possible	Good
Sandstone..........	Fair	Fair	Good	Good	Seldom	Good
Chert..............	Good	Poor	Poor	Fair	Likely	Poor
Conglomerate, breccia	Fair	Fair	Good	Good	Seldom	Fair
Shale..............	Poor	Poor	....	Good	Possible	Fair to poor
Metamorphic:						
Gneiss, schist........	Good	Good	Good	Good	Seldom	Good to poor
Quartzite...........	Good	Good	Good	Good	Seldom	Fair
Marble.............	Fair	Good	Good	Good	Possible	Good
Serpentinite........	Fair	Fair	Good	Fair to poor	Possible	Fair
Amphibolite........	Good	Good	Good	Good	Seldom	Fair
Slate..............	Good	Good	Good	Poor	Seldom	Poor

* From Ref. 6.

GEOLOGICAL AND MINERALOGICAL TERMS†

Acidic: A term applied to igneous rocks containing more than 65% silica (SiO_2).

Alluvial fan: A fan-shaped deposit of gravel or loose rock formed by a stream of running water at the mouth of a ravine. The coarsest material occurs at the high point (apex) of the fan or cone.

Amorphous: Having no crystal structure, such as opal or volcanic glass.

Arenaceous: Sandy or sandy-textured.

Argillaceous: Term applied to rocks containing clay. Clayey.

Basalt: An igneous, fine-grained, dense, volcanic rock, dark-colored or black. Commonly found in Northwestern states, but occasionally found in other areas of former volcanic activity. Also called "traprock." Some varieties have given trouble in gravel base courses.

*Procurable from the Superintendent of Documents, Government Printing Office, Washington, D.C. 20402 (20 cents).

† A much more comprehensive listing of terms and definitions is contained in "A Dictionary of Mining, Mineral and Related Terms," U.S. Bureau of Mines, 1968, pp. 1–1269, Superintendent of Documents, Government Printing Office, Washington, D.C. 20402 ($8.50).

Breccia: A rock formed of angular fragments of preexisting rock cemented together with bonding material such as silica or calcite compounds.

Calcareous: Term applied to rocks containing calcium carbonate.

Calcite: The mineral calcium carbonate, $CaCO_3$.

Carbonaceous: Containing organic matter.

Chalcedony: A submicroscopic variety of fibrous quartz SiO_2, generally translucent and containing variable amounts of opal. Reacts with alkalies in portland cement.

Chert: Very fine-grained siliceous rock containing cryptocrystalline quartz, chalcedony, or opal or a combination thereof. Porous varieties are usually light-colored and have splintery fractures. Dense varieties are hard, have conchoidal fracture, greasy luster, and occur in many colors including white, yellow, or brownish stained, or green. The colored varieties are sometimes designated "jasper." Dense, gray variety called "flint." All varieties will scratch glass and not be scratched by knife blade. Some of its constituents may be reactive with cement alkalies, and it should also be considered suspect as concrete aggregate for exposed concrete in northern climates.

Clay: Very fine particles consisting of hydrosilicates of aluminum or magnesium, or both. In concrete aggregates the term frequently refers to these materials occurring as coatings finer than the No. 200 (75 μm) sieve which may be removed by washing. Clay may also occur well dispersed in rocks between laminations so as to weaken the structure or cause the particle to be susceptible to freeze-thaw attack.

Conchoidal fracture: Breaks with a convex, curved face resembling a shell; typical of dense chert, flint, or obsidian.

Conglomerate: Rock consisting of rounded pebbles cemented together with finer material.

Cryptocrystalline: Rock texture too fine to be discernible with optical microscope.

Diabase: Same material composition as basalt but crystals slightly larger—just visible to unaided eye. Also called "traprock."

Diorite: Medium- to coarse-grained rock composed essentially of plagioclase feldspar and ferromagnesium minerals.

Dolomite: The mineral calcium–magnesium carbonate $CaMg(CO_3)_2$.

Feldspar: General name for a group of igneous minerals all of which are softer than quartz. Chemically they are calcium-aluminum silicates, potassium-aluminum silicates, or sodium-aluminum silicates.

Ferruginous: Containing iron and manifested by reddish-brown stains.

Foliated: More or less parallelism of mineral grains in metamorphic rocks as distinct from stratified structure of some sedimentary rocks.

Gabbro: Igneous rock similar to diorite predominantly composed of ferromagnesium minerals with crystals visible to the eye. Same mineral composition as basalt.

Glass: Component of some volcanic rocks resulting from such a rapid cooling from the molten state that no crystal structure is present. Obsidian is a natural glass.

Gneiss: A banded or foliated metamorphic rock (e.g., granite gneiss, diorite gneiss).

Granite: Rock with large grains easily visible to the eye and consisting predominantly of quartz and alkali feldspars.

Greenstone: Sometimes used by petrographers to designate rocks which defy ready identification or are a minor constituent which need not be identified for the purpose at hand.

Hardness: Measure of ability of one mineral to scratch another. Mohs' scale of hardness is as follows: 1, talc; 2, gypsum; 3, calcite; 4, fluorite; 5, apatite; 6, orthoclase; 7, quartz; 8, topaz; 9, sapphire; 10, diamond.

Hematite: Ferric oxide, Fe_2O_3.

Illite: One of the clay minerals of the type which swells and shrinks considerably upon wetting and drying.

Kaolinite: One of the clay minerals consisting of a hydrous aluminum silicate.

Limonite: A hydrated form of hematite. Also loosely applied to ferruginous sandstones, shales, and clay ironstones. Generally suspect as a constituent for exposed concrete use. Some forms used for heavyweight concrete.

Luster: Light-reflecting characteristic of a substance (e.g., dull, metallic, vitreous).

Megascopic: Term applied to gross examination by naked eye or hand lens.

Metamorphic: Altered from another rock by heat or pressure. The original rock may be igneous, sedimentary, or metamorphic.

Microcrystalline: Constituents too small to be seen without aid of a microscope.

Montmorillonite: One of the clay minerals which typically swells upon wetting and becomes soft and greasy. Very widespread.

Nodular: Occurring naturally as, or made into, small rounded lumps.

Opal: Hydrous form of silica containing 2 to 10% combined water. Amorphous. Softer than quartz. Reacts with cement alkalies and may be highly detrimental as a concrete-aggregate constituent.

Porphyritic: An igneous-rock texture in which larger crystals are set in a ground mass of smaller crystals or glass.

Pumice: Highly porous, frothy lava which is largely glass. Frequently of rhyolitic composition. May be reactive with cement alkalies. Sometimes used as aggregate for lightweight concrete.

Quartz: Most abundant form of the mineral silica (SiO_2). Very hard, will scratch glass but not be scratched by a knife. Colorless when pure, glassy luster, with conchoidal fracture.

Quartzite: Extremely hard, tough, and stable metamorphosed sandstone. Sand grains have been cemented together with secondary quartz. Excellent concrete aggregate but may crush to thin or elongated pieces.

Schist: May be formed from a number of igneous or sedimentary rocks. Characterized by crushing to thin, platy, flat fragments or crumbling to prismatic shapes. Weak parallel to plane of foliation.

Scoria: Clinkerlike material resulting from breakup of crust of lava flow or fragments thrown out by volcanic eruption. See Pumice.

Shale: Argillaceous sedimentary rock derived from silts or clays. Typically thinly laminated and weak along planes. Should be considered suspect as concrete aggregate unless proved otherwise.

Siderite: Iron carbonate, $FeCO_3$.

Slate: Fine-grained metamorphic rock, stratified and breaks easily, not necessarily parallel to laminations. Less suspect as concrete aggregate than shale.

Traprock: Common term applied to basalt and diabase or dark-colored rhyolite, dacite, fine-grained andesite, or trachyte.

Tuff: Consolidated volcanic ash.

Weathering: Rock decomposition or disintegration due to atmospheric exposure. Troublesome term for the concrete technologist who uses the same word for relatively short-exposure effects on concrete.

PROSPECTING AND EXPLORATION

Exploration or prospecting for suitable material for use as concrete aggregate may take one or more of several avenues depending upon the amount of material needed, location, and knowledge already acquired or readily procurable. In heavily populated areas where demand for concrete is large, exhaustion of present supplies of aggregate may occur, requiring major and continued attention by large suppliers in efforts to keep up with demand. Zoning restrictions, acquirement of property or mineral rights, investigating suitable haul roads, and attendant economic considerations make this an extremely complex undertaking even after successful location of suitable deposits. Avoidance of legal entanglements with property owners and various public agencies may be of overriding importance to all other considerations. Figure 3 shows a man-made lake developed by a producer subsequent to a gravel operation; good public relations result from such activities.

Clear distinction is to be made between methods for *searching out* deposits of granular materials or ledge rock and the methods used for *evaluating the suitability* of such materials. as concrete aggregates. The desired qualities are treated in Arts. 22 to 30. Devices or procedures capable of simultaneously evaluating both

Fig. 3. Attractive 80-acre man-made lake, part of land-rehabilitation project after gravel operation. (*Courtesy of American Aggregates Corp.*)

aspects are not yet available, and until both tasks are successfully accomplished undue optimism with respect to a potential aggregate source may be unwarranted.

14 Existing Geological Maps and Surveys

Much useful data pertaining to location of potential aggregate sources may be procured from Federal or state geological or agricultural agency maps. For instance, study of surface-geology maps available from the U.S. Geological survey, U.S. Department of Interior, Washington, D.C., or more detailed maps from state geologists may be rewarding as to location, for instance, of outwash plains and eskers where granular materials could be expected to be found. Similarly, state geology surveys are sometimes able to provide bedrock-formation maps which may give a clue as to potential quarry sources.

Much of the tillable land area of the United States has been covered by soil surveys, and soil maps are published by the Soil Conservation Service, U.S. Department of Agriculture, Washington, D.C. 20250. Maps of the immediate area can usually be obtained from the local soil-conservation service or agricultural agent. Such maps are normally thought of as pertinent only to the agriculturist since the weathered soil profile of about the depth that can be investigated with a 4-ft hand auger has actually been drilled. However, the pedological soil name assigned embodies a multiplicity of concepts—geology, topography, texture, and drainage characteristics.[7] Thus, ability to translate the soil-type names over to useful engineering information can be developed with practice—particularly when coupled with a study of surface-geology maps. Figure 4 reproduces a small (6 miles square) area covered by a 1901 (revised in 1915) surface-geology survey in southern Michigan.[8] Note a substantial size kame in the center of the area—a potential source of granular material. A soil-survey map of the same area made in 1930 is provided in Fig. 5.[9] The soil over the central part of the area is listed as "Bellefontaine sandy loam." The "Field Manual of Soil Engineering," Michigan Department of State Highways, January, 1970, gives in part the following characteristic for such soil:

. . . The surface relief ranges from undulating and smoothly rolling to rough and hilly. In places it is choppy, with high knolls, sharp ridges, steep slopes, and rounded depressions.

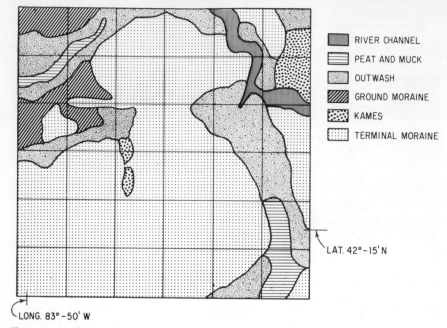

RIVER CHANNEL
PEAT AND MUCK
OUTWASH
GROUND MORAINE
KAMES
TERMINAL MORAINE

LAT. 42°–15' N

LONG. 83°–50' W

FIG. 4. Surface-geology map of 6-mile-square area in southern Michigan, same area as Fig. 5.[8]

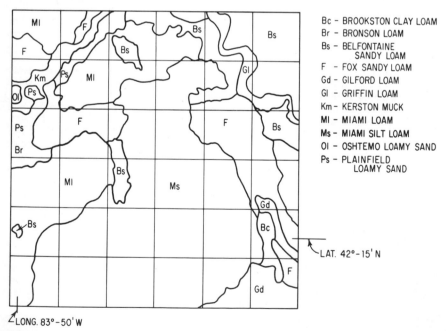

Bc – BROOKSTON CLAY LOAM
Br – BRONSON LOAM
Bs – BELFONTAINE SANDY LOAM
F – FOX SANDY LOAM
Gd – GILFORD LOAM
Gl – GRIFFIN LOAM
Km – KERSTON MUCK
Ml – MIAMI LOAM
Ms – MIAMI SILT LOAM
Ol – OSHTEMO LOAMY SAND
Ps – PLAINFIELD LOAMY SAND

LAT. 42°–15' N

LONG. 83°–50' W

FIG. 5. Soil-survey map of 6-mile-square area in southern Michigan, same area as Fig. 4.[9]

It is stony in places with large and small stones imbedded in the soil and scattered over the surface. Occasionally the stones may be nested. Layers and pockets of silt, clay, and very fine sand in conjunction with seepage may occur in the solum and substratum. Bellefontaine occurs on eskers, kames, and moraines in the southern part of the Lower Peninsula. *Gravel pits are common. . . .*

Thus, study of the soil-survey and surface-geology maps would surely predict granular material in this area. In fact, this kame has been the site of a commercial gravel and sand operation for many years, now worked so extensively that it is almost depleted. The promising kame in the northeast corner of the map is a residential area not available for development.

State highway departments are another source of information regarding ledge rock or granular-material deposits. A survey in 1957[10] showed 26 highway departments had made, or were in the process of making, detailed inventories of road-building materials in their states.

15 Aerial Surveys

Air photography interpretation, as contrasted with precise horizontal and vertical measurements derived from photogrammetry, is truly a specialized art, particularly with reference to searching for granular materials or rock deposits. Competence derives from intensive study of the geological and soil sciences, and in the hands of an expert a remarkable amount of information can be derived. Generally such information is interpreted from stereo pairs taken with black-and-white film but single, overlapping aerial photographs may also be employed. Color aerial photography is also being employed and appears to offer some advantages despite the greater cost of the film.

The interpreter of aerial photographs looks for air-photography pattern elements involving vegetation, drainage patterns, topography, and erosion as well as presence of the type of man-made features. An excellent treatment of air-photography interpretation is provided in Woods' "Highway Engineering Handbook" (McGraw-Hill), pp. 5-19 to 5-31 (1960).

The entire United States has been covered by aerial black-and-white photographs at a scale between 1:15,000 and 1:25,000, and such photographs may be procured from agencies of the Federal government at nominal cost, particularly from the Soil Conservation Service of the Department of Agriculture. The usefulness of such aerial photographs to the concrete engineer will depend entirely upon the experience and proficiency of the specialist engaged to interpret the photographs.[11]

16 Geophysical Exploration

Portable instruments are available which provide information, if coupled with supplementary knowledge, that may be useful to the prospector for mineral aggregates. These instruments operate on either of two principles: (1) electrical resistivity of the different underlying materials enabling distinguishing between bedrock, sand, gravel, groundwater level, or soil; and (2) seismic exploration by which the velocities of subsurface seismic or shock waves through these materials are measured or the waves may be observed to be reflected from underlying strata. Generally, dense materials such as bedrock or dry granular materials will offer greater resistance to passage of electric current employed in resistivity measurements, whereas shock waves will travel faster through dense rock or water.

Both these prospecting methods have been exploited by petroleum geologists for some 50 years for their special needs, but state highway departments have also found them of value in investigating highway cross sections, locating bedrock, and establishing borrow-pit areas. A survey in 1960 showed that, of 47 highway departments replying to a questionnaire, 29 used either or both methods.[12] Thus, a considerable body of information as to the potential use of these methods is being acquired upon which the aggregate prospector can draw.

Both the seismic and resistivity exploration methods require actually traversing the terrain with personnel and some reasonably portable equipment. Thus, before undertaking such a survey, supplementary knowledge should be acquired, such as from study of surface-geology maps, soil surveys, or aerial photography, that there is reason to believe the desired materials exist within the selected area. One of the most fruitful uses of these methods relative to aggregate prospecting is to delineate better the depth and horizontal extent of a known deposit without necessity of extensive boring.

Resistivity Surveys. In the resistivity method, four equally spaced brass electrodes (usually about ¾ in. diameter, 20 in. long) are driven in the ground and electric current is made to flow between the two outer electrodes (Fig. 6). Voltage

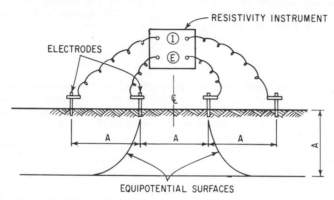

FIG. 6. Schematic diagram of resistivity-survey method.

drop is then measured between the two inner electrodes, from which the resistivity of the underlying material is measured according to the Wenner general equation

$$\rho = 191 \frac{AE}{I}$$

where ρ = resistivity of the underlying layer, ohm-cm
A = depth of layer and distance between electrodes, ft
E = voltage drop across the two inner electrodes, volts
I = current flow between two outer electrodes, amp

It is noted that the depth of material investigated A is numerically equal to the distance between electrodes; and by changing the amount of electrode separation, investigation of the electrical resistivity of the material at different depths can be made. Depending upon the character of the underlying material and sensitivity of the particular instrument, soundings up to depths of 50 to 100 ft are readily obtained.[13]

Skill, experience, and familiarity with the geology of the terrain are recognized necessities for successful predictions from resistivity surveys. For instance, moisture in the ground, such as occurs in early spring, is considered an asset to distinguish between sand, gravel, and loamy sand or silt. In dry seasons, resistivity values of such materials tend to merge closer together. Typical resistivity values for different materials are as follows:

Material	Resistivity, ohm-cm
Clay and silt...........	0–10,000
Sand-clay mixtures......	10,000–50,000
Sand.................	50,000–150,000
Gravel................	150,000–400,000
Rock.................	400,000 plus

Two methods of calculating the resistivity of individual underlying strata are available: (1) the Moore cumulative curve[14] and (2) the Barnes layer method.[15] Both methods derive from the fundamental Wenner equation and, under different subsurface conditions, offer advantages in assigning proper resistivity values to the materials encountered. The amount of calculation necessary for analysis of an extensive resistivity survey makes the digital computer extremely helpful.

Resistivity instruments now in use usually employ a low-frequency alternating current (19 to 97 cps) to help eliminate polarization effects and disturbance from stray 60-cycle currents in the ground. Battery power supply and necessary meters are contained in one portable instrument weighing about 20 lb. The "Michimho" instrument is calibrated to read directly in conductance (reciprocal of resistance, units "mhos") used in the Barnes layer method of interpreting resistivity results.

It is generally agreed that the resistivity survey is not complete until borings have been made at strategic locations to confirm or modify the survey findings.

Seismic Surveys. Seismic exploration employs devices by which the speed of travel of shock waves traveling through the subsurface material can be accurately measured. The shock waves are induced either by striking a sledge hammer on a steel plate resting on the ground or by detonating a small explosive charge lowered in an auger hole. Electrically connected to the sledge hammer or explosive charge is an electronic device capable of measuring intervals of time as small as $\frac{1}{4}$ msec (0.00025 sec). Also connected to the timing instrument is a pickup geophone which also rests on the ground a selected distance away from the impact or detonating point and responds to incoming signals transmitted through the subsurface material. The timing device measures the interval necessary for the induced shock wave to reach the geophone. By progressively moving the impact point farther away from the geophone, a time-distance graph can be plotted. Such a plot is generally neither a straight line nor a smooth curve but is observed to be a series of straight lines of diminishing slope indicating swifter pathways for the shock waves in the deeper and denser materials as distance increases between the impact point and pickup unit. Figure 7 gives a schematic diagram of the pathways of a typical ground profile. Thickness of the layers can be computed by the equation

$$D = \frac{\text{critical distance}}{2} \sqrt{\frac{V_2 - V_1}{V_2 + V_1}}$$

where the critical distance and velocities V are determined as in Fig. 7. In the example, the loam thickness is

$$D = \frac{12.5}{2} \sqrt{\frac{1,785 - 666}{1,785 + 666}}$$
$$= 4.2 \text{ ft}$$

Layer-depth computations are modified when explosives are detonated below ground level, and individual specialists also revise calculation methods to provide predictions better suited to the local terrain.

The first impulse to reach the pickup geophone will be the one whose travel distance and velocity through the different strata combine to produce the least elapsed time. By sounding the terrain on a grid pattern, the depths of strata providing discontinuities can be deduced as well as velocities within the individual strata. With judgment based on geological knowledge, observed velocities can be translated to those characteristic of specific materials such as topsoil, clay, glacial till, sand and gravel, and rock. By appropriate scales placed on the instruments, use of nomographs, etc., much useful information is available in the field as the survey progresses. This is contrasted with resistivity surveys where intelligible data are not readily obtainable until after office computation has been completed.

Portable equipment for conducting seismic surveys is on the market from several manufacturers. Truck-mounted equipment employing as many as **12** geophones simultaneously has also been used. Multichannel recording oscillographs provide a permanent record of the elapsed transit times for the pickup units.

As with resistivity surveys, authorities are agreed that the seismic exploration is not complete until confirmatory borings have been made in selected areas.

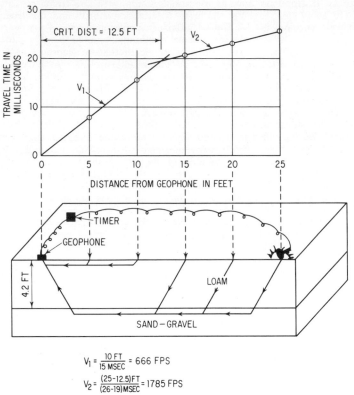

$$V_1 = \frac{10 \text{ FT}}{15 \text{ MSEC}} = 666 \text{ FPS}$$

$$V_2 = \frac{(25-12.5)\text{FT}}{(26-19)\text{MSEC}} = 1785 \text{ FPS}$$

FIG. 7. Schematic diagram of seismic-survey method. (*Adapted from Ref. 16.*)

SAMPLING PROSPECTIVE AGGREGATE SOURCES

After a potential source of concrete aggregate is located, it is necessary to determine if the desired quality and sizes of aggregates are present, or can be produced, so as to qualify the operation as an economical one. Failure to give proper attention to this step is foolhardy since substantial investment may be wasted upon a venture which could have been proved unwise in the first place if the proper investigational techniques had been utilized.

The reasons for sampling an undeveloped deposit are basically twofold: (1) to ensure the prospective operators that the deposit is of sufficient magnitude and contains the required sizes of aggregate or, if not, that the sizes needed can be acquired by crushing, so as to justify investment in processing equipment; and (2) to ascertain if the deposit contains aggregates which, if properly processed, will meet the quality requirements of the major consuming agencies within the market area. Public agencies or authorities will normally have the strictest quality requirements, and it is desirable to invite such agencies to participate in the testing of preliminary samples. The next best step is to submit samples to a testing organization whose qualifications for such work are acceptable to the public agency. Proved ability to furnish aggregate meeting stringent requirements of a respected agency in the area is a distinct business advantage.

Sampling of prospective aggregate sources is conveniently divided into the methods used for sand-gravel deposits and those used for prospective rock-quarrying operations. In either case, provision should be made for carefully identifying and preserving the samples in clean bags or tight boxes. Record should be made of both horizontal and vertical locations at which the samples were taken so that a detailed cross section can later be constructed.

17　Sampling Prospective Sand-Gravel Sources

The requisites of a satisfactory exploratory sample are that regardless of depth from which procured, it truly represents the material in place as to both gradation and quality and is of such size that the testing agency has sufficient material to conduct the required tests. It is wise to check with the testing agency before taking such samples, but generally such individual preliminary samples should weigh at least 200 lb. Comprehensive testing may be desirable at this stage, such as gradation, sulfate soundness, freeze and thaw, organic content, abrasion resistance, and petrographic analysis, thus requiring substantial amounts of aggregate to complete the tests.

The material encountered in a granular deposit may be so unpredictable and variable that the method first chosen for sampling is found unsuitable. Improvisation of equipment or changes in technique as sampling progresses is frequently necessary.

Locally obtained equipment operators may be unaware of the aims of the sampling procedure. Shovel operators, for instance, may be tempted to place the excavated material neatly into a single pile, thus defeating the aims of differentiating the materials found at different depths. A few minutes of prior briefing of such operators into the aims of the program is worthwhile. Instead of ending with a single pile, he may instead have many piles surrounding the excavated area, each representing a different stratum.

Steel-casing Method. Steel-cased holes from 6 to 60 in. diameter have been used in situations where boulders or large cobbles impeding downward progress of the casing are not encountered. Small clamshell buckets, posthole diggers, and soil augers have been used to recover samples as drilling progresses. Sledge-driven and rotary-driven casings have been employed, the latter with teeth cut in the leading edge of the casing. Cased test holes, although expensive, have great advantage in preventing cave-in and, with simultaneous removal of material as the casing sinks, enable procurement of thoroughly representative samples at each elevation. A larger shaft size obviously has great advantage in making it possible to recover more material for examination. The table below gives a rough estimate of the pounds of dry aggregate recoverable per foot of penetration for various sizes of casing.

Casing diam, in.	Estimated lb of dry aggregate per ft of penetration
6	20
10	55
15	130
20	230

When the casing reaches below groundwater level, a special bucket may be employed with rubber valve flaps over each end. The flaps are so arranged as to close when hoisting begins. Other equivalent devices may be employed to remove the material from below water elevation, but loss of fine material from the sand fraction by washing may distort the amount of fine-size fractions found in the subsequently tested samples.[17]

Excavated Test Holes or Trenches. Excavation of test holes is probably the most prevalent method of exploration of undeveloped granular deposits. Clamshell buckets, drag lines, bulldozers, and backhoes have been used when the excavation need not go deeper than about 15 to 20 ft. If the excavation proceeds rapidly and is completed before significant loss of ground moisture by evaporation, then sloughing or caving is inhibited and satisfactory samples at various depths are readily obtained. Again, the equipment operator should be forewarned to spread the material out in piles surrounding the excavation in an agreed-upon pattern for easy identification. If the deposit is on sharply sloping ground or is a narrow esker deposit, a trench or series of trenches may be cut which will intersect a substantial proportion of the intended working area. If the trench can be safely entered by personnel, it may be desirable later to sample directly from undisturbed exposed surfaces rather than from the material which has been excavated. This permits carefully selected samples and accurate representation of the profile. Sampling from intermixed material, particularly such as results from bulldozer operations, may not provide the desired detail in variations of the deposit.

Auger Borings. The limitations of the use of augers in investigating granular deposits usually make them not recommended for the purpose of obtaining representative samples. Cave-in of the sides of the bored hole is bound to occur in incoherent materials, thus preventing clear distinction of the material encountered

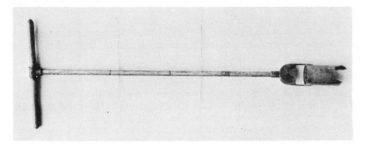

Fig. 8. Blade-type hand soil auger.

at a given depth of penetration. However, their usefulness in determining the general nature of the overall profile is unquestioned, and a few auger holes made in a potential site may establish that more formal sampling is not warranted.

Auger boring may then be considered an intermediate step between true exploration and sampling for laboratory testing.

Hand-driven blade augers are available in 2- to 12-in.-diameter sizes (Fig. 8). By rotating the auger handle, the head section penetrates the ground and must be withdrawn when filled and is then dumped. Penetration thus progresses in steps. Extension shafts, usually standard pipe 3 to 4 ft long with couplings, are threaded on, as desired, to dig deeper. When the water table is reached, only the most coherent material will remain in the auger head when it is withdrawn from the hole for dumping. Generally, digging to depths over about 15 ft is not attempted by hand means. Encountering large stones or cobbles precludes further penetration in any event.

Small, portable 2- to 9-hp gasoline-driven augers are available in diameters from 2 to 30 in. and with shaft extensions have penetrated to depths as much as 30 ft. Short, spiral cutting heads are usually employed. Penetration progresses in steps. When rotation has filled the spirals of the cutting head, it is quickly withdrawn and adhering material dislodged, and another bite taken. Disturbance of material on the sides of the hole during this operation precludes precise sampling at different depths. However, useful information may derive from later examining the sides of the hole with mirror and flashlight or light on an extension cord. Again, encountering cobbles or boulders may prevent further digging at any time.

Truck-mounted boring rigs are generally available from agencies performing soils exploration. Extendable, continuous-flight helical augers (Fig. 9) are most often used for sand and gravel exploration. Diameters range from 2 to 12 in. with extension bits from 3 to 10 ft long. Spindle (Kelly-bar) travel is usually 6 to 14 ft. The usual method of digging is to load the spiral by slow rotation, withdraw the head slightly, and then spin the spiral rapidly to bring up the excavated material and then take another bite. A skillful operator can sometimes "tease" his way around a fairly good-sized cobble or boulder by progressively "spinning" and raising or lowering the cutting head, thus forming a void into which the cobble can be pushed by the auger head. The excavated material is deposited in a concentric ring around the hole, and by "spinning" the spiral free and examining the material,

Fig. 9. Continuous-flight helical auger, truck-mounted.

a fairly good idea of the material encountered at that depth can be obtained. Because of possible contamination, however, the propriety of the material for formal testing can be questioned. Holes up to 100-ft depth have been reported excavated by continuous-flight augers although 50-ft depths are more usual. Hollow-stem augers are also available so that an undisturbed soil sample can be obtained through the center. These seem to be most appropriate for sampling fine-grained soils, but 12-in. augers are made with a 6-in.-ID sampling stem which can be utilized to take granular samples.

18 Sampling Prospective Quarry Sites

Quarries which have been once abandoned or exposed rock in mountainous regions may be potential sources of concrete aggregates, in which case samples should be procured from the exposed faces. Samples should represent the individual ledges if different strata are known to exist. Sheer faces may require sledging out samples by a man in a shovel bucket or "cherry picker" or may require descending from above in a sling. If weathering of the face is indicated, small-sized horizontal or oblique drilling for moderate blasting may be desirable to expose unweathered rock. Desirable features of the sampling are similar to those detailed below for unexposed rock.

Prospective quarry sites which have not been opened up so as to expose the rock faces to allow ready sampling are investigated with diamond or shot drills using hollow-core bits. Such drills utilize a swivel fitting at the top through which water is forced down inside the rotating bit for cooling and to carry out the

Fig. 10. Water swivel for rock-core drilling. Inset—head of diamond drill.

cuttings (Fig. 10). Small, portable gasoline- or electric-motor-driven drills are available for moderate depths and small-diameter diamond drilling, but vehicle or skid-mounted drills are more usual for quarry investigations. Mounted drills have the advantage of greater stability, thus decreasing wobble, and also allow better control of pressure exerted on the bit. Optimum pressure on a diamond bit is important to preserve the diamonds embedded in the cutting edge. Undue chatter or wobble will dislodge the diamonds and quickly destroy the bit. An experienced operator soon learns the operating characteristics of his drill rig and adjusts his technique to match the hardness of rock in order to minimize injury to the equipment. Drilling depths up to 100 ft are not unusual. Semipermanent gas and oil exploration rigs procure cores at much greater depths. Rotational speed of the drills is not high—usually of the order of 100 to 500 rpm.

Shot-core drills up to as large as 16 in. diameter have also been employed in quarry-rock investigations. The cutting head in this instance is plain steel pipe, and abrasive action derives from sifting chilled steel broken shot down the outside of the casing. Water is introduced at the top and forced down inside the casing through a swivel fitting to carry out the cuttings. In this case, the extracted rock-core surface is much rougher than obtained with diamond drills because of the irregular cutting action of the shot. Cutting speeds are much slower than with diamond drills in usual materials, but there is obvious advantage in not running the risk of suddenly spoiling a valuable diamond cutting head. Drilling speeds in average limestone with a 6-in.-diameter diamond drill may be of the order of 1 in. per min, whereas shot drilling will probably proceed at less than half this rate.

Before rock cores are secured, the prospective working area is first laid out on a grid pattern of 100- to 400-ft spacing to locate core holes accurately, and emphasis is first given to taking cores in the area where quarrying operations might first begin, i.e., where hauling and drainage conditions appear favorable. The cores recovered are then carefully labeled as to the depth and location and preserved in special long core boxes having curved indentations. Cores of the order of 2 to 4 ft, or even longer, are sometimes recovered intact, but breakage when encountering a new stratum is a frequent occurrence. An accurately maintained boring log should be kept reflecting the exact depth penetrated even if traversing a layer so soft that no coherent core was obtained. Presence of such a layer might well rule out the quarry as unsuitable for further exploration, particularly if such a layer was of appreciable thickness. On the other hand, if such a stratum is fairly deep, selective quarrying above it may prove profitable.

Choice of the proper size of core to be drilled is a perplexing problem. Small-diameter cores will not yield sufficient material to allow physical testing which may be demanded by consumer agencies to assess properly the quality of the rock. Large-diameter cores provide adequate material for testing but because of the greater expense of procurement do not encourage as extensive coverage of the prospective quarry area. Drilling is therefore sometimes accomplished in two stages: (1) exploratory cores of 1½ to 3 in. diameter are first diamond-drilled primarily for petrographic examination; if such exploration gives promise of producing material of acceptable quantity and quality, then (2) larger cores of 6- to 16-in. size are obtained at carefully selected locations upon which physical tests are conducted.

It is most desirable that a petrographer-geologist be continually present during the coring operation to identify the strata encountered and immediately make a rough plot of the horizontal and vertical position of the layers. Such information is helpful in planning drilling as the work proceeds. Certain strata may be tilted or pinch out entirely over rather short horizontal distances, and the services of a competent geologist at this stage of the investigation can be invaluable.

A word of caution regarding what the geologist can properly provide is here needed. He should not offer his opinions, derived from field observations, as a substitute for laboratory tests of the rock, or be asked to do so. His talents and training will be most effectively utilized in assessing similarities in ledges and in making educated guesses as to probable trends in the quarry layers.

SAMPLING PROCESSED AGGREGATE

It is highly recommended that ASTM Designation: D 75, Standard Methods of Sampling Stone, Slag, Gravel, and Sand for Use as Highway Materials, be studied for general instructions regarding sampling. This method is equally applicable to sampling aggregates for general concrete use as for highway purposes. Too much emphasis cannot be placed upon the necessity for careful and thoughtful sampling if the subsequent tests of the aggregate are to be meaningful.

Sampling should be entrusted only to those possessing the specialized knowledge regarding the amount of material needed for the schedule of tests to be performed and to those fully aware of the false indications which may derive from a sample not taken in such a manner as to be truly representative. It may well be that the sampler needs to exercise judgment superior to that of the personnel performing the more routine tests.

19 Sampling during Production

Sampling and testing aggregates during production has several important advantages: (1) The producer, and often also the consumer, are immediately informed as to the character of the production, i.e., whether it fails the specifications or is borderline or is well within the tolerance limits. (2) Corrective measures can be undertaken at once by the producer before a large quantity of aggregate is produced which may otherwise have to be disposed of or reprocessed, or before unacceptable material contaminates already approved material in a stockpile or bin. (3) Even more importantly, opportunity is provided to obtain periodic representative samples from the production stream in a manner which best characterizes the product.

Whenever possible, samples should be procured from the final conveyor belt or chute by passing a suitable container through the product stream so as to intercept the full cross section briefly. If it is feasible to stop the belt momentarily, an excellent method is to place across the belt two suitably spaced templates shaped to fit the belt convexity and then scrape and brush the entire belt load between the templates into a clean container.

Temptation to take samples which are too small must be resisted at all costs. They must be sufficiently ample in size so that chance inclusion or exclusion in the sample of one or two particles in the larger sizes does not make it appear that the material seriously departs from the grading limits. Coarse-aggregate samples to be examined for grading size, only, should be at least as large as the following:

Max particle size, in.	Min weight of sample, lb
$\frac{3}{8}$	5
$\frac{1}{2}$	10
$\frac{3}{4}$	15
1	25
$1\frac{1}{2}$	35
2	45
$2\frac{1}{2}$	55
3	100
$3\frac{1}{2}$	150

If the sample actually obtained chances to be considerably larger than prescribed above, it can be appropriately reduced in size by quartering or use of a mechanical splitter following techniques of Method for Reducing Field Samples of Aggregate to Testing Size, ASTM C 702. If tests other than grading are contemplated, the sampler should inform himself of the increased amount needed.

Freshly produced moist sand is much less subject to segregation and consequent sampling errors, and samples weighing 5 to 10 lb are adequate for routine testing and can be obtained in practically any convenient manner such as by intercepting the flow off a belt or chute. Again, if a larger schedule of tests is contemplated, larger samples may be needed.

Opinion is divided, and statistical theory is not well developed, as to whether it is better to composite several samples taken at shorter intervals or whether it is better to take one "grab" sample less frequently to represent a certain amount of production. For instance, some agencies require one test per 100 tons of production. If the plant is producing 100 tons per hr, three samples taken at 20-min intervals could be composited and the results be compared with a single grab sample taken sometime during the same hour. Some favor the latter since an examination of, say, 10 samples over a 10-hr day would give indication of the magnitude of the extremes in production whereas the compositing procedure tends to smooth out the results of each hour's production.

Whatever the frequency of sampling chosen, or whether "grab" or composite samples are elected to be taken, both producer and consumer benefit from choosing a rigorous method wherein opportunity for personal choices of the sampler is minimized. Sampling, splitting, and subsequent testing must be strictly devoid of personal whim. Complete randomness of sampling must be preserved wherein every grain of sand or coarse-aggregate particle has equal chance of being found in the selected test sample. This is the goal to be sought.

20   Sampling after Production

Sampling subsequent to production and at the destination has great advantage from the standpoint of best representing the aggregate actually being incorporated in the concrete. For instance, possibility of contamination, segregation of sizes, or generation of fines during handling may be suspected. Aggregates strictly conforming to specifications when produced may be subject to improper stockpiling procedures causing segregation or breakage, or may be contaminated by mud or clay scraped up or tracked onto the pile by bulldozers, or be handled so much as to generate fines. However, the obstacles to procuring thoroughly representative samples after delivery are many. Obtaining truly random samples is almost an insurmountable problem whether the aggregate be in trucks, railroad cars, or a stockpile. Practical considerations of the amount of manual labor involved preclude the probability that interior aggregate in the load or stockpile will be truly randomly sampled unless extraordinary effort is expended. It is almost mandatory that the aid of power equipment such as a front-end loader, power shovel, or clam be enlisted to aid sampling from large masses of aggregate.

Stockpiles.   Coarse-aggregate stockpiles, particularly if coned or tent-shaped, are subject to segregation of sizes unless formed in relatively thin layers. The larger particles tend to roll down to the toe of the slope. Therefore, whenever possible, sampling should be accomplished either as the pile is being formed or during loading out. If neither alternative is feasible, and it is recognized that such is the case in many instances, then some idea of the content of the pile can be obtained in the following manner: Individual shovelsful of coarse aggregate should be selected from the top, middle, and bottom of the pile and composited to form a single sample. Several such composite samples may be needed to represent adequately the material in the exterior surface of the pile. It is advisable to push a board into the pile just above the point of taking each shovelful so as to prevent segregation during sampling. If a fine-aggregate pile is being sampled, the exterior dry sand should first be scraped away before the individual shovelful is taken.

Railroad Cars.   Railroad cars of aggregate are manually sampled by first digging at least three equally spaced trenches across the car with the bottom of each trench at least 1 ft below the surface of the aggregate at the side of the car and the trench at least 1 ft wide at the bottom. Each trench is recommended to be sampled at seven equally spaced points by pushing the shovel downward into the trench without scraping horizontally. It is recommended that two of the seven points be directly against the side of the car.

Trucks, etc.   Trucks should be similarly sampled by trenching as prescribed above for cars, but only one or two trenches may be needed, depending upon the size of the load. Barges and other conveyances may require more trenches to provide adequately representative samples.

21   Identification of Samples

Figure 11 is a reproduction of a form successfully used to identify individual test samples. Persons submitting samples should be instructed to fill out meticulously such a form, or equivalent, to provide proper identification.

```
┌─────────────────────────────────────────────────────────────────────────────┐
│                                        Project _____   │
│                                                                               │
│            SAMPLE IDENTIFICATION       Pur. Order No. _____    │
│                                        Date Sampled _____     │
│          ═══════════════════════════   Lab. No.            Date Rec'd         │
│                                                                               │
│  Name of Material _____  │
│                                                                               │
│  Source _____ Manufacturer _____   │
│                                                                               │
│  Address _____ Address _____   │
│                                                                               │
│  Sampled from _____ Pit Name _____   │
│                                                                               │
│  Give car number and initial, if rail shipment.  If sampled at pit, give 1/4 section, town line and range. │
│                                                                               │
│  Quantity of Material Represented by Sample _____   │
│                                                                               │
│  Consigned to: _____   │
│             If sampled at source, state to whom and where material is to be shipped. │
│                                                                               │
│  Sampled by _____   │
│                       Name              Title                                 │
│  Submitted by _____   │
│                       Name              Title          Address                │
│  Intended Use _____   │
│                                                                               │
│  Specification _____ Sender's Identification _____   │
│  ══════════════════════════════════════════════════════════════════════════ │
│  Remarks _____   │
│  _____   │
│                                                                               │
│  Consign Sample to:  (Give specific address)                                  │
└─────────────────────────────────────────────────────────────────────────────┘
```

FIG. 11.   Suggested form to be filled out to accompany test samples.

AGGREGATE CHARACTERISTICS

Certain tests performed on concrete aggregates are for the purpose of establishing that minimum intrinsic *quality* requirements are fulfilled; others are more related to determining characteristics useful for selecting proportions for concrete, and still others may be a much abbreviated group of tests to assure routinely compliance with the job requirements. The first category includes such basic desirable qualities as toughness, soundness, abrasion resistance, etc., whereas specific gravity and absorption are included in the second category. Clean-cut distinction cannot always be made in a given case. For instance, the absorption value is a necessary item for the job engineer in calculating water-cement ratio of the concrete mix, but it may also, in some cases, reflect pore structure affecting freeze-thaw resistance of concrete in which the aggregate is placed. In most cases, tests applied to aggregates give an *index* to predicted behavior in concrete rather than evaluating a truly basic attribute.

22   Surface Texture

Satisfactory concrete is made containing aggregate of a great variety of surface characteristics ranging from very smooth to very rough and honeycombed. As a consequence, only recently have studies been initiated making rigorous examination of the matter.[18] Table 6 gives surface textures typical of a selected group of aggregates. Some specifications presently limit the amount of "glassy" pieces in slag coarse aggregate to a negligible amount, thus recognizing the poor bond between cement paste and such extremely smooth particles. Further discussion of surface texture is contained in Ref. 20.

Table 6. Surface Texture of Typical Aggregates*

Group	Surface texture	Characteristics	Examples
1	Glassy	Conchoidal fracture	Black flint, vitreous slag
2	Smooth	Water-worn, or smooth due to fracture of laminated or fine-grained rock	Gravels, chert, slate, marble, some rhyolites
3	Granular	Fracture showing more or less uniform rounded grains	Sandstone, oolite
4	Rough	Rough fracture of fine- or medium-grained rock containing no easily visible crystalline constituents	Basalt, felsite, porphyry, limestone
5	Crystalline	Containing easily visible crystalline constituents	Granite, gabbro, gneiss
6	Honeycombed and porous	With visible pores and cavities	Brick, pumice, foamed slag, clinker, expanded clay

* From Ref. 19, reproduced by permission of the British Standards Institution.

23 Aggregate Shape

As is the case of surface texture, satisfactory concrete has been made with aggregate consisting of particles of a great variety of individual shapes. Natural aggregate particles which have been subjected to wave and water action over geologic history may be essentially spherical; others broken by crushing may be cubical or highly angular with sharp corners. Of interest to the concrete technologist is that such changes in shape, without compensating changes in particle size, will be influential in altering the void characteristics of the aggregate. A highly angular coarse aggregate possessing larger void content will demand a greater amount of sand to provide a workable concrete. Conversely, well-rounded coarse aggregate tending toward spherical particles will require less sand. It is interesting to note, however, that concretes made with a great disparity in particle shapes at a given cement content per cubic yard of concrete will frequently have about the same compressive strength. Efforts at placing a numerical scale on particle shape so as to be able to characterize it better other than with the words "rounded," "subangular," "angular," etc., are showing promise. Highly polished, single-sized aluminum spheres have a "particle index" of zero and highly flaky, crushed limestone an index of 20 in a recently proposed test.[21] The latter test involves compacting the aggregate into a special rhombohedron-shaped mold.

Particle shape and texture of fine aggregate have also been measured by an orifice flow test[22] and by other means. Recent work indicates particle shape and surface texture of the fine aggregate may more importantly influence concrete strength than does the coarse aggregate.[23] The sand significantly influences the water required to provide a given concrete slump.

In view of the present uncertainty of the role of aggregate-particle shape in concrete technology, few specifications for coarse aggregate prescribe special requirements except possibly to limit the amount of thin or elongated particles to a maximum of about 10 to 15% by weight, to minimize harsh concrete mixtures whose surfaces may tear during finishing operations. Such particles are defined as those whose ratio of greatest dimension of a circumscribing rectangular prism to the least dimension is greater than 5. Exposed aggregate concrete sometimes uses entirely crushed or entirely rounded particles for pleasing aesthetic effects.

Specifications sometimes inadvertently influence particle shape as the result of controlling quality of gravel by requiring crushing of oversize. Decision may be made that the desired quality occurs only in the large-sized material in a gravel deposit. In this case, confirmation is made of crushing by observing that each particle has at least one fractured face resulting from the crushing process.

Summarizing, particle shape of coarse aggregate in concrete has not proved to be an important problem if increased sand content is chosen so as to compensate for aggregates tending otherwise to make harsh mixes such as can result from use of entirely crushed stone aggregate or blast-furnace slag. Information now being developed indicates particle shape of the fine-aggregate fraction may be more consequential than heretofore thought. Recognized standard tests have not yet been adopted for directly evaluating particle shape of either fine or coarse aggregate.

24   Structural Strength

High-strength concrete cannot be made containing aggregates which are structurally weak. For instance, insulating concrete containing vermiculite aggregate, which is itself a soft and friable material, rarely exceeds 750 psi compressive strength at 28 days, whereas carefully proportioned and cured concrete containing high-strength crushed limestone, crushed traprock, or quartzite gravel can be made to exceed 10,000 psi. Despite the seemingly obvious relation between concrete strength and aggregate strength, at least in extreme cases, other factors such as particle shape, surface texture, grading, and water-cement ratio of the concrete conspire against precise evaluation of the contribution of the structural strength of the aggregate itself. This is the case despite much research effort. For instance, compressive strengths of various rocks are shown in Table 7. The variability

Table 7. Compression Strength of Rocks Commonly Used as Concrete Aggregates*

Type of rock	No. of samples†	Compressive strength, psi		
		Avg‡	After deletion of extremes§	
			Max	Min
Granite............	278	26,200	37,300	16,600
Felsite............	12	47,000	76,300	17,400
Trap..............	59	41,100	54,700	29,200
Limestone........	241	23,000	34,900	13,500
Sandstone........	79	19,000	34,800	6,400
Marble............	34	16,900	35,400	7,400
Quartzite.........	26	36,500	61,300	18,000
Gneiss............	36	21,300	34,100	13,600
Schist............	31	24,600	43,100	13,200

* From Ref. 24.
† For most samples, the compressive strength is an average of 3 to 15 specimens.
‡ Average of all samples.
§ 10 % of all samples tested with highest or lowest values have been deleted as not typical of the material.

of compression values even for similar rocks does not give much encouragement for predicting concrete strengths. Elastic-modulus tests have likewise not been successful in making predictions. Crushing tests of graded aggregates under a piston, designated "attrition" tests, are sometimes prescribed in European specifications.

The test most often used in the United States to assess overall structural quality of coarse aggregate is the Los Angeles abrasion test (Fig. 12). In this procedure a carefully graded and weighed sample of the aggregate is placed in a hollow steel revolving drum along with a charge of steel balls. A shelf inside the rotating drum picks up the charge of balls and aggregate each revolution and drops them as the shelf approaches the high point of its travel. Thus, the aggregate experiences some scrubbing and tumbling action and considerable impact during the specified 500 revolutions of the drum. The aggregate is reexamined after the expiration of the required number of drum revolutions to determine the amount broken down finer than the No. 12 sieve. Except in the case of blast-furnace slag, the test appears to give a useful index of overall structural integrity of the aggregate as evidenced by the fact that so many organizations make use of it in their specifications. Uniquely enough, similar evidence is lacking with respect to the applicability of the test for blast-furnace slag; correlation of slag abrasion with concrete strength is poor or nonexistent.

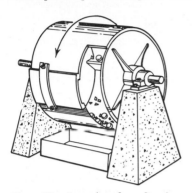

FIG. 12. Los Angeles abrasion machine. (*From T. D. Larson, "Portland Cement and Asphalt Concretes," McGraw-Hill Book Company, New York, 1963, used by permission.*)

Opinion is prevalent that the flexural strength of concrete used in design of pavements is importantly influenced by the structural quality of the coarse aggregate. No single routine test, or group of tests, is presently capable of reliably predicting development of high or low flexural strengths, and there is apparently no substitute for actually making trial batches of concrete from which flexural-strength specimens are tested.

25   Specific Gravity and Absorption

Specific gravity expresses the weight of an aggregate particle relative to that of an equal volume of water. As an example, granite was listed in Table 4 as having an average specific gravity of 2.65. Thus, since water weighs 62.4 pcf, a cubic foot of solid granite weighs approximately $2.65 \times 62.4 = 165$ lb. However, all aggregates are porous to a degree, allowing entrance of water into the pore or capillary spaces when placed in the concrete mixture or, as is more usual, they are already wet when entering the concrete. Careful definition of specific gravity must therefore properly account for both the weight and volume of the portion of water contained *within* the particles (see Fig. 13). Free water on the exterior surfaces of wet aggregate does not enter into computation of specific gravity but does contribute to water-cement ratio of the concrete. It is also noted that specific gravity by its definition disregards void spaces *between* aggregate particles, and average granite, for instance, will have

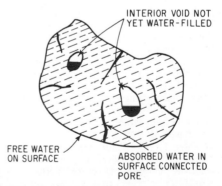

INTERIOR VOID NOT YET WATER-FILLED

FREE WATER ON SURFACE

ABSORBED WATER IN SURFACE CONNECTED PORE

FIG. 13.   Moist-aggregate particle showing distribution of exterior and interior water.

the same specific gravity regardless of whether it consists of angular material quarried from rock or of rounded particles from a gravel deposit. The same is true for absorption, and both specific gravity and absorption are therefore basic properties of the rock and are not influenced by the method of processing.

Absorption of an aggregate is arbitrarily expressed in terms of the water that enters the pores or capillaries during a soaking period of 24 hr and is calculated on the basis of the weight of the ovendry aggregate as follows:

$$\text{Absorption, \%} = \frac{B - A}{A} \times 100$$

where A = weight, g of ovendry sample in air
B = weight, g of saturated surface-dry sample in air
Average absorption values of rock types were shown in Table 4. It is observed that some of the softer, more porous sedimentary rocks typically have higher absorption values. Aggregates whose absorption of water has just been satisfied but whose surfaces are not visibly wet are said to be in a "saturated surface-dry" condition, frequently abbreviated "SSD."

Different agencies are not unanimous as to whether the quantities of ingredients for a concrete batch are most advantageously computed in terms of ovendry or saturated surface-dry aggregates, which leads, in part, to two definitions of specific gravity, i.e., "bulk specific gravity" or "bulk specific gravity (saturated surface-dry basis)." The differences can be exemplified by considering a coarse aggregate which has been soaked in water and then brought to a surface-dry condition, then weighed in air and again weighed when suspended in water:

$$\text{Bulk sp gr} = \frac{A}{B - C}$$

and

$$\text{Bulk sp gr}_{\text{SSD}} = \frac{B}{B - C}$$

where A = weight, g of ovendry sample in air
B = weight, g of saturated surface-dry sample in air
C = weight, g of saturated sample in water
The buoyant force $B - C$ in the denominator is the same in both cases and in the metric system can be considered equivalent to the volume of the aggregate in cubic centimeters as if each particle were enclosed in an infinitely thin sheath surrounding the bulk volume including the penetrable voids. It follows that, if the absorption value of the aggregate is known, bulk specific gravity (saturated surface-dry basis) can be calculated from the bulk specific gravity (sometimes the latter is referred to as "bulk specific gravity, dry basis" to reduce confusion) by the following relation:

$$\text{Bulk sp gr}_{\text{SSD}} = \text{bulk sp gr } (1 + \text{absorption})$$

where absorption is expressed as a decimal fraction. Thus, average dolomite is listed in Table 4 as having a bulk specific gravity of 2.70 and absorption value of 1.1%, from which it is calculated that its bulk specific gravity (saturated surface-dry basis) is $2.70(1 + 0.011) = 2.73$. The above equation can, of course, be used to solve for the bulk specific gravity by dividing the bulk specific gravity (SDD basis) by $(1 + \text{absorption})$, where again the absorption is expressed as a decimal fraction. Chapter 11 gives further details regarding use of specific gravity and absorption in concrete-mix computations.

Laboratory procedures for determining specific gravity and absorption are contained in Specific Gravity and Absorption of Coarse Aggregate, ASTM Designation: C 127, and Specific Gravity and Absorption of Fine Aggregate, ASTM Designation: C 128.

Specific gravity is *not necessarily related* to aggregate behavior. However, reference to Table 4 indicates three individual rock types, shale, sandstone, and chert, which may display poor performance in concrete, particularly in exposed concrete in northern climates. They are observed in the table to be of somewhat lower specific gravity than the others. It is for this reason that heavy-media beneficiation plants which sort materials on a specific-gravity basis are proving successful in many areas. An important exception to the rule that low-gravity materials are necessarily suspected is the case of air-cooled blast-furnace slag, which may have a specific gravity as low as 2.2 because of the high porosity of individual particles. Although not technically classified as a lightweight aggregate, concrete made containing such slag will average roughly 10 pcf lower than that containing natural aggregates.

26 Voids and Grading

Voids. The amount of space *between* aggregate particles, or voids, will be importantly influenced by the amount of compaction, aggregate shape, and surface texture, and by the amounts of the respective particle sizes, i.e., the grading of the aggregate. A graded aggregate is one which contains appropriate amounts of the progressively finer-size particles to fill in the apertures between the larger sizes and thus reduce the void content. However, excellently graded aggregate such as to provide minimum voids has not been found fundamental to acceptable concrete. In fact, "gap-graded" aggregates deficient in one or more sieve sizes have been successfully used and are even encouraged by some.

Voids in aggregates can be determined from the relation

$$\text{Voids, \%} = \left(1 - \frac{E}{D \times 62.4}\right) \times 100$$

where D = bulk specific gravity (see Art. 25)

E = weight of the aggregate, pcf

The value of E will depend upon the compactive effort expended to consolidate the aggregate. The "dry loose" and "dry rodded" are two standard conditions frequently used in concrete technology and are defined in Standard Method of Test for Unit Weight of Aggregate, ASTM Designation: C 29. Void contents of typical concrete aggregates will range from about 30 to 50%. Mixtures of coarse and fine aggregate will provide lower void content than either constituent measured separately because of intermingling of the sizes.

Grading. Concrete aggregates after excavation or quarrying are almost always subjected to a screening process to provide the proper sizes. Confirmation that the desired sizes are present in the product is made by the "mechanical analysis" or sieving test wherein a weighed sample of the aggregate is placed in the top of a set of nested testing sieves with progressively smaller openings from top to bottom. The nested set is then shaken by hand or by a mechanical shaker until practical refusal, i.e., when continued agitation causes essentially no more particles to pass through the respective sieves. The individual size fractions are then weighed and computation is made of the percentages retained or passing. ASTM Method C 136 gives details of the testing procedure.

Tables 8 and 9 display data from typical fine- and coarse-aggregate mechanical analyses, respectively. Computation has been made in the fourth column of each table of the *fineness modulus* of the aggregate, a number found convenient to characterize the overall "coarseness" or "fineness" of the aggregate. Definition of the fineness modulus requires that the sum of the cumulative percentages retained on a definitely specified set of sieves be determined and the result be divided by 100. The sieves specified to be used in determining fineness modulus (and no other) are No. 100, No. 50, No. 30, No. 16, No. 8, No. 4, ⅜-in., ¾-in., 1½-in., 3-in., and 6-in. Note in Table 9 that the sieves not in this particular series have been omitted in calculating fineness modulus. Fineness modulus of fine aggregate is often used as a uniformity requirement wherein successive shipments are required to not deviate from a base fineness modulus by more than 0.2. This

Table 8. Mechanical (Sieve) Analysis of a Concrete Fine Aggregate

Sieve	Fraction retained, g	Fraction retained, %	Cumulative retained, %	Cumulative passing, %	ASTM Specification C 33 grading requirements, % passing
⅜-in................	0	0	0	100	100
No. 4 (4.75-mm).......	22	4.0	4.0	96	95–100
No. 8 (2.36-mm).......	65	12.0	16.0	84	80–100
No. 16 (1.18-mm)......	103	19.0	35.0	65	50–85
No. 30 (600 μm).......	119	21.9	56.9	43	25–60
No. 50 (300 μm).......	157	29.0	85.9	14	10–30
No. 100 (150 μm)......	60	11.1	97.0	3.0	2–10
Pan.................	16	3.0			
Total..............	542	100.0	294.8		
Fineness modulus......			$\dfrac{294.8}{100} = 2.95$		

ensures that variations in grading will not be so large as to require change of concrete-mix proportions. It is often helpful to remember that coarser, larger-sized aggregate will have a larger numerical value of fineness modulus.

The last column in Tables 8 and 9 includes the grading limits given by ASTM Specification C 33 for concrete sand and for size 57 coarse aggregate. Coarse-aggregate grading requirements for most classes needed for concrete are shown in Table 10. The latter follow those issued by Simplified Practice Recommendation (SPR) R 163-48 of the U.S. Department of Commerce. Although specifications of local

Table 9. Mechanical (Sieve) Analysis of a Concrete Coarse Aggregate, Size 57

Sieve	Fraction retained, lb	Fraction retained, %	Cumulative retained, %	Cumulative passing, %	ASTM Specification C 33 grading requirements, size 57, % passing
1½ in...............	0	0	0	100	100
1 in................	1.2	4	...	96	95–100
¾ in................	9.3	30	34		
½ in................	6.8	22	...	44	25–60
⅜ in................	4.3	14	70		
No. 4 (4.75-mm)......	8.4	27	97	3	0–10
No. 8 (2.36-mm)......	0.6	2	99	1.0	0–5
No. 16 (1.18-mm).....	0.3	1	100		
No. 30 (600 μm)......	0	.....	100		
No. 50 (300 μm)......	0	.....	100		
No. 100 (150 μm).....	0	.....	100		
Pan.................	0				
Total..............	30.9	100.0	700		
Fineness modulus.....			$\dfrac{700}{100} = 7.00$		

agencies sometimes differ slightly from the SPR gradings, it is debatable whether such deviations are really justified. Concerted efforts are being made to encourage standardization to these gradings so as to reduce the necessity for producers to stock a multiplicity of sizes which differ only in insignificant details.

It is sometimes advantageous to plot aggregate gradings on special paper made for this purpose to enable rapid visualization of the grading. The coarse aggregate and sand in Tables 8 and 9 have been so plotted in Fig. 14, with shaded areas showing the allowable limits of grading for the respective aggregates as specified in ASTM C 33. Such plotting of gradings often reveals trends difficult to estimate from tabulated data. For instance, the plot clearly reveals that the coarse aggregate in Table 9 is very close to failing the intent of size 57 in the amount passing the ⅜-in. sieve since the actual grading curve almost falls outside the shaded area.

The horizontal spacings in Fig. 14 are proportional to the logarithm of the sieve opening. Paper is also available wherein the spacings are proportional to

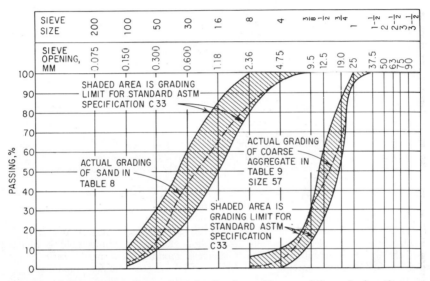

Fig. 14. Aggregate gradings plotted. Sand grading on left and size 57 coarse aggregate on right (data from Tables 8 and 9).

the sieve opening raised to the 0.45 power. The latter paper is often advantageous in plotting aggregate gradings for dense-graded aggregates for road bases or bituminous mixtures. Such gradings plotted on this paper will often approximate a straight line, making visual estimates of deviations in grading very easy.

A small amount of oversize must be allowed for in commercial production, and it is noted that the *maximum size* designated for the particular aggregate in the second column of Table 10 is always one size smaller than that through which 100% of the material is required to pass. Such a definition of maximum size becomes consequential in selecting proportions for concrete consistent with mixing-water requirements, form dimensions, and reinforcing-bar spacing, as further detailed in Chap. 11. Maximum size may also be a consequential item as to whether or not a particular granular deposit is really an economical one to operate. If major projects in the market area demand 1½-in. maximum-size aggregate of which the deposit is found deficient, then expensive importation of suitable larger sizes may be mandatory. On the other hand, if the deposit contains many good-quality cobbles and boulders, crushing down to the appropriate maximum size may entail additional expense. The latter eventuality may be offset by the sweetening effect of the larger sizes to remedy a deposit of poorer quality in the small sizes.

Blending of aggregates is undertaken for a variety of purposes, for instance, to "sweeten" an aggregate with that of better quality so as to make the combined aggregate acceptable, or to remedy deficiencies in grading. If interest centers on the fineness modulus of the blend or its resultant grading, both can be calculated

Table 10. Grading Requirements for Coarse Aggregates
(Simplified Practice Recommendations)

Size No.	Nominal size (sieves with square openings)	Amounts finer than each laboratory sieve (square openings), % by weight												
		4 in.	3½ in.	3 in.	2½ in.	2 in.	1½ in.	1 in.	¾ in.	½ in.	⅜ in.	No. 4 (4.75-mm)	No. 8 (2.36-mm)	No. 16 (1.18-mm)
1	3½ to 1½ in.	100	90–100	...	25–60	......	0–15	......	0–5					
2	2½ to 1½ in.	...	......	100	90–100	35–70	0–15	......	0–5					
357	2 in. to No. 4	...	......	...	100	95–100	......	35–70	......	10–30	......	0–5		
467	1½ in. to No. 4	...	......	...	......	100	95–100	......	35–70	......	10–30	0–5		
57	1 in. to No. 4	...	......	...	......	......	100	95–100	......	25–60	......	0–10	0–5	
67	¾ in. to No. 4	...	......	...	......	......	......	100	90–100	......	20–55	0–10	0–5	
7	½ in. to No. 4	...	......	...	......	......	......	......	100	90–100	40–70	0–15	0–5	
8	⅜ in. to No. 8	...	......	...	......	......	......	......	......	100	85–100	10–30	0–10	0–5
3	2 to 1 in.	...	......	...	100	90–100	35–70	0–15	......	0–5				
4	1½ to ¾ in.	...	......	...	......	100	90–100	20–55	0–15	......	0–5			

if the characteristics of the component aggregates are known. If two aggregates, designated as A and B, are mixed together having fineness moduli of FM_A and FM_B, respectively, the resultant blend will have the following fineness modulus:

$$FM_{\text{blend}} = FM_A \times \frac{P_A}{100} + FM_B \times \frac{P_B}{100}$$

where P_A and P_B are the percentages, by weight, of aggregates A and B in the blend. As an example, if the sand in Table 8 (fineness modulus 2.95) is blended with the coarse aggregate in Table 9 (fineness modulus 7.00) in the ratio of 40% sand to 60% coarse aggregate, the blend will have the fineness modulus

$$FM_{\text{blend}} = 2.95 \times 40/100 + 7.00 \times 60/100 = 5.38$$

If it is desired to determine in what proportion to combine materials A and B to achieve a certain fineness modulus of the blend, the amount of material A to be used can be calculated from

$$P_A = \frac{FM_{\text{blend}} - FM_B}{FM_A - FM_B} \times 100$$

Assume it was desired to determine how to combine the previously mentioned coarse and fine aggregates to achieve a blend with fineness modulus of 5.20; then

$$P_A = \frac{5.20 - 2.95}{7.00 - 2.95} \times 100 = 55\% \text{ coarse aggregate}$$

and the amount of fine aggregate, by weight, is $100 - 55 = 45\%$.

A problem often arising is that of determining in what proportion to blend two or more materials to meet a certain grading. For example, consider two hypothetical sands identified below as "fine" and "coarse," respectively. Their individual gradings below are compared with the requirements of concrete sand in ASTM C 33.

Sieve	% passing		
	Fine	Coarse	ASTM C 33 requirement
⅜-in.........	...	100	100
No. 4.........	...	95	95–100
No. 8.........	100	55	80–100
No. 16,,......	98	30	50–85
No. 30.......	75	15	25–00
No. 50.......	40	5	10–30
No. 100.......	15	1	2–10

Inspection of the gradings reveals that both sands seriously depart from the C 33 requirements, and question arises as to whether blending will be a successful remedy. Some prefer to attack the problem by first plotting the individual gradings on paper such as that used in preparing Fig. 14. Geometrical intuition is then used to guess how hard to the right or left it is necessary to pull the two lines to fall within the acceptable grading band.

Another method used is one of trial and error without plotting the data. For example, attention might be first given to the amount passing the No. 50 sieve in the above tabulation since many concrete technicians consider this amount as importantly influencing the workability of the concrete. A 50–50 blend might be first considered so as to provide $(40 + 5)/2 = 22.5\%$ passing, a value well within the 10 to 30% limit. However, it is quickly revealed that this is not an acceptable ratio for the No. 8 sieve since $(100 + 55)/2 = 77.5\%$, which violates the 80 to 100% requirement. A 60–40 ratio could then be tried, providing a little more of the fine sand, yielding the following:

Sieve		% passing	
		60–40 Blend	ASTMC 33 requirement
⅜-in...........	$0.6 \times 100 + 0.4 \times 100 =$	100	100
No. 4..........	$0.6 \times 100 + 0.4 \times 95 \ \ =$	98	95–100
No. 8..........	$0.6 \times 100 + 0.4 \times 55 \ \ =$	82	80–100
No. 16.........	$0.6 \times 98 \ \ + 0.4 \times 30 \ \ =$	71	50–85
No. 30.........	$0.6 \times 75 \ \ + 0.4 \times 15 \ \ =$	51	25–60
No. 50.........	$0.6 \times 40 \ \ + 0.4 \times 5 \ \ \ \ =$	26	10–30
No. 100........	$0.6 \times 15 \ \ + 0.4 \times 1 \ \ \ \ =$	9	2–10

The 60–40 blend successfully meets the C 33 grading requirements, and inspection likewise reveals that only a very little more of the fine sand could be used since the amount passing the No. 100 sieve is already close to the upper limit.

27 Deleterious Substances

Aggregates may contain mineral particles which in some exposure conditions of the concrete undergo excessive volume change, causing rupture of the concrete surface, or they may create sufficient interior stress so as to cause cracking and impair the structural integrity of the concrete. In other environments, these same mineral types may have negligible influence. Wetting and drying or substantial water saturation simultaneous with freezing and thawing will be destructive to some rock types. The latter is particularly true for lightweight, porous cherts, highly argillaceous limestones, and some shales. Water expands about 10% upon freezing and can be highly destructive when contained in aggregate pores unable to accommodate such expansion. Very soft particles such as ocher are detrimental if close to concrete surfaces subjected to abrasion since the thin mortar covering over the particles will be dislodged and the underlying soft fragments worn away, causing pitting of the surface.

Adherent clay coatings on aggregate particles which persist during the concrete-mixing process may impair bond with the cement paste, and specifications limit the amount of the very fine material (passing No. 200 sieve) which is removable by washing. Claylike materials, whether occurring as coatings or dispersed in such rocks as argillaceous limestones, are objectionable since the volume of the rock is then responsive to changes in moisture content. Shrinking and swelling are detrimental to the concrete. Shrinkage cracking of concrete in slabs or large masses or in otherwise restrained members, particularly that caused by some aggregates in the Southwest United States area, has been observed due to rocks of this type. Drying shrinkage of concrete is being considered for standard acceptance test as a means of eliminating such moisture-sensitive rocks for use.

Coal and lignite are detrimental to concrete. Lignite, a brownish-black substance intermediate between peat and coal, is particularly objectionable since it will cause unsightly brown stains on the concrete surface when it disintegrates.

Tables 11 and 12 give the deleterious substances permitted to be present in fine and coarse aggregates provided by Standard Specifications for Concrete Aggregates, ASTM C 33.

Table 11. Limits for Deleterious Substances in Fine Aggregate for Concrete[*]

Item	Max % by weight of total sample
Clay lumps and friable particles	3.0
Material finer than No. 200 (75-μm) sieve:	
Concrete subject to abrasion	3.0[†]
All other concrete	5.0[†]
Coal and lignite:	
Where surface appearance of concrete is of importance	0.5
All other concrete	1.0

[*] From ASTM C 33.

[†] In the case of manufactured sand, if the material finer than the No. 200 sieve consists of the dust of fracture, essentially free from clay or shale, these limits may be increased to 5 and 7%, respectively.

Still another class of rocks react deleteriously with alkalies of the cement, giving rise to the "alkali-aggregate reaction" or "alkali–carbonate rock reaction." Severe and highly destructive expansions in the first category have occurred for concretes in moist environments when the aggregates contain sufficient amounts of opal, chalcedony, tridymite, cristobalite, and certain rhyolites, andesites, or dacites. Freezing and thawing, although sometimes a complicating factor for a given structure, are not necessary to initiate this destructive action. Figure 15 gives locations in the United States known to have structures affected by alkali-aggregate reaction.[25] More recently, a few limestones have also been found which likewise react adversely with the cement alkalies. Those limestones which exhibit destructive reactions have a characteristic texture in which large crystals of dolomite are scattered in a fine grained matrix of calcite and clay.[26]

Assessment of the suitability of an aggregate which demonstrates susceptibility to either the aggregate-alkali reaction or the alkali-carbonate rock reaction for

Table 12. Limits for Deleterious Substances in Coarse Aggregate for Concrete*

Item	Max % by weight of total sample
Clay lumps and friable particles	5.0
Soft particles**	5.0
Chert as an impurity† that will disintegrate in 5 cycles of the soundness test, or 50 cycles of freezing and thawing (0 to 40°F);‡ or that has a specific gravity, saturated surface-dry, of less than 2.35:	
Severe exposure	1.0
Mild exposure	5.0
Material finer than No. 200 sieve	1.0§
Coal and lignite:	
Where surface appearance of concrete is of importance	0.5
All other concrete	1.0

* From ASTM C 33.

** This limitation applies only when softness of individual coarse aggregate particles is critical to performance of the concrete, e.g., in heavy-duty floors or other exposures where surface hardness is especially important.

† These limitations apply only to aggregates in which chert appears as an impurity. They are not applicable to gravels that are predominantly chert. Limitations on soundness must be based on service records in the environment in which they are used.

‡ Disintegration is considered to be actual splitting or breaking as determined by visual examination.

§ In the case of crushed aggregates, if the material finer than the No. 200 sieve consists of dust of fracture, essentially free from clay or shale, percentage may be increased to 1.5.

a given use requires a high degree of engineering judgment and careful examination of past performance under similar exposure. ASTM Specification C 33 gives useful criteria for evaluating potential alkali-aggregate reactivity and criteria for the alkali-carbonate rock reaction.

Positive identification and assignment of the degree of destructiveness of rock types which contribute to freeze-and-thaw destruction of concrete is probably the most troublesome matter presently facing both the aggregate producer and con-

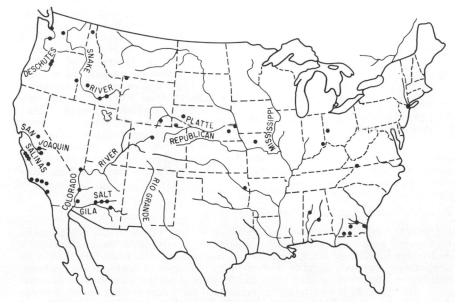

FIG. 15. Locations where alkali-aggregate reaction has been reported.[25]

sumer. Specifications of highway departments, etc., in a given locality frequently limit the percentages of certain rock types—e.g., chert, shale, argillaceous lime-stone—allowed in the aggregate for exposed concrete based upon past performance. It is often necessary to obtain local interpretation of such specifications in order to determine just what the impact on production will be. Efforts to establish criteria and tests of nationwide application have had only mediocre success, but the following have been or are being used: (1) Sulfate Soundness Test, ASTM C 88, (2) Freeze-Thaw Tests in Concrete, ASTM C 666, (3) Alcohol-Water Freeze-Thaw Test of Unconfined Aggregates,[27] and (4) Recommended Practice for Evaluation of Frost Resistance of Coarse Aggregates in Air-Entrained Concrete by Critical Dilation Procedures, ASTM C 682.

28 Organic Impurities

Organic impurities sometimes occur in natural fine aggregates which impair hydration of the cement and consequent strength development of the concrete. Such impurities are normally avoided by proper stripping of the deposit to remove topsoil completely and by vigorous washing of the sand. Problems with high-organic-content sand seem to be largely alleviated by the efficient washing equipment now provided by manufacturers. Detection of high organic content in sand is readily determined by the sodium hydroxide colorimetric test, ASTM C 40. Some impurities in sand may give indication of high organic content but not be really injurious. Assessment of this possibility can be made by ASTM Test C 87.

29 Bulking of Moist Sand

Sand which is moist such as fresh from a washing plant or even after prolonged stockpile storage has a considerably greater gross volume than that in the dry state because of the moisture films surrounding each particle. The thin films of moisture inhibit sliding of the particles upon each other to achieve a compact condition. This effect is the reverse of what the uninitiated tends to suspect would be the case; he mistakenly believes moisture to be an aid to compaction. Figure 16 shows typical volumes of a concrete sand plotted against moisture content. Most concrete sands will demonstrate maximum volume at about 5 to 6% moisture content. The sensitivity of sand volume to small changes in moisture content and resultant uncertainties as to how much sand is actually contained in a given volume of moist sand caused abandonment of volume proportions of job concrete. Likewise, the basis of purchase of sands has changed from a volume basis to weight basis in almost all areas.

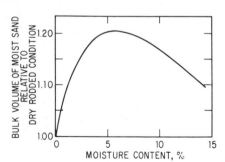

Fig. 16. Loose volume of concrete sand vs. moisture content.

30 Additional Aggregate Properties

Concrete projects sometimes require considerations of conditions not normally encountered such as unusually high- or low-temperature environments, potentially destructive alkalies or acids, or sulfates. Ice-removal salts have been known to contaminate aggregate supplies inadvertently. The role of aggregates in such circumstances may be questioned or it may be desired to find authoritative information. Reference 28 is particularly recommended for comprehensive coverage under such circumstances. Reference 29 may be helpful and gives many additional references.

AGGREGATE PROCESSING

Processing the raw product of a pit or quarry into an aggregate useful for concrete may be quite simple or very involved depending upon several factors. The reasons for choice of specific processing equipment acquired by the deposit operator do not lend themselves to precise engineering analysis because of variations in the deposit, experience of the operator, variations in equipment among different manufacturers, and a host of less definable factors. Economic considerations sometimes lead operators to continue using less efficient equipment simply to amortize their investment. "Package" installations wherein all individual units are assembled into a prearranged pattern for installation at the plant are becoming more frequent for concrete aggregates (Fig. 17). Preassembled portable gravel plants have long been common, for instance, for processing road-base aggregates.

Fig. 17. Rotary-screen "package" aggregate plant with screw classifier for sand. Plant feed (*A*) first goes to scrubber (*B*). Outer rotary screen (*C*) furnishes sand to screw classifier (*D*). Finished sand stored in bin (*E*). Pea gravel screened off to bin (*F*). Inner coarse-aggregate screens (*G*) provide three sizes with oversize stored in bin (*H*). (*Courtesy of Pioneer Engineering.*)

Concrete-aggregate-processing plants are often field-designed and assembled with a great variety of possible solutions. The individual processing operations shown in Fig. 17 are often physically separated some distance or even housed in separate structures. The References contain sources of information on processing equipment for concrete aggregates and manufacturer's catalogs are, of course, invaluable. Particular attention is called to the great wealth of information contained in the annual Pit and Quarry "Handbook and Purchasing Guide for the Non-metallic Mineral Industries," published by Pit and Quarry Publications, Inc., 105 W. Adams, Chicago, Ill. 60603. Reference 30 also provides much useful technical data relative to aggregate production.

31 Plant Layout

Initial plant layout of a pit or quarry operation must be given thorough consideration before embarking on developing a source. Some of the more important considerations are given below, assuming market-potential studies and quality tests of the aggregate have given indication of a successful venture.

1. *Water Supply.* Production of concrete aggregates without aid of water is not unknown, but an ample supply of clean water will undoubtedly be required. For instance, a rough estimate of water required is 1 gpm per cu yd of gravel washed per 10-hr day. A plant processing 2,500 cu yd per day would then require 2,500 gal of water per min. It is reported that some installations may need as much as five times this amount if very dirty aggregates are encountered. Driven wells, lakes, or rivers are usual sources of water. Disposal of the used water into settling ponds is employed and clean water reused after solids have settled if pumped from a source remote from the discharge end. Regardless of the source of the wash water, current emphasis is upon cleaning it before returning it to a stream or lake. The Rivers and Harbors Act of 1899 of the Federal Government and pollution controls of local governmental agencies may require extraordinary measures to ensure return of clean water meeting the regulations for allowable amounts of particulate matter and pH (a measure of acidity or alkalinity).[40]

2. *Power Supply.* Study will have to be made of whether it is more economical to provide electrical power by diesel-driven generators or procure electricity from a commercial source. Estimate of the power needed at the processing plant is hazardous without preparation of a flow sheet detailing the needed equipment. For instance, if heavy-media beneficiation is contemplated, 100- to 300-hp additional load can be anticipated.

3. *Traffic Planning.* Feed to the plant from quarry or pit may well be by rubber-tired trucks. Easy gradients and ample turn-around space should therefore be provided. Belt conveyors feeding the plant are also common and require careful planning of straight-line flights. Loading out of product by railroad car, barge, ship, or truck should likewise take advantage of the existing terrain if possible. Scales for weighing trucked finished product will be required. Provision must be made to enable circling back over the scales after removal of overload to meet public-highway legal-loading restrictions.[31]

4. *Storage Areas.* Storage of finished product demands ample space. Loading bins are rarely of sufficient capacity to provide adequate storage, and resort is made to stockpiling. Reclaiming tunnels in stockpiles are advantageous for subsequent conveyor loading, and stockpiles may be used in lieu of bins. However, in northern climates, blocking of reclaiming tunnels by lumps of frozen aggregate may be very troublesome and may make them impractical except for summer use. Probably the most vexing problem, and one whose magnitude can hardly be foreseen, is suitable disposal of waste sizes from a sand and gravel operation. A large order, or series of orders, may be filled which because of the nature of the deposit leaves certain screen sizes in excess. Beneficiation operations may likewise build up quantities of excess waste material. This waste must be disposed of either temporarily or even permanently if market for the material cannot be found. Some large plants after years of operation have built up waste piles covering several hundred acres. Costly double handling of this material when it encroaches on day-to-day operations should be avoided.

5. Generalizations as to the "ideal" plant layout are not here attempted. Topography, proximity to quarry or pit, and layout of nearby highways, railroads, or water transportation make estimate of such an ideal a fruitless undertaking. Each situation has its own proper and unique solution.

The following sections enumerate the processes used in aggregate production at various plants. Certain processes such as stripping, excavating, crushing oversize, and screening will be used in one form or another at almost all installations, whereas some processing equipment may be highly specialized. The latter may be unique to the geographical area because of mineralogical characteristics of the deposits or may be caused by consumer demand for a special product. Indeed, such consumer demand is forcing rapid developments in the processing industry and premium products are being sought to achieve special durability requirements or pleasing architectural effects. Producers accustomed to a high volume–low unit price market find it difficult to adjust to such special consumer desires. However, only about one-tenth of the cost of much structural concrete in place is attributable to the aggregates, and consumers are willing to make a larger contribution for aggregates if justified.

Ingenious plant personnel sometimes make modifications of commercially manufactured equipment to improve performance for their particular needs. The following sections cannot, of course, treat these matters since they are known only by plant operators. Too, such modifications ultimately are incorporated in the manufactured items if they prove generally applicable.

32   Stripping

Undesired overburden consisting of vegetation, trees, topsoil, clay, etc., must be removed to expose the desired material prior to beginning excavation. The processes do not differ essentially regardless of whether ledge rock or granular deposits are being developed.

Stripping should be accomplished by means that do not require rehandling the material. Power shovels or draglines loading into trucks are often used, with the trucks disposing of the material into the spent portion of the pit or quarry.

Fig. 18.   Gravel-stripping operation using twin-engine scraper. Struck capacity 24 cu yd, heaped capacity 30 cu yd. Average haul distance at this pit 1,500 ft or 3,000 ft round trip. Daily production based on double shifting (16 hr) is 3,500 cu yd. (*Courtesy of American Aggregates Corp.*)

With a small amount of overburden and a deep deposit, this may be only an intermittent operation to just keep ahead of the excavation and may be done with bulldozers. With shallow deposits and relatively more overburden, stripping may be a continuous operation. In some cases, lower costs are claimed for single-operator, push-pull, twin-engine, rubber-tired scrapers of 24 cu yd or more capacity (Fig. 18). These versatile machines can maintain their own haul roads and travel at high speeds under favorable conditions. Figure 19 gives some data on potential scraper production.

The depth of stripping that can be tolerated for an economical operation cannot be stated except to observe that as much as 50 ft has been removed in rare cases to expose desired material. Some gravel plants are able to utilize a portion of the clay overburden in making road-base material and reserve the deeper, clean gravel for manufacture of concrete aggregates.

Crawler-mounted rippers, with or without bulldozer blades, are sometimes used to break up softer rock or frozen ground in stripping operations.

33   Blasting Rock Quarries

Most rock in quarries is initially dislodged by blasting with explosives in order to reduce it to sizes which can be handled for subsequent crushing.

Spacing of drill holes and choice of explosive must be matched with the hardness of the rock. Such choices are made with close liaison between the explosive manufacturer and quarry operator. Rapid developments in drilling techniques and explosives behavior preclude formulating hard-and-fast rules, but the following general observations appear justified:

1. Blasting energy derives only from rapid expansion of *confined* gases. It is for this reason that charges initiated at the bottom of a hole with detonation progressing upward are more effective. Similar advantage derives from using a more powerful explosive at the hole bottom.

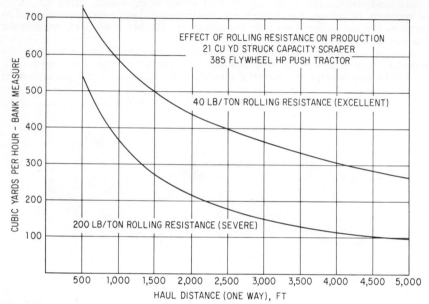

Fig. 19. Effect of "rolling resistance" on scraper production. Rolling resistance of 40 lb per ton is characteristic of hard, smooth roadway surface without penetration under load (concrete or bituminous pavement) and varies up to 200 lb per ton (or more) for soft, muddy, rutted roadway, or in sand. Appropriate efficiency factors should be applied to plotted curves, about 75% for daytime operation or 67% for night operation for rubber-mounted scrapers. (*Courtesy of Caterpillar Tractor Co.*)

2. Smaller-diameter closely spaced holes at a given distance back from the working face, or "burden," are most effective, but staggered rows are also used, with delayed blasting caps giving better fragmentation.

3. Harder rock requires closer spacing of drill holes.

4. Greater face heights enable greater spacing of drill holes. Spacings of about 10 to 30 ft are prevalent.

5. Blast holes are drilled by churn or well drills, percussion drills, rotary drills, and jet-piercing process. The well drill exerts a pile-hammer action on a steel shaft by alternately raising and dropping the shaft. The tip of the shaft is fitted with cutting edges. This technique is, of course, applicable only to vertical holes. Carbide-insert cutting heads are often used on percussion drills and compressed air is employed for removing cuttings. Rotary drills in the softer rocks are frequently used. Percussion and rotary drills can penetrate the rock at any desired angle assuming a suitable drill rig is used. A great variety of drill rigs are available,

usually crawler-mounted with hydraulic-operated booms for accurate control. Hole sizes drilled are of the order of 4 to 8 in. diameter with costs ranging from about $2 to $10 per foot. Jet-piercing drills employ an intense source of heat to shatter and fuse the rock, the heat deriving from a mixture of petroleum fuel and oxygen. Steam is used to aid the shattering process and blow the hole free of rock fragments.

6. A great variety of explosives are available for rock blasting. These explosives vary as to strength, rate of detonation, mode of placement in the hole, and sensitivity. Explosives are delivered in powder and stick form and more recently as a liquid slurry. The latter is delivered by bulk tank truck and placed in the holes by special transfer pump. Detonation is usually achieved with electric blasting caps, but blasts may also use dynamite with burnable fuses and caps. Manufacturers should be consulted regarding the use of their products and strict attention paid to all prescribed safety precautions. Blasting with nuclear explosives has recently been given serious consideration and, where feasible, has been estimated to cost from about 2 to 9 cents per ton of rock.[32]

In summary, blasting is a complex, specialized problem for the quarry manager and requires decision considering the following variables:

1. Height of face
2. Hardness and uniformity of rock
3. Dip of beds
4. Prevalence of bedding seams and joints
5. Dip of quarry floor
6. Size and depth of drill holes
7. Arrangement and spacing of drill holes
8. Number of holes shot at one time
9. Size of charge
10. Position of charges in drill holes
11. Type of explosive used
12. Method of firing shots
13. Method of loading rock
14. Size of shovel
15. Size of crusher

Comprehensive coverage of rock breaking is given in a series of four articles in Ref. 33.

34   Excavation

Power Shovels. Broken rock blasted from the quarry face is normally picked up by power shovels of from about 1 to 10 cu yd dipper capacity. Electric power shovels appear to be gaining popularity for the larger-capacity dippers and require current furnished through flexible cables. Rock has been reported to have been loaded using as little as 0.22 kwhr per ton with electric shovels. High-voltage alternating current is fed to the shovel and a-c–d-c converters drive d-c motors for the individual operations. With a 2½-yd dipper and 90° swing, rock loading is reported to progress at 200 to 275 cu yd per hr. The larger dippers do not appear to have commensurate increase in loading rate but obviously have greater power and reach for conditions where such is needed.

Sand-gravel deposits with a relatively shallow face are commonly excavated with power shovels. Loading rates are much higher than for rock and the corresponding 2½-yd dipper as mentioned above for rock may load as much as 400 cu yd per hr of the more easily handled gravel material. Higher working faces make shovels impractical because of danger of cave-in which may bury the shovel, and draglines are then used.

Draglines. Draglines (Fig. 20) with ever longer booms and larger scraper buckets are becoming available; 120-ft booms and 15-yd buckets are not uncommon. Some operators find it advantageous to drag *up* the bank, others *down*. Dragging down the bank may be desirable for a completely dry operation, and the operator can achieve blending of the material in the entire bank if such is desired—or he may exercise some selection. If part of the deposit worked is under water, dragging upward is mandatory. There is the attendant possibility of undermining the machine when dragging up the bank.

Dragline distances can be greatly increased by use of slackline techniques with a deadman. In one method, the scraper bucket rides on a sheave, or sheaves, attached to a cable which terminates at a fixed anchor some distance away and downgrade from the machine. To take a bite, the cable is tightened and the bucket rides down the cable by gravity a selected distance. The support cable

FIG. 20. Excavating raw material with walking dragline. 9-cu-yd bucket on 165-ft boom. Complete electric power, 4,100 volts a-c with 115-volt d-c conversion. 750 tons per hr capacity. Total weight of machine 500 tons. (*Courtesy of American Aggregates Corp.*)

is then slackened and the bucket drops down to the deposit. The separate load cable attached directly to the bucket then pulls the bucket toward the machine for loading. Repositioning the deadman is necessary for digging other than a single deep trench, or the machine can be moved radially around the deadman. By use of another cable with a sheave at an elevated deadman, dragging upgrade can be accomplished, or even excavated toward the deadman if desired. Slackline techniques allow a great variety of possibilities for loading or formation of surge piles and can cover considerable distances vertically and horizontally. Horizontal distances of 1,200 ft with a 240-ft-high bank have been reported. Hourly capacity drops rapidly for such distances and a 15-yd bucket is reported to have a production rate of 135 yd per hr at 1,000 ft and 1,350 yd per hr at 100 ft.

Dragline excavation under water is accomplished with perforated scraper buckets to release water from the load as it is brought up.

Hydraulic Excavation. Underwater deposits frequently lend themselves to dredging operations (Fig. 21). The dredging equipment is mounted on a floating barge or ship and either the dredged material is pumped ashore for processing or, in some cases, the same or adjacent barges house the processing machinery. Suction is supplied by centrifugal pumps specially lined for abrasion resistance, usually driven by 200-hp or frequently much larger engines.

The suction line, varying from 8 to 28 in. diameter, is lowered to the bottom by an A-frame arrangement with hoist and winch. In soft, sandy deposits suction alone is often adequate to bring up the material. In harder deposits containing gravel, mechanical means may be advantageous to dislodge the material (Fig. 22). Such mechanical devices employ rotary cutters of various types or chain-type digging ladders with special arrangement at the nozzle to prevent boulders from obstructing the suction line. The large amounts of water pumped simultaneously with the granular material tend to wash away fines desired in concrete sand. Construction of dikes and dewatering the enclosed area have been used with excavation proceeding in the dry.

Maximum digging depth for offshore hydraulic dredges is presently about 150 ft.

Barge-mounted dipper shovels and clamshells are also used to excavate materials underwater.

35 Transportation of Materials

Aggregate production is typically a high-volume business requiring handling large tonnages sometimes over quite a distance even within the plant area. Each cubic yard of aggregate weighs roughly 1½ tons (over 2 tons for solid ledge rock), and search for more economical means of handling these large tonnages is a continuing effort. Plants use a variety of means for conveying materials, ranging from railways, tramways, trucks, barges, and belts to open sand flumes. Others have chosen belts for practically all operations.

Railways. Some pit-and-quarry operations, particularly where the haul to plant exceeds about 1 mile, use diesel-electric locomotives with a variety of car types in use. Additional track is laid as excavation proceeds farther and farther from the plant, and this method seems to provide economical transportation despite necessary track laying and maintenance. Side-dump cars of various designs are most prevalent, enabling rapid discharge of their load at the plant, but bottom-dump cars are also used. Car capacities range up to 50 tons, but much smaller ones are also employed, particularly with narrow-gage track.

Motor Trucks. Motor trucks are certainly the most prevalent means of transporting material from the deposit to the processing plant. Their versatility for various uses is unquestioned. A great variety of sizes and types of dump bodies is available ranging up to 85-ton payloads without semitrailer. Better maneuverability seems to favor the single unit, without trailer, type of vehicle for intraplant hauling. Too, public-highway loading restrictions do not apply, and to the extent that plant haul roads can be maintained in good condition, these seem to provide economical transport. Major emphasis is given to condition of haul roads by many operators in view of vulnerability to injury of the large-sized, expensive rubber tires used on these vehicles. Driver fatigue and general maintenance costs are also reduced by maintaining excellent haul roads. Some plants operate a blade grader almost continuously to ensure smooth surfaces free of loose rocks.

Side-dump bodies are used but end-dump types are more prevalent, probably for reason of real or imagined greater stability.

Belt Conveyors. Materials are transported between operations within a plant almost universally by belt conveyors (Fig. 23), and many operators have also chosen this method for bringing material from the deposit to the plant itself even over considerable distances. A 5½-mile belt haul has been reported.[34] Belt conveyors can conveniently transport materials upward, downward, or on the level

Fig. 21. Excavating raw material with hydraulic dredges. Pumps have 24-in. intake and 20-in. discharge. Powered by 1,000-hp motors. Main pumps have 800 to 1,000 tons per hr capacity for solids. Dredges can excavate 65 ft below water. Discharge lines supported by pontoons sometimes attain length of 2,000 ft. (*Courtesy of American Aggregates Corp.*)

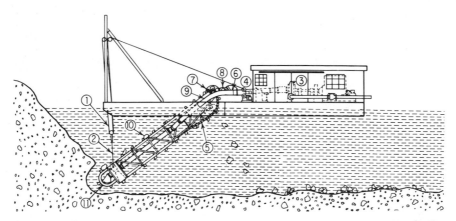

Fig. 22. Schematic view of "ladder" dredge. Front or nozzle end suspended on hoist line (1) and bail (2). Centrifugal dredge pump (3) is independently powered. Traveling-screen chain (5) driven by motor (4) through pulley (6), sprocket (7), and gear train (8). Feed enters nozzle (11) and is carried up through suction line (9). Boulder (10) too large for nozzle is carried up and dropped out of the way behind nozzle. (*Courtesy of Eagle Iron Works.*)

but individual flights do not normally exceed 20° from the horizontal. Excessive angle causes large rocks to roll back down the belt.

Raw feed from a quarry or pit is not recommended for belt conveyors without first removing oversize because of possible injury to the belt. This is often achieved by scalping and preliminary crushing at the deposit close to the point of excavation.

Table 13 gives conservative data on capacities of belt conveyors although different manufacturers vary in their recommendations.

Belts lend themselves to a great variety of applications such as stacking of materials, loading cars, bins, and trucks, and forming surge piles. Belts are sometimes

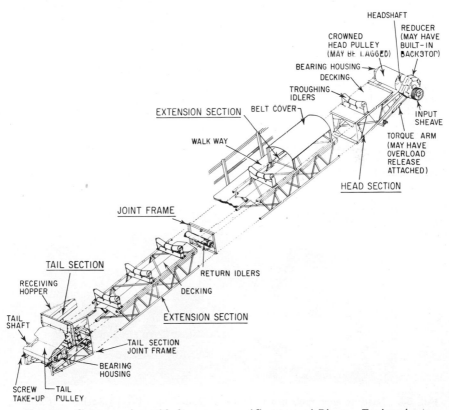

Fig. 23. Cutaway view of belt conveyor. (*Courtesy of Pioneer Engineering.*)

loaded at the center of the flight, and by reversing direction, material is piled at either end. Short flights (up to about 60 ft) are also mounted on wheels for portability and can quickly be transferred around the plant for a variety of loading purposes.

Devices are available for weighing the load traversing a belt in a given time interval using conventional weighing systems with special integrators and more recently one using nuclear-measurement principles has been developed.

A recently published report is not favorable to formation of coned or tent-shaped coarse-aggregate stockpiles because of segregation. Revision of stacker-belt procedures may be required.[35] Obviously, segregation of sizes cannot occur for single-size material and such tendencies are minimized for closely sized materials. Concrete technologists are therefore tending toward use of the latter in preference to the

Table 13. Characteristics of Belt Conveyors Carrying Concrete Aggregates Weighing 100 pcf

Width of belt, in.	Nominal belt speed, fpm	Tons/hr at nominal belt speed	Max aggregate size, in.	
			Uniform size	Mixed with 50% fines
18	300	150	3	5
24	300	300	4	8
30	300	450	7	12
36	300	700	8	15
42	350	1,100	10	18

"long" gradings. Even though handling closely sized aggregate, conveyor stockpiling may induce breakage of large-sized material when dropped several feet. In this case, rock ladders are sometimes used. This device contains steel steps or baffles which are alternately offset so that the rock does not freely fall more than about 4 ft before hitting another baffle, thus reducing breakage.

Front-end Loaders. Crawler and rubber-mounted front-end loaders (Fig. 24) have come into much use recently, particularly for loading out delivery trucks with finished product. Rubber-tired loaders are especially adaptable to circulating

FIG. 24. Loading shot rock using front-end loader with V-type bucket. (*Courtesy of Caterpillar Tractor Co.*)

rapidly among stockpiles for loading a variety of aggregates depending upon the needs of customers. Loaders having 3-yd buckets are popular, and all models have adequate lift height for many uses. They are well suited for loading grizzlies in a deposit and even perform some excavation on softer materials or broken stone. Indeed, they are rapidly displacing some of the time-honored methods of transport because of their increasing dependability, versatility, and speed.

Miscellaneous Modes of Transport. A great variety of other transport is used for aggregates, including waterborne vessels. Barges are used quite extensively in more protected waters and where shallow draft is needed, and self-unloading freighters are employed on the Great Lakes. The latter may have load capacities exceeding 2,000 long tons.

Bulldozers (Fig. 25) perform a variety of materials-handling and maintenance

FIG. 25. Preparing site for drill crew at quarry using track-type bulldozer. (*Courtesy of Caterpillar Tractor Co.*)

duties around aggregate plants, including clearing up stockpile areas and dozing materials toward tunnel elevators or depressed bins.

36 Crushing

Practically all quarried rock, slag, and oversize in a gravel deposit must be reduced to usable sizes for concrete by crushing. Such crushing is accomplished by a variety of commercially constructed machines classified as to type as follows: (1) jaw, (2) gyratory or cone, (3) roll (single, double, and triple), (4) hammer mill, and (5) impact. Diagrammatic sketches of three types are shown in Fig. 26.

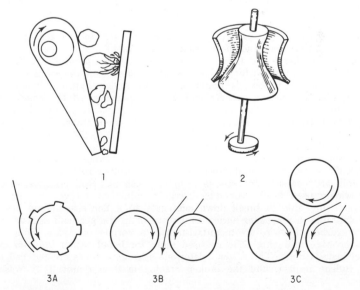

FIG. 26. Schematic sketches of (1) jaw crusher, (2) gyratory or cone crusher, and (3) roll crushers—(3*A*) single roll, (3*B*) double roll, and (3*C*) triple roll.

Choice of the type of crusher appropriate to a given application should consider at least the following:

1. Size and design should be such as to handle the largest-sized rock in the feed.

2. Capacity should be such as to handle the peak rate of feed contemplated.

3. Presence of fines in the feed should not jam and stall the crusher.

4. Wearing parts should have liners and be readily replaceable.

5. Provision should be made for stray iron or other uncrushable material which occasionally finds its way into the feed. Breakdown of the crusher should not result from such materials entering the crusher, but stalling or jamming is unavoidable.

Over the years of development of crushers, much effort has been expended attempting to evaluate the character of the product, i.e., whether it typically shatters into long, splintery pieces or into cubical, blocky fragments. The latter-shaped fragments are more desirable in concrete since the void content will be lower and less cement-sand-water mortar will be required. Opinion is prevalent that crushers depending more upon impact for breakage, such as the hammer-mill or impact types, will provide a more cubical product than those depending upon pure compression such as the jaw or double-roll type. It must be acknowledged, however, that thinly laminated stone, for instance, will tend to break along the laminations irrespective of the type of crusher employed, and superiority of performance of the crusher must be measured relative to the character of the particular stone itself. Gyratory crushers are considered intermediate in their tendency to impart pure compression and impact and perhaps for this reason, and for their compactness, have become very popular in the aggregate-production industry.

Jaw crushers are widely used in primary crushing of ledge rock particularly because of the large jaw openings available, up to 66 by 86 in. Replaceable special steel jaw liners, curved, smooth, or corrugated, are used and the different model crushers impart somewhat differing types of movement to the jaws. One popular model has an eccentric-mounted jaw such that movement at the top opening is quite large with relatively little movement at the bottom discharge. Jaw crushers invariably use a large flywheel to store energy for succeeding "bites." There is some opinion, if excessive reduction in one stage of crushing is not attempted, that jaw crushers tend to generate fewer fines than other types of crushers.

Gyratory crushers consist of a vertical shaft carrying a cone or "mantle." The cone does not rotate but is given a gyrating motion by an eccentric bearing arrangement on the lower end of the shaft. Movement of the cone is thus largest at the bottom discharge end (large end of cone) and minimum at the top. Gyrations of the cone occur within a bowl which is also cone-shaped and the space between the cone and bowl is therefore always wedge-shaped. Crushing action is quite similar to the jaw crusher except that curvature of the cone and bowl introduces some bending and shearing stresses to the stone. Stone is introduced through two semicircular openings in the top of the crusher which, because of the shaft housing and frame supports, do not accommodate stone larger than 72-in. size in the largest model now available.

Roll crushers are available in a great variety of types varying from single- to triple-roll types. Single-roll crushers with a studded roll, sometimes called a slugger roll, are occasionally used as primary crushers. Roll crushers, in general, are not widely used in the concrete-aggregate industry, however, but are advantageous in being able to break down aggregate to a fine size in one pass. This may be desirable in processing agricultural stone or in the portland-cement industry.

Impact crushers likewise are manufactured in a variety of styles. Some models, such as the cage mill (Fig. 27), are used more for beneficiation than for crushing in gravel installations. The softer deleterious aggregate types are reduced to fines by the violent impact, and the broken residue is then removed by washing or screening.

Crushing of aggregate for concrete is almost always done stepwise; that is, following primary crushing the undersize is removed and the oversize recirculated to a secondary crusher. Size reduction in a single crusher does not normally exceed about 1:4 for minimum production of fines.

Theories regarding crushing are extensive since it is of such importance in the field of mining-ore reduction as well as aggregate production. Rittinger, Kick,

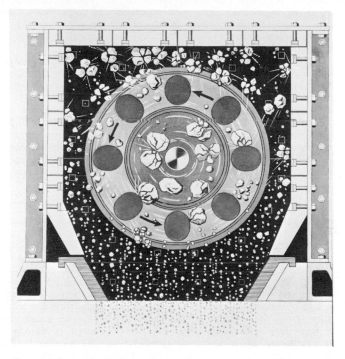

FIG. 27. Cage mill used for crushing and beneficiation. Aggregate is fed into center of spinning cage. (*Courtesy of Stedman Foundry and Machine Co., Inc., Aurora, Ind.*)

Bond, Charles, and others have pursued the matter. One general equation expressing the energy required is

$$dE = c\,\frac{dx}{x^n}$$

where dE = infinitesimal energy change
c = a constant
dx = infinitesimal size change
x = particle size
n = a constant

This equation proposes that the energy for a small change in particle size is inversely proportional to the particle size raised to some power n and directly proportional to the particle-size change. Reference 36 recently extends this theory to accommodate better the character of the rock itself.

37 Screening

As discussed in Art. 26, aggregates for use in concrete are furnished in particle sizes appropriate to the use of the concrete. This requires screening in commercial production to provide the necessary sizes.

Except for initial sand removal from bank material, the term "screening" in the aggregate industry applies exclusively to processing the coarser sizes, i.e., larger than about No. 10 (2-mm) sieve. Separation of the larger aggregate particle sizes is accomplished by various types of screens having square, round, diamond, or rectangular slotted openings. Sand-size materials are almost entirely processed by hydraulic classifying means since size separation of these finer materials by this method is so greatly facilitated. Sizing of the finer materials by sieves is confined to laboratory testing.

A few general comments apply to all screening operations:

1. Every particle traversing the screen must be given several chances to drop through if it is sufficiently small to do so. Irregularly shaped fragments may require several opportunities before their least dimension is so oriented as to allow dropping through.

2. It then follows that too steeply sloped screens or those given too violent agitation will cause the particles to hop rather than drop through, particularly if the screen is underloaded. Rolling action, not hopping, is desired.

3. If the screen is overloaded and agitation is insufficient, it may become "blinded" with particles slightly too large to pass through, thus blocking the openings.

4. Rounded gravel particles are easier to screen than angular crushed stone or slag.

5. Production rate in tons per hour is greater for screening materials having a considerable disparity in particle sizes. It is much easier to screen a material 90% of which passes the screen than one which is so closely sized that only 10% passes.

6. Wet materials screen more easily than dry material, particularly for the smaller gravel sizes. Dry screening is frequently used, however, in the crushed-stone and slag industries.

It is apparent from the above that efficient screening involves a host of variables. Manufacturers' literature is helpful in assessing screening capacity of their installations, but operators have found that there is no substitute for trial-and-error experimentation with their own materials and the particular sizing desired.

Grizzlies. Raw feed excavated from the pit or quarry is usually fed to a grizzly, which consists of a series of parallel bars, the spacing between which is adjusted to accommodate the largest rock which will be fed directly to the plant for further processing. The retained portion too large to pass the bars is fed to crushers to bring it down to usable sizes. The grizzly bars may be stationary or vibrating as the nature of the feed dictates. If the feed readily flows without sticking or hanging up on the bars, then vibration is not needed. Vibrating, sloped grizzlies act similarly to commercially manufactured mechanical *feeders* in that they level the material down and distribute it to a belt, crusher, or screen at a uniform rate.

Some gravel-pit grizzlies are job-made and simply consist of a hopper which is topped with an open horizontal grid of railroad-track iron to catch oversize material or clods (see Fig. 20). The latter are sledged down as necessary or wasted if only a few oversize boulders occur.

Sand-gravel operators sometimes feed directly by truck or shovel on a stationary grizzly opening and the material builds up, thus serving as a surge pile. Belt conveyors, trucks, etc., carry the material away at the bottom of the sloped bin underneath. Care must be exercised, of course, that the grizzly screen effectively prevents blocking at the bin discharge.

Commercially constructed vibrating grizzlies are available up to 6 ft wide by 20 ft long, and little trouble is encountered in obtaining sufficient tonnage capacity to match subsequent processing operations.

Revolving Screens. Rotating or revolving screens consist of wire or perforated-plate open-end cylinders revolving on a slightly inclined axis. Material is fed into the high end of the cylinder and if sufficiently small drops through the openings in the downhill tumbling action imparted by the rotating screen. Particles too large to pass the openings drop out the end of the cylinder into a chute, and oversize is crushed down and recirculated through the screens. By placing three cylindrical screens, for instance, concentrically, with successively smaller sieve openings in the outer cylinders, four sizes of materials can be obtained, i.e., oversize for the inner screen, that retained by each of the two outer screens, and that passing through the outer screen. With dry materials subject to abrasion, this may be a dusty operation, and water spray is introduced to remedy the situation. In other installations where the coarse aggregate has adherent clay coatings, larger volumes of spray water fed by nozzles inside the inner screen are introduced, and the screening operation acts simultaneously as a scrubber. In Fig. 17 a separate scrubber is shown used ahead of screening.

Fig. 28. Aggregate washing and flat-deck screening plant. Rotary scrubber at top followed by screening. Oversize fed to crusher at extreme left and then returned in closed circuit to main feed via belt conveyor. Flexibility of operation can be introduced at points (A), (B), and (C). For instance, coarser sand can be chuted to the screw at (B) and then blended with some or all of the finer sand, or it can be diverted to a separate screw as shown by the chute arrangement at (A). Any or all of the fine sand can be produced or wasted as desired. If only a small amount of fine sand is required, it can be produced by taking out the coarser sand at (C) and diverting the coarser sand to one of the bins as shown. If only a small amount of coarse sand is required to be blended with fine sand, the sliding gate at (C) can be used to meter out the required amount. (*Courtesy of Pioneer Engineering.*)

Rotating screens may be used at various points in the flow of materials through the plant, for instance, in sand-gravel operations sometimes immediately following removal of sand by vibrating screens and in quarry operation sometimes immediately ahead of primary crushing to remove oversize passed by the grizzly. In Fig. 17, however, initial separation of fine and coarse aggregate is accomplished by rotary screening.

Rotary screens are ordinarily used up to about 4 to 5 ft diameter and 16 to 20 ft long.

Deck Screens. Flat, vibrating screens are widely used either singly or stacked with up to three decks, for provision of closely sized materials (Fig. 28). Different manufacturers impart vibratory action to the screens by various eccentric or unbalanced weight drives. Some drives are so arranged as to induce a forward flow of material across the screen even when it is not sloped. Most screens are sloped a few degrees (about 20° is common) to provide gravity flow. Slopes up to 45° are used in rare cases for fine materials. The screens are steel-spring- or air-cushion-mounted to absorb vibration. A variety of screening areas are available for single screens ranging up to about 6 ft wide by 20 ft long. Likewise, the screening surface may be wire, bars, or perforated plate. Most wire screens require backup stiffeners to keep them from depressing from the superimposed load, and a 2-in. opening supporting perforated plate is also prescribed for the more fragile wire screens.

Computation of the tons per hour that a deck screen can handle presents uncertainties which are difficult to assess. One manufacturer suggests computation be based upon a certain "basic" capacity in tons per hour per square foot of screen to which a "correction factor" is applied whose value depends upon the product of four coefficients involving (1) number of screen decks above screen considered, (2) percentage of feed smaller than one-half screen opening. (3) presence or absence of water spray directly on screen, and (4) percentage of oversize fed to screen. The values of the individual coefficients are influenced in the following manner:

1. Efficiency of the lower decks is somewhat reduced from that of the top screen.

2. Greater percentages of feed whose size is less than one-half screen opening increase screening capability by a factor of 3 when percentage of small size increases from 10 to about 60%.

3. Water spray on screens increases screening capability almost three times as much for a $\frac{3}{16}$-in.-opening screen as for a 1-in.-opening screen.

4. Screening capacity is about halved for 90% oversize as compared with 20% oversize.

38   Washing and Sand Production

Most, but not all, aggregate production is carried on utilizing generous amounts of water. The purpose of the water may be for any one or a combination of the following: (1) to remove undesired adherent coatings such as clay, silt, or dust of fracture; (2) to lubricate the stone sizes to hasten screening operations; (3) to reduce dust nuisance created by dry screening and/or crushing or other handling; and (4) to accomplish sizing, or water "classification" of sand.

Some crushing and screening operations of slag or extremely hard stone such as quartzite are carried on entirely dry except for slight fogging to control dust. In such cases, proprietary wetting agents are sometimes added to the water for greater effectiveness. Such minimum amounts of treated water have been found desirable, particularly when making manufactured sand by crushing, in that balling or clumping is inhibited when roll-mill or pan-mill crushing is employed.

Washing of aggregates may consist of thorough drenching with high-pressure water by means of spray bars equipped with nozzles suspended over deck screens or inside rotary screens or may involve complete inundation with vigorous scrubbing by means of log washers or scrubbers. Stickiness of the coatings needed to be removed and other characteristics of the aggregates will dictate which method of washing is most advantageous. Additionally, many sand-gravel operations use both methods, for instance, spray-bar washing during preliminary screening followed later by log-washer scrubbing for only the coarse aggregate.

Log Washers and Scrubbers. Scrubbers illustrated in the top of Fig. 28 are used to wash aggregates when the nature of the adherent clay coatings is such that the more vigorous treatment of log washers is not needed. Scrubbers consist of cylindrical containers rotating on a horizontal or slightly inclined axis into which the aggregate is fed at one end along with large volumes of water. Lifters fastened to the inside of the drum bring up the aggregate and drop it at each revolution and churn the material violently. Suspended fines are washed away in the water. Some scrubber-screen combinations incorporate wet screening of the aggregate by having an interior cylindrical screen concentric with an outer solid jacket, thus making a separation of the coarse and fine aggregate simultaneously with scrubbing. Baffles impede flow through the scrubber so as to give a more thorough washing action than would be imparted by rotary wet screening alone.

Log washers derive their name from an old practice of putting wood logs inside a rotating drum along with the water-aggregate mixture to augment scrubbing action. Large-sized durable boulders were also used for a similar purpose, making the device akin to a ball mill. Present-day log washers more resemble pugmills and consist of a slightly inclined chamber into which aggregate is fed at the low end along with generous quantities of water. Screws or rotating shafts, similarly inclined, carrying hard-steel blades provide a violent churning action and work the material toward the high end for discharge. Adherent coatings are effectively removed from the aggregate, and the abrading action is sufficient to break down

Fig. 29. Double-shaft log washer used for removing adherent coatings and for disintegrating soft particles. (*Courtesy of Eagle Iron Works.*)

softer fragments such as soft sandstones, ocher, and soft shale. The suspended clay and abraded stone dust are carried away in the wash water. Figure 29 illustrates a blade-type log washer. This particular type is available up to 7 × 30-ft tub dimensions with a maximum capacity of 100 to 125 tons per hr. Power requirement is 75 hp, and it may require as much as 500 gpm of water.

Like so many other items of aggregate-processing equipment, choice of whether a scrubber, log washer, or both will be needed depends largely on the nature of the deposit. Operators sometimes experiment several seasons to determine just what equipment best fits their needs and to establish optimum operating characteristics. If the deposit is of a nature that these treatments will both wash and successfully reduce the deleterious materials down to an acceptably low percentage, the operator is indeed fortunate and may not be required to install more sophisticated and expensive beneficiation treatments such as jigs or heavy-media separation.

Hydraulic Sand Classification. Requirements for sand sizes for concrete were given in Art. 26. It almost invariably occurs that water classification of sand is needed in order to provide such graded sand and to furnish sand with an acceptably low amount finer than the No. 200 sieve. Too, many sand suppliers also stock clean, finer sand for mortar and plaster.

Hydraulic means are almost always used in transporting sand between different operations at the plant, the sand being carried in flumes along with high-velocity water.

Hydraulic classifiers operate on the basic principle that larger-sized particles fall faster in a water suspension than do the smaller sizes. Based upon the idealized assumption of spheres freely falling in a viscous medium, Stokes' law states the velocity of fall is proportional to the square of the radius of the sphere. Sand-classifying devices approximate these "ideal" conditions by various types of horizontal and rising-current classifiers such as screws, drags, rakes, wheels, and cyclones.

If a water suspension of sand is introduced into one end of a long, horizontal box or trough and the water overflows the far end, progressively finer and finer particles will settle out toward the overflow (weir) end. If the trough is lengthened and/or the water flow reduced, even finer particles will be deposited at the far

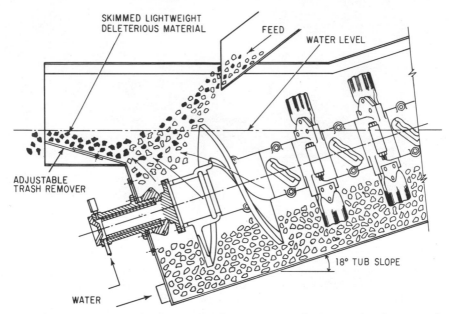

Fig. 30.   Coarse-material screw washer-dewaterer removing vegetation from gravel. (*Courtesy of Eagle Iron Works.*)

end since retention time is increased. The same concepts apply to rising-current classifiers in that the progressively finer particle size will be raised to greater heights in the pool, assuming lack of turbulence.

Commercial installations use the above principles in a variety of ways. Perhaps the simplest process is shown in Fig. 17, where the screw classifier is acting primarily to dewater sand before placing it in the delivery bin. The amount of water overflowing the weir at the low end of the screw classifier controls the amount of fines retained. The slowly turning helical screw tends to dewater the product as it raises the sand out of the pool. Control of even this apparently simple operation may be critical to preserve the desired fines. Some operators whose deposit runs low in fines or who use large volumes of water ahead of sand classification such as results from dredging, thus tending to lose fines, may experience difficulty in maintaining the product within the 10 to 30% requirement passing the No. 50 sieve.

Rakes accomplish essentially the same purpose as screw classifiers, but in this case the helical screw is replaced by an endless-chain arrangement with attached baffles which drag the sand up the slope of the container. Figure 28 shows a somewhat more complicated sand-production system wherein some, or all if desired, of the coarse size resulting from screening is blended back into the sand. Additionally, a "water scalper" is shown to dewater the sand to some extent prior to entering the screw classifier. This setup could be revised to produce finer mason's sand by introducing a second screw classifier, by blending back less coarse sand and adjusting the weir control to preserve more fines.

Figure 30 shows a variation of the use of a combined screw-type device and log washer to remove roots and other vegetation from the gravel. Such materials have proved very troublesome if of the proper size to clog screens. Others have used special cyclones to accomplish the same purpose, the principles of which are explained later.

FIG. 31. Water scalping–sand classifying tank. Different-sized sand is collected at seven stations and blended in appropriate amounts into collecting flumes below. Remote-control metering panels are available to control valves to provide uniform blends despite changes in feed. (*Courtesy of Eagle Iron Works.*)

Still other commercial sand classifiers use the principle of the horizontal long trough explained above but have as much as nine bins in the bottom of the trough to draw off the different sand sizes which settle out along the tank length (Fig. 31). A variety of sand sizes can thus be provided by blending these products. Such blending is achieved by wet blending under water by recirculating back through a screw classifier or rake. Belt blending of fine aggregate such as may be done with coarse aggregate does not successfully intermingle the sizes.

Another device used in concrete sand processing is the cyclone. This apparatus has found particular use in processing water from dredging operations where fines tend to be lost because of the large volume of water handled. Figure 32 illustrates the principles of operation of the cyclone. Spinning motion imparted to the feed tends to settle out the larger sand particles to the outside of the cone. By adjusting three flow rates, that of the slurry pumped into the cyclone, the amount of very fine material overflowing the top, and the amount of fines recovered at the bottom of the cone, the latter valuable fines can be recovered to blend back into the sand. Minus-50-mesh material has been reported to be successfully dewatered and recovered with this device, as much as 10 tons per hr with a 10-in. cyclone and a slurry feed of 300 gpm. Cyclones as large as 3 ft diameter are being used in some installations.

Dry Classification of Sand. In extremely arid regions with a scarcity of water, dry sizing of fine aggregate is undertaken using air separators. Crushed-stone-manufactured sand and slag sand are also sometimes dry-processed. Some air separators act similarly to the cyclone mentioned just previously except that air replaces water as the carrying medium. These are used more often for removal of the very fine material resulting from handling or the dust of fracture and resemble the cyclone classifiers used in portland-cement manufacture (see Chap. 1). In

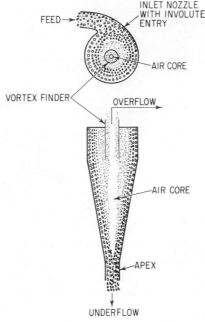

Fig. 32. Cyclone separator. Often used for recovering fines from large volumes of water resulting from dredging operations. Centrifugal force drives large particles to outside. Fines are caught in center overflow. (*Courtesy of Eagle Iron Works.*)

another device, the sand is projected out horizontally with a centrifugal impeller and subjected to a downward blast of air. Divider plates below catch the deflected sand thus separated as to size, centrifugal force casting coarse sizes the farthest. As in all other sizing operations (except laboratory sieving to refusal), clean-cut separation is not made or, in fact, necessary. Blending is later used to remedy deficiencies of grading.

Reference 37 provides a comprehensive series of three articles on sand classification giving the various processes in greater detail.

Pumps. Centrifugal pumps of large capacity are widely used in aggregate processing to handle water varying in solids content from that relatively clean for washing to sludges containing a high percentage of suspended solids. As mentioned in Art. 31 clean wash-water requirements of 2,500 to 5,000 gpm are not unusual for even moderately sized plants. Abrasive action on dredge and slurry pumps is severe, necessitating use of replaceable special steel or rubber liners.

Manufacturers should be consulted for capacities, installation instructions, etc., of their respective products. The prospective or present operator of pumping equipment should satisfy himself that answers are provided to the following types of questions:

1. Is the inlet properly screened and sufficiently deep to provide water to the pump at even peak demand and throughout the production season?

2. Does the inlet require a foot valve to hold prime on the pump during shutdown?

3. Are there depressions in the inlet line between inlet and pump which may encourage development of air lock or prevent complete drainage during freezing weather?

4. Are relief valves installed in the discharge line or shutoff valves so arranged that improperly trained workmen cannot close the discharge line, thus creating sudden pressure to burst or distort the pump housing?

5. Does the operation warrant installation of standby service?

39 Aggregate Beneficiation

Gradual depletion of existing aggregate sources, zoning restrictions preventing development of desired new sources, and tightening of specification requirements by some agencies are causing development and installation of special equipment to upgrade the quality of aggregate from present deposits or to allow development of sources heretofore considered to contain aggregate of marginal quality. Some equipment used in the past for beneficiation has proved ineffectual except under very unusual circumstances, and new techniques are constantly being proposed and undergoing trial. The following paragraphs will treat only those successful beneficiation techniques of which there are a substantial number of commercial installations presently in the United States. References 38 and 39 provide supplementary information to that given here.

Beneficiation techniques presently employ two basic principles for removal of deleterious materials: (1) specific-gravity separation and (2) mechanical impact or abrasion to break up softer fragments. Jigs and heavy-media plants separate the desired from the unwanted materials on the basis of specific gravity. Cage mills, roll mills with rubber liners, and to some extent log washers, break down the unwanted fragments by impact or abrasion. Obviously the nature of the deposit will dictate which of the beneficiation techniques is most advantageous in a given case. If the offending particles are predominantly soft, then impact beneficiation may be sufficient. On the other hand, if the deleterious particles are variously hard and soft, or predominantly hard as would be the case with chert, but are characterized by having a slightly lower specific gravity, then more expensive jig or heavy-media separation may be required. The latter seems true for many deposits in the North Central United States since installations are frequent in Ohio, Pennsylvania, Minnesota, and Michigan.

Cage Mill. Figure 27 illustrates the single cage mill wherein the rock is broken up or disintegrated when impacting against the spinning bars or when thrown against the steel housing. Aggregate is fed into the center of the cage and drops out of the bottom of the housing by gravity. Cage mills are also available with two and three concentric cages which rotate in opposite direction, thus imparting even greater disintegration action to the product.

Cage mills operating at high speed are used primarily for crushing with the claim that a highly cubical product is produced. However, some operators slow down the rotational speed, thereby reducing the crushing action but still giving sufficient impact to break down soft particles. Rotational speeds of about 600 rpm for a 3-ft-diameter cage have been employed for soft-stone reduction. In some installations, the product from the cage mill is then processed through a log washer for removal of the disintegrated fragments; in any event, the broken stone thus reduced to dust must be removed from the product by appropriate means.

As in all beneficiation procedures, too much emphasis cannot be placed on the necessity for prior careful study of the effectiveness of the cage mill in improving the product of a given deposit before installation is made. One of the best means of accomplishing this is to truck a load or two to the nearest cage-mill installation where appreciable quantities can be processed to determine benefits of the processing under full-scale operating conditions.

Power requirements for a cage mill processing 100 tons per hr are roughly 100 hp. No water is required.

Jig Beneficiation. Jig beneficiation uses the principle of "hindered settlement" wherein particles of different specific gravities tend to settle out in horizontal layers when placed in water subjected to rapid vertical pulsations. Heavy particles will sink faster than lightweight (low-specific-gravity) particles. If a directed continuous flow of water is maintained in the pulsating suspension within an enclosure

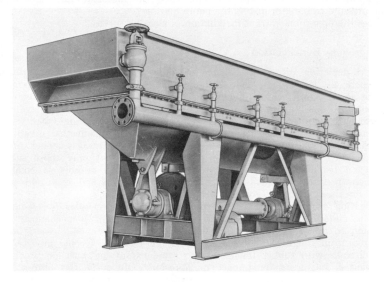

Fig. 33. Mechanically driven jig used for beneficiation to float off undesired lightweight shale, etc. Manifold valves allow adjustment of water flow for optimum operation. Speed and amplitude of jigging motion is also adjustable. (*Courtesy of Eagle Iron Works.*)

with an inclined bottom, the heavy particles will tend toward the bottom of the tank at the discharge end and the lightweight particles are effectively skimmed off in the water overflowing a dam whose height is adjusted to accomplish optimum beneficiation.

Figure 33 shows a photograph of one type of jig. Pulsations in this case are introduced by a mechanical eccentric drive; the lower, or hutch, section is attached to the sloping stationary tank by a rubber diaphragm. The tank bottom consists of perforated steel plate whose openings are restricted by a bed of loose steel balls. The balls serve to inhibit the materials from dropping into the hutch section and still allow transmission of water pulsations throughout the tank. Some jigs replace the mechanical drive with pulsating air.

Adjustment of the water-flow rate, rapidity and magnitude of jigging action, feed rate to the jig, and water overflow are critical and must be brought into careful balance by experimentation to establish optimum operation. Uniform rate of feed to the jig is essential to best operation to maintain essential "fluffing" action of the bed; surges are accompanied by loss of beneficiation effectiveness. Opinion is not unanimous as to how jigging efficiency is influenced by particle shape and size but both coarse and fine aggregates have been treated by the process.

Jig beneficiation has been most effective in gravel operations where there is considerable disparity in specific gravity between the wanted and unwanted material such as separating low-gravity wood, shale, or chert of less than 2.30 specific gravity, for example, from the desirable gravel having a gravity greater than about 2.60. Jigging does not provide entirely clean-cut gravity separations and some heavy material is bound to overflow the weir along with the lightweight material; similarly, some lightweight material will be trapped and end up in the product.

Heavy-media Separation. Heavy-media separation (HMS), despite its high initial cost, roughly five times that of a jig of equal capacity, has been adopted by the majority of operators where need for beneficiation is indicated for the reason that a precise specific-gravity "cut" can be made. In this process, aggregate is fed into a water suspension of finely divided magnetite and ferrosilicon. Aggregate particles whose specific gravity is lower than the media will float and those of higher gravity than the media will sink. By precise adjustment of the ratio of magnetite to ferrosilicon and the water dilution of the mixture, media gravity can be adjusted to an accuracy of about 0.02 anywhere in the range from about 2.00 to 3.00. Actual specific gravities used in gravel operations usually range from about 2.40 to 2.60.

Figure 34 shows a flow sheet of a typical heavy-media plant. The actual separation of sink and float takes place within the "separatory vessel." The latter may variously be of a stationary-tub type, rotating drum, spiral screw, or cone. In the tub type, the sink material is swept up from the bottom of the tank by a reciprocating paddle with the float material overflowing a weir along with the continuously circulating media. The drum type resembles a stationary mounted concrete mixer and has lifters attached to the inside perimeter of the drum which elevate the sink material from the bottom as the drum revolves and deposits it on a chute leading out of the drum. Again, float material overflows the lip of the drum along with the continuously circulating media. The spiral-screw type operates similarly to the screw dehydrator of Art. 38 except that heavy medium replaces water. Float material overflows a weir and sink gravel is carried up the incline by the spiral screw. The cone type uses a deep pool to accomplish separation with the sink product air lifted up the center from the bottom apex of the cone. Advantages are claimed for each of the four types of separating devices and all have been successfully used in gravel or coarse-sand fraction beneficiation.

The remaining devices in Fig. 34 are all concerned with reclaiming the relatively expensive heavy medium and returning it for reuse. Both the sink and float aggregate particles are coated with the black adhering media material when leaving the separating device. Some of the medium drains off at once and can be returned for immediate reuse. The adhering medium which is washed off by water sprays is magnetically recovered, screw-dewatered, and then demagnetized to inhibit clumping of the media particles. Both magnetite and ferrosilicon are magnetic, and a properly operating plant loses a minimum of these materials in the wash water despite their extreme fineness. Media losses amounting to 3 to 4 cents per ton of product have been reported. Total estimated cost of media beneficiation has ranged from 15 to 50 cents per ton of product.

The HMS plant does not distinguish deleterious aggregate particles as such but separates the material quite successfully into two specific-gravity groups, i.e., higher than the media gravity and lower than the media gravity. Insofar as the deleterious particles are of lower gravity, they will be successfully removed. However, some acceptable aggregate will inevitably also be of lower gravity and be wasted by the process. Similarly, high-gravity material such as limonite will not be removed. Thus, although the process is capable of making a clean-cut gravity separation, its success is dependent upon the character of the deposit and the type of particles necessary to remove to make an acceptable product. HMS gravel plants currently in use reject about 20% of the feed. It is not unusual to find that about one-half of this reject is acceptable material under prevailing specifications.

Disposal of the reject material has been a troublesome problem for some installations considerably removed from a metropolitan area. In the latter case, it can usually be sold for driveway gravel, porous backfill around drains, and such uses.

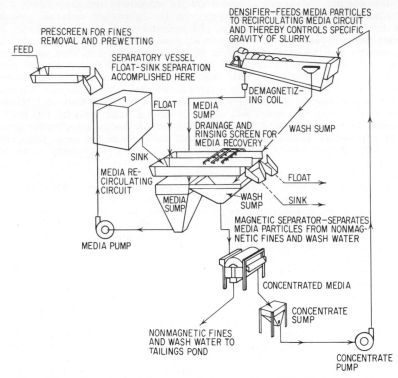

Fig. 34. Schematic view of heavy-media separation (HMS) beneficiation plant. (*Courtesy of Eagle Iron Works.*)

AGGREGATE TESTS AND SPECIFICATIONS

Large projects, particularly those supported by public funds, operate under written "specifications" in the contract document. The word "specifications" is loosely applied to (1) a document which lists the properties and appropriate numerical limits for each measured characteristic or (2) a document giving detailed procedures by which a particular test to evaluate a given property is to be conducted. Sometimes both are contained in a single "specification" but most organizations now attempt to separate the two. The contractor, inspector, or engineer should be fully aware of this distinction and be familiar with both test methods and truly designated "specifications" for the job. The unwary may discover, for instance, that despite unchanged specification limits, a really different material is being asked for because of a recent change in test method. Such change may be not too obvious in the contract document. Conversely, development of an improved testing technique may necessitate change of test limits for a certain aggregate when no revision of basic quality is really intended. In this case, concern by the aggregate producer that a new product is being required is uncalled for.

40 Specifying Agencies

Public agencies such as cities, counties, state highway departments, and Federal agencies provide specifications and, in some cases, test methods for concrete aggregates. Consulting engineering firms who prepare plans and specifications for public or private work may also furnish special requirements for concrete aggregates leading to even greater complexity in the overall number of provisions of which the aggregate producer must be aware if he is to furnish materials successfully for such projects. The specifications of local agencies will naturally reflect requirements applicable to locally available aggregates, local climate, and prevalent construction practices. On the other hand, aggregates are sometimes specified appropriate for a given locality but inappropriate for another.

Three organizations of national scope which undertake preparation of aggregate specifications are: (1) The American Association of State Highway Officials (AASHO), (2) The American Society for Testing and Materials (ASTM), and (3) Federal government agencies. As the name implies, the first listed organization confines its attention strictly to highway construction and maintenance. The test methods as well as many material specifications prescribed by AASHO are often identical to those of ASTM.

American Association of State Highway Officials. This organization currently publishes its specifications in two volumes:

Part 1. Standard Specifications for Highway Materials
Part 2. Standard Methods of Sampling and Testing

In addition, annual "Interim Specifications" are published which are ultimately provided in the formal publications. These publications may be obtained from American Association of State Highway Officials, 341 National Press Building, Washington, D.C. 20004.

American Society for Testing and Materials. This is a broadly based organization made up of producers, consumers, and "general-interest" groups. Society regulations of ASTM prohibit any particular one of these groups from exercising predominant voting influence in establishing the provisions of a test method or material specification. Additionally, ASTM establishes "Recommended Practices" which, because of the nature of the information provided, are not intended for insertion in a contract document. One of these, enumerated later, is of particular interest to the field of concrete aggregates.

Many of the test methods and specifications enumerated below can be purchased from ASTM as separate reprints at nominal cost. All are contained in the annually published book of ASTM Standards, currently Part 10, procurable from The American Society for Testing and Materials, 1916 Race St., Philadelphia, Pa. 19103.

1. Specifications for Concrete Aggregates, ASTM: C 33. Provides specifications for fine and coarse aggregate, other than lightweight aggregate, for use in concrete. Several provisions of this specification were given in Arts. 26 and 27.

2. Specifications for Wire-Cloth Sieves for Testing Purposes ASTM: E 11. Gives detailed requirements for laboratory testing sieves.

3. Method of Test for Resistance to Abrasion of Large Size Coarse Aggregate by Use of the Los Angeles Machine, ASTM: C 535. This gives method for testing coarse aggregate larger than ¾ in. See Art. 24 for additional details.

4. Method of Test for Resistance to Abrasion of Small Size Coarse Aggregate by Use of Los Angeles Machine, ASTM: C 131. Gives method of testing coarse aggregate smaller than 1½ in. See Art. 24 for further details. Aggregate otherwise identical except for size may not give the same abrasion

value when tested by this method and Method C 535 above. The latter procedure is inappropriate for concrete aggregates for most uses.

5. Method of Test of Graded Coarse Aggregate by Use of the Deval Machine, ASTM: D 289. This test has been largely superseded by Los Angeles abrasion test. See typical Deval abrasion values, Table 4.

6. Method of Test for Abrasion of Rock by Use of the Deval Machine, ASTM: D 2. As with Method D 289 above, this test has been largely superseded by Los Angeles test and was used for broken quarried rock.

7. Method of Test for Clay Lumps and Friable Particles in Aggregates, ASTM: C 142. This method provides a means, after soaking the sample, of determining the content of particles so soft and friable that they can be broken with the fingers.

8. Materials Finer than No. 200 Sieve in Mineral Aggregate by Washing, ASTM: C 117. This test evaluates the content of fine material passing No. 200 sieve which is either brought into suspension or dissolved when the aggregate is vigorously agitated with water. Has also been termed "elutriation test" or "loss by washing."

9. Lightweight Pieces in Aggregates, ASTM: C 123. This is a method by which the amount of material lighter than a selected specific gravity is determined by floating in heavy liquids. The coal and lignite determination, for instance, uses a liquid whose specific gravity is 2.00.

10. Organic Impurities in Sands for Concrete, ASTM: C 40. Potentially detrimental organic impurities in sands are detected by observing color, developed by the supernatant liquid when the sand is inundated in a 3% solution of sodium hydroxide. Glass color-comparison standards are available to facilitate assignment of color of liquid.

11. Effect of Organic Impurities in Fine Aggregate on Strength of Mortar, ASTM: C 87. Sands suspected of containing injurious amounts of organic matter detected by Method ASTM: C 40 above are evaluated for development of compressive strength in mortar by comparison with mortar strength of the same sand when thoroughly washed with sodium hydroxide solution to remove organic matter.

12. Potential Alkali Reactivity of Cement-Aggregate Combinations (Mortar Bar Method), ASTM: C 227. This is a long-time test wherein length changes are observed of mortar bars containing the sand under test when stored up to 1 year or longer. Bars are usually made with different cements having a range of alkali contents to establish sensitivity of the aggregate to different alkali-content cements. Warm, moist storage of the bars is used (100°F over water). Duration of the test over such long periods presents a serious obstacle to routine acceptance work but the test seems to be the most reliable now available to predict injurious expansions in concrete.

13. Potential Reactivity of Aggregates (Chemical Method), ASTM: C 289. This is the so-called "quick chemical method" for determining aggregates which may exhibit detrimental volume changes in concrete with high-alkali cements. The results of this test are not completely reliable but may be helpful when combined, for instance, with petrographic examination.

14. Potential Volume Change of Cement-Aggregate Combinations, ASTM: C 342. This is another long-time test involving exposing mortar bars to a series of moist storage and different temperature environments for 1 year or longer, to determine susceptibility of the particular aggregate-cement combination to detrimental volume change. The test appears to be particularly applicable to certain aggregates in Oklahoma, Kansas, Nebraska, and Iowa.

15. Potential Alkali Reactivity of Carbonate Rocks for Concrete Aggregates (Rock Cylinder Method), ASTM: C 586. This is a research screening method to determine observable length changes occurring during immersion of carbonate rocks in sodium hydroxide (NaOH), indicating general level of reactivity and whether tests should be made to determine the effect of aggregate prepared from the rocks upon volume change of concrete.

16. Sampling Stone, Slag, Gravel, and Sand, for Use as Highway Materials, ASTM: D 75. This method gives much helpful material on sampling aggregates for concrete as well as for highway use.

17. Sieve or Screen Analysis of Fine and Coarse Aggregates, ASTM: C 136. Gives detailed procedures for conducting sieve analyses of aggregates. An extremely important method for all concerned with concrete aggregates.

18. Scratch Hardness of Coarse Aggregate Particles, ASTM: C 235. The test evaluates the hardness of a particle by observing whether it is scratched or grooved by a sharp implement tipped with brass of specified hardness or whether the rock is so hard as to actually cause deposition of some of the brass on the rock when it is scratched. Often called the "brass-pencil test" since the lead in an ordinary pencil is sometimes replaced by the brass rod and this makes a convenient tool for field work.

19. Soundness of Aggregates by Use of Sodium Sulfate or Magnesium Sulfate, ASTM: C 88. The test is designed to simulate the destructive action of freezing and thawing to which some aggregates are vulnerable when water-soaked. In this case, the aggregates are alternately soaked in a saturated solution of either sodium or magnesium sulfate and then ovendried to drive off the water of crystallization. Reimmersion causes expansive action in the rock pores because of hydration of the desiccated crystals and is similar to the destructive action of formation of ice during freezing. Five cycles of the sulfate test is considered equivalent to many cycles of freezing and thawing.

20. Recommended Practice for Evaluation of Frost Resistance of Coarse Aggregates in Air-Entrained Concrete by Critical Dilation Procedures, ASTM: C 682. This details means of evaluating the frost resistance of coarse aggregate in air-entrained concrete using the slow-freezing method, ASTM: C 671, Critical Dilation of Concrete Specimens Subjected to Freezing. Nonlinear length changes occurring in a concrete specimen as it is slowly cooled through the freezing point indicate that expansion caused by the freezing water cannot be accommodated and that the contained aggregate is causing the concrete to be vulnerable to frost attack.

21. Specific Gravity and Absorption of Coarse Aggregate, ASTM: C 127. These methods were briefly discussed in Art. 25. Two precautions in technique are advised regarding these tests: (1) In order to attain a surface-dry condition following a 24-hr soak of the coarse aggregate, it is necessary to surface-dry each particle with a towel. It follows that it is a tedious operation to prepare more than a few hundred grams to a surface-dry condition. Furthermore, the surface-dry condition does not persist long unless the aggregate is sealed in a tight container to prevent further drying. (2) The specific-gravity determination requires that the aggregate be weighed when suspended in a basket in water. The buoyant force in water will be importantly influenced by air bubbles adhering to the basket and aggregate particles. Such bubbles must be dislodged by appropriate agitation to avoid a fictitiously low value of specific gravity.

22. Specific Gravity and Absorption of Fine Aggregate, ASTM: C 128. The same precautions as noted above for coarse aggregate apply to sand but are even more difficult to remedy. Some operators aid dislodgment of entrapped air in the volumetric flask by application of vacuum. Others prefer use of clean, dry, white kerosene as the liquid in the flask rather than water. Although such deviations from standard ASTM procedures are not recommended for continued use, a new operator will do well to check his techniques by such methods. Fictitiously low values of specific gravity are not unusual for an inexperienced operator.

23. Unit Weight of Aggregate, ASTM: C 29. This method gives a precise method of determining the unit weight of dry aggregates under three standard conditions of compaction: (1) rodding, (2) jigging, and (3) loose (by shoveling).

24. Voids in Aggregate for Concrete, ASTM: C 30. This method gives the procedure for calculating voids as detailed in Art. 26.

25. Recommended Practice for Petrographic Examination of Aggregates for Concrete, ASTM: C 295. This is an important "recommended practice," as distinguished from a formal contract document, for the concrete engineer as well as the trained petrographer. The petrographer is required to have had formal scholastic training, but the recommendations and techniques in this document are aimed specifically at the field of concrete aggregates. Likewise, the engineer required to examine the results of petrographic examinations will derive much benefit from study of this recommended practice to alert him more fully to the strengths and limitations of such examinations.

26. Definition of Terms Relating to Concrete and Concrete Aggregates, ASTM: C 125. Several of the terms were directly quoted in Arts. 3 to 9.

27. Standard Descriptive Nomenclature of Constituents of Natural Mineral Aggregates, ASTM: C 294. This gives a highly authoritative description of the minerals composing natural aggregates.

28. Method for Reducing Field Samples of Aggregate to Testing Size, ASTM: C 702. Describes techniques for reducing samples to size appropriate for testing using mechanical splitter or quartering. Suggests caution that, when only a few large particles are present, any reduction may impair proper representation.

Federal Government Agencies. Projects supported by funds from Federal agencies operate under a variety of specifications. The Federal Housing Authority relies quite heavily on ASTM specifications for aggregates. Some of the test methods employed by the U.S. Bureau of Reclamation are given in Ref. 17. The U.S. Corps of Engineers publishes the "Handbook for Cement and Concrete" with quarterly supplements available from the U.S. Army Engineer Waterways Experiment Station, Vicksburg, Miss. Some Federal agencies may use specifications procurable from the nearest district office of the General Service Administration. Other agencies rely entirely upon specifications prepared by the architect or engineering firm engaged to design the structure. The U.S. Bureau of Public Roads issues Standard Specifications for Construction of Roads and Bridges on Federal Highway Projects for construction supported solely by Federal funds.

NATIONAL AGGREGATE INDUSTRY ASSOCIATIONS

The four national organizations listed below are concerned with aggregate production, testing, specifications, and similar matters pertaining to their individual industries. Additionally, most of them actively support laboratories which undertake studies of interest to producers, consumers, and general-interest groups. Printed reports of such studies and reprints of those published in the technical literature are generally available to qualified persons upon request. Much of the progress in use of concrete aggregates derives from the devoted efforts of these organizations and their highly competent staffs.

1. National Crushed Stone Association, 1415 Elliot Place, N.W., Washington, D.C. 20007
2. National Limestone Institute, Inc., 1315 16th St., N.W., Washington, D.C. 20036
3. National Sand and Gravel Association, 900 Spring St., Silver Spring, Md. 20910
4. National Slag Association, 300 S. Washington, Alexandria, Va. 23219

REFERENCES

1. Powers, T. C.: Causes and Control of Volume Change, *J. Res. Develop. Lab., PCA,* vol. 1, no. 1, adapted from Fig. 4, p. 37, January, 1959.
2. The American Society for Testing and Materials: Standards, pt. 10, 1971.
3. Mineral Facts and Problems, *U.S. Bur. Mines Bull.* 585, p. 804, 1960.
4. Iron Blast-furnace Slag-production, Processing, Properties and Uses, *U.S. Bur. Mines Bull.* 479, pp. 1–304, 1949.
5. *Eng. News-Record,* vol. 188, no. 6, p. 44, Feb. 10, 1972.

6. Woolf, D. O.: "The Identification of Rock Types," rev. ed., U.S. Bureau of Public Roads, 1960.
7. Soil Mapping Methods and Application, *Highway Res. Board Bull.* 299, 1961.
8. Russell, I. C., and Frank Leverett: U.S. Geological Survey, Ann Arbor Folio, Original Survey 1901–02–03, revised 1915.
9. U.S. Department of Agriculture, Bureau of Chemistry and Soils, Michigan Agricultural Experiment Station, Soil Map of Washtenaw County, Michigan, 1930.
10. Material Inventories, *Highway Res. Board Record,* 1, January, 1963.
11. Maas, H. G.: It Can Pay Contractors to Prospect for Materials by Air, *Roads and Streets,* vol. 97, no. 5, pp. 42–43, May, 1954.
12. Subsurface Exploration: Organization, Equipment, Policies and Practices, *Highway Res. Board Bull.* 316, 1962.
13. Barnes, H. E.: Earth Resistivity Interpretation for Sand and Gravel Prospecting, *Pit and Quarry,* vol. 51, no. 11, pp. 92–96, May, 1959.
14. Moore, R. W.: Geophysical Methods of Subsurface Exploration in Highway Construction, *Publ. Roads Mag.,* vol. 26, no. 3, p. 53, August, 1950.
15. Barnes, H. E.: Soil Investigation Employing a New Method of Layer-value Determination for Earth Resistivity Interpretation, *Highway Res. Board Bull.* 65, pp. 26–36, 1952.
16. Juergens, R. A.: Subsoil Surveys Eliminate Pre-bid Guesswork, *Construct. Methods,* vol. 44, no. 9, pp. 123–130, September, 1962.
17. "Concrete Manual," 7th ed., pp. 107–128, U.S. Bureau of Reclamation, 1963.
18. Alexander, K. M.: Strength of the Cement-Aggregate Bond, *Proc. ACI,* vol. 56, pp. 377–390, November, 1959.
19. British Standards Institution: B.S. 812, Methods for Sampling and Testing of Mineral Aggregates, Sands and Fillers, 1960. (Copies of complete document may be obtained from the United States of America Standards Institute, 10 East 40th St., New York, N.Y. 10016.)
20. Neville, A. M.: "Properties of Concrete," John Wiley & Sons, Inc., New York, 1963.
21. Huang, Eugene Y.: A Test for Evaluating the Geometric Characteristics of Coarse Aggregate Particles, *ASTM Proc.,* vol. 62, pp. 1223–1242, 1962.
22. Rex, H. M., and R. A. Peck: A Laboratory Test to Evaluate the Shape and Surface Texture of Fine Aggregate Particles, *Publ. Roads Mag.,* December, 1956.
23. Gaynor, R. D.: "Aggregate Properties and Concrete Strength," reported at 49th Annual Convention, National Sand and Gravel Association, January, 1965.
24. Woolf, D. O.: Toughness, Hardness, Abrasion, Strength, and Elastic Properties of Concrete Aggregates, *ASTM Spec. Tech. Publ.* 169, p. 320, 1956.
25. Proceedings of the Fourth International Symposium on the Chemistry of Cement: Washington, D.C., National Bureau of Standards Monograph 43, vol. 2, p. 754, 1960.
26. Symposium on Alkali-Carbonate Rock Reactions, *Highway Res. Board Record* 45, pp. 1–244, 1964.
27. Brink, R. H.: Rapid Freezing and Thawing Test for Aggregate, *Highway Res. Board Bull.* 201, pp. 15–23, 1958.
28. Lea, F. M. (and Desch): "The Chemistry of Cement and Concrete," rev. ed., St Martin's Press, Inc., New York, 1956.
29. ACI Committee 621 Report, Selection and Use of Aggregates for Concrete, *J. ACI,* November, 1961, pp. 513–541.
30. Rockwood, N. C.: Production and Manufacture of Fine and Coarse Aggregates, Symposium on Mineral Aggregates, *ASTM Spec. Tech. Publ.* 83, pp. 88–116, 1948.

31. Meek, R. E.: Scales and the Mineral Aggregates Industry, *Pit and Quarry,* vol. 53, no. 3, pp. 102–106, September, 1960.
32. Hansen, S. M., and J. Toman: Rock Breaking Takes a Giant Step into Space Age, *Rock Prod.,* vol. 68, no. 6, pp. 53–59, June, 1965.
33. Ash, Richard L.: The Mechanics of Rock Breakage, *Pit and Quarry,* vol. 56, nos. 2, 3, 4, 5, August, September, October, November, 1963.
34. Peck, R. L.: "Concrete Industries Yearbook," p. 114, Pit and Quarry Publications, Inc., Chicago, 1965.
35. Effects of Different Methods of Stockpiling Aggregates—Interim Report, *Highway Res. Board Natl. Coop. Highway Res. Rept.* 5, pp. 1–48, 1964.
36. Moavenzadeh, F., and W. H. Goetz: Application of Statistical Mechanics to Analysis of Degradation of Aggregates, *Highway Res. Board Record* 51, pp. 112–123, 1964.
37. Golson, C. E.: Modern Classification Methods Applied to Fine Aggregates, *Pit and Quarry,* vol. 52, nos. 2, 3, 4, August, September, October, 1959.
38. Price, W. L.: Ten Years of Progress in Gravel Benefication—1948–1958, *Natl. Sand Gravel Assoc., Circ.* 71, pp. 1–19, March, 1958.
39. Legg, F. E., and W. W. McLaughlin: Gravel Beneficiation in Michigan, *J. ACI,* vol. 57, pp. 813–825, disc. pp. 1751–1760, January, 1961.
40. Smith, Charles A. Jr.: Pollution Control Through Waste Fines Recovery, National Sand and Gravel Association, circular no. 110, March, 1971.

Chapter 3

STEEL FOR REINFORCEMENT AND PRESTRESSING

CHARLES F. PECK, JR., AND
JOHN H. ITTEL

Concrete is an inherently brittle material, strong in compression but weak in tension and lacking ductility. Small steel bars, on the other hand, while strong in tension and quite ductile, cannot support sizable compressive loads. In 1861, Coignet set forth the concept of embedding metallic reinforcement through the tensile areas of a concrete structural member, so that the reinforcement carried the tensile loads while the concrete carried the compressive loads. Before the turn of the century, many investigators had provided a rational theoretical basis for reinforced concrete design. This concept of reinforced concrete construction led to structural members that were not only vastly stronger than those made with plain concrete, but also possessed the ductility that was lacking in the plain concrete members.

The widespread use of structural concrete in the last 75 years is, in large measure, due to the use of reinforcing steel acting with the concrete to create a new building material, reinforced concrete. With the capability of being cast in an endless variety of shapes and forms that cannot be obtained using any other common building material, reinforced concrete offers many advantages in creating structures that are esthetically and economically superior in comparison with other structural materials.

REINFORCING STEEL

Reinforced concrete's unlimited variety in shape and form can be safely and economically achieved only through the use of standardized materials. In the earlier days of reinforced concrete, an extremely wide variety of proprietary reinforcing material was available, but obvious advantages have led to a high degree of standardization in modern reinforcing materials. In the United States, the American Society for Testing and Materials (ASTM) has produced standards that govern both the form and materials of modern reinforcing.

The principal forms that standard concrete reinforcing takes are plain bars, deformed bars, and welded wire fabric. There are other miscellaneous forms, discussed later, which are more suitable for special situations. However, the bulk of concrete reinforcing is done with the above three types.

1 Plain and Deformed Bars

Plain round steel bars, the first reinforcing used, are still encountered as special-purpose reinforcing steel. All column spirals, most expansion-joint dowels, some light stirrups and column ties, and some very minor reinforcing use plain bars. The most commonly used size is a quarter-inch-diameter bar, but almost any size may be encountered. Plain bars are not used for any major tensile or compressive reinforcing since the bond or anchorage between the concrete and the steel is relatively poor. This makes the reinforcement ineffective unless special anchorage details are used to develop the strength of the bars. Even with special anchorage, the tensile cracking of the concrete covering plain bars will usually prove to be objectionable, since both control of the tensile cracking and the effectiveness of the reinforcing are largely determined by the bond between the steel bars and the concrete.

Standard reinforcing bars are rolled with protruding lugs or deformations. A typical deformed steel reinforcing bar is shown in Fig. 1. These deformations

Fig. 1. A typical deformed steel reinforcing bar. (*Ceco Corporation.*)

can serve to increase the bond and eliminate slippage between the bars and the concrete. They act much as the tread on a tire. The use of these deformed bars is an American invention, and in the early days of concrete construction there were literally thousands of different types of reinforcing bars available. By the 1930s the modern deformed reinforcing bar, as we know it, was established. With the publication of ASTM Specification A-305 in the 1940s, standardization of the sizes and deformations of modern reinforcing bars was essentially completed. This ASTM specification standardized the sizes of deformed bars and set certain specific minimum requirements for deformation spacing, height, and permissible gaps in deformations. The actual pattern of the deformations was not prescribed in the ASTM specification, however, and each producer was free to use his own

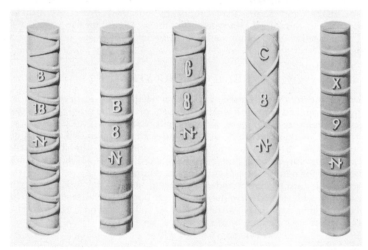

Fig. 2. Typical deformed reinforcing bars showing identification marks. (*Concrete Reinforcing Steel Institute.*)

distinctive pattern as long as he met the specification requirement. Figure 2 shows a few of the patterns used. Specification A-305 is now obsolete, since the size and deformation requirements have been incorporated into the newer ASTM reinforcing-bar specifications A-615, A-616, and A-617.

The sizes of the standard deformed reinforcing bar are designated by bar numbers. The bar number is based on the number of eighths of an inch in the nominal bar diameter. The nominal bar diameter of a deformed bar is the diameter of a plain round bar having the same weight per foot as the deformed bar. The actual maximum diameter is always larger than the nominal diameter. This increase is always neglected in design, except for the cases of sleeves or couplings that must fit over the bar when the actual maximum diameter, of course, must be used. Table 1 shows the nominal specification dimensions for deformed reinforcing

Table 1

Bar-size designation No.	Weight, lb/ft	Nominal dimensions, round sections		
		Diam, in.	Cross-sectional area, sq in.	Perimeter, in.
ASTM standard reinforcing bars				
2	0.167	0.250	0.05	0.786
3	0.376	0.375	0.11	1.178
4	0.668	0.500	0.20	1.571
5	1.043	0.625	0.31	1.963
6	1.502	0.750	0.44	2.356
7	2.044	0.875	0.60	2.749
8	2.670	1.000	0.79	3.142
9	3.400	1.128	1.00	3.544
10	4.303	1.270	1.27	3.990
11	5.313	1.410	1.56	4.430
ASTM large reinforcing bars				
14	7.65	1.693	2.25	5.32
18	13.60	2.257	4.00	7.09

bars. The special large bars, Nos. 14 and 18, are not commonly stocked and are normally encountered only as column reinforcing on projects where large quantities of these bars are used. In most construction, where isolated instances require only a few items of very heavy reinforcing, building codes permit the use of "bundles" of several smaller bars. The proper method of designating the size of a standard deformed bar is by its "bar number." On a drawing, bill of material, invoice, bar tag, etc., the bar number is preceded by the conventional number symbol (#). When more than one bar of the same size is indicated, the number of bars precedes the size marking; thus "6-#3" indicates 6 deformed bars of size 3 (approximately ¾ in. diameter), and "12-#8" would refer to 12 deformed bars of size number 8 (approximately 1 in. diameter).

2 Grade

Reinforcing bars are hot-rolled from a variety of steels in several different strength grades. Most reinforcing bars are rolled from new steel billets, but some are rolled from used railroad-car axles or railroad rails that have been cut into rollable shapes. Several strengths of reinforcing are available.

Reinforcing bars are produced to standards established by the American Society for Testing Materials. Table 2 lists the standard reinforcing-bar grades that are used today and a summary of the important physical property requirements. All the reinforcing standards are "mechanical property" specifications where the requirements for certain mechanical properties are stated. The specifications do not require any particular chemical analysis or composition of steel. Each manufacturer is

Table 2. Kinds and Grades of Reinforcing Bars as Specified in ASTM Standards

Type of steel and ASTM specification No.	Grade designation	Size Nos.	Yield point min, psi	Tensile strength min, psi	Elongation in 8 in. min, %	Bend test[a]
Billet steel A615	40	3	40,000	70,000	11	3t
		4, 5			12	3t
		6			12	4t
		7			11	4t
		8			10	4t
		9			9	5t
		10			8	5t
		11			7	5t
		14, 18			7	None
	60	3, 4, 5	60,000	90,000	9	4t
		6			9	5t
		7, 8			8	5t
		9, 10, 11			7	6t
		14, 18			7	None
	75	11	75,000	100,000	5	8t
		14, 18			5	None
Rail steel A616	50	2 to 11	50,000	80,000	1,000,000[b] tensile str.	None
	60	2 to 11	60,000	90,000		None
Axle steel A617	40	3	40,000	70,000	11	3t
		4, 5			12	3t
		6			12	4t
		7			11	4t
		8			10	4t
		9			9	5t
		10			8	5t
		11			7	5t
	60	3, 4, 5	60,000	90,000	8	4t
		6, 7			8	5t
		8			7	5t
		9, 10, 11			7	6t

[a] Diameter of test pin; t = diameter of specimen.
[b] Specification has size deductions and minimums.

free to use a wide range of chemical composition, so long as the specified mechanical properties are met. The critical mechanical properties are the yield point, tensile strength, elongation, and bendability. The yield point, the maximum elastic stress that the bar can withstand, is probably the most important mechanical property to the designer. The tensile strength, the maximum stress that the bar can withstand in tension without failing, is generally of lesser importance. Elongation is the amount of stretch that a bar undergoes when loaded. The specifications require that total stretch to failure measured over an original 8-in. length of bar be greater than a certain minimum percentage that varies with bar size and grade. Bendability is a measure of the ability of the bar to be bent to a minimum radius bend without cracking. In the ASTM specifications, it is the required diameter of the pin around which the specimen must be bent without cracking. The bend-test requirements differ with grade and bar size. It should be noted that there is presently no ASTM bend-test requirement for the 14 and 18 bars, although the American Concrete Institute's Building Code Requirements (ACI-318) requires, if these bars are shown bent, that the bar pass a 10t bend test. The higher design stress made possible by using the 60,000- and 75,000-psi yield-point reinforcing bars can often permit a substantial saving in size of members and quantity of materials required for a structure. There is, however, a practical limit on how strong the reinforcing steel should be in standard reinforced concrete construction. All strengths of steel have approximately the same amount of elongation for the same applied tensile stress. If one steel has twice the yield point of another, we can apply twice as much stress but we will get twice as much elongation. Under fairly moderate loads, the steel reinforcing will stretch about as much as the concrete surrounding it can stretch without severe cracking. If more load is applied, the steel may safely carry the load, but the concrete cover will crack. This is not only unsightly, but generally will permit corrosion of the reinforcing. Steels with yield points greater than 60,000 psi generally cannot use their greater strength in standard tensile reinforcing without causing cracking of the concrete, unless special provisions are made in the design of the member.

The present trend in reinforced-concrete construction is toward the use of higher-strength-grade reinforcing bars. Use of these high-strength bars results in a significant reduction of steel tonnage and size of concrete structural members, with resulting economy in labor and other material. With smaller columns, beams, and girders, the dead load is reduced, resulting in an accumulated saving for tall buildings. In addition, there are other advantages such as a gain in floor space and a reduction in story height.

3   Identification

With the wide range of strength available in otherwise identical reinforcing bars and the possible danger that lower-strength bars might find their way into locations where the design calls for high-strength bars, the ASTM has established a standard branding system for deformed reinforcing bars. There are two general systems of bar branding. Both systems serve the same basic purpose of identifying the maker, size, type of steel, and grade of each bar. In both systems an identity mark denoting the producer of the bar, the size number of the bar, and a symbol denoting the type of steel used are branded on every bar by engraving the final roll used to produce the bars so as to leave raised symbols between the deformations. Figure 3 shows the standard marking system for deformed bars. The producer identity mark which signifies the mill that rolled the bar is usually a single letter, or in some cases a symbol. The bar size follows the maker mark and is followed by a symbol indicating new billet steel (N), rerolled rail steel (I), or rerolled axle steel (A).

The lower-strength reinforcing bars will show only these three marks. High-strength bars must also show the grade marks. The two methods of showing grade marks are the continuous-line system and the number system. The American Bar Producers are about evenly divided in using the two systems. Figure 3 shows the two grade marking systems. In the line system, one continuous line is rolled into the 60,000-psi bars, and two continuous lines are rolled into 75,000-psi bars. The lines must run at least five deformation spaces. They fall between and are smaller than the two main ribs which are on opposite sides of all American-produced reinforcing bars. It is important that the main ribs, which are on all deformed bars, are not mistaken for grade marks. In the number system, a "60" is rolled into the bar following the steel type mark to denote the 60,000-psi bars, and a "75" is rolled into the 75,000-psi bars. Some producers place the identification

AMERICAN STANDARD BAR MARKS

LOWER-STRENGTH BARS SHOW ONLY 3 MARKS
(NO GRADE MARK) IN THE FOLLOWING ORDER:
1ST –PRODUCING MILL (USUALLY AN INITIAL)
2ND– BAR SIZE NUMBER (#3 THROUGH #18)
3RD –TYPE (**N** FOR NEW BILLET; **A** FOR AXLE;
I FOR RAIL)

HIGH-STRENGTH BARS MUST ALSO SHOW GRADE
MARKS:
60 OR ONE (1) LINE FOR 60,000-PSI STRENGTH
75 OR TWO (2) LINES FOR 75,000- PSI STRENGTH
(GRADE MARK LINES ARE SMALLER AND
BETWEEN THE TWO MAIN RIBS WHICH ARE ON
OPPOSITE SIDES OF ALL AMERICAN BARS.)
NUMBER GRADE MARKS ARE 4TH IN ORDER.)

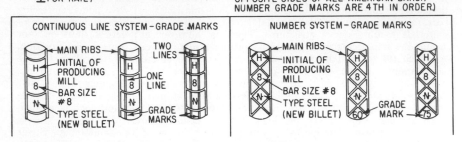

CONTINUOUS LINE SYSTEM – GRADE MARKS

NUMBER SYSTEM – GRADE MARKS

FIG. 3. The identification marking system for deformed reinforcing bars. (*Concrete Reinforcing Steel Institute.*)

marks rotated 90° from the position shown so that the brands read with the bar in the horizontal position.

4 Fabrication

The typical concrete reinforcing bar is a hot-rolled mill product. It begins as a steel billet or, in some cases, as a billet-shaped section cut from used railroad steel. The billet is heated to red heat and passed through successive stands of rolling-mill rolls. Each pass reduces the cross-sectional area and lengthens the resulting bar. The billet is reduced to the required bar size. The rolls of the last pass are generally deeply engraved to produce the standard deformations of the deformed reinforcing bar. The bars are then cropped to the standard mill length, usually about 60 ft long. After cooling, the bars are bundled in lifts of about 5 tons, tagged to permit identification of the heat of steel from which they were made, and shipped to the reinforcing-bar fabricator. Sufficient tests are conducted at the mill to ensure that each heat meets ASTM grade requirements. In the long mill length, the bars are not readily usable in very many structures. The basic job of the fabricator is to convert the long mill lengths to the usable reinforcing steel desired by the concrete structural designer.

An architect or engineer designing a concrete structure prepares drawings showing the complete design of the structure. From these design drawings a set of detailed placing drawings is prepared showing the number, size, length, and bending dimensions of each piece of reinforcing steel and its location in the structure. The detail drawings may be prepared by the designer or by the reinforcing-bar fabricator in accordance with the design drawings. In either case the fabricator will generally prepare the bar lists showing the full information of each mark of bent or straight bar that is shop-fabricated. In some areas, union regulations prohibit the shop fabrication of small-sized bent reinforcing, requiring that this be done on the jobsite. Where possible, it is preferred that all reinforcing steel be shop-fabricated for the fabrication can be performed with greater accuracy and less expense using the special machinery in the fabricating shop.

The first basic shop operation is cutting the reinforcing bars to length. Standard fabricating practice is to shear the bars with a tolerance of plus or minus 1 in. from the detailed length. Where the cut end of the bars must be exceptionally square, or where bevels or V grooves are required the bars must be saw-cut. Bars

FIG. 4. Reinforcing bars being sheared to length in a fabricating shop. (*Ceco Corporation.*)

that are to be provided as straight bars are then bundled, each size and length making a separate bundle. The bundle is then tagged to show the bar mark and the bundle is ready to ship. Much of the reinforcing will, however, have to be bent before it is bundled for shipping. Shop bending is generally divided into two classes—light and heavy bending. Light bending is more expensive per hundred pounds of bars than heavy bending. Light bending includes the small No. 2 and 3 bars, all stirrups and column ties, and all the larger bars 4 through 11 which are bent at more than 6 points, bent in more than one plane, radius-bent

FIG. 5. Shop bending of reinforcing bars. (*Ceco Corporation.*)

with more than one radius in any one bar, or a combination of radius and other bending. Heavy bending is the bending of bar sizes 4 through 11, which are bent at not more than six places, or the radius bending to one radius of these heavy bars. Figures 4, 5, and 6 show typical fabricating operations.

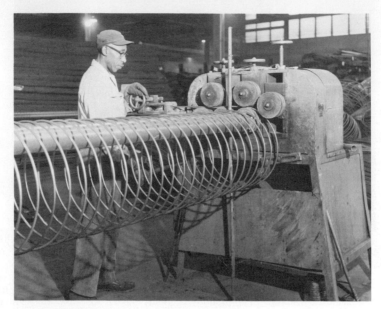

Fig. 6. Spiral reinforcing for a reinforced-concrete column being formed. (*Ceco Corporation.*)

5 Handling and Storage of Reinforcing Bars

Reinforcing bars should be handled and stored in such a manner that they will not be bent out of the planned shape. They should not be stored directly upon the ground. Open storage of reinforcing steel will in most cases result in rusting. The suitability of rusting reinforcing has been the subject of some concern in past years. Studies going back to 1920 have demonstrated that a rust film or tight mill scale, rather than harming the bond between the steel and concrete, actually causes an improvement in the bonding characteristics of the steel. Since the end of World War II, two major studies have confirmed the results of the earlier studies that rust on reinforcing that is to be placed has no harmful effect on the bond strength. The U.S. Bureau of Reclamation Concrete Laboratory conducted an extensive series of tests that led to the conclusion that normal handling was sufficient preparation even for extremely rusty reinforcing steel, and that sandblasting, wire brushing, or rubbing with burlap gave no better bond. Tests conducted at the West Virginia University confirm that rust on reinforcing bars has no adverse effect on bond. Where the reinforcing bars are very badly rusted, the cross-sectional area may have been reduced sufficiently so that the bars are unsuited for use. This can be checked by cleaning and weighing a length of bar to ensure that it will meet the specifications.

6 Welded-wire Fabric

Where light reinforcing and close spacing of bars are required for such items as concrete pavements, driveways, sidewalks, swimming pools, reservoirs, and thin floor slabs (such as those used in concrete-joist construction), it is usually more economical to use welded-wire fabric than to place individual reinforcing bars. Welded-wire fabric consists of longitudinal and transverse cold-drawn steel wires arranged to form a square or rectangular mesh. The wires are then electrically welded together at every intersection. The appearance of the resulting fabric is shown in Fig. 7. Using welded-wire fabric permits lightly reinforcing large areas with a minimum need for supervision and inspection. The wire used in making the fabric is generally produced in accordance with ASTM Specification A82. The plain cold-drawn wire sizes are designated by a size number consisting of a W followed by a number indicating the nominal cross-sectional area in hundredths of a square inch.

The wires are provided in sizes W-0.5 to W-31, as shown in Table 3. The fabric is designated by a "style" code which specifies the wire spacing and size. For example 4 × 6–W10 × W6 indicates that the longitudinal wires are W-10 on 4-in. centers and that the transverse wires are W-6 on 6-in. centers. The fabric is produced in accordance with ASTM Specification A185.

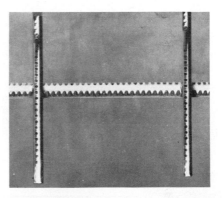

FIG. 7. A roll of welded-wire fabric. (*Ceco Corporation.*)

FIG. 8. Typical deformed wires used for welded-wire mesh. (*U.S. Steel Corp.*)

It is still common to find an earlier standard welded-wire fabric that uses the American Steel and Wire gauge numbers in place of the W-numbers. Thus 6 × 6–10 × 10 would refer to the equivalent of 6 × 6–W1.5 × W1.5. These previously used wire sizes are shown in Table 4.

Those styles of fabric in which the transverse wire provides the minimum steel area necessary for fabricating and handling are termed "one-way fabrics." Where significant reinforcement is provided in both transverse and longitudinal directions, the fabric is called "two-way."

Deformed-wire Fabric. A recent development is the use of deformed wires in producing welded-wire fabric. A cold-worked, deformed steel wire, as shown in Fig. 8, replaces the normal cold-drawn wire. The deformations along the wires are thought to improve the bonding characteristics for better concrete crack control. The deformed wires are manufactured to ASTM Specification A 496. The deformed-wire sizes are designated by a size number consisting of a D followed by a number indicating the nominal cross-sectional area of the wire in hundredths of a square inch. The wires are provided in sizes D-1 to D-31, as shown in Table 5. The deformed wires could also be used for reinforcing concrete structures in the same manner as deformed reinforcing bars. Since the A 496 deformations

Table 3. Cold Drawn Wire Sizes (ASTM A 82)

Size	Diam, in.	Area, sq in.	Weight, lb/ft
W-0.5	0.080	0.005	0.017
W-1	0.113	0.010	0.034
W-1.5	0.138	0.015	0.051
W-2	0.159	0.020	0.068
W-2.5	0.178	0.025	0.085
W-3	0.195	0.030	0.102
W-3.5	0.211	0.035	0.119
W-4	0.225	0.040	0.136
W-4.5	0.240	0.045	0.153
W-5	0.252	0.050	0.170
W-5.5	0.264	0.055	0.187
W-6	0.276	0.060	0.204
W-7	0.298	0.070	0.238
W-8	0.319	0.080	0.272
W-10	0.356	0.100	0.340
W-12	0.390	0.120	0.408
W-14	0.422	0.140	0.476
W-16	0.451	0.160	0.544
W-18	0.478	0.180	0.612
W-20	0.504	0.200	0.680
W-22	0.529	0.220	0.748
W-24	0.553	0.240	0.816
W-26	0.575	0.260	0.884
W-28	0.597	0.280	0.952
W-30	0.618	0.300	1.020
W-31	0.628	0.310	1.054

Table 4. A.S.&W. Wire Sizes

Gage	Diam, in.	Area, in.	Weight, lb/ft
0000000	0.4900	0.189	0.641
000000	0.4615	0.167	0.569
00000	0.4305	0.146	0.495
0000	0.3938	0.122	0.414
000	0.3625	0.103	0.351
00	0.3310	0.086	0.292
0	0.3065	0.074	0.251
1	0.2830	0.063	0.214
2	0.2625	0.054	0.184
3	0.2437	0.047	0.158
4	0.2253	0.040	0.136
5	0.2070	0.034	0.115
6	0.1920	0.029	0.099
7	0.1770	0.025	0.084
8	0.1620	0.021	0.070
9	0.1483	0.017	0.059
10	0.1350	0.014	0.049
11	0.1205	0.011	0.039
12	0.1055	0.009	0.030
13	0.0915	0.007	0.022
14	0.0800	0.005	0.017

Table 5. Deformed-steel Wire Sizes (ASTM A 496)

Size	Diam, in.	Area, sq in.	Weight, lb/ft
D-1	0.113	0.01	0.034
D-2	0.159	0.02	0.068
D-3	0.195	0.03	0.102
D-4	0.225	0.04	0.136
D-5	0.252	0.05	0.170
D-6	0.276	0.06	0.204
D-7	0.298	0.07	0.238
D-8	0.319	0.08	0.272
D-9	0.338	0.09	0.306
D-10	0.356	0.10	0.340
D-11	0.374	0.11	0.374
D-12	0.390	0.12	0.408
D-13	0.406	0.13	0.442
D-14	0.422	0.14	0.476
D-15	0.437	0.15	0.510
D-16	0.451	0.16	0.544
D-17	0.465	0.17	0.578
D-18	0.478	0.18	0.612
D-19	0.491	0.19	0.646
D-20	0.504	0.20	0.680
D-21	0.517	0.21	0.714
D-22	0.529	0.22	0.748
D-23	0.541	0.23	0.782
D-24	0.553	0.24	0.816
D-25	0.564	0.25	0.850
D-26	0.575	0.26	0.884
D-27	0.586	0.27	0.918
D-28	0.597	0.28	0.952
D-29	0.608	0.29	0.986
D-30	0.618	0.30	1.020
D-31	0.628	0.31	1.054

do not meet the requirements of the ASTM bar specifications, the common building codes do not allow sufficient bond stress to make the use of deformed wire practical.

The deformed-wire welded-wire fabric is produced to ASTM Specification A 497 and is similar to standard fabric except for the use of the deformed wire for either the longitudinal or both the longitudinal and transverse wires.

7   Other Reinforcing

There are other methods of reinforcing concrete structures. Although secondary in importance in terms of total tonnage used, these methods fill specialized needs and will be encountered in certain unique situations where they offer advantages over more customary forms of reinforcing. *Expanded metal mesh* is quite similar in purpose to welded-wire fabric. The mesh, trade-named Steelcrete, is made in flat sheets by slitting and stretching special low-carbon-steel sheets. It gives the appearance of being a "giant" size expanded metal lath. The cold work of stretching the material strengthens the strands. The strands are rectangular in cross section and are not mechanically or electronically fastened together to form the characteristic diamond shape, but are cut from one solid sheet of steel. The appearance and method of manufacture can be seen in Fig. 9. The size of the diamonds for structural reinforcing are about 3 in. wide by 8 in. long. A wide variety of weights and sheet sizes are available for differing design requirements. *Structural-steel shapes* in sizes ranging from small angles to large beams are used to reinforce some concrete structures. When they are used it is usually necessary to provide mechanically attached lugs or shear connectors to establish bond between the steel and the concrete. Figure 10 shows a typical beam with welded shear connectors. *Concrete-filled pipes* are commonly used for columns in light construction.

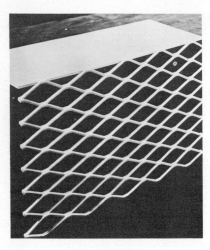

FIG. 9. A sheet of Steelcrete expanded metal mesh partially formed. (*Wheeling Corrugating Co.*)

FIG. 10. Nelson steel shear connector studs welded to a steel beam to permit composite action between the steel beam and the concrete bridge deck. (*Gregory Industries.*)

8   Accessories

Accessories, or bar supports, are used in placing the reinforcing bars to assure that the bars are correctly positioned in relation to the forms. The varied accessories all serve the basic function of supporting the reinforcing steel, in proper vertical position, enabling the bar setters to space and position the bars accurately. The bar supports should be heavy enough and numerous enough to support the bars properly. If too few supports are used, the bars will sag between supports and will not be correctly positioned.

Concrete bricks will usually be used for supporting the reinforcing bars in footings and in slabs that are placed on the ground. The reinforcing bars are tied together to form mats and set on concrete bricks which hold the mat at the correct height above the ground.

The necessary wire bar supports are furnished by the reinforcing fabricator for the reinforcing bars for beams, slabs, and joists. A wide variety of formed wire supports are available. Supports for the lower layer of reinforcing bars have legs that are straight or have a slight upturn. Supports for the upper layers of bars have runner wires that rest on the lower bars. The bar supports are identified by letter symbols. Figure 11 illustrates the variety of available bar supports.

The slab bolster (SB) has a corrugated top wire and legs spaced on 5-in. centers. The 1-in. corrugations are used on the top wire to help in spacing the bars. The slab spacer (SS) has a straight top wire and legs spaced on the exact centers

BAR SUPPORT SPECIFICATIONS AND STANDARD NOMENCLATURE

Wire Specifications—Cold drawn industrial quality basic wire

SYMBOL	BAR SUPPORT ILLUSTRATION	WIRE SIZE * TOP	WIRE SIZE * LEGS	DESCRIPTION
SB	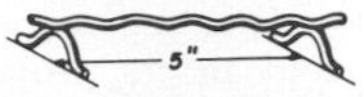 Slab Bolster	No. 4 Corrugated	¾" High, No. 6 Over ¾, No. 5	Legs spaced 5" centers—Vertical corrugations spaced 1" centers—Heights up to 2" **Stocked in ¾", 1", 1½" and 2" heights and 5 and 10 foot lengths.**
SBR	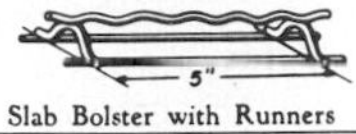 Slab Bolster with Runners	No. 4 Corrugated	Same as SB	Same as SB with No. 7 wire runners.
SS	 VARIABLE MIN. 4" Slab Spacer	No. 5 Smooth	Same as SB	Legs spaced to provide supporting leg under each bar. Minimum leg spacing 4"—Heights ¾", 1", 1½", 2". **Fabricated to order.**
BB	 Beam Bolster	No. 7 Smooth	No. 7	All legs spaced 2½" centers — Maximum height 3". **Stocked in 1", 1½", 2" heights, in 5 foot lengths.**
HBB	 Heavy Beam Bolster	No. 4 Smooth	No. 4	Same as BB except 1½" to 5" heights in increments of ¼".
UBB	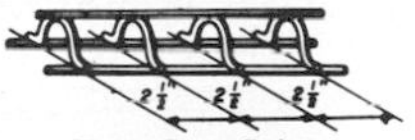 Upper Beam Bolster	No. 7 Smooth	No. 7	All legs spaced 2½" centers — Maximum height 3". **Stocked in 1", 1½", 2" heights, in 5 foot lengths.** Same as BB with No. 7 runner wire.
UHBB	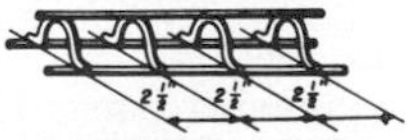 Upper Heavy Beam Bolster	No. 4 Smooth	No. 4	Heights same as HBB with No. 4 runner wire. **Fabricated to order.**
BC	 Individual Bar Chair	No. 7	No. 7	**Made and stocked only in ¾", 1", 1½" and 1¾" heights.**
JC	 Joist Chairs	No. 6	No. 6	**Made and stocked only in 4, 5, 6 inch widths and ¾", 1", 1½" heights.**
HC	 Individual High Chair	No. 4 from 2" to 6" No. 2 over 6" to 9" No. 0 over 9"		**Legs at 20° or less with vertical. When height exceeds 12 in., reinforce legs with welded cross wires, or encircling wire. No. 2 to 15" high in ¼" increments. Larger heights to special order.**
CHC**	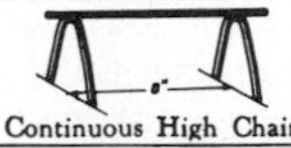 Continuous High Chair	No. 2 Smooth	Same as HC	All legs 8¼" centers (max.) with leg within 4" of end of chair. **Fabricated to order, heights 2" to 15" in ¼" increments.**
UCHC	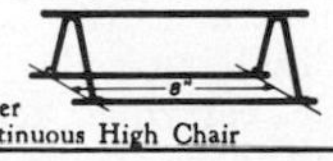 Upper Continuous High Chair	Same as CHC	Same as CHC	Same as CHC with No. 4 wire runners.

* AS&W wire gauges indicated in this table are the minimum sizes to be used.

**Continuous high chairs, CHC, may also be composed of individual high chairs, HC, spaced at 4'-0" with a #5 top bar. With this arrangement overlapped as required for a lap splice, the #5 bar may also be considered effective as reinforcement.

Types BC, JC, HC and CHC are supplied with straight legs as shown in the table, but can be furnished with up-turned legs, as shown at the right, if so ordered. All other types are supplied with up-turned legs.

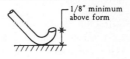

1/8" minimum above form

FIG. 11. Bar supports for placing reinforcing bars. (*Concrete Reinforcing Steel Institute.*)

of the reinforcing bars. When the reinforcing bars are placed directly over the legs they are correctly spaced.

Beam bolsters (BB), joist chairs (JC), and continuous high chairs (CHC) are used to support the bars in beams and joists. Quite often individual high chairs (HC) will be used to support a No. 4 or No. 5 reinforcing bar that will support the actual reinforcing.

The legs of the bar supports have often been the cause of rust streaking exposed concrete surfaces. Previously, galvanized bar supports were furnished to help eliminate this problem. A more recent solution is to furnish plastic-coated lower legs on bar supports for exposed surfaces, or molded plastic supports.

9   Placing

Accurate positioning of the reinforcing steel is of utmost importance, as discussed in Chap. 25. The reinforcing steel must be securely held in its proper position, as shown on the plans, so that it will not be dislodged before or during placement of the concrete. The steel is held in place in part by the accessories and in part by joining it together by welding, tying, or mechanical connectors. Often the joining is used to transmit stress from one bar to another in the completed structure.

Historically bars have been most commonly held together with tie wire. Usually No. 14 or No. 16 gage black soft-iron wire is used. Ten to fifteen pounds of wire are usually required per ton of reinforcing steel. Sufficient intersections of the reinforcing are tied to prevent shifting of the reinforcing. It is not necessary to tie every intersection. The tying adds nothing to the strength of the finished structure other than holding the bars in their proper position until the concrete has been placed. Although lapped splices are tied, the surrounding concrete forms the actual splice. A typical tie is shown in Fig. 12.

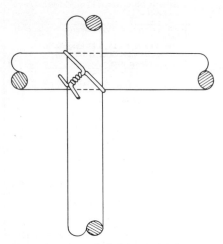

Fig. 12.   A typical "tie" holding cross-ing-reinforcing bars in place.

10   Welding

Welding has become a popular method for joining reinforcing bars. In the past, the majority of the welding was often tack welds, used in place of wire ties, to hold reinforcing bars together. This is an extremely poor practice. No tack welds should be made on reinforcing steel, since a single tack weld can reduce the strength of a reinforcing bar by more than half.

It has become common practice to eliminate standard lap splices of the larger sizes of reinforcing bars by butt-welding or mechanically coupling the bars. A properly welded butt joint can develop the full tensile strength of the bars that are joined. Most building codes require that welded splices carrying tension loads develop 125% of the yield strength of the reinforcing bars. There are many different welding methods and procedures, for steels of different chemical composition are quite differently affected by the heat of welding. The welding procedure necessary to produce sound crack-free welds depends upon the chemical composition of the reinforcing bars. A procedure that is suitable for one chemical composition can be totally unsuited for another composition of the same strength grade. It is essential that the composition of the steel to be welded be determined before a welding procedure is established. A basic rule that has been often stated is:

"Know the composition of the material that you are trying to weld. If you don't know, find out, and then adopt the most convenient and economical procedure that will give sound crack-free welds in steel of that composition."

Where reinforcing bars are to be ordered for new work, the fabricator of the bars should be told if welded splices are contemplated. Not only can the fabricator usually provide the chemical composition of the reinforcing bars, but in many cases he can supply bars that are more suited for welding. Where the reinforcing bars are of unknown composition, already in place, or of mixed lots, the only safe procedure is to cut a small sample from each bar that is to be welded and have the samples analyzed by a qualified laboratory.

It cannot be too strongly emphasized that the range of chemical compositions that can be encountered makes it impractical to specify a single welding procedure even for a single ASTM grade of reinforcing bars. Sound economical welds require that the actual steel composition be determined before establishing a welding procedure. The welding procedures used must conform to the American Welding Society specification D12.1. "Recommended Practices for Welding Reinforcing Steel, Metal Inserts, and Connections in Reinforced Concrete."

Welding Processes. Many different welding processes are capable of providing satisfactory reinforcing-bar splices. Thermit and electric-arc welding are most commonly used, but both oxyacetylene and resistance welding have been successfully used. *Thermit welding* has been used with considerable success in splicing the larger-sized bars, particularly when using hard-to-weld steels. Both the Erico "Cadweld" and and the Thermex Metallurgical "Thermit" welding systems are in wide use. Figure 13 shows thermit welding of vertical column bars. The welding heat is provided by burning thermit powder. The bars to be spliced are held clamped in position by a holding fixture. The ends of the bars should be relatively smooth, preferably saw-cut, and are spaced about ⅜ in. apart. A sand mold containing the powdered-steel thermit mixture is attached at the joint and ignited. The molten metal flows between the ends of the bars and is partially discharged into the mold on the other side. The bars and weld metal fuse together. The process is simple and yields excellent-quality welds. The sand molds and thermit powder must be protected from moisture and the bar ends must be free of oil, mud, or chalk marks. The thermit process can pose a serious fire hazard if not carefully handled.

The most commonly used welding processes are the electric-arc methods. The welding heat is provided by an electric arc drawn between the reinforcing bars and an electrode. The *shielded-metal-arc method* uses a consumable metallic rod electrode coated with a material that gives off an inert gas to prevent contamination

Fig. 13. Thermit welding column bars. (*Thermex Metallurgical.*)

of the molten weld metal by the atmosphere. The electrodes that are to be used should be clearly specified. The coating on low-hydrogen electrodes must be thoroughly dry. Electrodes that are exposed to air over 1 hr at relative humidities greater than 75% must be dried in accordance with the manufacturer's directions before using. At humidities below 75%, 4 hr exposure to air is usually permitted before redrying is required. It is vitally important that the manufacturer's recommendations be followed to the letter, and that no coated electrode that has been wet be used under any conditions. *Inert-gas shielded-electric-arc welding* is done in much the same manner as shielded-arc welding, except the shielding is accomplished by introducing a flow of inert gas (usually helium or argon) at the weld.

The electrode is usually semipermanent and the filler metal is introduced by automatically feeding a special welding wire. Inert-gas shielded-arc welding requires special equipment and is not so widely used as standard shielded-arc welding since the equipment requires more skill and experience to operate and maintain. It is capable of producing very high quality welds.

Oxyacetylene gas welding, where the welding heat is obtained by burning a mixture of oxygen and acetylene, is much slower and much more expensive than electric-arc welding. The inherent preheating effect of gas welding produces high-quality welds, but the process is little used except as an emergency measure.

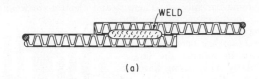

(a)

(b)

FIG. 14. A welded single lap splice.

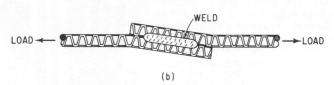

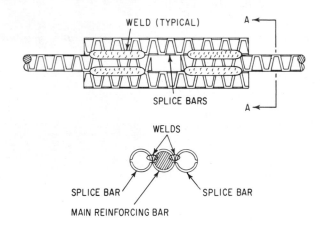

SECTION AA

FIG. 15. A welded double lap splice.

Resistance welding is used only in shop fabrication of the reinforcing, particularly welded-wire fabric and reinforcing mats. The welding is accomplished by a combination of heat and pressure. The welding heat is provided by the flow of a low-voltage current between two electrodes. Because of the equipment required, resistance welding is never used in field welding of reinforcing.

Types of Weld Joints. Probably the least desirable welded splice is the *lap joint* shown in Fig. 14a. When a lap splice is loaded the eccentricity of the bars causes a bending distortion, as shown in Fig. 14b. This distortion tends to split the concrete cover, producing a very unsatisfactory splice. Lap joints are particularly unsatisfactory for splicing the larger sizes of reinforcing bars. Where smaller bars are to be spliced and ties and stirrups are supplied to prevent splitting, satisfactory splices can be made as simple lap joints. The double lap joint of Fig. 15 is an improvement since it has less tendency to deform, but it is still not a completely satisfactory splice.

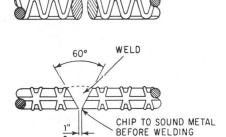

FIG. 16. A single-V welded splice.

The *butt-welded joints* shown in Figs. 16 to 19 are the preferred reinforcing-bar splice welds. The stress is transferred directly and concentrically across the joint, producing a splice that is compact and efficient. The actual detail of the butt-welded joint will vary somewhat, particularly with changes in bar sizes. For standard sizes of bars (No. 5 to No. 9) the single-V joint of Fig. 16 is generally used. The root of the weld must be back-chipped and welded. In many cases a small angle will be used as a backup to make welding of small bars easier. When

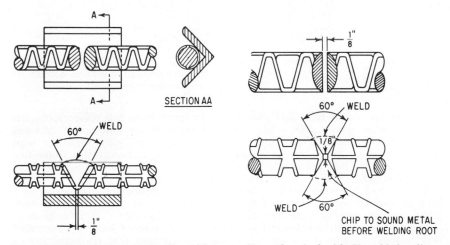

FIG. 17. A single-V welded splice with angle backup.

FIG. 18. A double-V welded splice.

the smaller-diameter reinforcing bars (No. 6 and smaller) are to be welded, an angle backup is almost essential for sound economical welds. The detail of Fig. 17 is the most customary weld for these smaller bars. The throat thickness should be about one-third of the bar diameter, and the length of the weld should be twice as long as required by normal weld allowable stresses.

The larger sizes of bars (No. 8 to 18) can best be welded using the double V of Fig. 18 for horizontal bars, or the double bevel of Fig. 19 for vertical bars. In both cases, care must be taken in back-chipping the root of the weld to avoid entrapping slag.

There are other weld details in common use. The *sleeve splice* of Fig. 20 has been used for splicing heavy column reinforcing where the bars carry only compressive loads. The bar ends must be saw-cut and the shop weld of the sleeve to

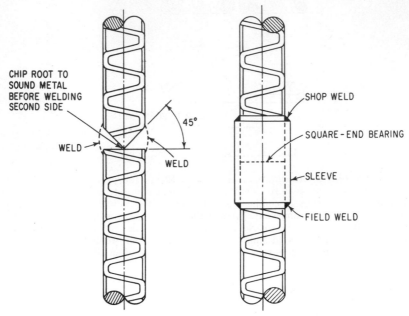

FIG. 19. A double-bevel welded splice.

FIG. 20. A welded sleeve splice for compression splices in column reinforcing bars.

upper bar should be a full fillet weld. In many cases this sleeve splice has Juced rather sloppy connections, where sleeves with too thin a wall thickness were used and attached only with tack welds. It is probably better to omit the field weld of the sleeve to a lower bar than to tack-weld this connection indiscriminately. Best practice is to use a full fillet weld in connecting the sleeve to the lower bar.

Welded cross joints are often encountered. Resistance-welded cross joints have been successfully made in many fabricating shops, but the field welding of cross joints to hold crossing bars in position is of dubious value. It is difficult in the field to deposit sufficient metal manually to make a sound weld. Uncontrolled tack welding can greatly reduce the strength of the bars that are being welded and should be avoided. In the same manner indiscriminate striking of arcs can also weaken reinforcing bars and should also be avoided.

A dangerous situation can arise where welding is being performed in the vicinity of the high-strength tendons for concrete prestressing. No welding should be allowed near these tendons, since even minor weld splatter can cause failure of a tendon during the stressing operation. The prestressing tendon should never be used as a welding ground. Best practice does not permit any cutting or welding of reinforcing steel around the prestressing tendons, once the tendons are in place.

It is often impractical to field-weld the reinforcing bars, particularly large bars of high-strength steel. Many patented mechanical devices are available for making both tension and compression splices that are as effective as welding for transmitting the stress. The G-Loc compression splice, shown in Fig. 21, is a mechanical device widely used in splicing column reinforcing bars.

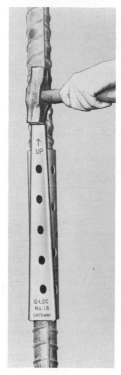

FIG. 21. A G-Loc mechanical splice for compression reinforcing bars. (*Gateway Erectors.*)

PRESTRESSED-CONCRETE REINFORCING

As discussed in the section on conventional reinforcing steel, concrete is a material that is weak in tension, and reinforcing steel must be supplied to carry the tensile forces which develop in normal structural behavior of the members. The concrete can, however, be prestressed in compression, so that the application of loads that would cause tension acts only to relieve the compression preload. An oversimplification of this is the simple case of lifting a stack of books as a beam. If sufficient pressure is supplied by pressing the ends of a stack of books together, the compression forces between the books counteract the tension forces set up by the bending action and the books will indeed carry their own weight as a beam. If we preload the standard reinforcing bars by applying tension loads to the bars and maintain the loads while the concrete that is placed around the bars sets, the prestressed-concrete member that results will be capable of carrying sizable "tensile" loads in the same manner as our stack of books.

To accomplish this in actual practice, an extremely high strength steel is required. Concrete is subject to shrinkage and creep, so that over a period of time in service the length of a concrete structure may shorten as much as 0.0006 to 0.0008 in. per in. of length. With steel reinforcing this would amount to a loss in preload of 18,000 to 24,000 psi. If the prestressing steel had originally been loaded to only 18,000 to 24,000 psi, this would amount to a complete loss of preload. Even with preloads of 30,000 to 40,000 psi, the irregular nature and high percentage of the preload loss would require a high factor of safety. Prestressing thus would not be economically competitive with standard reinforced concrete. The solution to the problem of the loss of preload by shrinkage and creep lies in the use of extremely high strength steels. If we can obtain steel preload stresses greater than 100,000 psi, the loss of preload from shrinkage and creep will be a small percentage of the total preload. Prestressing is the only method in which the extremely high-strength steels can be effectively used to reinforce a concrete structure. The high-strength steels must be elongated a great deal before their full tensile strength can be utilized. If such a high-strength steel is simply embedded in the concrete, the concrete surrounding the bar will have to crack very seriously before the steel can develop its full strength. By prestretching the high-strength reinforcing steel and anchoring it to the concrete, we produce the most desirable stresses and strains in both materials. The steel is pre-elongated, avoiding excessive lengthening under applied loads. The concrete is precompressed, avoiding cracking of the concrete under applied loads.

The methods of constructing a prestressed-concrete member are widely varied and are discussed in detail in Chap. 41. Many prestressed members are *precast*. The members are cast at a permanent plant or temporary plant at the site of the structure and then later erected. Precasting permits better control of the resulting member and is widely used for mass-producing relatively light sections, although many heavier sections have been precast. *Cast-in-place* concrete is, as the name implies, cast at its final location. This requires more forming and false-work than precasting and generally is not subject to the finer control that is possible at a precasting plant. There can, however, be a great savings in the cost of transporting and erecting extremely large and heavy members. In some cases a form of composite construction is used, with part of the structure precast and erected and the remainder of the structure then cast in place. This often permits some savings in forms and falsework.

Tensioning the prestressing steel strands or bars, which individually are called tendons, can be done before or after the concrete is placed. In *pretensioning*, the tendons are tensioned before the concrete is placed. This is done by stretching the tendons between abutments before casting the member. After the member is cast and the concrete is set, the prestress is transferred from the abutment to the member by cutting the tendons. The prestress is transferred to the concrete through bond between the prestressing steel and the concrete at the ends of the beam. Where the prestressing wires are over $\frac{1}{8}$ in. in diameter it is necessary to wave or corrugate the wires to increase the bond, since the effectiveness of the bond attachment decreases with increasing wire diameter. When the larger sizes of wire are used, a more positive bond can be assured by using metal clips (Dorland clips) that are crimped on the wire and tack-welded shut. These clips provide a positive end anchorage in pretensioned members. Pretensioning is gener-ally performed only at permanent precasting plants, although some pretensioned members have been field-cast where the abutments could be economically provided.

In *posttensioning*, the tendons are always tensioned after the concrete has been placed and has set. The tendons are prestressed against the hardened concrete and are then anchored to it. There is no need for heavy abutments for the tensioning operation, and posttensioning can be used on either precast or cast-in-place members. When a prestressed member is to be posttensioned, it is necessary to provide a means of preventing the prestressing steel from bonding to the concrete before tensioning. This protection is generally provided by conduits. The type of conduit is determined in large measure by whether the tendons are to be bonded or unbonded after prestressing. Bonded tendons are bonded throughout their length to the surrounding concrete, generally by grouting. For bonded ten-dons, the conduit can be made of metallic tubes or flexible metal hose. These form ducts through the member, permitting the prestressing steel to be posttensioned after the concrete hardens. It is also possible to form these ducts by withdrawing cores that are buried in the concrete. In some cases, metallic or fiber rods are embedded and then withdrawn before the concrete hardens. More frequently the core will be made of rubber, so that the core can be withdrawn with less effort. Since the rubber has a greater lateral contraction, it is more easily removed from the member. The rubber cores are often stiffened internally by steel pipes or rods to maintain their position during the concreting. In some cases, the core takes the form of an inflated rubber tube which can be deflated and then very easily withdrawn.

Cement grout is almost universally used where the tendons are to be bonded to the concrete after posttensioning. The cement grout also serves to protect the steel against corrosion. The grout is packed into the ducts through holes provided at each end of the member. The grout is forced into one end of the member until it is discharged from the other end. For very long members it is generally applied at both ends until it is discharged from a center vent. The best practice is to wash the cables with water before grouting, forcing the excess water out of the ducts with compressed air. The grout mix is determined by the space limitations in the ducts and around the prestressing tendons.

Where the tendons are to remain unbonded after posttensioning, the tendons are well lubricated to prevent corrosion and facilitate posttensioning and are then sheathed in plastic tube or heavy paper wrap. When the plastic tubes are used, the tubes should be properly lapped and sealed at the seams to prevent any leakage of mortar. When paper wrap is used, the paper should be spirally wrapped and double-wound to prevent any leakage of mortar. Where the tendons are unbonded, the tendons must be protected from corrosion. The usual methods of providing corrosion protection for the tendons are by galvanizing, greasing, or asphalt-coating the tendon.

11 Prestressing Steel

The steel used for prestressing concrete can be divided into four classes: bars, wires, small-diameter strands, and large-diameter strands. In every case, the prestressing steel will be considerably higher in strength than standard concrete-reinforcing bars.

High-strength Alloy Bars. For posttension prestressing use, high-tensile-strength, alloy-steel, smooth round bars are available in diameters ranging from ⅜ to 1½ in. by ⅛-in. increments and in lengths up to 84 ft. The bars are manufactured from high-alloy hot-rolled-steel rounds that are heat-treated and then cold-stretched by loading the bar to 90% of its ultimate tensile strength. The cold stretching cold-works the bar, producing a high yield strength. The cold stretching also acts as a proof load assuring that any defective bars will be eliminated in the manufacturing process. The bars are provided with either plain or specially threaded ends. The rigidity of the bars permits quick and easy placement, and they require less frequent securing in the form than other prestressing steel. The frictional losses due to tendon wobble and curvature are lower than cable-stressing systems.

The ultimate tensile strength of the high-tensile bars is considerably less than that obtained with prestressing wire or cable. Commercial high-tensile prestressing bars are produced in two grades, "regular" and "special." The "regular" grade has a minimum ultimate tensile strength of 145,000 psi, while the more expensive "special" grade has a minimum ultimate tensile strength of 160,000 psi.

Prestressing Wire. The prestressing wire is manufactured by cold-drawing high-carbon-steel bars. The bars are drawn into wire, using a series of tapered dies that reduce the diameter and increase the tensile strength of the wire. The strength of the wire is increased by cold work of each drawing, and consequently the smaller the diameter of the final wire the higher its ultimate strength. The as-drawn wire is very stiff and has very little ductility or elongation before failure. The as-drawn wire is very difficult to handle and takes a marked coil set, with the wire tending to retain the diameter of its shipping coil. The drawing process also results in nonuniform locked-in stresses. To relieve these locked-in drawing stresses and to improve the physical properties of the wire, the as-drawn wire is subjected to a closely controlled time-temperature treatment. This is done by passing the wire through a bath of molten lead or through a heated ceramic tube. The wire, in passing through the lead or air, will acquire a temperature of about 600°F. The wire will be exposed to the high temperature for about 30 sec. This mild annealing process produces only a slight reduction in ultimate strength, while more than doubling the ductility as shown by a greater elongation at fracture. The heat of the stress-relieving bath will also burn off any remaining drawing lubricant and leave the wire with a clean, oil-free surface.

When the wire is to be used in unbonded prestressed concrete where there is a danger of corrosion, the wires are often furnished galvanized. Where the wire is to be galvanized, the as-drawn wire is passed through a bath of molten zinc. The zinc bath has much the same effect as the lead bath used for stress relief.

Since the cold-drawing process tends to increase the strength of the wire with each additional drawing, the wire strength will vary with diameter. The smaller-diameter wires which have been drawn more often will show a higher ultimate tensile strength than the larger-diameter wires. The general variation of strength with diameter is shown in the curves of Fig. 22. Table 6 shows the physical properties of the commonly used prestressing wire.

Prestressing Strand. The time and labor required to place and tension many small prestressing wires can be extensive as the number of wires increases. To

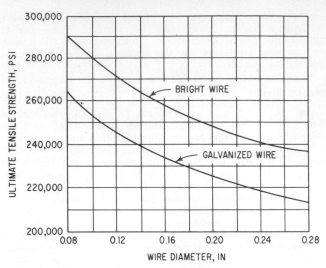

FIG. 22. A plot of the variation of strength with diameter for prestressing wire.

reduce the placing and tensioning expense, a strand of cable composed of several prestressing wires twisted together can be used. Although the prestressing strand is considerably more expensive than prestressing wire, the savings in placing and tensioning have made prestressing strand the most popular form of prestressing steel. The first strands were used for pretensioned concrete beams. The bond between the concrete and steel strands is generally all that is necessary to maintain the prestress of the beam. The strands used for pretensioning are almost universally seven-wire strands consisting of six wires helically twisted around a seventh straight wire. The straight central wire is made slightly larger than the other six wires to assure that the helical wires tightly grip the center wire. The seven-wire strands that are suitable for pretensioning use are classed as *small-diameter strands*. These small-diameter strands vary from ¼ to ½ in. in diameter

Table 6. Properties of Selected Sizes of Stress-relieved Wire

Diam, in.	Approx weight, lb/ft	Steel area, sq in.	Min breaking strength	
			lb	psi
0.1055 (12 gage)	0.0297	0.00874	2,440	279,000
0.1205 (11 gage)	0.0387	0.01140	3,110	273,000
0.135 (10 gage)	0.0486	0.01431	3,830	268,000
0.1483 (9 gage)	0.0587	0.01727	4,540	263,000
0.162 (8 gage)	0.0700	0.02061	5,340	259,000
0.177 (7 gage)	0.0836	0.02461	6,270	255,000
0.192 (6 gage)	0.0983	0.02895	7,260	251,000
0.196 (5 mm)	0.1025	0.03017	7,540	250,000
0.207 (5 gage)	0.1143	0.03365	8,380	249,000
0.2253 (4 gage)	0.1354	0.03987	9,770	245,000
0.250 (¼ in.)	0.1667	0.04909	11,780	240,000
0.2625 (2 gage)	0.1838	0.05412	12,880	238,000
0.276 (7 mm)	0.2032	0.05983	14,120	236,000

in $\frac{1}{16}$-in. increments. They are intended for use in pretensioned bonded prestressed concrete and are produced to ASTM Specification A 416. The physical properties of the seven-wire strand, which is always stress-relieved, are shown in Table 7. In addition to the ASTM A 416 prestressing strand, some manufacturers also provide a special 270K strand which is approximately 17% stronger than the standard ASTM 416 strand, permitting reductions in material and placement costs.

Table 7. Properties of Small-diameter Seven-wire Prestressing Strand

Diam, in.	Approx weight, lb/ft	Steel area, sq in.	Min breaking strength	
			lb	psi
$\frac{1}{4}$	0.122	0.036	9,000	250,000
$\frac{5}{16}$	0.198	0.058	14,500	250,000
$\frac{3}{8}$	0.274	0.080	20,000	250,000
$\frac{7}{16}$	0.373	0.109	27,250	250,000
$\frac{1}{2}$	0.494	0.144	36,000	250,000

Since there is a tendency for the helical wires to straighten under load, prestressing strand exhibits a somewhat greater elongation under load than prestressing wire. The ultimate strength is also somewhat less than the wire from which the strand was made.

Strands larger than $\frac{1}{2}$ in. in diameter are used only for posttension prestressing and are arbitrarily classed as *large-diameter strands*. They are capable of applying extremely large prestress forces, over a third of a million pounds per tendon. These strands are identical in construction to the cables used in suspension bridges and are often referred to as "bridge strands." The maximum sizes available run up to 2 in. in diameter. The large-diameter strands employ 7 to 91 wires and have breaking strengths from 54,000 to 520,000 lb for the bright wire strands as shown in Table 8.

Table 8. Properties of Large-diameter Bright Prestressing Strands

Diam, in.	No. of wires	Approx weight, lb/ft	Steel area, sq in.	Min breaking strength	
				lb	psi
$\frac{5}{8}$	7	0.82	0.238	54,000	227,000
$\frac{3}{4}$	19	1.16	0.337	80,000	237,000
$\frac{7}{8}$	19	1.61	0.470	108,000	230,000
1	19	2.06	0.599	134,000	224,000
$1\frac{1}{8}$	37	2.70	0.789	181,000	229,000
$1\frac{1}{4}$	37	3.35	0.976	220,000	225,000
$1\frac{3}{8}$	37	4.05	1.18	261,000	221,000
$1\frac{1}{2}$	37	4.84	1.41	306,000	217,000
$1\frac{9}{16}$	61	5.11	1.49	330,000	221,000
$1\frac{5}{8}$	61	5.52	1.61	335,000	220,000
$1\frac{11}{16}$	61	5.97	1.74	381,000	219,000
$1\frac{3}{4}$	61	6.45	1.88	407,000	216,000
$1\frac{13}{16}$	61	6.89	2.01	430,000	214,000
$1\frac{7}{8}$	61	7.37	2.15	460,000	214,000
$1\frac{15}{16}$	91	7.75	2.26	390,000	217,000
2	91	8.26	2.41	520,000	216,000

The large-diameter strands are furnished in bright (uncoated), stress-relieved, or non-stress-relieved, or in galvanized. Because of the greater number of wires, the larger-diameter strands tend to stretch more than the smaller-diameter strands. The large-diameter strands are generally furnished as strand assemblies with Roebling-type sockets attached at each end to form a specific length assembly.

12 Prestressing Systems

One of the unique features of prestressed concrete is that it is perhaps the only major structural field that is young enough that patents and patents pending encompass almost the entire field. Although the basic principle of prestressing cannot be patented, the actual details of accomplishing the prestressing can. The various systems involve either the details of end anchorage or the method of applying a prestress or both. There are hundreds of patents and dozens of workable prestressing systems. The profusion of patents and systems that exist appears to be quite confusing. There is, however, a great similarity between many of the systems. It is unnecessary and perhaps impossible to know the intimate details of every system in use in the United States today. The vast majority of prestressing work is being done with very few systems or variations of these systems.

Pretensioning. The simplest systems of prestressing concrete members are the various pretensioning methods. The Hoyer system is perhaps the most basic. The tendons are stretched between buttresses located at the ends of a casting bed. The concrete is placed and permitted to harden. The tendons are cut loose and the prestress transfers to the concrete by bond. The casting bed and buttresses must be strong enough to withstand the preload. By placing the buttresses several hundred feet apart several members can be cast at one time by providing shutters to separate the members along the casting bed.

Tapered conical wedges are generally used to grip the ends of the tendons at the loading buttresses. Figure 23 shows typical wedges that can be used for wires or strands. Quick-release grips, although more expensive, can be timesaving. A typical quick release for strands is the Strandvise of Fig. 24.

Usually no anchorage other than simple bond is used in pretensioned members. Larger wires or strands can be more reliable if a more positive anchorage is

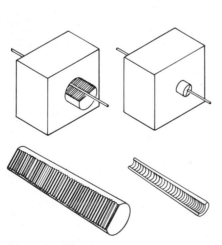

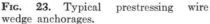

FIG. 23. Typical prestressing wire wedge anchorages.

FIG. 24. A Strandvise quick-release cable. (*Reliable Electric Co.*)

provided. Wires over ⅛ in. in diameter are often waved or corrugated or Dorland clips as shown in Fig. 25 are applied to each strand or wire.

Posttensioning Anchorages. The wide variety of anchorages available for use in posttensioning fall into several general classifications, first, by the type of tendon to be gripped; secondly, by the method of gripping. All forms of tendons, rods, wires, and strands are used in posttensioning. Anchorage may be by wedging action, by direct bearing, or by looping the tendon into concrete.

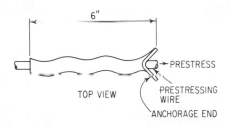
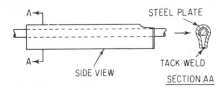

FIG. 25. A Dorland-clip anchorage for pretensioning prestressing wire.

In prestressing rods, the Lee-McCall or Stressteel system makes use of either wedging action or bearing of nuts or plates fastened to the ends of the bars. The Stressteel bars' anchorages are shown in Fig. 26. The threaded attachments use a special tapered thread that will develop the guaranteed minimum strength of the bars. On the unjacked end, only a short length of bar is threaded. On the jacked end, a longer thread is necessary. The Stressteel system also uses two styles of wedge anchorages that eliminate threading of the bars and require a less accurate determination of the final jacked length of the bar. In the threaded system, the nut must be turned to the very last thread. Split-washer shims must be inserted if the final bar length is greater than calculated.

Three systems for posttensioning prestressing strands are in general use. The Atlas system, shown in Fig. 27, uses a single seven-wire stranded cable ⅜ to ⅝ in. in diameter and a wedge grip similar in principle to those used in prestressing. The Atlas system is widely used for posttensioning thin slabs and decks, where the Atlas anchorages offer several construction economies in placing and in formwork requirements. An older system, still in wide use, is the Roebling system. The Roebling system anchorages are based on the fittings used for suspension-bridge suspenders. The fittings are based on a basket socket in which the ends of the strands are separated and then attached to the socket by filling the basket with molten zinc. The basket tapers down at the strand end, and the connection will develop the full strength of the strand. The sockets are provided in cast or machined steel. The machined steel is provided only in sizes up to 1-in.-diameter strand. The basic fitting has an internal thread at the large end of the socket. A stud bolt is threaded into the socket to permit jacking the strand and holding the preload after jacking. Rather than maintaining the preload with a nut on the stud bolt, the socket can be externally threaded to take a special nut which is used to maintain the preload. This permits the jacking stud to be removed after jacking.

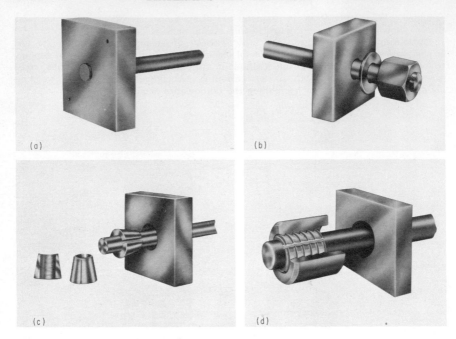

Fig. 26. Stressteel anchorage units. (*a*) Nonjacked threaded anchorage. (*b*) Nut and thread anchorage. (*c*) Wedge anchorage. (*d*) Howlet grip-nut anchorage. (*Stressteel Corporation.*)

The jacking stud and nut method is the more economical method. The Roebling system anchorages are shown in Fig. 28.

One of the most widely used posttensioning methods is the Freyssinet system, which is used for both prestressing wires and strands. The Freyssinet system uses a wedging principle. A Freyssinet-strand anchorage can preload six, eight, nine, or twelve ½-in.-diameter strands arranged in a circular fashion. Each anchorage consists of a two-piece conical wedge made of either steel or extremely high-strength concrete heavily reinforced with steel wire. A Freyssinet-strand anchor is shown in Fig. 29. The larger cylindrical piece contains a conical hole. The conical plug, which has the same taper as the conical hole, is fluted to receive

Fig. 27. A typical Atlas strand anchorage. The black wedges are used to create a void when the concrete is placed and are removed before prestressing. (*Atlas.*)

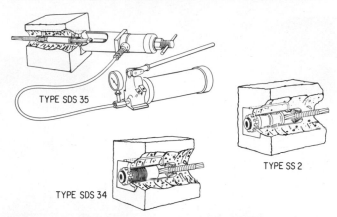

TYPE SDS 35

TYPE SS 2

TYPE SDS 34

FIG. 28. Typical Roebling prestressing strand anchorages. (*Colorado Fuel and Iron Corp.*)

the proper number of strands. The cyclinder is generally mounted flush in the concrete member, although it may be partially or completely exposed. The concrete member is reinforced with a small wire grid or steel spiral at the cylinder location. The cylinder transmits both the jacking reaction and the prestress to the member. A special double-piston jack is used to first tension the strands and then seat the conical plug. After the Freyssinet tendon has been tensioned and anchored, grout can be pumped into a central hole in the conical plug.

Posttensioning anchorages for wires take many forms. In almost every case, prestressing wires that are to be posttensioned are installed in the form of parallel wire cables, consisting of 2 to 60 or more wires attached to each anchorage block. The entire parallel wire cable is thus tensioned as a unit, simplifying the posttensioning operation. Most of the practical systems make use of either wedging action or direct bearing from button-shaped heads on the ends of the wires. The Freyssinet system, discussed under strand anchorages, is also used to pretension wires. As a wire anchorage the Freyssinet system can tension 12 or 18 wires each 0.196 in. in diameter or 12 wires each 0.276 in. in diameter. The details of the Freyssinet-wire anchorages are almost identical to those of the cable anchorages.

Buttonhead or direct-bearing methods are all similar in that the wires are tensioned and held by means of cold-formed rivet or buttonheads similar to the rivet head of Fig. 30. The ends of the wires are upset in a special press to cold-form the rivetlike head.

FIG. 29. Freyssinet prestressing strand anchorage. (*Freyssinet Co.*)

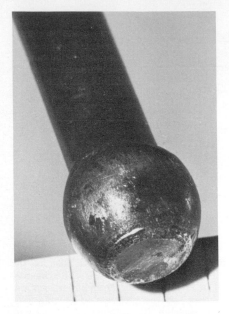

FIG. 30. A button or rivet head formed on the end of a prestressing wire. (*Prescon Corporation.*)

FIG. 31. A typical Prescon wire anchorage for posttensioning 43 prestressing wires. (*Prescon Corporation.*)

In the Prescon method, 1 to 60 wires cut to the exact length needed are passed through holes in a circular block and cold-headed. The holes in the block are small enough that the button end cannot pass through. The cylindrical surface of the blocks is grooved to permit the jack to grip the blocks. The jacking force is applied and shims are inserted between the circular block and a bearing plate as shown in Fig. 31.

The Ryerson BBRV method was of Swiss origin and now employs two separate systems for use with bonded or unbonded tendons. The Ryerson BBRV unbonded tendon anchorages, shown in Fig. 32, are quite similar to the Prescon anchorages.

FIG. 32. Ryerson BBRV unbonded wire posttensioning anchorages. (*a*) Jacked end. (*b*) Nonjacked end. (*Ryerson.*)

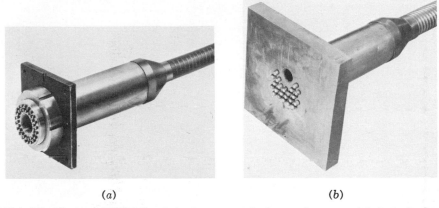

(a)   (b)

FIG. 33.   Ryerson BBRV bonded wire posttensioning anchorages.   (a) Jacked end.
(b) Nonjacked end.   (*Ryerson.*)

The Ryerson BBRV bonded tendon anchorages, shown in Fig. 33, are somewhat
different.   The circular block carrying the button ends is jacked by pulling a
holding rod that screws into the central threaded hole.   The nut is tightened
against the bearing plate and the jacking apparatus is removed.   The tendon
is then grouted.

Chapter 4

ADMIXTURES, CURING MEDIUMS, AND OTHER MATERIALS

RAYMOND J. SCHUTZ

ADMIXTURES

Admixtures are materials other than water, aggregates, and portland cement used as an ingredient of concrete and added to the batch immediately before or during mixing. Admixtures are used to modify the properties of concrete, such as to improve workability, increase strength, retard or accelerate strength development, and increase frost resistance. Generally, an admixture will affect more than one property of concrete, and its effect on all the properties of the concrete must therefore be considered.

Admixtures may increase or decrease the cost of concrete by lowering cement requirements for a given strength, changing the volume of the mixture, or lowering the cost of concrete placing and handling operations. Control of the setting time of concrete may result in economies such as decreasing waiting time for floor finishing or extending the time in which the concrete is plastic and therefore eliminating bulkheads and construction joints.

Admixtures have been used almost since the inception of the art of concreting. It is reported that the Roman builders used oxblood as an admixture in their concrete and masonry structures. Research has shown that oxblood is an excellent air-entraining agent. During the early part of this century it was common practice to add Gold Dust soap to concrete as a waterproofing agent. This soap was rich in stearates and acted as a combination air-entraining and dampproofing agent. It is doubtful whether these early uses of admixtures were carried out on a scientific basis; however, the users did realize their benefits.

1 Testing

Since the efficiency of an admixture may be affected by the type and amount of cement used, or modifications of the aggregate grading or mixture proportion, it is desirable to pretest all admixtures with the concrete to be used on the job. Since many admixtures affect more than one property of concrete, sometimes adversely, the admixture should be tested for more than one of the properties of concrete. To evaluate an admixture fully, tests should be carried out on the following properties and with the following test methods:

Slump. Method for test of slump of portland-cement concrete; ASTM Designation: C 143.

Air content. Method of test for air content of freshly mixed concrete by the pressure method; ASTM Designation: C 231.

Time of setting. Method of test for time of set of concrete mixtures by penetration resistance; ASTM Designation: C 403.

Compressive strength. Compressive strength of molded concrete cylinders; ASTM Designation: C 39. Concrete compression and flexure test specimens, making and curing in laboratory; ASTM Designation: C 192.

Flexural strength. Flexural strength of concrete; ASTM Designation: C 78.

Resistance to freezing and thawing. Either of the following test methods would be applicable: ASTM Designations: C 290-61T, C 291-61T.

Volume change. Method of test for volume change of cement mortar and concrete; ASTM Designation: C 157.

To test concrete containing admixtures only for compressive strength may lead to the choice of an admixture undesirable for the purpose intended. As an example, concrete with high compressive strength and poor durability to freezing and thawing would be undesirable for highway construction. A concrete with high compressive strength and high drying shrinkage would be undesirable for use in most structures. Evaluation should be made on the total effect of the admixture on all the characteristics necessary for the concrete in relation to its ultimate use.

2 Addition

Admixtures are generally used in relatively small quantities, as little as ⅓ fluid oz per sack of cement. It is therefore important that suitably accurate dispensing equipment be used in most cases. Liquid admixtures, including job mix solutions, may be added with the mixing water or added on nonabsorptive or saturated aggregates. Powdered admixtures should preferably be added onto the fine aggregate. All admixtures can be added after the concrete has been partially mixed. Under no circumstances should admixtures be added to portland cement prior to the addition of mixing water.

Liquid-admixture dispensers are available and have proved accurate and durable. These dispensers work on either a time-flow or a positive-displacement principle. A time-flow type of dispenser should be equipped with a transparent measuring tube, and each addition should be dispensed through this tube since any inaccuracy in flow rate with this type of dispensing equipment will change the volume delivered. A positive-displacement-type dispenser does not depend on flow rate for accuracy but is generally adaptable only to dispensing fixed quantities of the admixture.

It may be necessary to add two or more admixtures of different types to the concrete mixture to obtain desired characteristics in the concrete. As an example, a water-reducing admixture and an air-entraining admixture may be required. Most admixtures are compatible when mixed in concrete, but under no condition should two admixtures of different types be allowed to mix together prior to addition to the mixer. In most instances the admixtures will react, causing precipitation and loss of effectiveness. The incompatibility of two admixtures intermixed alone, or in water, does not indicate that such admixtures will be affected when combined in concrete.

3 Storage

Powdered admixtures generally have an indefinite shelf life if stored dry. Liquid admixtures may freeze or precipitate at low temperatures. Freezing may permanently damage some liquid admixtures; other liquid admixtures may be frozen and thawed without damage. The manufacturers' storage directions should be followed. Certain admixtures are shipped in powder form to be dissolved in water before addition to the concrete. In such cases only agitated storage tanks should be used to ensure that all the constituents of the admixture are added with each dose.

4   Types and Uses

Air-entraining Agents.   The purpose of any air-entraining admixture is to entrain small discrete air bubbles of proper size and spacing in the concrete mixture. Field and laboratory experience has demonstrated conclusively that proper air entrainment increases the resistance of concrete to disintegration by freezing and thawing by a very large factor.

As water freezes, it undergoes a volume change of approximately 9%. This increases in volume results in hydraulic pressure which can be sufficient to disintegrate the concrete. The evaporation of water and subsequent crystallization of deicing salts may also cause a similar phenomenon. Purposefully entrained air will provide voids spaced close enough to reduce pressures that tend to build up. It is evident that the efficiency of any air-entraining agent will depend on size and spacing of the voids induced by that agent. Numerous investigations have demonstrated that an air-void size of between 0.003 and 0.05 in. and a spacing factor of 0.004 to 0.008 in. will result in optimum durability. Expressed as a percentage of the volume of the concrete this range will be 4 to 7% (see Fig. 1).

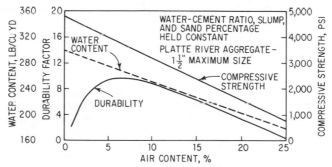

Fig. 1.   Effects of air content on durability, compressive strength, and required water content of concrete. (*From Ref. 1.*)

Air-entraining agents expand the volume of the cement matrix. Where a mixture is deficient in paste volume (usually 4 bags of cement per cu yd or less) concrete strength will be increased. With richer mixes where sufficient paste volume is present, the dilution of paste with entrained air will weaken the mixture. However, this strength penalty is more than offset by the increase in durability imparted by entrained air.

In the plastic concrete, the entrained air voids tend to block the capillaries, which are the natural paths for escaping bleed water; therefore, air-entrained concrete will tend to bleed less than non-air-entrained concrete. In the hardened concrete these air bubbles will also tend to interrupt the capillaries, resulting in lowered absorption. Entrainment of air greatly improves the workability of concrete and permits the use of aggregates deficient in fines or poorly graded. Since the workability of the concrete is improved by the addition of entrained air, for equal workability mixing water can be reduced 2 to 4% per percent of entrained air. Entrained air increases the volume of the concrete mixture, and therefore to maintain proportions it is necessary to reduce the sand content of the mix in an amount equal to the volume of the entrained air. Reduction in bleeding generally permits finishing of the concrete surfaces earlier.

Even though entrained air will increase the paste volume of the concrete, drying shrinkage is not significantly increased. Entrained air occurs only in the mortar fraction, and as the mortar is replaced by coarse aggregate with increasing maximum aggregate size, the air content for equivalent durability should be decreased ranging from 8% for ⅜-in. maximum-size aggregate to 3% for aggregate graded up to

6 in. maximum (see Fig. 2). Air entrainment should be used in all curbs, flat slab, or other concrete work which will be exposed to frost action or deicing salts. Air entrainment is not necessary in structural concrete made with well-graded sound and not harsh aggregates, not exposed to freezing and thawing.

Examples of air-entraining agents with proved satisfactory performance are those based on neutralized Vinsol resin and the sulfonated hydrocarbon salt of triethanolamine.

Control of Air Content. The efficiency of any air-entraining agent depends on temperature, slump, cement content and fineness, aggregate gradation, mixing time, and placing techniques. For a given quantity of air-entraining agent added, air content increases with greater slump and decreases with higher temperature, longer mixing time, greater cement content, and greater proportion of fines.

It can be seen that the quantity of air-entraining agent will have to be adjusted at the mixer to control air content under varying conditions.

Air content can be measured conveniently by the use of two types of air meters available on the market. These meters are based on either the volumetric method or the pressure method.

In the volumetric-type meter (Rollameter), a given quantity of concrete is added to the meter, generally $\frac{1}{20}$ cu ft, the air meter is closed, and a given quantity of water is shaken with the concrete displacing the air. The volume of water used to displace the air is read directly as entrained air. This type of meter may be used with lightweight or porous aggregates.

In the pressure-type meter (Acme), a given quantity of concrete is added to the meter, usually $\frac{1}{10}$ cu ft, the meter is closed, and air pressure is pumped into a small chamber. This air pressure is then released upon the plastic concrete, compressing any entrained air in the plastic concrete. The difference in pressure is then read directly as entrained air, the principle of Boyle's gas law

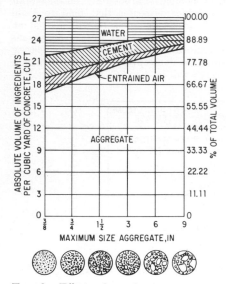

Fig. 2. Effect of maximum aggregate size on percentage of entrained air, cement, and water. Chart based on natural aggregates of average grading in mixes having a water-cement-ratio of 0.54 by weight, 3-in. slump, and recommended air contents. (*From Ref. 1.*)

applying; if there were no air in the concrete, no air could be released from the pressure chamber and the reading would be zero.

A small pocket air indicator is available (Chace air indicator) for indicating the percentage of air in mortar taken from the concrete. This indicator is handy to determine whether the concrete has entrained air or not but is subject to manipulation of the operator and should be considered a qualitative and not a quantitative instrument.

Accelerators. There are many chemicals that will accelerate the hardening of portland cement. Among these are the soluble chlorides, carbonates, silicates, fluosilicates, hydroxides, and organic compounds such as triethanolamine. For specialized purposes such as quick-setting mortars for sealing leaks there are proprietary compounds available with setting times as little as 15 sec; however, the most universally employed accelerator for portland cement is calcium chloride. Most proprietary compounds sold as accelerators are based on calcium chloride, although many of these compounds may contain other admixtures as well. Triethanolamine in combination with other admixtures is also used to a very limited extent.

Since calcium chloride is the most commonly used accelerator the following discussion refers only to this chemical or admixtures containing this chemical.

The addition of calcium chloride changes the complex reactions of portland cement and water. These changes are not fully understood, but calcium chloride can be considered a catalyst which triggers the hydration of portland cement. Calcium chloride is partially consumed during hydration, probably reacting with the tricalcium aluminate to form calcium chloroaluminate.

Calcium chloride is available in flake form 77% pure, or in granular form 94% pure. Calcium chloride is a hygroscopic salt; it should be stored in a dry place. Prior to use, calcium chloride should always be dissolved in water. A convenient solution can be prepared by dissolving 4 lb of calcium chloride in sufficient water to make 1 gal. Addition of this solution can be accomplished manually or by use of automatic dispensing equipment, 1 qt of solution being equivalent to 1 lb of calcium chloride.

Calcium chloride can be used safely in amounts up to 2% (2 lb per sack) by weight of the cement. Larger amounts may be detrimental and will provide little additional advantages. The addition of up to 2% calcium chloride by weight of the cement will increase compressive and flexural strengths at early ages of all types of portland cement. This increase will be in the area of 400 to 1,000 psi at both 70 and 40°F curing temperatures. This increase in strength reaches a maximum in 1 to 3 days and thereafter will generally decrease. At 1 year some increase may still be evident. Calcium chloride will not increase the flexural strength of concrete to the same degree as compressive strength, and decreases in flexural strength are generally obtained at or after 28 days.

Calcium chloride will also affect the following characteristics of concrete:

Drying Shrinkage. Calcium chloride will generally increase the drying shrinkage of concrete, this in spite of the fact that calcium chloride concrete will lose less water upon storage at low humidity.

Durability. Calcium chloride will lower the resistance of concrete to freezing and thawing and attack by sulfates and other injurious solutions.

Rate of Temperature Rise. Calcium chloride will increase the rate of temperature rise due to heat of hydration and in large sections will therefore increase stresses caused by thermal contraction.

Effect on Embedded Metals:

Prestressing Steel. Data are available indicating that calcium chloride may cause stress corrosion of prestressing steel.

Reinforcing Steel. Calcium chloride will not accelerate the corrosion of fully embedded reinforcing steel with adequate cover. Where large concentrations of stray currents are present such as concrete used in structures for electric railroads, powerhouses, or electrolytic-reduction plants, calcium chloride in the concrete can cause corrosion of adequately embedded steel.

Galvanized Metal. Galvanized metal embedded in concrete containing calcium chloride may be expected to corrode at an accelerated rate.

Combinations of Metals. Combinations of materials such as aluminum alloy conduit and steel reinforcing should not be used in concrete containing calcium chloride as electrolytic corrosion may take place.

Water-reducing Admixtures and Set-controlling Admixtures. Certain organic compounds are used as admixtures to reduce water requirements and to retard the set of concrete. In combination with other organic or inorganic compounds the characteristic of retarding the set of concrete can be offset and water reduction can be obtained without retardation. The chemicals generally used as water-reducing and set-controlling admixtures fall into four general classes:

1. Lignosulfonic acids and their salts
2. Modifications and derivatives of lignosulfonic acids and their salts

3. Hydroxylated carboxylic acids and their salts
4. Modifications and derivatives of hydroxylated carboxylic acids and their salts

Classes 1 and 3 are water-reducing set-retarding admixtures. Classes 2 and 4 are water-reducing admixtures in combination with other chemicals to offset the inherent retardation of these organic compounds. They therefore have no substantial effect on the rate of hardening and may achieve varying degrees of acceleration depending on the quantity of accelerator included in the formulation. These admixtures may in addition include an air-entraining agent.

The lignosulfonates are available as the calcium, sodium, or ammonium salts. They may extend the setting time of concrete 30 to 60%. In the amounts normally used the lignosulfonate retarders generally entrain 2 to 6% air in the concrete. The quantity of air may vary depending on all the factors influencing the air content of any air-entraining admixture. Data are available which indicate as much as 10% air has been entrained. Lignosulfonates generally reduce water requirements 5 to 10%. Compressive strength is usually less than reference concrete up to 2 or 3 days; 28-day strength may be 10 to 20% higher. Because of the air-entraining properties of the lignosulfonates they are generally used only in fixed proportions. Increase in the proportion of these materials may result in excessive air entrainment and/or erratic setting characteristics.

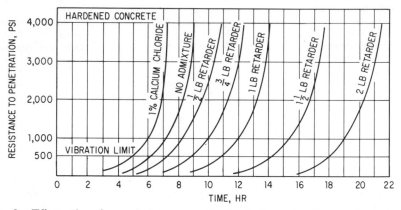

FIG. 3. Effect of various admixtures on the hardening rate of concrete (retarder of the hydroxylated carboxylic type). (*From Ref. 2.*)

Hydroxylated carboxylic acid salts act as water-reducing retarders; they do not entrain air. The rate of use can be adjusted to produce almost any degree of retardation desired, extended retardation as long as 42 days having been recorded (see Fig. 3). Used in the proportion needed to retard the set 30%, water content may be reduced 5 to 8%. The rate and capacity for bleeding are generally increased and drying shrinkage reduced with this type of admixture. Compressive strengths during the first 24 hr are generally lower, but after 3 days, strength increases of 10 to 20% may be obtained.

Both the salts of lignosulfonic acid and carboxylic acid or their modifications or derivatives can be mixed or reacted with other chemicals that entrain air or modify setting time. Characteristics of these modifications (classes 2 and 4) should be considered the same as the unmodified salt plus the particular admixture added. As an example, the salt of lignosulfonic acid, plus calcium chloride, will impart all the characteristics which these two chemicals will impart to concrete by themselves. In the combination of admixtures, effects of all the chemicals included in the admixture should be considered.

As with all admixtures, water-reducing and set-controlling admixtures will affect more than one characteristic of the concrete, and therefore to evaluate the effect of these admixtures on the properties of a given concrete the following tests should be performed:

1. Water reduction
2. Air content
3. Rate and capacity of bleeding
4. Rate of hardening
5. Compressive and flexural strength
6. Drying shrinkage
7. Resistance to freezing and thawing

ASTM Designation: C 494, Tentative Specifications for Chemical Admixtures for Concrete, is applicable and covers testing procedures for these characteristics.

Water-reducing admixtures can be used to improve the quality of concrete, lower cement content, increase workability without increase in water content, reduce drying shrinkage, and improve the properties of concrete made with poorly graded or harsh aggregates. They are desirable when placing concrete by means of pumping or when using a tremie. Set-retarding admixtures can be used to keep concrete plastic for sufficiently long periods to overcome the development of cold joints or discontinuities in a structural unit, to allow formwork to deflect while the concrete is still plastic, eliminating deflection cracks. Depending on the particular set-retarding admixture, the bleeding rate of concrete can . be changed so that bleeding can be accelerated to reduce or prevent plastic-shrinkage cracking. Set-retarding admixtures which do not entrain air may be used in varying proportions to offset the detrimental effect of temperature on concrete. As the concrete temperature increases, increased proportions of retarder can be added to obtain uniform water requirements and setting time over a large temperature range. Adjustment of the proportions of the combinations of a retarding and air-entraining admixture to give suitable retardation and air content may not be possible. Retarders are not normally effective in controlling false set. Their effects on rate of slump loss vary with the particular combinations of materials used.

Finely Divided Mineral Admixtures. Finely divided mineral admixtures are often added to the concrete to expand the paste volume or to offset poor gradation of the aggregates. Finely divided materials can be classified as either chemically inactive (inert), pozzolanic, or cementitious. All three classes affect plastic concrete in the same manner. The pozzolanic and cementitious materials may contribute to strength development in concrete and therefore usually require less cement to produce a given strength.

When added to concretes deficient in fines, the addition of finely divided mineral admixtures improves workability, reduces the rate and amount of bleeding, and increases strength. The deficiency in fines may be a deficiency in aggregate gradation or a mixture with insufficient cement paste to produce good workability. When mineral powders are added to concrete with sufficient fines, particularly concretes rich in portland cement, workability generally decreases for a given water content; therefore, water requirements and drying shrinkage are increased and strength decreased. These admixtures therefore have merit only in lean concrete or concrete manufactured with aggregates deficient in material passing a 200-mesh sieve.

The addition of fine mineral powders will decrease the efficiency of air-entraining agents, and the proportion of air-entraining agent generally has to be increased when these powders are added to the mix. The additional role of pozzolanic and cementitious admixtures is covered elsewhere. Examples of finely divided mineral admixtures are:

Inert materials:
  Ground quartz and limestone
  Crushed dusts
  Hydrated lime and talc

Cementitious types:
  Natural cements
  Hydraulic limes
  Granulated iron blast-furnace slag
Pozzolanic types:
  Covered under the following paragraphs.

Pozzolans. A pozzolan is a siliceous or siliceous and aluminous material which in itself possesses little or no cementitious value but will in finely divided form and in the presence of moisture chemically react with calcium hydroxide at ordinary temperatures to form compounds possessing cementitious properties. Examples of pozzolanic materials used in concrete are fly ash, volcanic glass, diatomaceous earth, calcined clays and shales, blast-furnace slag, and ground brick.

In plastic concrete, pozzolans will produce the same physical effects as finely divided materials; however, since pozzolans are chemically reactive, additional benefits are realized. In addition to improving the workability of concrete, pozzolans can improve quality and reduce heat generation, thermal volume change, and bleeding.

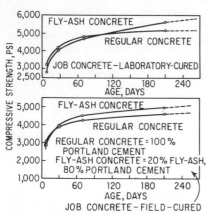

FIG. 4. Strength development of concrete with and without cement replacement with fly ash. (*From Ref. 3.*)

They can also be used to protect concrete from the destructive expansion caused by alkali-reactive aggregates. Certain pozzolans such as calcined clay and shale may increase water requirements resulting in increased drying shrinkage and resultant cracking; therefore, care must be exercised in their selection and use. Pozzolans will affect the following characteristics of concrete:

Cement Reduction. Since pozzolans will chemically react in the presence of water with the calcium hydroxide developed during the hydration process and form compounds with cementitious properties, part of the portland cement can be replaced or substituted for equal strength (see Fig. 4). The strength produced by pozzolanic admixtures is developed relatively slowly, particularly at low temperatures. Moist curing must therefore be continued for longer periods in order to develop the potential strength of such concrete. Under favorable curing conditions ultimate strengths of concrete containing pozzolans as a replacement for part of the cement will generally be higher than those obtained with portland cement alone.

Cement replacement by a pozzolan will range from 5 to 30% by weight. Because of the low bulk specific gravity of pozzolanic materials, the volume used may be equal to or greater than the volume of portland cement. As an example, structural concrete mixtures containing 4, 5, or 6 bags of portland cement per cubic yard may be reproportioned to replace 94, 83, and 71 lb of portland cement with 150 to 175, 125 to 137, and 100 lb of fly ash, respectively. In this particular test series, concrete of equal strength and workability was obtained with a general reduction in water requirements of 2 gal per cu yd.

Control of Alkali-Aggregate Reaction. The siliceous constituents of certain aggregates (alkali-reactive aggregates) will react with the alkalies of portland cement. The term alkali refers to the sodium and potassium present in small quantities and is expressed in mill reports and analyses as sodium oxide (sum of Na_2O and $0.658\ K_2O$ expressed in percent). This reaction produces an expansive alkali-silica gel which causes excessive expansion, cracking, and general deterioration of the concrete. Laboratory tests and field experience have indicated that the use of low-alkali cements (less than 0.6% Na_2O equivalent) and/or the use of pozzolans minimize alkali-aggregate reaction, thereby allowing the use of such aggregates. Pozzolans vary in their ability to control alkali-aggregate reaction; therefore, before a choice of a pozzolan is made for this purpose, tests should be carried out to determine their efficiency. An applicable test would be Method or Test for Effectiveness of Mineral Admixtures in Preventing Excessive Expansion of Concrete Due to Alkali Aggregate Reaction, ASTM Designation: C 441. Pozzolans that have proved effective in reducing alkali-aggregate reaction are some opals and highly opaline rocks, clays of the kaolinite type, some fly ashes, diatomaceous earth, and calcined clays of the montmorillonite type.

Heat Development. Pozzolans are generally used as cement replacements, thereby reducing the quantity of portland cement per cubic yard of concrete. This reduction in portland cement will reduce the total heat of hydration. This is very desirable when large masses of concrete are placed, as maximum temperature is reduced with subsequent reduction in thermal stresses and cracking on cooling.

This reduction in heat generation can be undesirable when relatively thin sections are placed during cold weather.

Coloring Admixtures. Inert pigments are often added to concrete to impart color. Coloring admixtures should be fast to ultraviolet light, stable in the presence of alkalies, and have no adverse effects on the characteristics of concrete. Insofar as their physical effect on concrete is concerned, coloring admixtures can be considered finely divided mineral admixtures. Coloring admixtures or pigments are available as natural or inert colors or as synthetic materials and are used in amounts of between 2 and 10% by weight of the cement. Coloring admixtures should be blended thoroughly with the dry cement or the dry concrete mixture before addition of water. The use of white portland cement in place of gray portland cement will always result in cleaner colors. White cement is much more effective than gray cement and a white pigment such as titanium dioxide.

Trial mixtures should be made to determine the type and concentration of pigment used. The dry hardened concrete will have a different color from the plastic concrete. The color of the concrete will also be affected by the water-cement ratio employed and the tools and procedures used in finishing and curing the surface.

The earth color or natural pigments available have a good service record for durability. The newer organic phthalocyanine blues and greens are satisfactory. Carbon black will impart a clear blue-gray color to concrete and mortar but has a tendency to float to the surface and requires more care in finishing than the black iron oxides.

The addition of coloring admixtures to concrete apparently does not affect durability, but a considerable increase over the normal amount of air-entraining agent may be required to produce the desired air content in concrete.

Acceptable Coloring Admixtures

Grays to black...............	Black iron oxide
	Mineral black
	Carbon black
Blue........................	Ultramarine blue
	Phthalocyanine blue
Bright red to deep red.........	Red iron oxide
Brown......................	Brown iron oxide
	Raw and burnt umber
Ivory, cream, or buff..........	Yellow iron oxide
Green......................	Chromium oxide
	Phthalocyanine green

Dampproofing Agents. Dampproofing agents generally are soaps based on stearates, oleates, or certain petroleum products. These admixtures may reduce penetration of the visible pores and may retard penetration of rain into porous concrete or block. Data have shown, however, that they do not resist transmission of moisture through unsaturated concrete. These admixtures cannot be considered waterproofing agents since they will not prevent passage of moisture. Concrete in itself is watertight and leaks occur only in faults such as honeycombs and cracks or other areas where a waterproofing admixture could not help. Commonly used chemicals are salts of fatty acids such as calcium, ammonium, stearate or oleate, and butyl stearate. These soaps tend to cause air entrainment during mixing, and many of the benefits obtained from their use are found in their air-entraining ability.

Other Admixtures. There are other admixtures available which are not in general use such as gas-forming, air-detraining, and grouting admixtures. These admixtures are generally used only in very specialized work, generally under the supervision of the manufacturer or a specialist in the particular field of concrete involved.

WATER

5   Quality

Water for mixing and curing concrete will be satisfactory if it is potable (fit for human consumption). Mixing and curing water should be reasonably clean and free from objectionable quantities of organic matter, silt, and salts. The maximum limit of turbidity should be 2,000 ppm. If clear water does not taste brackish or salty it can be generally used for mixing and curing of concrete without testing. Water that is apparently hard, or tastes bitter, may contain high sulfate concentrations and should be analyzed.

Experience and tests have shown that water containing sulfate concentrations of less than 1% can be used safely; however, a strength penalty may be incurred; 0.5% sulfate has been reported as reducing strength 4%, and 1% sulfate reduces strength 10%.

Ordinary salt (sodium chloride) in concentrations of 3½% may reduce concrete strength 8 to 10% but may produce no other deleterious effects. Highly carbonated mineral water may produce substantial reductions in strength. Water from swamps or stagnant lakes may contain tannic acid which may cause retardation of set and strength development. Where choice is available, the cleanest and purest source of water should be used.

6   Seawater

Seawater can be used successfully for mixing concrete. Seawater will contain an average salt content of 3.5%. Clean seawater has been used to manufacture concrete in many areas and the service record has been good. An 8 to 10% reduction in ultimate strength can be expected. Since concrete manufactured with seawater contains salts, mixing and placing techniques must be carried out with care to assure low porosity and full embedment, and cover of reinforcing steel, as the steel will rust if exposed to seawater penetrating through cracks and porosity. Many structures constructed with seawater may be exposed to freezing and thawing and alternate cycles of wetting and drying caused by waves and tides. Air entrainment should be employed and cement content should be high to ensure durability in this aggressive environment.

CONCRETE CURING MATERIALS

7   General

Concrete gains strength through the chemical process of hydration, a cement-water reaction. Drying of the concrete will cause this reaction to cease. Since the reaction of cement and water takes place over a prolonged period it is necessary to prevent loss of water from the concrete during this period to attain the inherent strength of the concrete mixture. As an example, concrete continuously moist-cured will attain approximately 80% of its 90-day strength in 28 days, 70% in 14 days, and 45% in 7 days. Drying of the concrete at these early ages will limit strength gain severely since after drying very little strength gain will take place. It can be seen that to obtain desired strength, curing of the concrete is of the utmost importance.

Concrete undergoes volume changes on alternate cycles of wetting and drying, shrinking on drying, and expanding on wetting. Flexural strength of the concrete is also developed slowly. If the concrete is alternately dried and wetted when it is young and low in flexural strength, cracking especially on the surface can occur. Therefore, curing must be continuous to prevent these detrimental volume changes.

Since any loss of moisture from the concrete will result in a strength penalty or surface cracking, curing regardless of the method should commence as soon as forms are removed or in the case of slabs as soon as the surface will not be damaged by the curing procedure.

Under certain conditions of low relative humidity, high wind, or radiant energy of the sun, or any combination, it is possible for the surface of concrete slabs to dry before hardening has taken place and while the interior is still plastic. The surface of the concrete will caseharden and shrink because of drying, causing plastic cracking to occur. Immediate application of curing, or any method that will lower wind velocity or raise relative humidity above the concrete, or will prevent the radiant energy of the sun from heating the surface, may solve this condition—fog sprays, wind breaks, and sun shades have proved effective. The application of a membrane curing compound immediately after the last finishing operation has under some conditions eliminated plastic cracking.

8   Types

Water. The application of water to concrete surfaces is the ideal curing medium as this cover of water prevents any loss of moisture from the concrete. Water can be supplied by continuous spraying or in the case of flat slabs by ponding with use of sand or other dikes. Water suitable for use in making concrete is suitable for curing.

Liquid Membrane-forming Compounds. Liquid membrane-forming curing compounds are generally solvent solutions of resins and/or waxes. Upon evaporation of solvent these compounds leave a membrane that reduces evaporation of water from the surface of the concrete. These compounds are available pigmented in clear, white, light gray, or black. The clear compound contains a fugitive dye that helps to indicate the areas covered but fades after a day or so. Clear or gray-pigmented membrane curing compounds are preferred where discoloration must be kept at a minimum; white membrane curing compounds where ambient temperatures are high, the reflective pigment preventing absorption of solar energy; and black membrane curing compounds on structural elements not exposed to the sun, or in cool weather where the solar energy will tend to warm the concrete. ASTM Designation: C 309, Liquid Membrane Forming Compounds for Curing Concrete, is an applicable specification. A test method for determining the efficiency of these compounds is covered by ASTM Designation: C 156. These specifications are based on a coverage of 200 sq ft per gal and a moisture loss of 0.055 g per sq in. under test conditions. Increasing this coverage will lessen the efficiency of these curing compounds, rendering them useless for the purpose. Care should be taken that the coverage of these curing compounds does not exceed this figure. Liquid membrane-forming curing compounds have the advantage that inspection is simplified. There is no chance that the concrete will dry out because of neglect such as is possible with water curing.

Paper and Plastic Sheets. Paper and plastic sheets are available for curing concrete. These sheets are applied to the concrete as soon as possible without disfiguring the surface of the concrete. The water-retentive qualities of paper and plastic sheets are the same as those required for liquid membrane curing compounds. Waterproof paper for curing concrete is covered by ASTM Designation: C 171. Plastic sheeting is available either reinforced with nylon threads or backed with paper. No ASTM test method or specification is presently available for plastic sheeting; however, the water-retaining properties should be the same as those acceptable for curing paper. Both paper and plastic sheets have the advantage that the concrete is not discolored by their use; however, they do have the disadvantage that they have to be fixed in position to prevent blowing away by wind. In hot climates a white pigmented sheeting is desirable. Reinforced sheeting has proved more economical since more reuses may be obtained.

Mats and Blankets. Wet curing can be accomplished by the use of burlap or cotton curing mats. Once wetted down with water these curing mats will supply moisture to the concrete for a long period of time, eliminating the need for continuous spraying of the concrete by a workman. These mats consist of a burlap, jute, or cotton cloth covering filled with linters of raw cotton or cotton waste. A minimum of 12 oz per sq yd of filling should be required. ASTM Designation: C 440 applies to this item. These mats are bulky to use and store but have proved an excellent curing medium.

For winter concreting, reusable commercial blankets are available. Ordinary commercial bat insulation may also be used. The object of these materials is to insulate the concrete from cold. To prevent loss of their insulating value these blankets must be protected from wind, rain, snow, or other wetting by use of moistureproof cover material. To be effective they must also be kept in close contact with the concrete or concrete-form surface. Table 1 lists the insulation requirements of concrete walls and floor slabs placed above ground as listed in the ACI Recommended Practices for Winter Concreting, Publication 604–56.

JOINT SEALANTS

Joint sealants are flexible materials used to seal joints subject to movement between adjacent sections of concrete or between concrete and other construction materials. Joint sealants are available as field-molded sealants such as the common poured- or troweled-in-place sealants, and as premolded sealants such as the plastic and rubber water stops and gaskets.

9　Field-molded Sealants

Sealants falling into this class are either thermosetting elastometers, thermoplastics, or mastics. The shape of these sealants in use is determined by the joint slot cast or sawed into the concrete. All sealants are solids and their volume cannot be changed by application or release of pressure; only the shape of the sealant will be changed upon movement of the joint. Therefore, the shape factor (depth-to-width ratio) of the sealant is critical and will affect the performance of any field-molded sealant. As a joint slot is compressed the sealant will have to bulge up and/or down. As the joint slot is extended the sealant will have to neck down in the center as the sealants in themselves cannot expand or compress. The strain S_{max} on the outer fiber of the sealant caused by this movement will determine whether the sealant will perform successfully. The significance of S_{max} is that it is the outer fiber of the sealant that has the greatest strain, and failure of a sealant in extension or compression will occur in this outer fiber. Therefore, knowing the strain on the outer fiber that a given sealant can withstand allows one to design a joint with any depth-to-width ratio. The flatter the cross-sectional

Table 1. Insulation Requirements for Concrete Walls and Floor Slabs above Ground Concrete placed at 50°F (10°C)

Wall thickness, ft (m)	Min air temp allowable for these thicknesses of commercial blanket or bat insulation, °F (°C)			
	0.5 in. (1.3 cm)	1.0 in. (2.5 cm)	1.5 in. (3.8 cm)	2.0 in. (5.1 cm)
Cement content, 300 lb/cu yd (178 kg/cu m)				
0.5 (0.15)	47 (8.3)	41 (5.0)	33 (0.56)	28 (−2.2)
1.0 (0.30)	41 (5.0)	29 (−1.7)	17 (−8.3)	5 (−15)
1.5 (0.46)	35 (1.7)	19 (−7.2)	0 (−18)	−17 (−27)
2.0 (0.61)	34 (1 1)	14 (−10)	− 9 (−23)	−29 (−34)
3.0 (0.91)	31 (−0.56)	8 (−13)	−15 (−26)	−35 (−37)
4.0 (1.2)	30 (−1.1)	6 (−14)	−18 (−28)	−39 (−39)
5.0 (1.5)	30 (−1.1)	5 (−15)	−21 (−29)	−43 (−42)
Cement content, 400 lb/cu yd (237 kg/cu m)				
0.5 (0.15)	46 (7.8)	38 (3.3)	28 (−2.2)	21 (−6.1)
1.0 (0.30)	38 (3.3)	22 (−5.6)	6 (−14)	−11 (−24)
1.5 (0.46)	31 (−0.56)	8 (−13)	−16 (−27)	−39 (−39)
2.0 (0.61)	28 (−2.2)	2 (−17)	−26 (−32)	−53 (−47)
3.0 (0.91)	25 (−3.9)	− 6 (−21)	−36 (−38)	
4.0 (1.2)	23 (−5.0)	− 8 (−22)	−41 (−41)	
5.0 (1.5)	23 (−5.0)	−10 (−23)	−45 (−43)	
Cement content, 500 lb/cu yd (296 kg/cu m)				
0.5 (0.15)	45 (7.2)	35 (1.7)	22 (−5.6)	14 (−10)
1.0 (0.30)	35 (1.7)	15 (−9.4)	− 5 (−21)	−26 (−32)
1.5 (0.46)	27 (−2.8)	− 3 (−19)	−33 (−36)	−65 (−54)
2.0 (0.61)	23 (−5.0)	−10 (−23)	−50 (−46)	
3.0 (0.91)	18 (−7.8)	−20 (−29)		
4.0 (1.2)	17 (−8.3)	−23 (−31)		
5.0 (1.5)	16 (−8.9)	−25 (−32)		
Cement content, 600 lb/cu yd (356 kg/cu m)				
0.5 (0.15)	44 (6.7)	32 (0)	16 (−8.9)	6 (−14)
1.0 (0.30)	32 (0)	8 (−13)	−16 (−27)	−41 (−41)
1.5 (0.46)	21 (−6.1)	−14 (−26)	−50 (−46)	−89 (−67)
2.0 (0.61)	18 (−7.8)	−22 (−30)		
3.0 (0.91)	12 (−11)	−34 (−37)		
4.0 (1.2)	11 (−12)	−38 (−39)		
5.0 (1.5)	10 (−12)	−40 (−40)		

Insulating material	Equivalent thickness, in. (mm)
1 in. (31 mm) of commercial blanket or bat insulation	1.000 (25.4)
1 in. (31 mm) of loose-fill insulation of fibrous type	1.000 (25.4)
1 in. (31 mm) of insulating board	0.758 (19.2)
1 in. (31 mm) of sawdust	0.610 (15.5)
1 in. (31 mm) (nominal) of lumber	0.333 (8.45)
1 in. (31 mm) of dead-air space (vertical)	0.234 (5.94)
1 in. (31 mm) of damp sand	4.023 (102.2)

shape of a given sealant, the greater the extension or compression of the joint. Expressed another way, at the same extension, the flat joint will induce a smaller strain in the joint material than the narrow, deep joint. Figure 5 illustrates the effect of shape factor (ratio of depth to width) upon the strain on the outer fiber of a sealant. This relationship will hold true regardless of the type of sealant under consideration. The strain on the outer fiber of the sealant for a given movement and shape factor can be determined mathematically or graphically from Fig. 6. For example, for a joint 1 in. deep and ½ in. wide $D_x/W_{max} = 1/0.5 = 2$, and for a linear expansion of 100%, $S = 160\%$. To prevent bond of the sealant on the bottom of the joint slot a bond breaker must be used to allow free movement of the sealant. This bond breaker can be silicone-treated paper, polyethylene tape, or even strips of newspaper. In order to maintain the proper shape factor

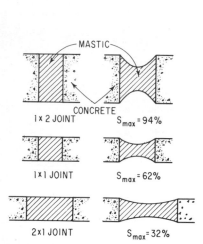

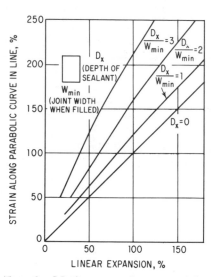

Fig. 5. Strain on the extreme fiber of a sealant for a ½-in. extension of different joint designs. (*From Ref. 5.*)

Fig. 6. Maximum strain on a joint plotted against percentage of linear expansion, applicable to any joint and any sealant. (*From Ref. 5.*)

it is often necessary to fill the bottom of the joint slot with a compressible, nonabsorptive backup material. Materials which have proved good for backing up field-molded joint sealants are closed-cell neoprene, polyurethane, or polystyrene foam. Materials composed of wood fibers and other organic materials that will rot or saturate with water have proved unsatisfactory.

The choice of a particular sealant will depend upon the movement anticipated in a given joint, physical abuse which might be encountered in service, and the physical environment of the installation (injurious solutions, temperature, water pressure, etc.).

Properties and uses of several types of field-molded sealants are summarized in Table 2.

10   Types of Field-molded Sealants

Mastics. Recommended only for use in joints subject to little strain or no movement. The group of sealants classified as mastics are composed of viscous

Table 2. Field-molded Sealants

Base compound	Uses	General comments
Self-leveling and nonsag thermosetting elastomers—chemical-curing type—little or no shrinkage on curing		
Epoxy.............	For caulking purposes	Relatively high modulus as compared with other sealants. Limited elongation
Polysulfide...........	Caulking, glazing, expansion and construction joints	Good weathering and ozone resistance. High elongation. Low modulus. Resistant to solvents, fuels, and other chemicals
Polyurethane.........	General caulking and glazing	Good weathering properties. Good resistance to fuel and chemicals. High modulus. Limited elongation
Silicones.............	Caulking and glazing	Good weathering, limited elongation, maintains properties over wide temperature range
Polysulfide coal tar.....	Highway, airport, and bridge joints	Low modulus. Good fuel resistance Good elongation
Thermosetting sealants—solvent-release type—shrink on curing		
Neoprene.............	General caulking and glazing	Good weathering. Moderate elongation. Small cross sections
Chlorosulfonated polyethylene Hypalon	General caulking and glazing	Good weathering. Nonstaining, exhibits shrinkage, for joints with small cross sections. Moderate elongation
Butyl rubber.........	General caulking and glazing	Good weathering characteristics. Limited elongation. Good in all colors
Thermoplastic sealants—hot-melt type		
Lead................	Joints in floors having heavy traffic	Joints subject to no movement. Good impact resistance. No elongation
Rubber asphalt........	Concrete highway joints	Good for use in nonfuel areas. Limited elongation
Rubber coal tar.......	Concrete highway joints	Good for use in nonfuel areas. Limited elongation
Thermoplastic sealants—solvent-release type—shrink on curing		
Acrylic..............	General caulking and glazing	Limited elongation
Vinyl...............	General caulking and glazing	Limited elongation
Rubber asphalt........	Joints in canal linings	Limited elongation
Rubber asphalt emulsion	Narrow cracks	Limited elongation
Mastics—trowel or gun grade		
Oleoresinous...........	Caulking and glazing	Very limited elongation. Short service life
Polybutene............	Sealing of butt and lap joints	Good weathering properties. Good adhesion. Use only for compression joints and where material is concealed. Limited elongation
Asphalt..............	Construction joints, tanks, and canals	Limited elongation

liquids rendered immobile by the addition of fibers and fillers. The vehicle in mastics may be a nondrying oil, a polybutene, a low-melting-point asphalt, or any combination of these materials. The filler used may be asbestos fiber, fibrous talc, or a finely divided calcareous or siliceous material.

Thermoplastics. Recommended maximum strain on the outer fiber 5 to 10% depending on the formulation. Thermoplastics include such materials as asphalts, rubber asphalts, pitch, and coal tar. They will become soft upon heating and will harden when cooled. For design purposes these are considered elastomers with limited extensibility.

Thermosetting Elastomers. Recommended maximum strain on the outer fiber 10 to 25% depending on the formulation. The thermosetting plastic sealants can be two-component liquid systems, which react after installation to form elastomers, or thermosetting plastics in solvent solution which set or cure by evaporation of solvent. These materials will not soften appreciably in applications up to 225°F.

Preparation of Joint Slot. The surfaces of the joint slot must be clean and free of oil or dust, and dry in order that the sealant may bond properly. Generally speaking, even joints cast in new concrete will be dirty; form oil, loose mortar, and dust may be present. Sawed joints will be dusty and should be blown out with oil-free air. Cast joints can best be cleaned by mechanical abrading. This can be accomplished by sandblasting, power or hand wire brushing, or power-driven mechanical routers. If any honeycombs or faults exist on the side of the joint slot these should be patched prior to application of the sealant.

11 Premolded Joint Sealants

Premolded joint sealants are either embedded in the concrete, termed water stops, or installed by compressing an extruded shape into a joint slot, commonly termed compression gaskets.

Water Stops. Water stops are either sheets of metal, Z- or N-shaped, or extruded natural or synthetic rubber or plastic installed in the plastic concrete by use of split forms.

Metal water stops may be of steel, stainless steel, copper, or lead. Copper is the most expensive metal water stop; it has high resistance to corrosion but is easily damaged in construction. Lead water stops enjoyed popularity in the past but are seldom used today. Steel is the stiffest of the metal water stops; if plain carbon steel is used it is subject to corrosion.

Splicing of steel water stops can be accomplished by welding. Copper water stops can be spliced by brazing or soldering.

Flexible water stops are made of natural or synthetic rubber, polyvinyl chloride, or other plastics. They are quite often extruded with a hollow bulb in the center so that movement can be accommodated with minimum strain on the sealant. Complex shapes are available to assure better mechanical grip to the concrete. Although natural- and synthetic-rubber water stops are available the most commonly used flexible water stops are manufactured of polyvinyl chloride.

Splicing of rubber water stops can be accomplished by use of special cements. Polyvinyl chloride water stops can be spliced by heating the ends and pressing them together when they are in the molten state. Many grades of polyvinyl chloride are available, and a grade should be chosen that will be flexible at the lowest anticipated service temperature. Figure 7 illustrates some of the shapes available in flexible water stops.

Installation. Water stops, both metal and flexible, are installed by the use of split forms as shown in Fig. 8. Care must be exercised when placing the concrete so that the water stop is not bent or misplaced. Compaction and vibration of the concrete adjacent to the water stop must be thorough so that no honeycombs or voids exist on the underside of the water stop. Since the water stop will be inaccessible after the concrete has hardened, the success for the installation will be dependent upon the position of the water stop and the thoroughness of the compacting of the concrete.

Compression Gaskets. Compression gaskets are premolded shapes of either neoprene foam, extruded cellular neoprene shapes, or bitumen-impregnated polyurethane foam. They are installed by compressing the gasket into the joint slot. The joint must be so designed that the gasket will always be under compression

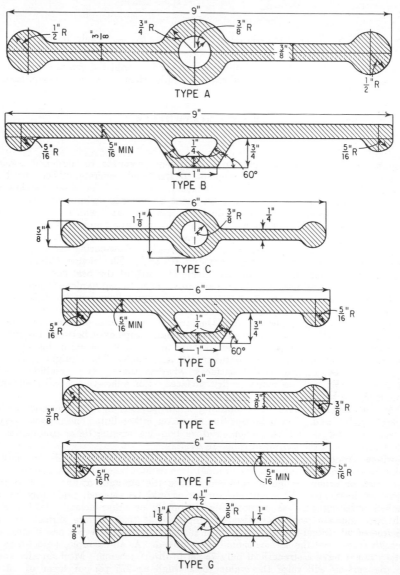

Fig. 7. Typical shapes available in flexible water stops. (*From Ref. 5.*)

even at the widest opening anticipated. Sealing is accomplished only if the gasket is under compression. Compression gaskets are useful where little or no water pressure is anticipated such as in slabs on grade or above grade joints. Neoprene is the most suitable elastomer for compression gaskets as it has the greatest resistance

PLACING CONCRETE

FORMWORK

FIG. 8. Typical installation of water stops by use of split forms.

to compression set. Quite often adhesives are used to hold the compression gaskets in place. The use of adhesives is always desirable when using compression gaskets.

OTHER MATERIALS

12 Adhesives

Organic adhesives are available to bond plastic concrete to hardened concrete or other materials, and hardened objects to hardened concrete. Those used for structural purposes are based on epoxy resins; those used for small patches and other nonstructural purposes can also be based on various organic latexes.

Structural Adhesives. Structural adhesives available and which have a proved service record are all based on epoxy-resin systems. For bonding plastic concrete to hardened concrete the only proved systems are based on epoxy-resin polysulfide polymer systems. These systems are, of necessity, slow-curing as they cannot cure faster than concrete or mortar gains strength. The bonds obtained using these systems are far in excess of the tensile strength of the best concrete. These systems can also be used to bond hardened concrete to hardened concrete, or metal devices, anchor bolts, and plates to concrete; however, for this purpose the slow cure of the epoxy polysulfide systems may be a disadvantage. For bonding hardened objects to hardened concrete, the straight epoxy-resin systems are generally more suitable. Where the concrete is rough or large gaps exist between the surfaces to be bonded, the straight epoxy-resin systems are available in gel or high-viscosity form to facilitate filling the gap between the adherends. The straight epoxy-resin systems must be applied to dry surfaces, whereas the epoxy-polysulfide systems will work equally well on damp or dry surfaces. For adhesive use, all epoxy-resin systems should be 100% solids containing no solvent.

Nonstructural Adhesives. For bonding small patches or thin overlays, organic latexes can be used. These latexes are based on either butadiene styrene, acrylic, or polyvinyl acetate resins. When used as bonding agents, these compounds are generally mixed in the mortar or concrete overlay.

Surface Preparation. The strength of any adhesive system will be only as strong as the weakest link. Surface preparation cannot be overemphasized; laitance, dust, dirt, or other foreign matter will reduce the strength of the bond of these adhesives drastically. The concrete surface should be cleaned, preferably by mechanical abrading such as sandblasting, chipping, or other means. Acid washing with 10% muriatic acid may also be employed if the concrete surface is sound and free of oil. Steel-troweled or formed surfaces are usually weak, and if structural strength is required the surface should be removed. As an example, a steel-troweled surface might have a strength of only 75 psi in direct tension. Mechanically abrading the surface will raise this value to as high as 275 psi. A layer of oil can reduce this bond value to zero.

Application—Structural Adhesives. For bonding plastic concrete to hardened concrete the two components of the epoxy-polysulfide adhesive must be thoroughly mixed and applied as soon as possible after mixing. Concrete or mortar can then be placed at any time during which the adhesive remains tacky. This may vary anywhere from 30 min to 2 hr depending upon the particular system. During hot weather, the curing time of these resins is faster and the contact time will be reduced. If thin portland-cement concrete or mortar overlays are being applied, great care should be exercised that the thin overlay is kept moist, as drying will result in shrinkage, curling, and loss of bond of the overlay. The epoxy-polysulfide adhesives are vapor barriers before cure and the overlays will dry only on the top surface. This drying will tend to curl the thin overlay away from the adhesive; wet curing or the application of a curing compound should be carried out as soon as possible.

When bonding new concrete to old concrete which has high absorption, the old concrete should be wet with water to prevent the adhesive from being absorbed into the old concrete. The concrete should be wet but have no free puddles standing on the surface.

Application of the bonding compound can best be carried out by the use of brushes or brooms, although automatic spraying and mixing equipment is available for this purpose.

For bonding hardened concrete to hardened concrete or other objects to concrete, straight epoxy systems are employed. If the surfaces to be bonded do not mate, the epoxy adhesive can be mixed with fine sand to form a mortar, although for this purpose the high-viscosity or gelled epoxy-resin adhesives are easier to use.

Nonstructural Adhesives for Bonding Concrete or Mortar Overlays with the Organic Latexes. The latexes are generally diluted with water and added to the cement and aggregate. The proportion of latex should be approximately 15% resin solids based on the weight of the cement. The mixed mortar should be placed on the area to be patched, and this mortar should be scrubbed into the concrete with a stiff brush. The remainder of the mortar should then be screeded into position before the scrub coat has dried.

13 Floor Hardeners

Floor hardeners currently employed in the industry are based on either fluosilicates or organic resins. Fluosilicates combine chemically with the free lime in the concrete and reduce dusting and increase hardness slightly. The fluosilicates may be salts of zinc or magnesium or a combination thereof. Some proprietary compounds contain wetting agents and/or organic acids to improve their efficiency. Fluosilicate hardeners should be applied as aqueous solutions at the rate of approximately 60 to 150 sq ft per gal. The fluosilicate content should be between 1 and 1½ lb per gal.

Combination curing and hardening compounds are marketed. These compounds are usually chlorinated-rubber or butadiene-styrene based coatings. They do not actually harden the floor but apply a protective lacquer to the concrete. Since these materials are classified and used as curing compounds they should be applied at a rate not exceeding 200 sq ft per gal as covered by ASTM Specification C 309. Higher coverages will result in little or no curing of the concrete. Neither the fluosilicate nor the resinous-type floor hardeners will overcome basic faults in the concrete such as the results of overfinishing, high water content, or carbonation.

REFERENCES

1. "Concrete Manual," 7th ed., U.S. Bureau of Reclamation, 1963.
2. Schutz, R. J.: Setting Time of Concrete Controlled by Use of Admixtures, *J. ACI*, January, 1959.
3. Pearson, A. S.: Fly Ash Improves Concrete and Lowers Its Cost, *Civil Eng.*, September, 1953.
4. ACI Committee 306 Report, Recommended Practice for Cold Weather Concreting, (ACI 306-66).
5. Schutz, R. J.: Shape Factor in Joint Design, *Civil Eng.*, October, 1962.
6. ACI Committee 212 Report, Guide for Use of Admixtures in Concrete, *J. ACI*, September, 1971.

Section 2

PROPERTIES OF CONCRETE

Chapter 5

WORKABILITY

GEORGE W. WASHA

ELEMENTS OF WORKABILITY AND CONSISTENCY

1 General

All concrete mixtures must be properly proportioned to have satisfactory economy, workability, required strength, and durability properties. These different objectives require differences in proportioning, and consequently most concrete mixtures are compromises rather than mixtures having the best workability, or the highest compressive strength at a given age or the greatest economy. Excellent workability, for example, normally requires high cement and fine-aggregate contents with a low coarse-aggregate content and a relatively high water content. Such a mixture would certainly not be economical, and its properties would not be optimum. Consequently as the proportions of a given mixture are changed to provide greater workability the effects on the other properties must be considered, and the desired improvement in workability should be obtained by means of such changes as have the least harmful effect on the other desired properties. In other words, the concrete mixture must have sufficient workability to enable satisfactory placement under job conditions, sufficient strength to carry design loads, sufficient durability to allow satisfactory service under expected exposure conditions, and necessary economy not only in first cost but in ultimate service.

Workability of a concrete mixture may be defined as the ease with which it may be mixed, handled, transported, and placed into its final position with a minimum loss of homogeneity. Workability is dependent on the proportions and physical characteristics of the ingredients, but there is no general agreement on the properties of a concrete mixture that may be used as measures of workability. Workability also depends on the mixing, transporting, and compacting equipment used, on the size and shape of the mass of concrete to be formed and on the spacing and size of reinforcement. Further, workability is relative because a concrete may be considered workable under some conditions and unworkable under others. For example, concrete having satisfactory workability for a pavement slab would be difficult to place in a heavily reinforced concrete column. Special considerations are usually involved for concrete to be transported by pump or placed under water.

Consistency or fluidity of concrete is an important component of workability which relates to the degree of wetness of concrete. It is measured by ball penetration, slump, and flow tests. Normally wet concretes are more workable than dry concretes, but concretes of the same consistency may vary greatly in workability. Other components of workability may be determined from a trowel test that will allow evaluation of such properties as harshness, stickiness, cohesiveness, and ease of manipulation. Another measure of workability may be obtained from the bleeding test, which determines the tendency of the water to separate from the other constituents of the concrete and to rise to the top of the concrete mass.

Since all the properties needed to determine the workability of a concrete mixture correctly are either not known or are impossible to measure, systematic visual inspection must be used along with the results of the consistency, trowel, and bleeding tests to ensure the use of concrete with satisfactory workability. Constant vigilance on the part of the inspector is necessary to avoid the undesirable effects resulting from the use of a concrete with improper workability.

MEASUREMENT OF WORKABILITY

2 Slump Test

Despite its limitations the slump test is widely used to measure the consistency of concrete. It is commonly the first test made on a sample of freshly mixed concrete and frequently determines whether the batch will be accepted or rejected. Details of the slump test are given in ASTM Standard Method of Test for Slump of Portland Cement Concrete (C 143). The test is conducted with a standard slump mold 12 in. high having a base diameter of 8 in. and a top

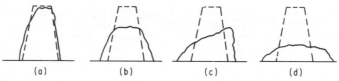

Fig. 1. Types of slump. (*a*) Near-zero slump. (*b*) Normal slump. (*c*) Shear slump. (*d*) Collapse slump.

diameter of 4 in. The dampened mold is placed on a smooth, horizontal, nonabsorbent surface and is filled in three layers, each approximately one-third of the volume of the mold, while the mold is held firmly in place. Each layer of concrete is tamped 25 times with a round, straight steel rod ⅝ in. in diameter and 24 in. long, having one end rounded to a hemispherical tip. The bottom layer is rodded throughout its depth and each remaining layer is rodded with strokes that just penetrate the underlying layer. Strokes should be uniformly distributed over the cross section of the mold. As the third layer is rodded additional concrete, if needed, should be added to keep the level of concrete above the top of the mold at all times. After completion of rodding, any excess concrete should be removed by means of a rolling and screeding motion of the tamping rod. The slump cone is then lifted vertically in about 5 sec with a steady upward lift, with care being taken to avoid any lateral or twisting motion. The slump mold is then placed gently near the settled concrete mass and the slump is measured by placing the tamping rod across the top of the mold and reading the vertical distance from the center of the settled concrete mass to the bottom of the tamping rod with ruler as shown in Fig. 1, Chap. 12. Slump is usually determined to the nearest ¼ in.

The slump test may be quickly performed and provides some answers that are helpful in evaluating the workability of the concrete. However, at best, only partial answers are obtained and in some situations they may be of doubtful value. The concrete cone after removal of the metal cone will subside into one of the four forms shown in Fig. 1. The near-zero slump shown in Fig. 1*a* may be the result of a concrete that has all the requirements of good workability except that the water content is too low, or it may be the result of a harsh concrete mixture that is free-draining and allows the water to run out of the concrete mass without causing any significant change in subsidence. Special care must be used when lightweight aggregates are used because many are angular and interlocking and produce a harsh concrete, unless special precautions are taken to increase workability by additions of fine sand, air-entraining agents, or special admixtures. When the normal slump cone, as shown in Fig. 1b, is obtained, the concrete should usually have good to excellent workability. The slump used for placing most structural concrete ranges between 2 and 7 in. The shear slump shown in Fig. 1c indicates that the concrete lacks plasticity and cohesiveness and that the results of the slump test are of questionable value. The collapse slump shown in Fig. 1d is obtained with lean, harsh or extremely wet mixes. Concrete of this type normally has slump values ranging between 7 and 10 in., and in the slump test the grout or mortar tends to run out of the concrete, leaving the coarser material in the center. Generally, concrete exhibiting shear or collapse slumps has poor or unsatisfactory placeability.

The slump test is not considered applicable to nonplastic and noncohesive concrete, and is further limited to concrete containing coarse aggregate not over 2 in. in size.

3 Ball-penetration Test

A measure of consistency may also be determined by means of the ASTM Standard Method of Test for Ball Penetration in Fresh Portland Cement Concrete

FIG. 2. Measuring ball penetration of concrete.

C (360). Essentially, this test consists of placing a 30-lb metal cylindrical weight, 6 in. in diameter and 4⅝ in. in height, having a hemispherically shaped bottom, on the smooth level surface of the concrete and determining the depth to which it will sink when released slowly. During penetration the handle attached to the weight slides freely through a hole in the center of the stirrup which rests on large bearing areas set far enough away from the ball to avoid disturbance when penetration occurs. The depth of penetration is obtained from the scale reading on the handle attached to the weight using the top edge of the independent stirrup as the line of reference. Penetrations are measured to the nearest ¼ in., and each reported value should be the average of at least three penetration tests. The apparatus is shown in Fig. 2.

This test can be quickly made, is less subject to personal errors, and does not require molding a specimen as does the slump test. The test may be performed on concrete as placed in the forms prior to any manipulation, or in suitable containers such as pans, hoppers, or wheelbarrows. The depth of the concrete should not be less than 8 in. or three times the maximum size of aggregate depending on which of the two values is larger. The centerline of the handle should be more than 9 in. horizontally from the nearest edge of the level surface.

The ratio of slump to penetration is usually reported between 1.3 and 2.0. The variations are apparently due to differences in mix proportions, size and type of aggregate, and test conditions. In the absence of comparable field data the ratio of slump to penetration may be approximated as 1.67. However, where possible, the ratio should be determined under actual job conditions (see Chap. 12).

4 Trowel Test

This is not a standard test, but it may be used to provide valuable information about the workability characteristics of any concrete mixture. The test consists

(a)

(b)　　　　　　　　　　　　　　　　(c)

FIG. 3. Trowel workability. (a) Insufficient mortar. (b) Excess mortar. (c) Correct amount of mortar.

of subjecting the concrete mixture to troweling action and observing the behavior of the mixture under this action. The cohesiveness may be judged by noting if concrete sticks together and also if it tends to stick to the trowel under the troweling action. The magnitude of bleeding, if any, may be visually observed around the perimeter of the concrete batch. The appearance of the surface after troweling may suggest changes in the proportions of the concrete mixture. Figure 3a to c shows surface conditions of three different concrete mixtures. In Fig. 3a the mixture contains insufficient mortar to fill the voids in the coarse aggregate. This mixture would be difficult to place and would result in a porous concrete with honeycombed surfaces. The concrete mixture in Fig. 3b has an excess of mortar and is plastic and workable. It would produce smooth surfaces, but the concrete may be porous. In addition, this concrete would be uneconomical because of a low yield. In Fig. 3c the concrete mixture contains the correct amount of mortar. The voids between the coarse-aggregate particles have been filled with mortar under light troweling, but no excess mortar is evident. This workable mixture will provide maximum yield of concrete for a given amount of cement.

5   Bleeding Test

The tendency for water to rise to the surface of freshly placed concrete, known as bleeding, may be determined from the ASTM Standard Method of Test for Bleeding of Concrete (C 232). The test consists of placing concrete in a ½cu. ft container to a height of $10 \pm \frac{1}{8}$ in. using standard procedures. The surface is troweled smooth and level under minimum action, and the concrete is then allowed to stand at a temperature between 65 and 75°F on a level floor free from vibration. At specific intervals the water collecting on the surface is drawn off with a pipette and measured. The rate of bleeding and the accumulated bleeding water, expressed as a percentage of the net mixing water in the test specimen, may be obtained.

Bleeding might be regarded as desirable because the decrease in water content must lead to a decrease in water-cement ratio. However, bleeding disturbs the internal homogeneity of the concrete and causes other results that are not desirable. Because of the water gain in the top portion of freshly placed concrete it tends to be weak and porous and subject to disintegration by freeze-and-thaw action or by percolation of water. The water rising to the surface may carry fine inert particles of cement that weaken the top portion and form a scum called laitance which must be removed if a new layer of concrete is to be placed over the original layer. In addition, as the water rises through the concrete flow channels are formed in the concrete mass and water accumulates under the coarse-aggregate particles and under horizontal reinforcing bars. This action results in a weaker concrete structure because of the lack of bond between the paste and coarse aggregate and between the concrete and reinforcing steel. As a result concrete with a large amount of bleeding water may be very permeable and the reinforcing steel may be subject to corrosion.

Bleeding may be largely controlled by the proper choice of constituents and the proportions of the concrete mixture. Richer mixtures made with finely ground cements having normal setting properties, minimum amounts of mixing water, smooth natural sands with an adequate percentage of fines, air-entraining admixtures, or admixtures consisting of fine particles are all helpful in decreasing bleeding of concrete mixtures. Some improvement of the properties of concrete that has had appreciable bleeding may be obtained by vibrating or tamping the concrete at the end of bleeding and at the beginning of setting.

EFFECTS OF CONSTITUENTS

6   Cement

The workability of a concrete mixture depends on the amount of cement, the fineness of the cement, and its chemical composition. Very lean mixtures are apt to be harsh and have poor workability because of the lack of sufficient material of cement fineness. Generally, other things being equal, workability increases as the amount and fineness of the cement increase. Extremely rich mixtures, however, may be too cohesive or sticky.

In some instances the necessary period for proper workability before stiffening may not be available. This premature stiffening or false set is believed due to unstable gypsum in the cement resulting from high finishing temperatures in the manufacture of the cement. It is usually possible to rework the stiffened concrete to its original workability without the addition of extra mixing water. Remixing after development of false set will prevent further stiffening until the start of true setting and hardening action. Improperly proportioned or manufactured cements may cause a flash set accompanied by an appreciable liberation of heat. This condition is permanent and cannot be improved by additional mixing or agitation.

7　Aggregates

The grading and shape of the fine and coarse aggregates and the maximum size of the coarse aggregate have important effects on workability. Both the fine and coarse aggregates should be uniformly graded from fine to coarse and should be free from an excessive amount of any one size fraction that would tend to cause particle interference and would result in poor workability. It appears that for any given aggregate a large number of gradings may be used which will be about equally satisfactory for workability provided that reasonable limits and uniformity of grading are maintained. Gap gradings in which one or more of the intermediate size fractions have been eliminated should be checked under job conditions before adoption.

Natural sands with rounded grains produce more workable concrete than crushed sands made of angular, flat, or elongated pieces. The latter types usually have a high percentage of voids and may cause excessive bleeding of the concrete. The coarse-aggregate particles should preferably be spherical in form. Crushed coarse aggregates cubical in shape will, if properly graded and combined with the proper amount of workable mortar, produce workable concrete. Flat, disk-shaped pieces and long, thin wedgelike particles are objectionable because they cannot be easily and closely compacted.

The maximum size of coarse aggregate must be selected for each specific construction condition. The choice will usually involve consideration of such factors as spacing of reinforcing bars, minimum width of form, and methods of placing and compacting the concrete mass.

8　Admixtures

Workable concrete mixtures made with satisfactory aggregates, sufficient cement, and the correct amount of water to produce the required slump do not normally require any additions of admixtures for satisfactory workability. However, admixtures are helpful in lean, harsh concrete mixtures of poor workability and where difficult placement conditions are involved.

In order to increase workability of a lean concrete it is necessary to increase the surface area of the solids per unit volume of water. This may be accomplished by adding finely divided admixtures such as hydrated lime, diatomaceous earth, bentonite, or fly ash or by increasing the cement content. The amount of the finely divided admixture added to the concrete mixture must be carefully controlled since large quantities tend to require a higher water-cement ratio with undesirable effects on strength, durability, and drying shrinkage unless appropriate adjustments are made.

A large variety of proprietary compounds are available that affect various workability characteristics. These compounds are classified as wetting agents, dispersing agents, densifying agents, retarders, accelerators, and air-entraining agents. When these compounds are properly used to overcome certain deficiencies of a concrete mixture they may be very effective. Indiscriminate use of these compounds, however, may produce harmful rather than desirable effects.

Air-entraining agents are usually added to increase the durability properties of concrete. In addition, the billions of disconnected small air bubbles that they produce increase workability, decrease bleeding and segregation, decrease density, and may decrease strength. Air entrainment permits a reduction of the sand content in a concrete mixture approximately equal to the volume of the entrained air. It also permits a reduction of the mixing water, about 2 to 4% for each percent of entrained air, with no loss of slump and some gain in workability. While purposeful air entrainment improves workability of concrete mixtures it may cause problems related to finishing of horizontal surfaces. The marked reduction in bleeding of air-entrained concrete usually requires that finishing be accomplished much more quickly than for concrete containing no entrained air. This is especially true where the water evaporation from the surface is rapid as in the case of a slab exposed to hot, dry winds. Under such conditions the water that is very rapidly evaporated from the surface is not replaced by bleeding water. Concrete containing entrained air can be properly finished, but special care and understanding are required.

9 Temperature

Extreme temperatures produce additional problems in placing concrete and may cause undesirable changes in properties. The difficulties due to placement of concrete in cold or hot weather must be anticipated and provision must be made to avoid or to lessen undesirable effects. In general this requires that mixing temperatures be kept between 55 and 80°F and that the concrete be protected during its early life.

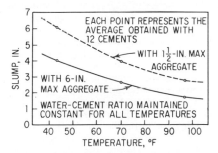

Fig. 4. Relation between slump and temperature of concrete made with two maximum sizes of aggregate.[1]

Concrete placed at low temperatures above freezing and kept from freezing will develop higher ultimate strength and greater durability than similar concrete placed at higher temperatures. The advantages of placing concrete under cold-weather conditions should be realized along with the limitations. Generally, the temperature of freshly placed concrete in thin sections should not be less than 55°F, while the minimum temperature for mass concrete should be at least 40°F. In order to obtain these temperatures under very cold conditions it may be necessary to heat the mixing water or the aggregate. Overheating of these constituents should be avoided because of a possible increase in the water requirement, a loss in slump, and the possibility of causing flash set. The relationship between slump and temperature of concrete is shown in Fig. 4. Frequently air-entraining agents and accelerators may be added to concrete that is to be mixed and placed under cold temperatures. The most commonly used accelerator, calcium chloride, is added in amounts of 1 to 2% by weight of the cement.

Hot weather also presents special problems in manufacture and placement of concrete. High temperatures usually accelerate the setting rate, increase evaporation of mixing water, increase the required amount of mixing water, reduce ultimate strength, and increase the tendency to crack. Undesirable effects of hot weather may be minimized by cooling the ingredients, to keep the temperature of freshly placed concrete in the range 60 to 80°F, and to protect and cure the concrete promptly after placement. Water-reducing retarders are used to counteract the accelerating effects of high temperatures and to decrease the need for increased mixing water. See Sec. 8 for detailed discussion of hot- and cold-weather concreting.

SEGREGATION

10   Factors Influencing Segregation

Segregation of the various constituents of fresh concrete is detrimental, and every effort should be made to minimize it. Imperfections in hardened concrete such as rock pockets, sand streaks, weak and porous layers, crazing, pitting, and surface scaling are usually related to segregation. Repair of the damaging effects due to segregation is difficult and costly. It is much better to avoid segregation by using well-designed mixtures and by placing the concrete properly under competent supervision.

Segregation of concrete is always a problem because concrete is not a homogeneous material but an aggregation of constituent materials that vary widely in size and specific gravity. Special care to avoid defects due to segregation must be taken when the concrete mixture is very lean, or very wet, or contains rough-textured aggregate not of cubical or spherical shape or if the maximum aggregate size is large in comparison with the dimensions of the member to be formed. Segregation is not limited to the solids in a concrete mixture. The mixing water tends to rise as the heavier solid particles of aggregate and cement settle through it. This type of segregation may be apparent during placing but it is most evident after placement.

Proper proportioning is only one of the important factors that influence segregation. Every step in the manufacturing and placing procedures must observe time-proved methods in order to minimize segregation. Accurate batching and thorough mixing are essential. After mixing, every operation involved in transporting and placing the concrete affords further opportunity for loss of uniformity. Filling and discharging concrete from hoppers, buckets, or cars and discharging from chutes, belt conveyors, or buckets, and placing concrete in the forms by hand methods or vibration provide further opportunity for segregation. Special problems are encountered when the concrete is to be pumped, when it is to be placed under water, when the reinforcement is closely spaced, or when the mold is complex and has sharp corners.

Segregation must be avoided. It will not be corrected automatically in the succeeding operations required in the construction of any concrete structure. Every operation involved in handling, placing, and consolidating the concrete mixture must be carefully planned and controlled to avoid segregation. In general, the concrete mixture should be no wetter than necessary, it should be allowed to fall vertically in a continuous flow, and it should be placed as near to its final location as possible to avoid excess lateral motion. Correct and incorrect methods of handling, placing, and consolidating concrete are discussed in detail in Chap. 24.

REFERENCES

1. "Concrete Manual," 7th ed., U.S. Bureau of Reclamation, 1963.
2. ACI Committee 614 Report, Proposed Recommended Practice for Measuring, Mixing, and Placing Concrete (ACI 614-72), *J. ACI,* July, 1972.

Chapter 6

STRENGTH AND ELASTIC PROPERTIES

GEORGE W. WASHA

STRENGTH TESTS

1 General Considerations

Since concrete is a hardened mass of heterogeneous materials its properties are influenced by a large number of variables related to differences in types and amounts of ingredients, differences in mixing, transporting, placing, and curing, and differences in specimen fabrication and test details. Because of the many variables, methods of checking the quality of the concrete must be employed. The usual procedure is to cast strength specimens at the same time that the structure is placed and to regard the specimen strength as a measure of the strength of the concrete in the structure. The reliability of this assumption should always be questioned because of different curing conditions for the specimen and the structure, because poor workmanship in placing concrete in the structure may not be reflected in tests of specimens, and because poor testing procedures may provide false results. A pattern of tests should be used rather than placing reliance on only a few tests to check uniformity and other characteristics of concrete. Statistical methods as given in ACI Standard 214 should be used where large quantities of concrete are involved.

Most concrete is proportioned for a given compressive strength at a given age, and consequently a compression test is most frequently used. A 6 × 12-in. cylinder is most commonly required but larger cylinders are frequently used with mass concrete to be placed in dams. Compressive strengths may also be determined from modified cube tests made on beam specimens remaining after flexural tests and on cores of various sizes cut from hardened concrete. The details of all strength tests are given in ASTM standards.

Concrete is not normally required to resist direct tensile forces because of its poor tensile properties. However, tension is important with respect to cracking due to restraint of contraction caused by chemical activity, drying shrinkage, or decrease in temperature. Tension tests also provide an approximate measure of the compressive strength. Direct tension tests of concrete are infrequently made because of difficulties in applying the tensile forces. The splitting tensile strength of concrete is determined by applying increasing diametral compressive loads on a 6 × 12-in. cylinder placed on its side until splitting failure occurs and then calculating the tensile strength from the maximum load and known dimensions (ASTM C 496).

Bending tests on beams are usually required when unreinforced concrete is to be subjected to flexural loading, as in the case of highway pavements. The beams are usually end-supported and loaded with a concentrated force at the center or with concentrated forces at the third points. The stress in the fiber farthest from the neutral axis, calculated from the flexure formula $S = Mc/I$, is called modulus of rupture. While this is a fictitious stress value because the assumptions on which the flexure formula is based are not valid near failure, it is useful for calculation and comparison purposes. It is usually from 60 to 100% higher than the tensile strength.

A standardized test procedure to determine the effects of variations in the properties of concrete on the bond strength between concrete and reinforced steel is provided by ASTM C 234. Tests to determine bond strength are infrequently performed except for research purposes. Tests to determine fatigue strength and strength under various types of combined loading are not standardized and usually are performed only for research purposes.

FACTORS AFFECTING STRENGTH

2   Mix Proportions; Water-Cement Ratio

If satisfactory materials are combined into a workable concrete mixture that is allowed to age under satisfactory curing conditions, the strength of the hardened concrete at a given age will be greatly influenced by the water-cement ratio of the mixture. The exact position of the strength vs. water-cement ratio curve will be dependent on the properties of each of the ingredients, the proportions of the ingredients, the mixing and placing methods, and the curing methods.

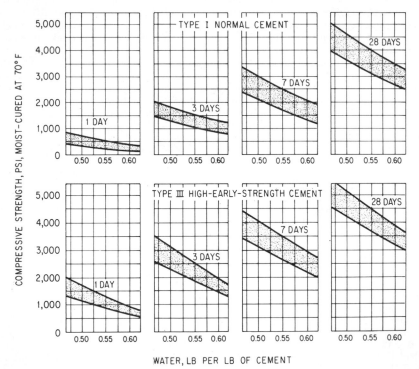

FIG. 1. Age-compressive strength relations for air-entrained concrete (concrete with air content within recommended limits and maximum aggregate size of 2 in. or less) made with Types I and III portland cements.[1] These relationships are approximate and should be used only as a guide in lieu of data on job materials.

Proportions of concrete mixtures are specified in many ways. Ready-mixed concrete producers are frequently asked to provide concrete with a given cement content, pounds of cement per cubic yard of concrete, and a given slump. In some instances concrete proportions may be specified by weight or volume as 1:2:4, 1 part of cement to 2 parts of fine aggregate to 4 parts of coarse aggregate. In other instances concrete proportions will be dictated by a required strength at a given age and a given slump. Regardless of the method of specifying proportions the most important consideration for good strength is the use of a workable mixture

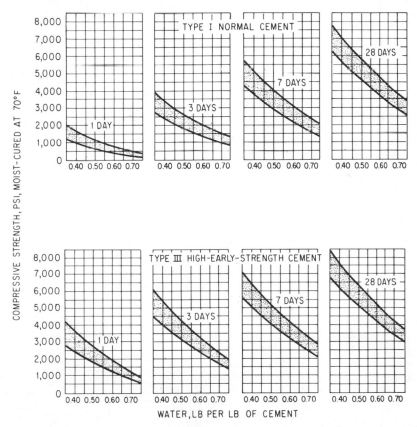

FIG. 2.   Age-compressive strength relations for non-air-entrained concrete made with Types I and III portland cements.[1] These relationships are approximate and should be used only as a guide in lieu of data on job materials.

with a water-cement ratio that is as low as possible. A concrete that does not have its water-cement ratio specified may have a wide range in strength. A 1:2:4 mixture, by weight, may have relatively good or poor strength depending on the amount of water added. If a desired strength is required the amount of water (or the slump) must be given along with the proportions of dry materials.

Relations between compressive strength and water-cement ratio for air-entrained concrete made with Types I and III portland cement for ages of 1, 3, 7, and 28 days are given in Fig. 1. In each instance a band of values is shown rather than a single curve in order to cover variations in materials and testing procedures. On a given job with reasonably uniform materials and procedures a single curve for each age can usually be obtained. Relations between compressive strength and water-cement ratio for non-air-entrained concrete made with Types I and

III portland cement for ages of 1, 3, 7, and 28 days are given in Fig. 2. When the compressive strength is known the flexural strength may be approximated from the equation

$$R = K \sqrt{f_c'}$$

where f_c' = compressive strength, psi
       R = flexural strength (modulus of rupture), psi
       K = a constant, usually between 8 and 10

3   Cement Content and Type

With materials, consistency, and density constant, the strength of concrete increases with the proportion of cement in the mixture until the strength of the cement or aggregate, whichever is the weaker, is reached. The data in Fig. 3 represent tests on moist-cured, workable concretes having the same slump, 4 to 6 in., and made from the same aggregate mixture, 38% sand and 62% gravel. For workable mixtures of drier consistencies, the corresponding curves would be higher and for those of wetter consistencies the curves would be lower. Also the use of higher-strength cements tends to move the curves higher.

The effect of cement fineness (Wagner), expressed as specific surface in square centimeters per gram of cement, on the compressive strength of concrete at four different ages is shown in Fig. 4. Finely ground cements are desirable in that they increase strength, especially at early ages, and also increase workability. They may be undesirable because they contribute to cracking and have lower resistance to freezing and thawing. The ASTM mini-

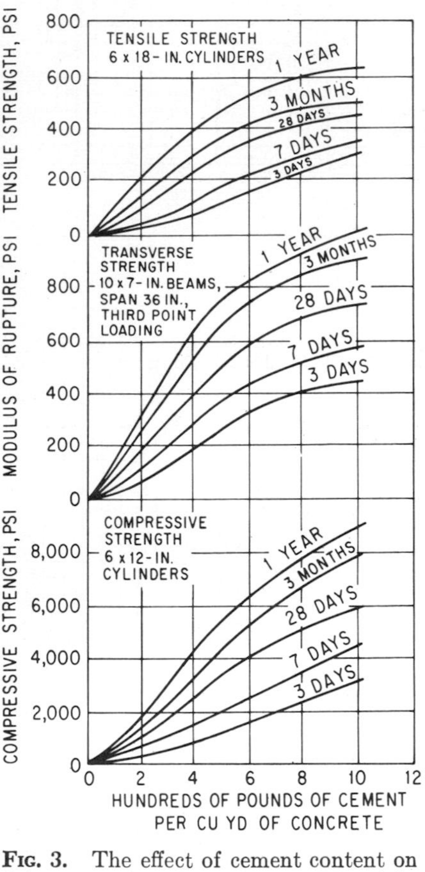

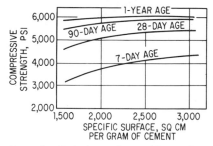

FIG. 3. The effect of cement content on the strength of concrete. Materials, Elgin sand and gravel graded from 0 to 1½ in. Fineness modulus of mixed aggregate = 5.5.[2]

FIG. 4. Relation of specific surface of cement to compressive strength of concrete.[3]

mum fineness requirements for portland cement are 1,500 sq cm per g (Wagner) and 2,600 sq cm per g (Blaine).

The curves in Fig. 5 show the change in compressive strength of concrete with age for the five standard types of portland cement specified by ASTM. All curves cross at an approximate age of 3 months. The concrete made with high-early-strength cement (Type III) has relatively high strength at ages up to 3 months but after that its strength is slightly lower than that of concrete made with normal

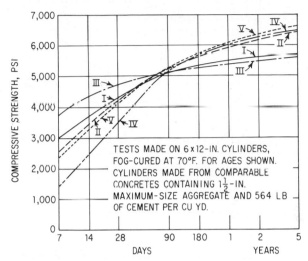

FIG. 5. Rates of strength development for concrete made with various types of cement.[4]

cement (Type I) and considerably lower than the strengths of concretes made with modified (Type II) or low-heat (Type IV) or sulfate-resisting (Type V) cement.

4 Aggregate Types and Characteristics

Heavy aggregates such as magnetite, barite, and scrap iron are used for making concrete (180 to 220 pcf) for radiation shielding and concrete counterweights. Light-weight aggregates such as expanded shales and clay, expanded slag, vermiculite, perlite, pumice, and cinders are used to make insulating, masonry unit, and light-weight structural concrete weighing between 25 and 120 pcf.

The principal materials used in normal-weight concrete, usually 140 to 155 pcf, include the sands and gravels, crushed rock, and air-cooled blast-furnace slag. The more commonly used crushed rocks are granite, basalt, sandstone, limestone, and quartzite. Aggregate characteristics that affect concrete strength include particle shape, texture, maximum size, soundness, grading, and freedom from deleterious materials. Usually the effect of type of normal-weight aggregate with satisfactory properties and gradation on the strength of concrete is small because the aggregates are stronger than the cement paste.

As the maximum size of aggregate in a concrete mixture of a given slump is increased the water and cement contents in pounds per cubic yard of concrete are decreased. The influence of maximum size of aggregate on compressive strength of concrete is shown in Fig. 6. The curves show that the compressive strength varies inversely with the maximum size of aggregate for a minimum cement, that in the lower-strength ranges maximum size is less important, and that at the higher-strength ranges concretes containing the smaller maximum-size aggregates generally develop the greater strengths.

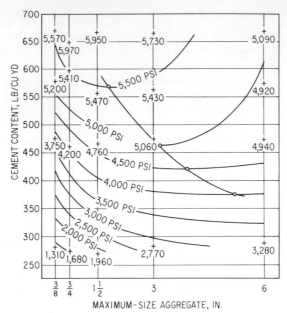

Fig. 6. Variation of cement content with maximum size of aggregate for various compressive strengths. Each point represents an average of four 18 × 36-in. and two 24 × 48-in. concrete cylinders tested at 90 days for both Clear Creek and Grand Coulee aggregates. Mixes had a constant slump of 2±1 in. for each maximum-size aggregate.

5   Admixtures

The effect of admixtures on the strength of concrete varies greatly with the properties of the admixture and with the characteristics of the concrete mixture. The admixtures with important strength effects include the accelerators, the water reducers, the air entrainers, and the finely divided minerals.

The most widely used accelerator is calcium chloride. Added in proper amounts it provides increased early strength during cold weather and also provides better protection against damage due to freezing temperatures, by shortening the time that the concrete requires protection. The amount of calcium chloride should be restricted to that necessary to produce the desired results and should not exceed 2% by weight of the cement. Frequently 1% will be sufficient to meet the requirements. The effects of calcium chloride additions, up to 3% on the compressive strength of concrete made and cured at temperatures of 40 and 70°F, are shown in Fig. 7. Under some conditions the desirability of calcium chloride additions for strength purposes must be considered in terms of possible undesirable effects on other properties such as drying shrinkage, resistance to sulfate attack, resistance to alkali-aggregate reaction, and corrosion of reinforcing steel.

Certain organic compounds usually available in various proprietary forms are used to reduce the water requirement of concrete mixtures. In some cases they are also used to retard set. The usual effect of these materials is to improve compressive strength and impermeability of hardened concrete. The decrease in the water content of a concrete mixture produces a reduction in the water-cement ratio which results in an increase in strength, but the increase in strength is frequently greater than that indicated by the reduction in the water-cement ratio alone.

In order to increase workability and durability, concrete mixtures are frequently designed to contain between 3 and 6% air. The inclusion of this air may normally result in some loss of strength. When the water-cement ratio is held constant the compressive strength is reduced about 5% for each percentage of air entrained. However, when the cement content is kept constant and advantage is taken of the opportunity to reduce the water-cement ratio for a given workability the strength losses are smaller and may actually be slight increases for the leaner mixtures. This is true because relatively large decreases in water-cement ratio in the leaner mixtures offset the strength loss due to entrained air. In the richer mixes the relatively small decreases in water-cement ratio do not offset strength loss due to entrained air. The effect of air content on compressive strength is shown in Fig. 8. The curves in the top portion of Fig. 8 were obtained for

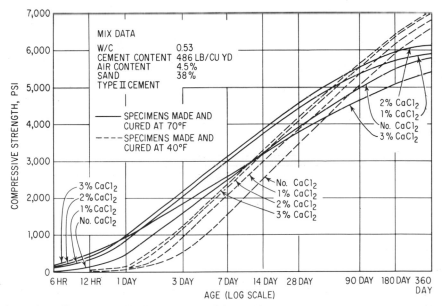

Fig. 7. Effect of small additions of calcium chloride on the compressive strength of concrete.[3]

mixtures with and without entrained air when the cement content was kept constant; the curves in the bottom portion were obtained for mixtures with and without entrained air when the water-cement ratio was kept constant. Flexural-strength losses due to entrained air are usually not so great as the losses in compressive strength.

The finely divided mineral admixtures are usually classified into three types: (1) relatively chemically inert materials such as ground quartz, ground limestone, bentonite, hydrated lime, and talc; (2) cementitious materials including natural cements, hydraulic limes, slag cements, and granulated blast-furnace slag; (3) pozzolans including materials such as fly ash, volcanic glass, diatomaceous earth, and some shales and clays. In many instances these admixtures are added primarily to improve workability but strength changes are also obtained. Generally, the strengths of lean mixtures are improved while the strengths of rich mixtures may be only slightly affected and may actually decrease.

Strength contributions caused by pozzolanic and cementitious admixtures are relatively slow, especially at low temperatures. Under continued moist curing the strengths at later ages will normally be higher than those obtained with portland cement alone. When fly ash is used as a replacement for up to 30% of the cement the compressive strength of concrete is usually reduced at 7 and 28 days and may be increased after 3 months. However, compressive strengths at 7 and 28 days for concrete containing fly ash may be made about equal to the strength of concrete without fly ash by proper redesign of the mixture.[5]

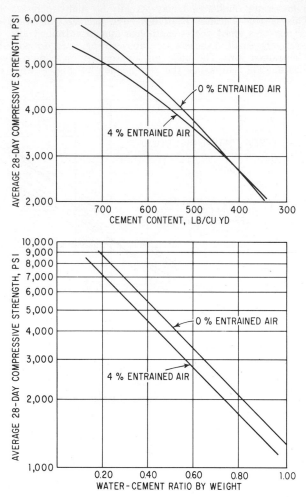

Fig. 8. Effect of air content on compressive strength of concrete.[3]

6 Curing Method

Proper curing requires a satisfactory moisture content and a favorable temperature during the early life of the concrete until satisfactory properties have been developed. When satisfactory curing is not provided serious loss of desirable potential properties occurs.

The effect of moisture during curing is shown in Figs. 9 and 10. Both figures show that compressive strength increases at a decreasing rate as the moist-curing period increases and that strength development stops at an early age if the concrete

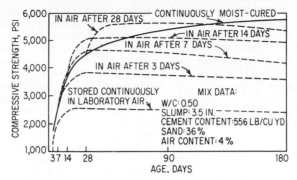

FIG. 9. Effect of air drying on the compressive strength of moist-cured concrete.[4]

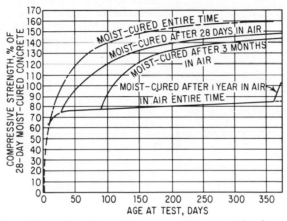

FIG. 10. Effect of curing on the compressive strength of concrete.[1]

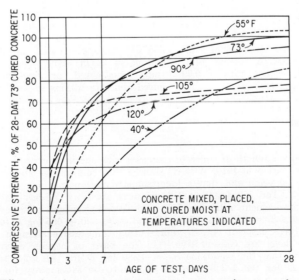

FIG. 11. Effect of curing temperature on the compressive strength of concrete.[1]

is kept in air. Figure 9 shows that when moist curing is discontinued the compressive strength increases for a short period but remains constant or decreases thereafter. Figure 10 shows that when moist curing is resumed after a period in air, strength increases are also resumed.

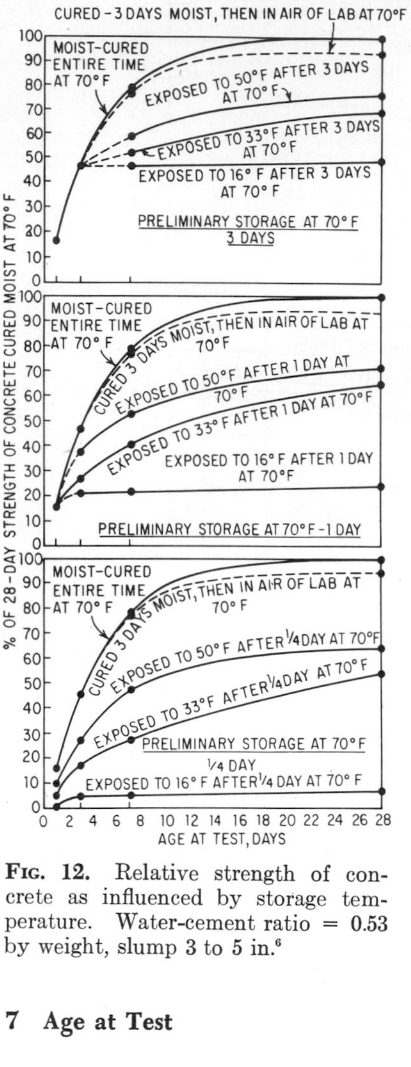

FIG. 12. Relative strength of concrete as influenced by storage temperature. Water-cement ratio = 0.53 by weight, slump 3 to 5 in.[6]

The effect of temperature on the compressive strength of concrete is represented in Figs. 11, 12, and 13. Figure 11 shows that higher strengths are obtained at early ages with higher curing temperatures and that the strengths at 28 days for temperatures above 55°F are inversely related to the curing temperature. A curing temperature of 40°F develops low strengths at all ages, but the rate of increase at 28 days is large. The additional information on the effect of temperature on compressive strength in Fig. 12 shows the desirable effects of maintaining a good initial curing period for as long a period as possible. The extremely harmful effects of curing temperatures below freezing are clearly evident.

While structural concrete is usually subjected to moist curing, concrete products are frequently cured in low-pressure steam or in high-pressure steam (autoclaved). Low-pressure-steam curing is usually carried out in saturated steam at temperatures between 135 and 195°F. The effect of various steam temperatures on compressive strength is shown in Fig. 13. The greatest acceleration in strength gain and a minimum loss in ultimate strength are obtained at temperatures between 130 and 165°F. High-pressure-steam curing is usually performed at a maximum steam pressure of 140 to 150 psig (about 360 to 365°F). Concrete cured in this manner attains in 24 hr about the same strength that moist-cured or low-pressure-steam-cured concrete attains in 28 days.[7]

7 Age at Test

The compressive strength of concrete increases with age if moisture is present. In a series of tests started in 1910, Withey[8] showed that compressive strength increases with age up to 50 years and that a logarithmic equation may be used to express the relationship. The 1:2:4 and 1:3:6 mixtures used by Withey and shown in Fig. 14 had water-cement ratios of 0.62 and 0.90, by weight. Withey suggested that the equations $S_c = 350 + 1{,}195 \log D$ and $S_c = 220 + 665 \log D$ for the 1:2:4 and 1:3:6 mixtures, respectively, S_c representing compressive strength in psi and D the age in days, are conservative. On the basis of certain assumptions he further suggested that the equations $S_c = 1{,}380 \log D$ for the 1:2:4 mixture and $S_c = 810 \log D$ for the 1:3:6 mixture appear to be satisfactory for estimating

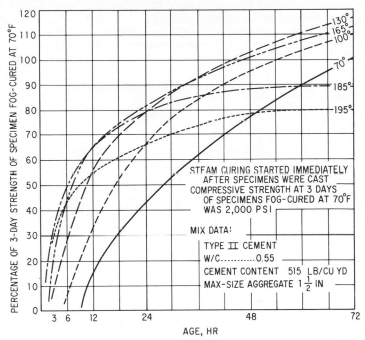

FIG. 13. Effect of steam curing at temperatures below 200°F on the compressive strength of concrete at early ages.[4]

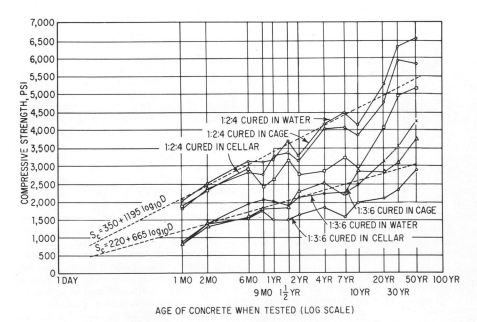

FIG. 14. Relation of concrete compressive strength and age.[8]

strength gain with age of similar concrete mixtures made with present Type I cements when cured under comparable conditions.

The rates at which compressive, tensile, and flexural strengths are developed under moist-curing conditions for concrete made with different types of aggregates and with a water-cement ratio of 0.532, by weight, are given in Table 1. The

Table 1. Rate of Strength Development*

Type of test	Age at test				
	3 days	7 days	28 days	3 months	1 year
Compression........	35	59	100	135	161
Flexure............	53	71	100	126	143
Tension............	46	68	100	121	150

* From Report of the Director of Research of the Portland Cement Association, 1928.

28-day values are taken as 100%, and the values at all other ages are based on the 28-day values.

STRENGTH RELATIONSHIPS

8 Relations between Various Types of Strength

Compressive strength, tensile strength, flexural strength, shearing strength, and bond strength are all related, and usually an increase or decrease in one is reflected similarly in the others, but not necessarily in the same degree. Relations between compressive strength, tensile strength, and modulus of rupture are shown in Fig. 15. The tensile strength usually ranges between 8 and 12% of the compressive strength and averages about 10%. The modulus of rupture usually ranges between 12 and 20% of the compressive strength and averages about 15% for concrete with a compressive strength of 3,500 psi.

The relations between splitting tensile strength and flexural strength and between splitting tensile strength and compressive strength are shown in Figs. 16 and 17 for gravel concrete, crushed-stone concrete, and lightweight-aggregate concrete. The curves in Fig. 16 show that the ratio of splitting tensile strength to the flexural strength was 67% for the concrete made with crushed stone, 62% for

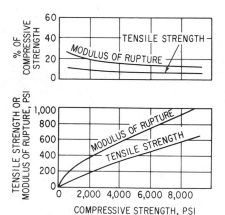

Fig. 15. Relations between compressive strength, tensile strength, and modulus of rupture.[9]

the gravel concrete, and 76% for the lightweight-aggregate concrete. The curves in Fig. 17 show nonlinear relations between splitting tensile strength and compressive strength. Average ratios of splitting tensile strength to compressive strength were about 10.7% for the concrete made with crushed stone, 10.8% for the gravel concrete, and 8.0% for the lightweight-aggregate concrete.

Satisfactory bond between reinforcing steel and concrete is necessary if rein-forced-concrete structures are to perform satisfactorily. Bond may be the result of adhesion, friction, lug action, or end anchorage. Bond-strength values determined

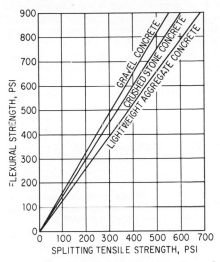

FIG. 16. Relation between flexural strength and splitting tensile strength for concrete made with three different aggregates.[10]

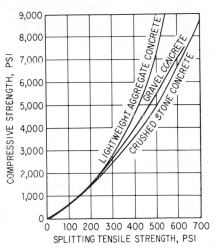

FIG. 17. Relation between compressive strength and splitting tensile strength for concrete made with three different aggregates.[10]

from pullout tests are usually noted at initial slip of the free end and at a slip of 0.01 in. at the loaded end. They are significantly influenced by the type of reinforcement, the properties of the concrete, and test procedures. The curves in Fig. 18 show large differences between plain and deformed bars. It is also apparent that the bond stress at a given slip increased as the compressive strength increased up to a compressive strength of about 3,000 psi, and that above 3,000 psi the increase in bond stress continued, but at a decreasing rate, up to a compressive strength of about 6,000 psi.

Bond strength is greatly affected by the quality of the paste that is used. It is not significantly affected by air entrain-ment if the air content is the normal amount of 3 to 6%. It is increased by delayed vibration that is properly timed and applied. It is less for horizontal bars than for vertical bars because of the water gain on the underside of horizontal bars. It is smaller for horizontal bars

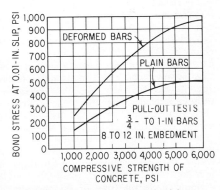

FIG. 18. Variation of bond with com-pressive strength of concrete.[3]

1 ft or more from the bottom than for horizontal bars near the bottom. In general, bond strength is reduced by alternations of wetting and drying, of freezing and thawing, and of loading.

The importance of shearing strength is evident from the fact that standard concrete cylinders tested in axial compression usually fail by shearing along an inclined plane. The failure is actually due to a combination of normal and shearing stresses on the plane. Similarly, when the term "shear failure" is applied to the diagonal failure in the web of a concrete beam, it is misleading because the cause of such failure is a tensile stress resulting from a combination of tensile and shearing stresses. The resistance of concrete to pure shearing stress has not been directly determined because such a stress condition causes principal tensile and compressive stresses, equal in magnitude to the shearing stresses, acting on other planes. Since concrete is weaker in tension than in shear, failure occurs as a result of the tensile stresses. Reported shearing strengths vary widely because of difficulties and differences in test procedures. However, on the basis of available test data it appears that the strength of concrete in pure shear is about 20% of the compressive strength and that the shear strength may increase when compressive stresses also occur on the failure plane.

CAUSES OF STRENGTH VARIATIONS

9 Materials

In order to maintain production of concrete having nearly constant properties, strict control of all materials and manufacturing processes is necessary. Careful control and adequate inspection will help avoid many of the difficulties.

One of the commonest reasons given for a poor concrete is the use of a cement with presumed unsatisfactory properties. While cements are variable and have a large effect on concrete properties they are only one of a large number of possible causes of poor concrete. All the other factors that contribute to the properties of the concrete must also be carefully examined before the reason for the poor concrete can be established. Lack of uniformity of portland cement is an important factor relating to strength. There are large differences in the strength-producing properties of cements of a given type from different sources, and also significant differences in the day-to-day production from a given cement mill. Obviously, also, different types of portland cement have significantly different strength properties, and variations in the amount of the cement in a given mixture will greatly influence strength properties. Consequently, while variations in portland cement must be carefully considered, they should not always be blamed for poor concrete properties. All other factors involved must also be considered.

Variations in the aggregate from batch to batch may cause significant strength changes. Aggregate from a given source should be used and the grading should be carefully controlled. Variations in grading may require changes in the amount of mixing water and result in strength changes. Changes in the moisture content of the aggregates, especially the fine aggregate, must be determined and allowed for or serious strength changes may occur. In addition, undesirable materials such as clay lumps, soft particles, organic matter, silt, mica, lignite, and easily friable pieces should be kept at a minimum to decrease harmful effects on strength.

Water for concrete mixtures should not contain any material that can have an appreciably harmful effect on strength, durability, or time of set. Normally water that is acceptable for drinking purposes has been considered satisfactory for concrete. However, relatively small amounts of sugar in water may have serious effects on time of set and strength. Substances that have harmful effects if present in sufficient quantity include silt, oils, acids, alkalies and their salts, organic material, and sewage. General limits of tolerance for the degree of contamination have not yet been developed. The possible effects of contaminated water on the properties of concrete may be evaluated by making comparative tests for time of set and soundness of the cement and for strength and durability of mortars with the contaminated water and with satisfactory water. Contaminated waters that have had no significant effect on time of set and soundness and have not produced strength decreases greater than 15% have been frequently used with good success. The presence of salt in mixing water and its possible effect on corrosion of the reinforcing steel must also be considered.

Satisfactory admixtures that are properly used will have beneficial effects on concrete properties. However, admixtures should not be expected to compensate automatically for improperly proportioned concrete mixtures or for poor construction practices. Great care must be taken to use the admixtures properly, and constant checks on the amounts added must be made since small changes in the very small quantities used may seriously influence properties. Overdosages must be avoided.

10 Production

Uniformity of concrete strength depends on careful attention to all the factors involved. Without such care wide variations in strength will occur. All materials should be accurately weighed, and special precautions must be taken in ensuring the correct amounts of cement, water, and admixtures. Water in the aggregates must be taken into consideration in determining the amount of water to be added. After the correct quantities of materials have been added care should be taken to prevent any losses, especially cement or water. A written record of all quantities in a given batch should be obtained as a check and for possible future reference in the event of unusual results.

The necessary mixing procedure and period for a given mixture should be kept constant. The period of mixing must be long enough to obtain a homogeneous concrete, but it should not normally be longer than necessary because of changes due to loss of water by evaporation and to grinding action resulting in an increase of fine material. Figure 19 shows that when water was not added to restore slump the strength increased progressively until the concrete became too dry to allow satisfactory molding of specimens. When water was added to restore slump the strength decreased progressively with time

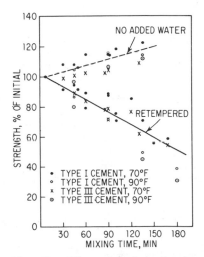

Fig. 19. Effect of prolonged mixing on compressive strength of concrete.[11]

to about one-half of the original value after a 3-hr mixing period.

Every precaution must be taken to avoid segregation while handling and placing concrete to avoid variations in strength from one portion of a concrete structure to another. In addition, equal compaction of concrete batches from a given mixture is required if equal strengths are to be obtained.

The effect of curing on the strength of concrete has been discussed in Art. 6. The importance of curing has frequently been overlooked. It is unfortunate that in many instances great effort and expense have led to proper mixture proportions and satisfactory placement procedures, but poor concrete properties have been obtained because of improper curing. In general, concrete will continue to gain strength as long as moisture is available and the temperature is satisfactory. Lack of moisture or excessively low or high temperatures will stop strength gain.

11 Sampling and Testing Procedures

Compression tests, usually on 6 × 12-in. cylinders, are commonly made to evaluate properties of concrete. Tests may also be made on cores taken from hardened concrete. Flexural tests of beams, frequently 6 × 6 in. in cross section tested over a span of 18 in., are made on concrete for highway use. Details of test procedures are given in ASTM Standards. While the tests of relatively small specimens do not necessarily provide information on the strength of concrete in a structure, they are sufficiently accurate for purposes of design and job control.

Concrete specimens may be subjected to standard curing, moist at 73.4 ± 3°F, and tested to determine potential strength, or they may be cured in the field under the conditions of moisture and temperature to which the structure is exposed. Field-cured specimens are usually tested to determine when forms may be removed and when the structure may be placed in service. Many factors affect the strength values determined from tests, and consequently great care should be used in evaluating the results.

The number of specimens that should be tested for a given structure is frequently specified. Usually the requirement is based on a given number of specimens per day, or a given number of specimens for a given volume of concrete, or a given number of specimens for a given area of paved surface. Generally a minimum of three specimens per test result are required. On large structures statistical methods are used to assess strength results and to refine design criteria. The methods used to evaluate compression-test results of field concrete are provided in detail in ACI Standard 214. Test ages most commonly specified are 7 and 28 days, although tests may also be required at other ages such as 3, 60, 90 days and at 1 year.

12 Bearing Conditions

Concrete strengths obtained by tests are significant only if all proper precautions are taken in conducting the tests. In all compression tests the specimen axis should be vertical, the specimen should be accurately centered in the testing machine, the bearing surfaces should be plane and should be perpendicular to the axis, and a spherically seated bearing block should be used.

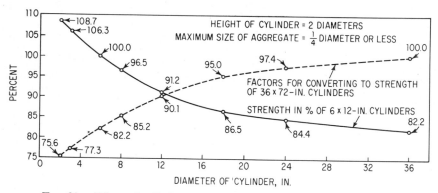

FIG. 20. Effect of cylinder size on compressive strength of concrete.[3]

All bearing surfaces that depart from a plane by more than 0.001 in. should either be ground smooth or be capped with a material that has as nearly as possible the same strength and elastic properties as the concrete specimen. Materials frequently used for capping concrete cylinders include neat high-alumina cement, neat portland cement, mixtures of sulfur and granular materials, and high-strength gypsum plasters used with specimens having a compressive strength less than 5,000 psi. All capping materials must be properly applied and be given sufficient time to harden before the specimen is tested.

The capping material used may have considerable influence on the compressive strength of concrete cylinders. High-strength concrete cylinders, above 5,000 psi, are more affected by the capping material, with strength losses as large as 40% reported for plaster of paris caps. Low-strength concrete cylinders around 2,000 psi or less are not greatly affected by the capping material, with strength losses usually less than 10% even when poor capping materials are used.

13 Form and Size of Specimen

Concrete strengths are dependent on the type of specimen, cube or cylinder, and on the specimen dimensions. The effect of specimen size, for concrete specimens made and cured under the same conditions and having a fixed ratio of height to diameter of 2, is shown in Fig. 20. The values are based on the average of tests at 28 and 90 days. The decrease in strength with an increase in cylinder diameter is believed to be due to a faster strength gain of the smaller-diameter cylinders and other reasons. Further, it is believed that age may equalize the difference due to difference in specimen diameter.

The effect of the ratio of length of cylinder to the diameter is shown in Fig. 21. It is evident that decreasing values of L/D below 2.0 result in large strength changes. The actual values of the changes are dependent on many factors such as concrete strength, type of aggregate, amount

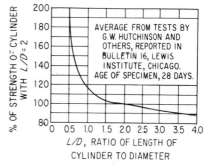

FIG. 21.   Relation of length and diameter of specimen to compressive strength.[3]

of entrained air, and method of curing. The effect of concrete strength on correction factors required for various L/D ratios is shown in Fig. 22a and b for normal-weight air-entrained concrete and for lightweight concrete.

The method of determining compressive strength of a concrete using portions of

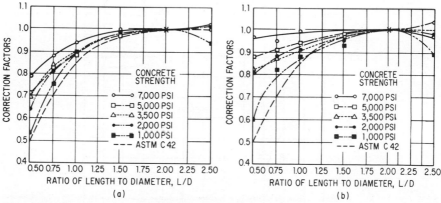

FIG. 22.   Correction factors for 6-in. concrete specimens. (a) Normal-weight air-entrained concrete. (b) Lightweight concrete.[12]

beams broken in flexure (modified-cube method) is given in ASTM Standard C 116. The modified-cube compressive strength is usually somewhat higher than the cylinder strength, especially for low-strength concrete. The compressive strength determined from tests of 6-in. cubes is normally taken as ⅘ times the compressive strength of a cylinder with an L/D of 2.0. However, the actual correction is affected by the concrete strength and other variables.

14 Moisture Content at Test

The moisture content of a concrete specimen at the time of test has an appreciable influence on the strength. Compression specimens tested in an air-dry condition usually have strengths about 20 to 40% higher than similar specimens tested in a saturated condition. Consequently most test specifications require that all specimens be in a saturated condition at the time of test to eliminate indefinite effects due to partial drying. Cores taken from a structure should be soaked for about 48 hr prior to test. The effect of drying on compressive strength is also evident in Fig. 9.

The effect of moisture on tensile and flexural strength appears to be more variable, probably because of nonuniform drying. Generally, drying appears to reduce both tensile and flexural strength, with reported losses up to 25%. In any event, care should be taken to avoid partial drying of specimens prior to test.

15 Loading Rate

An increase in the rate at which a concrete specimen is loaded causes an increase in strength. Consequently ASTM Standard C 39 requires that compression cylinders be tested at a rate of 35 ± 15 psi per sec if a hydraulic machine is used and that the loading head operate at an idling speed of 0.05 in. per min if a screw-gear machine is used. Flexural tests specified in ASTM Standard C 293 should be performed at a rate such that the increase in the extreme fiber stress does not exceed 150 psi per min.

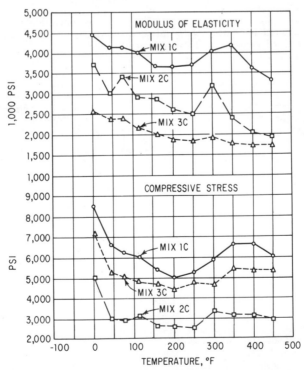

Fig. 23. Variation of compressive strength and modulus of elasticity with testing temperature.[15]

Tests[13] have shown that loading at 1 psi per sec reduced indicated compressive strength by about 12% when compared with tests at a standard rate of 35 psi per sec, and that tests at 1,000 psi per sec increased the indicated compressive strength by about 12% when compared with the standard. Other tests[14] have shown that the compressive strength of concrete tested at stress rates between 3.0×10^6 and 20.0×10^6 psi per sec was about 85% greater than that obtained from tests requiring an approximate failure time of 30 min. Investigators have also shown that concrete may be expected to sustain indefinitely only about 70% of the ultimate load determined from tests at a standard loading rate.

16 Test Temperature

Tests of concrete specimens are usually made at room temperature, and the effect of temperature on strength may be overlooked. In Fig. 23 mixtures 1C and 2C were made with sand and gravel aggregate and had water-cement ratios of 0.48 and 0.84, by weight, respectively. Mixture 3C was made with an expanded shale aggregate and had a total water-cement ratio of 0.81. The curves in Fig. 23 show that compressive strength and modulus of elasticity decreased as the specimen temperature increased to approximately 200°F, and then increased before finally decreasing again. The decrease appeared to be greatest as the temperature increased from 0 to 50°F. At about 450°F the strengths were about equal to those at 70°F, but the moduli of elasticity were considerably lower. Sand-and-gravel concrete and expanded-shale concrete responded similarly to variations in test temperature.

17 Lateral Restraint

Concrete that is restrained on all sides will support a much greater load than unrestrained concrete. Figure 24 shows that as the lateral restraint increased the axial stress also increased. While concrete in normal use is not subjected to the large lateral restraints shown in Fig. 24, the effect of lateral restraint in concrete is frequently encountered in bearing pressures. Test results have shown that when the ratio of total concrete area to the loaded area exceeds 40 the maximum bearing pressure may be about five times the cylinder compressive strength. The effect of lateral restraint is recognized in the ACI Building Code by the fact that the allowable bearing stress is $0.25f_c'$ when the bearing pressure is applied to the full area, and is $0.375f_c'$ when the

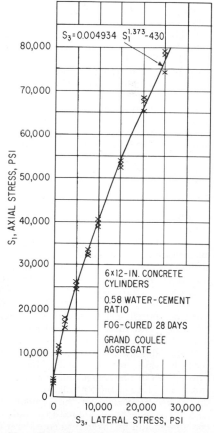

FIG. 24. Relation of axial stress to lateral stress at failure in triaxial compression tests of concrete.[3]

bearing pressure is applied on one-third or less of the total area and when the least distance between the edges of the loaded and unloaded areas is at least one-fourth of the parallel side dimension of the loaded area.

STRENGTH GAIN

18   Retarded Strength Gain

Seriously retarded strength gains of concrete may be due to a wide variety of causes usually associated with poor concrete materials or poor concrete practices. In order to prevent serious retardation of strength it is necessary to pay particular attention to the following:

1. Aggregates and mixing water must be clean and free from detrimental amounts of silt, oil, acids, organic matter, alkalies, sewage, and alkali salts. Very small amounts of some impurities such as sugar may greatly delay the rate of setting and the rate of strength gain.

2. The effects of retarding and accelerating admixtures should be determined for the concrete used. The correct amount of admixture needed to obtain the desired result should be ascertained and used in all batches. Suitable and accurately adjusted dispensing equipment is necessary for this purpose. The effect of temperature on the required amount of the admixture must be taken into account.

3. The correct design proportions in each batch must be used. Special care must be taken to make certain that the specified type of cement is consistently used and that the correct amounts of all materials are always used. Special care must be taken to avoid low cement contents and/or high water contents.

4. The air content of each batch should be determined to avoid low strengths due to very high air contents.

5. The concrete must be properly cured. This requires proper temperature and moisture conditions for the required length of time to obtain the desired properties. The concrete should not be allowed to freeze until it has developed the required strength. Under winter concreting conditions care must be observed to vent space heaters properly to avoid surface carbonation effects.

6. Compressive-strength tests should be made in accord with ASTM standards.

19   Accelerated Strength Gain

Under some circumstances it is highly desirable that concrete develop high strength at early ages. In some instances economical reuse of forms may be required or it may be necessary to use the concrete within a few days after manufacture or placement, or placement in cold weather may require fast initial strength gains to resist freeze-and-thaw damage. High early strength may be obtained by using richer concrete mixtures made with Type I or IA portland cement; high-early-strength cement Type III; accelerating admixtures such as calcium chloride, water-reducing admixtures, high-alumina cement, high steam-curing temperatures at normal pressures, or high curing temperatures at elevated pressures (autoclaving). The method or combination of methods to be used depends on the strength requirements and economical considerations.

If special cements or admixtures are used extra care is required to make certain that they are correctly used. High-alumina cement for example, requires special precautions in placement and curing, and the instructions provided by the manufacturer should be carefully followed. In cold weather, additions of calcium chloride, up to 2% by weight of the cement, will help in increasing the rate of strength gain. The effects of admixtures and special cements on other properties of the concrete should be taken into consideration.

High temperatures during moist curing or low-pressure-steam curing accelerate the early rate of strength gain, but if the temperatures are too high they may have an adverse effect on strengths at later ages. Optimum temperatures are dependent on the materials used, the mix proportions, and the details of the curing cycle. Low-pressure steam curing widely used for precast structural elements usually includes a presteam period at room temperature in excess of 3 hr, followed by a steaming period of 12 to 18 hr during which time the temperature is increased at a rate not exceeding 40°F per hr up to a maximum between 150 and 175°F, and finally return to room temperature.[16]

High-pressure-steam curing of precast-concrete products greatly accelerates early strength and is generally considered to give 28-day strengths in 24 hr. In addition this type of curing is generally considered to increase chemical resistance, decrease efflorescence, and increase volume stability. When this curing method is used additions of 10 to 40% of reactive silica, by weight of the cement, are made to allow reaction between the calcium hydroxide, liberated during hydration of the cement, and the reactive silica to form new fairly insoluble compounds. The curing cycle usually consists of a 1- to 4-hr presteam period at room temperature followed by a buildup of steam pressure in about 3 hr to 150 psig (350°F), then a period of 5 to 8 hr at full pressure, and finally a pressure release in a period of 20 to 30 min. The rapid pressure release allows the final product to emerge in a relatively dry condition.

20   Predicting 28-day Strength

Concrete specifications are usually based on required compressive strength at an age of 28 days. Improved methods of scheduling concreting operations require methods of anticipating the age at which concrete will be strong enough for all subsequent operations. Methods of predicting strengths at later ages from values obtained at early ages have been used with fair success.

Several investigators have attempted to provide equations relating 3- and 7-day compressive strengths with the 28-day strength and have concluded that predictions based only on 3-day strengths were not reliable. On the basis of a study made on a large amount of test data for moist-cured concrete, W. A. Slater (*Proc. ASCE*, January, 1925) proposed the following equation relating the 7-day strength S_c to the 28-day compressive strength S_c':

$$S_c' = S_c + 30 \sqrt{S_c}$$

This equation generally provides satisfactory predictions.

Rough approximations of the 28-day compressive strength of moist-cured concrete may be made by assuming that the strength at 3 days in one-third of the 28-day strength and that the strength at 7 days is two-thirds of the 28-day strength. However, it should be noted that large errors are possible because of variations in materials, proportions, manufacture, and curing and testing methods.

A method reported by P. Smith and B. Chojnacki[17] provides a means of predicting the 28-day compressive strength from a test made at an age of about 1 day. The method consists of placing the cylinders in their molds in boiling water 20 min after the concrete reaches a pin-pullout bond strength of 12 psi or a Proctor-needle penetration resistance of 3,500 psi. The water temperature should again be brought up to boiling in 2 hr after the cylinders are placed in the water. The cylinders in their molds should be kept in the water for a period of 16 hr. The cylinders are then removed from the water, the molds are removed, and the cylinders are capped with a sulfur-granular material after cooling for ½ hr. They are tested in compression within 1 hr of the time that they were removed from the boiling water. The equation for determining the 28-day compressive strength is

$$R_{28} = \frac{26,160 \, R_a}{R_a + 11,620}$$

in which R_{28} is the 28-day compressive strength of standard moist-cured cylinders and R_a is the accelerated strength determined at an age of about 1 day. It is stated that the method takes into account variations due to different cements and admixtures, and that low-strength aggregates should be avoided.

ASTM Standard C 684 covers three procedures (warm water, boiling water, and autogenous curing) for making accelerated compression tests of concrete specimens at ages between 1 and 2 days.

MODULUS OF ELASTICITY

21 General

Concrete is not a truly elastic material, but concrete that has hardened thoroughly and has been moderately preloaded has essentially a straight-line compressive stress-strain curve within the range of usual working stresses. The modulus of elasticity, defined by the equation E = unit stress/unit strain, is a measure of stiffness or of the resistance of the concrete to deformation. The modulus of elasticity of structural concrete normally varies between 2 and 6 million psi.

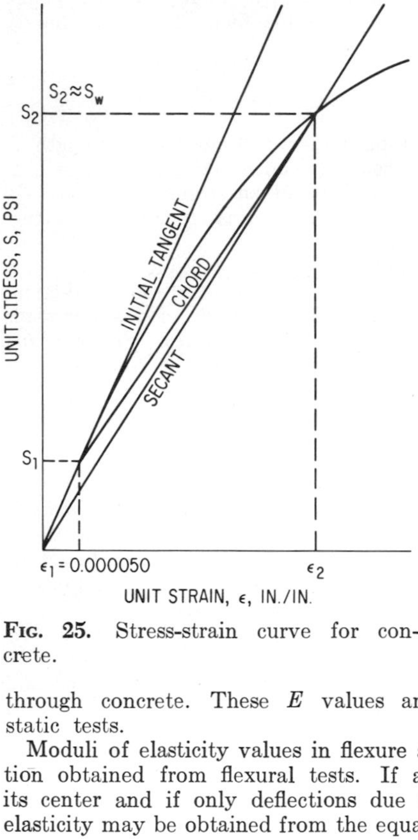

22 Method of Determining Modulus of Elasticity

Generally moduli of elasticity are determined from compression tests of concrete cylinders. Different values that may be determined from one test include initial tangent modulus, secant modulus, and chord modulus. Each of these may be represented by the slope of the appropriate line shown in Fig. 25. The secant or chord modulus is usually determined and used in calculations. Details for the determination of the chord modulus of elasticity are given in ASTM Standard C 469. It is apparent that if the curve approaches a straight line up to the working stress S_w the different values of modulus of elasticity E tend to become equal.

Values of the dynamic moduli of elasticity may be obtained by measuring the natural frequency of vibration of specimens of known dimensions, ASTM Standard C 215, or by measuring the velocity of sound waves as they travel through concrete. These E values are larger than the values obtained from static tests.

FIG. 25. Stress-strain curve for concrete.

Moduli of elasticity values in flexure are calculated from load-deflection information obtained from flexural tests. If a simple end-supported beam is loaded at its center and if only deflections due to bending are considered the modulus of elasticity may be obtained from the equation

$$E = \frac{PL^3}{48ID}$$

where E = modulus of elasticity, psi
   P = applied central load (in the working range), lb
   D = deflection at midspan due to P, in.
   L = distance between supports, in.
   I = moment of inertia of the section with respect to the centroidal axis, in.[4]

Appreciable effects on the flexural modulus of elasticity may result from the neglect of shear deflections, especially for beams that are relatively deep and have a short span.

Modulus of elasticity in shear, also called modulus of rigidity, may be determined from torsion tests by means of the equation

$$E_s \text{ or } G = \frac{TL}{\theta J}$$

where $G = E_s$ = modulus of elasticity in shear, psi
T = applied torque, in.-lb
L = length over which the angle of twist is developed, in.
θ = angle of twist in length L, rad
J = polar moment of inertia of the cross section, in.[4]

23 Effects of Variables on Modulus of Elasticity

The same factors that cause strength variations in concrete also cause variations in modulus of elasticity, but there does not appear to be any direct relationship between the two. In general, factors that cause strength increases also cause increases in modulus of elasticity. Thus lower water-cement ratios, richer mixtures, lower air contents, longer curing periods, and greater ages at time of test result in improved strength and modulus of elasticity. Test variables are also important. The modulus of elasticity decreases as the rate of testing is decreased because creep effects increase with time; as the stress S_2 at which E is determined is increased because creep effects increase at higher stresses; and as the moisture content at test is decreased. The amount, type, and grading of the aggregate have important effects on E. Under comparable conditions concrete made with limestone (E = 4,000,000 psi) has a much lower modulus of elasticity than concrete made with traprock (E = 13,300,000 psi).

24 Modulus of Elasticity Relationships

Under comparable conditions and for loads less than 50% of the ultimate the static moduli of elasticity for tension, compression, and flexure are approximately equal. The dynamic moduli of elasticity are greater than the static values. The modulus of elasticity in shear appears to vary between 0.4 and 0.6 of the modulus of elasticity in compression.

A widely used equation for calculating modulus of elasticity, given in the ACI Building Code (ACI 318), relates the modulus of elasticity to the ultimate compressive strength f_c' and the unit weight of the concrete w. This equation

$$E = 33w^{1.5} \sqrt{f_c'}$$

is satisfactory for values of w between 90 and 155 pcf.

OTHER PROPERTIES

25 Poisson's Ratio

The ratio of lateral strain to longitudinal strain, within the elastic range, for axially loaded specimens is called Poisson's ratio. Values of Poisson's ratio are required for structural analysis and for design of many types of structures. The method of determining Poisson's ratio is detailed in ASTM Standard C 469. Consistent relations between the values of Poisson's ratio and the usual mixture variables have not been demonstrated. There is some evidence that suggests that Poisson's ratio increases with age up to about 1 year. Most reported values for Poisson's ratio, up to an age of 50 years, fall within the range of 0.10 to 0.20. In the absence of experimental data Poisson's ratio is frequently assumed as ⅙.

26 Fatigue Strength

Information on fatigue properties of plain and reinforced concrete is still meager and effects of various factors on fatigue strength have not been reliably established. The following facts appear to be generally accepted.[18]

1. The fatigue strength of air-dry plain concrete subjected to several million cycles of repeated compressive loading, for a stress range between zero and a maximum, is between 50 and 55% of the static ultimate compressive strength.

2. The fatigue strength of plain concrete in tension under similar loading is about 55% of the modulus of rupture.

3. The fatigue strength in flexure of plain concrete under several million cycles of repeated loading is about 55% of the static ultimate flexural strength for a range in load from zero to a maximum.

4. Plain concrete does not appear to have an endurance limit at least within 10 million load repetitions.

5. The age of the test specimens, conditions of curing, moisture content and range of stress may significantly affect the results.

6. Most fatigue failures of reinforced-concrete beams appear to be due to failure of the reinforcing steel, associated with severe cracking and possible stress concentration and abrasion effects. Beams with critical longitudinal reinforcement appear to have a fatigue strength of 60 to 70% of the static ultimate strength for about 1 million cycles.

7. Most fatigue failures of prestressed-concrete beams are due to fatigue failure of the stressing wires or strands and are related to the extent and severity of cracking. There is some evidence to suggest that prestressed beams are superior to conventional beams for resisting fatigue loading.

27 Toughness

Plain concrete has poor toughness properties, with the energy of rupture in compression generally under 10 in.-lb per cu in. and the energy of rupture in bending generally under 0.1 in.-lb per cu in. The energy of rupture of steel in tension, for comparative purposes, is about 15,000 in.-lb per cu in. The toughness of concrete may be slightly increased by increasing the concrete strength. A significant increase in toughness (shock resistance) may be obtained by the addition of proper amounts of randomly placed reinforcement consisting of appropriately sized nylon fibers and/or steel wires.[19] The randomly placed steel wires are also effective in increasing the tensile and bending strengths of mortar and concrete.[20]

Concrete used for structural purposes, as in bridges and buildings, is reinforced with tensile and shear steel and may also have compressive and temperature steel. The design of these reinforced-concrete structures must take into account shock loading resulting from earthquakes and other types of dynamic loading. Properly designed reinforced-concrete structures are very resistant to blasts produced by bombing and to dynamic loadings caused by earthquakes.

Toughness, strength, and durability properties of fresh and hardened concrete have been significantly improved by impregnating the concrete with a monomer and polymerizing by either radiation or thermal-catalytic techniques. Radiation polymerization of dry concrete specimens containing from 4.6 to 6.7%, by weight, of methyl-methacrylate increased the strength almost four times, drastically reduced absorption and permeability, and greatly increased freeze-thaw durability and resistance to chemical attack.[21]

NONDESTRUCTIVE TESTS

28 Types and Significance of Nondestructive Tests

Nondestructive tests are used on specimens and concrete structures to determine information on certain properties of concrete. These tests may be subdivided into the following groups:

1. Indention tests in which the rebound of a spring-driven hammer is measured and related to ultimate strength.

2. Sonic tests usually involving determination of the resonant frequency of longitudinal, transverse, or torsional vibration of small concrete specimens. These tests provide information on dynamic modulus of elasticity, damping constant and logarithmic decrement.

3. Pulse-transmission tests at sonic and ultrasonic frequencies. These tests measure the velocity of a compressional pulse traveling through the concrete and provide information on the presence or absence of cracking in monolithic concrete. They also may be used to measure the thickness of slabs one face of which is inaccessible.

4. Radioactive tests involving absorption of gamma and x-rays. These tests provide information on density or quality of concrete and also on the presence or absence of reinforcing steel.

5. Penetration probe tests require measurement of the depth that a carefully made steel pin is driven into the concrete by a special gun fired with a precisely measured powder charge. In addition the Moh's hardness of the aggregate must be determined. With these known values, an estimate of the compressive strength of the concrete may be obtained from given tables or charts.

The most widely used indentation device is the Schmidt concrete test hammer. It is a valuable tool when properly used, but careless and indiscriminate use may lead to incorrect conclusions. The results of the hammer tests are affected by many variables including position of hammer (vertical, horizontal, or intermediate angle), smoothness of test surface, internal and surface moisture content, size, shape and rigidity of specimen, aggregate type, aggregate size and concentration near the surface, air pockets, age, temperature, and previous curing conditions. Standardization of test conditions will reduce or eliminate some of the variations, but the others require special calibration curves. Since the manufacturer's calibration curves provided with the hammer apply only to specific conditions, it is apparent that strength values obtained from test-hammer readings and the given calibration curve cannot be accurate or reliable except by chance. More reliable and accurate strength results may be expected when the hammer is calibrated for the specific conditions relating to the given job. The principal advantages of the indentation tests are their rapidity, simplicity, and economy. However, the results must be carefully interpreted and should be considered qualitative rather than quantitative unless a job calibration curve has been developed. The indentation tests should not be considered as a replacement for standard compression tests but as a complement to provide additional information.

Sustained sonic frequencies are primarily used for laboratory testing of beams and cylinders to follow the course of deterioration as the specimens are subjected to weathering or exposure tests. Pulse-velocity tests are usually made on concrete structures to establish information on degree of uniformity, especially with regard to the amount and position of cracking. On the basis of available test results it appears that concrete with a longitudinal wave velocity in excess of 15,000 fps is in excellent condition and that concrete with a longitudinal wave velocity less than 7,000 fps is in very poor condition.

The penetration probe test is a valuable addition to the available test procedures used to determine concrete compressive strength. However, the test must be carefully made and the significance of the final numerical results must be carefully evaluated. Correlation of penetration probe tests with cylinder or core tests may be very helpful in evaluating final results. The penetration probe test is not completely nondestructive since it may leave craters similar to popouts in the concrete surface.

REFERENCES

1. "Design and Control of Concrete Mixtures," 11th ed., and "Design of Concrete Mixtures," Portland Cement Association.

2. Report, Director of Research, Portland Cement Association, 1928 Series 210.
3. Price, W. H.: Factors Influencing Concrete Strength, *J. ACI,* February, 1951, pp. 417–432.
4. "Concrete Manual," 7th ed., U.S. Bureau of Reclamation, 1963.
5. Lovewell, C. E., and G. W. Washa: Proportioning Concrete Mixtures Using Fly Ash, *J. ACI,* June, 1958, pp. 1093–1101.
6. Timms, A. G., and N. H. Withey: Temperature Effects on Compressive Strength of Concrete, *J. ACI,* January, February, 1934, pp. 159–180.
7. ACI Committee 516 Report, High Pressure Steam Curing: Modern Practice, and Properties of Autoclaved Products, *J. ACI,* August, 1965, p. 869.
8. Withey, M. O.: Fifty Year Compression Test of Concrete, *J. ACI,* December, 1961, pp. 695–712.
9. Gonnerman, H. F., and E. C. Shuman: Compression, Flexure, and Tension Tests of Plain Concrete, *ASTM Proc.,* vol. 28, pt. II, pp. 527–573, 1928.
10. Grieb, W. E., and G. Werner: Comparison of Splitting Tensile Strength of Concrete with Flexural and Compressive Strengths, *ASTM Proc.,* vol. 62, pp. 972–995, 1962.
11. Gaynor, R. D.: Effects of Prolonged Mixing on the Properties of Concrete, *NRMCA Publ.* 111, June, 1963.
12. Kesler, Clyde E.: Effect of Length to Diameter Ratio on Compressive Strength— An ASTM Cooperative Investigation, *ASTM Proc.,* vol. 59, pp. 1216–1228, 1959.
13. Jones, J. P., and F. E. Richart: The Effect of Testing Speed on Strength and Elastic Properties of Concrete, *ASTM Proc.,* vol. 36, pt. II, pp. 380–391, 1936.
14. Watstein, David: Effect of Straining Rate of the Compressive Strength and Elastic Properties of Concrete, *J. ACI,* pp. 729–744, April, 1953.
15. Saemann, J. C., and G. W. Washa: Variation of Mortar and Concrete Properties with Temperature, *J. ACI,* pp. 385–395, November, 1957.
16. ACI Committee 517 Report, Recommended Practice for Atmospheric Pressure Steam Curing of Concrete (ACI 517-70).
17. Smith, P., and B. Chojnacki: Accelerated Strength Testing of Concrete Cylinders, *ASTM Proc.,* vol. 63, pp. 1079–1104, 1963.
18. Nordby, G. M.: Fatigue of Concrete—A Review of Research, *J. ACI,* pp. 191–220, August, 1958.
19. Fibrous Reinforcements for Portland Cement Concrete, U.S. Army Engineer Division, Ohio River Corps of Engineers, Cincinnati, Ohio, 45227, *Tech. Rept.* 2-40, May, 1965.
20. Romualdi, J. P., and J. A. Mandel: Tensile Strength of Concrete Affected by Uniformly Distributed and Closely Spaced Short Lengths of Wire Reinforcement, *J. ACI,* pp. 657–672, June, 1964.
21. Dikeou, J. T., L. E. Kukacka, J. E. Backstrom, and M. Steinberg: Polymerization Makes Tougher Concrete, *J. ACI,* pp. 829–839, October, 1969.

Chapter 7

DURABILITY

GEORGE W. WASHA

GENERAL CONSIDERATIONS

1 Weathering

Good concrete is a relatively durable material under a wide variety of exposures. Ordinary weathering conditions, however, may have harmful effects and may cause disintegration of poor concrete. Weathering effects are due to the disruptive action of freezing and thawing, to alternate wetting and drying, to undesirable chemical activity, and to temperature variations in the concrete mass. Many laboratory tests have been proposed and used for determining the durability of concrete under various types of exposure conditions, but correlation between laboratory tests and field service records is difficult, if not impossible.

Major factors that may influence concrete durability include the physical properties of the hardened concrete, the constituent materials of which the concrete is composed, the manufacturing and construction methods used, and the nature of the deteriorating influences. In order to have good durability concrete should have a low water-cement ratio, it should be made with properly selected sound materials, it should be dense and well made, it should be properly cured, and where freezing and thawing are involved it should contain between 4 and 6% entrained air. Evaluation of the durability of concrete may also involve a study of the elastic, plastic, and thermal properties of the constituents and possible incompatibilities.

FREEZING AND THAWING

2 Effect of Entrained Air

All concrete contains some entrapped air, usually between 0.5 and 1.5%, which is relatively ineffective in increasing resistance to freezing and thawing. However, additions of 4 to 6% purposefully entrained air in concrete, with a 1½-in. maximum-size aggregate, greatly increase resistance to freezing and thawing. As the maximum coarse-aggregate size is decreased the amount of entrained air should increase, and as an example concrete with a maximum aggregate size of ¼ in. should have an entrained-air content between 8 and 10%. While the total air content is important, other factors such as size and spacing of bubbles are equally significant. Purposefully entrained air bubbles are usually very small and are closely spaced throughout the mass, while entrapped air generally consists of a smaller number of larger voids spaced at greater distances. Methods of determining air content and bubble characteristics are given in ASTM Standard C 457. Entrained air bubbles usually have diameters varying between 0.003 and 0.05 in. and should have a maximum spacing factor (index related to the maximum distance of any point in the cement paste from the periphery of an air void, in inches) about 0.008 in. Estimates have placed the number of entrapped air bubbles in a cubic yard of non-air-entrained concrete, 1.5% entrapped air, at about 5 billion, and the number of bubbles in a cubic yard of air-entrained concrete (658 lb of cement per cu yd, 7% air, ⅜-in. maximum-size aggregate) at about 130 billion.

Air entrainment affects the properties of freshly mixed and hardened concrete. Air-entrained concrete is more plastic and workable, it has less tendency to bleed, and it may be placed with less segregation than non-air-entrained concrete. Hardened concrete containing entrained air is more uniform, has less absorption and permeability, and is much more resistant to the action of freezing and thawing. Additions of air to a given concrete mix will result in lower strength of the hardened concrete. The compressive strength of concrete will decrease about 5% for each percent of entrained air when concretes of a given water-cement ratio are compared. However, if comparison is made on the basis of a fixed cement content and if the mixture containing entrained air is redesigned by reducing the water and sand contents to maintain the slump and volume of mortar, the strength decreases will be less and may in fact be strength increases. The latter may be secured with lean concrete mixtures containing up to about 470 lb of cement per cu yd of concrete. Richer mixes that have been redesigned may suffer strength losses between 10 and 15%.

During freezing and thawing either the cement paste or the aggregate or both may be damaged by dilation. In this process stresses beyond the proportional limit may be produced which may cause permanent enlargement or actual disintegration. Dilation and associated stresses are believed to be due to (1) hydraulic pressure generated when growing ice crystals displace unfrozen water and cause it to flow against resistance to unfrozen portions of the mass, (2) growth of capillary ice crystals, (3) osmotic pressures brought about by local increases of alkali concentration caused by the separation of pure ice from the solution. The large number of closely spaced small air bubbles in air-entrained concrete are believed to serve as reservoirs for the relief of pressure developed within the freezing concrete and thus to relieve or avoid the high tensile stresses that lead to failure.

Some of the important factors involved in freezing and thawing concrete are (1) the size, homogeneity, and characteristics of the aggregates, especially the porosity and related absorption and permeability characteristics; (2) the proportions of the concrete mixture with special reference to the water-cement ratio and the amount and characteristics of the entrained air; (3) the amount and distribution of the moisture available to the cement paste and to the aggregate; (4) the degree of saturation of the paste and the aggregate in relation to the critical saturation coefficient of 0.917 (in unsaturated concrete the constituent with the finest texture tends to become more nearly saturated and consequently cement paste tends to be more saturated than the aggregate); (5) the rate of freezing and thawing; (6) the details of the freezing and thawing cycles including the medium in which freezing and thawing take place; (7) the age of the concrete when freezing and thawing start; (8) the total number of freezing and thawing cycles; and (9) the presence or absence of calcium chloride, sodium chloride, or other salts during freezing and thawing.

In summary, good durable concrete will be obtained if sound aggregates with proven records are used, if a properly proportioned dense concrete with a relatively low water-cement ratio is used, and if the proper amount and type of entrained air are present in the concrete.

3   Freeze-and-Thaw Tests

Freezing and thawing tests are costly and are difficult to standardize and to evaluate. ASTM Standard C 666 provides two procedures for conducting tests to determine the resistance of concrete specimens to rapid freezing and thawing in water, and to rapid freezing in air and thawing in water. These procedures require that the test be continued until the specimens have sustained 300 cycles of freezing and thawing or until the dynamic modulus of elasticity has reached 60% of the initial modulus. A measure of the durability, the durability factor DF, may then be calculated from the equation

$$DF = \frac{PN}{M}$$

where P = relative dynamic modulus of elasticity at N cycles, %
  N = number of cycles at which P reaches the specified minimum value for discontinuing the test or the specified number of cycles at which the exposure is to be terminated, whichever is less
  M = specified number of cycles at which the exposure is to be terminated

The standard states that the methods are not intended to provide a quantitative measure of the length of service that may be expected from a specific type of concrete under field conditions.

Results of freezing and thawing tests have usually been evaluated by one or more of the following measures: (1) reduction in dynamic modulus of elasticity, (2) loss in compressive or flexural strength, (3) loss in weight, (4) change in visual appearance, and (5) expansion of specimen. The reduction in dynamic modulus is widely used and provides a good means of evaluating freeze-and-thaw tests. Characteristic behaviors of a good and poor concrete are shown in Fig. 1. Reasonably good correlation has also been obtained between changes in dynamic modulus and changes in flexural strength. Flexural-strength loss is a good indication of deterioration but compressive strength is not. In addition, a large number of specimens are required for method 2. Methods 3 and 4 may provide little information during the early stages of deterioration, and information obtained during the later stages may be largely dependent on the interpretation of a single individual. Expansion tests satisfactorily indicate deterioration but are largely limited to research programs because of the experimental details involved.

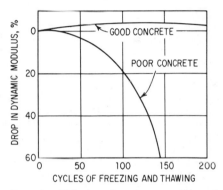

FIG. 1. Relation between number of cycles of freezing and thawing and drop in dynamic modulus of elasticity.

Ultrasonic tests are used to determine deterioration of concrete structures such as dams, piles, and bridge piers in the field. The method requires the determination of the longitudinal wave velocity in the concrete structure. On the basis of previous tests it appears that if the velocity through the concrete is between 13,500 and 16,000 fps the concrete is in good to excellent condition, and that if the velocity is in the range between 10,000 to 13,500 fps the concrete is in fair or questionable condition, and velocities below 10,000 fps generally indicate a questionable to poor condition.

4 Deicing Agents

In areas where climatic conditions are severe concrete pavements are frequently subjected to direct application of flake calcium chloride or rock salt to remove ice, or to repeated applications of granular materials impregnated with these salts. Under these conditions surface scaling has developed, usually in those locations where heavy or frequent applications of salts have been used.

Concrete that is to be subjected to freezing and thawing and to the action of deicing salts should normally contain between 4 and 7% entrained air. In addition, since concrete is more vulnerable to surface scaling damage during its early life, surface coatings are frequently applied to the concrete prior to the first winter season. Applications of two coats of either boiled or raw linseed oil to the concrete surface have produced superior resistance to scaling.

Properly proportioned concrete containing the necessary amount and type of entrained air will resist satisfactorily scaling due to the use of calcium chloride or sodium chloride for ice removal but it will not resist the action of all salts. Deicing agents consisting primarily of either ammonium sulfate or ammonium nitrate should not be used because of their chemical attack on concrete. Solutions of these salts will disintegrate concrete under freezing or normal room-temperature conditions. The chemical composition of a deicing agent and its possible effects on concrete properties should be considered before the agent is used.

CHEMICAL ATTACK

5 Effect of Sewage

Except in the presence of certain industrial wastes there is little direct attack of the sewage since the usual sewage and waste water is nearly neutral, varying only slightly above or below a pH of 7.0. Whether concrete or mortar can be used in sewer constructions depends consequently on the character of the sewage and operating characteristics. Many concrete sewers have been in successful operation for long periods, but there are also published accounts of others that have failed after short periods in service. If the sewage is of such nature that a strong odor of hydrogen sulfide is evolved, sulfuric acid of sufficient strength will be formed to attack the lime compounds of the concrete and produce disintegration above the water level. The hydrogen sulfide, formed by bacterial decomposition of the sulfur compounds in the sewage, rises and combines with oxygen and moisture to form sulfuric acid on the upper portions of the sewer.

In some instances changes of the operation of the sewer may be helpful. Since the flow of sewage is not normally itself aggressive, deterioration of the crown may be prevented by keeping the sewage at lower temperatures, running the sewers full, ventilating at high velocities the space between the sewage and the crown of the sewer, or providing protective coverings or coatings. The protective coating frequently used is polyvinyl chloride, applied as a liner by placing the polyvinyl in the form when the pipe is cast, or subsequently as a painted-on coating. If proper operating conditions are not possible or if protection cannot be provided, portland-cement concrete should not be used.

Some improved resistance to the action of sewage may be obtained with the use of high-alumina cement. Improved resistance may also be obtained by the use of a patented process called Ocrate concrete. In this process hardened and dried concrete in a vacuum is subjected to the action of silicon tetrafluoride gas in accord with the equation

$$2Ca(OH)_2 + SiF_4 = 2CaF_2 + Si(OH)_4$$

In a given time the gas penetrates the concrete to some depth and provides a shell resistant to acid attack.

6 Effect of Other Acids

Basically, no portland-cement concrete is resistant to acids. When the most resistant cements are used in sound impermeable concrete that is properly placed in well-designed structures, resistance to mild acids may be satisfactory. Where strong acids are involved other materials should be used or a protective surface treatment or covering must be provided. Common acid attacks of concrete are due to lactic and acetic acids in food-processing plants. These attacks are comparatively mild but persistent, result in softening of the working floors, and may require repair or replacement at periodic intervals. Acetic acid attack of concrete silos due to silage is also fairly common.

Bacteria and fungi find good conditions for growth on the floors and walls of food-processing plants. These agencies cause damage by mechanical action and by secretion of organic acids. Antibacterial cements are available for decreasing these actions by permanently sublimating the microbiological metabolism.[1] The active component in these cements usually is a toxic material such as arsenic or copper. Antibacterial cements are effective in decreasing odors, development of slimes, and the rate of deterioration of concrete and mortar surfaces. They have been satisfactorily used in dairies, kitchens, food-processing plants, food warehouses, pharmaceutical plants, breweries, and chemical plants. They have also been used in "wet-foot traffic areas" in locker rooms, shower rooms, bathhouses, and around swimming pools to reduce the possibility of contact contagion.

7 Effect of Leaching

In the process of cement hydration soluble calcium hydroxide is formed. This material is easily dissolved by water that is lime-free and that contains dissolved carbon dioxide. Snow water in mountain streams is particularly aggressive because it is relatively pure, cold, and contains carbon dioxide which produces a mild carbonic acid solution with a greater capacity for dissolving calcium hydroxide than pure water. As a result of this action surfaces of water-carrying structures develop a rough sandy appearance and may suffer reduction in capacity.

In some hydraulic structures leakage of water through cracks or joints or porous concrete may carry the calcium hydroxide in solution through the concrete structure. At the surface, reaction between the calcium hydroxide and carbon dioxide will cause precipitation of a white deposit of calcium carbonate. Generally this type of leaching does not result in any serious problem, but it may over a long period of time cause serious disintegration.

The problems relating to leaching may be minimized by the use of properly proportioned dense concrete mixtures, careful attention to the proper design and placement of contraction and construction joints, provision for satisfactory drainage, and provision of effective and durable coatings where necessary. High-alumina cements, portland blast-furnace slag cements, and pozzolan cements are effective in minimizing leaching.

8 Effect of Sulfates

In regions where alkali is present in soil and groundwaters deterioration of concrete structures may take place. The harmful effects are primarily due to the sulfates of magnesium and sodium. These salts react with the hydrated calcium aluminate to form crystals of calcium sulfoaluminates accompanied by considerable expansion that may result in eventual disintegration. An example of serious disintegration due to sulfate attack that took place in a few years is shown in Fig. 2. The rate and severity of the sulfate attacks increase as the concentration of the sulfates in the groundwater increases and as the temperature increases. Dry concrete in dry sulfate-bearing soils is not attacked. Continuous saturation in strongly sulfate-bearing water will produce rapid and severe effects. Frequently alternated saturation and drying conditions appear to produce the most harmful effects.

The deposition and growth of sulfate crystals in the surface pores of concrete may also cause disintegration. During periods of drying the alkali waters may evaporate and deposit salt crystals which will grow with periods of alternate wetting and drying until they fill the pores and ultimately develop pressures great enough to cause pitting and scaling.

The resistance of concrete to the action of sulfates may be greatly improved in several ways.

1. Sulfate-resisting cement Type V (C_3A less than 5%) should be used where concentrations greater than 0.2% water-soluble sulfates in the soil or 1,000 ppm

Fig. 2. Concrete slab (20 by 100 ft by 4 in. thick) disintegrated by sodium-sulfate-bearing soil.[2]

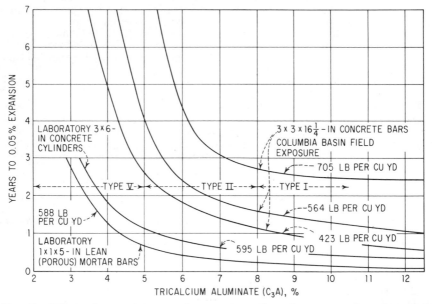

Fig. 3. Effect of cement type and amount on the rate or expansion of concrete exposed to sulfate waters.[2]

sulfates in the water are present. When the concentrations are 0.1 to 0.2% water-soluble sulfates in the soil or 150 to 1,000 ppm sulfates in the water Type II cement (C_3A less than 8%) should be used.

2. The water-cement ratio should not exceed 0.50 and the concrete should have an ample cement content. The effect of quantity and type of cement content on the rate of expansion of concrete exposed to sulfate waters is shown in Fig. 3.

3. Substitution of 15 to 30% of the cement, by weight, of an active pozzolanic material may be very effective.

4. Autoclaving concrete products at 350°F or higher greatly improves sulfate resistance.

9 Effect of Seawater

Concrete in seawater is subjected to many different effects but if it is properly proportioned, mixed, and placed it should resist these effects almost indefinitely. Wetting and drying, leaching, temperature variations, corrosion of reinforcing steel, battering of waves and tides, sulfate attack, and freezing and thawing action may all be involved.

Many of the potentially harmful effects can be largely controlled by the use of normal cement along with sound nonreactive aggregates that are properly proportioned to provide strong impermeable concrete. Reinforcing steel should be properly protected from corrosion by means of a thick protective shell of concrete. In northern climates where freezing and thawing may be important the concrete mixture should contain from 4 to 6% of entrained air. Since sulfate attack may be an important consideration it may be desirable to use cements low in C_3A such as Types V and II. The use of portland blast-furnace cements or high-alumina cements or additions of active pozzolans should also be considered.

10 Effect of Carbonation

Carbonation of concrete is a chemical combination of carbon dioxide with the hydration products of portland cement. Carbon dioxide reacts principally with calcium hydroxide but also with calcium silicate and calcium aluminate and combines with the calcium portion of these compounds to form calcium carbonate. When fresh concrete is placed in cold weather in rooms heated by improperly vented space heaters a concentration of carbon dioxide results. The carbon dioxide reacts with the fresh concrete near the surface and produces a soft crumbly surface layer between 0.10 to 0.30 in. thick. The action takes place readily at a temperature between 30 and 50°F in an atmosphere that provides both carbon dioxide and moisture. If it is not possible to keep the carbon dioxide content of the air at a low level, the fresh concrete may be protected with a membrane curing compound or surface seal applied as promptly as possible to protect the concrete during the first 24 hr. A carbonated surface will not respond to the action of chemical hardeners and the only way to remove the poor surface is by surface grinding.

Carbonation also causes important changes in hardened concrete, especially lightweight porous concrete. The reaction of the atmospheric carbon dioxide with the hydration products of portland cement produces significant weight gains and irreversible shrinkages. The shrinkage values due to carbonation may be about as large as the shrinkage obtained by air-drying saturated concrete. Carbonated products may possess improved volume stability to subsequent moisture changes.

Carbonation proceeds slowly, produces little direct shrinkage at relative humidities near 100 or below 25%, and appears to be most active at about 50% relative humidity. Size of specimen, density of specimen, concentration of carbon dioxide, previous drying and carbonation history, and method of curing all appear to have significant effects on carbonation shrinkage.

11 Effect of Free CaO and MgO

Excessive expansion of concrete or masonry construction due to hydration of the uncombined lime and magnesia present in cement is called unsoundness. Quantitative information on this type of expansion may be obtained from the autoclave test as given in ASTM Standard C 151. If, as is usually the case, the maximum autoclave expansion for a given cement is lower than the value allowed by the appropriate ASTM standard there should be no large expansions due to the hydration of the uncombined lime and magnesia when the cements are used. Problems due to this type of expansion are most frequently associated with the use of limes, or masonry cements containing limes, that have not been properly hydrated prior to use. High-magnesium limes, because they hydrate much more slowly, should be given special care. There are many descriptions in the literature of the serious consequences resulting from the use of materials containing incompletely hydrated lime.[3] Many relate to the use of these materials as the cementing material in brick walls.

12   Effect of Various Substances on Concrete and Protective Treatments, Where Required[4]

Concrete of a suitable quality must be assumed in a discussion of the effect of various substances on concrete, and protective treatments. In general, this means a properly proportioned, carefully placed, and well-cured concrete resulting in a watertight structure. This requires:

1. Low water-cement ratio, not to exceed a water-cement ratio of 0.50 by weight.

2. Suitable workability, to avoid mixes so harsh and stiff that honeycomb occurs, and those so fluid that water rises to the surface.

3. Thorough mixing, at least 1 min after all materials are in mixer, or until mix is uniform.

4. Proper placing, spaded or vibrated, to fill all corners and angles of forms without segregation of materials—avoid construction joints.

5. Adequate curing, protection by leaving forms in place, covering with wet sand or burlap, and sprinkling. Concrete to be kept wet and above 50°F for at least the first week. Not to be subject to hydrostatic pressure during this period.

Many solutions, such as brines and salts, which have no chemical effect on concrete, may crystallize upon loss of water. It is especially important that concrete subject to alternate wetting and drying of such solutions be impervious. When the free water in the concrete is saturated with salts, the salts crystallize in the concrete near the surface in the process of drying and this crystallization may exert sufficient pressure to cause surface scaling. Salt solutions corrode steel more rapidly than plain water. In structures which are to be subject to frequent wetting and drying by these solutions, it is essential to provide impervious concrete and sufficient coverage over the steel, and it may be advisable to provide some surface coating such as sodium silicate, linseed oil, or one of the varnishes as an added precaution.

Surface Treatments. Materials are available for almost any degree of protection required on concrete. The best material to use in a given case will depend on many factors in addition to the substance to be protected against. These include concentration of solution, temperature, taste and odor, and abrasive action. High temperatures usually accelerate any possible attack and therefore better protection is required than for normal temperatures. Bituminous materials soften at elevated temperatures, may even melt and become ineffective. Grades are available for a fairly wide temperature range and manufacturers should be consulted as to grade required for given conditions. Where taste or odor is important it should be determined whether proposed treatment will be satisfactory. As a rule, thin coatings are not so durable as heavier coatings, where there is considerable abrasion.

The more common treatments are indicated in Table 1, the numbers corresponding to the descriptions given below. For most substances, several treatments are suggested. These will provide sufficient protection in most cases, but any of the other treatments designated by a number higher than the highest shown would be equally suitable and often may be advisable. In making a selection, economy should be considered as well as the factors discussed above. Where continuous service over long periods is desirable it may be more economical to use the more positive means of protection rather than those of lower first cost which may be less permanent.

Protective coatings usually require dust-free, surface-dry concrete for satisfactory application.

1. *Magnesium fluosilicate or zinc fluosilicate.* The treatment consists of two or more applications. First, a solution of about 1 lb of the fluosilicate crystals per gal of water is used. For subsequent applications about 2 lb of crystals per gal of water is used. Large brushes are convenient for applying on vertical surfaces, and mops on horizontal areas. Each application should be allowed to dry; after the last one has dried, the surface should be brushed and washed with water to remove crystals which have formed. The treatment hardens the surface by chemical action and makes it more impervious. Fluosilicates are available from chemical dealers.

2. *Sodium silicate* (commonly called water glass). This is quite viscous and must be diluted with water to secure penetration, the amount of dilution depending on the quality of the silicate and permeability of the concrete. Silicate of about 42.5° Baumé gravity diluted in proportions of 1 gal with 4 gal of water makes a good solution. It may be applied in two or three or more coats, allowing each coat to dry thoroughly. On horizontal surfaces it may be poured on and then spread evenly with brooms or brushes. Scrubbing each coat with water after it has hardened provides a better condition for application of succeeding coats. For tanks and similar structures, progressively stronger solutions are often used for the succeeding coats.

3. *Drying oils.* Boiled or raw linseed oil may be used, but the boiled oil dries more rapidly. China wood oil or tung oil and soybean oil are also effective. Applied hot, better penetration is secured. The oil should be applied immediately after heating, however, as it will become more viscous if allowed to stand. Two or three coats may be applied, allowing each to dry thoroughly before the next application. Diluting the oil with turpentine, up to a mixture of equal parts, gives better penetration for the first coat. The concrete should be well cured and seasoned before the first application. The oil is sometimes applied after the magnesium fluosilicate treatment, providing a good coating over a hardened surface.

4. *Cumar.* Cumar is a synthetic resin soluble in xylol and similar hydrocarbon solvents. A solution consisting of about 6 lb of cumar per gal of xylol with ½ pt boiled linseed oil makes a good coating. Two or more coats should be applied. Concrete should be fairly dry. The cumar should be powdered to aid dissolving. It is available in grades from dark brown to colorless, and sold through paint and varnish trades.

5. *Varnishes and paints.* Any varnish can be applied to dry concrete. High-grade varnishes of the spar, china wood oil, or bakelite types and synthetic resin paints and coatings, or paints consisting largely of chlorinated rubber or synthetic rubber give good protection against many substances. Two or more coats should be applied. Some manufacturers can provide specially compounded coatings for certain conditions.

6. *Bituminous or coal-tar paints, tar and pitches.* These are usually applied in two coats, a thin priming coat to ensure bond and a thicker finish coat. Finish coat must be carefully applied to secure continuity and avoid pinholes. Surface should be touched up where necessary.

7. *Bituminous enamel.* This is suitable protection against relatively strong acids. It does not resist abrasion at high temperatures. Two materials are used, a priming solution and the enamel proper. The priming solution is of thin brushing consistency and should be applied so as to completely cover, touching up any uncoated spots before applying the enamel. When primer has dried to slightly tacky state, it is ready for the enamel. The enamel usually consists of a bitumen with a finely powdered siliceous mineral filler. The filler increases the resistance to flowing and sagging at elevated temperatures, and to abrasion. The enamel should be melted and carefully heated until it is fluid enough to brush. The temperature

Table i

Material	Effect on concrete	Surface treatment*
Acids		
Acetic...................	Disintegrates slowly	5, 6, 7
Acid waters.............	Natural acid waters may erode surface mortar, but usually action then stops	1, 2, 3
Carbolic................	Disintegrates slowly	1, 2, 3, 5
Carbonic...............	Disintegrates slowly	2, 3, 4
Humic..................	Depends on humus material, but may cause slow disintegration	1, 2, 3
Hydrochloric...........	Disintegrates	8, 9, 10, 11, 12
Hydrofluoric...........	Disintegrates	8, 9, 11, 12
Lactic.................	Disintegrates slowly	3, 4, 5
Muriatic...............	Disintegrates	8, 9, 10, 11, 12
Nitric..................	Disintegrates	8, 9, 10, 11, 12
Oxalic.................	None	None
Phosphoric.............	Attacks surface slowly	1, 2, 3
Sulfuric................	Disintegrates	8, 9, 10, 11, 12
Sulfurous..............	Disintegrates	8, 9, 10, 11, 12
Tannic.................	Disintegrates slowly	1, 2, 3
Salts and alkalies (solutions)†		
Carbonates of ammonia, potassium, sodium....	None	None
Chlorides of calcium, potassium, sodium, strontium	None unless concrete is alternately wet and dry with the solution, when it is advisable to treat with	1, 3, 4
Chlorides of ammonia, copper, iron, magnesium, mercury, zinc...	Disintegrates slowly	1, 3, 4
Fluorides..............	None except ammonium fluoride	3, 4, 5
Hydroxides of ammonia, calcium, potassium, sodium..............	None	None
Nitrates of ammonia, calcium, potassium, sodium..............	Disintegrates	8, 9, 10, 11, 12
	None	None
Potassium permanganate	None	None
Silicates...............	None	None
Sulfates of ammonia, aluminum, calcium, cobalt, copper, iron, manganese, nickel, potassium, sodium, zinc	Disintegrates	6, 7, 8, 9
	Disintegrates; however, concrete products cured in high-pressure steam are highly resistant to sulfates	1, 3, 4
Petroleum oils		
Heavy oils below 35° Baumé‡.............	None	None
Light oils above 35° Baumé‡	None—require impervious concrete to prevent loss from penetration, and surface treatments are generally used	1, 2, 3, 5, 9
Benzine, gasoline, kerosene, naphtha	None—require impervious concrete to prevent loss from penetration, and surface treatments are generally used	1, 2, 3, 5, 9
High-octane gasoline..		12

Table 1 (*Continued*)

Material	Effect on concrete	Surface treatment*
	Coal-tar distillates	
Alizarin, anthracene, benzol, cumol, paraffin, pitch, toluol, xylol	None	None
Creosote, cresol, phenol.	Disintegrates slowly	1, 2, 5, 9
	Vegetable oils	
Cottonseed............	No action if air is excluded. Slight disintegration if exposed to air	None 1, 2, 5, 9
Rosin.................	None	None
Almond, castor, china wood,§ coconut, linseed,§ olive, peanut, poppy-seed, rapeseed, soybean,§ tung,§ walnut.................	Disintegrates surface slowly	1, 2, 5, 9
Turpentine............	None—considerable penetration	1, 2, 5, 9
	Fats and fatty acids (animal)	
Fish oil...............	Most fish oils attack concrete slightly	1, 2, 3, 5, 9
Foot oil, lard and lard oil, tallow and tallow oil..................	Disintegrates surface slowly	1, 2, 3, 5, 9
	Miscellaneous	
Alcohol...............	None	None
Ammonia water (ammonium hydroxide)......	None	None
Baking soda...........	None	None
Beer..................	Beer will cause no progressive disintegration of concrete, but in beer storage and fermenting tanks a special coating is used to guard against contamination of beer	Coatings made and applied by Turner Rostock Co., 420 Lexington Ave., New York, and Borsari Tank Corp. of America, 60 E. 42d St., New York
Bleaching solution......	Usually no effect. Where subject to frequent wetting and drying with solution containing calcium chloride provide	1, 3, 4
Borax, boracic acid, boric acid.................	No effect	None
Brine (salt)...........	Usually no effect on impervious concrete. Where subject to frequent wetting and drying of brine provide	1, 3, 4
Buttermilk............	Same as milk	3, 4, 5
Charged water.........	Same as carbonic acid—slow attack	1, 2, 3
Caustic soda...........	None	None
Cider.................	Disintegrates (see acetic acid)	5, 6, 7
Cinders...............	May cause some disintegration	1, 2, 3

Table 1 (*Continued*)

Material	Effect on concrete	Surface treatment*
Miscellaneous—*Continued*		
Coal..................	Great majority of structures show no deterioration. Exceptional cases have been coal high in pyrites (sulfide of iron) and moisture showing some action but the rate is greatly retarded by deposit of an insoluble film. Action may be stopped by surface treatments	1, 2, 3
Corn syrup...........	Disintegrates slowly	1, 2, 3
Cyanide solutions......	Disintegrate slowly	7, 8, 9, 10, 12
Electrolyte...........	Depends on liquid. For lead and zinc refining and chrome plating use	7, 8, 9, 10, 11
	Nickel and copper plating	None
Formalin.............	Aqueous solution of formaldehyde disintegrates concrete	5, 9, 10, 11, 12
Fruit juices...........	Most fruit juices have little if any effect as tartaric acid and citric acid do not appreciably affect concrete. Floors under raisin-seeding machines have shown some effect, probably due to poor concrete	1, 2, 3
Glucose..............	Disintegrates slowly	1, 2, 3
Glycerin.............	Disintegrates slowly	1, 2, 3, 4, 5, 9
Honey...............	None	None
Lye..................	None	None
Milk.................	Sweet milk should have no effect, but if allowed to sour the lactic acid will attack	3, 4, 5
Molasses.............	Does not affect impervious, thoroughly cured concrete. Dark, partly refined molasses may attack concrete that is not thoroughly cured. Such concrete may be protected with	2, 5, 9
Niter................	None	None
Sal ammoniac.........	Same as ammonium chloride—causes slow disintegration	1, 3, 4
Sal soda.............	None	None
Saltpeter.............	None	None
Sauerkraut...........	Little, if any, effect. Protect taste with	1, 2
Silage...............	Attacks concrete slowly	3, 4, 5
Sugar................	Dry sugar has no effect on concrete that is thoroughly cured.	None
	Sugar solutions attack concrete	1, 2, 3
Sulfite liquor..........	Attacks concrete slowly	1, 2, 3
Tanning liquor........	Depends on liquid. Most of them have no effect. Tanneries using chromium report no effects. If liquor is acid, protect with	1, 2, 3
Trisodium phosphate...	None	None
Vinegar..............	Disintegrates (see acetic acid)	5, 6, 7
Washing soda.........	None	None
Whey................	The lactic acid will attack concrete	3, 4, 5

Table 1. (*Continued*)

Material	Effect on concrete	Surface treatment*
Miscellaneous—*Continued*		
Wine..................	Many wine tanks with no surface coating have given good results but taste of first batch may be affected unless concrete has been given tartaric acid treatment	For fine wines the concrete has been treated with 2 or 3 applications of tartaric acid solution (1 lb tartaric acid in 3 pt water). Sodium silicate is also effective. In a few cases tanks have been lined with glass tile
Wood pulp...........	None	None

* Treatments indicated provide sufficient protection in most cases but any of the other treatments designated by a number higher than the highest shown would be equally suitable and often may be advisable. See discussion.

† Dry materials generally have no effect.

‡ Many lubricating and other oils contain some vegetable oils. Concrete exposed to such oils should be protected as for vegetable oils.

§ Applied in thin coats the material quickly oxidizes and has no effect. Results indicated above are for constant exposure to the material in liquid form.

should not exceed 375°F. When fluid it should be mopped on quickly, as it sets and hardens rapidly.

8. *Bituminous mastic.* This is used chiefly for floors on account of the thickness of the layer which must be applied, but some mastics can be troweled on vertical surfaces. Some mastics are applied cold. Others must be heated until fluid. The cold mastic consists of two compositions—the priming solution and the body coat or mastic. The primer is first brushed on. When the primer has dried to a tacky state, a thin layer—about $\frac{1}{32}$ in.—of the mastic is troweled on. When this has dried, successive $\frac{1}{32}$-in. coats of the mastic are applied until the required thickness has been built up. The mastic is similar to the primer but is ground with sufficient asbestos and finely powdered siliceous material fillers to make a very thick, pasty, fibrous mass.

The hot mastics are somewhat similar to the mixtures used in sheet asphalt pavements, but contain more asphaltic binder so that when heated to fluid condition they can be poured and troweled into place. They are satisfactory only when applied in layers 1 in. or more in thickness. When ready to lay, the mixture usually consists of about 15% asphaltic binder, 20% finely powdered siliceous mineral filler, and the remainder is sand graded up to $\frac{1}{4}$-in. maximum size.

9. *Vitrified brick or tile.* These are special burnt clay products which possess high resistance to attack by acids or alkalies. They must, of course, be laid in mortar which is also resistant against the substance to which they are to be exposed. A waterproof membrane and a bed of mortar are usually placed between the brick or tile and concrete. Some of the acid-resistant cements are melted and poured in the joints. Only materials suitable for the conditions should be used and the manufacturer's directions for installation must be followed. Silica brick and cement are not resistant to hydrofluoric acid and the hydroxides, but special brick and cement for these substances are available.

10. *Glass.* May be cemented to the concrete.

11. *Lead.* May be cemented to the concrete with an asphaltic paint.

12. *Sheets of synthetic resin, rubber, and synthetic rubber.* Thin sheets of synthetic resin, rubber, or synthetic rubber resistant to many acids, alkalies, and

other substances are available. These are cemented to the concrete with special adhesives.

REACTIVE AGGREGATES

13   Alkali-Aggregate Reaction

Reactions between certain highly siliceous constituents of aggregates and cements having high alkali contents result in strength loss, excessive expansion, cracking, and disintegration. Cracks resulting from this action develop an irregular pattern as

FIG. 4. Typical pattern cracking on the exposed surface of concrete affected by alkali-aggregate reaction.[5]

shown in Fig. 4. The expansions associated with the action may close expansion joints, cause structural members to shift with respect to others, cause machinery to be dislocated and other undesirable effects. The reaction takes place when the alkalies Na_2O and K_2O are present in amounts exceeding 0.6% by weight, expressed as Na_2O equivalent, when aggregates contain significant amounts of reactive substances such as opal, chalcedony, tridymite, heulandite, zeolite, some phyllites, cristobalite, dacites, andesites and some rhyolites, and when moisture is present. The expansive deterioration is believed to be caused by osmotic swelling of the alkali silica gels produced by the chemical reaction between the reactive siliceous material and the alkalies released by hydration of the cement and from other sources. All siliceous materials in concrete aggregates do not expand excessively when combined with high-alkali cement as shown in Fig. 5. Also it is

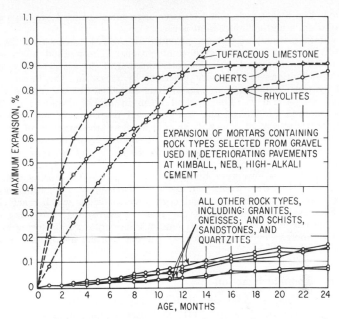

Fig. 5. Relation between expansion, age, and type of aggregate when a high-alkali cement is used.[6]

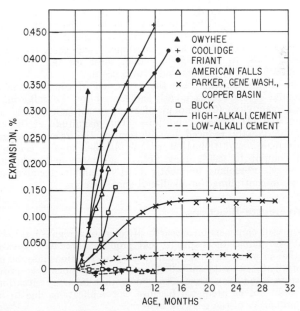

Fig. 6. Relation between expansion, age, and alkali content of cement of mortars made with various aggregates.[6]

evident from Fig. 6 that even when reactive aggregates are used excessive expansion occurs only when a high-alkali cement is used.

Two standard methods of test are provided by ASTM to check on the potential alkali reactivity of cement-aggregate combinations. The method given in ASTM Standard C 227 is concerned with the determination of the susceptibility of cement-aggregate combinations to expansive reactions by measuring the linear expansions developed by the combination in mortar bars stored above water in covered containers kept at $100 \pm 3°F$. The method given in ASTM Standard C 289 covers a chemical method of determining the potential reactivity of an aggregate with alkalies in portland-cement concrete as indicated by the amount of the reaction during a 24-hr period at 80°C between $1N$ sodium hydroxide solution and the aggregate crushed and sieved to pass a No. 50 sieve and be retained on a No. 100 sieve.

Alkali-aggregate reactions may be controlled or reduced by using cements with an alkali content less than 0.60% expressed as Na_2O equivalent; using nonreactive aggregates; using certain finely ground pozzolanic materials which react chemically with the alkalies before they attack the reactive aggregates; providing exposures that cause the concrete to become and remain relatively dry; and using air-entraining agents to increase the void space within the mortar, not considered effective with large deteriorations.

14   Alkali-Carbonate Reaction

The study of the problems of alkali-carbonate reactivity is a relatively new development in concrete research but the large amount of reported research is a good indication of its importance. Although most of the dolomitic aggregates give excellent service in concrete, certain types of fine-grained argillaceous dolomitic aggregates may react with the cement alkalies and produce undesirable expansions.

Tests for alkali-carbonate reactivity are made by exposing concrete prisms made with the questionable aggregate and a high-alkali cement to an atmosphere at 73°F and 100% relative humidity and noting the amount of expansion for a given period. Another test used to detect alkali-carbonate reactivity consists in storing prisms of rock in a $2M$ alkali hydroxide (equal molar parts of NaOH and KOH) solution. Expansions of the questionable material are then compared with those for companion prisms of known sound limestone. Aggregates having linear expansions in excess of 0.03 to 0.04% greater than those for the control aggregate at ages up to 10 months should be regarded with concern.

Experimental results[7] indicate that the expansion of concrete containing alkali-reactive carbonate aggregates is greatly influenced by the following factors: (1) degree of rock reactivity, (2) amount of reactive component, (3) alkali content of the cement, and (4) storage conditions. The magnitude of the expansion and cracking may be greatly reduced by dilution of the reactive aggregate with a nonreactive aggregate and reduction of the alkalies in the cement. Reduction of the alkali content to 0.60% may not be adequate and reductions to 0.40% may be necessary. Pozzolans frequently used to inhibit the alkali-silica reaction are not effective in controlling the alkali-carbonate reaction.

15   Other Aggregate Reactions

With an increased emphasis on research of cement-aggregate reactions it becomes apparent that the actual mechanisms that produce excessive expansion are very complex and that in many instances combinations of reactions may be involved. Investigations of alkali-aggregate and alkali-carbonate reactions have provided answers to some questions, and they have also shown that in many situations the actual performance could not be predicted because of complications due to other physical and chemical properties of the reacting materials.

Under certain conditions concrete made with sand-gravel aggregates from Kansas, Nebraska, and portions of several adjacent states may suffer large expansions and possible deterioration. Apparently alkali-aggregate reaction to a varying degree, delayed hydration of free magnesia, rupture of the cement-aggregate bond, and physical and chemical phenomena that have not yet been defined may be involved. Replacement of 30% or more, by weight, of the questionable aggregate with crushed limestone may be used for partial control of the undesirable expansion.

Large expansions have also been associated with aggregates that undergo excessive drying shrinkage in air and large expansion in water. In the Union of South Africa certain sandstones containing clay minerals, and in Scotland certain basic igneous rocks of the basalt and dolerite type have given trouble. The cause of the dimensional instability of the aggregate itself under wetting and drying conditions is uncertain. The presence of clay minerals with expansive lattices such as montmorillonite offers only a partial explanation since large dimensional changes have also been observed in aggregates that do not contain such clay minerals. Further, it appears that the shrinkage and expansion of the mortar is not simply related to the dimensional change of the rock and that complex phenomena are involved.

WEAR

16 Abrasion

The information on the resistance of concrete to abrasion is limited but it appears that a high degree of resistance to abrasion can be obtained with a dense concrete that has a low water-cement ratio and a minimum of fine aggregate compatible with good workability, and that has been properly placed and cured. Overworking and early troweling of the concrete surface should be avoided in order to prevent formation of an oversanded, very wet mortar at the surface. The abrasion resistance increases roughly as does the compressive strength up to about 6,000 psi. Entrained air influences abrasion in a manner similar to its effect on strength and consequently the amount of entrained air should generally not exceed 4% when good resistance to abrasion is desired.

Wear of concrete surfaces has been classified as follows:[8] (1) attrition, wear on concrete floors due to normal foot traffic, light trucking, and sliding of objects; (2) attrition combined with scraping and percussion, wear on concrete road surfaces due to vehicles with and without chains; (3) attrition and scraping, wear on underwater construction due to the action of abrasive material carried by moving water; and (4) percussion, wear on hydraulic structures where a high hydraulic gradient is present—generally known as cavitation erosion.

Cavitation effects have caused severe damage to hydraulic structures. The action is caused by the abrupt change in direction and velocity of the water so that the pressure at some points is reduced to the vapor pressure and vapor pockets are created. These pockets collapse with great impact when they enter areas of higher pressure, producing very high impact pressures over small areas which eventually cause pits and holes in the concrete surface. It has been estimated that the impact of the collapse may produce pressures as high as 100,000 psi. Apparently cavitation damage is not common in open conduits at water velocities below 40 fps. In closed conduits velocities as low as 25 fps may cause pitting.

A large variety of abrasion and impact tests have been developed to provide information for special purposes. Tests to determine abrasion properties of concrete aggregates by means of the Los Angeles machine are given in ASTM Standards C 131 and C 535, and by means of the Deval machine in ASTM Standards D 2 and D 289. A test to determine the abrasion resistance of concrete is provided by ASTM Standard C 418. In this test the abrasion resistance of concrete is determined by subjecting it to the impingement of air-driven silica sand and determining weight loss under given test conditions.

The use of absorptive form linings and the vacuum process reduce wear as a result of the lowered water content. Properly applied combinations of dry cement and special aggregates decrease wear of unformed surfaces. Application of pressure through hard troweling to the concrete surface after it begins to set is also effective. Abrasion properties of hardened concrete floors may be improved by the use of surface hardeners or by painting.

CORROSION

17   Steel

Sufficient concrete coverage normally provided protects the reinforcement against corrosion. Special protection may be required when concrete is submerged in or exposed to water containing salt alkalies or chlorides. Some work has been reported on the use of corrosion-inhibiting admixtures, but agreement on the effectiveness of these materials over a wide variety of exposure conditions has not yet been reached. Under some circumstances additions of calcium chloride to accelerate rate of strength gain are considered harmful because of adverse effects on corrosion of the reinforcement. The use of calcium chloride should not be permitted in prestressed concrete, in concrete containing galvanized reinforcing steel, or in concrete that may be subjected to electrolytic action.

Disintegrations of certain reinforced-concrete structures have apparently been due to electrolytic action of stray currents from neighboring power circuits. This type of action is accelerated when the concrete is wet and when calcium chloride is present. The action results in a loss of bond between the steel and concrete due either to the oxidation of the steel and the development of high radial stress, or to the softening of the concrete around the steel.

18   Aluminum, Copper, Lead, and Zinc

Since some nonferrous metals are frequently used in contact with portland-cement concrete the possibilities of corrosion contact must be considered. Aluminum may be used for coils, pipelines, and linings for concrete vats; copper may be used for flashings, roofing, and diaphragms in joints; lead may be used for cable sheathing and for lining tanks; zinc may be used for flashings, roofing, and leaders.

Aluminum is corroded by caustic alkalies. It is attacked by fresh and unseasoned concrete, reacting with calcium hydroxide to form calcium aluminate. The action is progressive because the corrosion products are nonadherent and do not protect the underlying metal. Aluminum should be protected with a coating of asphalt, varnish, or pitch when it is to be embedded in fresh concrete or is to come in contact with wet concrete. It has also been shown that severe corrosion of aluminum conduit may occur when the conduit is embedded in concrete containing calcium chloride and steel that is electrically connected to the aluminum. The pressures developed by the corrosion product are great enough to crack the concrete or to collapse the conduit.[9]

Copper is almost immune to the action of caustic alkalies. Copper may be safely embedded in fresh concrete and will not normally react with hardened concrete in either the wet or dry state. However, if copper is to be in contact with wet concrete, admixtures containing chlorides should not be used because of their corrosive action.

Lead in contact with fresh concrete will react with calcium hydroxide and will corrode. If it is to be in contact with fresh or green concrete it should be given a protective coating of asphalt, varnish, or pitch. Electrolytic action may also cause gradual disintegration of embedded lead under moist conditions when part of the lead is exposed to the air and the other part is embedded in concrete.

Zinc is attacked by caustic alkalies such as calcium hydroxide. Zinc in fresh concrete will react with calcium hydroxide to form calcium zincate. This corrosion product appears to form a dense, closely adherent film that protects the underlying material from further attack. If all action between the calcium hydroxide and the zinc is to be avoided the metal should be given a protective coating of asphalt, varnish, or pitch. Zinc does not corrode in contact with dry seasoned concrete.

REFERENCES

1. Levowitz, D.: *Food Eng.*, June, 1952, p. 43.
2. Higginson, E. C., and D. J. Glantz: The Significance of Tests on Sulfate Resistance of Concrete, *ASTM Proc.*, vol. 53, pp. 1002–1020, 1953.
3. McBurncy, J. W.: Cracking in Masonry Caused by Expansion of Mortar, *ASTM Proc.*, vol. 52, p. 1228, 1952.
4. "Effect of Various Substances on Concrete and Protective Treatments, Where Required," Portland Cement Association, Concrete Information ST 4.
5. "Concrete Manual," 7th ed., U.S. Bureau of Reclamation, 1963.
6. Blanks, R. F., and H. L. Kennedy: "The Technology of Cement and Concrete," vol. 1, "Concrete Materials," John Wiley & Sons, Inc., New York, 1955.
7. Newlon, Howard H., Jr., and W. Cullen Sherwood: Methods for Reducing Expansion of Concrete Caused by Alkali-Carbonate Rock Reactions, *Highway Res. Board Record*, no. 45, p. 134.
8. Kennedy, H. L., and M. E. Prior: Abrasion Resistance, *ASTM Spec. Tech. Publ.* 169, pp. 163–174, 1956.
9. Monfore, G. E., and Borje Ost: Corrosion of Aluminum Conduit in Concrete, *J. Res. Develop. Lab., PCA,* vol. 7, no. 1, p. 10, January, 1965.

Chapter 8

PERMEABILITY AND ABSORPTION

GEORGE W. WASHA

GENERAL CONSIDERATIONS

1 Definitions

Water may enter a porous body as either a liquid or vapor through capillary attraction, it may be forced in under pressure, or it may be introduced by a combination of pressure and capillary attraction. Motion of water through the body may also involve osmotic effects. Absorption refers to the process by which concrete draws water into its pores and capillaries. Permeability of concrete to water or vapor is the property which permits the passage of the fluid or vapor through the concrete.

All concrete mixtures absorb some water and are permeable to a certain extent. Tests under hydrostatic heads of 100 ft have indicated that neither portland cement nor mixtures made from it are absolutely impervious. However, there is abundant evidence which shows that concrete and mortar can be made so impermeable that no leakage or dampness is visible on the surface opposite that at which the water enters. Apparently, even when the humidity is high, the frictional resistance to flow prevents the water from leaving the free surface of the concrete at a rapid enough rate to escape evaporation. In hydraulic structures where watertightness is of great concern, permeability may be more important than strength. Permeability and absorption may also be important because of their relation to various disintegration agencies that damage concrete.

2 Pore Structure of Concrete

Concrete contains pores in all its constituents. Pores in the aggregates undergo only small changes with time but pores in the cement paste are subject to great changes, especially during the early life of the paste. In a freshly mixed neat cement paste the water-filled space is available for the formation of hydration products. This space, originally a function of the water-cement ratio of the paste, is continually reduced by the volume of the precipitated hydrated gel. The capillary system at any time is that part of the original water-filled space not filled with hydrated gel. It is thus apparent that hydration reduces the size and volume of capillary pores and increases the gel volume, and that the process is continuous as hydration progresses. It has been stated that if the original capillary space is low (water-cement ratio less than 0.40 by weight) the gel will eventually fill all the original water space and leave a paste with no capillary pores.[1] As the water-cement ratio is increased and as the degree of hydration is decreased the volume of capillary pores is increased. The capillary pores are of submicroscopic size, interconnected, and randomly distributed.

The pores in the gel are very numerous and are much smaller than the capillary pores. Water in the gel pores does not behave as normal free water because of the very small pore size. This is also true for water in capillary pores, but to a lesser degree, because of the larger pore size. The permeability of paste is most closely associated with the capillary pores because water in these pores is more responsive to changes in hydrostatic pressure than the water in the gel pores.

The aggregates which make up from 70 to 75% of the total concrete volume have porosities that range from nearly 0 to about 20% by solid volume. The pores vary considerably in size, are larger than the gel pores, and may be at least as large as the largest capillary pores. Associated with the aggregates are the voids in contact with the bottom surfaces of the coarse aggregates that result from the upward flow of water through the plastic concrete. In addition, the tendency of the cement paste to settle during the plastic state results in voids between the sand particles.

Concrete usually contains entrapped air and may in addition contain purposefully added entrained air voids. The entrained air voids are generally noncoalescing and separated spheroids that reduce bleeding, tend to reduce the channel structure, and decrease permeability. The air voids usually constitute from 0.5 to 6.0% of the concrete volume.

In summary, concrete contains a wide variety of pores between and within its various components. They may exist as separated spheroids or they may be interconnected and randomly distributed. Further the pore structure will change as hydration continues and causes a reduction of the capillary pores and an increase in gel pores. Since capillary pores are directly related to permeability increased hydration decreases the permeability.

ABSORPTION AND PERMEABILITY TESTS

3 Test Methods

Absorption tests are most frequently performed by immersing concrete in water for 24 or 48 hr, weighing after surface drying, ovendrying, and again weighing. The absorption may then be calculated by dividing the loss in weight by the ovendry weight. In some instances the concrete may be boiled for 5 hr and then dried. If rate of absorption is required the weight after a short period of immersion, as 30 min, may be compared with that after the total 24- or 48-hr period. Total absorption is considered to be a criterion of concrete durability but correlation is usually not very satisfactory. Rate of absorption appears to provide a better correlation.

Although the principal objective of permeability testing is to determine the permeability characteristics of concrete, the tests may have little direct relation to the imperviousness of the structure made with the concrete because of the presence of cracks and poor joints. However, besides providing information on the permeability properties of the concrete the test is also useful in determining the corrosive effects of percolating waters that leach out the free lime and gradually attack the lime in the tricalcium silicate. Tests also provide information on the relative efficiencies of cements, on the use of integral and surface waterproofing agents, and provide information on the basic pore structure which is related to other properties such as absorption, capillarity, uplift, resistance to freezing and thawing, and others.

The permeability of concrete is usually obtained by determining the amount of water under pressure that is forced to flow into a specimen during a given time interval or by determining the amount that flows out of the opposite surface, exposed to air, in a given time interval. Each cylindrical specimen is usually sealed along its curved surface in metal containers and subjected to water under pressure, hydrostatic heads up to 1,000 ft, on one of its flat surfaces. Means for measuring the water inflow or outflow, protected from evaporation losses, are provided. Initially outflow will be less than inflow because of absorption, but ultimately the values will be the same. The leakage, frequently given in gallons or cubic feet of water per square foot of area per hour, is usually determined at a given time after the start of the test. This is done to eliminate absorption effects during the early portion of the test and to provide a standard for comparison purposes.

The leakage of water through concrete can be calculated from Darcy's law for viscous flow

$$\frac{Q}{A} = K_c \frac{H}{L}$$

where Q = rate of flow, cfs

A = area of cross section under pressure, sq ft

$\frac{H}{L}$ = ratio of head of water to percolation length

K_c = permeability coefficient, unit rate of discharge in cfs due to a head of 1 ft acting on a specimen of 1 sq ft cross section and 1 ft thick

Other factors influence the results of permeability tests. Percolating water of an aggressive nature will increase permeability due to the leaching of lime from the cement. Percolating water that contains sediment or bacteria will act to close the pores and decrease permeability. The direction of water flow through the concrete in relation to the direction of placement may be important. End effects

Table 1. Tabulation of Typical Permeability Coefficients for Varying Materials

Material	$K \times 10^{12}$	Class
Granite specimen	2–10	12
Slate specimen	3–7	12
Concrete and mortar, W/C 0.5–0.6	1–300	12–10
Breccia specimen	20–	11
Concrete and mortar, W/C 0.6–0.7	10–650	11–10
Calcite specimen	20–400	11–10
Concrete and mortar, W/C 0.7–0.8	30–1,400	11–9
Limestone specimen	30–50,000	11–8
Concrete and mortar, W/C 0.8–1.0	150–2,500	10–9
Dolomite specimen	200–500	10
Concrete and mortar, W/C 1.2–2.0	1,000–70,000	9–8
Biotite gneiss in place, field test	1,000–100,000	9–7
Sandstone specimen	7,000–500,000	9–7
Cores for earth dams	1,000–1,000,000	9–6
Slate in place, field test	10,000–1,000,000	8–6
Face brick	100,000–1,000,000	7–6
Concrete, unreinforced canal linings, field test	100,000–2,000,000	7–6
Steel sheetpiling—junction open 1/1,000 in. with ½ in. of contact—18-in. sections*	500,000–	7
Concrete, restrained slabs with ¼ to ½ % reinforcing—30° temp. change*	1,000,000–5,000,000	6
Water-bearing sands	1,000,000,000	3

Coefficient represents quantity of water in cubic feet per second, per square foot of surface exposed to percolation, passing through 1 ft of substance with 1-ft head. $Q = \dfrac{K_c H}{L}$.

* Flow through 1 ft length of crack, 1 ft deep, $Q = 60,000 \dfrac{H}{L} D^3$.

are important because of the concentration of mortar and paste near the surface. Consequently short specimens are less permeable per unit length than long specimens.

Tests made on Boulder Dam mass concrete containing 9-in. gravel and 4 sacks of low-heat cement per cu yd of concrete showed that it was relatively impermeable and that flow through its pore structure was negligible.[2] The report on these tests also included information on permeability coefficients and classes of permeability for various materials as shown in Table 1. An increase in the class number of unity indicates an increase of impermeabilty of ten times.

FACTORS AFFECTING WATERTIGHTNESS

4 Materials

Test results[3] indicate that permeabilities of cement pastes made with portland cements differing in chemical composition are similar if the initial water-cement

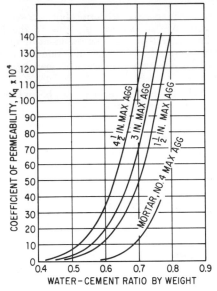

Fig. 1. Relation between coefficient of permeability and water-cement ratio for mortar and for concrete with three different maximum coarse-aggregate sizes. K_q is a relative measure of the flow of water through concrete in cubic feet per year per square foot of area for a unit hydraulic gradient.[4]

ratios, corrected for bleeding, are equal and equal fractions of the cements have hydrated. However, at a given age and given water-cement ratio, pastes containing slow-hydrating cements are more permeable than those made with fast-hydrating cements. Cement fineness is also important, with watertightness decreasing as fineness increases.

Permeability as well as strength is related to the water-cement ratio as shown in Fig. 1. The permeability of concrete increases at a rapid rate as the water-cement ratio exceeds about 0.65, by weight. For given materials and conditions an increase in the water-cement ratio from 0.40 to 0.80 may increase permeability about 100 times.

The effect of aggregate size on permeability is also shown in Fig. 1. With a given water-cement ratio the permeability increases as the maximum size of the aggregate increases, probably because the water voids present on the underside of the coarse aggregate increase as the maximum aggregate size is increased. Sound dense aggregates having a low porosity along with proper grading are essential to the development of concrete with low permeability. Sufficient fine aggregate must be used to ensure placement of concrete free from honeycombing.

Purposely entrained air generally reduces permeability because of the increased workability, decreased bleeding, and effect of the separated voids in reducing water-channel structure. Substitution of fly ash for a part of the portland cement generally reduces the permeability of the concrete.

5 Proportions

The exact proportions of aggregates and the cement content are dependent on the grading and shape of the aggregate particles, on the conditions of placement,

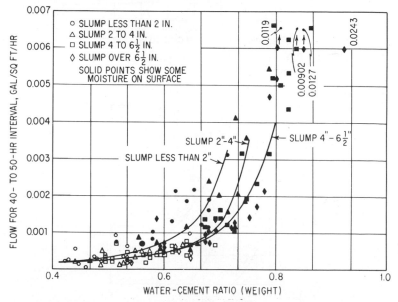

Fig. 2. Effect of water-cement ratio and slump on permeability.[5]

and on the desired permeability characteristics. The correct amount of water must be determined and used. The use of excess water reduces density and increases flow. Hand placement of a dry mix leads to pinholes and large leakages, especially with lean concrete. Generally, a slight excess of water produces less leakage than a slight deficiency. For most hand-rodded concrete the minimum slump that can be successfully used varies from 2 to 6 in., depending on the richness of the mix, the maximum size of aggregate, intricacy of reinforcing, width of forms, and amount of puddling. The relation between water-cement ratio and slump is shown in Fig. 2. It is apparent that the slump is especially important in mixtures with higher water-cement ratios. With vibrated concrete, slumps of ½ to 2 in. can often be used effectively to produce concrete with low permeability.

6　Placing and Curing

Handling, placing, and compacting properly proportioned concrete in forms is an important step in the production of watertight concrete. Every attempt must be made to avoid segregation, which may cause honeycombing or a porous structure.

The need of continued hydration of the cement to reduce the volume of pores through development of gel has been previously stated. The effect of the length of the curing period on the permeability is shown in Fig. 3. It is apparent that

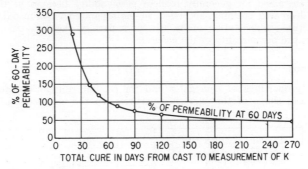

Fig. 3. Effect of length of curing period on permeability.[2]

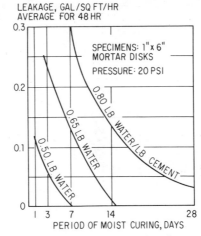

Fig. 4. Effect of water-cement ratio and curing on permeability.[6]

curing during the early life of the concrete is especially effective in reducing permeability. The relation between water-cement ratio and length of curing period on permeability is shown in Fig. 4. The curves show that concrete with a water-cement ratio of 0.50 became essentially impermeable after 7 days moist curing, concrete with a water-cement ratio of 0.65 required 14 days, and concrete with a water-cement ratio of 0.80 did not reach that condition even after curing 28 days.

7　Construction Practices

Since construction joints are a frequent source of water-leakage problems they should be avoided whenever possible. In massive concrete structures, where expansion joints are provided, concrete placement should be continuous between joints. If construction joints are necessary, precautions must be taken to ensure good bond between the old and the new concrete. A properly proportioned mixture with a minimum of bleeding is needed to avoid formation of a weak and porous layer on the surface which will provide poor bond with the next layer. The bond between two layers of concrete may be improved by:

1. Careful cleaning of the surface of the first layer of concrete, including wire brushing to expose the coarse aggregate before the first layer has become thoroughly hard and before the second layer is placed

2. Dampening the surface of the first concrete

3. Providing a slush coat of neat portland-cement grout on the surface of the first concrete layer

4. Placing, before the cement grout has attained its initial set, several inches of new concrete with less coarse aggregate to provide enough mortar to avoid coarse-aggregate pockets at the bottom of the new layer

5. Continuing placement of regularly designed concrete to complete the immediate construction (vibration of the fresh concrete is desirable)

Galvanized-iron or copper stops may be placed in construction joints of tanks. These stops are usually made of about 20-gage sheet metal and are about 7 or 8 in. wide. Half of the stop is embedded in the first layer of concrete and the other half in the second layer. Various types of rubber and plastic water stops are effectively used to make watertight joints in concrete structures.

WATERPROOFING AND DAMPPROOFING

8 Integral Mixtures

The terms waterproofing and dampproofing are frequently used interchangeably, but, in general, dampproofing refers to any method of making concrete impervious to water vapor or to liquid water under low pressure, while waterproofing refers to any method of making concrete impervious to liquid water that may be under high pressure.

If concrete is properly proportioned, placed, and cured it is quite impermeable under heads as high as 100 ft, and integral mixtures or surface treatments should not be necessary. Integral mixtures cannot be expected to offset poor workmanship or improper curing. They will not be effective if the concrete cracks.

A large number of proprietary materials are available as integral mixtures. They should be used in accord with the manufacturer's directions in order to be as effective as possible. The extra expense involved with the use of these materials should be considered in relation to the use of concrete made without integral mixtures but with higher cement contents. A wide variety of available integral mixtures may be classified as (1) calcium chloride; (2) water solutions of inorganic salts; (3) water suspensions of pore-filling substances; (4) soaps containing fatty acids, largely as stearates; (5) soap solutions with an evaporable solvent which react with hydrated lime; (6) finely subdivided materials; (7) combinations of solutions in successive applications which react chemically; (8) solutions of solid hydrocarbons in oil or paraffin; (9) bituminous coatings; and (10) portland cement with an added water repellent. As a result of an extensive series of tests[7] it was reported that the finely subdivided fillers generally decreased permeability, and that other types gave variable results. Some integral mixtures actually increased permeability.

9 Surface Applications

Materials used as surface treatments for concrete may be classified as those which penetrate the surface of the concrete and fill the pores or those which form surface films. The penetrating materials may be inert materials that simply fill the pores, or they may react with constituents in the concrete surface to form compounds of greater volume and with greater pore-filling capacity. These materials are also used as hardeners to prevent surface dusting and disintegration.

A large variety of proprietary materials are available. These materials should be used in accord with the manufacturer's recommendations to secure the best results. Surface waterproofing materials are usually more effective if they are applied on the face of the concrete in contact with water, but frequently this is not economically feasible for structures in existence.

Test results[8] have shown that surface waterproofing agents are more or less effective in decreasing permeability of concrete. Properly applied and cured portland-cement mortars, asphalt emulsions, heavy petroleum distillates dissolved in volatile solvents, and transparent coatings of linseed or china wood oil were among the most effective. In general, exposure to the elements decreases the effectiveness of surface treatments. Essentially absolute imperviousness can be obtained by the use of alternate layers of fabric, such as the better grades of roof felt, and hot asphalt or coal tar. While this method of waterproofing is costly it is universally acknowledged as satisfactory.

10   Air and Vapor Permeability[9,10,11]

Problems involving air and vapor permeability of concrete are important in various types of tanks and pressure vessels. The flow of air and water vapor depend on the air or vapor pressure; the thickness of the concrete; the properties of the concrete, and the properties of the air, gas, or vapor. Concrete made with a low water-cement ratio, with well-graded aggregates, with a pozzolan addition, and properly cured for an adequate time will be relatively impermeable to the flow of air, gases, or water vapor. Severe drying increases air permeability, probably because of the formation of shrinkage cracks, but vapor permeability is not apparently increased. There appears to be no significant relation between the air, gas, and water permeabilities of concrete.

REFERENCES

1. Verbeck, George: Pore Structure, *ASTM Spec. Tech. Publ.* 169, p. 136, 1956.
2. Ruettgers, A., E. N. Vidal, and S. P. Wing: An Investigation of the Permeability of Mass Concrete with Particular Reference to Boulder Dam, *J. ACI,* March–April, 1935, p. 382.
3. Powers, T. C., L. E. Copeland, J. C. Hayes, and H. M. Mann: Permeability of Portland Cement Paste, *J. ACI,* November, 1954, p. 285.
4. "Concrete Manual," 7th ed., U.S. Bureau of Reclamation, 1963.
5. Norton, P. T., and D. H. Pletta: Permeability of Gravel Concrete, *J. ACI,* May, 1931, p. 1093.
6. "Watertight Concrete," PCA Concrete Information ST-33.
7. Jumper, C. H.: Tests of Integral and Surface Waterproofings for Concrete, *J. ACI,* December, 1931, p. 209.
8. Washa, G. W.: The Efficiency of Surface Treatments on the Permeability of Concrete, *J. ACI,* September–October, 1933, p. 1.
9. "Air Permeability of Concrete," *Concrete and Constructional Engineering,* vol. 60, May, 1965, p. 166.
10. Henry, R. L., and G. K. Kurtz: "Water Vapor Transmission of Concrete and Aggregates," *Tech. Rept.* R244, U.S. Naval Civil Engineering Laboratory, Port Hueneme, Cal., June, 1963.
11. Graf, O.: "Die Eigenschaften des Betons," 2nd ed., Springer-Verlag, Berlin-Wilmersdorff, 1960.

Chapter 9

VOLUME CHANGES

GEORGE W. WASHA

1 General

Volume changes in concrete due to variations in stress, temperature, and humidity are partly or completely reversible, but volume changes due to destructive chemical and mechanical action are usually cumulative as long as the action continues. Unrestrained volume changes due to variations in temperature, moisture, or stress are usually not important, but volume changes that are restrained by foundations, reinforcement, or connecting members may lead to stresses which may cause distress and even failure. Restrained contractions are usually more important than restrained expansions because concrete is much weaker in tension than in compression. The magnitude of volume changes is usually given in linear rather than volumetric units because linear changes can be easily measured and because they are of primary concern to engineers. A length change may be given, for example, as 900 millionths of an inch per inch or 0.09% or 1.08 in. per 100 ft.

The causes of volume changes and the reactions of concrete to these changes are generally known, but it is not yet possible to build concrete structures such as buildings, bridges, pavements and dams with assurance that they will not crack. However, if proper attention is given to all the factors that are involved it is possible to build these structures so that they are relatively crack-free and resistant to the action of destructive agents frequently associated with cracking.

FRESH CONCRETE

2 Bleeding and Setting Shrinkage

Fresh concrete in the plastic state undergoes significant volume changes. They are due to water absorption, sedimentation (bleeding), cement hydration, and thermal changes and are influenced by the temperature and humidity of the surrounding atmosphere. Absorption of water by the aggregates and bleeding of the free water to the top, where it may be lost by drainage from the forms or by evaporation, cause shrinkage. Bleeding starts shortly after the concrete has been placed and continues until maximum compaction of the solids, particle interference, or setting stops the action. Usually a large portion of the shrinkage due to absorption and bleeding takes place during the first hour after placement. Profuse bleeding along with rapid evaporation or leaky or absorbent forms will result in excessive shrinkage, which may be as high as 1%. Since the concrete is in a plastic or semiplastic state during bleeding, there are no appreciable stresses associated with the large shrinkage. Setting shrinkage may be kept low by the use of saturated aggregates, low cement content, properly designed concrete mixtures, moist and cool casting conditions, tight and nonabsorbent forms, and shallow lifts in placing.

3   Plastic Shrinkage

Plastic strinkage and plastic cracking occur in the surface of fresh concrete soon after it has been placed and while it is still plastic. The principal cause of this type of shrinkage is rapid drying of the concrete at the surface. With the same materials, proportions, and methods of mixing, placing, finishing, and curing, cracks may develop one day but not on the next. Changing weather conditions that increase the rate of evaporation from the surface are usually involved. The highest evaporation rates are obtained when the relative humidity of the air is low, when the concrete and air temperatures are high, when the concrete temperature is higher than the air temperature, and when a strong wind is blowing over the concrete surface. This combination of circumstances, which may remove surface moisture faster than it can be replaced by normal bleeding, is most frequently obtained in summer. However, rapid drying is possible even in cold weather if the temperature of the concrete is high compared with air temperature.

Plastic-shrinkage cracks may have considerable depth and are usually straight lines without any symmetry or pattern, but they may have a crowfoot pattern. Corrective measures to prevent plastic-shrinkage cracking are all directed toward reducing the rate of evaporation or the total time during which evaporation can take place.

HARDENED CONCRETE

4   Caused by Cement Hydration

Volume changes that are due to cement hydration and do not include changes due to variations in moisture, temperature, or stress are called autogenous volume changes. These changes may be either expansions or contractions, depending on the relative importance of two opposing factors—expansion of new gel due to the absorption of free pore water or shrinkage of the gel due to extraction of water by reaction with the remaining unhydrated cement. Generally the initial expansions obtained during the first few months do not exceed 0.003% while the ultimate contractions obtained after several years usually do not exceed 0.015%. This type of volume change is especially important in the interior of mass concrete where little or no changes in total moisture content may take place.

Test results indicate that autogenous volume changes are influenced by the composition and fineness of the cement, quantity of mixing water, mixture proportions, curing conditions, and time. The magnitude of the autogenous volume change appears to increase as the fineness of the cement and the amount of cement for a given consistency are increased. Ultimate contractions seem to be somewhat greater for low-heat cements (Type IV) than for normal portland cement. The most significant autogenous shrinkage usually takes place within 60 to 90 days after placement of the concrete.

5   Caused by Thermal Changes

Unrestrained concrete expands as the temperature rises and contracts as it falls. The average value of the coefficient of thermal expansion, the rate at which thermal volume change takes place, is 5.5 millionths per degree Fahrenheit, which is fortunately close to the value for steel. While the coefficient may frequently be close to the average value, it may also vary between 2.5 and 8.0 millionths per degree Fahrenheit, depending on the richness of the mix, the moisture content, and primarily on the thermal coefficient of the aggregate.

The coefficient of thermal expansion of hardened cement paste usually varies between 5.0 and 12.5 millionths per degree Fahrenheit. Ovendried and vacuum-saturated specimens have similar coefficients, but specimens with intermediate moisture contents may apparently have coefficients almost twice as large.

Thermal expansion of concrete is greatly influenced by the type of aggregate because of the large differences in the thermal properties of various types of aggregates and because the aggregate constitutes from 70 to over 80% of the total solid volume of the concrete. Siliceous aggregates such as chert, quartzite, and sandstone have thermal coefficients of expansion between 4.5 and 6.5 millionths per degree Fahrenheit, while the coefficients for pure limestone, basalt, granite, and gneiss may vary between 1.2 to 4.5 millionths per degree Fahrenheit. Further single mineral crystals may have different coefficients along three different axes. As an example, feldspar has values of 9.7, 0.5, and 1.1 millionths per degree Fahrenheit along three different axes. An estimated value of the coefficient of thermal expansion for concrete may be computed from weighted averages of the coefficients of the aggregate and the hardened cement paste.

It has been suggested that significant internal stresses due to temperature variations may develop because of differences in the thermal coefficients of the cement paste and the aggregate. The relative importance of this thermal incompatibility and its effect on durability is in some doubt.

Thermal changes in mass concrete are kept as low as possible by the use of low-heat portland cements, by artificial refrigeration, and by other special procedures in order to avoid cracking as the concrete cools from the maximum to the stable temperature. Thermal changes in pavements are also important. Differential temperature gradients at night may cause the top surface to shorten in relation to the bottom surface, tending to lift the slab ends above the subgrade and decreasing the ability of the slab to support traffic loads without cracking.

6 Caused by Continuous Moist Storage

Moist-cured concrete starts to expand after the setting shrinkage has taken place and continues at a decreasing rate under continuous moist storage. The ultimate expansion is usually less than 0.025% and shows little change after 10 years of moist storage. The maximum expansion due to moist storage is usually about one-fourth to one-third of the shrinkage due to air drying. Expansions obtained with Type I cement appear to be higher than those with Types II, III, IV, and V. Replacement of cement with pozzolanic material usually increases slightly the expansion under continuous moist curing. However, the amount of the cementing material is much more important than the type, since neat cement pastes expand about twice as much as average mortars and the mortar expands about twice as much as an average concrete.

7 Caused by Drying

Drying shrinkage and carbonation shrinkage (due to the reaction between carbon dioxide and the constituents of cement) take place concurrently. Normally the two types of shrinkage are not separated in reported data and the total is usually designated as drying shrinkage. Continuously wet or dry structural concrete made with sand and gravel aggregates will have only a small amount of·carbonation shrinkage, but the carbonation shrinkage of porous concrete dried to equilibrium in air at 50% relative humidity may approach the drying shrinkage. Drying shrinkage of hardened concrete is usually caused by the drying and shrinking of the cement gel which is formed by hydration of portland cement. It is affected primarily by the unit water content of the concrete. Other factors affecting drying shrinkage include cement composition; cement content; quantity and quality of paste; characteristics and amounts of admixtures used; mineral composition and maximum size of aggregate; mixture proportions; size and shape of the concrete mass; amount and distribution of reinforcing steel; curing conditions; humidity of surrounding air during the drying period; and the length of the drying period.

Materials. The type of cement used influences drying shrinkage. In general, finer cements exhibit slighty greater shrinkages. The tricalcium aluminate contributes most and the tricalcium silicate least to drying shrinkage. Gypsum has a large effect on drying shrinkage, and for a given cement there appears to be an optimum amount that produces the smallest shrinkage.

Aggregate particles embedded in cement paste restrain drying shrinkage. Well-graded aggregates with a large maximum size reduce shrinkage because they allow low water contents, require small quantities of paste, and encourage cracking between particles. Concrete of the same cement content and slump with ⅜-in. maximum-size aggregate usually develops from 10 to 20% greater drying shrinkage than concrete with ¾-in. maximum-size aggregate, and from 20 to 35% greater drying shrinkage than concrete containing 1½-in. maximum-size aggregate. The actual amounts are dependent on many variables. The effect of aggregate size is also evident in the following drying shrinkages for neat cement, mortar, and concrete of 0.25 to 0.30, 0.06 to 0.12, and 0.03 to 0.08%, respectively. Aggregates having a high modulus of elasticity and those having rough surfaces offer greater restraint to shrinkage. Concrete made with sandstone, slate, hornblende, and pyroxene may shrink up to two times as much as concrete made with granite, quartz, feldspar, dolomite, and limestone. Dirty sands and unwashed coarse aggregates containing detrimental clay cause increased shrinkage.

Drying shrinkage of lightweight concrete can be relatively high, but careful selection of materials and proper attention to proportioning may provide lightweight concrete with about the same drying shrinkage obtained with normal-weight concrete. Drying-shrinkage values of structural lightweight concrete normally vary between 0.04 and 0.15% and are likely to be more pronounced for concrete containing aggregates that have high absorptions and that require high cement contents for strength. Moist-cured cellular products made with neat cement weighing between 10 and 20 pcf may have drying-shrinkage values between 0.30 and 0.60%. Autoclaved cellular products that contain fine siliceous material may weigh about 40 pcf and have drying-shrinkage values in the range 0.02 to 0.10%.

Admixtures appear to have a variable effect on drying shrinkage. When drying shrinkage is important admixtures should be used with care, preferably after evaluation under job conditions. Admixtures that increase the unit water content of concrete will usually increase drying shrinkage, but many admixtures that reduce the unit water content do not reduce drying shrinkage. Shrinkage is usually increased by replacing some of the cement with pozzolanic materials such as pumicite or raw diatomaceous earth. However, replacement of cement with a low-carbon, high-fineness fly ash results in about the same or slightly lower shrinkage. Compounds frequently used to increase the rate of strength development, calcium chloride and triethanolamine, usually significantly increase drying shrinkage. Within normal limits, entrained air has little effect on drying shrinkage.

Reinforcing steel restricts but does not prevent shrinkage. Shrinkage of reinforced unrestrained structures produces tension in the concrete and compression in the steel. Increasing amounts of reinforcing steel reduce the contraction but increase the tensile stress in the concrete. If sufficient reinforcement is used the restraint may be great enough to cause cracking of the concrete. Reinforced-concrete structures with normal amounts of reinforcement may have drying shrinkages in the range of 0.02 to 0.03%.

Proportions. The importance of the quality and quantity of cement paste on the drying shrinkage of concrete is evident from the fact that cement paste in concrete, if not restrained by the aggregate, shrinks from five to fifteen times as much as the concrete. However, the single most important factor affecting drying shrinkage of concrete is the amount of water per unit volume of concrete. Aggregate size and grading, mix proportions, cement content, slump, temperature of fresh concrete, and other factors affect drying shrinkage principally as they influence the total amount of water in a given volume of concrete. Figure 1 gives the interrelation of shrinkage, cement content, and water content and shows clearly that drying shrinkage is primarily governed by the unit water content.

Size and Shape. In large concrete members, differential volume changes take place, with the largest drying shrinkages at and near the surface. Because of the large moisture variations from the center to the surface, tensile stresses are present at and near the surface while compressive stresses are present in the interior. If the tensile stresses are very high, surface cracks may appear. However, creep may act to prevent cracking and may cause permanent elongation of the fibers in tension and shortening of the fibers in compression. The rate and ultimate shrinkage of a large mass of concrete are smaller than the values for small concrete specimens, although the action continues for a longer period for the large mass.

Structural concrete slabs reinforced only in tension tend to warp, because the concrete near the top compressive surface shrinks more because of the water accumulation at that surface during placement and also because the steel on the tensile side acts to resist shrinkage.

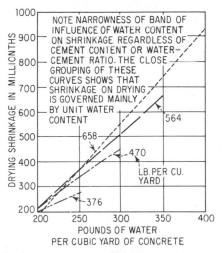

FIG. 1. Relation between drying shrinkage, cement content, and water content.[2]

Exposure Conditions. Moist curing of concrete beyond that required to develop the required strength has little effect on drying shrinkage. High-pressure steam curing at about 350°F is effectively used in block and precast-concrete products plants to reduce greatly subsequent drying shrinkage.

The length of the drying period and the humidity of the surrounding air have an important effect on drying shrinkage. Small specimens of neat cement paste have shown that drying shrinkage increases as the logarithm of the age, at least out to 20 years. The shrinkage of mortars and concretes is usually small after drying for 3 years. The rate and magnitude of drying shrinkage increase as the relative humidity of the surrounding air decreases and the rate increases as air movement past the member increases. Pastes, mortars, and concretes that have reached equilibrium under given drying conditions will shrink more if the relative humidity is then decreased, or will expand if it is increased. Shrinkage of concrete in dry air will be greatly retarded by coating the concrete with an impervious material which tends to prevent escape of the contained moisture.

Combined Effect of Unfavorable Factors. It has been experimentally shown[3] that the cumulative effect of the individual factors that increase drying shrinkage can be very large, and that the combined effect is the product rather than the sum of the individual effects. As shown in Table 1, the use of less favorable construction practices—concrete discharge temperature of 80°F rather than 60°F, a 6- to 7-in. rather than a 3- to 4-in. slump, a ¾-in. maximum-aggregate size rather than 1½-in., and too long a mixing and waiting period—could be expected to increase shrinkage 64%. If, in addition, a cement with high shrinkage characteristics, dirty aggregates of poor inherent shrinkage quality, and admixtures that increase shrinkage are used the final shrinkage could be about five times as large as the shrinkage that would be obtained with the best choice of variables.

Table 1. Cumulative Effect of Adverse Factors on Shrinkage*

Factor	Equivalent increase in shrinkage, %	Cumulative effect
Temperature of concrete at discharge allowed to reach 80°F, whereas with reasonable precautions, temperature of 60°F could have been maintained....................................	8	$1.00 \times 1.08 = 1.08$
Used 6- to 7-in. slump where 3- to 4-in. slump could have been used......................	10	$1.08 \times 1.10 = 1.19$
Use of ¾-in. maximum size of aggregate under conditions where 1½-in. size could have been used..	25	$1.19 \times 1.25 = 1.49$
Excessive haul in transit mixer, too long a waiting period at jobsite, or too many revolutions at mixing speed.............................	10	$1.49 \times 1.10 = 1.64$
Use of cement having relatively high shrinkage characteristics............................	25	$1.64 \times 1.25 = 2.05$
Excessive "dirt" in aggregate due to insufficient washing or contamination during handling.....	25	$2.05 \times 1.25 = 2.56$
Use of aggregates of poor inherent quality with respect to shrinkage.......................	50	$2.56 \times 1.50 = 3.84$
Use of admixture that produces high shrinkage...	30	$3.84 \times 1.30 = 5.00$

* Based on effect of departing from use of best materials and workmanship. From Ref. 3.

8   Caused by Alternate Wetting and Drying

Storage of concrete in moist air or in water produces expansion while storage in dry air produces shrinkage. If the storage consists of alternate wetting and drying at room temperature alternate swelling and shrinkage will result in a residual shrinkage. This shrinkage usually increases during the initial cycles and then tends to become constant. Consequently the shrinkage after the initial cycles may be considered as completely reversible. When wetting and drying cycles are combined with alternations of high and low temperatures, residual expansions that increase with the number of cycles are caused. These expansions may be quite large. In a series of tests,[4] 120 cycles consisting of 9 hr of oven drying at 180°F followed by 48 hr immersion in water at 70°F and then by 15 hr of air storage at 70°F caused expansions of 0.10 to 0.25% for concrete mixtures made with different cements, water-cement ratios, and methods of placement.

9   Caused by Carbonation

When concrete is exposed to air containing carbon dioxide it increases in weight and undergoes irreversible carbonation shrinkage that may be about as large as the shrinkage due to air drying at 70°F and 50% relative humidity from a saturated condition. Apparently under ideal conditions all the constituents of cement are subject to carbonation. The rate and extent of carbonation are dependent on variables such as moisture content and density of concrete, temperature, concentration of carbon dioxide, time, size of member, method of curing, and the sequence of drying and carbonation. Carbonation proceeds slowly and usually produces little shrinkage at relative humidities below 25% or near saturation.[5] The maximum effect appears to take place when the relative humidity is about 50%. Less dense concrete products, such as building block made with lightweight aggregate, are more susceptible to carbonation than dense concrete products.

Since carbonation-shrinkage values may be large and since they are irreversible, efforts have been made to carbonate concrete building blocks before they are to be placed in a wall. Some success with precarbonation treatments involving exposure of block to hot flue gases after curing has been obtained.[6] It also appears that precarbonation may improve volume stability to subsequent moisture changes.

10 Caused by Undesirable Chemical and Mechanical Attack

Many types of chemical and mechanical attack act to shorten the useful life of concrete. While the mechanism of destruction varies and may be quite complicated, signs of its action are usually first evident as expansions. As the action continues the expansions increase and ultimately may lead to disintegration. Some of the commoner destructive agents or actions are sewage of high acid or sulfide content, sulfate waters, electrolysis, seawater, fire, freezing and thawing, hydration of uncombined lime or magnesia or both, alkali-aggregate reaction, and reaction of alkalies in the cement with other types of aggregates. These undesirable volume changes should be avoided as much as possible by careful control of the materials used, proper proportioning, desirable construction practices, and control of exposure conditions. Prevention of undesirable reactions is much more effective than questionable remedies applied after expansions have started.

11 Cracking

Cracks in concrete due to restrained volume changes are largely dependent on the degree to which shrinkage is resisted by internal and external forces. The ability of concrete to resist cracking is dependent on the degree of restraint; the magnitude of the shrinkage due to carbonation, drying, and thermal effects; the stress produced in the concrete; the amount of stress relief due to creep; and the tensile strength of the concrete. In order to have good resistance to cracking, concrete should have values of sustained modulus of elasticity, thermal coefficient of expansion, drying shrinkage, and carbonation shrinkage as low as possible and a tensile strength as high as possible. Properly placed expansion joints in long concrete walls help in decreasing the amount and severity of cracking.

NONSHRINK OR EXPANSIVE CONCRETE

12 Prepacked Concrete

This type of concrete is made by first filling the form with clean coarse aggregate, usually graded from ½ in. to the largest practicable maximum size, and then compacting and wetting the coarse aggregate. A mortar usually consisting of water, sand, cement, and frequently a workability agent is then pumped into the forms until it fills the voids in the coarse aggregate. Because the coarse-aggregate particles are in contact and not separated by cement paste the hardened concrete has exceptionally low shrinkage. This type of concrete is widely used in resurfacing dams, repair of tunnel linings, piers, spillways, and for miscellaneous patchwork of hardened concrete. Preplaced aggregate concrete is discussed in Sec. 10.

13 Expansive Materials

Shrinkage may be minimized or prevented when concrete must be placed under special or difficult conditions, such as bedments under machinery or around a congestion of reinforcing steel when vibration is not possible, by the addition of small amounts of superfine unpolished aluminum powder. The aluminum powder, usually added in the range of 0.005 to 0.02% by weight of cement, reacts with the hydroxides in fresh cement paste to produce small bubbles of hydrogen gas. This gas if properly controlled causes a small expansion of the fresh concrete or mortar and reduces or eliminates settlement.

Another method of producing expanding or shrinkage-compensating mortars and concretes involves the use of finely divided or granulated iron in combination with an oxidizing material. The expansion produced is due to the increase, in solid volume of the iron as it is converted to iron oxide. Careful control of the proportion of the oxidizing catalyst is needed to obtain the desired expansion. The possibility of staining the concrete when moisture is present should be considered.

Expansive cements that compensate for shrinkage by controlled expansion of the cement are also used. In most of these cements the expansions are due to the controlled formation of hydrated calcium sulfoaluminates. These cements are also known as self-stressing cements since they are used to induce stress in the restraining steel. The potential uses of expanding cements are still under development. The chief difficulty associated with the use of expansive materials is the careful control needed to obtain the desired expansion. The production of concrete with desired predetermined volume-change characteristics requires careful consideration and control of all factors involved in the production, with special attention to compositions, proportions, and fineness of the components of the expansive cement; proportions of the mixture; time and temperature during the curing period; and the degree of restraint.

Expansive materials are used for limited special conditions and are not used for normal placement of structural concrete because of the difficulties of proper control and possible undesirable variable effects. When expansive materials are used the instruction of the manufacturer should be carefully followed.

CREEP

14   Nature of Creep

Elastic deformations occur immediately when concrete is loaded. Nonelastic deformations increase with time when the concrete is subjected to a sustained load. Consequently since concrete is frequently subjected to dead loading it usually is subjected to both types of deformations. The nonelastic deformation, creep, increases at a decreasing rate during the period of loading. It has been shown that significant creep takes place during the first few seconds after loading and that creep may increase out to 25 years. Approximately one-fourth to one-third of the ultimate creep takes place in the first month of sustained loading, and about one-half to three-fourths of the ultimate creep occurs during the first half year of sustained loading in concrete sections of moderate size. When the sustained load is removed there is some recovery but the concrete does not usually return to its original state. Figure 2 gives the deformation record of a specimen loaded at an age of 1 month and unloaded 6 months later. The elastic and creep recoveries are less than the deformations under load because of the increased age of the concrete at the time of unloading.

Creep may be due partly to closure of internal voids, viscous flow of the cement-water paste, crystalline flow in aggregates, and flow of water out of the cement gel due to external load and drying. The last cause is generally believed to be the most important. The magnitude and rate of creep for most concrete structures are intimately related to the drying rate, but creep is also important in massive structures where little or no drying of the concrete takes place. In these structures, most of the creep is believed due to the flow of the absorbed water from the gel (seepage) caused by external pressure.

The ultimate magnitude of creep on a unit-stress basis, psi, usually ranges from 0.2 to 2.0 millionths and ordinarily is about 1 millionth per unit of length. This value is about three times the elastic deformation of concrete having a secant modulus of elasticity of 3 million psi.

In order to visualize the effects of both elastic and creep deformations at a given time the sustained modulus of elasticity, defined as the ratio of the constant sustained stress to the sum of the elastic and creep deformations at a given time, may be used. Tests have shown that the modulus of resistance after 2 years of sustained loading varied between one-fifth and one-half of the initial secant modulus of elasticity. A reduced value of short-time secant or chord modulus of elasticity is frequently used in design to take creep into account.

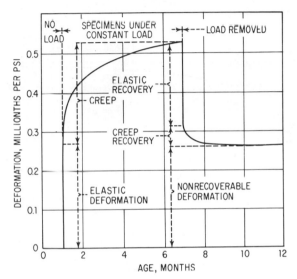

Fig. 2. Elastic and creep deformations of mass concrete under constant load followed by load removal.[2]

Creep, unlike shrinkage, which is generally undesirable, may be either desirable or undesirable depending on the circumstances. It is desirable in that it generally promotes better distribution of stresses in reinforced-concrete structures. It is undesirable if it causes excessive deformations and deflections that may necessitate costly repairs or if it results in large losses of prestress in prestressed-concrete members.

15 Effects of Constituents, Proportions, and Manufacture

Concrete made with low-heat cement creeps more than concrete made with normal cement, probably because of its influence on the degree of hydration. This desirable characteristic explains, at least in part, the relative freedom from cracking of mass concrete structures as they cool to normal temperatures. The effect of cement fineness on creep appears to be variable. Pozzolanic additions to cements generally increase creep. Approved air-entraining agents added in the proper amounts appear to have no appreciable effect on creep. If creep is an important factor, proprietary compounds should not be used unless their effects have been previously evaluated.

Under comparable conditions creep and shrinkage generally decrease when well-graded aggregates with low void contents are used and when the maximum size of the coarse aggregate is increased. Hard, dense aggregates with low absorption and a high modulus of elasticity are desirable when low shrinkage and creep are wanted. The mineral composition of the aggregate is important, and generally increasing creep may be expected with aggregates in the following order: limestone, quartz, granite, basalt, and sandstone.

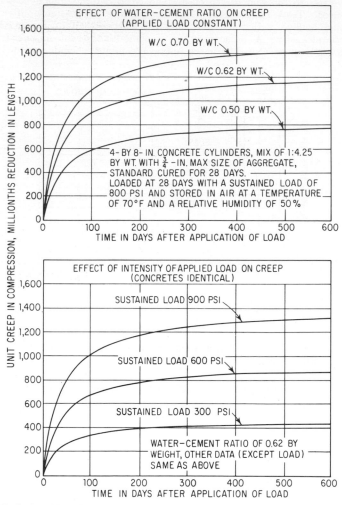

Fig. 3. Relationship of creep of concrete to water-cement ratio and to intensity of applied load.[2]

Creep of concrete increases as the water-cement ratio increases as shown in Fig. 3. In addition it appears that if two concrete mixtures have the same water-cement ratio, the mixture having the greatest volume of cement paste will creep the most. In general, lean concrete mixtures exhibit considerably greater creep than do rich concrete mixtures because the water-cement-ratio effect of increasing creep for the lean mixtures is more important than the paste effect of decreasing creep.

The tendency of concrete to creep decreases as cement hydration increases, and consequently water-cured concrete should creep less than air-cured concrete. However, the shrinkage or swelling produced during the initial curing period also influences creep. Under compressive load preswelled specimens (moist-cured) creep more than preshrunk specimens (cured in dry air). Size effects are also important in curing because small specimens respond more rapidly to moisture changes than large concrete members.

16 Effects of Exposure and Loading Conditions

The rate and ultimate magnitude of creep increase as the humidity of atmosphere decreases. The relation between relative humidity and creep is not linear. Concrete under sustained load in air at 70% relative humidity will have an ultimate creep about twice as large as concrete in air at 100% relative humidity. The ultimate creep in air at 50% relative humidity will be about three times as large. Protection of concrete members against rapid drying is very beneficial in reducing the rate and ultimate amount of creep.

The rate and magnitude of creep generally decrease as the size of the concrete member increases. It has been estimated that the creep of mass concrete may be about one-fourth of that obtained with small specimens stored in moist air.

With a given material and sustained load the rate and magnitude of creep decrease as the age at which the sustained load is applied increases, as long as hydration continues. Creep is also approximately proportional to the sustained stress within the range of usual working stresses as shown in Fig. 3. Sustained stresses above the working stresses produce creep that increases at a progressively faster rate as the magnitude of the sustained stress is increased.

Most creep information for plain concrete has been obtained for sustained compressive loading. However, test information for creep under sustained tension, bending, torsion, and biaxial- and triaxial-stress conditions generally shows the same behavior pattern. The ultimate tensile creep reduced to a 1-psi stress basis is about the same as the ultimate compressive creep on the same basis.

A test method to determine the creep characteristics of molded concrete cylinders subjected to sustained longitudinal compressive load is provided in ASTM Standard C 512. The method is intended to compare concrete specimens tested under controlled conditions, but it does not provide a means for calculating deflections of reinforced-concrete members in structures. The equation for total strain per psi ϵ is given by the equation

$$\epsilon = \frac{1}{E} + F(K) \log_e (t + 1)$$

where ϵ = total strain per psi
E = instantaneous elastic modulus, psi
$F(K)$ = creep rate, calculated as the slope of a straight line representing the creep curve on a semilog plot (logarithmic axis represents time)
t = time after loading, days

The quantity $1/E$ is the initial elastic strain per psi, and the second term in the equation represents the creep strain per psi for any given age.

17 Effects of Reinforcement

Reinforced-concrete columns subjected to sustained compressive loads also exhibit creep. Under the action of shrinkage and creep there is a tendency for additional stress to be transferred to the steel, with a consequent decrease in the concrete stress. In some tests of reinforced-concrete columns in air the compressive stress in the steel increased from three to five times the design value, while the compressive stresses in the concrete decreased materially and in some instances they changed to low tensile stresses. Concrete columns in water showed much smaller stress changes with time.

Beams and slabs that are reinforced in tension only and that are not loaded will usually warp because shrinkage is resisted by the steel near the bottom and because the top concrete shrinks more than the bottom concrete because of bleeding. Consequently beams and slabs subjected to sustained loading will be deflected and strained by warping as well as by loading. In general, the deflections and compressive strains of beams and slabs under sustained load increase at a decreasing rate while the strain at the tensile-steel level shows relatively small changes with time. Creep deflections and deformations increase rapidly as the ratio of span length to total depth increases. Inclusion of arbitrary amounts of compressive steel, not required for design strength, at the section of maximum moment in simply supported reinforced-concrete beams is effective in reducing creep. Tests have shown that inclusion of compressive steel equal in amount to the tensile steel reduces creep deflections and strains by about one-third to one-half. Consequently it appears that when a combination of high length/depth ratio and a large sustained load (causing steel and concrete stresses approaching the limiting design value of the ACI Code) are involved, compressive reinforcement may be used to control creep. In the case of framed structures or continuous beams creep frequently, but not always, may cause a more favorable stress redistribution.

Great care is required in the manufacture of prestressed reinforced-concrete beams to minimize drying shrinkage and creep in order to avoid large reductions in the amount of prestress. Under poor conditions the loss in prestress may be as high as 50,000 psi.

REFERENCES

1. Washa, G. W.: Volume Changes, Significance of Tests and Properties of Concrete and Concrete Aggregates, *ASTM Spec. Tech. Publ.*, 1965.
2. "Concrete Manual," 7th ed., U.S. Bureau of Reclamation, 1963.
3. Tremper, B., and D. L. Spellman: Shrinkage of Concrete—Comparison of Laboratory and Field Performance, *Highway Res. Board Record* 3, *Publ.* 1067, 1963.
4. Washa, G. W.: Comparison of the Physical and Mechanical Properties of Hand Rodded and Vibrated Concrete Made with Different Cements, *J. ACI*, June, 1940, p. 617.
5. Verbeck, G. J.: Carbonation of Hydrated Portland Cement, *ASTM Spec. Tech. Publ.* 205, 1958.
6. Toennies, H. T., and J. J. Shideler: Plant Drying and Carbonation of Concrete Block, NCMA-PCA Cooperative Program, *J. ACI*, May, 1963, p. 617.
7. Neville, Adam M.: "Hardened Concrete Physical and Mechanical Aspects," ACI Monograph no. 6, 1971.
8. "Expansive Cement Concretes—Present State of Knowledge," Report by ACI Committee 223, *J. ACI*, August, 1970.
9. "Temperature and Concrete," *ACI Publ.* SP-25, 1971.

Chapter 10

THERMAL, ACOUSTIC, AND ELECTRICAL PROPERTIES

GEORGE W. WASHA

THERMAL PROPERTIES

1 General

The thermal properties, as well as the strength properties, of concrete may be varied considerably by changes in the materials, proportions, and manufacturing methods. Knowledge of the thermal properties of concrete is necessary to design and to predict the performance of a large variety of concrete structures. Although concrete is generally superior to metals and natural stones in its insulating ability it is surpassed at room temperature by such materials as asbestos, powdered magnesia, mineral wool, and pulverized cork. At high temperatures, materials such as powdered magnesia and infusorial earth are much better insulating materials. The protective value of concrete at high temperatures, proved in many large conflagrations, is due to its high resistance to fire in conjunction with relatively low conductivity and high strength.

The thermal properties of hardened concrete that are important to the engineer are thermal conductivity, specific heat, thermal diffusivity, coefficient of thermal expansion, and adiabatic temperature rise. In addition, the effect of temperature on the strength properties must be known.

2 Thermal Conductivity

Thermal conductivity is the rate at which heat passes through a material of unit area and thickness when there is a unit temperature change between the two faces of the material. It is important in connection with temperature variations in mass concrete, and also with insulation and condensation properties of walls and slabs. Definitions and numerical values for the various coefficients are given in the "American Society of Heating, Refrigerating and Air-conditioning Guide and Data Book."[1]

The various coefficients used to calculate heat losses are given below:

k = thermal conductivity of a homogeneous material between the surface of the warmer side and the surface of the cooler side, Btu per hr per sq ft of area per degree difference in temperature per in. of thickness

C = thermal conductance of an obstruction (wall) between the surface of the warmer side and the surface of the cooler side, Btu per hr per sq ft per degree difference in temperature for a stated thickness (frequently given, for example, for 4-in., 8-in. and 12-in. concrete masonry units)

f = film or surface conductance, time rate of heat flow between a unit area of a surface and the surrounding air; f_i designates the inside surface and f_o the outside surface, Btu per hr per sq ft per degree difference in temperature

a = thermal conductance of an air space, Btu per hr per sq ft per degree difference in temperature

R = thermal resistance, the reciprocal of conductance, as $1/k$, $1/C$, $1/U$, etc; the overall transmission coefficient of a composite wall may be obtained by determining the total resistance by adding the reciprocals of the various conductivity coefficients for the individual parts of the composite wall:

$$R = \frac{1}{f_i} + \frac{x_1}{k_1} + \frac{1}{a} + \frac{x_2}{k_2} + \frac{1}{C} + \frac{1}{f_o}$$

where x_1, x_2 = thicknesses of various materials

U = overall coefficient of heat transmission, Btu per hr per sq ft per degree difference in temperature between the air on the warmer side of an obstruction and the air on the cooler side:

$$U = \frac{1}{R}$$

Methods of calculating the loss through a given wall construction and also the temperature variation between the cold and warm sides of a wall are given

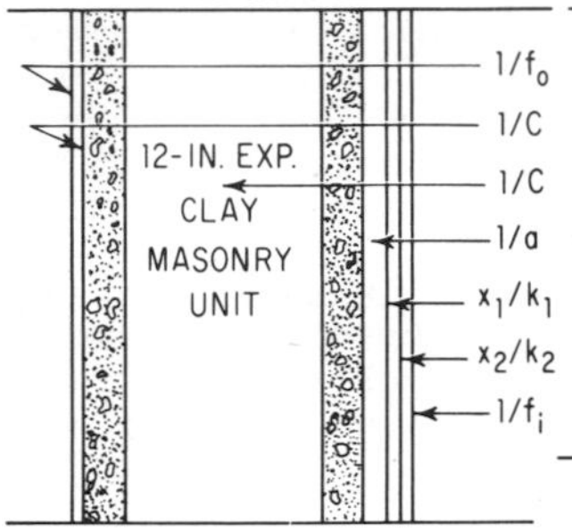

WALL SECTION		PART	RESISTANCE TO HEAT TRANSMISSION
	$1/f_o$	OUTSIDE SURFACE	0.17
	$1/C$	PORTLAND CEMENT PAINT	0.19
12-IN. EXP. CLAY MASONRY UNIT	$1/C$	12-IN. EXP. CLAY MASONRY UNIT	2.27
	$1/a$	AIR SPACE (FURRING)	0.91
	x_1/k_1	1/2-IN. RIGID INSULATION	1.51
	x_2/k_2	1/2-IN. PLASTER	0.15
	$1/f_i$	INSIDE SURFACE	0.68
		TOTAL RESISTANCE R_T =	5.88

$U = \dfrac{1}{R_T} = \dfrac{1}{5.88} = 0.17$ BTU/HR/SQ FT/°F

BTU LOSS PER 1,000 SQ FT OF WALL PER 24 HR PER 90°F TEMPERATURE DIFFERENTIAL = 0.17 x 1,000 x 24 x 90 = 367,000

Fig. 1. Calculation of heat loss through wall.[2]

in Figs. 1 and 2. The various numerical coefficients have been obtained from the "Heating, Refrigerating and Air-conditioning Guide."

The mineralogic composition of the aggregate has a large effect on conductivity. Basalt and trachyte have low conductivities, quartz has high conductivity, and dolomite and limestone have relatively high conductivity. Lightweight aggregates have conductivities roughly proportional to their densities. The approximate relation between thermal conductivity and ovendry density is shown in Fig. 3.[3] The air content of concrete has a pronounced effect in reducing thermal conductivity. Tests[4] have shown that the thermal conductivity for both sand and gravel and lightweight-aggregate concrete increases as the moisture content of the hardened concrete increases. These tests also showed that as the temperature of the hardened

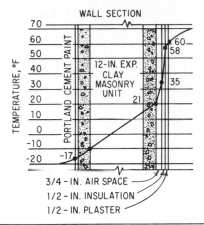

Part	Resistance	Temp change, °F
Outside-surface resistance..........................	0.17	(0.17 ÷ 5.88)90 = 3
12 in. expanded clay masonry unit and 2 coats portland-cement paint............................	2.46	(2.32 ÷ 5.88)90 = 38
Air space, furring................................	0.91	(0.91 ÷ 5.88)90 = 14
½ in. rigid insulation............................	1.51	(1.51 ÷ 5.88)90 = 23
½ in. plaster....................................	0.15	(0.15 ÷ 5.88)90 = 2
Inside-surface resistance.........................	0.68	(0.61 ÷ 5.88)90 = 10
		90

Fig. 2.   Temperature variation in wall.[2]

concrete was increased from −250 to 75°F the thermal conductivity of sand and gravel concrete decreased, and that a similar increase in temperature caused only little change in the conductivity of lightweight aggregate concrete.

3   Thermal Conductivity and Condensation

The elimination of condensation on or within walls and floors may be as important as the reduction of heat loss through them. Condensation of moisture on the inside surface of an exterior wall of a building may be prevented if the overall coefficient of heat transmission is low enough to maintain the inside surface at a temperature above the dew-point temperature t_d. The maximum heat-transmission coefficient which will prevent condensation may be computed from the equation

$$U = f_i \frac{t_i - t_d}{t_i - t_o}$$

where U = coefficient of heat transmission
f_i = inside surface-film conductance (average value for still air is 1.46)
t_i and t_o = inside and outside temperatures, respectively
t_d = dew-point temperature (available in hygrometric tables)

Condensation on a wall may be avoided by keeping the temperature of the inside surface above the dew-point temperature (as suggested above), by reducing the relative humidity of the air inside the room, or by increasing the circulation of the air passing over the inside surface. The first of these methods is usually used. In addition, in order to reduce the possibility of condensation or dampness

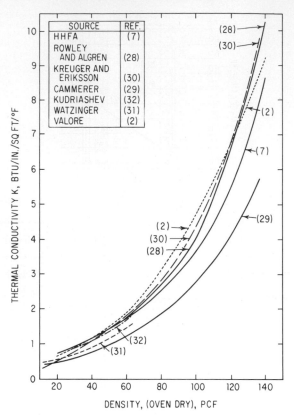

FIG. 3.   Relation of density and thermal conductivity. References shown here to identify the curves, are given in the original report.[3]

within the wall vapor barriers should be placed as near to the warm side of the wall as possible.

4   Specific Heat and Thermal Diffusivity

Specific heat is defined as the amount of heat required to change the temperature of 1 lb of material 1°F. The specific heat of hardened concrete usually varies between 0.20 and 0.28 Btu per lb per °F. The mineral composition of the aggregate has little effect, and for calculation purposes the specific heat of the dry materials in concrete is frequently assumed between 0.20 and 0.22 Btu per lb per °F. The specific heat of water is 1.0 Btu per lb per °F.

When calculations involving the flow of heat in concrete masses are required, thermal conductivity k, specific heat S, and density d (pcf) must be considered These three values may be used to provide the thermal diffusivity D by means of the equation

$$D = \frac{k}{Sd}$$

Thermal diffusivity is a measure of the rate at which temperature changes will take place within a mass of hardened concrete. Its value normally varies between 0.020 and 0.060 sq ft per hr. The value of D may also be determined experimentally.

5   Coefficient of Thermal Expansion

The thermal coefficient of expansion is a length change (expansion due to temperature increase and contraction due to a temperature decrease) in a unit length per degree of temperature change. It is expressed in inches per inch per degree Fahrenheit, or more commonly as millionths per degree Fahrenheit. An average value for hardened concrete is 5.5 millionths per degree Fahrenheit, although it may normally vary between 3½ and 7 millionths per degree Fahrenheit. The thermal coefficient of expansion of concrete varies mainly with the character and amount of the coarse aggregate. Quartz has a coefficient about 7 millionths per degree Fahrenheit while some limestones have a value as low as 3 millionths per degree Fahrenheit. Usual values of the coefficient for well-cured neat cement pastes in either a dry or saturated condition vary between 5 and 8 millionths per degree Fahrenheit, although values between 5 and 12 millionths per degree Fahrenheit have been obtained. The coefficient for concrete may be estimated as the weighted average of the coefficients of the various constituents. It appears that the coefficients of thermal expansion of ovendry and water-saturated concretes are approximately equal but that partly dried concrete has a higher coefficient.

6   Temperature Rise in Mass Concrete

In concrete structures that are less than a few feet in maximum thickness the heat of hydration normally does not need to be considered because it is dissipated rapidly. The temperature rise in mass concrete is important because the heat evolved in setting is dissipated so slowly that a marked rise in temperature occurs. Eventually when the interior cools and contracts tensile stresses are induced in the interior of the mass. If these stresses are sufficiently high they may cause cracks which in turn may unite with surface cracks and thus increase leakage and subsequent disintegration. The magnitude of the adiabatic temperature rise can be controlled by proper selection of the type of cement, proportions of the mixture, use of pozzolanic materials, rate of placement, temperature of fresh concrete, artificial cooling, and proper design and manufacturing procedures. These items are detailed in Sec. 13.

The heat generated is closely related to cement composition, with the greatest contribution per unit of compound due to C_3A and followed in decreasing order by C_3S, C_2S, and C_4AF. Low-heat portland cement, ASTM Type IV, is effective in keeping the adiabatic temperature rise at a low level. Pozzolanic materials usually are more effective in decreasing the temperature rise than the cement which they replace.

Lean mixtures made with low-heat cementing materials may produce mass concrete having the required strength and an adiabatic temperature rise less than 40°F. In massive structures where high cement factors have been used temperature increases in excess of 100°F have been noted. The maximum temperature from which the mass concrete must cool to its final stable temperature may also be decreased by placing it at relatively low temperatures.

The rate of heat dissipation may be increased by using concrete of high conductivity, placing small rather than large units, placing units with a high ratio of exposed surface to volume, providing a long exposure period for the unit before it is covered with concrete, avoiding insulation such as backfill for as long as possible, and providing refrigeration soon after placement of the concrete.

In large mass concrete structures the thermal properties—heat of hydration, specific heat, thermal conductivity, and thermal diffusivity—are important because specifications for precooling procedures, placing temperatures, construction schedules, and design of refrigeration system are all dependent on them. In concrete structures where the thickness is less than a few feet the thermal properties other than the coefficient of expansion and the thermal conductivity are usually not considered.

7 Effect of Temperature on Properties

The physical properties of concrete vary with temperature.[4,5,6] Generally the physical properties are greater at subnormal than at room temperature and lower at high temperatures than at room temperature. Tests[5] have shown that the compressive and splitting strengths appear to reach a maximum value between −75 and −250°F. These tests indicate that the compressive strength of normal sand and gravel concrete more than tripled as the temperature decreased from 75 to a temperature around −150°F. The splitting strengths more than doubled under the same circumstances. Moduli of elasticity usually increased as the temperature decreased from 75 to −250°F. The amount of the increase varied from 0 for dry concrete to about 50% for moist concrete. Poisson's ratio remained essentially constant as the temperature decreased from 75 to −250°F. The properties of moist concrete appeared to vary much more with temperature, between 75 and −250°F, than the properties of dry concrete.

In another group of tests[6] conducted between −75 and 450°F it was noted that at subnormal temperatures mortar and concrete properties generally increased as the temperature decreased. Above room temperature as the temperature increased the properties usually first decreased, then increased, and finally again decreased. At 450°F the strengths of the mortars and concretes were about the same as those at room temperature, but the moduli of elasticity were considerably lower. Mortar properties investigated included energy of rupture, modulus of rupture, tensile strength, compressive strength, and modulus of elasticity. Concrete properties investigated were compressive strength and modulus of elasticity.

Test results[4] between 75 and 1500°F are given for thermal expansion, density, and dynamic modulus of elasticity. These tests show that the weight loss due to loss of water was largely complete at 800°F and that weight changes at higher temperatures were related to the chemical composition of the aggregates. The coefficient of thermal expansion was significantly greater above 800°F because it is not affected by drying shrinkage of the paste at those temperatures. Moduli of elasticity at 1400°F appeared to be roughly one-third of the values at 75°F. After dehydration, for a given water-cement ratio, the modulus of elasticity was essentially independent of age or curing conditions.

8 Fire Resistance

Concrete strength and stiffness properties decrease significantly as the temperature is increased much above 200°F. Consequently concrete load-bearing members should not be exposed continuously to temperatures above 500°F. Continuous exposure to temperatures above 900°F may result in spalling. At high temperatures, constituents of concrete may undergo important changes. Quartz, for example, changes state and expands about 0.85% at 1063°F, causing a severe disruptive effect. In a fire, high temperatures are initially obtained on one side or in a small portion of the structure. Under these conditions differential expansion between the hot concrete and the cold concrete will take place. The cement paste tends to shrink because of the loss of moisture and to expand because of the increase in temperature, while the aggregate expands continuously with the rise in temperature. The opposing actions lead to cracking and spalling, and in reinforced concrete to exposure of the reinforcement to the fire. The unprotected reinforcement then rapidly loses strength as the temperature rises.

The fire endurance of a concrete wall is primarily dependent on wall thickness, type of construction, type of aggregate, and quality of concrete. Tests[7] have shown that for a given aggregate type the fire-resistance period generally doubled when the thickness (weight per square foot of wall) increased 35 to 40%. Differences in construction such as solid concrete wall as against a hollow masonry-unit wall, full bedding of masonry units as against bedding of the face shells only, and thickness of cover for reinforcing steel all have important effects. Natural aggregates given in descending order of their fire resistance are (1) calcareous; (2)

feldspathic, such as basalt; (3) granites and sandstones; and (4) siliceous, such as quartz and chert. The lightweight aggregates such as the inflated clays, shales, and slags are superior to the natural aggregates in producing fire-resistant concrete. Greater cement content is effective in increasing· the fire-resistance period and increasing the load capacity of a wall, both before and after fire exposure.

The fire resistance of a wall is determined by subjecting a loaded test panel (frequently with a 9-ft minimum dimension and with an area of at least 100 sq ft) on one side to temperatures rising in a prescribed manner from 1000°F at 5 min to a maximum of 2300°F at 8 hr (ASTM E 119). A similar procedure is followed in testing floors and roofs except that a larger test unit is usually required. During the test period the temperature of the unexposed surface is taken at a number of positions and the behavior of the test specimen is noted. Generally the specimen is considered to have passed the test for the required classification period, usually between 1 and 4 hr, if it has sustained the applied load without permitting passage of a flame or gases hot enough to ignite cotton waste, and if the temperature on the unexposed surface has not increased more than 250°F above the initial temperature. In some instances a hose-stream test may be required in which the exposed face is subjected at a given time to the action of water of given pressure for a given period. The cooled specimen may then be subjected to a load test. Fire tests of columns are made by subjecting columns to loads that should produce design stresses and then exposing all four sides of the columns to fire. The test is considered successful if a column sustains the applied load during the fire test for the desired classification period.

Concrete exposed to high temperatures present in a fire and subsequently cooled exhibits decreased strength and stiffness. Tests[5] of the compressive strength of walls, 6 ft high, made with 8-in. hollow masonry units, showed after a 3- to 3½-hr fire exposure that the ratio of the wall strength to the original strength of the unit averaged about 25% for units made with sand and gravel aggregates and about 35% for units made with inflated clay aggregates. The moduli of elasticity for walls prior to exposure to fire ranged between 200,000 and 750,000 psi, and after exposure the values varied between 100,000 and 300,000 psi on a gross-area basis. Other tests[7] have shown that solid concrete walls, 6 and 8 in. thick and 10 ft high, carried satisfactorily uniformly distributed working loads of 400 psi during and after severe fire exposure.

ACOUSTIC PROPERTIES

9 Sound Absorption

The control of sound in a room must be considered with respect to the origin of the sound, whether originating within or outside the room. Consideration must also be given to whether the sound is air- or solid-borne.

Control of sound originating within a room requires that the walls, ceiling, flooring, and furnishings in the room have good sound-absorption qualities. The most commonly used coefficient of sound absorption is a number that expresses the ratio of sound energy absorbed by a surface to the amount of energy incident upon the surface. The *absorption coefficient* is generally given for a sound frequency of 500 cps. The *noise-reduction coefficient* (NRC) is the average of the coefficients at 250, 500, 1,000, and 2,000 cps expressed to the nearest 0.05.

Tests have shown that a porous surface with interconnected pores from the surface to the interior is the chief characteristic needed for good sound absorption. With a porous surface the energy of sound is converted into heat. If the surface pores are sealed by painting or plastering the sound absorption is markedly reduced.

The sound-absorption coefficient for plain cast concrete is about 0.02, which indicates that 98% of the incident sound energy is reflected by the surface. Average values of the sound-absorption coefficients for various materials are unpainted cinder block—0.45; painted cinder block—0.40; vermiculite acoustic plaster—0.55; cork—0.70; sprayed asbestos—0.90.

10   Sound Transmission

Sound originating outside a given room may be transmitted as solid-borne as well as airborne sound. The solid-borne sound should be suppressed at its source. A concrete floor in a hall, for example, should be either broken at the partitions or covered by a carpet or other resilient material in order to suppress sound transmission into the room. Airborne sound transmission from outside the room may be suppressed by barriers such as masonry walls and attention to details of construction.

Transmission loss expressed in decibels (db), a measure of sound insulation, is given by the equation

$$\text{Transmission loss (TL)} = 10 \log \frac{E_i}{E_t}$$

Where E_i is the level of sound energy incident upon a wall and E_t is the level of sound energy transmitted through the wall. A transmission loss of 10 db indicates

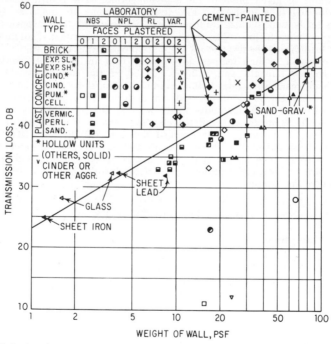

FIG. 4. Relation between transmission loss and weight of wall for airborne sound.[3]

that one-tenth of the incident sound energy is transmitted, and 20 db indicates that one-hundredth is transmitted, etc. Sound intensities vary from 0, the threshold of audibility, to about 130 db, the threshold of painful sound or the limit of the ear's endurance. The sound intensity on an average busy street is about 60 db.

Values of transmission loss (TL) frequently vary between 40 and 55 db for walls between rooms in apartments, hospitals, and schools. Various studies[3,9] have shown that the weight per unit wall area is a very important factor influencing transmission loss of airborne sound. Figure 4 shows average transmission-loss data for airborne sound for various types of walls obtained from tests conducted at the National Bureau of Standards, National Physical Laboratory of Great Britain, Riverbank Laboratories, and other laboratories. The TL values are averages for frequency ranges of approximately 100 to 2,000, 3,000 or 4,000 cps, and the heavy line is an average TL-weight relation.[9] Porous rigid materials, such as concrete masonry, follow the TL-weight relationship only if the pores are not continuous. Transmission of airborne sound through continuous air paths in the material greatly reduces TL. Consequently painting or plastering lightweight-concrete masonry units greatly increases the TL.

The factors which provide good sound absorption tend to provide poor sound insulation. Porous concrete absorbs sound but has a poor sound-insulation value (low TL). Plastering or painting porous concrete greatly reduces sound absorption but increases sound insulation.

ELECTRICAL PROPERTIES

11 Electrolysis, Resistivity, Dielectric Strength

Disintegration of some reinforced concrete structures has apparently been due to electrolytic action of stray currents from nearby power circuits. The action is much greater for wet concrete than for dry concrete, especially if calcium or sodium chloride are present in the mixing water. It has been observed that, when the current flows from the concrete to a steel cathode, the concrete around the reinforcement softens because of a gradual concentration of sodium and potassium alkalies and the bond is destroyed. When the steel acts as the anode, splitting of the concrete surrounding the steel is caused by the oxidation and resultant volume increase of the steel.

Increasing use of concrete crossties has emphasized the need for more information on the electrical resistivity of concrete. Studies[10] including direct and alternating current have shown that moist concrete behaves essentially as an electrolyte having a resistivity of the order of 10^4 ohm-cm, which is in the range of semiconductors. When concrete has been oven-dried, its resistivity is of the order of 10^{11} ohm-cm and it is considered to be a reasonably good insulator.

The dielectric strength of an oven-dried concrete of a 1:2:4 mix made with Type I cement and with a water-cement ratio of 0.49 tested with direct current has been reported[11] as 15.9 kv per cm for the first breakdown and 12.5 kv per cm for the third breakdown. Under similar conditions, concrete made with high alumina cement showed slightly higher values; and concrete made with Type II cement showed slightly lower values. When tests were made with 50 cycles per sec alternating current, generally slightly lower values were obtained. The dielectric strength of air-dry concrete was approximately the same as that of the oven-dried concrete.

REFERENCES

1. "American Society of Heating, Refrigerating and Air-conditioning Guide and Data Book, Fundamentals and Equipment" (1965–1966), American Society of Heating, Refrigerating and Air-Conditioning Engineers, Inc., New York.
2. Stone, A.: Thermal Insulation of Concrete Homes, *J. ACI*, May, 1948, pp. 849–874.
3. Valore, R. C., Jr.: Insulating Concretes, *J. ACI*, November, 1956, pp. 509–532.
4. Philleo, R.: Some Physical Properties of Concrete at High Temperatures, *J. ACI*, April, 1958, p. 857.
5. Monfore, G. E., and A. E. Lentz: Physical Properties of Concrete at Very Low Temperatures, *J. Res. Develop. Lab., PCA*, vol. 4, no. 2, May, 1962.
6. Saemann, J. C., and G. W. Washa: Variation of Mortar and Concrete Properties with Temperature, *J. ACI*, November, 1957, p. 385.

7. Menzel, C. A.: Tests of the Fire Resistance and Thermal Properties of Solid Concrete Slabs and Their Significance, *ASTM Proc.*, vol. 43, 1943.

8. Menzel, C. A.: The Strength of Concrete Masonry Walls after Standard Fire Exposure, *J. ACI*, November, 1932, pp. 113–142.

9. Knudsen, V. O., and C. M. Harris: "Acoustical Designing in Architecture," John Wiley & Sons, Inc., New York, 1950.

10. Monfore, G. E.: The Electrical Resistivity of Concrete, *J. Res. Develop. Lab., PCA,* vol. 10, no. 2, May, 1968.

11. Hammond, E., and T. D. Robson: The Comparison of Electrical Properties of Various Cements and Concretes, *The Engineer,* vol. 199, January 21, 1955, pp. 78–80, and January 28, 1955, pp. 114–115.

Section 3

PROPORTIONING MIXES AND TESTING

Chapter 11

PROPORTIONING MIXES

J. F. ARTUSO

CRITERIA

1 Specifications

Project specifications differ widely from one to another in the manner in which the concrete mix or mixes are described. In general these may be divided into two classes, "prescription specifications" and "performance specifications," with some including features of both.

A straight prescription specification will spell out the exact proportions of all ingredients, one to the other. Typical of this type is the now largely outmoded but still occasionally encountered volume-proportion requirement such as 1:2:3, meaning 1 part cement, 2 parts sand, and 3 parts coarse aggregate, by volume. In these volume-proportion specifications, quantity of water is seldom limited, except indirectly by means of a maximum-slump requirement.

Prescription-type specifications more often cover only minimum cement content in pounds per cubic yard of concrete and maximum water-cement ratio usually by weight.

In either of such prescriptions, the specification writer is taking full responsibility for sufficiency of the mix design. The contractor or concrete supplier merely has the responsibility to prepare his concrete in the proportions specified, assuming no liability for the strength developed. Many people feel that this is proper. However, probably more feel that the concrete producer, be it contractor or ready-mix plant operator, should be responsible for the properties of the finished product.

Performance specifications, which tend to place the responsibility on the concrete producer, are those which spell out the strength requirement, often accompanied by certain other limiting factors. These others may include minimum cement content, or maximum water-cement ratio, and they practically always include a requirement on consistency as measured by slump, and usually include some limitation on aggregate sizes and properties. Too many limitations tend to convert the specifications back to a prescription type, and if included without full knowledge of local conditions and materials, they may make it impossible to produce concrete without violating one or more of the requirements.

If one accepts the theory that the contractor should at least share the responsibility for the final product, the specification should cover the desired strength, air entrainment, maximum-size aggregate, and slump range. Quality of aggregate, consistent with local conditions to be encountered, and with due consideration of the economics of locally available material, should be covered.

With these few limiting factors the contractor, working with his laboratory, can submit for consideration of the specifying agency a suitable and economical concrete mix best adapted to the local conditions.

MIX CHARACTERISTICS

2   Consistency

Consistency is the most easily defined method of expressing workability of plastic concrete. Consistency is measured by means of the slump test. Although the slump test does not exactly indicate other workability characteristics such as finishing qualities, bleeding, and segregation tendency, it is the best presently known test for predicting the ease of placing and consolidating concrete.

After a slump test is made, the side of the slumped specimen should be tapped lightly to investigate degree of cohesiveness. A specimen of a cohesive mix will sag and the aggregates will not fall away from it. This mix would be more easily consolidated in formed and heavily reinforced structures. A harsh mix with fewer fines but of equal slump may be satisfactory in slabs or mass concrete that can be readily consolidated by strong mechanical vibration. In order to assure the highest quality, concrete must be placed at the lowest slump that can be handled practically, and be adequately consolidated.

Consistency can also be measured by means of the penetration apparatus (Kelly ball) described in Chap. 12.

3   Aggregate Size

Generally, the largest maximum size of aggregate should be used in concrete mixtures to develop the optimum properties of strength, durability, shrinkage, etc. This is primarily achieved because the larger aggregate enables the use of minimum unit water content.

For many years there has been a generally accepted rule of thumb that the largest-sized aggregate available and usable in the section to be placed would be the most desirable. It has been felt that smaller coarse aggregate would result in higher sand requirement with consequent higher water demand with all its accompanying undesirable effects. This view remains prevalent with most concrete engineers, and there is much in experience to defend it. However, some recent data published by the National Sand and Gravel Association have indicated that, for a given cement factor and slump, the compressive strength of rich concrete mixtures increases as size aggregate decreases from 2½ to ⅜ in. This is shown in Fig. 1. It also indicates that from a strength standpoint, there is no significant improvement from the use of aggregate larger than ¾ in. It is noted that the range of aggregate under consideration was limited, sizes of 2½ to ⅜ in. were used, and cement factors of 375, 565, and 750 lb per cu yd were evaluated. In addition to the above considerations, the maximum size aggregate should not exceed one-fifth of the narrowest dimension between sides of forms, three-fourths of the minimum clear spacing between reinforcing bars, or one-third the depth of slabs. Also consideration must be given to the maximum size that is economically available and consistent with the above criteria.

4   Air Entrainment

The benefits of air entrainment in concrete mixtures have made its use very common. These advantages include improved resistance to freezing and thawing damage, reduced permeability, improved workability, and reduced bleeding. The entrainment of air in concrete mixtures requires special consideration in proportioning, because of the volume aspects and the interrelationships with the other ingredients. When compared with non-air-entrained concrete mixtures, the additional volume produced by the entrained air is compensated for by both a reduction in water content and a decrease in sand quantity. Air entrainment will increase strengths of lean mixes, which may allow a slight reduction of cement. However, in rich mixes, that is, those over 600 lb of cement per cu yd, there is usually a significant strength reduction. This requires the use of additional cement or proper water-reducing admixtures.

The amount of air entrained by a given agent or air-entraining cement is affected by many factors. Generally the air content is increased by an increase in slump, water-cement ratio, or sand content. An increase in the amount of sand size between the No. 30 and No. 50 sieve increases the air entrainment. It is decreased by an increase of fines in sand, higher cement content, higher temperature, longer mixing time, and the addition of pozzolans. The reduction in entrained air that accompanies addition of fly ash is commonly attributed to the carbon content of the fly ash and amount retained on the No. 325 sieve.

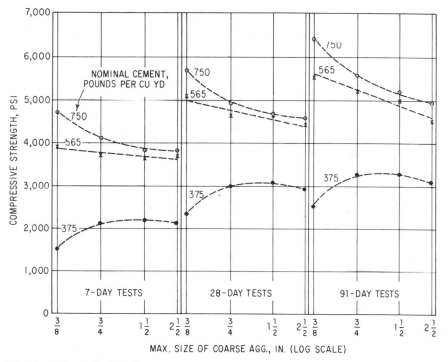

Fig. 1. Effect of size of coarse aggregate on compressive strength of air-entrained concrete.[5]

The optimum amount of air entrainment for a given concrete mixture is governed by the applicable specification. The amount will vary with the maximum-size aggregate and for concrete in buildings will generally follow the limits given in Table 1. Slabs on ground, such as highway pavements, are often specified to have 6 to 8% entrained air, regardless of maximum size of coarse aggregate.

Although the air contents are given in relation to aggregate sizes, the desirable amount is actually dependent directly on the volume of air in the cement-sand mortar part of the concrete. The optimum appears to be 9% in this mortar at all times. This is therefore reflected in the various volumes because of the reduced mortar content in the concrete with larger-sized aggregates.

5 Structural Requirements

The basic factors affecting proportioning are strength and durability. The concrete must be proportioned to develop sufficient strength to sustain adequately the loads to be imposed upon it in service. The specific strength of the concrete mixture is established by the designer and is usually predicated on the allowable stresses used in the structural analysis. For adequate durability, the ACI Building Code[8] provides that normal weight aggregate concrete subject to freezing and thawing while wet shall be properly air-entrained and shall have a water-cement ratio of 0.53 by weight. When lightweight aggregate concrete is used, the compres-

Table 1. Concrete Air Content for Various Sizes of Coarse Aggregate*

Nominal maximum size of coarse aggregate, in.	Total air content, % by volume
⅜	6 to 10
½	5 to 9
¾	4 to 8
1	3.5 to 6.5
1½	3 to 6
2	2.5 to 5.5
3	1.5 to 4.5

* From Table 4.2.5 ACI Standard 318-71.

Table 2. Maximum Permissible Water-Cement Ratios for Concrete in Severe Exposures*

Type of structure	Structure wet continuously or frequently and exposed to freezing and thawing†	Structure exposed to sea water or sulfates
Thin sections (railings, curbs, sills, ledges, ornamental work) and sections with less than 1 in. cover over steel	0.45	0.40‡
All other structures................	0.50	0.45‡

* Based on report of ACI Committee 201, "Durability of Concrete in Service." From Table 5.2.4(b), ACI 211.1-70.
† Concrete should also be air-entrained.
‡ If sulfate resisting cement (Type II or Type V of ASTM C 150) is used, permissible water-cement ratio may be increased by 0.05.

sive strength f_c' shall be 3,000 psi minimum. It has been established that durability is a function of both air entrainment and water-cement ratio.

Concrete used under various degrees of exposure should meet minimum requirements as given in ACI Standards on proportioning concrete mixtures. This requires the use of the proper type of cement as shown in Chap. 1, and the use of proper water-cement ratios shown in Table 2.

The ACI Standard 318-71 Building Code covers other specific conditions. When normal weight aggregate concrete is intended to be watertight, it should have a maximum water cement ratio of 0.48 for exposure to fresh water and 0.44 for exposure to sea water. With lightweight aggregate concrete, the specified compressive strength f_c' should be at least 3750 psi for exposure to fresh water and 4000 psi for exposure to sea water. Concrete that will be exposed to injurious concentrations of sulfate-containing solutions should conform to above watertight concrete quality requirements and should be made with sulfate resisting cement.

MATERIALS

6 Cement

Several types of cement are commonly available to meet specific needs of various projects as described in Chap. 1. The most commonly used cement is Type I. Types II, IV, or V are specified to cover the need for special properties such as sulfate resistance and low heat development. The 28-day strength of concretes made with these cements is normally lower than that of concretes made with equal contents of Type I, but strengths at later ages will equal those of Type I. Type III, high-early-strength cement, is usually made to achieve in 7 days the comparable 28-day compressive strength of Type I cement.

Air-entraining Cement. This cement contains an interground air-entraining agent, and is used for air-entrained concrete in lieu of a separate air-entraining agent added at the mixer. Adjustments in proportions to accommodate the air entrainment must be made, as shown later. Air-entraining cements are designated as Types IA, IIA, IIIA, IVA, VA, or ISA.

Cement Variations. It is recognized that strength-development capabilities of a given cement type may vary appreciably from plant to plant, and in some cases from shipment to shipment within the same plant. It is therefore necessary to use in trial batches for establishing design the brand of cement intended for the project, obtained from current production of the specific plant that will furnish the project.

7 Aggregates

Coarse Aggregates. The most important properties of coarse aggregates that affect the strength and proportions of a concrete mixture are grading, size, shape, and surface texture. These properties vary with the various conventional coarse aggregates available. The conventional aggregates are primarily gravels, crushed stone, and blast-furnace slag. (Open-hearth slags are, in general, considered to be unsatisfactory as concrete aggregates.) The gravels are available in natural rounded form, crushed, or a combination of both. The crushed stones are angular in shape and usually have significantly different surface textures from gravels. The crushed air-cooled blast-furnace slag is similar in some respects in angularity to the crushed stones. In addition it has a vesicular quality and texture that affect proportioning. The unit weight of air-cooled blast-furnace slag is usually less than that of other aggregates, and slag mixes will therefore appear quite different. In all cases, the method of determining proportions given here will automatically make allowance for the different particle shapes by use of the fixed dry-rodded volume method of ACI Standard 211.1-70[1]. The strength-development capabilities as affected by texture, structural strength, and water demand of each aggregate will appear in the trial-mix results. Aggregates consisting primarily of particles having rough texture or many flat or elongated pieces require different mix proportions from those consisting of relatively smooth, rounded, or cubical particles. These qualities will require greater proportions of sand, which requirements will develop in the calculations of mix design by the dry-rodded volume method. Grading of coarse aggregates will also be reflected in the fixed dry-rodded volume method by the void content. The variations in grading of coarse aggregates within different maximum-size limits are not reflected; however, this omission is considered of little practical importance when limits fall within conventional grading specifications.

Fine Aggregates. Fineness modulus is a convenient method of expressing the overall grading of the sand. It is a single figure determined by adding together the cumulative percentages retained on sieves Nos. 4, 8, 16, 30, 50, and 100, and dividing the total by 100. This fineness modulus of the sand is used in determination of dry-rodded volume for mix designs in ACI Standard 211.1-70. Permissible limits of fineness are given in ASTM C 33, limiting the fineness modulus between 2.3 and 3.1. Generally, the ratio of sand to coarse aggregate in the mix is lower when the sand is fine than when it is coarse. A grading of sand in which one or two particle sizes predominate should be avoided. This condition would result in a large void content and would therefore require a greater amount of cement-water paste to produce a workable mixture. The greatest workability is achieved when the individual sizes of a given sand form a smooth curve within the grading limits. Compliance with the numerical limits of some specifications does not always result in a smooth curve.

Manufactured Sand. Sands produced by crushing rock are often made up of particles having rough and angular surfaces. When this quality is coupled with flat and elongated shapes, it will produce concrete mixtures that are harsh and not so workable as concrete made with rounded sands, in which case it is necessary to increase the proportion of sand. A greater content of cement paste and greater water content then are generally necessary. Some manufactured sands, because of their angularity and toughness, produce greater concrete compressive strength for a given cement factor, even with higher water content, than natural sand. Not all sands manufactured by crushing rock are "harsh." Some, produced with modern equipment in modern plants, behave almost the same as natural sands.

Lightweight Aggregates. There are many varieties and types of lightweight aggregates, each of which may behave differently in concrete from the others. Many are angular and have rough surface textures producing harsh mixes. (Air entrainment helps tremendously to overcome this tendency.) Natural lightweight aggregates include diatomite, pumice, scoria, volcanic cinders, and tuff; however, these are not widely used except in a few areas where they are easily available. Pumice, most widely used of these, is a frothlike volcanic glass with a bulk density of 30 to 55 pcf.

The manufactured aggregates are marketed under a variety of names. The most common and structurally superior are expanded clay, shale, slate, and specific types of sintered fly ash. These produce concrete of excellent strength characteristics. Some are manufactured by a process that forms spherical particles with an impervious surface, in effect lowering absorption and improving workability. Another type commonly used is expanded blast-furnace slag and industrial cinders. These are more vesicular and usually are softer and more absorptive and thus result in lower compressive strength and higher shrinkage than the manufactured type. Perlite and vermiculite are much lighter in weight and consequently produce low strength and high shrinkage. They are seldom used for purposes other than insulation.

Heavy Aggregates. These include magnetite and limonite iron ores, barites, ferrophosphorus, steel punchings, steel shot, and lead. Specifications covering quality of these materials is contained in ASTM C-637, High Density Aggregate. Also see paragraph 30. Their use in concrete is primarily for nuclear shielding, or in high-density concrete for such usage as movable-bridge counterweights.

The high density and angularity of the natural ores usually produce harsh mixes. These ores tend to dust, have high absorption, and may break down during handling and mixing. This complicates mix design, but satisfactory concrete can be made with densities of about 200 to 300 pcf. The artificial heavy aggregates such as steel punchings, shot, and lead produce concrete with a density as high as 350 pcf, but with a tendency to segregation during handling and placing. This is due to the considerably higher density of the metals as compared with that of the fine aggregates and cement used in the mixtures. The steel must be free of oil to develop proper bond. It is advisable to corrode the surfaces prior to use or to acid-etch the steel for improved bond.

Although boron frit is not a heavy aggregate, mention is made of it here because of its use in shielding concrete. It is efficient in the absorption of thermal neutrons when the boron content is about ½% by weight of the concrete. Care must be exercised in the type of boron frits that are used. A type of low solubility is necessary to prevent retarding effects on the cement paste.

8 Water

Water used in manufacturing concrete including that free water on the aggregates should be clean and free from injurious amounts of oils, acids, alkalis, salts, organic materials, or other substances that may be deleterious to concrete or steel. Also, mixing water for prestressed concrete or for concrete which will contain aluminum embedments should not contain deleterious amounts of chloride ion. See also Chap. 4.

9 Admixtures

Air-entraining Admixtures. These admixtures can be interground in portland cement to achieve air-entraining cement or can be added independently to concrete mixtures. The effects for specific limits are the same. Air entrainment is one of the most important single developments in concrete technology since the invention of portland cement. Its primary purpose is to increase durability, which it does manyfold. It also improves workability. There are also side effects on strength, water requirements, and sand requirements, all of which are taken into consideration in the design of the concrete mix.

Accelerators. Accelerating admixtures are used, as is indicated by the name, to speed hardening and strength gain of concrete. They are used primarily in winter concreting to permit early finishing and to hasten the progress of cement hydration to the point that frost damage will not occur. They *cannot* be considered to be antifreeze mixtures. In permissible quantities they cannot lower the freezing point of the concrete more than 1 or 2°. They *can* and *do* hasten the set and strength gain to the point that the fresh concrete is susceptible to frost damage for a shorter length of time.

The commonest accelerator is calcium chloride, and this is the active constituent of most of the accelerators sold under many trade names. Some of the accelerators sold under various trade names contain alkali silicates and carbonates, fluosilicates, and triethanolamine.

It is recommended that calcium chloride content not exceed 2% by weight of cement. It is most effective in rich mixes in accelerating setting time and strength development. Calcium chloride generally does not affect unit water content; however, it may reduce air entrainment and thus require adjustment of the amount of air-entraining agent used. Some proprietary water-reducing admixtures contain an accelerator to counteract the retarding action of the admixture.

Retarders and Water Reducers. The common retarders available are generally a form of lignosulfonic acid or hydroxylated carboxylic acid. Both types retard the setting time about 1 to 4 hr over normal setting time but do not affect normal strength gain after set has occurred. The lignosulfonic acid types tend to entrain some air and must be modified with air-entraining agents to meet levels of air entrainment for durability. Exclusive of the air entrained, normal amounts of these retarders will reduce the water requirement about 8 to 25 lb per cu yd and increase compressive strength about 10 to 15%. Adjustments must be made in mix proportions to accommodate these effects.

Pozzolans. The pozzolans develop cementitious properties in concrete mixtures by their reaction with calcium hydroxide and water. Some pozzolans also inhibit expansion resulting from alkali-aggregate reaction. In general they increase workability of lean or harsh mixtures.

Pozzolans are sometimes used as a replacement of a part of the portland cement. Such replacement seldom exceeds 25% of the cement content. In cases of replacement, strength gain of the concrete will usually be delayed somewhat, with 90-day strengths of the cement-pozzolan mixes probably being about equal to the 28-day strengths of straight portland-cement mixes. When pozzolans are used as an addition to the portland cement, they frequently increase the 28-day concrete strength and will almost always increase the strengths at later ages. Some pozzolans will also affect air entrainment and will require adjustment in quantity of agents. The specific gravity of most common pozzolans is usually less than that of conventional aggregates and cement, often falling between 2.35 and 2.50. Specific gravity must be determined for proper proportioning of concrete mixtures.

Integral Waterproofers and Dampproofers. Many of the admixtures sold specifically for this purpose are stearates or oleates, though some probably contain other chemicals. The effect or value of any should be corroborated by trial concrete mixes to determine effect on compressive strength, durability, shrinkage, and other properties of the concrete, as well as waterproofing.

Mix proportioning to adjust for a change in absolute volume of concrete is generally not necessary unless the admixture contains an air-entraining agent. Air entrainment in itself will generally reduce absorption and permeability. Most of the proprietary types which do not include an air-entraining agent do not have a significant effect on water content or compressive strength when used in normal quantities.

Other Admixtures. Other admixtures are used for a wide variety of purposes and have widely different effects on concrete mixtures. These include materials designated as air-detraining, gas-forming, expansion producers, bonding, chemicals to reduce cement-aggregate expansion, corrosion-inhibiting, fungicidal, germicidal, insecticidal, flocculating, and coloring admixtures. The individual effect on concrete properties should be verified by concrete mixes and tests. Predicted effect on proportioning can be made prior to mixing and refinements made on subsequent trial mixes.

SELECTING MIX PROPORTIONS

Numerous tables have been prepared and presented in various publications of the American Concrete Institute which are of valuable assistance in the planning of proportions and quantities of materials for the production of concrete to meet certain requirements. If no advance laboratory checks of designs are planned, the recommendations of these tables may be combined to produce a calculated design which will be on the side of safety in both strength and durability. In many cases preliminary laboratory designs, based fundamentally on the data of these tables, but with the laboratory's knowledge of local materials, may result in appreciable economies with even greater margins of safety.

10 Water-Cement Ratio Determinations

The water-cement ratio may be established numerically by definite project specifications. An alternative found in some specifications is specific reference to ACI Standard 211.1-70 (Table 2). In this table the water-cement ratio is considered primarily from the standpoint of durability under various exposure conditions. In addition to this standard, ACI Building Code Requirements for Reinforced Concrete, ACI 318-71, states that the water-cement ratio shall in no case exceed 0.53 for concrete exposed to freezing while wet. In addition to the establishing of water-cement ratios for exposure conditions, the other principal criterion used to establish a water-cement ratio is the compressive strength at 28 days f_c'. The ACI Building Code gives allowable water-cement ratios for specific strengths of concrete mixtures when preliminary tests are not made, as shown in Table 3.

It must be noted that the ACI Building Code does not cover, in Table 3, strengths in excess of 4,500 psi air-entrained or 5,000 psi non-air-entrained, or lightweight concrete. Under these conditions the proposed concrete mixture must be established by previous reliable tests or by a water-cement ratio curve using at least three different water-cement ratios (or cement contents in the case of lightweight concrete).

Table 3. Maximum Permissible Water-Cement Ratios for Concrete
(When strength data from trial batches or field experience are not available)*

Specified compressive strength f_c', psi†	Maximum permissible water-cement ratio			
	Non-air-entrained concrete		Air-entrained concrete	
	Absolute ratio by weight	U.S. gal. per 94-lb bag of cement	Absolute ratio by weight	U.S. gal. per 94-lb bag of cement
2,500	0.65	7.3	0.54	6.1
3,000	0.58	6.6	0.46	5.2
3,500	0.51	5.8	0.40	4.5
4,000	0.44	5.0	0.35	4.0
4,500	0.38	4.3	0.30	3.4
5,000	0.31	3.5	‡	‡

* From Table 4.2.4, ACI Standard 318-71.

† 28-day strengths for cements meeting strength limits of ASTM C 150 Type I, IA, II, or IIA and 7-day strengths for Type III or IIIA; with most materials, the water-cement ratios shown will provide average strengths greater than indicated in Table 4 as being required.

‡ For strengths above 4,500 psi with air-entrained concrete, proportions should be selected by the methods of paragraph 10.

The proportions should be selected to produce an average strength at the designated test age exceeding f_c' by the amount indicated below, when both air content and slump are the maximum permitted by the specifications.

Where the concrete production facility has a record, based on at least thirty consecutive strength tests representing materials and conditions similar to those expected, the strength used as the basis for selecting proportions should exceed the required f_c' by at least:

400 psi if the standard deviation is less than 300 psi
550 psi if the standard deviation is 300 to 400 psi
700 psi if the standard deviation is 400 to 500 psi
900 psi if the standard deviation is 500 to 600 psi

If the standard deviation exceeds 600 psi or if a suitable record of strength test performance is not available, proportions should be selected to produce an average strength at least 1,200 psi greater than the required f_c'. Table 3 is generally conservative by necessity to cover the more extreme conditions. More realistic water-cement ratios for various compressive strengths are given in Table 4 and are normally used in conjunction with laboratory trial mixes. This table gives an approximation, is conservative, and covers a wide band of conditions. It does not distinguish other factors which affect a resultant compressive strength. Strengths for air-entrained concrete are 20% lower than for non-air-entrained concrete of the same cement factor. It is usually found that this is conservative for the lean mixes and optimistic for the rich mixes.

11 Estimate of Total Water

The quantity of water required for a cubic yard of concrete to provide a certain slump within the fine aggregate–coarse aggregate proportions given here will depend primarily on the size, shape, and grading of the aggregates and amount of entrained air. The quantity of cement will generally not affect water demand for normal structural-concrete mixes. Table 5 will provide an adequately accurate unit water content for preliminary designs.

Table 4. Relationships Between Water-Cement Ratio and Compressive Strength of Concrete*

Compressive strength at 28 days, psi†	Water-cement ratio, by weight	
	Non-air-entrained concrete	Air-entrained concrete
6,000	0.41	....
5,000	0.48	0.40
4,000	0.57	0.48
3,000	0.68	0.59
2,000	0.82	0.74

* From Table 5.2.4(a), ACI Standard 211.1-70.

† Values are estimated average strengths for concrete containing not more than the percentage of air shown in Table 5.2.3. For a constant water-cement ratio, the strength of concrete is reduced as the air content is increased.

Strength is based on 6 × 12 in. cylinders moist-cured 28 days at 73.4 ± 3 F (23 ± 1.7 C) in accordance with Section 9(b) of ASTM C 31 for Making and Curing Concrete Compression and Flexure Test Specimens in the Field.

Relationship assumes maximum size of aggregate about ¾ to 1 in.; for a given source, strength produced for a given water-cement ratio will increase as maximum size of aggregate decreases; see paragraph 33.

The selection of slump should be based upon the recommendations given by ACI Committee 301. Unless otherwise permitted or specified, the concrete shall be proportioned and produced to have a slump of 4 in. or less if consolidation is to be by vibration, and 5 in. or less if consolidation is to be by methods other than vibration. A tolerance of up to 1 in. above the indicated maximum shall be allowed for individual batches, provided the average for all batches or the most recent 10 batches tested, whichever is fewer, does not exceed the maximum limit. Concrete of lower than usual slump may be used provided it is properly placed and consolidated. The slump shall be determined by the Test for Slump of Portland Cement Concrete (ASTM C 143). Recommendations to cover special situations and varying from those of the table include a suggestion that the slump of concrete made with lightweight aggregate never exceed 3 in. and that concrete for steep ramps, folded plates, or similar sloping surface may, when necessary, have a slump of less than 1 in. Other conditions which will affect total water requirements are given in Table 6.

12 Cement-factor Determination

The cement content is determined by utilizing the water-cement ratio specified, or it is taken from Table 2 for exposure conditions or Table 3 for compressive-strength requirements, whichever is lower.

The unit water content given in Table 5 selected for the given conditions is divided by the water-cement ratio to produce the cement factor in pounds per cubic yard.

13 Aggregate-quantity Determination

The quantity of coarse aggregate is determined by the use of Table 7. The maximum-size aggregate contemplated and the fineness modulus of the sand must be first determined. The volumes from this table are taken from an empirical relationship and will automatically provide for different types of aggregates by

**Table 5. Approximate Mixing Water and Air Content
Requirements for Different Slumps and Maximum
Sizes of Aggregates***

Slump, in.‡	Water, lb per cu yd of concrete for indicated maximum sizes of aggregate†						
	⅜ in.	½ in.	¾ in.	1 in.	1½ in.	2 in.	3 in.
Non-air-entrained concrete							
1 to 2	350	335	315	300	275	260	240
3 to 4	385	365	340	325	300	285	265
6 to 7	410	385	360	340	315	300	285
Approximate amount of entrapped air in non-air-entrained concrete, percent	3	2.5	2	1.5	1	0.5	0.3
Air-entrained concrete							
1 to 2	305	295	280	270	250	240	225
3 to 4	340	325	305	295	275	265	250
6 to 7	365	345	325	310	290	280	270
Recommended average total air content, percent.......	8	7	6	5	4.5	4	3.5

* From Table 5.2.3, ACI Standard 211.1-70.
† These quantities of mixing water are for use in computing cement factors for trial batches. They are maxima for reasonably well-shaped angular coarse aggregates graded within limits of accepted specifications.
‡ The slump values for concrete containing aggregate larger than 1½ in. are based on slump tests made after removal of particles larger than 1½ in. by wet-screening.

Table 6. Adjustments to Water Content for Changed Conditions*

Changes in conditions	Effect on unit water content, %
Each 1-in. increase or decrease in slump............................	±3
Each 1% increase or decrease in air content.........................	±3
Each 1% increase or decrease in sand content.......................	±1
To well-graded rounded aggregate from angular aggregate..............	−7 to −10

* Adapted from "Concrete Manual," 7th ed., Table 14, U.S. Bureau of Reclamation, 1963.

making allowances for differences in mortar contents as reflected by the void content. These values are so designed to provide for variations in gradations within acceptable standards. This frequently results in greater than necessary quantities of sand. Therefore, prior knowledge of the character of aggregate may warrant an increase in coarse aggregate, usually about 10% higher than the tabular value.

14 Absolute-volume Computations

The final determination of the proportions of each material is made by the absolute-volume method. This is shown in the subsequent examples. In order to utilize ACI Standard 211.1-70, the following data must be established: (1) gradation, (2) specific gravity, (3) dry-rodded unit weight of the coarse aggregate.

Table 7. Volume of Coarse Aggregate Per Unit of Volume of Concrete*

Maximum size of aggregate, in.	Volume of dry-rodded coarse aggregate† per unit volume of concrete for different fineness moduli of sand			
	2.40	2.60	2.80	3.00
⅜	0.50	0.48	0.46	0.44
½	0.59	0.57	0.55	0.53
¾	0.66	0.64	0.62	0.60
1	0.71	0.69	0.67	0.65
1½	0.75	0.73	0.71	0.69
2	0.78	0.76	0.74	0.72
3	0.82	0.80	0.78	0.76
6	0.87	0.85	0.83	0.81

* From Table 5.2.6, ACI Standard 211.1-70.

† Volumes are based on aggregates in dry-rodded condition as described in ASTM C 29 for Unit Weight of Aggregate.

These volumes are selected from empirical relationships to produce concrete with a degree of workability suitable for usual reinforced construction. For less workable concrete such as required for concrete pavement construction they may be increased about 10 percent. For more workable concrete, such as may sometimes be required when placement is to be by pumping, they may be reduced up to 10%.

The specific gravity of the cement can be taken as 3.15 with reasonable accuracy. The fineness modulus of the sand is computed from the gradation. The absorption and total moisture content of each aggregate must be established in order to compute the batch weights for trial mixes and field use.

15   Estimated Unit Weight Basis

The ACI 211 Recommended Practice for Selection Proportions for Normal Weight Concrete describes an alternate method to that of the absolute-volume method. It is based on an estimated weight of the concrete per unit volume. The selection of water-cement ratio, estimation of mixing water and air content, calculation of cement content, and estimation of coarse aggregate content are the same as in the absolute-volume basis. The fine aggregate content then may be estimated by the weight method. The unit weight of concrete is obtained from Table 8 or can be estimated from experience. The required weight of fine aggregate is the difference between the weight of fresh concrete and the total weight of the other ingredients.

16   Examples

The use of typical examples will best explain the calculation of proportions by the design criteria of ACI Standard 211.1-70.

1." **Non-air-entraining.** Required: Mix proportions of concrete of the following characteristics:

1. Not exposed to weather or sulfate water.

2. Reinforced walls 6 to 11 in. thick with minimum clear reinforcing spacing of 2 in.

3. Concrete design strength f_c' is 4,000 psi.

Table 8. First Estimate of Weight
of Fresh Concrete*

Maximum size of aggregate, in.	First estimate of concrete weight, lb per cu yd†	
	Non-air-entrained concrete	Air-entrained concrete
⅜	3,840	3,690
½	3,890	3,760
¾	3,960	3,840
1	4,010	3,900
1½	4,070	3,960
2	4,120	4,000
3	4,160	4,040
6	4,230	4,120

* From Table 5.2.7.1, ACI 211.1-70.

† Values calculated for concrete of medium richness (550 lb of cement per cu yd) and medium slump with aggregate specific gravity of 2.7. Water requirements based on values for 3 to 4 in. slump in Table 5. If desired, the estimated weight may be refined as follows if necessary information is available: for each 10-lb difference in mixing water from the Table 5 values for 3 to 4 in. slump, correct the weight per cu yd 15 lb in the opposite direction; for each 100-lb difference in cement content from 550 lb, correct the weight per cu yd 15 lb in the same direction; for each 0.1 by which aggregate specific gravity deviates from 2.7, correct the concrete weight 100 lb in the same direction.

Laboratory data given:

1. Dry-rodded unit weight of the angular coarse aggregate is 100 pcf.

2. Fineness modulus of sand is 2.60.

3. Saturated surface-dry bulk specific gravity of both fine and coarse aggregate is 2.65.

4. Absorption of fine aggregate is 1.0%, and that of coarse aggregate is 0.5%.

Computation of proportions:

1. Since exposure is not critical and non-air-entrained concrete is specified, the water-cement ratio will be established for compressive strength only.

2. The compressive-strength requirement will be 4,000 psi plus 1,200 psi because standard deviation is not known. This is 5,200 psi. From Table 4, the water-cement ratio required to produce 5,200 psi in non-air-entrained concrete is 0.47.

3. From paragraph 11, the slump shall be 4 in. Therefore, 4 in. is used in design.

4. From the spacing of reinforcement, maximum-size aggregate is 1½ in.

5. From Table 5, the approximate mixing water for 4-in. slump under test conditions is 300 lb per cu yd.

6. From the information in items 2 and 5, the required cement content is 300/0.47 = 638 lb per cu yd.

7. From Table 7, the volume of coarse aggregate per unit volume of concrete for 1½-in. aggregate and sand with a fineness modulus of 2.60 is 0.73. For a cubic yard the quantity of coarse aggregate will be 0.73 × 27 = 19.7 cu ft. For aggregate of 100 pcf, the dry weight is 1,970 lb per cu yd. The saturated surface-dry (SSD) weight is the dry weight multiplied by 1 plus absorption of the aggregate: 1,970 × 1.005 = 1,980 lb cu yd.

8. The sand content per cubic yard is determined by calculating the solid volumes of all other ingredients and subtracting this quantity from the total volume of 1 cu yd. The calculations are usually conducted in terms of cubic feet as in the following:

Solid volume of cement:

$$\frac{\text{wt of cement}}{\text{sp gr of cement} \times 62.4} = \frac{638}{3.15 \times 62.4} = 3.26 \text{ cu ft}$$

Volume of water:

$$\frac{\text{lb used}}{\text{lb/cu ft}} = \frac{300}{62.4} = 4.80 \text{ cu ft}$$

Solid volume of coarse aggregate:

$$\frac{\text{wt of coarse aggregate}}{\text{sp gr} \times 62.4} = \frac{1,980}{2.65 \times 62.4} = 11.95 \text{ cu ft}$$

Volume of air (estimated 1% entrapped):

$$0.01 \times 27 \text{ cu ft} = 0.01 \times 27 = \underline{0.27 \text{ cu ft}}$$
$$\text{Total solid volume, all ingredients except sand} = \overline{20.28 \text{ cu ft}}$$
$$\text{Solid volume of sand} = 27 - 20.28 = 6.72 \text{ cu ft}$$

Required weight of sand:

$$\text{Volume} \times (\text{sp gr} \times 62.4) = 6.72 \times (2.65 \times 62.4) = 1,110 \text{ lb}$$

9. From the above trial batch, quantities per cubic yard of concrete are

Cement	=	638 lb
Water	=	300 lb
Sand (SSD basis)	=	1,110 lb
Coarse aggregate (SSD basis)	=	1,980 lb
	Total =	4,028 lb/cu yd

$$\text{Calculated unit weight of plastic concrete} = \frac{4,028 \text{ lb}}{27 \text{ cu ft/yd}} = 149.0 \text{ pcf}$$

2. Air-entraining. Required: Mix proportions of concrete of the following characteristics:

1. Exposed moderate sulfate water and severe freezing and thawing.

2. Reinforced walls 12 to 24 in. thick with minimum clear reinforcing-bar spacing of 3 in.

3. Concrete design strength f_c' is 3,000 psi.

Laboratory data given:

1. Dry-rodded unit weight of the angular coarse aggregate is 100 pcf.

2. Fineness modulus of sand is 2.60.

3. Saturated surface-dry bulk specific gravity of both fine and coarse aggregate is 2.65.

4. Absorption of fine aggregate is 1.0%, and that of coarse aggregate is 0.5%.

Computation of proportions:

1. From Table 2, moderate section in contact with sulfate requires water-cement ratio of 0.45. Type II moderate sulfate-resisting cement will be used, and this permits increase of water-cement ratio to 0.50. Air-entrained concrete will be required.

2. The compressive-strength requirement will be 4,200 psi (3,000 psi plus 1,200 psi because standard deviation is not known). From Table 4, the water-cement ratio required to produce 4,200 psi air-entrained concrete is 0.46. However, the water-cement ratio based on exposure (from Table 2), being lower, must govern. Therefore, the trial batch will be calculated at 0.45.

3. From paragraph 11, the maximum slump shall be 4 in., and therefore 4 in. is used in the design.

4. Based on clear spacing, the maximum-size aggregate shall be 2¼ in. Since only 1½- and 3-in. sizes are available, 1½-in. shall be used.

5. From Table 5, the approximate mixing water for 4 in. slump under these conditions is 275 lb per cu yd. In this table the desired air content is indicated as 4.5%. However, the maximum given in Table 1 is 6%. Therefore 6% shall be used in the design.

6. From the information in items 1 and 5, the required cement content is 275/0.45 = 612 lb per cu yd.

7. From Table 7, the volume of coarse aggregate per unit volume of concrete for 1½-in. aggregate and sand with a fineness modulus of 2.60 is 0.73. For 1 cu yd the quantity of coarse aggregate will be $0.73 \times 27 = 19.7$ cu ft. For coarse aggregate of 100 pcf, the dry weight is 1,970 lb per cu yd. The saturated surface dry weight is the dry weight multiplied by 1 plus absorption of the aggregate: $1,970 \times 1.005 = 1,980$ lb per cu yd.

8. The sand content is determined by the following absolute-volume method:

Solid volume of cement:

$$\frac{\text{wt of cement}}{\text{sp gr of cement} \times 62.4} = \frac{612}{3.15 \times 62.4} = 3.12 \text{ cu ft}$$

Volume of water:

$$\frac{\text{lb used}}{\text{lb/cu ft}} = \frac{275}{62.4} = 4.40 \text{ cu ft}$$

Solid volume of coarse aggregate:

$$\frac{\text{wt of coarse aggregate}}{\text{sp gr} \times 62.4} = \frac{1,980}{2.65 \times 62.4} = 11.95$$

Volume of air (estimated 6% entrained):

$$0.06 \times 27 = \underline{1.62}$$

$$\text{Total solid volume, all ingredients except sand} = 21.09 \text{ cu ft}$$

$$\text{Solid volume of sand } 27 - 21.09 = 5.91 \text{ cu ft}$$

Required weight of sand:

$$\text{Volume} \times (\text{sp gr} \times 62.4) = 5.91 \times (2.65 \times 62.4) = 980 \text{ lb}$$

9. The estimated batch quantities per cubic yard of concrete are

$$
\begin{array}{rr}
\text{Type II cement} = & 612 \text{ lb} \\
\text{Water (31 gal)} = & 275 \text{ lb} \\
\text{Sand (SSD basis)} = & 980 \text{ lb} \\
\text{Coarse aggregate (SSD basis)} = & 1,980 \text{ lb} \\
\text{Total} = & 3,847 \text{ lb}
\end{array}
$$

Air entrainment is obtained by use of air-entraining agent added in accordance with recommendations of the manufacturer. Adjustment should be made as found necessary by the trial mix.

$$\text{Calculated unit weight of plastic concrete} = \frac{3,847}{27} = 142.0 \text{ pcf}$$

MAKING TRIAL MIX

17   Procedure

After the mix proportions are determined by the designs as shown in the preceding examples, a trial mix is made. This is to check the assumptions and establish the effects of the variables on water requirement and air entrainment.

1. For our purposes we select a trial mix of sufficient capacity to establish a 7-day and 28-day compressive strength, each with four specimens. This will require a $\frac{1}{10}$ cu yd batch as shown in the following tabulation for Example 2 above.

	Wt/cu yd, lb	Wt per $\frac{1}{10}$ batch, lb	% moisture	Wt of water in aggregate	Corrected batch wt, lb
Type II cement...	612	61.2	...	...	61.2
Water...........	275	27.5	...	(4.9)	22.6
Fine aggregate...	980	98.0	5.0	4.9	102.9
Coarse aggregate.	1,980	198.0	SSD	0	198.0
Total..........	3,847	384.7	...	0	384.7
Approved air-entraining agent..	5.6 fluid oz	0.56 oz or 17 cu cm	...	...	17 cu cm

Calculated volume of batch = 2.7 cu ft at 142.0 pcf.

2. In order to compensate for moisture in the aggregates, the amount of free moisture is determined. Free moisture is that water over the absorbed water in the aggregate which is available as mixing water. The method of test is given in Chap. 12. The coarse aggregate was found to be saturated dry and therefore contained only the absorbed water. The fine aggregate contained 5% free moisture. Therefore, the batch weight of the sand was increased to allow for this water and the mixing water was reduced accordingly. The calculations are

$$98 \times 0.05 = 4.9 \text{ lb}$$

This weight of water must be added to the design weight of saturated surface-dry fine aggregate.

$$98 + 4.9 = 102.9 \text{ lb}$$

3. The materials at room temperature (68 to 77°F) are batched for the trial mix as follows:
   a. Weigh aggregates cumulatively beginning with the smallest size on a scale accurate to 0.03 lb or $\frac{1}{2}$ oz. A sufficient size of container to hold either a full batch or a half batch may be used with tare corrections made. The fine aggregate should be moist to avoid segregation when it is weighed.
   b. Weigh cement separately.
   c. Weigh a greater quantity of water than required. Add air-entraining agent to a portion of the water. Use for this portion about two-thirds of the total amount that will be required in the trial batch.
4. The mixing procedure is as follows:
   a. Capacity of the mixer should be equal to or slightly greater than the batch size.
   b. The mixer must be "buttered" by initially mixing a small amount of mortar of the same composition as the mortar of the batch. The interior of the mixer is thus coated. Any excess should be discharged.
   c. Place the coarse and fine aggregate in the mixer with about two-thirds of the expected water which includes all the air-entraining agent. Add the cement, and mix for about $\frac{1}{2}$ min. Then add water from the measured amount until the concrete in the mixer appears to have attained the desired slump.

d. Mix the batch for 3 min, rest for 3 min, and then remix for 2 min.

e. Dump the concrete into a watertight and nonabsorbent receptacle large enough to hold the entire batch and to permit remixing with a shovel. (A clean damp floor may be used, but a large pan is preferred.) Remix to eliminate any segregation that may have occurred during discharge of the mixer.

f. Determine the amount of water actually used by subtracting the weight of that remaining from the weight of the amount originally weighed.

5. The following tests must be made after the above remixing:

a. Slump by ASTM Method C 143.

b. Air content by ASTM Method C 231 (Pressure Method) or Method ASTM C 173 (Volumetric Method). Use Volumetric Method only on concrete made with porous aggregate.

c. Unit weight, yield by ASTM Method C 138.

d. Strength specimens following procedures of ASTM Method C 192.

e. Concrete used for air-content tests must be discarded. Concrete from other tests can be recombined and remixed with balance of concrete batch for making the strength specimens.

6. Concrete from the trial mix will, in practically all cases, vary from the anticipated conditions slightly because of the variables of water demand and air content and possible slight differences in specific gravities and test deviations. As an illustration, the following tabulation represents the actual results of the trial mix of Example 2:

Type II cement..............	61.2 lb
Water.....................	27.9 lb
Fine aggregate..............	98.0 lb
Coarse aggregate...........	198.0 lb saturated surface-dry
Total.....................	385.1 lb
Slump....................	4.0 in.
Air content................	6.0%
Unit weight...............	144.5 pcf
Yield.....................	385.1/144.1 = 2.67 cu ft

$$W/C = 27.9/61.2 = 0.46$$

In order to have provided the planned yield, the original batch weights should have been $2.70/2.67 = 1.011$ times those used, namely,

	Batch wt used, lb	Corrected for yield (original × 1.011), lb
Cement (Type II).........	61.2	61.9
Water total.............	27.9	28.2
Fine aggregate (SSD)....	98.0	99.1
Coarse aggregate (SSD)...	198.0	200.2
Total................	385.1	389.4

However, this calculation and the higher water content required by the mix leave a water-cement ratio of 0.46, while the specification permits a maximum of only 0.45. The water cannot be reduced without adversely affecting the desired workability; therefore the cement must be increased by the factor of 0.46/0.45, which is 1.02. Thus the cement for the trial batch should be 1.02×61.9 lb, which is 63.1 lb. To verify the calculations, we check:

$$W/C = 28.2/63.1 = 0.45$$

The increase of 1.2 lb of cement to the trial-batch weights in order to adjust the water-cement ratio has added $1.2/(3.15 \times 62.4) = 0.006$ cu ft of absolute volume to our trial-batch volume, giving a calculated volume of 2.706 cu ft. To correct for this, reduce the fine-aggregate content by 0.006 cu ft absolute volume, which is $(2.65 \times 62.4) \times 0.006 = 1.0$ lb. Thus the final corrected trial batch weights would be

Type II cement...................	63.1 lb
Water..........................	28.2 lb
Fine aggregate (SSD).............	98.1 lb
Coarse aggregate (SSD)..........	200.2 lb
Total........................	389.6 lb

Theoretically a new trial batch should now be prepared using these weights, but inherent variations in concrete materials and test procedures are such that to do so would again give experimental results indicating that new corrections should be calculated, and still another trial batch be prepared, and so on, indefinitely.

It is normally considered sufficient to make the calculated corrections as shown and to use these corrected figures as the basis for the recommended field mix. The concrete test specimens made from this trial mix are almost certainly representative of the recommended field mix within the normal concrete test variations.

Thus the batch weights per cubic yard recommended to the field in this particular example would be

Ingredient	Quantity/cu yd
Type II cement...................	631 lb
Water..........................	282 lb
Fine aggregate (SSD).............	980 lb*
Coarse aggregate (SSD)..........	2,000 lb*
Air-entraining admixture..........	5.8 oz

* These figures have been rounded off to normal batch-plant scale graduations.

18   Strength Curves for Trial Mixes

The water-cement ratio law can be utilized to determine mixes for specific compressive strengths by means of a series of trial mixes. This is required in the ACI Building Code for reinforced concrete when reliable test data is not available or when proportioning new materials of unknown properties.

The usual procedure is as follows:

At least three and preferably four different trial mixtures are made with different cement contents and with constant slump. The total water per cubic yard will be practically the same for all mixes. This will result in three or four water-cement ratios spaced such that a range of compressive strengths will be achieved that encompasses those required for the project. The data obtained from each mix must include water-cement ratio, percent air, slump, and workability characteristics. The yield determinations will enable calculations of water, cement, fine aggregate, and coarse aggregate per cubic yard for each mix. This calculation will be the same as the foregoing trial-mix example. Of course, it is not necessary to calculate corrections, since the results are to be plotted at their actual points on the curve. The cylinders (and beams if required) must be cured in accordance with ASTM C 192. Three cylinders should be tested at 7 days and three at 28 days. Beams should be tested as required by the specifications or in a manner similar to the cylinders.

A typical compressive-strength curve developed from four trial mixes is shown in Fig. 2. A water-cement ratio and corresponding mix design can be taken from this curve for a given set of materials. As an example of a method of preparing and using, see the curve and notes of Fig. 2, which gives the following results:

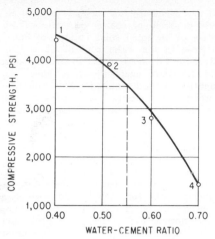

Fig. 2. Trial mix curve. For a described compressive strength of 3,450 psi, enter the curve at 3,450 psi, intersect the curve, then drop down and read 0.55.

If other considerations of durability do not require a lower water-cement ratio, use 0.55. Use this water-cement value to interpolate quantities of materials from the values of the two nearest mixes plotted.

	Mix 2	Mix 3	Design mix, 3,450 psi
W/C.......................	0.5	0.6	0.55
Water......................	306 lb	300 lb	303 lb
Cement.....................	612 lb	500 lb	556 lb
Fine aggregate (SSD)..........	1,120 lb	1,240 lb	1,180 lb
Coarse aggregate (SSD)........	1,950 lb	1.950 lb	1,950 lb

Compressive strengths are related to flexural strengths; however, the relationship is not exactly proportional. Characteristics of various materials affect each to different degrees. Therefore, trial mixes with beam tests are necessary to establish proportions for a given flexural-strength requirement. No attempt should ever be made to predict flexural strength from compressive-strength results.

19 Adjustments in Field

Laboratory-designed mixes should always be refined in the field. This is necessary because of the many variables that affect mix proportions. Temperature will affect slump and therefore will cause a change in the water-cement ratio for equal slump. Figure 5, Chap 5, shows that as the temperature increases the slump decreases. It should be noted that the use of retarding admixtures will retard setting time but generally will not affect the loss of slump resulting from temperature rise. The use of a water-reducing retarder as a means of water reduction can offset increased water when higher temperatures are encountered.

Variations in aggregate gradings and shape will affect water requirement. A change of ±0.20 in the fineness modulus of the sand will usually require a change in aggregate proportions. Generally an increase of 0.20 in the fineness modulus will require a decrease of 50 lb of coarse aggregate per cu yd of concrete, with a corresponding increase of an equal solid volume of fine aggregate. This might in turn cause an increase in the water content of as much as 4%. The above would be conversely true for a decrease of 0.20 in the fineness modulus.

Changes in grading of coarse aggregates may affect the void content, which would change the quantity of sand required. This would be reflected by a change in dry-rodded unit weight. For an aggregate of given specifications a decrease in unit weight means an increase in the void content, and consequently the fine-aggregate requirement increases. A change in particle shape consisting of a larger proportion of flat and elongated pieces would require more sand and water for equal slump. The converse is equally true. It has been shown that each 1% increase in air content will decrease the unit water requirement by about 3%. This is conversely true for each 1% decrease in air content. In order to maintain yield, the fine aggregate should be decreased ½% for each 1% increase in air content. This is conversely true for each 1% decrease in air content. Generally the mix proportions need not be changed for ±1% air-content variation, since this much fluctuation is normal over a period of operations. However, a change should be made for a consistently high or low air content by either a change in air-entraining agent or a change in proportions. Each 1-in. slump increase will cause an increase of 3% in water required.

It is generally acceptable to have variations of about 0.02 in water-cement ratios without changing proportions. However, it is prudent to adjust proportions for a consistent change in one direction of the water-cement ratio. Any adjustments of proportions in the field should be made under the guidelines of the trial-mix water-cement ratio unless otherwise permitted by consistent compressive-strength development.

PREPARED MIXES

20   Charts

The use of previously established mixes may be desirable on small projects when the user can use judgment in selecting an overdesign for the extra factor of safety. These mixes may also be used as a guide in establishing trial designs. Tables 9 and 10 can be used for this purpose. These mixes are designed for natural sand and well-graded stone or gravel aggregates using a specific gravity of 2.65 for both aggregates. The respective compressive strengths may be approximated by the use of Table 4. If the use of slag is desired, compute the solid volume of the coarse aggregate and replace it with an equal solid volume of slag. Increase or decrease water per cubic yard by 3% for each respective increase or decrease of 1 in. slump; then increase or decrease the fine aggregate by an equal change in the solid volume of water. If manufactured sand is used, increase the sand by 3% and increase the water by 15 lb per cu yd, and make a respective change in the solid volume of coarse aggregate. When a less workable concrete for use in pavements is desired, decrease the sand by 3% and decrease the water by 10 lb per cu yd with a corresponding change in solid volume. The solid volume of water can be taken from Table 11, and the solid volume of aggregates can be computed from Table 12. In all cases the method of trial-mix adjustments given should be used for field adjustments.

21   Useful Tables

Several tables and curves, taken from Ref. 2, are reproduced to facilitate computations in mix designs. Table 12 is used for absolute-solid-volume determinations of aggregates and cement, Table 13 to convert bags of cement to both weight and absolute solid volume, and Table 11 for conversion of gallons of water to pounds and cubic feet. It is sometimes convenient to interchange water-cement ratio by weight with gallons per bag. A table for this purpose is given in Table 14.

LIGHTWEIGHT CONCRETE

22 Mix Characteristics and Application

This discussion is limited to structural lightweight concrete obtained primarily through the use of lightweight aggregates discussed earlier in this chapter and does not cover lightweight insulating concretes obtained by the use of foaming agents such as aluminum powder that produce insulating concretes with little or no structural significance. Structural lightweight concrete is used primarily to reduce dead load and occasionally for insulating applications.

Selection of the type of lightweight aggregate for strength characteristics can be generally guided as follows:

1. Expanded shale, clay, and sintered fly ash, produce high-strength concrete. Some of the better grades can be equivalent in strength to some common natural aggregate for equal cement factors.

2. Some expanded slags, scoria, and pumice produce intermediate-strength concrete.

3. Perlite, vermiculite, diatomite, and certain cinders produce low strength.

4. The incorporation of hard natural sand in lieu of lightweight fine aggregate will usually enhance the strength and other properties. Its use, of course, will increase unit weight.

23 Selecting Mix Proportions

Most specifications for lightweight structural concrete include a requirement for maximum unit weight and minimum compressive strength.

The inherent weight characteristic of the lightweight aggregate proposed is the primary controlling factor in the unit weight of the concrete. Appreciable variations in the unit weight of the concrete can seldom, if ever, be accomplished with a given weight of aggregate. Decreasing cement content will decrease the weight of concrete but will also decrease the strength of concrete. If the aggregate being considered is a mixture of lightweight coarse and natural sand, considerable variations in weight of concrete can sometimes be accomplished by varying proportions of the two aggregates and of cement, while still maintaining reasonable workability and specified strength. If a lightweight coarse- and fine-aggregate combination is involved and the maximum unit weight is exceeded, a different aggregate source capable of producing lighter material will probably be mandatory. If the strength can be maintained at a satisfactory level, it is sometimes effective to increase the amount of entrained air.

Proportioning of lightweight concrete is best accomplished by the cement-factor basis. The method used in regular-weight concrete is not easily applicable because of the difficulty in establishing specific gravity and absorption values of most lightweight aggregate. ACI Standard 211.2-69[2], Selecting Proportions for Structural Lightweight Concrete, has been found to be one of the best methods of determining proportions. Table 15 gives the approximate relationship between strength and cement content. It should be used only as a guide for the preliminary proportions for a trial mix, because of variations in cement and aggregate quality.

The optimum proportion of fine to coarse aggregate has been established at between 40 and 60% by a study of void content. Therefore, preliminary proportioning of aggregates is usually based on 50% fine aggregate and 50% coarse aggregate. When experience so dictates, initial proportioning may deviate from the above method of establishing cement factor and aggregate proportions. Also it may be desirable to utilize conventional fine aggregate when permitted by specified unit weight limits to improve structural strength. Various trials or experience will govern the quantities that should be used. The amount of fines to be used is proportional to the cement content. The aim of the trial mix is to determine

Table 9. Suggested Trial Mixes for Air-Entrained Concrete of Medium Consistency*
(3- to 4-in. slump†)

Water-cement ratio, lb/lb	Maximum size of aggregate, in.	Air content, %	Water, lb/cu yd of concrete	Cement, lb/cu yd of concrete	With fine sand (fineness modulus = 2.50)			With coarse sand (fineness modulus = 2.90)		
					Fine aggregate, % of total aggregate	Fine aggregate, lb/cu yd of concrete	Coarse aggregate, lb/cu yd of concrete	Fine aggregate, % of total aggregate	Fine aggregate, lb/cu yd of concrete	Coarse aggregate, lb/cu yd of concrete
0.40	3/8	7.5	340	850	50	1,250	1,260	54	1,360	1,150
	1/2	7.5	325	815	41	1,060	1,520	46	1,180	1,400
	3/4	6	300	750	35	970	1,800	39	1,090	1,680
	1	6	285	715	32	900	1,940	36	1,010	1,830
	1 1/2	5	265	665	29	870	2,110	33	990	1,990
0.45	3/8	7.5	340	755	51	1,330	1,260	56	1,440	1,150
	1/2	7.5	325	720	43	1,140	1,520	47	1,260	1,400
	3/4	6	300	665	37	1,040	1,800	41	1,160	1,680
	1	6	285	635	33	970	1,940	37	1,080	1,830
	1 1/2	5	265	590	31	930	2,110	35	1,050	1,990
0.50	3/8	7.5	340	680	53	1,400	1,260	57	1,510	1,150
	1/2	7.5	325	650	44	1,200	1,520	49	1,320	1,400
	3/4	6	300	600	38	1,100	1,800	42	1,220	1,680
	1	6	285	570	34	1,020	1,940	38	1,130	1,830
	1 1/2	5	265	530	32	980	2,110	36	1,100	1,990
0.55	3/8	7.5	340	620	54	1,450	1,260	58	1,560	1,150
	1/2	7.5	325	590	45	1,250	1,520	49	1,370	1,400
	3/4	6	300	545	39	1,140	1,800	43	1,260	1,680
	1	6	285	520	35	1,060	1,940	39	1,170	1,830
	1 1/2	5	265	480	33	1,030	2,110	37	1,150	1,990

0.60	3/8	7.5	340	565	54	1,490	1,260	58	1,600	1,150
	1/2	7.5	325	540	46	1,290	1,520	50	1,410	1,400
	3/4	6	300	500	40	1,180	1,800	44	1,300	1,680
	1	6	285	475	36	1,100	1,940	40	1,210	1,830
	1½	5	265	440	33	1,060	2,110	37	1,180	1,990
0.65	3/8	7.5	340	525	55	1,530	1,260	59	1,640	1,150
	1/2	7.5	325	500	47	1,330	1,520	51	1,450	1,400
	3/4	6	300	460	40	1,210	1,800	44	1,330	1,680
	1	6	285	440	37	1,130	1,940	40	1,240	1,830
	1½	5	265	410	34	1,090	2,110	38	1,210	1,990
0.70	3/8	7.5	340	485	55	1,560	1,260	59	1,670	1,150
	1/2	7.5	325	465	47	1,360	1,520	51	1,480	1,400
	3/4	6	300	430	41	1,240	1,800	45	1,360	1,680
	1	6	285	405	37	1,160	1,940	41	1,270	1,830
	1½	5	265	380	34	1,110	2,110	38	1,230	1,990

* From "Proportioning Concrete Mixtures," Portland Cement Association.

† Increase or decrease water per cu yd by 3% for each increase or decrease of 1 in. in slump, then calculate quantities by absolute volume method. For manufactured fine aggregate, increase percentage of fine aggregate by 3 and water by 17 lb per cu yd of concrete. For less workable concrete, as in pavements, decrease percentage of fine aggregate by 3 and water by 8 lb per cu yd of concrete.

Table 10. Suggested Trial Mixes for Non-Air-Entrained Concrete of Medium Consistency* (3- to 4-in. slump)

Water-cement ratio, lb/lb	Maximum size of aggregate, in.	Air content (entrapped air), %	Water, lb/cu yd concrete	Cement, lb/cu yd concrete	With fine sand (fineness modulus = 2.50)			With coarse sand (fineness modulus = 2.90)		
					Fine aggregate, % of total aggregate	Fine aggregate, lb/cu yd of concrete	Coarse aggregate, lb/cu yd of concrete	Fine aggregate, % of total aggregate	Fine aggregate, lb/cu yd of concrete	Coarse aggregate, lb/cu yd of concrete
0.40	3/8	3	385	965	50	1,240	1,260	54	1,350	1,150
	1/2	2.5	365	915	42	1,100	1,520	47	1,220	1,400
	3/4	2	340	850	35	960	1,800	39	1,080	1,680
	1	1.5	325	815	32	910	1,940	36	1,020	1,830
	1 1/2	1	300	750	29	880	2,110	33	1,000	1,990
0.45	3/8	3	385	855	51	1,330	1,260	56	1,440	1,150
	1/2	2.5	365	810	44	1,180	1,520	48	1,300	1,400
	3/4	2	340	755	37	1,040	1,800	41	1,160	1,680
	1	1.5	325	720	34	990	1,940	38	1,100	1,830
	1 1/2	1	300	665	31	960	2,110	35	1,080	1,990
0.50	3/8	3	385	770	53	1,400	1,260	57	1,510	1,150
	1/2	2.5	365	730	45	1,250	1,520	49	1,370	1,400
	3/4	2	340	680	38	1,100	1,800	42	1,220	1,680
	1	1.5	325	650	35	1,050	1,940	39	1,160	1,830
	1 1/2	1	300	600	32	1,010	2,110	36	1,130	1,990
0.55	3/8	3	385	700	54	1,460	1,260	58	1,570	1,150
	1/2	2.5	365	665	46	1,310	1,520	51	1,430	1,400
	3/4	2	340	620	39	1,150	1,800	43	1,270	1,680
	1	1.5	325	590	36	1,100	1,940	40	1,210	1,830
	1 1/2	1	300	545	33	1,060	2,110	37	1,180	1,990

0.60	3/8	3	385	640	55	1,510	1,260	58	1,620	1,150
	1/2	2.5	365	610	47	1,350	1,520	51	1,470	1,400
	3/4	2	340	565	40	1,200	1,800	44	1,320	1,680
	1	1.5	325	540	37	1,140	1,940	41	1,250	1,830
	1 1/2	1	300	500	34	1,090	2,110	38	1,210	1,990
0.65	3/8	3	385	590	55	1,550	1,260	59	1,660	1,150
	1/2	2.5	365	560	48	1,390	1,520	52	1,510	1,400
	3/4	2	340	525	41	1,230	1,800	45	1,350	1,680
	1	1.5	325	500	38	1,180	1,940	41	1,290	1,830
	1 1/2	1	300	460	35	1,130	2,110	39	1,250	1,990
0.70	3/8	3	385	550	56	1,590	1,260	60	1,700	1,150
	1/2	2.5	365	520	48	1,430	1,520	53	1,550	1,400
	3/4	2	340	485	41	1,270	1,800	45	1,390	1,680
	1	1.5	325	465	38	1,210	1,940	42	1,320	1,830
	1 1/2	1	300	430	35	1,150	2,110	39	1,270	1,990

* From "Proportioning Concrete Mixtures," Portland Cement Association.

Table 11. Water Conversion Factors*

Gal	Lb	Cu ft
0.12	1.0	0.01607
1.0	8.33	0.1338
7.48	62.3	1.0
26.0	216.67	3.48
26.5	220.83	3.54
27.0	225.00	3.61
27.5	229.17	3.68
28.0	233.33	3.75
28.5	237.50	3.81
29.0	241.67	3.88
29.5	245.83	3.95
30.0	250.00	4.01
30.5	254.17	4.08
31.0	258.33	4.15
31.5	262.50	4.21
32.0	266.67	4.28
32.5	270.83	4.35
33.0	275.00	4.41
33.5	279.17	4.48
34.0	283.33	4.55
34.5	287.50	4.61
35.0	291.67	4.68
35.5	295.83	4.75
36.0	300.00	4.82
36.5	304.17	4.88
37.0	308.33	4.95
37.5	312.50	5.02
38.0	316.67	5.08
38.5	320.83	5.15
39.0	325.00	5.22
39.5	329.17	5.28
40.0	333.33	5.35

* From Ref. 2, Table 16-8.

Table 12. Density and Volume Relationships*

Table for unit weight

Specific gravity S	Density (D)	$\dfrac{1}{\text{Density}}$
2.45	152.64	0.006552
2.46	153.26	0.006525
2.47	153.88	0.006498
2.48	154.50	0.006472
2.49	155.13	0.006446
2.50	155.75	0.006420
2.51	156.37	0.006395
2.52	157.00	0.006370
2.53	157.62	0.006344
2.54	158.24	0.006319
2.55	158.86	0.006295
2.56	159.49	0.006270
2.57	160.11	0.006246
2.58	160.73	0.006222
2.59	161.36	0.006197
2.60	161.98	0.006174
2.61	162.60	0.006150
2.62	163.23	0.006126
2.63	163.85	0.006103
2.64	164.47	0.006080
2.65	165.10	0.006057
2.66	165.72	0.006034
2.67	166.34	0.006012
2.68	166.96	0.005989
2.69	167.57	0.005967
2.70	168.21	0.005945
2.71	168.83	0.005923
2.72	169.46	0.005901
2.73	170.08	0.005880
2.74	170.70	0.005858
2.75	171.33	0.005837
3.10	193.13	0.005178
3.15	196.24	0.005096
3.20	199.36	0.005016

Notation

S = specific gravity
D = density, pcf
W = weight, lb
V = volume, cu ft

BASIC FORMULAS

Weight of water is 62.3 pcf
$D = 62.3S$
$W = DV$
$\quad = 62.3SV$

$$V = \frac{W}{62.3S} = W \times \frac{1}{\text{density}}$$

* From Ref. 2, Table 16-6.

optimum amount of fines that must be added to overcome harshness for a selected air content.

24 Estimating First Trial Mix Proportions

1. Determine dry loose unit weight, moisture content of fine- and coarse-aggregate fractions, and specific gravity factors at various moisture contents of such aggregates and coarse lightweight aggregate. See Chap. 12.

PROPORTIONING MIXES AND TESTING

Table 13.   Cement Conversion Factors*

Weight and solid-volume equivalents
per sack from 0.1 to 10.9 sacks

Sacks						Weight and solid volume, cu ft							
		0	0.1	0.2	0.25	0.3	0.4	0.5	0.6	0.7	0.75	0.8	0.9
0	Wt	0	9.4	18.8	23.5	28.2	37.6	47.0	56.4	65.8	70.5	75.2	84.6
	Vol	0	0.048	0.096	0.120	0.144	0.192	0.240	0.287	0.335	0.359	0.383	0.431
1	Wt	94	103.4	112.8	117.5	122.2	131.6	141.0	150.4	159.8	164.5	169.2	178.6
	Vol	0.479	0.527	0.575	0.599	0.623	0.671	0.719	0.766	0.814	0.838	0.862	0.910
2	Wt	188	197.4	206.8	211.5	216.2	225.6	235.0	244.4	253.8	258.5	263.2	272.6
	Vol	0.958	1.006	1.054	1.078	1.102	1.150	1.198	1.245	1.293	1.317	1.341	1.389
3	Wt	282	291.4	300.8	305.5	310.2	319.6	329.0	338.4	347.8	352.5	357.2	366.6
	Vol	1.437	1.485	1.533	1.557	1.581	1.629	1.677	1.724	1.772	1.796	1.820	1.868
4	Wt	376	385.4	394.8	399.5	404.2	413.6	423.0	432.4	441.8	446.5	451.2	460.6
	Vol	1.916	1.964	2.012	2.036	2.060	2.108	2.156	2.203	2.251	2.275	2.299	2.347
5	Wt	470	479.4	488.8	493.5	498.2	507.6	517.0	526.4	535.8	540.5	545.2	554.6
	Vol	2.395	2.443	2.491	2.515	2.539	2.587	2.635	2.682	2.730	2.755	2.778	2.826
6	Wt	564	573.4	582.8	587.5	592.2	601.6	611.0	620.4	629.8	634.5	639.2	648.6
	Vol	2.874	2.922	2.970	2.994	3.018	3.066	3.114	3.161	3.209	3.234	3.257	3.305
7	Wt	658	667.4	676.8	681.5	686.2	695.6	705.0	714.4	723.8	728.5	733.2	742.6
	Vol	3.353	3.401	3.449	3.473	3.497	3.545	3.593	3.640	3.688	3.713	3.736	3.784
8	Wt	752	761.4	770.8	775.5	780.2	789.6	799.0	808.4	817.8	822.5	827.2	836.6
	Vol	3.832	3.880	3.928	3.952	3.976	4.024	4.072	4.119	4.167	4.192	4.215	4.263
9	Wt	846	855.4	864.8	869.5	874.2	883.6	893.0	902.4	911.8	916.5	921.2	930.6
	Vol	4.311	4.359	4.407	4.431	4.455	4.503	4.551	4.598	4.646	4.671	4.694	4.742
10	Wt	940	949.4	958.8	963.5	968.2	977.6	987.0	996.4	1005.8	1010.5	1015.2	1024.6
	Vol	4.790	4.838	4.886	4.910	4.934	4.982	5.030	5.077	5.125	5.150	5.173	5.221

* From Ref. 2, Table 16-7.
Weight = 94 × sacks.
Volume = weight × 0.00509567, based on specific gravity of 3.15.
1 bbl = 4 sacks = 376 lb = 1.916 cu ft.

2. Assume a satisfactory cement factor from Table 15 or experience.

3. Assume total water required for 1 cu yd including absorbed water. This will usually fall between 400 and 550 lb per cu yd for a 3-in. slump.

4. Assume that 16 cu ft of both fine and coarse lightweight aggregate is required. These proportions may vary as found by experience and if some natural aggregate is used.

5. Compute the weights for 1 cu yd and make a trial mix with a convenient-sized batch in the same proportions.

6. Determine the resultant quantitative and qualitative properties. Make adjustments in proportions as required to attain proper workability and unit weight. Workability can usually be changed by an increase or decrease in fine aggregate, which may also affect unit weight.

7. Utilize the specific gravity factors for the aggregates and make additional mixes at various cement factors to establish a cement-strength curve to develop a mix of any specific strength.

Table 14.   Conversion of Water-Cement Ratio by Weight to Gallons per Sack
(Gallons per Sack = W/C by Weight × 11.28)*

Lb water per lb cement	Lb water per lb cement									
	0	0.01	0.02	0.03	0.04	0.05	0.06	0.07	0.08	0.09
0.20	2.2560	2.3688	2.4816	2.5944	2.7072	2.8200	2.9328	3.0456	3.1584	3.2712
0.30	3.3840	3.4968	3.6096	3.7224	3.8352	3.9480	4.0608	4.1736	4.2864	4.3992
0.40	4.5120	4.6248	4.7376	4.8504	4.9632	5.0760	5.1888	5.3016	5.4144	5.5272
0.50	5.6400	5.7528	5.8656	5.9784	6.0912	6.2040	6.3168	6.4296	6.5424	6.6552
0.60	6.7680	6.8808	6.9936	7.1064	7.2192	7.3320	7.4448	7.5576	7.6704	7.7832
0.70	7.8960	8.0088	8.1216	8.2344	8.3472	8.4600	8.5728	8.6856	8.7984	8.9112
0.80	9.0240	9.1368	9.2490	9.3024	9.4752	9.5880	9.7008	9.8136	9.9264	10.0392
0.90	10.1520	10.2648	10.3776	10.4904	10.6032	10.7160	10.8288	10.9416	11.0544	11.1672

Note: Water-cement ratio by weight equals pounds of water per pound of cement.
EXAMPLE: A W/C of 0.55 by weight equals 6.2 gal per sack.   Six gal per sack equals 0.532 by weight.
* From Ref. 2, Table 16-10.

**Table 15.   Approximate Relationship Between
Strength and Cement Content***

Compressive strength		Cement content	
Psi	Kg/cm²	Lb per cu yd	Kg/m³
2,500	175	425 to 700	250 to 420
3,000	210	475 to 750	280 to 450
4,000	280	550 to 850	330 to 500
5,000	350	650 to 950	390 to 560

* From Table 3.3, ACI 211.2-69.

25 Weight Method

Lightweight aggregate concrete may also be proportioned by the weight method similar to that given earlier in this chapter for proportioning normal weight concrete. The sum of the weights of all materials in a mix is equal to the total weight of the same mix. Therefore, when the unit weight of the concrete is known (by experience or assumed), the weight of cement and water (known or assumed) is deducted from the given unit weight. This subtraction gives the weight of the total lightweight aggregates SSD. The volume can be determined from the specific gravity factor, and proportion of fine to coarse is determined later by experience or the method given above.

26 Example of Lightweight-concrete Proportioning

Required: Lightweight concrete with compressive strength of 3,500 psi, and a maximum air-dry unit weight of 105 pcf.

Given: Dry loose unit weight of available lightweight fine aggregate is 60 lb per cu ft, and that of lightweight coarse aggregate of ¾ in. nominal size is 50 lb per cu ft. Moisture content of the coarse aggregate is 2% and of the fine aggregate is 4%.

Solution:
1. From Table 15 assume 564 lb cement per cu yd required for 3,500 psi.
2. Assume 16 cu ft of each aggregate required.
3. Assume total water including absorbed water will be 500 lb per cu yd.
4. Use air-entraining agent for 5% entrained air.
5. Computations—batch quantities:

	Dry basis		Damp basis	
	Wt/cu yd, lb	Wt/$\frac{1}{10}$ cu yd batch	Wt/cu yd, lb	Wt/$\frac{1}{10}$ cu yd, lb
Cement......................	564	56.4	564	56.4
Fine aggregate 16 × 60.....	960	96.0	960 × 1.04 = 998	99.8
Coarse aggregate 16 × 50.....	800	80.0	800 × 1.02 = 816	81.6
Water.......................	500	50.0	500 − 65 = 446	44.6
Approved air-entraining agent..	Follow manufacturer's directions		Follow manufacturer's directions	
Total weight..............	2,824	282.4	2,824	282.4

6. The aggregates should be soaked for 24 hr and the absorbed water and free water calculated to determine total moisture content. Dry basis of aggregates is determined for making adjustments.

7. The tests of the trial batch of plastic concrete resulting from the above calculated mix might be as follows:

Water required......................	50 lb
Unit weight, ASTM C 138...........	103 pcf
Slump, ASTM C 143................	3 in.
Air content, ASTM C 173...........	5%
Yield = 282.4 ÷ 103................	2.74 cu ft

8. Quantities per cubic yard are obtained by multiplying batch quantities by the ratio of cubic feet in a yard to cubic foot of batch:

$$\frac{27}{2.74} = 9.86$$

	Dry basis	Damp basis
Cement...............	9.86 × 56.4 = 556	556
Fine aggregate.........	9.86 × 96.0 = 947	947 × 1.04 = 985
Coarse aggregate.......	9.86 × 80.0 = 789	789 × 1.02 = 805
Water................	9.86 × 50 = 493	493 − 54 = 439
Total weight.........	2,785	2,785

Water actually changes with changes in moisture content of aggregate. See following example for illustration of changes in water content and specific gravity factor.

9. Adjusting mix proportions of structural lightweight concrete. The specific gravity factor of the aggregate method is used to provide means for mix adjustments and proportioning.

The test method is described in Chap. 12, paragraph 17. The specific gravity factor is a function of the moisture content and therefore must be computed for various moisture contents and each aggregate type. An example of a typical relationship is shown in Fig. 3.

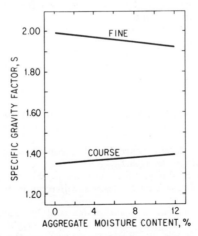

Fig. 3. Example of the relationship between pycnometer specific gravity and moisture content for a lightweight aggregate. (*From Fig. 4.2.2, ACI 211.2–69.*)

10. A trial batch is made and the procedure shown from items 5 through 7 is performed with the necessary adjustments for yield.

11. Additional mix designs and trial mixes should be made to provide at least three and preferably four points on a cement-strength curve so that the mix for the desired strength can be obtained. (An experienced laboratory technician can frequently calculate a trial mix of conventional-weight concrete so as to produce concrete with a given cement factor or water-cement ratio or both, to sufficient accuracy as to require only minor and insignificant corrections. In such case a

second trial batch with the minor correction is usually unnecessary. With light-weight concrete there is much less probability of being able to calculate a single trial batch and have it produce a concrete so close to the desired requirements that additional mixes are unnecessary. Almost always a series of mixes will be required before a project mix for lightweight concrete can be recommended.)

27 Adjustment

Even more than in regular concrete, the many variables encountered during production and construction may require adjustment of the mixes in the field. It is essential that when the parameters are established in the trial mixes, these should be maintained in the field. Generally the tolerances used for regular concrete may not be completely applicable to lightweight concrete. For example, lower slumps can be used in lightweight concrete for equal placeability. The mix may appear stiff but if properly designed will flow and consolidate readily under mechanical vibration, and slumps of 1 to 2 in. can be readily used. However, when it is found necessary to increase the water content, the cement content must be proportionally increased to maintain the design strength level. The water-cement ratio law applies. When the moisture content of the aggregate changes, care must be taken to maintain the weight of dry aggregate constant by making the moisture correction. If the density of the aggregate has changed, the volume of dry aggregate must be kept constant to maintain yield. A completely new mix may be required if the unit weight of the concrete varies appreciably from the design mix weight. A unit weight change due to air-content variation may be corrected by changing the quantity of air-entraining agent. A unit weight change due to a change in unit weight of aggregate may possibly be corrected if the aggregates involve lightweight coarse and natural sand by changing proportions; but if the aggregate is all lightweight, a change in the aggregate being furnished will almost certainly be required. Generally the most important factors in field adjustment of a mix made of high and consistent quality materials are to maintain constant cement content, slump, air content, and volume of dry aggregate per cubic yard. A constant volume of dry aggregate requires continuous moisture determinations and necessary corrections.

The following rules of thumb will serve as a guide for approximate mix adjustments of structural lightweight concrete:

1. A 1% increase in fine aggregate increases water by 3 lb per cu yd.
2. An increase of 3 lb per cu yd in water will require an increase of about 1% in cement content.
3. An increase of 1% in air content will decrease water content by 5 lb per cu yd. (May also decrease strength.)
4. An increase of 1 in. in slump will require about 10 lb per cu yd water increase.
5. An increase of 10 lb per cu yd in water will require increased cement content of approximately 3%.

Mix adjustments for cement, air, and water content, using the absolute-volume method, are calculated in the same manner as normal weight concrete proportioning. An example is shown below to illustrate the use of the specific gravity factor and effective-displaced-volume method which can be used for adjustments in proportions. In this case, the method is used to increase the fine-aggregate proportion of the above mix, assuming it was necessary for an increase in workability.

28 Example of Adjustments

Total effective displaced volume of aggregates in original mix computed above was about 50% coarse aggregate and 50% fine aggregate. Experience tells us that we should change to 55% fine and 45% coarse. New effective displaced volume of aggregates is now $27.0 - 2.98 - 1.35 - 8.15 = 14.52$. (See adjusted mix dry basis

Table 16. Mix Adjustments, Dry Basis

	Weight, lb.	Effective displaced volume, cu ft
	Original mix	
Cement.............	556	2.83
Air 5%.............	....	1.35
Coarse aggregate......	$\dfrac{805}{1.02} = 789$	$\dfrac{789}{62.4 \times 1.70} = 7.44$
Fine aggregate........	$\dfrac{985}{1.04} = 947$	$\dfrac{947}{62.4 \times 2.03} = 7.48$
Added water..........	$7.90 \times 62.4 = 493$	$27.00 - 19.10 = 7.90$
Total..............	2,785	27.00
	Adjusted mix	
Cement.............	$556 + 28 = 584$	$\dfrac{584}{62.4 \times 3.15} = 2.98$
Air 5%.............	....	1.35
Coarse aggregate......	$6.53 \times 62.4 \times 1.70 = 695$	$0.45 \times 14.52 = 6.53$
Fine aggregate........	$7.99 \times 62.4 \times 2.03 = 1,010$	$0.55 \times 14.52 = 7.99$
Added water..........	$493 + 15 = 508$	$\dfrac{508}{62.4} = 8.15$
Total..............	2,797	27.00

	Coarse aggregate	Fine aggregate
Moisture content, %...........	0	0
Specific gravity factor..........	1.70	2.03

in Table 16.) Volume of water increases, and volume of cement also increases because of increase in fine aggregate. Quantities of materials that change are calculated from rules of thumb 1 and 5. These changes are illustrated in Table 16 for the dry basis and Table 17 for the damp basis.

29 Field Testing and Production Control

The purpose of routine tests of lightweight concrete is to maintain uniform quality, thus assuring proper values for unit weight, cement content, and water content. It is usually quite difficult to produce a uniform mixture consistently because of the greater variations in absorption, specific gravity, moisture content, and grading of lightweight aggregate. Therefore, lightweight concrete must be thoroughly tested in the field during production and use. Unit weight, slump, and air-content tests must be made frequently. ACI Standard 301 suggests unit-weight tests of every load. Variations in unit weight will signal air-entrainment variations or improper aggregate proportioning, grading, or specific gravity. The Volumetric Method, ASTM C 173, should be used for air-entraining tests. Although the design is usually based on dry weights of aggregates, it is most advantageous to use wet aggregate during production. This avoids segregation and avoids the subsequent absorption of water from the mix that causes rapid stiffening and segregation. It also permits better control of total water content. Periodic adjust-

Table 17.   Mix Adjustments, Damp Basis

	Weight, lb	Effective displaced vol., cu ft
	Original mix	
Cement...............	556	2.83
Air 5%................	.....	1.35
Coarse aggregate.......	805	$\dfrac{805}{62.4 \times 1.71} = 7.52$
Fine aggregate.........	985	$\dfrac{985}{62.4 \times 1.97} = 8.05$
Added water...........	452	$\dfrac{452}{62.4} = 7.25$
Total...............	2,798	27.00
	Adjusted mix	
Cement...............	584	2.98
Air 5%................	....	1.35
Coarse aggregate.......	$695 \times 1.02 = 710$	$\dfrac{710}{62.4 \times 1.71} = 6.64$
Fine aggregate........	$1{,}010 \times 1.04 = 1{,}025$	$\dfrac{1{,}010}{62.4 \times 1.97} = 8.23$
Added water.........	$7.80 \times 62.4 = 486$	$27.00 - 19.20 = 7.80$
Total...............	2,805	27.00

	Coarse aggregate	Fine aggregate
Moisture content, %............	2	4
Specific gravity factor..........	1.71	1.97

ments may be necessary as predicated by the field tests that deviate from established relationships. It should always be the aim in adjustments to keep the volume of the dry aggregate constant (which is the basis of the design) unless a check reveals a change in density.

The placeability of lightweight concrete can be significantly improved by the use of air entrainment, which will also assist in improved bleeding characteristics, reduced segregation, and lowered unit weight. The amount of air-entraining agent will vary widely for a specific amount of air with the many varieties of lightweight aggregates that are used. This characteristic demands that frequent air-entrainment tests be made.

Concrete made with lightweight aggregate that has been shown to absorb less than 2% by weight during the first hour after inundation, based on test of a sample from the field-conditioned supply, should be batched and mixed in the same manner as normal weight concrete.

Concrete made with lightweight aggregates not conforming with this absorption limit of 2% maximum should be batched and mixed by adding 80% of the mixing water to the aggregate in the mixer, then mixed for a minimum of 1½ min (15 revolutions in a truck mixer). The balance of the materials can then be added in the normal manner.

Acceptance of lightweight concrete in the field should be based on fresh unit weight measured in accordance with ASTM C567 as described in Chap. 12. When the normal fresh unit weight varies more than 2 lb per cu ft from the required weight, the mixture must be adjusted as promptly as possible.

HEAVY CONCRETE

30 Application

The greatest use of heavy concrete is in nuclear shielding. Gamma rays and x-rays can be shielded against by a mass material. Concrete has proved effective and economical. The use of high-density or heavy concrete enables a reduction in thickness for the same degree of shielding. In order to shield against neutrons, it is necessary to have a mass material such as concrete containing elements whose atoms can both thermalize and capture neutrons. The hydrous atom in the form of water in the concrete thermalizes the fast neutrons, which may then be captured by atoms of elements such as cadmium or boron which have high neutron-capture cross sections. Care must be exercised in the form in which boron is introduced into the concrete because of the high retardation effect on concrete of some forms of boron. The ACI Symposium of Heavy Concrete[4] is a good reference for information on design. The heavy concrete usually is designed to have a unit weight of 180 to 300 pcf. This is predicated upon the degree of shielding necessary and thickness of structure.

31 Aggregate Characteristics

High-density concrete is achieved by the use of heavy aggregates, including iron ores such as magnetite and limonite; quarry rock such as barite, steel punchings, and iron shot; or synthetic materials such as ferrophosphorus. The aggregates can be made to conventional gradings for both coarse and fine sizes. Conventional sand may be incorporated in the mixture when permitted by the allowable unit weight. All the many aggregates tend to produce a harsh mix because of their shape and weight. Air entrainment should be avoided because of its detrimental effect on density, and it therefore cannot be used for improved workability. Compressive strength attainable with these materials can range between 3,000 and 6,000 psi at 28 days. Some of the heavy aggregates are not suitable for exposure to weathering, and they should therefore be protected from the elements.

ASTM Standard C-637, Aggregates for Radiation Shielding Concrete, covers important properties of heavy aggregates and should be used to evaluate this type of material.

The iron ores, ilmenite, hematite, and magnetite, range in specific gravity from 4.1 to 4.8 and can produce concrete with densities from 200 to 300 pcf. Iron ores are usually readily available and can be processed with little difficulty to conventional fine and coarse gradings. The hydrous iron ores, limonite and geothite, have specific gravities of 3.4 to 4.8 and contain water as part of the compound; therefore they are well suited for use in helping to thermalize neutrons.

Barite is a quarry rock composed of 90 to 95% barium sulfate, with a specific gravity of 4.1 to 4.3. It can be processed with little additional effort to both conventional fine and coarse gradings. The optimum density concrete possible is about 230 pcf. The limited source of barite makes it more expensive in most locations than iron ores. Serpentine is a quarry rock with fixed water and is suitable for shielding but it has a low specific gravity of 2.4 to 2.7.

Ferrophosphorus has a very high specific gravity of 5.8 to 6.3 and therefore is capable of producing concrete with densities of about 250 to 350 pcf. However, caution must be exercised in its use because, when it is mixed with portland cement concrete, a flammable and possibly toxic gas is generated which can develop high pressure if confined.

Iron and steel punchings and shot have the highest specific gravities of about 6.2 to 6.7. This permits concrete density of about 300 pcf. In order to develop maximum densities, the punchings and shear bars are used as coarse aggregate. However, iron ores and barites usually make a concrete with better placeability and should be used if suitable from unit-weight consideration. Steel aggregate is undesirable because of normal poor gradation and extreme problems of segregation. It is essential that the steel particles be free of oil coatings in order to develop proper bond and compressive strength. Bond can be improved by corroding or pickling in acid, which improves surface texture and also assures cleanliness. Acid must be removed by washing the aggregate before it is used in concrete.

32   Mix Proportioning

The procedure used for regular-weight concrete is used to proportion mixes for heavy concrete using the solid-volume method. The cement contents will usually range from 470 to 750 lb per cu yd, with strength and other qualities similar to those of corresponding water-cement ratios of regular concrete. The method of proportioning is best shown by example in which the method of establishing a required density is incorporated.

Required: Heavy concrete with a minimum compressive strength of 3,500 psi and minimum plastic unit weight of 220 pcf.

Given: Magnetite iron ore with a specific gravity of 4.1 in both fine and coarse gradation. Dry-rodded unit weight of 1½ in. to No. 4 gradation is 150 pcf.

Fineness modulus of fine gradation is 2.70.

Iron punchings are available of 6.5 specific gravity.

Previous experience indicates 333 lb of water per cu yd required for 4-in. slump and 564 lb of cement per cu yd required for strength desired.

1. From Table 7, volume of dry-rodded coarse aggregate per unit volume of concrete is 0.72 for 1½-in. aggregate, and 2.80 fineness modulus of sand.

2. Weight of 1½-in. size magnetite, $0.72 \times 27 \times 150 = 2,920$ lb.

3. Solid volumes:

	Weight, lb	Volume, cu ft
Cement..........................	564	2.87
Water..........................	333	5.35
Coarse aggregate $\dfrac{2,920}{4.1 \times 62.4}$........	2,920	11.41
Volume of entrapped air, 1%........	.....	0.27
Total, except fine aggregate........	3,817	19.90

Solid volume of fine aggregate $27 - 19.90 = 7.10$ cu ft

Required dry wt of fine aggregate $7.10 \times 4.1 \times 62.4 = 1,820$ lb

Unit weight of plastic concrete:

$$\frac{3,817 + 1,820}{27} = \frac{5,637}{27} = 208.8 \text{ pcf}$$

4. In order to increase the unit weight to 220 pcf, the 1 cu yd must be $27 \times 220 = 5,940$. Therefore, the coarse-aggregate weight must be increased $5,940 - 5,637 = 303$ lb with the same solid volume of 11.41.

5. Let

$$M = \text{weight of magnetite}$$
$$P = \text{weight of iron punchings}$$

Then

$$11.41 = \frac{M}{4.1 \times 62.4} + \frac{P}{6.5 \times 62.4}$$

or

$$405.6M + 255.8P = 1,183,816$$

since

$$M + P = 2,920 + 303 = 3,223$$

Substitute and simplify:

$$P = 824 \text{ lb}$$
$$M = 2,399 \text{ lb}$$

$$\text{Volume of magnetite} = \frac{2,399}{4.1 \times 62.4} = 9.38 \text{ cu ft}$$

$$\text{Volume of punchings} = \frac{824}{6.5 \times 62.4} = 2.03 \text{ cu ft}$$

The revised mix in quantities per cubic yard is

Material	Weight, lb	Volume, cu ft
Cement..................	564	2.87
Water...................	333	5.35
Coarse aggregate:		
Magnetite.............	2,399	9.38
Punchings.............	824	2.03
Fine aggregate..........	1,820	7.10
Entrapped air, %........	.....	0.27
Total................	5,940	27.00

$$\text{Unit weight } \frac{5,940}{27} = 220 \text{ pcf}$$

6. A trial mix is made similar to that of regular concrete as shown in paragraph 17, with the same tests. The proportions are adjusted in the same manner.

33 Adjustments

The variables encountered in the field with heavy concrete are similar to those of regular concrete. It is usually necessary to make field adjustments in the amount of water to compensate for variations in the aggregate. It may be necessary to adjust quantities of fine aggregate for workability, unit weight, and yield. Because the iron ores, barite aggregates, and ferrophosphorous tend to break down in sizes during mixing, it is advisable to avoid prolonged mixing; and if it is found that excessive fines are produced during mixing, an adjustment should be made in the ratio of fine to coarse aggregate. A wet mix must be avoided because of the likelihood of segregation; however, the concrete must be consolidated to avoid honeycomb. The workability of heavy concrete is not easily determined by slump. Therefore the degree of compaction to avoid segregation and honeycomb must be visually determined to establish the consistency desired. Moisture corrections for surface water on the aggregates must be made with cement ratio control maintained since the water-cement ratio law applies to heavy concrete mixtures.

REFERENCES

1. ACI Committee 211 Report, Recommended Practice for Selecting Proportions for Normal Weight Concrete, (ACI 211.1-70).
2. Waddell, Joseph J.: "Practical Quality Control for Concrete," McGraw-Hill Book Company, New York, 1962.
3. ACI Committee 211 Report, Recommended Practice for Selecting Proportions for Structural Lightweight Concrete (ACI 211.2-69).
4. "Concrete for Radiation Shielding," ACI Compilation 1, American Concrete Institute, 1962.
5. Walker, Stanton, and Delmar L. Bloem: Effects of Aggregates Size on Properties of Concrete, *J. ACI,* September, 1960.
6. "Concrete Manual," 7th ed., U.S. Bureau of Reclamation, 1963.
7. ACI Committee 301 Report, Specifications for Structural Concrete for Buildings, (ACI 301-72).
8. ACI Standard 318-71, Building Code Requirements for Reinforced Concrete, American Concrete Institute, 1971.

Chapter 12

TESTING CONCRETE

J. F. ARTUSO

ASTM TEST METHODS

1 Significance

The American Society for Testing and Materials[1] provides the most widely accepted standards for determining the properties of concrete materials. This is particularly true for the test methods of concrete and its components. In correlation with the determination of properties, ASTM has developed specifications denoting the minimum properties that the material must possess for a certain use. The development of both specifications and methods is not a static situation, as there is continual progress in these areas through research and application. Therefore, it is necessary to utilize the latest revision of the ASTM standards. This can be determined by noting the number following the dash in the designation listed in the latest ASTM Index. This number indicates the year of the latest revision. The test method may often be divided into tests usually performed in the laboratory and tests usually performed in the field, particularly in the case of concrete. Some of the tests are performed under both circumstances. In addition to ASTM test methods, recognition is given to other outstanding agencies such as the Bureau of Reclamation, the Corps of Engineers, the American Association of State Highway Officials, and organizations such as the American Concrete Institute. All these make tremendous contributions to the science of testing concrete.

LABORATORY TESTS

2 Cement

The specifications for all kinds and types of cement contain detailed limits on chemical compounds and physical properties. Conformance to these limits is generally checked by tests in the laboratory as detailed in applicable ASTM test methods. This can be accomplished by sampling an entire bin or by testing samples from each shipment for a given project. In many cases mill test reports are accepted; in others, independent-laboratory verification is required. In addition to the routine chemical and physical tests, cement may be tested for special properties such as false or unusual setting or excessive bleeding.

3 Water

Most specifications covering water used in the manufacture and curing of concrete state that it must be clean and free of injurious impurities. In general it is felt that water that is potable will be acceptable for use in concrete. This may not be indisputable since water contaminated with sugars could be potable but would significantly reduce the quality of concrete. When water is obtained from a treated source, it usually contains a total dissolved solids content of 200 to 1,000 ppm, with only a rare source as high as 2,000 ppm. With the exception of carbonates and bicarbonates, individual impurities of 1,000 ppm will not affect concrete quality. It is advisable to determine the total solids content, and when concentrations exceed 1,000 ppm, particularly of carbonates and bicarbonates, strength tests are in order. Excessive solids content may also warrant concrete tests for durability, volume change, setting time, flexural strength, and bond to steel. The usual physical tests performed for a check of water quality are setting time, soundness, and compressive strength as given in AASHO Standard T-26. Sea water may reduce strength only slightly but is not recommended for reinforced concrete because of the corrosive effect on the steel. Nevertheless it is sometimes used in some areas because of a scarcity of fresh water. Water used for prestressed concrete should not contain more than 250 ppm chloride content.

4   Aggregate Tests

ASTM Specification C 33 indicates the acceptable level of physical properties for general construction. The test methods given there cover each property. Routine field tests usually include:

	ASTM Designation
Sampling	D 75
Grading	C 136
Amount of material finer than No. 200 sieve	C 117
Organic impurities	C 40
Clay lumps	C 142
Unit weight	C 29
Fineness modulus	C 125
Coal and lignite (lightweight pieces)	C 123

The detailed tests required for qualification prior to use in the field are more extensive and usually require laboratory facilities. These include:

Mortar-making properties	C 87
Soundness	C 88
Abrasion of coarse aggregate	C 131
Soft particles	C 235
Reactive aggregates	C 227, C 289, C 295, C 342, C 586
Freezing and thawing	C 666

5   Concrete

Routine laboratory tests of concrete are usually confined to trial mixes to determine economical proportions necessary for strength, durability, and workability, followed by strength tests of concrete from the project. More exhaustive examinations are sometimes required for specific projects including, for example, volume change and creep. After the proportions are established, they must be checked in the field, and when changes occur in materials or conditions, additional laboratory trial mixes for new designs may be necessary. Strength tests of concrete produced on the project are the most frequent and common of all laboratory tests of concrete.

FIELD TESTS

6   Sampling

The procedure used in taking a sample of concrete depends upon the equipment from which it is sampled. The following procedures have been summarized from

ASTM Designation: C 172. Conformance to the proper sampling method is essential for reliable results and is considered the most important single factor in concrete testing. The test can be only as accurate as the sample.

Transit-mix Trucks. (a) Take a portion of the sample from two or more regularly spaced intervals during discharge of the middle portion of the batch.

(b) Obtain each portion either by passing a receptacle through the entire discharge stream or by diverting the stream completely into a container.

(c) Take the composite samples to the place where the test specimens are to be molded.

(d) Composite the samples into one sample to ensure uniformity, and protect it from the sun and wind while testing.

(e) The time required from taking the sample to using it must not exceed 15 min.

(f) Take care not to restrict flow of concrete from the mixer container or transportation unit so as to cause segregation.

Paving Mixers. From paving mixers, the entire batch should be discharged and the sample collected from at least five different portions of the pile. Compositing, protection, and time limit are as above.

Stationary Mixers. All the conditions given under transit-mix trucks apply. If discharge of the concrete is too rapid to divert the complete discharge steam, discharge the concrete into a container or transportation unit sufficiently large to accommodate the entire batch and then accomplish the sampling in the same manner as for transit-mix trucks.

7 Consistency

Slump. The procedure used in making the slump test is given in ASTM Designation: C 143, and is summarized as follows:

(a) Moisten inside of slump cone and place on a flat, moist, nonabsorbent surface at least 1 by 2 ft (plank, piece of heavy plywood, concrete slab, etc.). Surface must be *firm* and *level*. Hold slump cone in place by standing on foot pieces.

(b) Fill cone one-third full and rod the concrete exactly 25 times with tamping rod, distributing the rodding evenly over the area. *Caution:* Use standard steel tamping rod, ⅝ in. diameter by 24 in. long with one end rounded to a hemispherical tip. Do not use a piece of reinforcing steel.

(c) Fill cone with second layer until two-thirds full; rod this layer 25 times as before with rod penetrating into but not through first layer. Note: one-third or two-thirds full means one-third or two-thirds of the volume, not depth of cone.

(d) Overfill cone slightly and rod this layer 25 times, making sure the rod penetrates into but not through second layer. Distribute the rodding evenly over entire area.

(e) Use the tamping rod to scrape off excess concrete from top of cone, and clean spilled concrete from around bottom of cone.

(f) Lift the cone vertically and slowly in 5 to 10 sec. Avoid torsional movement, jarring or bumping the concrete.

(g) Set slump cone on surface next to but not touching slumped concrete—lay tamping rod across top of cone. Measure amount of slump from bottom of tamping rod to top of slumped specimen over the original center of the base of the specimen (see Fig. 1). Discard this concrete after slump has been measured. Do not use it for making test cylinders.

(h) The entire operation from the start of filling should be performed within 2½ min. If there is a decided falling away from one side or a shearing off of a portion of the sample in two consecutive tests, the concrete probably lacks the necessary plasticity and cohesiveness for the slump test to be applicable.

Ball Penetration. The consistency of concrete may be determined by the use of the ball-penetrating (Kelly ball) method in accordance with ASTM Designation: C 360, summarized as follows:

(a) The concrete sample used shall be in suitable containers or forms and the test shall be made without preliminary manipulation. Suitable containers may be wheelbarrows, Georgia buggies, or the forms themselves.

(b) The depth of the sample shall be at least three times the maximum size aggregate, but not less than 8 in. The minimum horizontal distance from the center of the ball to the free edge of the concrete shall be 9 in.

(c) Strike off the concrete to a level surface and place the Kelly ball on the surface.

(d) Release handle without twisting or jolting the ball.

(e) Read the penetration directly on the handle.

(f) The control nut should be adjusted before the test starts so that the guide will be on zero.

(g) Three or more tests should be taken for a representative reading.

(h) The ball-penetration method should be calibrated for use in controlling consistency. This is done making both the slump test and ball test on mixes of various consistencies and each class of concrete. The relationship will vary with different aggregates, cement contents, air contents, and other variables.

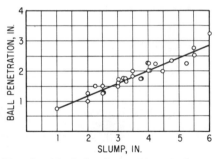

Fig. 1. Measurement of slump of con-crete.

Fig. 2. Typical relationship of slump to ball penetration must be determined individually for variations in ingredients.

A curve may be made similar to Fig. 2, with at least 10 tests made. As an alternate, a conversion factor can be compiled to be used within restricted limits of slump. The correlation should be periodically checked.

8 Air Content

The measurement of the amount of air entrained in fresh concrete during place-ment is essential to maintain the quality desired. It should be checked at regular intervals and at least at any change in conditions such as weather or consistency. The part used for the air test must be discarded and must not be used for any other test. There are three procedures covered by ASTM, namely, Designation: C 138, Gravimetric Method; C 173, Volumetric Method; and C 231, Pressure Method.

The gravimetric method (C 138) can be made in conjunction with the yield test described here. It is calculated from the knowledge of mix proportions and specific gravities of all materials. The theoretical air-free weight can be used as a constant when the mix consistency remains the same. It is computed by the formula

$$\% \text{ air } = \frac{\text{theoretical unit weight} - \text{measured unit weight}}{\text{theoretical unit weight}} \times 100$$

It is generally not recommended for field control because of likely inaccuracies. An error of 2% in moisture content of aggregate can cause an error of 1% in the indicated air, and an error of 0.02 in specific gravity of the aggregate can cause an error of ½% in computed air content.

The volumetric method (C 173), commonly called the Roll-A-Meter method, is necessary for concrete made of lightweight aggregates, slag, and any other vesicular aggregates but can also be used for any type aggregate concrete. It utilizes the principle of direct determination of air by displacement in water. The procedure is as follows:

(a) Place concrete sample in the bowl in three equal layers and rod each layer in a manner similar to that described in the slump test.

(b) Tap the exterior surface of the measure 10 or 15 times after each rodding, or until no large bubbles of air appear.

(c) Strike off the surface flush using sawing motion with the straightedge and wipe the flange clean.

(d) Clamp the top section, add the water to the zero mark, and attach screw cap.

(e) Invert the unit and agitate until the concrete is free of the base, and rock and roll the unit until the air is displaced. Roll and rock the unit with neck elevated. Repeat this until there is no further drop in the water column.

(f) Add isopropyl alcohol to dispel foam, using the measuring cup which is furnished with the equipment.

(g) Make a direct reading to the bottom of the meniscus. The air content of the concrete is the reading plus 1% for each measureful of alcohol added.

The pressure method (C 231) is the commonest and most accurate for all concrete except lightweight, which requires the volumetric method described above. It utilizes the principle of Boyle's law to determine air content by the relationship of pressure to volume. The manufacturer of each meter provides detailed instructions for operation and calibration for variations in atmospheric pressure. ASTM Designation: C 231 provides a detailed procedure. Briefly the procedure, for concrete with aggregate of less than 2 in. nominal size, is (see Fig. 3):

(a) Place the sample in the bowl in three layers, tamp, tap, and strike off the top in similar manner to that given in the volumetric method.

(b) Thoroughly clean the flanges, clamp on the cover, and add water to the zero mark. Tap the sides after adding part of the water to remove entrapped air.

(c) Operate the valves in accordance with the manufacturer's instructions and apply desired test pressure to read the air content directly on the graduated gage glass. The air content is the difference in height at specified pressure and reading at zero pressure after release of pressure.

(d) Release pressure and read level at zero pressure.

(e) The aggregate correction factor may be significant and should be made in accordance with ASTMC-231 when the characteristics are not known. This is applied to the reading made on the concrete.

The "Washington meter," originally developed by the Washington State Highway Department, is a pressure meter that is commercially available. It determines air content by measuring the change in pressure of a known volume of air under an established pressure when released in contact with the concrete in a sealed container.

The Chace air meter is not covered by ASTM but actually utilizes the same principle as the volumetric method. Only the mortar portion of the concrete

Fig. 3.   Determination of entrained air by the pressure method.

is used. The meter consists of a small graduated glass tube and a small brass base. The mortar fraction of the concrete is placed in the base, and the air content is determined by displacement with isopropyl alcohol. Detailed instructions and mortar factors are given by the manufacturer. The Chace meter should be used only as a guide and in conjunction with a standard meter. It requires less time than any of the standard methods and is generally considered to be accurate to about ½% of air.

9   Strength Specimens

The ACI Building Code (ACI 318) recommends that a test consist of two specimens made for test at 28-day age. Most specifications require one test for each 150 cu yd, or fraction thereof, of each class of concrete placed in any one day. The samples of concrete should be taken in accordance with ASTM Designation: C 172 described earlier, and the specimens made and laboratory-cured in accordance with ASTM Designation: C 31 when the strength is determined as a basis for acceptance of the concrete. Additional test specimens cured entirely under field conditions may be required to check adequacy of curing and protection of the concrete for stripping strength, or for prestressing strength. A modified method of casting cores in place with special methods of extraction for the compression test has been used for the latter purpose.

Compression Specimens.   The method for casting cylinders is described in ASTM Designation: C 31, and the following procedure is a summary of the procedure when applied to the standard 6-in.-diameter by 12-in.-high cylinders.

1. The sample should be taken as described in Art. 6 and should preferably be taken to the point of casting.

2. The molds may be reusable steel or single-use coated cardboard or sheet metal, as covered by ASTM Designation: C 470.

3. Place the molds on a smooth, firm, and level surface.

4. Fill the molds in three equal layers and rod each layer uniformly 25 times with a ⅝-in. hemispherical-tipped rod. Rod the upper layers so that the rod just extends into the layer underneath about ½ in. When making two or more cylinders for one test, place and rod the bottom layer of all cylinders, then the second layer of all cylinders, and finally the third layer. Tap the sides of the molds after rodding each layer if necessary to close any voids.

5. The last layer should contain a slight excess of concrete. After rodding and tapping, strike off the excess concrete with a straightedge.

6. In the case of low-slump concrete (less than 3 in.), internal or external vibration may be used for the consolidation of concrete in the cylinder molds. The mold is filled in two layers with sufficient vibration to consolidate each layer without segregation. When internal vibration is used, the vibrating element, which should be 1½ in. in diameter or less, should be inserted at three different points for each layer and must not rest on the bottom or sides of the mold and, when vibrating the second layer, should extend about 1 in. into the first layer. When external vibration is used, the mold must be securely held against the vibrating element or surface. The excess concrete should be struck off in the same manner as in rodded specimens.

During the first 24 hr the specimens must be maintained at temperatures between 60 and 80°F, and specimens must be covered to prevent evaporation. Specimens formed in cardboard molds should not be stored for the first 24 hr in contact with wet sand or wet burlap or by other methods that would cause the cardboard

Fig. 4. Moist-curing room. Cabinets on wall at right are for curing cement test cubes (normal fog condition has been dissipated for photographic purposes).

mold to absorb water. They can be covered with polyethylene sheeting and then with wet burlap. Cylinders must not be moved or disturbed during the first 24 hr. Insulated storage boxes should be used to maintain the desired temperature, using a light bulb in cool weather and damp sand or wet burlap in hot weather. A minimum-maximum thermometer can be used to check the temperature inside the box. When strength tests are made for acceptance of concrete, the specimens must be removed from their molds at the age of 16 to 24 hr and stored in a moist room at a temperature of 73.4 ± 3.0°F, with free water on the surface at all times, but not exposed to running water, as shown in Fig. 4. If a fog room is not available, the cylinders can be immersed at the above temperature, in which case 'the water should be saturated with lime. Transportation to a quality laboratory will ensure that these curing conditions are maintained. Specimens should be adequately protected from breakage during transportation to the laboratory.

The cylinders should be properly tagged as they are made, and all the data concerning the concrete recorded to correlate with the cylinder marking. It is advisable to use metal tags attached with light wire to the specimens with the tags numbered consecutively. This numbering system provides for a check of all cylinders on a given project. Pertinent data recorded separately and correlated with the cylinder numbers should include project, mix proportions, date, slump, air content, temperature, location in the structure, ages to be tested, and identity of person making the cylinders.

The compressive-test specimens must have bottom and top surfaces that are plane within 0.002 in. and with each surface perpendicular to the axis within 0.5° (approximately 1/8 to 12 in.). Capping of both surfaces of single-use molds is usually necessary to achieve this, as shown in Fig. 5. The caps should conform to the requirements covered in detail in ASTM C 617:

1. They should be as thin as practical, 1/8 to 5/16 in., and made of material with a compressive strength greater than that of the concrete. Cap should not be off center on specimen by more than 1/16 in.

FIG. 5. Capping cylinders with sulfur compound showing capping stands and heating ovens.

2. Capping materials can be neat portland cement, high-alumina cement, high-strength gypsum plaster such as Hydrocal or Hydrostone, or sulfur compounds. There are a number of good proprietary compounds available. Ordinary plaster of paris should not be used.

3. The top surfaces of specimens cast in metal molds may be capped with a thin layer of neat portland-cement paste before the concrete has hardened after settlement has ceased.

4. Hardened specimens should be capped with any of the above materials, against plane surfaces to assure the tolerance requirements. In all cases the cap must be allowed to gain strength of 5,000 psi when tested. Sulfur compounds usually require 2 hr.

The method of testing specimens for compressive strength, as shown in Fig. 6, is given in ASTM Designation: C 39. Some of the pertinent points include:

1. Cylinders must be tested as soon as practical after removal from curing and while in a damp condition.

2. The diameter of the cylinders must be measured to the nearest 0.01 in. by averaging two diameters taken at right angles near the center of the specimen.

3. The specimen must be aligned in the center of the top bearing block of the testing machine. The bearing faces of the blocks must be plane to within 0.001 in. and the spherically seated bearing block must be on top and must have a face diameter at least slightly larger than the cylinder. The bottom block must be at least that large but may be larger. The top bearing block must be able to rotate freely and to tilt slightly in any direction. Loading must be continuous at specified rates (34,000 to 85,000 lb per min for 6×12-in. cylinders on hydraulic machines, or at 0.05 in. per min on screw-type machines). Hand-operated machines do not comply with ASTM C 39.

FIG. 6. Compression test of concrete cylinder.

4. The report of compression-test results must give the maximum load and the unit strength calculated to the nearest 10 psi. It should also include information mentioned above, including age at test, specified strength, and any abnormalities observed. Some specifications require the report to include type of break, i.e., "cone," "shear," or other. There is very little evidence that this information has any appreciable significance.

Splitting Tensile Strength. The splitting tensile strength of concrete cylinders is considered to be an accurate determination of the true tensile strength of concrete. The American Concrete Institute Standard 318-71, Building Code Requirements for Reinforced Concrete, permits use of actual splitting tensile strength in the computation of allowable stresses when structural lightweight-aggregate concrete is used.

The method for determination of this splitting tensile strength is given in ASTM Designation: C 496. Some of the pertinent points include:

1. The test specimens conform to the size, molding, and curing requirements given here for the compressive-strength specimens. When used for evaluation of lightweight concrete in accordance with American Concrete Institute Standard 318-71, six specimens tested at 28 days shall be in an air-dry condition after 7 days of moist curing followed by 21 days of drying at $73 \pm 3°F$ and $50 \pm 5\%$ relative humidity, as given in ASTM C 330.

2. The testing machine shall conform to the requirements given here for the compressive-strength test of concrete cylinders. A supplementary bearing bar is required when the lower bearing block or upper bearing face is less than the length of the cylinder to be tested. The bar shall be machined to within ± 0.001 in. of planeness. The width shall be at least 2 in., and the thickness not less than the distance from the edge of the spherical or rectangular bearing block to the end of the cylinder. The bar shall be used in such a manner that the load will be applied over the entire length of the cylinder.

3. Two bearing strips of nominal ⅛-in.-thick plywood, approximately 1 in. wide and of length equal to or slightly larger than the cylinder, shall be provided adjacent to the top and bottom of the specimen during testing. Bearing strips shall not be reused.

4. Prior to positioning in the testing machine diametral lines shall be drawn on each end of the specimen using a suitable apparatus that will ensure that they are in the same axial plane.

5. Measurements shall be made of the diameter of the test specimen to the nearest 0.01 in. by properly averaging three diameters, and measurements shall be made of the length of the specimen to the nearest 0.1 in. by proper averaging of two lengths.

6. Positioning of the specimen, bearing strips, and bearing bar shall be made so as to ensure that the projection of the plane of the two diametral lines on the cylinder intersects the center of the upper bearing bar and bearing strips, and is directly beneath the center of thrust of the spherical bearing blocks.

7. The rate of loading shall be continuous without shock at a constant rate within the range of 11,300 to 22,600 lb per min for a 6 × 12-in. cylinder.

8. The calculation of splitting tensile strength of the specimen shall be as follows:

$$T = \frac{2P}{\pi l d}$$

where T = splitting tensile strength, psi
P = maximum applied load indicated by the testing machine, lb
l = length, in.
d = diameter, in.

9. The report of splitting-tensile-strength test must give the strength calculated to the nearest 5 psi. It should also include identification number; diameter and length, in inches; maximum load, in pounds; estimated proportion of coarse aggregate fractured during test; age of specimen; curing history; defects in specimen; and type of fracture.

Flexural Specimens. The same ASTM Designation: C 31 covers making and curing flexure-test specimens. The procedures are briefly as follows:

1. Beam molds must be rigid, watertight, and nonabsorptive. The size of beam for concrete containing aggregates up to a nominal size of 2 in. is 6 × 6 × 20 in.

2. The beam molds should be filled in two equal layers and consolidated by rodding, for slumps of 3 in. or more. They may be consolidated by vibration or rodding for slump between 1 and 3 in. Vibration must be used for slump less than 1 in.

3. Rodding frequency for each layer is one stroke for each 2 sq in. of surface. After rodding, the sides and ends are spaded with a suitable tool such as a mason's trowel. Tapping the sides may be used to close voids if necessary.

4. When vibration is used, the vibrator is applied uniformly to assure consolidation without segregation. Internal vibrator is spaced at intervals not exceeding 6 in.

5. After consolidation, the top surface is struck off with a straightedge and finished with a wood or magnesium float. Overfinishing must be avoided and proper identification made.

6. Loss of moisture must be prevented by covering the beams with a plate, plastic, or double layer of wet burlap. All other conditions as described for making compressive-strength specimens must be followed.

7. For beams for checking a laboratory mix or for basis of acceptance, remove specimens from mold between 20 and 48 hr after molding and cure the same as cylinders except that a storage for a minimum of 20 hr in saturated lime water at 73.4° ± 3 prior to testing is required.

ASTM Designation: C 78 covers the procedure for making flexural tests. The following general requirements are necessary:

1. Loading is on the third points of the beam at a span of three times the depth of the beam. Hence, for 6 × 6-in. beams, this would be 18-in. span with the points of load application 3 in. each side of the center.

2. The loading device must be able to apply force to the beam vertically without eccentricity, and with the load and reactions remaining parallel.

3. Specimens must be soaked in water for at least 40 hr before testing. It is absolutely essential that the specimens be kept moist while testing. Uniformity of moisture conditions throughout is essential. One dry side may affect results drastically.

4. Beams must be loaded on their sides as cast. If the load points are not in full contact, capping or leather shims must be used. Loading rate for a 6 × 6-in. beam is 1,800 lb per min. Calculations are based on measurements of the dimensions at place of failure to the nearest 0.1 in.

5. Flexural strength is calculated as modulus of rupture, when fracture occurs within the middle third of the span length, by the formula

$$R = \frac{Pl}{bd^2}$$

where P = maximum applied load
l = span length, in.
b = average width of specimen, in.
d = average depth of specimen, in.

Breaks outside the middle third of span length should be disregarded.

10   Unit Weight and Yield

To assure conformance to specifications, field tests are made to determine yield and cement factor of the concrete being placed. ASTM Designation: C 138 covers this method by utilizing the unit weight of the plastic concrete. The method, generally summarized, is as follows:

1. Equipment required is a scale accurate to 0.3% of the test load, standard tamping rod, strike-off plate, and a measure of proper capacity. The size depends upon the size of aggregate in the concrete. Use the 1 cu ft measure when the concrete contains aggregate larger than 2 in. A ½ cu ft measure is usually used for aggregates up to 2 in. nominal size. Figure 7 shows the necessary equipment.

2. Fill the measure in three equal layers and rod 25 or 50 strokes per layer for the ½ cu ft and 1 cu ft measures, respectively. The method of rodding is similar to that described for the air-content test and for casting concrete cylinders. For measures of 0.4 cu ft or greater, consolidation may be performed by internal vibration in the same manner as for cylinders.

3. After consolidation, strike off the top carefully with a rigid flat plate large enough to cover the measure, by a sawing action, while maintaining contact with the edges of the measure at all times. Striking off with a straightedge is not satisfactory as it usually yields falsely high results.

Fig. 7.   Equipment for unit-weight test of concrete or aggregates.

4. Weigh the measure to the scale accuracy and calculate the weight per cubic foot.

5. Total the weight of all the ingredients in the batch and divide it by the unit weight as determined above. This results in the calculated cubic feet of concrete which, when divided by 27, gives the actual number of cubic yards per batch.

6. Determine the cement factor by dividing the total quantity of cement in the batch by the number of cubic yards.

7. Determine yield in cubic feet of concrete per bag of cement by dividing 27 cu ft per cu yd by cement factor determined above.

8. A practical consideration is the yield in actual cubic feet per cubic yard ordered. This is determined by dividing the number of cubic feet of concrete by the number of cubic yards ordered. Any amount under 27 cu ft would indicate a low yield; anything over 27 cu ft, a high yield. A reasonable tolerance would be ±1%.

Example.   Given: The 6-yd-batch weights of the first example in Chap. 11.

$$
\begin{aligned}
\text{Cement} &= 3{,}156 \text{ lb} \\
\text{Sand, including 5\% moisture} &= 7{,}056 \text{ lb} \\
\text{Coarse aggregate} &= 12{,}060 \text{ lb} \\
\text{Mixing water added} &= 1{,}230 \text{ lb} \\
\hline
\text{Total} &= 23{,}502 \text{ lb}
\end{aligned}
$$

Weight per cubic foot from above test is 145.9 pcf

$$
\text{Volume of batch, cu ft} = \frac{23{,}502}{145.9} \qquad = 161.08
$$

$$
\text{Volume of batch, cu yd} = \frac{161.08}{27} \qquad = 5.97
$$

$$
\text{Cement factor, lb per cu yd} = \frac{3{,}156}{5.97} \qquad = 529
$$

$$
\text{Cement factor, bags per cu yd} = \frac{529}{94} \qquad = 5.63
$$

$$
\text{Yield, cu ft per bag} = \frac{27}{5.63} \qquad = 4.80
$$

$$
\text{Relative yield, cu ft per cu yd designed} = \frac{161.08}{6} \qquad = 26.85
$$

$$
\text{Variations, \%} = \frac{-0.15}{27} \times 100 = 0.56 \text{ shortage}
$$

11 Unit Weight of Structural Lightweight Concrete

Density of the lightweight concrete at the age of 28 days for design control is determined by ASTM Method C 567, Unit Weight of Structural Lightweight Concrete. The unit weight of the freshly mixed concrete is determined in accordance with ASTM C 138, Unit Weight of Concrete, except that vibration of specimens as described in ASTM C 192, Making and Curing Concrete Test Specimens in the Laboratory, shall be permitted. This is used for placement control.

The air-dry unit weight is determined on 6×12 in. cylinders, generally in the following manner. The cylinders are made in acordance with ASTM C 192.

The cylinders are covered and cured at 60° to 80°F in a wrapped or sealed condition for 7 days. On the seventh day, the cylinders are removed from cylinder molds or plastic covering and dried for 21 days in $50 \pm 2\%$ relative humidity. Determine the weight of the cylinders at 28-day age as dried; then completely immerse the cylinders in water for 24 hr. Determine the immersed weight and the saturated surface dry weight. Calculate the weight per cubic foot of concrete in accordance with the following equation:

$$\text{Wt/cu ft} = \frac{62.3A}{B - C}$$

where A = 28-day weight of concrete cylinder, as dried, lb
 B = saturated, surface, dry weight of cylinder, lb
 C = immersed weight of cylinder, lb

12 Uniformity Tests of Truck Mixers

ASTM Designation: C 94, Specifications for Ready Mixed Concrete, stipulates that when the concrete is mixed completely in a truck mixer, 70 to 100 revolutions at manufacturer's specified mixing speed, it is necessary to produce the uniformity of concrete indicated in Table 1. Concrete uniformity tests may be made to determine if the uniformity is satisfactory and at least five of the six tests shown in Table 1 must be within the specified limits. Slump tests of individual samples taken after discharge of approximately 15% and 85% of the load may be made for a quick check of the probable degree of uniformity. These two samples shall be obtained with an elapsed time of not more than 15 min. If these two slumps differ more than that specified in Table 1, the mixer shall not be used unless the condition is corrected or when operation with a longer mixing time, a smaller load, or more efficient charging sequence will permit the requirements of Table 1 to be met.

13 Uniformity Tests of Stationary Mixers

ASTM Designation: C 94 specifies mixing time of 1 min for 1 cu yd mixers and an additional 15 sec for each cubic yard, or fraction thereof, of additional capacity.

It is considered that these minimum time requirements will produce uniformity, but they are known to be unnecessarily time-consuming for many of the large-capacity mixers. Therefore, provisions have been incorporated in the specification to permit reduction in this mixing time, provided tests are conducted to show that uniformity can be achieved in less time.

A 6 cu yd mixer under the time restriction alone would be required to mix each full batch of concrete 2¼ min. If the owner has reason to believe that uniformity can be achieved in less time he may have tests made after any predecided time of mixing, and if samples taken from near the two ends of the batch meet specified uniformity requirements, the mixing time can be reduced accordingly. The uniformity requirements are shown in Table 1.

Table 1.   Requirements for Uniformity of Concrete*

Test	Requirement, expressed as maximum permissible difference in results of tests of samples taken from two locations in the concrete batch
Weight per cubic foot (weight per cubic meter) calculated to an air-free basis...	1.0 lb (16 kg/m³)
Air content, % by volume of concrete.......................	1.0
Slump:	
If average slump is 4 in. (10.2 cm) or less..................	1.0 in. (2.5 cm)
If average slump is 4 to 6 in. (10.2 to 15.2 cm).............	1.5 in. (3.8 cm)
Coarse aggregate content, portion by weight of each sample retained on No. 4 (4.76 mm) sieve, %......................	6.0
Unit weight of air-free mortar† based on average for all comparative samples tested, %.................................	1.6
Average compressive strength at 7 days for each sample,‡ based on average strength of all comparative test specimens, %....	7.5§

* ASTM C 94–71, Table A1.

† Test for Variability of Constituents in Concrete, Designation 26, p. 565, U.S. Bureau of Reclamation, 1963. "Concrete Manual," 7th ed.   Available from Superintendent of Documents, U.S. Government.

‡ Not less than 3 cylinders will be molded and tested from each of the samples.

§ Tentative approval of the mixer may be granted pending results of the 7-day compressive-strength tests.

14   Method of Test for Coarse-aggregate Content

One measure of uniformity of a concrete batch is the percentage of coarse aggregate in two different portions of the batch (see Table 1). The procedure for this determination is:

(a) Weigh a sample of plastic concrete. In order to combine this determination partially with that for the air-free unit weight of mortar, the base of an air meter (¼ cu ft capacity for nominal-size aggregates up to 1½ in. or ½ cu ft capacity for aggregate up to 3 in.) is used for unit-weight determination. The sample is then tested for air content.

(b) After the air test, wash and sieve over a No. 4 screen.

(c) Weigh the aggregate while immersed in water.

(d) Compute the saturated surface-dry weight using the known bulk saturated surface-dry specific gravity G of the aggregate by use of the formula given in ASTM Designation: C 127, Specific Gravity and Absorption of Coarse Aggregate.

$$\text{SSD weight of coarse aggregate} = \frac{\text{immersed weight} \times G}{G - 1}$$

W = coarse-aggregate content (% by weight of sample)

$$= \frac{\text{SSD weight of coarse aggregate}}{\text{weight of sample of concrete}} \times 100$$

(e) In lieu of the immersed-weight method, the washed coarse aggregate may be towel-dried and weighed. However, the immersed-weight method is recommended.

15   Method of Test for Air-free Unit Weight of Mortar

The method for air-free unit weight of mortar is described under Variability of Constituents in Concrete, Designation 26, p. 565, in the "Concrete Manual."[2]   The procedure is summarized as follows:

(a) Perform the weight tests as described in Art. 13 (a) through (d).

(b) Determine the weight of the mortar in the sample by subtracting the SSD weight of the coarse aggregate from the weight of the concrete sample.

(c) The volume of the air-free mortar is obtained by subtracting the volumes of the coarse aggregate and volumes of air.

(d) The air-free unit weight of mortar is its weight divided by volume and calculated by the following formula:

$$M = \frac{b - c}{V - \left(A + \dfrac{c}{G \times 62.3}\right)}$$

where M = unit weight of air-free mortar, pcf

b = weight of concrete sample in air meter, lb

c = SSD weight of aggregate retained on No. 4 sieve, lb

V = volume of sample, cu ft

A = volume of air computed by multiplying the volume of container V by % of air divided by 100

G = SSD specific gravity of coarse aggregate

(e) The following example is given to illustrate the method of unit weight of air-free mortar determinations and coarse-aggregate content test as described above.

Given: Concrete sample containing 1½ in. maximum nominal size aggregate.

G = bulk specific gravity of coarse aggregate SSD = 2.65
b = weight of concrete sample in ¼ cu ft V measure of air meter = 35.0 lb
Air content on this sample =   5.0%
Submerged weight of aggregate retained on No. 4 sieve = 12.0 lb

Calculation:

c = SSD weight of coarse aggregate = $12.0 \times \dfrac{2.65}{2.65 - 1}$ = 19.27

W = coarse-aggregate content (% by weight of sample), $\dfrac{19.27}{35} \times 100 = 55.0\%$

A = volume of air = $\dfrac{0.25 \text{ cu ft} \times 5.0}{100}$ = 0.0125 cu ft

M = unit weight of air-free mortar = $\dfrac{35 - 19.27}{0.25 - \left(0.0125 + \dfrac{19.27}{2.65 \times 62.3}\right)}$ = 130.22 pcf

In order to determine compliance of a truck or stationary mixer, one must first make identical determinations on each of two samples, one from near each end of the batch. Following the determination and calculations of the second sample, conducted as above, let us assume that the results of this second sample were 60% coarse-aggregate content and 132 pcf unit weight of air-free mortar. Then we compare results as follows:

Test	Front	Back	Variation	Specification requirements expressed as maximum permissible difference between samples, ASTM C 94
Coarse-aggregate content, % of each sample retained on No. 4 sieve....	55	60	5.0	6.0
Unit weight of air-free mortar.......	130.2	132.0	0.7%, 1.8 pcf	1.6%*

* Percent variation based on average of comparative samples calculated as follows:

$$\text{Average unit weight of air-free mortar} = \frac{130.2 + 132.0}{2} = 131.1 \text{ pcf}$$

$$\text{Variation in unit weight from average} = \frac{131.1 - 130.2}{131.1} \times 100 = 0.7\%$$

Table 2. Example of Computation for Unit Weight Air-free Mortar

Plant B.J.A.   Project FBA   Mixer JEA   Date 8–12–65
Inspector R.J.A.   Time of test 1:45 P.M.   Slump 2 in.
Mixing time 1 min   Batch No. 77   Mix No. 7
Maximum size of aggregate 1½ in.   Specific gravity of coarse aggregate 2.65

	Sample from front of mixer or first portion of batch as discharged from mixer		Sample from back of mixer or last portion of batch as discharged from mixer	
	Weight, lb	Volume, cu ft	Weight, lb	Volume, cu ft
Weight and volume of sample in air meter........	35.00	0.2500	35.16	0.2500
Air content by air meter...	5.00%		4.75%	
Volume of air............		0.0125		0.0119
Weight and air-free volume of sample..............	35.00	0.2375	35.16	0.2381
Submerged weight of sample retained on No. 4 screen..	12.00		11.50	
Computed SSD weight and solid volume of plus 4 material*..............	19.27	0.1170	18.49	0.1120
Weight and volume representing mortar in sample	15.73	0.1205	16.67	0.1261
Computed unit weight of air-free mortar, pcf......	130.22		132.00	

NOTE: Comparison of unit weight of air-free mortar for compliance with the given specification is the same as shown previously when computed by the formula.

$$* \text{ SSD weight of coarse aggregate} = \frac{\text{immersed weight} \times G}{G - 1}.$$

Methods of comparison of other tests as given in Table 1 are self-evident. Test results conform to uniformity requirements of the specifications.

The above example illustrates the method of uniformity determinations by use of the given formulas. A tabular method adapted from the U.S. Bureau of Reclamation "Concrete Manual" Designation 26 may be preferred. An example of this determination of unit weight of air-free mortar is shown in Table 2.

16   Determination of Modulus of Elasticity

The modulus of elasticity is the ratio of stress to strain. Unlike most metals, concrete has no true "straight-line" portion of a stress-strain curve, although for most modern concretes the lower portion of the curve is very nearly straight. Since there is no true straight-line portion, there can be no true proportional limit and the modulus will be different at any selected stress point.

FIG. 8.   Modulus of elasticity test setup for determination of static Young's modulus.

Three methods of expressing and calculating the modulus of elasticity have been proposed. These are known as "chord modulus," "secant modulus," and "tangent modulus." "Chord modulus" is probably the one most used.

ASTM Designation: C 469 covers the procedure for this determination. The method generally consists of measuring the strain at various loads to establish data for a stress-strain curve. This is accomplished by clamping two metal rings on a 6×12-in. cylinder, one near the top and one near the bottom, not over 8 in. apart, as shown in Fig. 8.

The movement of the rings is measured by a dial gage and the strain is calculated by dividing the total measured movement by the gage length. Electric strain gages of at least 6-in. lengths are sometimes used in lieu of rings and a dial gage for the purpose of measuring the strain.

The chord modulus is the slope of a straight line originating at the point on the curve corresponding to 50 μin. per in. strain and extending to the stress point selected. ASTM C 469 recommends that the chord modulus be calculated at

40% of the ultimate strength. Figure 9 illustrates determination of chord modulus at 1,600 psi.

The secant modulus and the tangent modulus are sometimes denoted in engineering texts. The secant modulus is similar to the chord modulus except that it is the slope of a line from the origin to any selected point on the curve. The tangent modulus is the slope of a line tangent to any selected point on the curve.

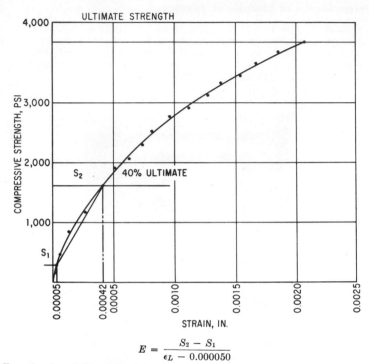

$$E = \frac{S_2 - S_1}{\epsilon_L - 0.000050}$$

where E = chord modulus of elasticity, psi
S_2 = stress corresponding to maximum applied load, psi
S_1 = stress corresponding to a longitudinal strain of 50 μin. per in., psi
ϵ_L = longitudinal strain produced by stress S_2

$$E = \frac{1,600 - 300}{0.00042 - 0.00005}$$

$$E = 3,513,500$$

Fig. 9.   Typical stress-strain curve for calculation of modulus of elasticity.

The modulus of elasticity of concrete will usually increase with age similar to strength development. Although modulus of elasticity is not directly proportional to strength, concretes of higher strengths have higher moduli of elasticity. This is recognized by the ACI Building Code. The determination of modulus of elasticity is not necessary for normal concrete testing but is sometimes required for special study.

The modulus of elasticity can also be calculated at any desired stress, although it is usually at 40% of the ultimate strength, using the conventional formula

$$E = \frac{\text{unit stress}}{\text{unit strain}}$$

17 Determination of Specific Gravity Factor of Structural Lightweight Aggregate

This specific gravity factor, as given in ACI 211.2-69, is the relationship of weight to displaced volume for lightweight aggregates and is determined either dry or moist for use in making adjustments in proportions of lightweight concrete. The basic apparatus is a pycnometer consisting of a narrow-mouth 2-qt mason jar with a spun brass pycnometer top. Two representative samples are obtained for each size of lightweight aggregate to be tested. The first is weighed, and then dried in an oven at 105° to 110°C or on a hot plate to constant weight.

The dry aggregate weight is recorded, and the aggregate moisture content (percent of aggregate dry weight) is calculated. This value is used to compute the specific gravity factor and moisture content curve shown in Fig. 3, Chap. 11. The second aggregate sample is weighed (weight C in grams). The sample is then placed in the empty pycnometer; it should occupy one-half to two-thirds of the pycnometer volume.

Water is added until the jar is three-fourths full, and the entrapped air is removed by rolling and shaking the jar. After all the air is dispelled, the jar is filled to full capacity and weighed. The pycnometer specific gravity factor is calculated by the formula

$$S = \frac{C}{C + B - A}$$

where A = weight of pycnometer charged with aggregate and then filled with water, grams
B = weight of pycnometer filled with water, grams
C = weight of aggregate tested, moist or dry, grams

The larger test samples of coarse aggregate that cannot be evaluated in the pycnometer can be tested for specific gravity factor by the equivalent weight in air and water procedure of ASTM C 127. The top of the container used for weighing the aggregates under water must be closed with a screen to prevent light particles from floating.

Specific gravity factors by this method are calculated by the formula

$$\text{Specific gravity factor } S = \frac{C}{C - E}$$

where C = weight in air at particular moisture content, grams
E = weight of coarse aggregate sample under water, grams
S = specific gravity factor at particular moisture content

The specific gravity determinations are made at various periods of immersion dependent upon conditions contemplated in the field.

18 Durability

The usually accepted method of evaluating durability is by freezing and thawing in accordance with ASTM Designation: C 666, Test of Resistance of Concrete to Freezing and Thawing, Procedure A, Rapid Freezing and Thawing in Water, and Procedure B, Rapid Freezing in Air and Thawing in Water, using equipment similar to that depicted in Fig. 10. These methods will not yield an accurate correlation with months or years of natural exposure, but are valuable for comparison purposes. Concrete specimens of specific size are cast and cured for a specified time and method. These are then placed in freezing and thawing apparatus equipped as necessary with refrigerating units, heating units, fans, and water pumps to produce reproducible cycles with specified temperature and time requirements. This is usually attained by use of automated equipment utilizing electronically

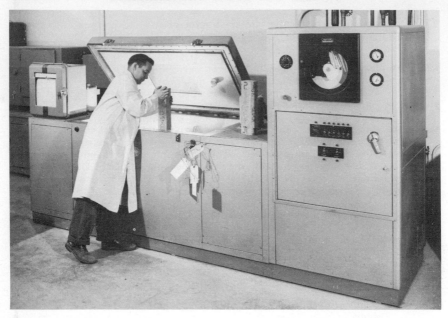

Fig. 10. Automatic freeze-and-thaw unit, 12 cycles per day. (Recorder on left is for monitoring temperatures in bar with thermocouple wires shown in front. Recorder in cabinet is part of control mechanism and records cycles and cabinet temperatures.)

programmed cycling apparatus. Approximately 8 cycles per day can be achieved in many of the units. Fundamental transverse frequency determinations are made periodically on each specimen, and from these the relative dynamic modulus of elasticity is calculated after 300 cycles, or until the relative dynamic modulus of elasticity of the specimen drops to 60% of the initial modulus. The durability factor is then computed for each specimen.

19   Cores from Hardened Concrete

The compressive strength of concrete is sometimes determined from cores extracted from a structure. This is usually performed when the compressive strength of a given concrete is not known or because of questionable concrete cylinder test results. The cores are secured by use of a core drill normally utilizing diamond drill bits of the type pictured in Fig. 11. Care must be exercised not to cause overheating and to avoid obvious defects such as rock pockets and joints. When the core is used for compressive-strength determinations, the length should preferably be as near as possible to twice the diameter, and the diameter should be at least three times the maximum size of aggregate in the concrete. Cores having lengths less than the diameter are unsuitable for compression testing. The ends should be sawed if necessary to produce an even surface, and the core should be ground or capped to enable proper compression test. Moisture condition at time of test will affect the results. Concrete tested while dry will have a higher indicated compressive strength than concrete tested in the saturated condition. For many years the only recognized moisture condition was that obtained by immersion in lime-saturated water for at least 40 hr immediately prior to testing. However, a note has now been added to the standard method of testing cores to the effect that other moisture conditions may be used at the option of the agency for

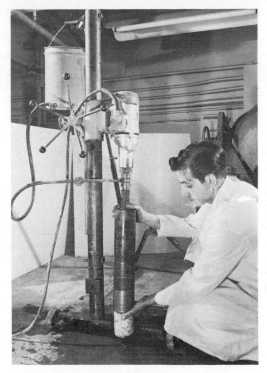

Fig. 11. Core drilling inside building with portable equipment and using diamond bit.

which the test is being conducted. The details of this test are covered in ASTM C 42. The correlation between indicated core strengths and cylinder strengths of the same concrete is the subject of much controversy. Some excellent investigators have found excellent agreement; others have found core strengths higher; others have found them lower.[3,4,5]

20 Impact Hammer

An approximate or relative compressive-strength determination can be made by the impact hammer, sometimes called a "Swiss" hammer. This hammer utilizes a spring-driven plunger, and the rebound of the plunger after impact with the concrete surface is recorded on a built-in scale. The instrument carries a chart which correlates the rebound number to compressive strength. This correlation should be accepted with caution and skepticism. It is recommended that determinations made with this instrument be used exclusively for comparative purposes on concrete of the same composition, age, and moisture content, and that judgment be used in the interpretation. This is critical because the rebound is affected by many factors such as moisture content of the concrete, type of surface finish, type of aggregate, and age of the concrete. Although there are many data purporting to show the accuracy of the instrument, data shown in Table 3 taken from several projects, comparing impact-hammer results with immediately adjacent cores, show the potential fallibility of accepting the correlation given on the instrument.

It is not intended to suggest that the extreme lack of correlation shown in Table 3 is typical. These are unusual cases, but they do show the danger of accepting impact-hammer test results with full confidence.

Table 3.   Comparison of Core Strengths per ASTM C 42 and Indicated Strengths as Determined by Impact Hammer

Core No.	Strengths as determined on cores per ASTM C 42, psi	Indicated strength of immediately adjacent concrete based on impact-hammer tests, psi	Core No.	Strengths as determined on cores per ASTM C 42, psi	Indicated strength of immediately adjacent concrete based on impact-hammer tests, psi
W 1	3,400	4,000	1	3,890	6,500
W 2	3,150	4,200	2	3,840	6,800
W 3	2,700	3,750	3	5,040	7,400
S 4	2,480	4,400	4	4,650	8,000
S 5	3,450	3,250	5	4,630	7,400
E 6	2,190	3,000	6	4,490	7,600
E 7	2,280	4,200			
E 8	2,200	3,250			
E 9	2,340	3,250			
N 10	3,130	3,250			

Each impact-hammer "indicated strength" is based on the average of 10 most uniform of 15 readings and on the curve that accompanies the hammer.

Probably in the majority of cases, results taken from the curve which is attached to the hammer are reasonably accurate, but there are enough wide variations from this relation to introduce doubt into any specific investigation.

If the hammer is calibrated against actually determined strengths of a given mix of concrete for a given project, under approximately comparable moisture conditions and approximately uniform age, it can be an excellent tool. It is of great value—when used with proper understanding and discretion—for checking a specific project to investigate and isolate an area of suspected low strength.

21   Penetration Probe Gage

A proprietary method for the nondestructive determination of compressive strength of concrete in the field is known as the Windsor probe and utilizes the principle that the penetration of the probe gage is inversely proportional to the compressive strength of the concrete being tested. Generally, three probe gages are driven into the concrete using a powder-actuated device of a specific concrete muzzle energy-producing force (Fig. 12).

The compressive strength is determined by correlation of the kinetic energy to compressive strength. An approximate relationship of probe penetration to compressive strength is given in a table furnished by the manufacturer. The moh hardness rating (scratch test) of the coarse aggregate is required for use of the table. The average of three probe values (exposed height of probe gage) is used with the table.

Intensive study of this method was made in the laboratories of the National Ready Mix Concrete Association. The conclusions of this study are that this probe method is comparable to the concrete test hammer in accuracy. Therefore this test should not be used as an alternate to ASTM C-39, Test of Compressive Strength of Molded Cylinders, but is suitable for in-place, nondestructive approximation of strength.

22   Other Tests

Other vital tests on concrete include volume change, bond to steel, bleeding, and setting time. These are usually performed in the laboratory under applicable

Fig. 12. The Windsor probe is based on the principle that the penetration of a powder-driven probe into the concrete is inversely proportional to the strength of the concrete.

ASTM designations and primarily as qualification tests. Structural load tests are usually performed in accordance with the ACI Building Code Specifications and are generally performed when there is a question of the concrete strength in a specific structure.

Pulse-velocity tests can be used on structures to detect progressive deterioration, concealed cracking, honeycombing, and other properties. Some investigators have attempted to determine compressive strengths by pulse-velocity determinations. Few, if any, authentic data are now available indicating that this can be done with any degree of accuracy. Other things being equal, strong concrete will show a higher pulse velocity than weak concrete; but that is about as far as the correlation can be extended on the basis of present published information.

ANALYSIS OF STRENGTH RESULTS

23 Terminology

The discussion of analysis of strength results involves the mathematics of statistical analysis. Some of the terms and their customary symbols used may be somewhat unfamiliar to many engineers, scientists, and technicians. We are threfore providing a series of definitions of these terms with the symbols customarily used in connection with them:

$\bar{X}$ average strength—the numerical average of all strengths under consideration
    deviation—the numerical difference between any given strength and the average strength

σ standard deviation—the rms deviation of the strengths from their average. The standard deviation is found by extracting the square root of the average of the square of deviation of individual strengths from their average; thus

$$\sigma = \sqrt{\frac{(X_1 - \bar{X})^2 + (X_2 - \bar{X})^2 + \cdots + (X_n - \bar{X})^2}{n}}$$

n total number of tests
V coefficient of variation—the standard deviation expressed as a percentage of the average strength

$$V = \frac{\sigma}{\bar{X}} \times 100$$

R range of strengths—the numerical difference between the highest and lowest strengths of tests under consideration
f_c' specified strength—self-explanatory
f_{cr} required average strength—the average strength, determined mathematically, that will be required to produce not more than a specified number of test results below the specified strength of f_c'

$$f_{cr} = \frac{f_c'}{1 - tV}$$

where t = a constant dependent on proportion of tests permitted below f_c' and on number of tests used to establish V (see Table 4)

σ_1 within-test standard deviation—standard deviation between individual specimens of a given test, preferably calculated on at least 10 sets of specimens

$$\sigma_1 = \frac{1}{d_2} \bar{R}$$

where $\dfrac{1}{d_2}$ = a constant depending on number of samples (see Table 5)

$\bar{R}$ average range—the average range R of the sets of specimens used to determine σ_1
V_1 within-test coefficient of variation—the variation between results of cylinders in a given batch expressed as the coefficient (see V above)

$$V_1 = \frac{\sigma_1}{\bar{X}} \times 100$$

Test: A group of specimens (usually two or three) all made from the same sample and all tested at the same age.

24   Quality-control Requirements

The control of concrete quality encompasses the wide spectrum of conformance of each component of concrete to a specification, as well as to the mixing, placing, and curing techniques. Generally, the efficiency of concrete manufacture, testing, and material control will determine the degree of uniformity and quality of concrete. In order to determine the degree of concrete uniformity and quality, it is necessary to apply statistical methods of analysis. The American Concrete Institute Recommended Practice for Evaluation of Compression Test Results of Field Concrete[6] was developed for this purpose. In order to develop this type of study, a sufficient number of field compression tests must be made in accordance with the acceptable standards. Concrete strength-test results will fall in the normal frequency-distribution curve as shown in Fig. 13. The primary conclusion from this analysis is the fact that variations will occur and an absolute minimum compressive-strength specification is unrealistic.

Table 4. Values of t*

No. of samples minus 1†	Percentage of tests falling within the limits $\bar{X} \pm t\sigma$							
	50	60	70	80	90	95	98	99
	Chances of falling below lower limit							
	2.5 in 10	2 in 10	1.5 in 10	1 in 10	1 in 20	1 in 40	1 in 100	1 in 200
1	1.000	1.376	1.963	3.078	6.314	12.706	31.821	63.657
2	0.816	1.061	1.386	1.886	2.920	4.303	6.965	9.925
3	0.765	0.978	1.250	1.638	2.353	3.182	4.541	5.841
4	0.741	0.941	1.190	1.533	2.132	2.776	3.747	4.604
5	0.727	0.920	1.156	1.476	2.015	2.571	3.365	4.032
6	0.718	0.906	1.134	1.440	1.943	2.447	3.143	3.707
7	0.711	0.896	1.119	1.415	1.895	2.365	2.998	3.499
8	0.706	0.889	1.108	1.397	1.860	2.306	2.896	3.355
9	0.703	0.883	1.100	1.383	1.833	2.262	2.821	3.250
10	0.700	0.879	1.093	1.372	1.812	2.228	2.764	3.169
15	0.691	0.866	1.074	1.341	1.753	2.131	2.602	2.947
20	0.687	0.860	1.064	1.325	1.725	2.086	2.528	2.845
25	0.684	0.856	1.058	1.316	1.708	2.060	2.485	2.787
30	0.683	0.854	1.055	1.310	1.697	2.042	2.457	2.750
∞	0.674	0.842	1.036	1.282	1.645	1.960	2.326	2.576

* From ACI Standard 214-65, Table 4. Values of t extracted from table originally produced by Fisher and Yates, "Statistical Tables for Biological Agriculture and Medical Research."
† Degrees of freedom.

Other values of t for $n - 1 = \infty$

Percentage within $\bar{X} \pm t\sigma$	Chances of falling below lower limit	
40	3 in 10	0.524
68.27	1 in 6.3	1.000
95.45	1 in 44	2.000
99.73	1 in 741	3.000

Values of t increase for small samples because of the unreliability of small samples to establish a true estimate of σ. The advantage of establishing V from a large number of tests becomes apparent in the reduction of t and f_{cr}.

It can be noted that the law of probability indicates that one strength out of every six tests will be more than standard deviation σ below the average; one out of every 44 will be more than twice the standard deviation below average, and one out of every 741 will be more than thrice the standard deviation below average.

25   American Concrete Institute Building Code Strength Requirements

The American Concrete Institute Building Code criteria for evaluation of concrete are based on the condition that the average strength of concrete produced must always exceed the specified value of f_c' that was used in the structural design phase.

Table 5. Factors for Computing Within-test Standard Deviation*

No. of specimens	d_2	$1/d_2$
2	1.128	0.8865
3	1.693	0.5907
4	2.059	0.4857
5	2.326	0.4299
6	2.534	0.3946
7	2.704	0.3698
8	2.847	0.3512
9	2.970	0.3367
10	3.078	0.3249

* From ACI Standard 214-65, Table 3.

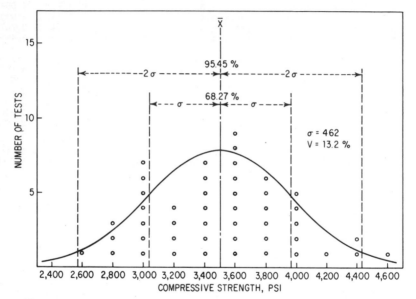

Fig. 13. Normal frequency distribution of strength data from 46 tests.[6]

The probability concept is used to provide for single test values that are lower than f_c'. It specifies, "The concrete strength is considered to be satisfactory as long as averages of any three consecutive tests remain above the specified f_c' and no individual test falls below the specified f_c' by more than 500 psi."

The ACI Building Code requires that samples for strength tests of each class of concrete shall be taken not less than once a day nor less than once for each 150 cu yd of concrete or for each 5,000 sq ft of surface area placed. The samples for strength tests shall be taken in accordance with ASTM C 172, Method of Sampling Fresh Concrete. Cylinders for acceptance tests shall be molded and laboratory cured in accordance with ASTM C 31, Method of Making and Curing Concrete Compressive and Flexural Strength Test Specimens in the Field and tested in accordance with ASTM C 39, Method of Test for Compressive Strength of Molded Concrete Cylinders. Each strength test result shall be the average of two cylinders from the same sample tested at 28 days or the specified earlier age.

If individual tests of laboratory-cured specimens produce strengths more than 500 psi below f_c', and computations indicate that the load-carrying capacity may have been significantly reduced, tests of cores drilled from the area in question may be required in accordance with ASTM C-42, Method of Obtaining and Testing Drilled Cores and Sawed Beams of Concrete. Three cores shall be taken for each case of a cylinder test more than 500 psi below f_c'. Concrete in the area represented by the core tests will be considered structurally adequate if the average of the three cores is equal to at least 85% of f_c' and if no single core is less than 75% of f_c'. To check testing accuracy, locations represented by erratic core strengths may be retested. If these strength acceptance criteria are not met, the responsible authority may order load tests conducted in accordance with the ACI Building Code.

20   Quality-control Charts

A convenient and reliable method of graphically analyzing the trend of concrete strengths is by the use of a control chart. Figure 14 is an example of a control

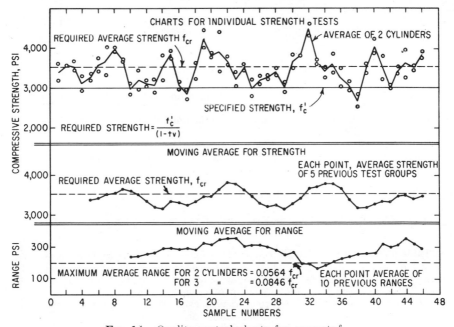

FIG. 14.   Quality-control charts for concrete.[6]

chart for structural concrete showing individual 28-day compressive-strength tests. A curve of this type shows how many cylinders fall below the specified strength and thus indicates whether there is compliance with the specification. The daily plotting of this curve conveniently indicates whether a detrimental trend has occurred. If so, prompt remedial action such as a change in mix design, materials, mixing, etc., should be taken. The specified strength f_c' (design data) and the required average strength f_{cr} are included in this curve. The f_{cr} is calculated from the following formula:

$$f_{cr} = \frac{f_c'}{1 - tV}$$

where t is taken from Table 4, and V is the coefficient of variation.

Another curve which is helpful in this type of analysis is the moving-average curve, also shown in Fig. 14. The moving average for strength is plotted for the previous five sets of companion cylinders, and by specification the lower limit is the specified strength f_c'. The moving average for range is plotted for the previous 10 groups of companion cylinders.

It is considered that the within-test variation V_1 should be a maximum 5% for good control. Therefore, the maximum for average range in control becomes

$$\bar{R}m = (0.05 \times 1.128)f_{cr} = 0.0564f_{cr} \quad \text{(for two cylinders in a group)}$$
$$\bar{R}m = (0.05 \times 1.693)f_{cr} = 0.08465f_{cr} \quad \text{(for three cylinders in a group)}$$

27 Statistical Evaluation

Statistical data are applied to concrete strength of each class or type of concrete separately. In addition to the use of statistical formulas for establishing control charts, prescribed statistical methods are used for computing standard deviation and coefficient of variation. The procedure used is:

(a) Compute $\bar{X}$, the average 28-day strength of all test specimens (preferably after at least 30 tests are available), by the following formula:

$$\bar{X} = \frac{X_1 + X_2 + X_3 + \cdots + X_n}{n}$$

where $X_1, X_2, X_3, \ldots, X_n$ are strengths of individual tests and n is the total number of tests.

(b) Compute the standard deviation σ (see the normal-frequency curve for graphic illustration) by the following formula:

$$\sigma = \sqrt{\frac{(X_1 - \bar{X})^2 + (X_2 - \bar{X})^2 + \cdots + (X_n - \bar{X})^2}{n}}$$

(c) Compute the coefficient of variation V by the formula

$$V = \frac{\sigma}{\bar{X}} \times 100$$

Table 6 gives ratings that are commonly assigned to various coefficients of variation. As can be noted from Fig. 15, the higher the coefficient the higher the average strength must be and, therefore, the more costly the design mix will be. This emphasizes the value of good quality-control operations.

(d) As an adjunct to the above statistical evaluation of compressive strengths, it is desirable to determine the uniformity of companion cylinders. This will, as previously discussed, indicate the efficiency of fabrication, curing, and testing. The within-test standard deviation σ_1 and within-test coefficient of variation V_1 are computed as follows:

$$\sigma_1 = \frac{1}{d_2} \bar{R}$$

$$V_1 = \frac{\sigma_1}{\bar{X}} \times 100$$

in which $\bar{X}$ = average strength

$\bar{R}$ = average range of groups of companion cylinders

$\dfrac{1}{d_2}$ = a constant depending on number of cylinders in each group, taken from Table 5

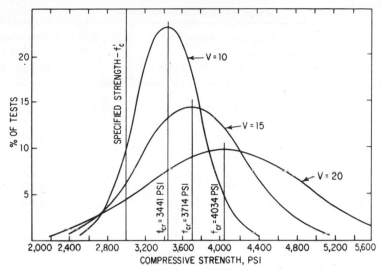

Fig. 15. Normal frequency curves for coefficients of variation of 10, 15, and 20%. Required average strength f_{cr} based on a probability of 1 in 10 that a test will fall below a specified strength f_c' of 3,000 psi.[6]

Table 6. Standards of Concrete Control*

Class of operation	Coefficients of variation for different control standards			
	Excellent	Good	Fair	Poor
Overall variations:				
General construction.........	Below 10.0	10.0–15.0	15.0–20.0	Above 20.0
Laboratory control..........	Below 5.0	5.0–7.0	7.0–10.0	Above 10.0
Within-test variations:				
Field control...............	Below 4.0	4.0–5.0	5.0–6.0	Above 6.0
Laboratory control..........	Below 3.0	3.0–4.0	4.0–5.0	Above 5.0

NOTE: These standards represent the average for 28-day cylinders computed from a large number of tests. Different values for other than average concretes can be expected.
* From ACI Standard 214-65, Table 2.

In the case of within-test computations, a coefficient V_1 of 5% is considered satisfactory for field control (see Table 6).

An example of the calculations of Art. 24 (*a*), (*b*), (*c*), and (*d*) is given in Table 7.

28 Frequency-distribution Curve

It may be desirable in the statistical evaluation of a given project to determine the skewness of data. This is achieved by plotting a histogram from the data. If the resultant curve does not follow the normal-distribution curve, some influencing factor has developed requiring further study. The histogram is constructed by arranging all the strength results into groups or cells. (A cell as used in this sense is a strength bracket of a given maximum and minimum. Each test between

Table 7. Sample Calculations

Test No.	Cylinder 1			Cylinder 2			Avg X 1 and 2	Range R	Moving avg	
	X	$X - \bar{X}$	$(X - \bar{X})^2$	X	$X - \bar{X}$	$(X - \bar{X})^2$			Strength	Range
1	3,300	170	28,900	3,100	370	136,900	3,200	200		
2	3,700	230	52,900	3,800	330	108,900	3,750	100		
3	3,200	270	72,900	3,300	170	28,900	3,250	100		
4	3,600	130	16,900	3,400	70	4,900	3,500	200		
5	3,200	270	72,900	3,000	470	220,900	3,100	200	3,360	
6	3,800	330	108,900	3,500	30	900	3,650	300	3,450	
7	4,000	530	280,900	3,800	330	108,900	3,900	200	3,480	
8	3,300	170	28,900	3,600	130	16,900	3,450	300	3,520	
9	3,200	270	72,900	3,000	470	220,900	3,100	200	3,440	
10	3,000	470	220,900	3,200	270	72,900	3,100	200	3,440	200
11	3,000	470	220,900	3,100	370	136,900	3,050	100	3,320	190
12	3,200	270	72,900	3,300	170	28,900	3,250	100	3,190	190
.	.	.	.	.	.	.	.	.	.	.
.	.	.	.	.	.	.	.	.	.	.
30	3,400	70	4,900	3,500	30	900	3,450	100	3,280	150
	103,500		3,581,600	104,700		5,360,400		5,400		

$n = 60$ (30 tests, 2 cylinders each)

$$X = \frac{103,500 + 104,700}{60} = 3,470$$

$$\sigma = \sqrt{\frac{3,581,600 + 5,360,400}{60}} = 386 \text{ psi}$$

$$V = \frac{386}{3,470} \times 100 = 11.1\%$$

Within-test variation:

$$\bar{R} = \frac{5,400}{30} = 180$$

$$\sigma_1 = 180 \times 0.8865* = 160$$

$$V_1 = \frac{160}{3,470} \times 100 = 4.6\%$$

$$\bar{R}m = (0.05 \times 1.128) \times 3,470 = 196$$

* Factor from Table 5.

these figures is assigned to that cell.) There should be at least 10 cells. The cell boundary selected must be such that any single value belongs definitely in a specific cell. Since compressive strengths are computed to the nearest 10 psi this can easily be accomplished by having the cell boundaries end in 5, such as 2,405–2,605–2,805, etc.; or by having one boundary end in 0 and the other in 9, such as 2,400–2,599; 2,600–2,799. Table 8 shows a convenient method of tabulation.

The normal curve is then plotted by computing the ordinate at $\bar{X}$, at $\bar{X} \pm \sigma$, and at $\bar{X} \pm 2\sigma$, and then drawing a smooth curve through these points with the points of inflection at $\bar{X} \pm \sigma$ as shown in Fig. 16. The equations for computing these points are

At $\bar{X}$, $\qquad\qquad\qquad y = 0.3989 \dfrac{nc}{\sigma}$

At $\bar{X} \pm \sigma$, $\qquad\qquad\qquad y = 0.242 \dfrac{nc}{\sigma}$

At $\bar{X} \pm 2\sigma$, $\qquad\qquad\qquad y = 0.054 \dfrac{nc}{\sigma}$

in which $\bar{X}$ = average strength
n = total number of specimens
σ = standard deviation
c = cell size or range in strength of each cell

Figure 16 is a graphical comparison of the data from Table 8 and the corresponding normal-distribution curve.

Table 8. Frequency of Tests Tabulation

Cell boundary	No. of test results in cell	Total No.
2,405		
	/	1
2,605		
	/	1
2,805		
	⊬⊬⊢	4
3,005		
	⊬⊬⊢ ///	8
3,205		
	⊬⊬⊢ ⊬⊬⊢ ////	14
3,405		
	⊬⊬⊢ ⊬⊬⊢ ///	13
3,605		
	⊬⊬⊢ //	7
3,805		
	⊬⊬⊢ /	6
4,005		
	///	3
4,205		
	//	2
4,405		
	/	1
4,605		
	Total	60

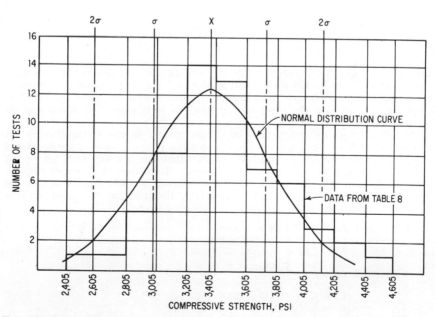

Fig. 16. Normal-frequency-distribution curve with example of job data (Table 8) superimposed.

29 Significance of Statistical Analysis

The ACI Building Code utilizes the statistical concept of classification by attributes. This is based on minimizing the frequency of strength tests below the specified strength, f_c'. If this limiting number is not exceeded, the concrete is acceptable. In order to satisfy this requirement, the average strength f_{cr} must obviously be in excess of f_c'.

The amounts by which the average strength f_{cr} should exceed the specified strength f_c' have been calculated by procedures outlined in the report of ACI Committee 214, Recommended Practice for Evaluation of Compressive Strength Tests of Field Concrete. The listed values are required to meet all three of the following criteria, using the maximum standard deviation from the range shown in each case:

1. A probability of less than 1 in 10 that a random individual strength test will be below the specified strength f_c'.

2. A probability of 1 in 100 that an average of three consecutive strength tests will be below the specified strength f_c'.

3. A probability of 1 in 100 that an individual strength test will be more than 500 psi below the specified strength f_c'.

Using values of t from Table 4, formulas for calculating the required average strengths reduce to the following for the respective criteria above:

1. $f_{cr} = f_c' + 1.282\sigma$

2. $f_{cr} = f_c' + \dfrac{2.326\sigma}{\sqrt{3}} = f_c' + 1.343\sigma$

3. $f_{cr} = f_c' - 500 + 2.326\sigma$

where f_{cr} = average strength to be used as the basis for selecting concrete proportions, psi

$\quad\ f_c'$ = strength level used in the design of the structure, psi (specified f_c')

$\quad\ \sigma$ = standard deviation of individual strength tests, psi

If probabilities other than those given in the ACI requirements are desirable, replacement of the values of t given in the above values can be obtained from Table 4.

It can be seen that criterion 2 always produces a required average strength higher than criterion 1. Criterion 2 will produce a higher required average strength than will criterion 3 for low to moderate standard deviations; however, criterion 3 governs, i.e., limiting the expected frequency of tests more than 500 psi below the specified f_c' to 1 in 100.

The use of these criteria permits a change in mix design whenever sufficient data (normally 30 or more successive tests of a given class of concrete) are developed on a specific project. As an example, assume that the original 5,000 psi design was overdesigned to 6,200 psi for a given project. Development of proper strength data indicated that the standard deviation of this mix was 500 psi. In accordance with the above stated probability requirements, the mix may be adjusted from the water-cement ratio curve to produce the following strength level:

1. Criterion 1: $f_{cr} = f_c' + 1.282\sigma$
$\qquad\qquad f_{cr} = 5,000 + (1.282 \times 500)$
$\qquad\qquad f_{cr} = 5,640$

2. Criterion 2: $f_{cr} = f_c' + \dfrac{2.326\sigma}{3}$

$\qquad\qquad f_{cr} = f_c' + (1.343 \times 500)$
$\qquad\qquad f_{cr} = 5,670$

3. Criterion 3: $f_{cr} = f_c' - 500 + 2.326\sigma$
$\qquad\qquad f_{cr} = 5,000 - 500 + (2.326 \times 500)$
$\qquad\qquad f_{cr} = 5,660$

Therefore, $f_{cr} = 5,670$ governs. This strength is used to enter the original water-cement ratio curve to develop the water-cement ratio requirement for a 5,670 psi strength instead of 6,200 psi. This would permit a more economical mix for the balance of the project. A typical water-cement ratio-strength curve used for this purpose is shown in Chap. 11, Fig. 2.

30 Mix Design Strength Requirements

The ACI Building Code emphasizes a statistical basis for establishing the average strength required to assure attainment of the strength level f_c' that was used in the structural design. If an applicable standard deviation for strength tests of the concrete is known, the required average strength can be established. If this standard deviation is not known, the proportions must be selected to produce an excess of 1,200 psi over the required f_c'. These average strength or overdesign factors are given in paragraph 10 of Chap. 11.

In addition to strength requirements, ACI Building Code stipulates maximum water-cement ratios for various exposures together. These are given in paragraph 10 of Chap. 11.

31 Simplified Method of Evaluation

ACI 214 now contains a method of evaluation utilizing cell values to eliminate the need of lengthy computations. This method is approximate but is nearly correct for evaluation of large numbers of strength tests and is sufficiently accurate for as few as 30 strength tests.

Table 9. Frequency Distribution of Strength Prepared for Simplified Method of Evaluation

Cell boundaries	Tallied frequencies	Midcell value (1)	Coded midcell Z (2)	Frequency f (3)	Zf (4)	Z^2 (5)	Z^2f (6)	$(Z+1)^2f$ (7)*
2,300–2,499		2,400	12					
2,500–2,699		2,600	13					
2,700–2,899	//	2,800	14	2	28	196	392	450
2,900–3,099	卌	3,000	15	5	75	225	1,125	1,280
3,100–3,299	卌 /	3,200	16	7	112	256	1,792	2,023
3,300–3,499	卌 卌	3,400	17	10	170	289	2,890	3,240
3,500–3,699	卌 卌 //	3,600	18	12	216	324	3,888	4,332
3,700–3,899	卌 卌 卌	3,800	19	15	285	361	5,415	6,000
3,900–4,099	卌 卌 卌 //	4,000	20	17	340	400	6,800	7,497
4,100–4,299	卌 卌 卌 //	4,200	21	18	378	441	7,938	8,712
4,300–4,499	卌 卌 卌 卌	4,400	22	20	440	484	9,680	10,580
4,500–4,699	卌 卌 卌 ///	4,600	23	18	414	529	9,522	10,368
4,700–4,899	卌 卌 卌 /	4,800	24	16	384	576	9,216	10,000
4,900–5,099	卌 卌 ///	5,000	25	13	325	625	8,125	8,788
5,100–5,299	卌 卌 /	5,200	26	11	286	676	7,436	8,019
5,300–5,499	卌 //	5,400	27	7	189	729	5,103	5,488
5,500–5,699	卌 /	5,600	28	6	168	784	4,704	5,046
5,700–5,899	///	5,800	29	3	87	841	2,523	2,700
5,900–6,099	/	6,000	30	1	30	900	900	961
6,100–6,299		6,200	31			961		
6,300–6,499		6,400	32					
Totals......				181	3,927		87,449	95,484

* Column 7 is only for the purpose of verifying calculations of the table. It is computed by multiplying f by Z^2 of the next higher cell; i.e., for 2,700 to 2,899, f is 2; use 225; then $(Z + 1)^2f = 2 \times 225 = 450$.

To check calculations:

Total column 3 + twice total column 4 + total column 6 must equal total of column 7
181   +   2 × 3,927   +   87,449   =   95,484

Table 10. Squares of Numbers*

z	z^2	z	z^2	z	z^2
1	1	18	324	35	1,225
2	4	19	361	36	1,296
3	9	20	400	37	1,369
4	16	21	441	38	1,444
5	25	22	484	39	1,521
6	36	23	529	40	1,600
7	49	24	576	41	1,681
8	64	25	625	42	1,764
9	81	26	676	43	1,849
10	100	27	729	44	1,936
11*	121	28	784	45	2,025
12	144	29	841	46	2,116
13	169	30	900	47	2,209
14	196	31	961	48	2,304
15	225	32	1,024	49	2,401
16	256	33	1,089	50	2,500
17	289	34	1,156	51	2,601

* From ACI Standard 214-65, Table A6.

Table 11. Coefficient of Variation V*

$\dfrac{y}{\text{avg } z}$	$V, \%$	$\dfrac{y}{\text{avg } z}$	$V, \%$
1.0002–1.0006	2	1.0342–1.0380	19
1.0006–1.0012	3	1.0380–1.0420	20
1.0012–1.0020	4	1.0420–1.0462	21
1.0020–1.0030	5	1.0462–1.0506	22
1.0030–1.0042	6	1.0506–1.0552	23
1.0042–1.0056	7	1.0552–1.0600	24
1.0056–1.0072	8	1.0600–1.0650	25
1.0072–1.0090	9	1.0650–1.0702	26
1.0090–1.0110	10	1.0702–1.0756	27
1.0110–1.0132	11	1.0756–1.0812	28
1.0132–1.0156	12	1.0812–1.0870	29
1.0156–1.0182	13	1.0870–1.0930	30
1.0182–1.0210	14	1.0930–1.0992	31
1.0210–1.0240	15	1.0992–1.1056	32
1.0240–1.0272	16	1.1056–1.1122	33
1.0272–1.0306	17	1.1122–1.1190	34
1.0306–1.0342	18	1.1190–1.1260	35

* From ACI Standard 214-65, Table A7.

To follow this simplified method, tabulate all single values into cells with cell range r normally of 200 psi. Calculate a coded midcell "value Z" by the equation

$$Z = \frac{\text{midcell value}}{r}$$

With cell range of 200 psi and cell boundaries of 2,300 to 2,499, 2,500 to 2,699, 2,700 to 2,899, etc., the coded midcell value becomes a simple figure with no fractions or decimals.

Table 9 shows the tabulation used in the simplified method of evaluation. Table 10 is a tabulation of squares of numbers from 1 to 51 which will be convenient for calculating the Z^2 column of Table 9. Table 11 gives the values of the coefficient of variation as determined from the values of Table 9.

Calculation of coefficient of variation from Table 9:

$$\text{avg } Z = \frac{\text{sum of column 4}}{\text{sum of column 3}} = \frac{3,927}{181} = 21.70$$

$$\text{avg } Z^2 = \frac{\text{sum of column 6}}{\text{sum of column 3}} = \frac{87,449}{181} = 483.15$$

$$y = \frac{\text{avg } Z^2}{\text{avg } Z} = \frac{483.15}{21.70} = 22.27$$

Then

$$\frac{y}{\text{avg } Z} = \frac{22.27}{21.70} = 1.027$$

Coefficient of variation V from Table 11 = 16%.

REFERENCES

1. ASTM Standards, Parts 9 and 10, American Society for Testing and Materials, 1973.
2. "Concrete Manual," 7th ed., U.S. Bureau of Reclamation, 1963.
3. Bloem, D. L.: Concrete Strength Measurements—Cores vs. Cylinders, *ASTM Proc.*, 1965.
4. Wagner, Walter K.: Effect of Sampling and Job Curing Procedures on Compressive Strength of Concrete, *Mater. Res. Std.*, August, 1963.
5. Investigation of Compressive Strength of Molded Cylinders and Drilled Cores of Concrete, U.S. Army Engineers Waterways Experiment Station, *Tech. Rept.* 6-522, Vicksburg, Miss.
6. ACI Committee 214 Report, Recommended Practice for Evaluation of Compression Test Results of Field Concrete (ACI 214-57), *J. ACI*, July, 1957, revised September, 1964.

Chapter 13

INSPECTION AND LABORATORY SERVICE

J. F. ARTUSO

INSPECTION REQUIREMENTS

1 ACI Recommendations

A recommended practice has been developed by the ACI to encourage and enable more effective inspection of concrete construction.[1,2] Some of the pertinent recommendations include:

1. The responsibility for inspection must remain with the architect-engineer.
2. Inspectors must be well qualified and properly supervised.
3. The types of inspection that must be provided are given as:

   a. Inspection and approval of batching and mixing facilities, as well as batching inspection when the size of job or type of concrete warrants it.
   b. Inspection, testing, and approval of materials.
   c. Inspection of forms, reinforcing steel, shoring, bracing, embedded items, joints, etc.
   d. Inspection of concrete handling and placing equipment such as buckets, chutes, buggies, hoppers, vibrators, and pumps.
   e. Inspection of concrete handling, placing, consolidation, finishing, curing, protection, and repair or patching.
   f. Inspection at the plant of precast items including prestressed work for strength, dimensions, and special properties.
   g. Inspection of stripping forms and removal of shoring.
   h. Preparation and testing of concrete specimens, testing of consistency, air entrainment, unit weight.
   i. Daily reporting of all these items. Figures 1 and 2 are examples of typical reports for field inspection.

2 Batch-plant Inspection

A vital requirement for concrete control is the inspection of the batching operation, including tests of aggregates for grading and moisture content.

Briefly, the usual routine duties of batch-plant inspectors include witnessing the weighing and measurement of all constituent materials, including cement, aggregates, water, and admixtures, if any, and the checking of aggregates for size and moisture content. Adjustment of aggregate weights and of amount of added water to compensate for the free moisture in the aggregates is an important part of the control at this point. Periodically the scales and other measuring devices should be checked for accuracy by calibration with standards furnished by the National Bureau of Standards.

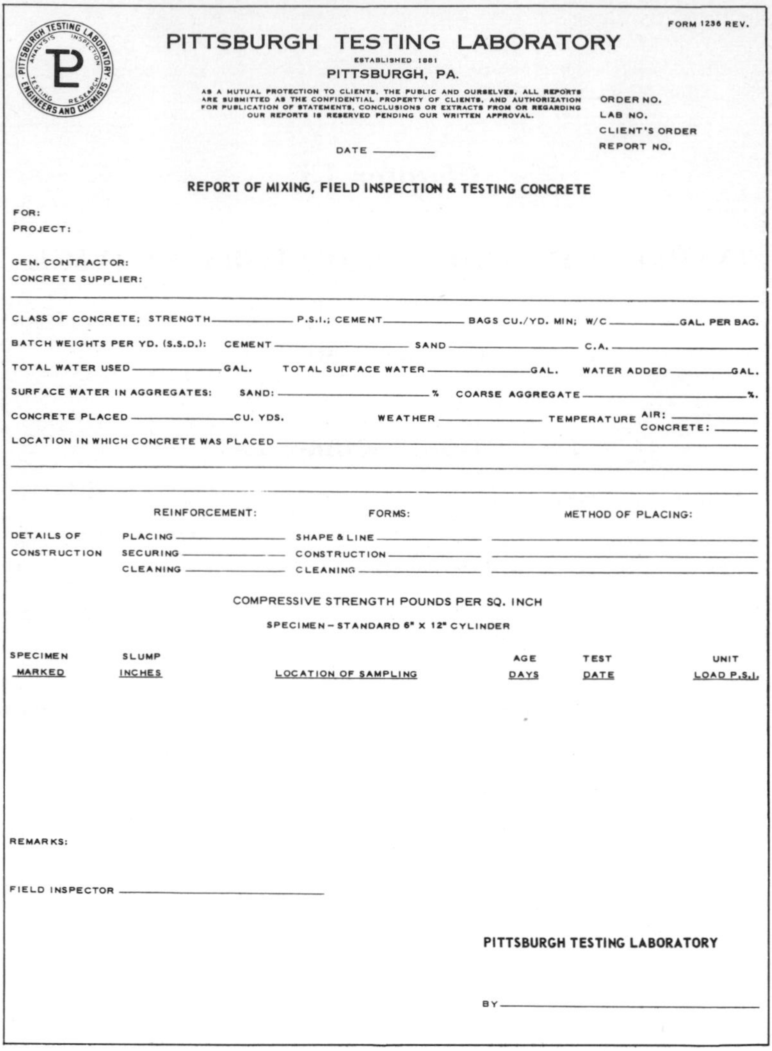

FORM 1236 REV.

PITTSBURGH TESTING LABORATORY

ESTABLISHED 1881

PITTSBURGH, PA.

AS A MUTUAL PROTECTION TO CLIENTS, THE PUBLIC AND OURSELVES, ALL REPORTS ARE SUBMITTED AS THE CONFIDENTIAL PROPERTY OF CLIENTS, AND AUTHORIZATION FOR PUBLICATION OF STATEMENTS, CONCLUSIONS OR EXTRACTS FROM OR REGARDING OUR REPORTS IS RESERVED PENDING OUR WRITTEN APPROVAL.

ORDER NO.

LAB NO.

CLIENT'S ORDER

REPORT NO.

DATE _____

REPORT OF MIXING, FIELD INSPECTION & TESTING CONCRETE

FOR:

PROJECT:

GEN. CONTRACTOR:

CONCRETE SUPPLIER:

CLASS OF CONCRETE; STRENGTH _____ P.S.I.; CEMENT _____ BAGS CU./YD. MIN; W/C _____ GAL. PER BAG.

BATCH WEIGHTS PER YD. (S.S.D.): CEMENT _____ SAND _____ C.A. _____

TOTAL WATER USED _____ GAL. TOTAL SURFACE WATER _____ GAL. WATER ADDED _____ GAL.

SURFACE WATER IN AGGREGATES: SAND: _____ % COARSE AGGREGATE _____ %.

CONCRETE PLACED _____ CU. YDS. WEATHER _____ TEMPERATURE AIR: _____
CONCRETE: _____

LOCATION IN WHICH CONCRETE WAS PLACED _____

REINFORCEMENT: FORMS: METHOD OF PLACING:

DETAILS OF PLACING _____ SHAPE & LINE _____

CONSTRUCTION SECURING _____ CONSTRUCTION _____

CLEANING _____ CLEANING _____

COMPRESSIVE STRENGTH POUNDS PER SQ. INCH

SPECIMEN – STANDARD 6" X 12" CYLINDER

SPECIMEN MARKED	SLUMP INCHES	LOCATION OF SAMPLING	AGE DAYS	TEST DATE	UNIT LOAD P.S.I.

REMARKS:

FIELD INSPECTOR _____

PITTSBURGH TESTING LABORATORY

BY _____

FIG. 1. Typical report of mixing, field inspection, and testing concrete.

The batch plant should be so equipped that cement can be weighed on a scale entirely separate from the aggregate scales. Various-sized aggregates may all be weighed in a single scale hopper on the cumulative basis. Details of batch plants are covered in Sec. 5 of this handbook.

The accuracy to which measurement of various ingredients should be made is shown in Table 1.

Table 1. Accuracy to Which Concrete Ingredients Should Be Measured in a Batch Plant

Ingredient	Accuracy tolerance, %
Cement	±1
Aggregates	±2
Water	±1
Admixtures	±3

Figure 2 is an example of a typical report form for batch-plant inspection. Adequate means for sampling the materials must be provided. The samples must be taken in such a manner as to ensure positive representation. When samples are taken from a bin they should be taken from the entire cross section of the

PITTSBURGH TESTING LABORATORY FORM 9003-A

ESTABLISHED 1881

PITTSBURGH, PA.

AS A MUTUAL PROTECTION TO CLIENTS, THE PUBLIC AND OURSELVES, ALL REPORTS ARE SUBMITTED AS THE CONFIDENTIAL PROPERTY OF CLIENTS, AND AUTHORIZATION FOR PUBLICATION OF STATEMENTS, CONCLUSIONS OR EXTRACTS FROM OR REGARDING OUR REPORTS IS RESERVED PENDING OUR WRITTEN APPROVAL.

Order No.

Report No.

REPORT OF BATCH PLANT OPERATIONS   Clients No.

At _____ * Date _____
Project _____
Contractor _____
Reported to _____
Location of Concrete Placed (per information from job site) _____

SPECIFICATION REQUIREMENTS

Strength _____ psi Min. @ 28 days; Slump _____ In. Max.; Entr. Air _____ % to _____ %
Cement Type _____ ; Amt. _____ bags/cu. yd. Min.; W/C _____ Gal./bag Max.
Aggregate: (Kind and Size Range) Coarse _____ Fine _____
Admixture: _____

SOURCE OF MATERIALS

Cement _____ Admixture _____
Fine Aggregate _____
Coarse Aggregate _____

DESIGN BATCH QUANTITIES per CUBIC YARD – MIX DESIGN NO. _____

Cement _____ lbs. _____ bags; Fine Aggregate (S.S.D.) _____ lbs.
Coarse Aggregate (S.S.D.) _____ lbs. _____ lbs.
Admixture (Kind) _____ ; Amount _____ ; Total Water _____ Gals.
Admixture (Kind) _____ ; Amount _____

SURFACE MOISTURE IN AGGREGATES			**GRADING OF AGGREGATES**	
Fine Aggregate _____ % _____ %		Sieve	Percent (Passing) (Retained)	
Coarse Aggregate _____ % _____ %			Coarse Agg.	Fine Agg.
ACTUAL BATCH WEIGHTS per CUBIC YARD		2"		
(Adjusted for Surface Moisture on Aggs.)		1½"		
Cement _____ lbs.		1"		
Fine Agg. _____ lbs.		¾"		
Coarse Agg. _____ lbs.		½"		
Coarse Agg. _____ lbs.		⅜"		
Admix. _____		No. 4		
Admix. _____		No. 8		
Free Water in Agg. _____ Gallons		No. 16		
Added Water _____ Gallons		No. 30		
Total Water _____ Gallons		No. 50		
		No. 100		
Batching Started _____ Finished _____		Mat'l. finer		
Concrete batched this report _____ cu. yds.		than No. 200 _____		
		Inspector:		

PITTSBURGH TESTING LABORATORY

Joseph F. Artuso, Manager
Cement-Concrete Dept.

FIG. 2. Typical report of batch-plant operations.

flow of material as it is being discharged. Sufficient material should be permitted to pass at the beginning of the discharge to ensure normal uniformity.

When samples of coarse aggregate are taken from stockpiles, they should be taken at or near the top, base, and an intermediate point. A board placed above the point of sampling will prevent segregation. Samples taken from railroad cars, trucks, barges, or boats should be taken from three or more trenches dug across the vehicle at representative areas. The trench should be at least 1 ft deep and

1 ft wide at the bottom. At least 7 shovelfuls should be taken by digging down into the bottom of the trench. Two of these should be against the sides of the vehicle. Fine aggregate can be sampled in a similar manner as above, or a sampling tube 1¼ in. in diameter by 6 ft long may be repeatedly inserted to provide a 10- to 20-lb sample.

The minimum amount of coarse aggregate required to be reasonably confident of having a representative sample is dependent on the maximum size of the material. Table 2 gives the suggested minimum quantity for each of various sizes from ⅜ to 2½ in. maximum.

Table 2. Minimum Quantities for Coarse-aggregate Samples*

Nominal max size of particle, in.	Min weight of field sample, lb
⅜	10
½	20
¾	30
1	50
1½	70
2	90
2½	100
3	125
3½	150

* From ASTM Standard D 75-59.

3 Aggregate-moisture Determination

An important function of aggregate inspection at the batch plant is to ensure that correct moisture corrections are made. Several methods are available.[3]

1. Sample dried in an oven or hot plate. The sample is completely dried and a correction is made for absorption to determine amount of free moisture. An approximate formula of sufficient accuracy is given by

$$\% \text{ moisture} = 100 \left(\frac{\text{wet weight} - \text{dry weight}}{\text{dry weight}} \right) - \% \text{ absorption}$$

Figure 3 may be used for convenience. A 400-g sample is dried and weighed. The percent moisture is taken from the intersection with the curve of the appropriate absorption.

2. A convenient method for free moisture in sand is that based on U.S. Bureau of Reclamation Designation 11-B for use with Table 3. The specific gravity must be known.

Apparatus. Chapman-type volumetric flask of 500 ml capacity accurate to 0.15 ml, balance of 2 kg capacity accurate to 0.1 g, and pipette, ¼-in.-diameter glass tube, of sufficient length to enable water-level adjustment.

Procedure. Fill the flask with water at room temperature to the calibrated 200-ml mark. Use the pipette to adjust the water level (lower part of the meniscus) to the 200-ml mark. Introduce 500 g of representative sand into the flask. Roll to dispel all entrapped air. Break foam. A drop or two of ether, alcohol, or commercial wetting agent will accomplish this readily. Stirring with a fine wire is tedious and time-consuming, although usually a successful method. Read the volume of the combined water and fine aggregate directly on the graduated scale. The percentage of free moisture can be taken directly from Table 3.

An expedient method that is used in commercial practice utilizes a pressure vessel similar to a bomb calorimeter in which sand and calcium carbide react and produce acetylene gas. The resultant pressure is indicated on a scale calibrated to read directly the percent of free moisture in the sample.

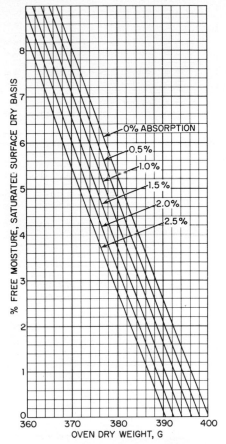

Fig 3. Chart for determining moisture content of aggregate. (1) Use 400-g wet sample. (2) Dry to oven dryness. (3) Enter chart at dry weight, intersect curve for correct absorption, and read percent moisture on scale at left.

$$\%M = 100 \frac{400 - A(1 + X)}{A(1 + X)}$$

where A = ovendry weight
     X = absorption expressed as a decimal
(*From Ref. 3.*)

LABORATORY FACILITIES

4   Permanent Laboratories

There are two conventional types of testing facilities. One is a Federal, state, county, or city governmental agency designed to serve the needs of government-sponsored construction. The other type is the commercial independent laboratory serving all types of private and commercial and, frequently, of governmental construction. The permanent laboratory of either type is normally equipped to perform detailed investigations and tests for evaluation of concrete materials which include physical and chemical testing. Concrete is routinely tested for strength,

Table 3. Percent Moisture in Sand*

V	\-\- Bulk specific gravity, SSD basis \-\-																				
	2.55	2.56	2.57	2.58	2.59	2.60	2.61	2.62	2.63	2.64	2.65	2.66	2.67	2.68	2.69	2.70	2.71	2.72	2.73	2.74	2.75
382																					0.1
383																				0.2	0.4
384																		0.1	0.3	0.5	0.7
385																	0.2	0.4	0.6	0.8	1.0
386																0.3	0.5	0.7	0.9	1.1	1.3
387														0.1	0.4	0.6	0.8	1.0	1.3	1.5	1.7
388													0.2	0.4	0.7	0.9	1.1	1.4	1.6	1.8	2.0
389											0.1	0.3	0.5	0.8	1.0	1.2	1.5	1.7	1.9	2.1	2.3
390									0.0	0.2	0.4	0.6	0.9	1.1	1.3	1.6	1.8	2.0	2.2	2.4	2.6
391									0.3	0.5	0.7	1.0	1.2	1.4	1.7	1.9	2.1	2.3	2.5	2.7	3.0
392							0.1	0.4	0.6	0.8	1.1	1.3	1.5	1.8	2.0	2.2	2.5	2.6	2.8	3.1	3.3
393						0.2	0.5	0.7	0.9	1.2	1.4	1.6	1.9	2.1	2.3	2.6	2.8	3.0	3.2	3.4	3.7
394				0.1	0.3	0.6	0.8	1.0	1.3	1.5	1.7	2.0	2.2	2.4	2.7	2.9	3.1	3.3	3.5	3.8	4.0
395		0.0	0.1	0.4	0.6	0.9	1.1	1.4	1.6	1.8	2.1	2.3	2.5	2.8	3.0	3.3	3.4	3.7	3.9	4.1	4.3
396		0.2	0.5	0.7	1.0	1.2	1.5	1.7	1.9	2.2	2.4	2.6	2.9	3.1	3.4	3.6	3.8	4.0	4.2	4.4	4.7
397	0.3	0.6	0.8	1.1	1.3	1.5	1.8	2.0	2.3	2.5	2.7	3.0	3.2	3.5	3.7	4.0	4.1	4.3	4.6	4.8	5.0
398	0.7	0.9	1.1	1.4	1.6	1.9	2.1	2.4	2.6	2.8	3.1	3.3	3.6	3.8	4.1	4.3	4.5	4.7	4.9	5.1	5.4
399	1.0	1.2	1.5	1.7	2.0	2.2	2.5	2.7	2.9	3.2	3.4	3.7	3.9	4.2	4.4	4.6	4.8	5.0	5.3	5.5	5.7
400	1.3	1.6	1.8	2.1	2.3	2.6	2.8	3.0	3.3	3.6	3.8	4.0	4.3	4.5	4.7	4.9	5.2	5.4	5.6	5.8	6.1
401	1.7	1.9	2.2	2.4	2.6	2.9	3.1	3.4	3.7	3.9	4.2	4.4	4.6	4.8	5.1	5.3	5.5	5.8	6.0	6.2	6.4
402	2.0	2.2	2.5	2.7	3.0	3.2	3.5	3.8	4.0	4.3	4.5	4.7	5.0	5.2	5.5	5.7	5.9	6.1	6.3	6.5	6.8
403	2.4	2.6	2.8	3.1	3.3	3.6	3.9	4.1	4.4	4.6	4.8	5.1	5.3	5.6	5.8	6.0	6.2	6.5	6.7	6.9	7.1
404	2.7	2.9	3.2	3.4	3.7	4.0	4.2	4.5	4.7	4.9	5.2	5.4	5.7	5.9	6.2	6.4	6.6	6.8	7.0	7.3	7.5
405	3.0	3.3	3.5	3.8	4.0	4.3	4.6	4.8	5.1	5.3	5.5	5.8	6.0	6.3	6.5	6.7	7.0	7.2	7.4	7.7	7.9

406	3.4	3.6	3.9	4.2	4.4	4.7	4.9	5.2	5.4	5.7	5.9	6.1	6.4	6.6	6.9	7.1	7.3	7.5	7.8	8.0	8.2
407	3.8	4.0	4.2	4.5	4.8	5.0	5.3	5.5	5.8	6.0	6.3	6.6	6.7	7.0	7.2	7.5	7.7	7.9	8.1	8.4	8.6
408	4.1	4.3	4.6	4.9	5.1	5.4	5.6	5.9	6.1	6.4	6.7	6.9	7.1	7.3	7.6	7.8	8.1	8.3	8.5	8.7	9.0
409	4.4	4.7	5.0	5.2	5.5	5.7	6.0	6.3	6.5	6.8	7.0	7.3	7.5	7.7	8.0	8.2	8.4	8.6	8.9	9.1	
410	4.8	5.1	5.3	5.6	5.8	6.1	6.4	6.6	6.9	7.1	7.4	7.7	7.8	8.1	8.3	8.6	8.8	9.0	9.3		
411	5.2	5.4	5.7	6.0	6.2	6.5	6.7	7.0	7.2	7.5	7.8	8.0	8.2	8.5	8.7	8.9	9.2				
412	5.6	5.8	6.1	6.3	6.6	6.8	7.1	7.4	7.6	7.9	8.1	8.3	8.6	8.8	9.1	9.3					
413	5.9	6.2	6.4	6.7	7.0	7.2	7.5	7.7	8.0	8.3	8.5	8.7	9.0	9.2							
414	6.3	6.5	6.8	7.1	7.3	7.6	7.8	8.1	8.4	8.7	8.8	9.1									
415	6.7	6.9	7.2	7.4	7.7	8.0	8.2	8.5	8.8	9.0	9.2	9.5									
416	7.0	7.3	7.6	7.8	8.1	8.4	8.6	8.9	9.1	9.4	9.6										
417	7.4	7.7	7.9	8.2	8.5	8.7	9.0	9.2													
418	7.8	8.0	8.3	8.6	8.8	9.1	9.4														
419	8.2	8.4	8.7	9.0	9.2	9.5															
420	8.6	8.8	9.1	9.4																	
421	8.9	9.2																			
422	9.3																				

$$\% \text{ moisture} = \frac{V - (500/\text{sp gr}) - 200}{200 + 500 - V}$$

where V = flask reading
  sp gr = bulk specific gravity, SSD basis
See "Concrete Manual," 7th ed., p. 531, U.S. Bureau of Reclamation, 1963.
*From Ref. 3.

often for other properties such as durability, volume change, modulus of elasticity, and permeability.

5 Jobsite Laboratories

Governmental agencies and commercial laboratories often establish a field laboratory on a major construction project. The size will depend upon the magnitude of construction and may range from one equipped to check gradations of aggregates and to make compressive-strength tests of concrete only, to elaborate facilities such as those found in permanent laboratories. Many large private and commercial projects will require field laboratories because of distance from permanent laboratory installations. Generally these laboratories are equipped only for routine compression testing of specimens and for checking gradations and deleterious substances of aggregates. The primary equipment required for this scope of testing includes the following:

Cylinder molds
Identification tags and data cards
Slump cone
Tamping rod
Platform scales of 100 or 200 lb capacity accurate to 0.1 lb
Laboratory scales of 100 and 2,000 g capacity accurate to 0.2 g
Air meters (pressure type for stone and gravel concrete—volumetric types for slag or lightweight-aggregate concrete)
Thermometers (steel-dial type and maximum-minimum type)
Pocket rule
Calibrated measures (¼, ½, or 1 cu ft capacity dependent on aggregate size)
Buckets, scoops, shovels, pans, marking pens, sample bags
Curing tanks and lime
Motor-driven adjustable-speed campression-testing machine
Project specifications and reference methods of test

CEMENT AND CONCRETE REFERENCE LABORATORY

6 Functions and Scope

The Cement and Concrete Reference Laboratory is an agency jointly sponsored by the American Society for Testing and Materials and the National Bureau of Standards. The primary function of the Cement and Concrete Reference Laboratory is the inspection of cement and concrete testing laboratories. This inspection is available to any interested testing laboratory and is performed only by request. A nominal fee is charged. Although there is no direct regulatory action, many agencies, architects, and engineers require that a laboratory engaged in their work be regularly inspected by the Cement and Concrete Reference Laboratory.

The inspection includes examination of the apparatus used in testing concrete, aggregates, and cement to ASTM procedures. Procedures are also evaluated and instructions are given regarding any deviations from standards that are detected. A detailed report is submitted to the inspected laboratory which contains an evaluation of every piece of apparatus and of every method inspected. This can be used as a guide for any corrections that should be made. The report may also be shown as evidence of compliance with ASTM methods and use of approved apparatus to any party using the facilities of the laboratory. This is one of the most reliable methods of determining the capability of a specific laboratory.

The present published scope of the work of the Cement and Concrete Reference Laboratory is: "To promote uniformity and improvement in the testing of cement and concrete through field inspection of testing laboratories; the study of special problems evolving from such inspections; instructions in methods of testing; sponsorship of comparative test programs, and contributions to the work of ASTM Technical Committees."

APPROVAL OF MATERIALS

7 Sources

The engineer-architect has the privilege and should exercise the right to approve or disapprove all materials and sources proposed for use in the concrete for the specific structure. It is desirable to have the contractor submit the proposed source of each material well in advance of start of concreting. In the case of established manufacturers of cement, reinforcing steel, and prestressing items, approval can be granted on the basis of experience and knowledge of the source. This can also apply to aggregates. However, the many variables encountered in the production of aggregates and sometimes cement warrant current evidence in the form of recent test data. This is particularly important, usually essential, for the product of new deposits. The approval of the source does not obviate the need for continuing evidence of conformance to the applicable specifications. During construction, mill test reports of cement, supplemented by tests by an independent testing laboratory, provide a high level of assurance of quality. Aggregates should be continuously tested for grading and cleanliness and periodically for inherent properties of durability and abrasion.

8 Off-site Testing and Inspection

It is often desirable to perform inspection and testing of specific materials at the point of production. This expedites their use in construction with the assurance of compliance with the applicable contract specifications and drawings. This is accomplished by use of a testing agency that can have inspectors at the plant at the time of production. The engineer should arrange for this inspection and the contractor should indicate on his purchase order that inspection will be made by the specified testing agency in sufficient time to have this inspection and testing performed prior to shipment. Individual items must be properly tagged, stamped, or identified in some manner for shipment and acceptance at the project. In the case of cement, the inspecting agency can sample the cement as a bin is loaded for the exclusive use of a given project. The bin is sealed and the tests are performed by the testing agency to verify compliance with the applicable specifications. When conformance is established, shipments can be made to the project. The testing-agency inspector seals and appropriately identifies each shipment with tags and supporting documents for proper acceptance at the project. Aggregate stockpiles can be tested in the same manner. It is essential that a reliable method of identification be maintained for acceptance of the material at the project.

9 Acceptance on Basis of Manufacturer's Certifications

If agreeable to the engineer, materials of standard manufacture may be accepted for a project on the basis of certification by the manufacturer. This is accomplished by the submission of certified test results which show complete compliance to the applicable specifications on each lot of consignment. In addition, a statement certifying that the process used in the manufacture has remained constant should be included. Proper identification must be shown on all items to correlate respective lots. It is prudent to supplement acceptance by this method with check tests by an independent testing agency. It has been found that such check has disclosed errors in classifications of types of steel, such as between intermediate-grade and high-strength grade. It has also revealed variations in strengths and air contents of cement which have explained in part fluctuations in concrete strength. Variations in admixtures may cause appreciable difficulties, but the standard laboratory tests of these products are too time-consuming to be of much practical value in controlling uniformity of shipments.

10   General Sampling Methods

The first step in testing is often one of the most important, and this is certainly true in the case of sampling. It is essential that every effort be made to secure a sample which is representative of the material. The test, obviously, can only be as accurate as the sample. Sampling is often based on the mathematics of probability and a statistical analysis is applied. In sampling of concrete and concrete materials, methods are covered by appropriate standards of the American Society for Testing and Materials. The method for sampling aggregates is covered by ASTM D 75 and has been described previously here. The method for sampling portland cement is given in ASTM C 183. Liquids such as air-entraining agents, chemical admixtures, and curing compounds must usually be sampled from large drums. In this event, appropriate measures must be taken to ensure that the contents are thoroughly mixed by mechanical agitation, preferably using motorized agitators, before the sample is taken. When individual units are sampled, the number of samples should equal the cube root of the number of units in the lot that is inspected. As an example, 10 units would represent 1,000 units in a lot.

The American Association of State Highway Officials and various other organizations also publish many applicable standards, many of which are very similar to and often identical with those of ASTM.

11   Shipping and Handling Samples

When samples are shipped to a laboratory, complete information should be included in such a manner that it is securely attached or identified with the sample. The information should include project, sample number, complete material identification as to type and grade, source, date of sampling, quantity represented, location to be used in the structure, tests required, and shipper's identification. The samples must be adequately packed to prevent damage in transit.

When concrete cylinders are shipped, they should be packed in moist sand with sufficient covering to prevent drying or damage from impact. Green specimens, of an age of a day or two, must be handled carefully, as they can easily sustain serious breakage or damage. However, the prevalent belief that rough handling of a hardened concrete specimen is likely to result in a lower test result is erroneous. Certainly if parts of the specimen are actually broken away, or if it is broken into two or more major parts, it is unsuitable for test. Specimens should not be dropped, or hauled around loose in the bed of a truck or the trunk of a car. The principal damage that a specimen might sustain under such conditions would be chipping and spalling of edges, making proper capping difficult or impossible. However, if the specimen shows no obvious physical damage, the results of the strength test will not be affected. These comments should not be taken to apply to that period of time during which the concrete in the specimen is passing from the plastic state into the hardened state. Even slight disturbance during this period may well result in internal microfractures that cannot be seen but will seriously affect the test result.

REFERENCES

1. "Manual of Concrete Inspection," American Concrete Institute Special Publication SP-2, 1967.
2. ACI Committee 311 Report, Recommended Practice for Concrete Inspection (ACI 311-64), *J. ACI*, July, 1964.
3. Waddell, Joseph J.: "Practical Quality Control for Concrete," McGraw-Hill Book Company, New York, 1962.

Section 4

FORMWORK AND SHORING

Chapter 14

FORMWORK REQUIREMENTS AND MATERIALS

DAVID R. HARNEY

In order to build a concrete structure it is necessary to have molds by which the plastic concrete can be shaped. These molds are called concrete forms. Because the cost of molding concrete to the required shape is often equal to the combined cost of both the concrete and reinforcing steel, forming makes up a significant portion of the cost of the whole structure. It is therefore necessary to look into all aspects of the design and construction of concrete forms so the costs can be minimized without sacrificing either safety or quality. In other words, sufficient

FIG. 1. The beauty of architectural concrete can be realized by careful formwork.

planning is necessary to get the best available system for the least expenditure of money. It is almost as shortsighted to overdesign a concrete form as it is to underdesign it.

Forms are temporary structures; even so they must be built accurately so the desired dimensions are maintained, and built with sufficient strength so as to be self-supporting while bearing the plastic concrete and any other live load that will be imposed upon them (Fig. 1). To eliminate form failures it is also important that the rate of concrete placing stay within the design limits. Because concrete formwork is only temporary, the method of form removal must be given close consideration lest the form-stripping cost equal or exceed the form-erection cost.

1 Architect-Engineer Specifications and Requirements

Both architects and engineers can help cut the cost of formwork for most structures without losing the individuality they wish to create, by tailoring formwork specifications to the particular project at hand, rather than attempting to apply a general specification to all structures, and also by considering form requirements when finalizing their design of the structure, keeping in mind that the cost of concrete is cheap compared with that of formwork. It is wise to try to standardize beam, column, and pilaster sizes on most structures, thus making it possible to obtain the greatest reuse factor. Reuse of a form is usually the fastest way to cut forming costs. For example, if the initial cost to construct a form is 60 cents per square foot, two uses will reduce the per-use cost to 30 cents per square foot and additional reuses will make the form even more economical.

In most multistory buildings the beam and column sizes are reduced as the load on these members decreases in the upper floors, saving concrete but increasing forming costs. Remodeling (cutting down) forms is four to eight times more expensive than the concrete savings.

For example, assume that in a certain building the column sizes for four floors are as follows:

Fourth floor, 14 × 14 in.
Third floor, 18 × 18 in.
Second floor, 22 × 22 in.
First floor, 24 × 24 in.

A 24 × 24-in. column 12 ft high contains 48 cu ft of concrete, and a 22 × 22-in. column 12 ft high contains 40 cu ft of concrete. Difference 8 cu ft.

Concrete cost, 50 cents per cubic foot, material
10 cents per cubic foot, labor to place
Total, 60 cents per cubic foot

Then the saving on 8 cu ft reduction in quantity of concrete is $4.80. Cost of column-form fabrication is 60 cents per square foot, labor, and 15 cents per square foot, material, for a total of 75 cents per square foot. A 24 × 24-in. form 12 ft high has 96 sq ft of contact area. If the cost to remodel (cut down) a column form from 24 × 24 in. to 22 × 22 in. is 25 cents per square foot, then the remodeling cost is 96 × 0.25 = $24. In this case, then, it actually cost $19.20 to "save" 8 cu ft of concrete ($24.00 minus $4.80).

Comprehensive formwork specifications that clearly state the results that are expected and the tolerances that are allowed rather than how to obtain them are of great value to all concerned with any particular project. Specifications should not be written so loosely that they leave interpretation of their intent to personal preferences. By stating results and tolerances architects and engineers will encourage bidders to use their imagination in bidding jobs, allowing the contractor in many cases to give a better job for less money, thus making a saving for the owner.

It is frequently advantageous for the architect-engineer to consult independent form engineers who can assist by explaining how forms can be designed for a problem structure. For example, by adjusting dimensions one way or another, or by sloping corners or surfaces one degree or so, forming costs can sometimes be reduced substantially. Also these engineers can give advice on sources and types of material that will give the architect the results he desires.

The naming of proprietary items should be avoided as much as possible. Naming of a specific type of form or form hardware narrows down the source of supply and greatly restricts the contractor. It also tends to increase the job cost, as the cost of exclusive items usually goes up. If the architect-engineer designs for a particular finish that can be obtained only by a specific product, the plans and specifications should point out very clearly the proprietary item, how it is to be used, and where it is obtainable (name and address of supplier). He should also check with the supplier as to the availability before plans are released for bids.

Since the in-place concrete is the finished product, the tolerances for any structure should be set for it, expecting the formwork to be constructed with sufficient strength and accuracy so the finished work is within the allowable limits. Suggested tolerances for various structures are given in Table 1 and in Chap. 25.

Tolerances in construction are usually prescribed in the specifications or on the drawings, based on the ACI Recommended Practice for Concrete Formwork[1] or similar standards. Whether or not these limits can be violated depends on the experience and judgment of the engineer, contractor, and form designer. For example, the limits shown in Table 1 are workable and practical, although some of them exceed the ACI recommendations, and others are smaller.

Table 1. Suggested Tolerances for a Reinforced-concrete Building*

Footings:	
Variation from plans (size)	Plus 6 in.†
	Minus ½ in.
Variation in thickness	Plus 10% of thickness†
	Minus 5% of thickness
Variation from plans (location)	5% of footing width in direction of mislocation
Pits and sumps:	
Variation from plans (size)	Plus 2 in.†
	Minus ½ in.
Variation from plans (location)	Either direction (pits) ½ in.
	Either direction (sumps) 2 in.
Vertical surfaces: walls, columns, piers:	
Variation from plumb	In 10 ft, ¼ in.
	In one story, 20 ft max, ⅜ in.
	In 40 ft, ½ in.
Horizontal surfaces: decks, beam soffits	In 10 ft, ¼ in.
	In 20 ft, ⅜ in.
	In 40 ft, ½ in.
Variation from true location for columns and walls	20 ft, ¼ in.
	40 ft, ¼ in.
Variation from true thickness or cross section for walls, pilasters, columns, slabs, beams	Plus ½ in.†
	Minus ¼ in.

* Proposed tolerances should be checked against engineer's drawings to determine if they can be used safely. These tolerances apply to the concrete, and not to location of reinforcement.

† Tolerance is allowable only if there will be no interference with other building elements.

MATERIALS FOR FORMWORK

The type of concrete finish required usually determines the type of forming material selected. Materials used to construct concrete forms and the form accessories that are used in conjunction with forming have become varied in recent years to keep pace with the development of new techniques in concrete construction, such as precast concrete for all parts of structures, thin shells, and folded plate roofs, to mention a few. Lumber, in one form or another, is used with almost every forming system no matter how ingenious the system might be. Steel, aluminum, fiber glass, plastic, styrofoam, and fiberboard are other materials used in forming and in connection with form-tying devices or shoring. The commonest materials and standards are listed in Table 2.

When the forming material and form-tying devices are selected for a forming system, careful consideration should be given to their properties so that they are equal to what is required from the standpoint of strength and economy with a sufficient safety factor. For example, when a form (using 2 × 4 studs and wales*) is built to withstand only low concrete pressures (300 to 400 psf) a tie with tensile strength of 6,000 lb is incongruous and uneconomical, as its capacity is far in excess of any load plus factor of safety that might be imposed on it.

Table 2. Forming Materials

Material	Application	Standard	Structural data
Lumber..............	Form framing Form facing Shoring and bracing	National Lumber Manufacturers Association	"Wood structural design data"
Pressed wood (hard- board).............	Form facing (liners)	Manufacturers' data	Manufacturers' data
Plywood.............	Form facing	U.S. commercial standards	Manufacturers' data
Steel...............	Form framing Form facing Falsework	American Institute of Steel Construc- tion	AISC Handbook
	Individual post shores	Manufacturers' data	Manufacturers' data
	Scaffold shoring	Manufacturers' data	Manufacturers' data
	Horizontal shores	Manufacturers' data	Manufacturers' data
	Prefabricated patent forms	Manufacturers' data	Manufacturers' data
	Sheet-metal forms	Manufacturers' data	Manufacturers' data
Aluminum...........	Form framing Form facing Horizontal shores	Manufacturers' data	Manufacturers' data
Fiber glass (RIP).....	Form facing and body	Manufacturers' data	Manufacturers' data
Laminated fiber.......	Column forms and concrete voids	Manufacturers' data	Manufacturers' data
Corrugated paper.....	Voids in slabs Voids under beams (soil conditions re- quire)	Manufacturers' data	Manufacturers' data
Styrofoam...........	Relief forming in pre- cast	Manufacturers' data	Manufacturers' data
	Voids in beams, slabs, columns	Manufacturers' data	Manufacturers' data

NOTE: Industry standards and data are given where available for standard materials. Special materials and applications are based on manufacturers' limits and are described in the various manufacturers' catalogs and handbooks.

2 Wood and Hardboard

Lumber is necessary for most jobs regardless of what other form material is used. It is wise to check local stocks to find what type and in what quantities lumber products are readily available. Lumber is classified by size into boards (1 to 1½ in. thick, 2 in. wide and wider), dimension (2 to 4 in. thick, any width), and timbers (5 in. thick and thicker, 5 in. wide and wider). Lumber comes from

* The words "wale" and "waler" are synonymous, describing a horizontal support-ing timber outside the studs (see Fig. 5).

the mill either rough or dressed; rough is saw-cut to specific cross-sectional sizes, for example, 2×4 in. or 4×8 in. Rough lumber becomes dressed lumber when it is put through a planing machine that reduces the cross section by $\frac{1}{8}$ to $\frac{1}{4}$ in. per side so that a nominal (S4S) 2×4 is actually about $1\frac{5}{8} \times 3\frac{5}{8}$ in. in size. Lumber can be ordered dressed with any of the four sides or edges finished as desired. The designation for the various surfacing is as follows: S1S (surface one side), S2S (surface two sides), S2E (surface two edges), S4S (surface four sides), or any combination of sides and/or edges is available on special order,

Table 3. Standard Lumber Sizes and Volumes

Nominal size, in.	Board feet per lin. ft	Board feet for standard lengths						
		8 ft	10 ft	12 ft	14 ft	16 ft	18 ft	20 ft
1×2	0.166	1.33	1.67	2.0	2.33	2.67	3.0	3.33
1×3	0.25	2.0	2.5	3.0	3.5	4.0	4.5	5.0
$1 \times 4; 2 \times 2$	0.333	2.67	3.33	4.0	4.67	5.33	6.0	6.67
$1 \times 6; 2 \times 3$	0.5	4.0	5.0	6.0	7.0	8.0	9.0	10.0
$1 \times 8; 2 \times 4$	0.667	5.34	6.67	8.0	9.34	10.67	12.0	13.34
$1 \times 12; 2 \times 6; 3 \times 4$	1.0	8.0	10.0	12.0	14.0	16.0	18.0	20.0
$2 \times 8; 4 \times 4$	1.333	10.66	13.33	16.0	18.66	21.33	24.0	26.66
2×10	1.667	13.34	16.67	20.0	23.34	26.67	30.0	33.34
$2 \times 12; 3 \times 8; 4 \times 6$	2.0	16.0	20.0	24.0	28.0	32.0	36.0	40.0
2×14	2.333	18.66	23.33	28.0	32.66	37.33	42.0	46.66
3×10	2.5	20.0	25.0	30.0	35.0	40.0	45.0	50.0
4×10	3.333	26.66	33.33	40.0	46.62	53.33	60.0	66.66
3×6	1.5	12.0	15.0	18.0	21.0	24.00	27.0	30.0
$3 \times 12; 6 \times 6$	3.0	24.0	30.0	36.0	42.0	48.00	54.0	60.0
3×14	3.5	28.0	35.0	42.0	49.0	56.00	63.0	70.0
$3 \times 16; 4 \times 12;$ 6×8	4.0	32.0	40.0	48.0	56.0	64.00	72.0	80.0
4×8	2.667	21.34	26.67	32.0	37.34	42.67	48.0	53.34
4×14	4.667	37.34	46.67	56.0	65.34	74.67	84.0	93.34
$4 \times 16; 8 \times 8$	5.333	42.66	53.33	64.0	74.66	85.33	96.0	106.66
6×10	5.0	40.0	50.0	60.0	70.0	80.0	90.0	100.0
6×12	6.0	48.0	60.0	72.0	84.0	96.0	108.0	120.0
6×14	7.0	56.0	70.0	84.0	98.0	112.0	126.0	140.0
$6 \times 16; 8 \times 12$	8.0	64.0	80.0	96.0	112.0	128.0	144.0	160.0
8×10	6.667	53.34	66.67	80.0	93.34	106.67	120.0	133.34
8×14	9.333	74.66	93.33	112.0	130.67	149.33	168.0	186.66
8×16	10.667	85.34	106.67	128.0	149.34	170.67	192.0	213.34

such as S1S1E. When ordering lumber even though it is dressed, the specific piece of lumber is requested by cross-sectional dimensions before it is dressed; this is termed nominal dimension. Lumber is purchased by the board foot unit of measure, 1 board foot being 144 cu in., or 1 sq ft of lumber 1 in. thick nominal dimension (Table 3). Lumber is priced by the thousand board measure. Lumber is commercially stocked in lengths of even feet in 2-ft modules (6, 8, 10, etc.). Because the pieces of lumber come in even 2-ft modules, thought should be given to the lengths in which the material will be used, so as to have the least waste in cutoffs. If 7 ft-0 in. lengths are needed, 14-ft material should be ordered rather than cutting the required 7-ft lengths from 8-ft mill lengths, or perhaps one 7-ft and one 9-ft required length can be cut from one 16-ft mill length. Sheathing

(1-in.-thick boards) is required for some jobs. This material is available in S4S (surfaced four sides), shiplap, or tongue and groove (T & G). Tongue and groove is also called D & M, which is dressed and standard matched boards (Fig. 2).

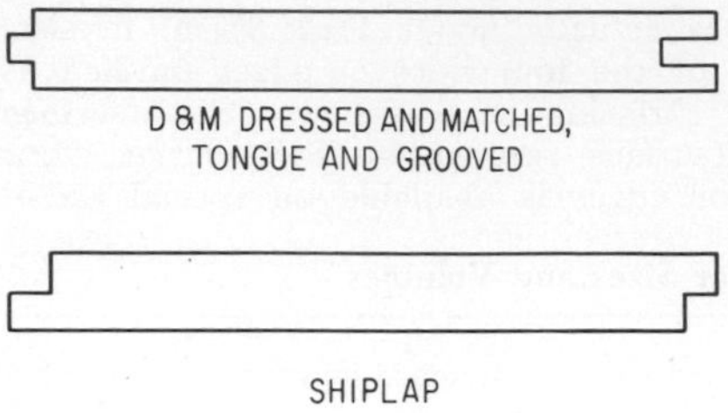

Lumber is graded by characteristics and usability as select, construction, standard, utility, and economy grades. Select grade is the highest-priced, with reduction for each grade down to economy as the cheapest. Construction grade is usually used for most formwork, select for all scaffold plank and for falsework that is to support a particularly heavy load. Grade of lumber is normally stamped or branded on every piece. Allowable unit stresses for construction grade are shown in Table 4.

D & M DRESSED AND MATCHED,
TONGUE AND GROOVED

SHIPLAP

FIG. 2. Cross section of commonly used form boards. Thickness is nominal 1 in.

Lumber in general is divided into two classes, hardwood and softwood, derived from the tree from which the lumber is cut. Hardwood trees have broad leaves; softwood trees have needlelike leaves (conifer). The division of hardwood and softwood trees does not always apply to the wood itself, as some hardwoods are softer than some softwoods, while there are softwoods that are as hard as medium-density hardwood. For concrete forming the so-called softwoods are most commonly used. The type used depends upon the area of the country where the work is being performed and the availability. The most widely used kinds of lumber for formwork are Douglas fir, southern pine, western hemlock, and eastern spruce.

Table 4. Suggested Allowable Stresses for Formwork Lumber

Species (construction grade)	Allowable unit stress, psi*				Modulus of elasticity, psi
	Extreme fiber bending	Compression parallel to grain	Compression perpendicular to grain	Horizontal shear	
Douglas fir.......	1,500	1,200	390	120	1,760,000
Southern pine....	1,500	1,350	390	120	1,760,000
Eastern spruce...	1,200	900	300	95	1,320,000
Western hemlock.	1,500	1,100	365	100	1,540,000

* Unit stresses may be increased up to one-third when short-term loading is involved.

Plywood. The use of plywood in concrete forming for form facing has improved the quality of finished concrete. The relatively large sheets of plywood have reduced the cost of building and maintaining forms, at the same time giving smooth surfaces that reduce the cost of hand-finishing concrete.

Plywood is a manufactured wood product consisting of an odd number of veneer sheets, or plies, placed crosswise to each other and bonded together with an adhesive under hydraulic pressure, and then put through a sanding machine, sanded to uniform thickness. The odd number of veneer sheets gives the manufactured panel strength to minimize warping and twisting. The strength of plywood depends upon the number of plies and the direction they run in relation to their support. Plywood is strongest when the exterior grain is parallel with the span and perpendicular to the supports (Table 5). Plywood is manufactured in standard sizes up to 48 in. wide and 96 in. long and in standard thicknesses from $\frac{3}{16}$ to $1\frac{1}{8}$ in. Panels up to 5 ft wide and up to 16 ft long can be produced on special order. Two types of plywood, exterior and interior, are manufactured. Both types are used to form concrete, but the exterior type is more durable because the adhesive bonding between the plies is waterproof and gives maximum reuse where needed. The grade of plywood is determined by the quality of the surface veneer. The grades range from A to D: A grade (sound) is sound, tight veneer free of knots and open defects; B grade (solid) has solid plugs with only minor sanding defects; C grade (exterior back) has knotholes limited to 1 in. maximum and splits limited to $\frac{3}{16}$ in.; and D grade (utility) contains solid knots not more than $2\frac{1}{2}$ in. maximum.

The U.S. Department of Commerce establishes standards[2] for the quality of plywood, including a requirement that plywood labeled "concrete form grade" shall be edge-sealed and mill-oiled. If particular standards are necessary for a specific application, the panels purchased should be so marked and certified that they have been inspected and tested to comply with the applicable U.S. Commercial Standard.

Table 5. Properties of Plywood Commonly Used for Formwork

Thick-ness, in.	No. of plies	Weight, lb per 1,000 sq ft	Weight, lb per 4 × 8-ft sheet	Parallel to face grain		Perpendicular to face grain	
				Moment of inertia I, in.^{4}	Section modulus S, in.^{3}	Moment of inertia I, in.^{4}	Section modulus S, in.^{3}
$\frac{1}{4}$	3	790	25	0.0143	0.114	0.0014	0.0247
$\frac{1}{2}$	5	1,525	49	0.0926	0.370	0.0324	0.1995
$\frac{5}{8}$	5	1,825	58	0.1670	0.534	0.0771	0.3520
$\frac{3}{4}$	7	2,225	71	0.2510	0.670	0.1710	0.6080
$1\frac{1}{8}$	7	3,350	107	0.7710	1.371	0.6530	1.3950

Exterior-type solid-core plywood with a resin-impregnated fiber face permanently fused under heat and pressure into one or both sides is called overlaid plywood. The overlay is available in either medium density or high density. The high-density overlay is commonly called plastic-coated, and it has a hard, smooth, grainless surface that resists abrasion and moisture penetration. Plastic-coated plywood requires only a light coat of oil between uses to ensure prolonged life (up to 200 uses). Medium-density overlay is similar to the high-density overlay, but it is substantially less resistant to abrasion and moisture penetration. It requires a normal application of form oil or bond breaker between each use to aid stripping and ensure added uses.

Plywood with striated surfaces is available with a variety of patterns to give the exposed concrete a textured finish. This type of plywood surface does not resist wear and therefore will endure fewer reuses than conventional plywood.

To form curved surfaces, plywood can be bent to the required radius, as shown in Table 6. If the radius is short, steaming or wetting the plywood will make it flexible. These aids to bending should be done only to exterior-type plywood, manufactured with waterproof glue. The curves can often be made better by using a double thickness of thin plywood rather than one thick sheet. Saw kerfing can also facilitate the bending of thick plywood. Care should be exercised when using this method of curving plywood so as not to cut through more than two plies of the plywood, as to do so would substantially weaken it.

Hardboard. Partly refined wood particles are impregnated with a special liquid, pressed together, and baked to form fiberboard or hardboard. This material makes a good form liner for certain types of work. It is particularly good for obtaining smooth concrete surface when used to face circular forms made of sheathing. It is usually used in ¼-in. thickness and is dependent upon backing for the strength to withstand concrete pressures.

Table 6. Minimum Recommended Bending Radii for Plywood

Thickness, in.	Min bending radius in plane of curvature, in.	
	Across face grain	Parallel to face grain
¼	15	24
⅜	36	54
½	72	96
⅝	96	120
¾	120	144

3 Steel

Standard and special rolled structural shapes (angles, H and I beams, channels, T sections) and various gages of flat rolled and corrugated sheet steel are used to make and support concrete forms. Steel allows longer spans and heavier loads than comparable-sized lumber and, of course, by comparison is more expensive at initial purchase. However, when many uses of a form are anticipated, it is more reasonable to choose the more durable, more expensive material. When steel, either rolled sections or sheet steel (flat corrugated), is to be used for support only (requiring no fabricating), it can be ordered from a steel warehouse. The warehouse will have cutting facilities so they can cut structural sections to length and trim sheets to width as well as cut them to length. Corrugated sheet metal is often used as a permanent form for concrete decks: that is, it is left in place permanently in the structure. Corrugated sheet metal comes in widths of 26 and 27½ in. When ordering corrugated sheets, the practice is to request the length required, and figure a net width of 24 in.

Because it takes special equipment to shear, bend, and weld steel, few construction projects are equipped to fabricate their own steel forms, so that the work of building these forms is done in a metal-fabricating shop. Building a good-quality steel form takes experience for both design and fabrication. The steel handbook[3] of the American Institute of Steel Construction has tables available that will assist the steel-form designer to select form elements that will best suit his requirements. The tables list the dimensions, weights, and properties of rolled steel.

Flat sheet steel is shaped into pans for forming pan joists, dome pans for forming waffle slabs, flat pan for flat slabs, round forms for column capitals, as well as form facing for steel ribbed forms.

4 Aluminum

Aluminum is the second most common metal used for concrete-form construction. The principal reason that contractors are using aluminum more today than in the past is that it is much lighter in weight than steel and therefore easier to transport and handle. Contractors are aware that if a product is easier for a man to handle, it will usually get better care, and the cost to use the lighter product will be less. The cost of an extrusion die to obtain a specific cross-sectional shape of aluminum is much less than that of a set of rolls to roll a steel section.

Plain aluminum, considered a soft metal when compared with steel, would wear fast and dent easily; hence it is necessary to use an aluminum alloy that has a hardness that compares with that of structural steel, which has a Brinell hardness rating of 150.

5 Fiber Glass

The use of fiber-glass-reinforced plastic (FRP) for concrete forms has greatly extended the versatility of forming for both precast and cast-in-place concrete. The use of fiber-glass-reinforced plastic has spread throughout the country. This rela-

FIG. 3. Circular beam forms fabricated of fiber-glass-reinforced plastic. (*Interform Division, La Mesa Industries, Inc.*)

tively new technique of building forms for concrete should give greater freedom of design to architects and engineers and reflect savings to both contractors and owners. The initial cost of the mock-up, or master mold, is high, but this cost is reduced substantially as many parts (forms) are made from the original mock-up.

It is important to deal with a reliable and experienced manufacturer when purchasing fiber-glass-reinforced plastic concrete forms. Almost any qualified fiber-glass manufacturer can build a mold, but it takes one with engineering and concrete-forming knowledge to build a mold that will act as a concrete form. The number of successful reuses of correctly built fiber-glass-reinforced plastic molds is over 200 on large molds and more on small molds, without a single failure. The shock of placing and vibrating concrete, the heat of steam curing, and the repetition of assembly and disassembly over many uses put a strain on any form. For these reasons

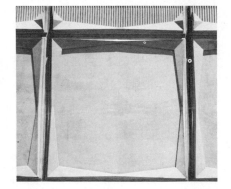

FIG. 4. Architectural details cast against fiber-glass molds. (*Plastic Engineering & Mfg. Co.*)

it is necessary to purchase a quality set of molds if the finished product is to meet the specifications and reflect a labor saving to the contractor. One method of comparing the quality and strength of one form with another is by weight. Usually the form that weighs more is stronger because it has more layers of fiber-glass buildup in the body of the form itself. Fabrication of a fiber-glass mold for a simple rectangular column is described in Chap. 33. Uses of fiber glass are shown in Figs. 3 and 4.

FORM TIES AND ACCESSORIES

6   Form Ties

The unit that holds concrete forms together to withstand the hydrostatic pressure of fluid concrete on the forms is called a form tie. Installation of form ties and alignment of wall forms comprise between 40 and 60% of the total form-erection cost for most applications. Because these two operations are so closely associated with each other, careful consideration should be given to the selection of the form tie. The tie should have sufficient strength to develop the strength of the forming material when it is loaded to its working capacity, with a large enough safety factor to ensure against failure. The safety factor varies with different types of forming requirements. A good rule of thumb is to use a tie with a 50% safety factor with light formwork, and use a tie that has a 100% safety factor for heavy formwork and where there may be any unusual hazard to life or limb.

Selection of the correct form tie for use on almost any form job will reflect a saving in the cost of the formwork. The savings will be nominal on a small job but may be substantial on a large project. The type of form tie chosen should be determined by the type of forming system selected, as discussed in Chap. 15.

Form ties fall into two general categories: those which are both form spreaders and form ties, and those which are only form ties. There are also form ties that are expendable (left in concrete) and some that are not expendable (removable for reuse) and still others in which part of the tie is expendable and the balance of it, called working parts, is reusable.

Form ties over the years have changed from job-fabricated ties made from tie wire and banding strap to factory-manufactured ties. The reasons for the change from "job-fabricated" to factory-manufactured ties are many, but the most significant is that factory-fabricated ties reduce labor costs. The banding strap and tie wire have relatively low tensile strength and so require the installation of many more units than a factory-built tie for a similar application. As the construction industry has developed means of placing concrete at faster rates, the demands for form ties to keep pace have been met by the tie manufacturers. Today, factory-manufactured ties are available with working capacity of from 1,000 lb for light forming to an ultimate strength of 72,000 lb for heavy-duty forming. The loads on form ties and tie spacing are discussed in Chap. 15.

Form Clamp. The form clamp, or button, as it is sometimes called, is used with a plain round rod to hold the forms together. The form-clamp-type tie affords a simple, flexible tying system, but it affords no spreader action. The round rod is cut to the necessary length and put through holes drilled in the forms. A form clamp is fastened to one end of the rod by means of a setscrew about 4 in. from the end of the rod. The second clamp is slipped onto the other end of the rod on the other side of the wall form and forced into position by means of a tightening wrench, and secured by a setscrew (Fig. 5).

Form clamps are available to fit over ¼-, ⅜-, ½-, and ⅝-in. rod stock. Working loads shown in Table 7 should not be exceeded. If it is necessary to remove

Table 7. Safe Working Loads for Plain Steel Rod

Rod diam, in.	Safe working load,* lb
¼	1,500
⅜	3,000
½	5,000
⅝	9,000

*Based on 32,000 psi for mild steel.

the rod from the concrete it can be accomplished by two means. The rod can be encased in paper tubing to prevent bonding with the concrete and be removed after the concrete has hardened, or it can be removed from the concrete (within 2 days) by means of a rod puller. If so desired, and if allowed by the specifications, the rod may be cut off flush with the wall face. Most specifications prohibit the use of a form tie that leaves a hole completely through the wall, or exposure of metal on the surface that might subsequently rust and stain the concrete except in concrete that is to be backfilled.

The **snap tie** (Fig. 6) is the most commonly used form tie in conjunction with wall forming. The snap tie gets its name from the fact that it is notched so that it can be broken or snapped off inside the wall face without spalling the hardened concrete. Snap ties are available in either 3,000- or 5,000-lb working-load strength. The ultimate strength of the 3,000-lb tie is assumed to be 4,500 lb and for the 5,000-lb tie it is 6,250 lb. These ties are often termed 3M and 5M snap ties.

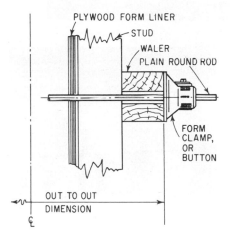

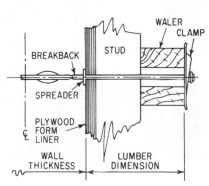

FIG. 5. One commonly used form clamp.   FIG. 6. The snap tie is designed to twist off at the "break-back" when it is removed from the concrete.

The standard snap tie is manufactured from a single piece of medium- or high-carbon wire, cut to length, notched to the required setback or, in this case, break-back, deformed in the center so it will not turn within the wall when the ends are being snapped off, and headed on each end to hold the snap-tie clamp. There are a variety of devices used with snap ties to provide spreader action and/or facilitate the break-back action. Plastic, wood, or metal cones, used as spreaders, also provide for a more positive break-back as well as furnishing a hole of sufficient size that can be dry-packed or pointed to get a better-appearing concrete wall finish. Flat metal washers (large or small) are also used as spreaders on snap ties; some are loose and rotate freely, while others are fixed by swaging to the wire. The snap tie is installed in the form by feeding the ends of the tie through holes drilled in the form face and through the wall, then secured at each end by a snap-tie clamp (slotted wedge-shaped pieces of steel), sometimes called a hairpin, that fits over the tie and wedges between the tie head and the wale.

When ordering snap ties, it is necessary to specify (1) the strength of tie (3,000 or 5,000 lb), (2) the type of spreader (fixed or loose washer, wood, plastic, or metal cone), (3) the depth of break-back necessary (1, 1½, or 2 in.), and (4) the tie size, which is lumber and wedge dimension plus wall thickness plus lumber and wedge dimension.

To figure the lumber and wedge dimension, add up the lumber that makes up the form thickness (from face of form to back of wale) plus ½ in. for the wedge.

Example: 2×4 studs and wales with ¾-in. plywood require 8½-in. lumber and wedge dimension for snap tie $(3\frac{5}{8} + 3\frac{5}{8} + \frac{3}{4} + \frac{1}{2})$. Used for an 8-in.-thick wall the tie would be ordered by length of $8\frac{1}{2} \times 8 \times 8\frac{1}{2}$ in.

Snap ties are available with neither break-back nor spreaders, in which case they are called tip-to-tip ties.

There is another type in which fiber-glass-reinforced nylon is injection-molded to form plastic snap ties that overcome the corrosion and rust staining often caused by metal ties left in the concrete. The plastic snap tie system uses plastic tie clamps and waler brackets (Fig. 7).

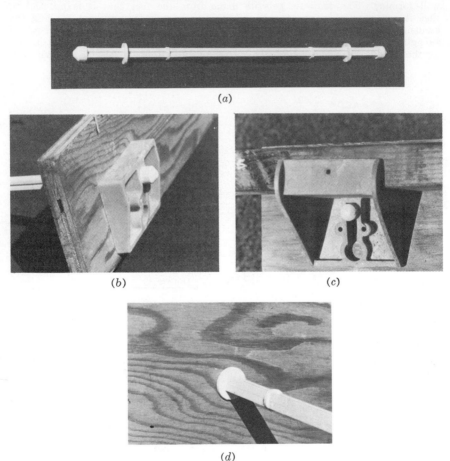

(a)

(b)   (c)

(d)

FIG. 7. Plastic snap ties and working parts, showing: (a) the snap tie unit; (b) form clamp for plastic snap tie; (c) combination plastic snap tie clamp and 2×4 waler bracket; (d) plastic snap tie secured in place to wall form. (*P.D.H. Photo Service.*)

The **coil tie** is a simple yet versatile type of form tie. Two helical wire coils are welded to opposite ends of two or four steel rods, called struts, to form the basic element or inner portion of the coil-tie assembly. The rest of the standard assembly consists of two coil bolts and two flat washers. The tie is assembled in place, by threading one of the coil bolts through a flat washer and then through the form, and is screwed into the open end of the helical coil until tight (Fig. 8).

The same procedure is followed when the second form face is put in place and the tie assembly is completed.

The coil tie assembled as described above serves to hold the form together but does not have any spreader action, nor does it provide the setback of the coils (break-back). The coil tie can be equipped with plastic, metal, or wood cones to provide spreader action with setback. To accomplish this, cones can be slipped on loose over bolts (between the form face and the inner portion of tie) or can be screwed to the coil itself when so ordered (Fig. 8). Only the inner section of the coil-tie assembly is expendable (left in the concrete), but the coil bolts and plate washers are reusable (removed) and are termed working hardware. Working hardware is available for rental or purchase with a returnable option.

Coil ties with two struts are available for use with ½- ¾- and 1-in.-diameter bolts, and four struts for use with 1- and 1¼-in.-diameter bolts; working loads

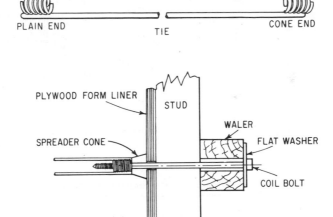

FIG. 8. The coil tie is designed for either plain or cone spreader installation.

vary from 6,000 to 36,000 lb. Water-seal ties are also available with washers swaged to struts, designed to eliminate seepage along the tie caused by water pressure.

When ordering coil ties the following information should be given: diameter of the coil bolt required, working-load strength of the tie, and overall length of the coil tie required. If screw-on cones are to be used or if the tie requires a water seal this should be specified. Example: ½-in. 9,000-lb 8-in. water-seal tie for screw-on cones. Cones are ordered by the nominal bolt diameter (½, ¾, 1 in.), material that the cone is made of (wood, plastic, metal), and screw-on or slip-on type. Metal cones are more durable. However, they are castings and are not so uniform as plastic cones; as a result the plastic cones are removed more easily.

She-bolts. The she-bolt assembly (Fig. 9) consists of two nut washers, two waler rods, and an inner section called the center or tie rod, which is a piece of smooth round rod stock threaded on both ends (threads are U.S. Standard). It is job-assembled in much the same manner as the coil tie, with the exception that the nut washer threads onto the waler rod, and the center rod screws into the waler rods. Spreader action is provided by incorporating adapter cones in the assembly.

Strength of the she-bolt assembly is primarily determined by the diameter and type of steel in the center rod, as shown in Table 8. Cold-rolled steel gives a stronger tie rod than rods made of hot-rolled stock, and, of course, the larger-diameter rod is capable of taking higher tensile loads. As the diameter of the tie rods increases for more tensile capacity, so also is it necessary that the diameter

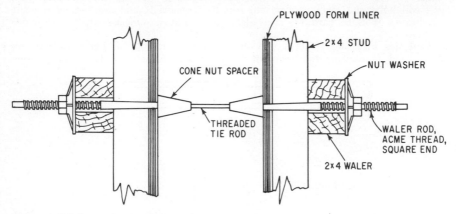

FIG. 9. Waler rods for the she-bolt are tapered to facilitate removal from the concrete.

Table 8. Properties of Threaded Tie Rods

Rod diam, in.	Hot-rolled stock		Cold-rolled stock		Special high tensile strength	
	Safe working load, lb	Ultimate strength, lb	Safe working load, lb	Ultimate strength, lb	Safe working load, lb	Ultimate strength, lb
⅜	3,000	4,500	5,000	8,100	7,500	8,300
½	5,000	7,200	9,000	13,800	15,000	19,000
⅝	.....	.....	14,000	20,500	20,000	25,000
¾	.....	.....	20,000	29,500	28,000	46,000
⅞	.....	.....	......	......	46,000	58,000
1	.....	.....	......	......	57,000	72,000

of the waler rod be increased. The bearing area of the nut washer can also be critical when figuring load capacity.

Threads on tie rods that are made of hot-rolled steel are either cut or rolled; threads on tie rods of cold-rolled or high-tensile steel are rolled. One advantage of rolled threads is that they are likely to be smoother than cut threads, as cold working hardens the threads so that they are not easily damaged. The cold working also develops compressive stresses in the threads, making them more resistant to fatigue from stresses in bending or tension. Tie rods that have cut threads can be pulled out of the hardened concrete, but roll-threaded tie rods cannot be withdrawn. The center section of a tie rod that is not to be removed should be deformed so that it will not rotate when waler rods are removed.

When ordering tie rods the supplier should be given the tie-rod diameter and length, type of steel and strength required, and the length of thread required on each end. To develop the tie rod's strength the rod must have at least as much thread in length engaged with the waler rod as the rod is in diameter. For example, a ½-in.-diameter rod must be engaged at least ½ in. in the waler rod.

7 Miscellaneous Form Accessories

Void formers serve the purpose of creating voids within the concrete either for utility ducts, or for weight reduction by displacing concrete. Two general types are available, expendable and removable.

Expendable formers remain in place after the concrete has been cast. Square and rectangular voids are available (Fig. 10), obtaining their rigidity by means

Fig. 10. The rectangular void former for a prestressed box girder has been placed on the initial layer or bottom membrane of fresh concrete. Note the piece of ¼-in. plywood on top to protect the former during placing of steel and concrete. This former, with the plywood, is completely surrounded by concrete and remains in the girder.

of their interior "eggcrate" reinforcement. Square and rectangular formers are usually shipped flat and are easily assembled on the job by stapling, taping, or gluing. Another common type is the fiberboard tube described in Art. 10, which can be sawed to any desired length on the site (Fig. 11).

Removable formers are frequently of rubber, similar to heavy hose that can be inflated with air or water. When inflated, the former becomes quite rigid, expands slightly, and decreases in length as much as 10%. Another type is reinforced so that it is stiff and does not expand when inflated. During installation, the former can be bent or curved to enable it to pass around obstructions. Diameters range from about ¾ to 120 in. Smaller sizes are frequently used for conduits for electrical wiring, and to form ducts for posttensioning cables. Inflation pressures may be as high as 30 psi for the small sizes, ranging down to 5 psi for the large sizes which are used for water and ventilation conduits and shafts.

Normally no parting compound is used on these formers as they release from the concrete when deflated.

Miscellaneous form accessories that are available for use with concrete forms are too numerous to be discussed in detail here, but some of the most useful will be considered.

Chamfer strip is made of plastic, metal, rubber, or wood. The manufactured chamfer strip is more expensive in first cost than mill-run wooden strip but can

FIG. 11. Two cylindrical void formers tied in place with heavy wire. These formers were set in place on the fresh concrete and wired to the reinforcing steel. Formers must be rigidly fixed in place, as any movement after concrete is in place is apt to cause a crack.

be reused many times. In addition it is moistureproof, so it does not dry out and warp after being wet. Sheet metal comes in straight lengths; plastic and rubber are available in either straight lengths or coils, with either rounded or straight chamfer face, as shown in Fig. 12. The manufactured chamfer has a tail that is pinched between the forms at their junction to hold the strip in place as well as to ensure a tight joint. The plastic and rubber can also be nailed, screwed, or glued to the form if so desired.

FIG. 12. Chamfer strips come in a variety of sizes and shapes. This shows a triangular cross section with a tail.

Rustication strips are made of wood, rubber, plastic, or metal (see Fig. 15, Chap. 25). The advantage of plastic, rubber, or metal strips as compared with wood strips is that they do not absorb water and expand as the wood does; thus they come loose from the concrete more easily when forms are removed. Rustication strips made of wood cost about one-fourth as much as manufactured ones. They are usually milled from soft knot-free lumber such as white pine, and are most economical when only a small amount is required for only a few reuses. The wood strips are nailed in place when the form is to be used not more than three times; otherwise the strips are either bolted or screwed to the form.

Cork stoppers provide one of the best methods for plugging relatively small holes (up to 1 in. in diameter) in wood form faces. The cork is pushed or driven, small end first, into the hole from the form face to the outside until it fits snugly. It is then cut off flush with the form face with a very sharp flexible blade. The texture of the cork is similar to that of the plywood; so the patch does not show in the finished concrete surface. There are two sizes, $9/16$ and $13/16$ in. diameter, of plastic plugs available for plugging form holes. The $9/16$-in. plug usually fits snap-tie holes and the $13/16$-in. size fits she-bolt holes.

Hole covers made of light-gage sheet metal are also available. These covers are easily applied by use of a magnetic applicator that holds the cover in place while it is struck with a hammer. Sheet-metal hole covers are not recommended for use on forms where high-quality finished concrete is desired because the contrast in texture between the plywood and the sheet metal causes unsightly blemishes on the concrete. For forms where concrete is not considered "finished" concrete, the sheet-metal covers are economical in both initial cost and installation cost.

A time- and laborsaving method for aligning wall or column forms is to use an adjustable brace. A turnbuckle-type bracket attached either between the brace and its anchor or between the brace and the form will make the brace adjustable. The form, when initially erected, is plumbed close to line and the brace is secured in place. The form can then be set to exact alignment by adjusting the turnbuckle to shorten or lengthen the brace.

8 Form Oils and Parting Compounds

In order to facilitate stripping or removal of forms after concrete has been placed, the form faces should be treated to prevent adhesion or bonding between the concrete and the form. Many kinds of form sealers, form coatings, and release agents are available for this purpose. The type of treatment that should be used depends upon the form material, the concrete surface required after forms have been removed, and the prevailing weather.

Prevention of adhesion of concrete to the form can be accomplished by:

1. Applying a liquid to the form that leaves an oily coating on the form
2. Applying a resin or varnish that dries and leaves a smooth, glasslike surface
3. Applying a liquid retardant to the form that prevents hydration of a thin layer of cement paste immediately adjacent to the form
4. Applying a liquid that dries, but saponifies when the concrete comes in contact with it
5. Using dense, smooth form-lining material, such as fiber glass or some plastic-coated plywoods.

Some of the treatments are applied to the plywood at the factory, some can be applied either at the factory or on the jobsite, and some can be applied only on the job immediately prior to placing concrete.

Plain lumber and untreated plywood should be sealed to inhibit moisture penetration from the fresh concrete, or should be treated with a combination product that acts both as a form sealer to prevent moisture penetration and as a parting agent that aids form removal. Plywood can be coated with a resin-base product, plastic compound (rolled on at the mill or job-applied), shellac, or lacquer that seals it from absorbing moisture. If the form-facing material is to be sealed as described, it is recommended that the treatment be given to both faces of the wood so it will weather uniformly. Sealing of plywood in this manner prevents wet concrete from raising the grain, giving a smoother finish than would be obtained with untreated plywood. When plywood is to be sealed on the jobsite, it may be advisable to use a dip-tank and drainboard system to apply sealer. The plywood is submerged in the sealer and then set up to drain and dry. This assures an even coat of sealer on both faces as well as the edges. Sealed plywood requires only a light coat of form oil before each use.

Steel-faced forms do not absorb moisture; so a releasing-agent type of form coating is most suitable. However, steel forms rust and concrete bonds well to oxidized steel; so to prevent this situation from occurring a form-releasing agent with a built-in rust inhibitor is advantageous. If a steel form has any rough spots where concrete would bond, these should be ground or wire-brushed smooth and a heavier form coating applied to the questionable areas.

Reinforced fiber-glass plastic forms, like those made of steel, do not absorb moisture from concrete and hence do not need a form sealer. Some manufacturers of plastic forms advise that no form coating is necessary to strip their forms. This is true, but stripping is greatly facilitated when a very light coat of parting agent is used.

Forms that are a combination of wood and metal require a form-releasing agent that will work compatibly well with both wood and metal.

For good economical care of forms, cleaning and oiling should be done as soon as forms are stripped. Green concrete can easily be removed and the oiling will protect the forms from the weather (hot and dry or cold and wet).

In some situations the only time the forms can be coated is when they are in place. If that is the case, care should be exercised to see that the form coating is not slopped on reinforcing steel or on areas of existing concrete where bonding is necessary.

When choosing a form coating, it is important to be sure that the product selected will not interfere with curing and/or sealing the concrete surface or with the adhesion of paint or plaster if either is to be applied to the hardened concrete. Form coatings are constantly being improved by the manufacturers, and the user should rely on the advice of a reputable manufacturer for specific instructions.

Special care is necessary when white cement is being used in the concrete, as even slight discoloration or texture variation in the white surface is objectionable. The cement manufacturer should be consulted to ascertain which compounds have been found to be acceptable.

PREFABRICATED FORMS AND SHORING

The many types of prefabricated forms that are available to mold concrete can be divided into two general classes: those made specifically for one particular form job, and those which are modular in size and are adapted to many different forming jobs. Both these types may be either job-built or factory-manufactured. Figure 13 shows factory assembly-line production of form panels. Some of the

Fig. 13. Assembly-line fabrication of factory-built forms, showing attachment of plywood sheathing to metal frames. (*Symons Mfg. Company.*)

factory-built modular forms are discussed here, and job-built and special-application forms are discussed in Chap. 15.

9   Factory-built Forms

One proprietary forming system, the Universal, consists of three basic elements; the form panel, the form tie, and the tie key. The form is a plywood-faced panel within a steel frame constructed of specially rolled structural-steel sections that act as built-in studs. The edge member is T-shaped and offset so that it cradles and protects the plywood edge to assure long plywood life and maximum reuse, as shown in Fig. 14. It is slotted to receive ties and punched for the attachment of form accessories. Exterior-grade plywood is attached to the frame by means of split rivets, allowing the plywood to be removed and either replaced or reversed. The tie is made of flat strap steel with loops on each side of each end to hold the tie to the form. With a 3,000-lb working-load capacity, the form tie acts also as form spreader and form lock. It is placed in the slot formed between the panel frames when two panels are brought together. The tie key (similar to the old-style cut nail) is inserted in the tie loop, thus securing the form tie to the form.

This type of form can be erected to whatever height is required by stacking the form units one on top of the other. The panels are held to line by alignment (horizontal and vertical) and bracing lumber, which is required on only one wall of a double-wall form. Horizontal aligning members are single 2×4's attached to the back of each row of forms near the top of the panel. A second brace is added near the bottom of the first row of forms by means of liner clamps when the forms are erected more than one panel high. Vertical alignment is accomplished by using 4×4's or double 2×4's approximately 8 ft on center. Bracing lumber is usually attached to these vertical members. Inside corners are assembled by means of prefabricated-steel inside corner forms; outside corner forms are held in place by use of an outside corner angle and clamps, which eliminates the need to "log-cabin," or stagger, the boards at the corner. By using two filler angles and plywood, job-built wood fillers can be made to any required width. To compensate for the circumferential difference between inside and outside surfaces metal filler forms of 1-, 1½-, and 2-in. widths are used with the outside form when forming circular structures (Fig. 15). Scaffold brackets can be attached directly to the back of the form's steel frame.

Standard panels are 24 in. wide and range in height from 1 to 8 ft in 1-ft increments. Fractional panels are available in 12- and 18-in. widths and in heights from 2 to 8 ft in 1-ft increments.

The Symons system consists of plastic-coated plywood supported and edge-encased in a special steel frame. These panels are equipped with handles to assist in handling. The forms, as shown in Fig. 16, are assembled by means of the connecting wedge bolts, which also serve to secure the ties, wales, and outside corner angles in position. The basic form panels are 24 in. wide and 3, 4, 5, 6, 7, and 8 ft high; plywood-faced filler forms from 4 through 20 in. wide in the same heights are available, making this system very adaptable to column forming. Inside corners are formed by using a steel inside corner form which connects to standard panels. Two types of form ties are used. One is made of wire with loop welded to each end; the other is a flat strap tie slotted at both ends. The tie is a combination form spreader and form tie that has a safe load capacity of 3,000 lb.

The form panels are held in horizontal alignment by means of a double 2×4 and secured in position by the waler plate, waler tie, and the steel wedge. Vertical alignment is handled in a similar manner. Pilasters can be easily formed by using the steel pilaster forms and the correct-width filler form.

These forms have been adapted for use as a suspended slab-forming system (Fig. 17). By using I beams with adjustable angle seats as stringers and steel-post shores that can be adjusted from either below or above, the form panels become deck forms. This deck-forming system was developed so shores and stringers could

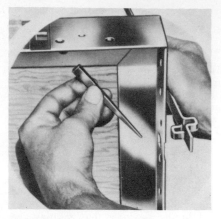

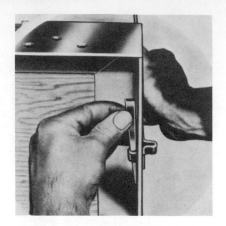

FIG. 14. (a) Tie loop is placed in the square tie hole; then the tapered tie key is set into the tie loop to lock the tie to the panel.

FIG. 14. (b) Bringing the next form panel into position and inserting a tie key in the second loop ties next form in place.

FIG. 14. (c) Repeating these steps on the opposite side closes the wall form.

FIG. 14. Three steps in assembly of the Uni-form tie key. (*Universal Form Clamp Co.*)

be left in place to support the concrete slab that has not fully developed its strength, but permitting removal of the deck forms for use elsewhere.

A third forming system is the Economy, using an all-steel form as its principal element. The rigid form panels are constructed by welding an alloy-sheet-steel face to a steel frame specially designed for strength and the attachment of accessories. The panels are clamped and locked together by means of plate clamps, commonly called "dogs" on most jobs. Alignment both horizontally and vertically is accomplished by the use of single 2 × 4's fastened to the back of the forms

Fig. 15. Concentric circular tanks cast in factory-built forms. Forms can be aligned from one side only, thus simplifying erection. (*Universal Form Clamp Co.*)

by cam-action aligner clamps. The form tie is a combination form spreader and form tie made from flat strap steel designed to withstand a minimum breaking load of 8,000 lb, secured to the panel frame by spreader tie pins. Standard rigid panel lengths are 48, 36, 24, and 12 in. and 24, 16, 12, 8, 7, 6, and 5 in. wide. Flexible panels are the same length as the rigid panels and 4, 3½, 3, 2¾, 2½, 2¼, and 2 in. wide. Steel inside corner forms are made to form 90° corners, and flexible angle form panels are available to form obtuse-angle corners.

Using specially designed trusses that attach to the side of adjustable post shores and act as stringers, this system can be used for suspended-slab forming. The form panels are placed between the trusses to form the deck. When the forms are ready to strip, the trusses and form panels are lowered 4 in. by removing pins. The shores can remain in place to support the deck, thus eliminating the need for reshoring. All other equipment may then be removed for use elsewhere.

The Atlas Compo Form system consists of a metal-frame plywood-faced form using two different types of metal frames. One frame is made of a special rolled-steel section; the other is made of a high-strength aluminum-alloy extrusion.

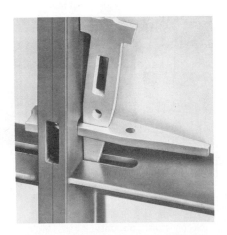

Fig. 16. Connecting hardware, consisting of two wedge bolts, for holding form tie and connecting two prefabricated panels together. (*Symons Mfg. Company.*)

Both frames are designed to protect the edges of the plastic overlay plywood to allow it to be reused many times. The steel-frame form weighs about 5 psf and is available in widths from 12 to 24 in. in increments of 2 in., and in lengths from 3 to 8 ft in increments of 1 ft. The aluminum-frame forms are manufactured in three widths, 12, 18, and 24 in., and in 3-, 4-, 5-, 6-, 7-, and 8-ft lengths, weighing just a little more than 3 psf (Figs. 18 and 19). Both types of form frames are notched for the insertion of form ties which act as form spreaders, eliminating the need for wales. Alignment of the forms is assured by use of one row of double 2×4's on only one side of the wall for each lift of forms erected. Special fabricated steel forms are available in two sizes, 6×6 in. and 8×8 in., to form inside corners. A punched angle is used to lock forms together at outside corners.

Yet another system is one in which lightweight aluminum panels 2 ft square can be assembled into wall forms for foundation and patio walls, carports, planters,

FIG. 17. Prefabricated form panels in use for deck forming. (*Symons Mfa. Company.*)

and retaining walls, as well as for garages and commercial and industrial buildings. The face of the panels is patterned so that concrete cast against them resembles brick or stone masonry (Figs. 20 and 21). The machined edges of the panels have small recesses through which the spacer ties fit, and panels are held rigidly in alignment by a system of pins and wedges. The machined edges permit assembly of the panels to close tolerances so as to obscure the joint between panels. One man can handle a 2×8-ft panel without difficulty.

10 Fiberboard and Corrugated Paper

The popularly called "paper forms" are made of laminations of paper, impregnated with waxes and resins to bond the layers together and to make the finished product

(a)

(b)

FIG. 18. (a) Basement wall forms consisting of prefabricated metal frame panels; (b) stripping prefabricated form panels. Note that one man can handle the panel. (*Western Forms, Inc.*)

water-repellent. Thickness of this solid cardboard material may be as much as ½ in. Corrugated paper similar to the type commonly used for shipping cartons, consisting of layers of paper separated by a layer of corrugated paper, is used for void formers of various shapes, such as rectangular or square. Cylindrical void formers and column forms are made of heavy cardboard.

Fiberboard tube forms, developed to form circular concrete columns, are available in lengths to 20 ft long. Range of inside diameter is from 2 to 48 in. Most manufacturers make the tube forms in two wall thicknesses, regular and lightweight. Ten feet is the maximum length of the lightweight tube available from most manufacturers. Table 9 gives general information on standard-wall round tube forms. Figures 22, 23, and 24 show applications.

FIG. 19.　Atlas prefabricated panels forming a square column.　(*Hico Corp. of America.*)

FIG. 20.　Panels of Con-tech forms assembled into 2 ft wide by 8 ft high.　Those on the right have been erected and aligned.　(*International Concrete System Co.*)

FIG. 21.   The lower wall of this house was cast in place against brick-patterned aluminum forms.   (*International Concrete Systems Co.*)

FIG. 22.   Bridge columns cast in fiberboard tube forms.   (*Sonoco Products Co.*)

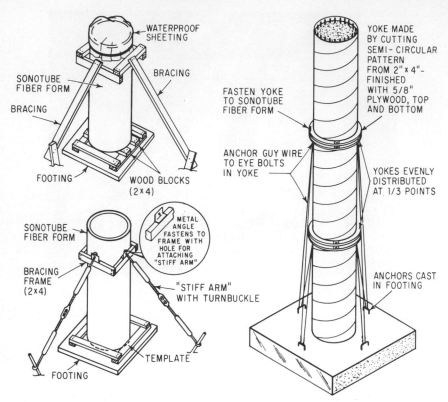

FIG. 23. Methods of bracing fiberboard column forms. (*Sonoco Products Co.*)

Table 9. Cylindrical Fiberboard Forms

Inside diam, in.	Volume/lin ft		Light-wall tube		Regular-wall tube	
	cu yd	cu ft	Wall thickness, in.	Weight, lb/lin ft	Wall thickness, in.	Weight, lb/lin ft
6	0.0073	0.197	0.125	0.79	0.200	1.27
8	0.0129	0.348	0.125	1.05	0.200	1.62
10	0.0202	0.545	0.150	1.55	0.225	2.27
12	0.0291	0.786	0.150	1.85	0.225	2.71
14	0.0396	1.096	0.175	2.48	0.250	3.48
16	0.0517	1.396	0.175	2.83	0.300	4.73
18	0.0654	1.766	0.200	3.61	0.300	5.31
24	0.1164	3.143	0.225	5.35	0.375	8.75
30	0.1818	4.909	0.300	8.77	0.400	11.61
36	0.2618	7.069	0.300	10.52	0.400	14.24

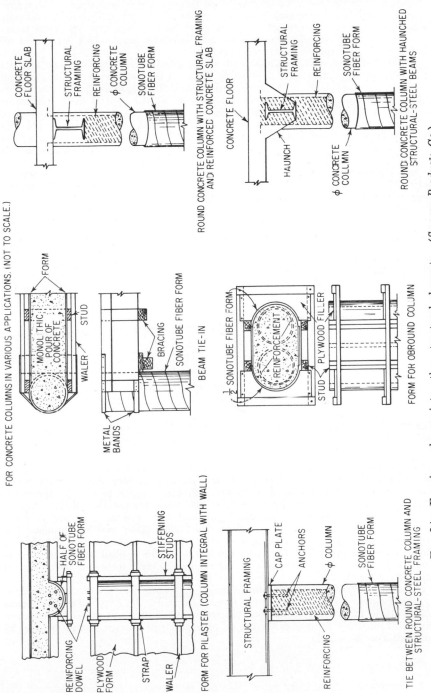

FIG. 24. Framing columns into other structural elements. (*Sonoco Products Co.*)

Fiberboard forms should be stripped from concrete columns as soon as permitted by job specification. Stripping is best accomplished by cutting the form vertically with a portable electric saw with the blade set slightly less than the form thickness. The balance of the form can be cut with any sharp blade and pried off with a wide-blade tool.

Round fiberboard voids equipped with either plastic, metal, or fiber end closures are available from 2¼ to 49 in. outside diameter and up to 50 ft long to form voids within concrete slabs or other members, such as precast piles. The voids reduce weight by displacing low-working concrete at the neutral axis. The voids also present a problem to the installer, that of holding them in place while concrete is placed below the hollow voids. One method of holding the void in place is to strap it in place with steel banding material. However, the best method for medium-sized voids seems to be a combination spacer and void tie-down, as shown in Fig. 11.

One company makes a conical fiber tube for use as a sleeve over an anchor bolt. The small end of the sleeve goes over the bottom of the bolt; so the large end permits horizontal adjustment at the top of the bolt.

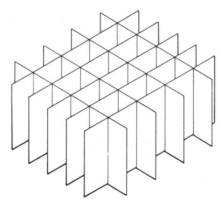

FIG. 25. The so-called "eggcrate" type of fiberboard reinforcement for void formers and paper pan molds.

Rectangular-shaped corrugated fiberboard voids are used to form a void between the soil and the structural element, such as a grade beam or a slab on grade. The void supports the concrete until it reaches sufficient strength to be self-supporting. Moisture and soil acids gradually collapse the void, leaving the element free of pressure from underneath, eliminating beam and slab cracking from soil pressure. A similar type of former is used in fabricating hollow prestressed box girders. The "eggcrate," as it is called, consists of a skin of heavy corrugated paper forming the periphery of the void. The space inside the former is filled with three-dimensional corrugated paper, interlaced at intervals of several inches to make a rigid support for the skin.

Corrugated fiberboard is made into forms for forming pan joist and waffle slabs by the Lawrence Paper Co. These forms are called Jay-pans. The joist pans are available with either straight or tapered ends. Both these types of forms are supported on fully formed decks by an eggcrate type of core shown in Fig. 25 (see also Figs. 26 and 27).

11 Shoring

A temporary substructure, called shoring, is necessary to hold formwork for decks, beams, and girders in place until the concrete has attained sufficient strength

FIG. 26. Jay-hawk corrugated-paper pan forms in place, ready for installing reinforcing steel. (*Lawrence Paper Co.*)

FIG. 27 Underside of suspended slab cast in place on corrugated-paper pan forms. (*Lawrence Paper Co.*)

to be self-supporting. The temporary support may be either vertical, horizontal, or, in most cases, a combination of both. There are many types and sizes of shores and shoring systems available from which to choose for each particular application.

Vertical Shores. Lumber, at one time the only material used for shoring, and still an important component, has been replaced by steel on many jobs. Lumber used for shores varies from 3 × 4's to large timbers, either as individual posts or built up into shoring bents. Some advantages of wood shores are their low initial cost and high load capacity in proportion to weight as compared with metal, but they have the disadvantages of higher labor installation cost and lack of versatility. For instance, wood shores cut to a short length for use in one area cannot be easily adapted to longer lengths for use elsewhere without expensive reworking. By using a screw jack at either the top or bottom of a wood shore, however, it can be easily adjusted to exact height, within a range of about 4 to 6 in. Ellis clamps, used in pairs, give adjustability to two-piece vertical wood shores (Fig. 28). Standard clamps are available for 3×4's, 4×4's, and 6×6's (S4S). Two clamps are used in the makeup of each shore, one clamp being slipped over the lower shore member and nailed securely about 14 in. from the top of this member; then the second clamp is secured about 2 in. from the top of the member, making the clamps about 12 in. apart. The clamps are held at 90° to the lower member while the upper member is threaded into the clamps. For shores up to 20 ft high the lower member should be 7 ft long, giving an adjustment of over 5 ft.

Fig. 28. Two Ellis shore clamps are nailed onto the bottom post; then the upper extension is fitted in and adjusted to the required elevation. (*Ellis Mfg. Co., Inc.*)

These shores are recommended for a maximum load of 3,000 lb per shore using two standard clamps on each shore.

Baker, Roos, Inc., makes two types of shores that incorporate both wood and steel in their construction. The standard shore, available in three sizes, has a wood head for the easy attachment of T heads. The extension-type shore is available in two sizes; it has a flat steel head through which a 4×4 can be inserted to extend the length of the shore but still maintain its adjustability.

The Symons shore is an adjustable composite of sizes and capacities shown in Table 10. Standard extensions are available in lengths of 2, 4, 6, and 8 ft. The extension consists of a 4×4 with a $1\frac{1}{4}$-in. pipe bolted in one end.

All-steel post shores, available from a number of manufacturers, have a load-carrying capacity substantially greater than other types of individual shores, which tends to offset the disadvantages of heavier weight and difficulties in lacing and bracing because of a lack of nailing surfaces. Most steel post shores telescope

Table 10. Symons Composite Steel and Wood Shores

Nominal length, ft	Length of shores				Weight, lb
	Telescoped		Extended		
	ft	in.	ft	in.	
4	4	6	7	6	40
6	6	6	11	6	48
7	7	6	13	0	54
8	8	6	15	0	60

Safe Load Capacity, Lb Unbraced to Heights Shown

Nominal length, ft	Extended to following heights, ft							
	7½	9	10	11	12	13	14	15
4	6,000							
6	6,000	6,000	5,500	4,500				
7	6,000	6,000	6,000	5,500	3,500	3,000		
8	.....	6,000	6,000	6,000	4,000	3,500	2,500	2,000

NOTE: The above ratings may be increased 100% by proper bracing.

one steel tube within another. The upper section, called the mast, has a series of holes punched through the tube wall to receive a pin that connects it to the lower section of the shore, called the base, the top of which is threaded. Rough adjustment is accomplished by placing the pin in the desired hole, thus fixing the mast in position. Then, by rotating a threaded collar on top of the base section, the shore is brought into final adjustment (Fig. 29).

Tables 11 and 12 give statistics on two commonly known shoring systems. Others are available, and the user should consult the manufacturer's current catalog for data on any particular shore, whether composite or all-steel.

Horizontal Shores. The adjustable horizontal shore, developed in Western Germany, has proved to be an asset for supporting deck forming by reducing both labor and material costs. Most adjustable horizontal shores consist of two basic units, a plate section and a lattice section. Adjustability is realized by telescoping the plate section into the lattice section, thus allowing the shores to be made into a variety of lengths. Each member has camber rolled into it to compensate for loading deflection, so that it is capable of supporting heavy concrete loads without suffering negative deflection.

Spanall adjustable shores are available in two categories, standard and lightweight (Fig. 30). The standard beam (No. 730) is particularly good for long spans and heavy loads. It spans from 6 ft 11 in. to 29 ft 7 in. and has a number of members that can be used in combination with other members to make various lengths of shores, as shown in Table 13. Spanall Series 610 lightweight beams, weighing 49 lb, are adjustable from 6 ft 1 in. to 10 ft 2 in. and the Spanall

FIG. 29.   The Waco all-steel adjustable post shore.   (*Waco Mfg. Company.*)

FIG. 30.   Spanall horizontal shores, fully extended.   Adjustable steel post shores are also shown.   (*Spanall Inc., Div. of Patent Scaffold Company.*)

Table 11. Waco Steel Post Shores

Shore No.	Height			
	Min		Max	
	ft	in.	ft	in.
701-P and J8..............	6	2	11	1
701-J14....................	6	3½	11	1
702 Series.................	8	2	13	1
703 Series.................	10	9	15	7

Recommended Working Loads

Height of shore, ft	Max load, lb	
	Unbraced	Braced
6	9,000	9,000
7	9,000	9,000
8	8,000	8,500
9	7,000	7,750
10	6,000	7,250
11	5,000	6,750
12	4,250	6,250
13	3,500	5,750
14	2,900	5,250
15	2,500	4,750

1015, weighing 100 lb, is adjustable from 9 ft 11 in. to 15 ft 5 in. Spanall beams are locked into position with each other by means of bolts permanently attached to each end of the lattice or outer section.

The Hico Corporation manufactures adjustable horizontal shores consisting of two-member units made from high-strength aluminum-alloy extrusions, as depicted in Fig. 31. The inside member is an I section that telescopes into the outside member, which is a box-beam section. Sliding the I section within the box section varies the length of the beam as designed. Hico beams are usually in pairs and named for their minimum and maximum spans. For example, Beam 59, weighing 28 lb, is adjustable from 5 ft 9 in. to 9 ft 9 in., and Beam 610, weighing 47 lb, is adjustable from 6 ft 6 in. to 10 ft 6 in. Other popular sizes are No. 915, 67 lb, and No. 1119, 82 lb.

Global Industries, Inc., offers three sets of adjustable horizontal shores, the Globalite 9-15, the Globalite 7-12, and the Globalite 6-10. This equipment is all considered lightweight, so it can be handled by one man. The Globalite units are held in position by use of a wedge that slides on the bottom chord of the plate or inner section. Units range in length from 5 ft 11 in. to 15 ft 2 in., weighing from 40 to 88 lb each.

American Pecco Corporation has available both standard and lightweight horizontal shores. The standard beams consist of three basic units that can be made up to various lengths from 8 ft 6 in. to 27 ft 7 in. The basic units are a P-8 plate section 8 ft 6 in. long weighing 76 lb, an L-8 lattice section 8 ft 6 in. long weighing 63 lb, and an L-11 lattice section 11 ft 8 in. long weighing 86

Table 12. All-steel Post Shores, Patent Scaffold Co.

Height of shore	Safe working capacity, lb	Height of shore	Safe working capacity, lb
Shore No. 1		**Shore No. 2**	
5 ft 7 in. fully closed	10,000	6 ft 7 in. fully closed	9,200
5 ft 8 in. to 5 ft 11 in.	9,200	6 ft 8 in. to 6 ft 11 in.	8,000
6 ft 0 in. to 6 ft 5 in.	7,500	7 ft 0 in. to 7 ft 5 in.	7,200
6 ft 6 in. to 6 ft 11 in.	5,700	7 ft 6 in. to 7 ft 11 in.	6,800
7 ft 0 in. to 7 ft 5 in.	5,600	8 ft 0 in. to 8 ft 5 in.	6,000
7 ft 6 in. to 7 ft 11 in.	5,400	8 ft 6 in. to 8 ft 11 in.	5,600
8 ft 0 in. to 8 ft 11 in.	5,200	9 ft 0 in. to 9 ft 11 in.	5,500
9 ft 0 in. to 9 ft 10 in.	5,100	10 ft 0 in. to 10 ft 10 in.	5,450
Shore No. 3		**Shore No. 4**	
8 ft 2½ in. fully closed	8,300	11 ft 0 in. fully closed	5,300
8 ft 3½ in. to 8 ft 11 in.	7,500	11 ft 1 in. to 11 ft 5 in.	4,500
9 ft 0 in. to 9 ft 5 in.	6,600	11 ft 6 in. to 11 ft 11 in.	4,000
9 ft 6 in. to 9 ft 11 in.	6,200	12 ft 0 in. to 12 ft 5 in.	3,600
10 ft 0 in. to 10 ft 5 in.	6,000	12 ft 6 in. to 12 ft 11 in.	3,200
10 ft 6 in. to 10 ft 11 in.	5,750	13 ft 0 in. to 13 ft 5 in.	2,800
11 ft 0 in. to 11 ft 5 in.	5,000	13 ft 6 in. to 14 ft 5 in.	2,600
11 ft 6 in. to 12 ft 5½ in.	4,700	14 ft 6 in. to 16 ft 0 in.	2,400

NOTE: Safe working-load capacities are based upon a safety factor of 3.
Proper bracing should be installed between shores to assure adequate stability.

FIG. 31.   Aluminum horizontal shores, consisting of a box section and telescoping I section.   (*Hico Corp. of America.*)

Table 13. Spanall Horizontal Shores, 730 Series

Individual Members

Lattice member	Length		Wt, lb	Plate member	Length		Wt, lb
	ft	in.			ft	in.	
L-4	4	2	32	P-6	6	8	59.0
L-8	8	4	59	P-8	8	4	73.0
L-11	11	8	78.5	P-9	9	11	87.5

Combinations

Member combination	Min span		Max span	
	ft	in.	ft	in.
L4 + P6	6	11	9	0
L8 + P8	8	7	14	10
L8 + P9	10	2	16	5
L11 + P8	11	11	18	2
L11 + P9	11	11	19	9
L8 + P8 + L8	16	11	21	4
L8 + P9 + L11	20	3	26	3
L11 + P9 + L11	23	7	29	7
P6 + L11 + P6	13	7	21	4
P6 + L11 + P8	15	3	23	0
P8 + L11 + P8	16	11	24	8
P8 + L11 + P9	18	6	26	3
P9 + L11 + P9	20	1	27	10

Fig. 32. Adjustable horizontal shores consisting of plate and lattice sections wedge-locked together ready to receive slab formwork. (*American Pecco Corporation.*)

Table 14. Steel-frame Shoring Safety Rules°

Following are some common-sense rules designed to promote safety in the use of steel-frame shoring equipment. These rules are illustrative and are intended to deal only with some of the many practices and conditions encountered in the use of steel-frame shoring. The rules do not purport to be all-inclusive or to supplant or replace other additional safety and precautionary measures. They are not intended to conflict with, or supersede, any state or local statute or regulation; reference to such specific provisions should be made by the user (see Rule 1).

1. Follow local codes, ordinances, and regulations pertaining to shoring.

2. Inspect all equipment before using. Never use any equipment that is damaged or deteriorated in any way.

3. A shoring layout should be available on the jobsite at all times.

4. Inspect erected shoring and forming: Immediately prior to placing, during, and immediately after placing concrete.

5. Use manufacturer's recommended safe working loads consistent with the type of shoring frame and the height from supporting sill to formwork.

6. Do not exceed the shore frame spacings or tower heights as shown on the shoring layout.

7. Shoring load should be carried on legs. Consult your shoring supplier for shoring frames that are designed for taking loads on top horizontal.

8. If motorized concrete equipment is to be used, be sure that the shoring layout has been designed for use with this equipment and such fact is noted on the layout.

9. Provide and maintain a solid footing to distribute maximum loads properly.

10. Use adjustment screws to adjust to uneven grade conditions.

11. Use adjustment screws to level off, to position the falsework accurately, and for easy stripping.

12. Keep screw extensions to a minimum for maximum load-carrying capacity (follow manufacturer's recommendation on screw extension).

13. Make certain that all adjustment screws are firmly in contact with sills, formwork, and frame legs.

14. Plumb and level all shoring frames as the erection proceeds. Do not force braces on frames to fit—level the shoring towers until proper fit can be made easily. Check plumb and level of shoring towers just prior to pour.

15. Fasten all braces securely.

16. Tie high towers of shoring frames together with sufficient braces to make a rigid, solid unit (see manufacturer's recommendations).

17. Exercise caution in erecting or dismantling free-standing shoring towers to prevent tipping.

18. Do not climb cross braces.

19. Avoid eccentric loads on U heads, top plates, and similar members by centering stringers on those members.

20. Use special precautions when shoring from or to sloped surfaces.

21. Use lumber stresses as shown on layout and consistent with age, type, and condition of the available lumber to be used. Use only lumber that is in good condition.

22. Reshoring procedure should be approved by a qualified engineer.

23. Do not remove braces or back off on adjustment screws until proper authority is given.

* As recommended by the Steel Scaffolding and Shoring Institute. Adapted from *Concrete Construction*, July, 1962.

lb. The Pecco lightweight beam system consists of four units which, used in various combinations, can be made up into lengths from 7 ft 9 in. to 17 ft 4 in. using two units and up to 25 ft 3 in. using three units. Pecco Quick-Beam I and II are lightweight but strong enough to handle shorter spans. The Quick-Beam weighs 48 lb and spans from 6 ft 1 in. to 10 ft 6 in. The Quick-Beam II weighs 71 lb and spans from 7 ft 1 in. to 13 ft 9 in. Pecco adjustable horizontal shoring units (plate and lattice sections) are locked together and camber is automatically set by means of the Wedgelock (Fig. 32).

Scaffold-type Shoring. Tubular-steel frame scaffold was originally designed and developed to support workmen and relatively light loads of material for doing work off the ground, and was used for this purpose until someone discovered that it could be used as shoring to support suspended concrete formwork. By assembling two frames with two cross braces into an individual shoring tower, the contractor has the equivalent of four individual post shores completely braced for stability. This modular method of assembling shores has become so advantageous that many new accessories have been developed to make scaffold shoring a very versatile means of furnishing vertical support.

Loads imposed by fluid concrete are usually heavier and less stable than those the scaffold was originally designed to carry safely when used strictly as a working platform. To accommodate these potentially heavier shoring loads, scaffold-type frames were designed principally for shoring. Readily available now from stock,

Fig. 33. Shoring for a highway bridge structure, using 6-in.-square tubing with a leg-carrying capacity of 100,000 lb. (*Waco Scaffold & Shoring Co.*)

for rental or purchase, are both light-duty and heavy-duty shoring frames. The basic difference between the light-duty and heavy-duty frame is that the latter is manufactured from larger-diameter thicker-wall tubing, which gives it greater leg-carrying capacity. Leg-carrying capacity varies from 4,000 lb per leg for light-duty up to 11,000 lb per leg for heavy-duty shoring. To support even greater loads than the round tube, 11,000 lb per leg capacity, some manufacturers are now using heavy-wall square tubing. Six-inch square tubing $\frac{5}{16}$ in. thick is being fabricated into a system with a capacity of 100,000 lb per leg. A capacity of 25,000 lb per leg is obtained with $3 \times 3 \times \frac{3}{16}$ in. tubing (Fig. 33).

With the development of scaffold frames for shoring came the demand for a variety of sizes of cross braces to allow the frames to be modularly spaced to obtain maximum safe capacity on each frame. Originally the two sizes of crosses that were available were for 7- and 10-ft frame spacing so standard lengths of 12- and 16-ft scaffold planks could be used.

As the loads imposed on this shoring and its number of applications increased, so too did the necessity of providing the users with more accurate load-carrying information so that steel scaffold shoring can be put to its maximum use without exceeding its limitations, thus leading to possible failures. The Steel Scaffolding and Shoring Institute has conducted physical tests on ladder-type frames to determine failure patterns and safe capacity of shoring frames. The test results as well as their recommended safety practices and erection procedures are available from most shoring manufacturers. One bit of advice that should always be adhered to is: Never exceed the manufacturer's recommended safe working loads consistent with the type of shoring frame and height from supporting base to formwork. Other safety rules are given in Table 14.

REFERENCES

1. ACI Committee 347-68 Report, "Recommended Practice for Concrete Formwork," ACI Book of Standards.
2. Douglas Fir Plywood Commercial Standards, CS45, U.S. Department of Commerce, Washington, D.C.
3. "Manual of Steel Construction," American Institute of Steel Construction, New York.

Chapter 15

DESIGN, APPLICATION, AND CARE OF FORMS

DAVID R. HARNEY

FORM DESIGN

Concrete form design begins with the selection of a forming system. It may be a simple decision to use the forms on hand, or it may be the result of completely analyzing total job requirements. The latter method has proved more economical for concrete jobs, both large and small. Naturally the larger, more complex jobs take more planning but have greater potential savings than the small jobs.

Fig. 1. Mechanized equipment, such as these tower cranes, expeditiously handles forms, concrete, and other materials with a minimum of labor.

The availability of material and equipment too often wrongly influences the selection of a forming system. It can be a costly mistake to choose a system principally because the material is available, a decision that is especially costly on a large job. One positive way to reduce forming costs is by increasing the number of reuses of a form; also frequently overlooked is the method of handling forms. Mechanical equipment of the type shown in Fig. 1 is available that will do the work of many men at a fraction of their cost.

1 Form Layout

The form layout must be simple, clear, and consistent. It is always less costly to work out solutions to forming problems on paper than it is to have it done in the field. Mistakes that can be easily corrected with an eraser on paper can be very expensive to correct in the field. For this reason it is important to make a general layout of work to be performed. Use a scale that is sufficiently large; ¼ in. = 1 ft should be the minimum; ⅜ or ½ in. = 1 ft is better for layout. Detail should be ¾ or 1 in. = 1 ft. Notes should be kept to a minimum, as drawings are best. Care should be exercised to avoid cramming too much on one sheet. Each sheet should carry a title and a number. Form assembly drawings show how fabricated form sections fit together. Sometimes an isometric view is helpful.

The fabrication drawing should show the information necessary to build each individual section. It is often helpful for the fabrication drawings to include the bill of material. The information contained on the detail sheets must be adequate enough so the workmen doing the fabrication need not consult the contract drawings for clarification. A useful place to lay out the construction joints is on the general layout drawings. Embedded items should be shown on the drawings where the information is most advantageous for the easy flow of work.

2 Selecting a Forming System

In order to evaluate what forming system will do the best job for a particular project, forming cost must be analyzed. Form costs can be separated into three general categories: (1) fabrication (form manufacture), (2) erection (form installation), and (3) stripping (form removal). Each category is subdivided into labor, material, and handling, and each is interdependent upon the other. On some jobs the categories for fabrication and erection may overlap, when forms are constructed in place.

Answers to the following questions are necessary in order to select a forming system intelligently:

1. How much contact area is to be formed and how long is it going to take to build the structure? This information varies from job to job. Contact area should be readily available from the bid estimate, and the job duration would normally be determined by the projected job schedule.

2. What type of concrete finish is required? The type of concrete finish is normally spelled out in the individual job plans and specifications.

3. How many reuses can be obtained from the forms? A careful inspection of plans and the sequence of construction as well as consulting the specifications for stripping time in conjunction with the job schedule will assist in compiling this information.

4. What will be the rate of placing concrete? The rate of placing concrete is controlled by a number of factors including type and size of structure, rate of concrete delivery, concrete-mixing facilities, transportation, type of handling and placing equipment, labor force, and weather, to mention a few, as well as the specifications.

5. What mechanical equipment is available for this job? No type of equipment, large or small, that will reduce labor costs should be overlooked without carefully evaluating its potential. More will be discussed on this subject under form handling.

When the basic forming system has been chosen, the materials selected, and the type of handling equipment determined, the form designer then proceeds to design the forms in detail so they will support themselves, hold their shape, provide adequate safety when fully loaded, and strip economically.

3 Form Pressures

The load or forces imposed upon a form that must be considered by the designer are classified as live load and dead load. Live load is the weight of workmen and equipment such as pipelines and buggies, and other temporary loads that are supported by the form during concrete placing and finishing. Dead load consists of the weight of the form itself plus the concrete and reinforcing steel. However, in most cases the form weight is negligible; so the concrete weight is of prime importance to the designer. The weight of reinforced concrete varies from 50 to over 400 pcf depending on the type of aggregate used. However, these very light or very heavy concretes are designed for special applications, and the form designer is usually concerned with normal-weight structural concrete at 150 pcf or lightweight concrete at 110 pcf. The same principles of form design apply in any case, regardless of the concrete density

When concrete is placed in or on forms it is in a quasi-liquid state, exerting hydrostatic pressure on its confining surfaces in the same manner that any liquid does. Because of this, the pressure in all directions at any given point is calculated the same as for other liquids. That is, the pressure, in pounds per square foot, at any point equals the depth, measured in feet, below the fluid head (top of liquid) times the density (150 pcf for standard concrete or 110 pcf for lightweight concrete). Two important factors that influence the hydrostatic pressure of fresh concrete are the rate of placement, measured in feet of depth per hour, and the ambient temperature, measured in degrees Fahrenheit, unless the concrete itself is heated or cooled to a controlled temperature, in which case the controlled temperature is used.

As concrete hardens it changes from a liquid to a solid state, eliminating the lateral pressure. Temperature of the concrete influences this rate of hardening; thus temperature must be taken into consideration when calculating the hydrostatic pressure of concrete. Concrete hardens faster in warm weather than it does in cold (assuming there is no retarder in the concrete). The rate of placement (feet per hour) controls the height of the fluid head of concrete, making it necessary to use this rate when calculating hydrostatic pressure.

There are many other factors that can affect the pressure of concrete on the forms, some of which are the type of vibration used to consolidate the concrete (external vibration, internal vibration); impact loading caused by free fall of concrete when it is discharged into or on forms; and slump of the concrete. The degree of influence that each of these variables has upon the lateral pressures that concrete exerts upon forms, and the limit of hydrostatic pressure to be expected, are apt to be assigned differing degrees of importance by different persons. However, ACI Committee 347 has developed workable formulas for calculating the lateral pressures exerted by fluid concrete on wall and column forms.[1] These formulas take into consideration the many variable factors that are involved. For wall forms with rate of placement not exceeding 7 ft per hr,

$$p = 150 + \frac{9,000R}{T} \qquad \text{(max 2,000 psf or 150}h\text{, whichever is least)} \qquad (1)$$

and in walls, with rate of placement greater than 7 ft per hr,

$$p = 150 + \frac{43,400}{T} + \frac{2,800R}{T} \qquad \text{(max 2,000 psf or 150}h\text{, whichever is least)} \qquad (2)$$

For columns,

$$p = 150 + \frac{9,000R}{T} \qquad \text{(max 3,000 psf or 150}h\text{, whichever is least)} \qquad (3)$$

in which p = maximum lateral pressure, psf
R = rate of concrete placement, ft per hr
T = temperature of concrete in the forms, °F
h = maximum height of fresh concrete in form, ft

The lateral pressure of fluid concrete (dead load) acts as a force against the walls of its mold. This force must be held static by the mold to ensure that the required shape can be obtained. To do this the concrete form must be constructed of form members, selected by the form designer, with physical properties adequate to withstand the stresses of compression, bending, and shearing that are exerted upon them.

4 Basic Formulas

For many years form members have been selected and spaced on the basis of the designer's experience, by adapting materials that worked satisfactorily in the past to similar applications for the job at hand, often called "design by assumption." This practice fortunately has given way to design by application. Tables are readily available to the form designer giving the physical properties of plywood, lumber, and steel, these being the materials most commonly used for form members. Information regarding plywood properties is available from the American Plywood Association.[2] There are tables in the "Wood Structural Design Data" book[3] that give directly the required beam sizes, loadings, shear, fiber stresses, and other data for almost any loading and construction condition. The "Manual of Steel Construction" gives the physical properties of structural shapes for design and detailing. Tables and references are given in Chap. 14 also.

(a)   **(b)**

Fig. 2. (a) Bending strength is a measure of the load-carrying capacity of a beam. (b) Fiber stress in bending results when the beam is subjected to a load that causes deflection.

Deflection. When a beam is subjected to an external moment the fibers on one side elongate, while the fibers on the other side shorten (Fig. 2), causing the beam to deflect. Design of beams may be based on strength requirements (load-carrying capacity) or on stiffness, when only a small deformation, or deflection, is acceptable. The modulus of elasticity, which controls stiffness, varies with different types of lumber and, in any one kind of lumber, varies as the moisture content varies, as shown in Table 4, Chap. 14.

Form members, including walers, studs, and sheathing, all act as beams under load when supporting fresh concrete vertically, horizontally, or in some intermediate direction.

Maximum deflection for a simple beam uniformly loaded is given by the equation

$$D_{max} = \frac{5wl^4}{384EI} \tag{4}$$

in which D_{max} = maximum deflection, in.
w = uniformly distributed load, lb per lin ft
l = length of beam between supports, ft
E = modulus of elasticity of beam material, psi
I = rectangular moment of inertia, in.[4]

Other beam formulas used in design of formwork are shown in Table 1.

When loads acting perpendicular to the longitudinal axis of a form member cause deflection, the member is subjected to fiber stress in bending. The stress is calculated by the equation

$$f = \frac{M}{S} \tag{5}$$

in which f = extreme fiber stress, psi
M = induced bending moment, in.-lb
S = section modulus, in.[3]

Table 1. Properties of Beams

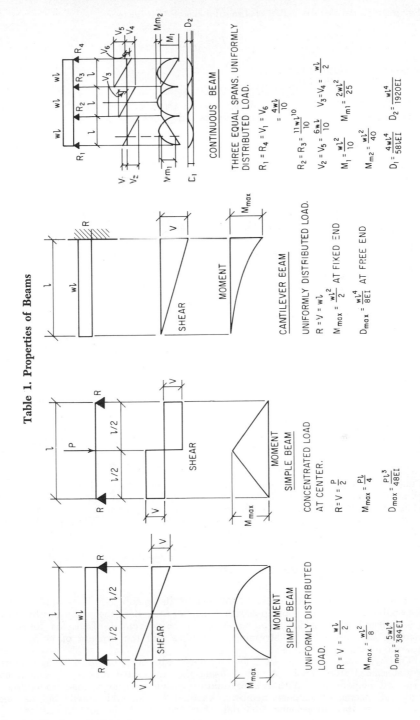

CONTINUOUS BEAM

THREE EQUAL SPANS. UNIFORMLY DISTRIBUTED LOAD.

$$R_1 = R_4 = V_1 = V_6 = \frac{4wl}{10}$$

$$R_2 = R_3 = \frac{11wl}{10}$$

$$V_2 = V_5 = \frac{6wl}{10} \qquad V_3 = V_4 = \frac{wl}{2}$$

$$M_1 = \frac{wl^2}{10} \qquad M_{m1} = \frac{2wl^2}{25}$$

$$M_{m2} = \frac{wl^2}{40}$$

$$D_1 = \frac{4wl^4}{58lEI} \qquad D_2 = \frac{wl^4}{1920EI}$$

CANTILEVER BEAM

UNIFORMLY DISTRIBUTED LOAD.

$$R = V = wl$$

$$M_{max} = \frac{wl^2}{2} \quad \text{AT FIXED END}$$

$$D_{max} = \frac{wl^4}{8EI} \quad \text{AT FREE END}$$

SIMPLE BEAM

CONCENTRATED LOAD AT CENTER.

$$R = V = \frac{P}{2}$$

$$M_{max} = \frac{Pl}{4}$$

$$D_{max} = \frac{Pl^3}{48EI}$$

SIMPLE BEAM

UNIFORMLY DISTRIBUTED LOAD.

$$R = V = \frac{wl}{2}$$

$$M_{max} = \frac{wl^2}{8}$$

$$D_{max} = \frac{5wl^4}{384EI}$$

For a simple uniformly loaded rectangular beam, this becomes

$$f = \frac{3wl^2}{4bh^2} \qquad (6)$$

in which b = width of beam, in.
h = depth of beam, in.

Computed values of f should not exceed those in Table 4, Chap. 14, which gives allowable values of f for various types and grades of wood.

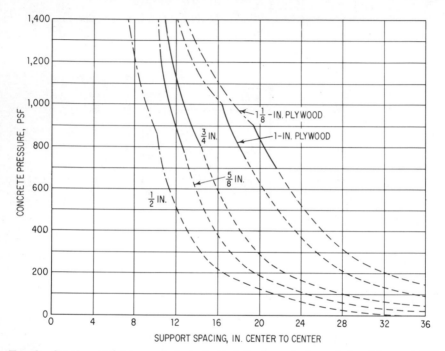

FIG. 3. Spacing of joists or studs to limit deflection of plywood sheathing to 1/270 of span between studs or joists, based on flexural stress of 2,000 psi and shear in plane of panel of 94 psi. For Douglas-fir interior plyform or exterior all-sanded grades. Panels are continuous across two or more spans. When panel is used on a single span, stiffness is only half that shown on chart. For face grain of plywood across supports. When face grain is parallel with supports, use the following percentages of the load given in the chart for each thickness: ½ in., 40%; ⅝ in., 51%; ¾ in., 73%; 1 in., 90%; 1⅛ in., 90%.

If a form is designed so the deflection of its members is maintained within reasonably close limits, poor-appearing concrete, as well as the problems that often arise when it is time for form removal, can be prevented. Deflected forms can increase the cost of form stripping as well as add the expense of chipping off unwanted hardened concrete.

The hydrostatic pressure that fluid concrete will exert on form-facing material is determined from Tables 2 and 3, using the anticipated rate of placement and expected temperature. With this information, calculations can be made to determine size and spacing of support members, whose deflection limit should be the same as that for the form-facing material. For ordinary walls not subject to close

Table 2. Maximum Lateral Pressure for Design of Wall Forms

Rate of placement R, ft/hr	p, maximum lateral pressure, psf, for temperature indicated					
	90°F	80°F	70°F	60°F	50°F	40°F
1	250	262	278	300	330	375
2	350	375	407	450	510	600
3	450	488	536	600	690	825
4	550	600	664	750	870	1,050
5	650	712	793	900	1,050	1,275
6	750	825	921	1,050	1,230	1,500
7	850	938	1,050	1,200	1,410	1,725
8	881	973	1,090	1,246	1,466	1,795
9	912	1,008	1,130	1,293	1,522	1,865
10	943	1,043	1,170	1,340	1,578	1,935

NOTE: Do not use design pressures in excess of 2,000 psf or 150 × height of fresh concrete in forms, whichever is less.

Table 3. Maximum Lateral Pressure for Design of Column Forms

Rate of placement R, ft/hr	p, maximum lateral pressure, psf, for temperature indicated					
	90°F	80°F	70°F	60°F	50°F	40°F
1	250	262	278	300	330	375
2	350	375	407	450	510	600
3	450	488	536	600	690	825
4	550	600	664	750	870	1,050
5	650	712	793	900	1,050	1,275
6	750	832	921	1,050	1,230	1,500
7	850	938	1,050	1,200	1,410	1,725
8	950	1,050	1,178	1,350	1,590	1,950
9	1,050	1,163	1,307	1,500	1,770	2,175
10	1,150	1,275	1,435	1,650	1,950	2,400
11	1,250	1,388	1,564	1,800	2,130	2,025
12	1,350	1,500	1,693	1,950	2,310	2,850
13	1,450	1,613	1,822	2,100	2,490	3,000*
14	1,550	1,725	1,950	2,250	2,670	
16	1,750	1,950	2,207	2,550	3,000*	
18	1,950	2,175	2,464	2,850		
20	2,150	2,400	2,721	3,000*		
22	2,350	2,625	2,979			
24	2,550	2,850	3,000*			
26	2,750	3,000*				
28	2,950					
30	3,000*					

NOTE: Do not use design pressures in excess of 3,000 psf or 150 × height of fresh concrete in forms, whichever is less.

* 3,000 psf maximum governs.

scrutiny, and the underside of most slabs, a deflection of $\frac{1}{270}$ of the span is acceptable. Spacing of support members can be computed with the aid of Fig. 3. Vertical-facing support members, commonly called studs, should be modularly spaced to fit the size of the facing material. For example, if computations indicate

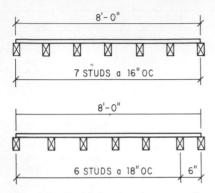

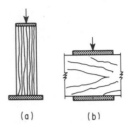

FIG. 4. Careful planning permits modular spacing of studs, making a better-looking form that is more economical.

FIG. 5. (*a*) Compression parallel to the grain is a measure of the capacity of a short column to withstand loads acting in a direction parallel to the long axis. (*b*) Compression perpendicular to the grain is a measure of the bearing strength of wood across the grain.

that a stud spacing of 18 in. on centers is adequate to support an 8-ft length of plywood, we can see that 16 in. on center requires no more studs (Fig. 4) but does give uniformity to the form. Therefore, the latter spacing should be used.

Compression. Wood form members are apt to be subjected to compression

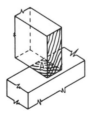

FIG. 6. The end bearing area A should be such that the reaction V divided by A will not exceed the allowable unit stress in compression perpendicular to the grain. For bearings shorter than 6 in. located 3 in. or more from the end of the member, the stress can be increased as follows:

Length of bearing, in.	$\frac{1}{2}$	1	$1\frac{1}{2}$	2	3	4	6 or more
Factor..............	1.75	1.38	1.25	1.19	1.13	1.10	1.00

in either of two directions, compression parallel to the grain of the member, and compression perpendicular to the grain (Fig. 5). Compression parallel to the grain is most commonly encountered when lumber is used as shoring, in which case the member is acting as a column and the load capacity is calculated by using the slenderness ratio, which is defined as the relationship of the unsupported length of the shore (acting as a column) to the cross-sectional dimension of the face under consideration, usually the narrowest of the two faces. It is expressed as l/d, in which l = unsupported length, in inches, and d = net dimension in inches of face under consideration. The slenderness ratio for simple solid columns should not exceed 50.

Allowable unit stresses for compression perpendicular to the grain are of concern to the form designer when form members are subjected to loads that would cause the material to fail by being crushed by its support. An area of form members where this situation can occur is shown in Fig. 6.

Values for compression perpendicular to grain are adjusted for duration of loading the same as other strength properties. No allowance need be made for the fact that as members deflect the pressure on the inside edge of the bearing area is greater than it is on the open side. When the bearing is not more than 6 in. in length and more than 3 in. from the end of the member, an increase in the allowable unit stress for compression perpendicular to the grain is permitted, as shown in Fig. 6. For round bearing areas the diameter is used for the length.

(a)   (b)

Fig. 7. Shearing strength parallel to the grain is a measure of the ability of the wood to resist slipping of one part upon another along the grain.

Fig. 8. (a) Horizontal shear is the resistance to the tendency of the upper half of the beam to slide upon the lower half when the beam is loaded. (b) Loads and reactions at the supports are vertical forces tending to shear or cut the beam across. This is vertical shear.

Shear. Vertical and horizontal form members must be selected so they are capable of resisting failure in shear (Fig. 7). Vertical shear is the tendency of the load forces to cut the member into two pieces perpendicular to its longitudinal axis, while in horizontal shear the load forces tend to split the member into two pieces parallel with the longitudinal axis (Fig. 8). Vertical and horizontal shear are of equal magnitude at any point on the member acting perpendicular to each other. These terms, horizontal and vertical, were originated for viewing beams in the horizontal position. However, if their relationship with the beam axis is remembered, they can be applied to beams at any angle, such as studs being used to form either vertical or inclined surfaces. Horizontal shear stress is common to fibrous material such as lumber.

Horizontal shear is a maximum at the neutral axis and is given by the equation

$$H = \frac{3V}{2bd} \tag{7}$$

and vertical shear by

$$V = \frac{wl}{2}$$

in which b = width of beam, in.
$\quad\quad d$ = depth of beam, in.
$\quad\quad l$ = length of beam, ft
$\quad\quad w$ = uniformly distributed load, lb
$\quad\quad V$ = total vertical shear at supports

APPLICATION OF FORMS

5   Forms for Footings and Foundations

Forms for wall footings (continuous footings) or for columns and piers (individual footings) are rather simple forms. This is because footings for most jobs are

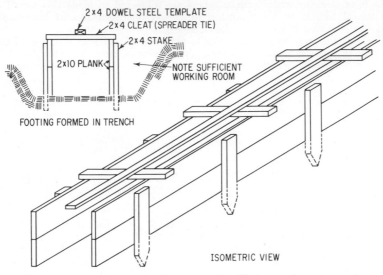

FIG. 9. Footing forms, quite simply made, should have adequate working space to facilitate construction.

FIG. 10. Top of form for a basement wall, showing forming for corbel. Note adequate width of excavation for working space.

not very thick, and therefore the fresh concrete does not develop much in the way of lateral pressure on the forms. Usually when footings are thicker they are also larger in cross-sectional area and the rate of concrete placing (feet per hour) is low; therefore, again very little hydrostatic pressure is generated.

Concrete footings are normally placed in excavated soil and not exposed to view, so that the quality of form finish is not important; what is important is that the concrete forms be so constructed that the concrete will be in the location called for by the job plans and will be ready to receive the walls, piers, or columns that are supported by the footing. On jobs where the excavated soil is sufficiently stable concrete can be placed directly against the earth, using it for the form. If this is the case the soil should be excavated slightly smaller than the footing and then hand-trimmed to the correct size. If a trencher or similar machine is used for excavation, by machine cutting a little oversize most of the hand labor can be eliminated. When the earth is used as the form it is commonly called "pouring neat." This is a very good method of reducing costs, whenever possible, as no forming is the cheapest forming.

If it is necessary to form footings rather than "pouring neat," the excavation, whenever space is available, should be large enough to provide sufficient working room behind forms, so they can be installed with the minimum amount of hindrance,

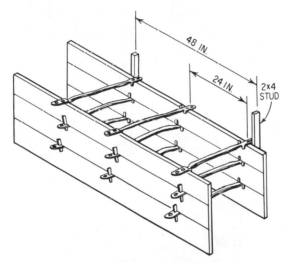

Fig. 11.   Low-tensile-type ties can be used for forms for low foundation walls.

as shown in Figs. 9 and 10. It is false economy to excavate so tight that a man does not have working room to install ties and braces, and then later excavate so forms can be removed. If the excavation is made large enough to start with, it will be cheaper than coming back for a second cut.

Two-inch-thick planks, 8 or 10 in. wide, are most commonly used for footings less than 24 in. high. When footings are more than 24 in. thick, old or used forms can be used. Patented prefabricated forms can be particularly advantageous to form thick footings (30 in. or more). Their modular sizes make them adaptable to almost any length and width of spread footing or pile cap.

For low foundation walls without much repetition, sheathing can be used with sheathing-type ties, or plywood and snap ties with single waler brackets (Fig. 11).

Forms for vessel bases or equipment foundations usually have to be both strong and very accurately located, so that the finished concrete will provide adequate clearances for piping that services whatever is being supported, and so that the equipment itself will be serviceable. Usually this type of structure has cast-in-place anchor bolts that need to be positioned accurately. These anchor bolts are best set and held in place by a template that is strong enough to withstand bumping or other shocks when concrete is being placed. It is wise to place a sleeve around the top half of the bolt so it will have some horizontal adjustment when the equipment is set in place over it, so that as soon as the concrete has been placed the accuracy of the template's location can be checked and any adjustment made while the concrete is still plastic.

Too often extra concrete is put into footings, increasing their thickness and swelling forms only because the concrete foreman is reluctant to discard a wheelbarrow or two of concrete that is left. Care should be exercised when casting footings to see that they are level and true to line. Straight and level footings will eliminate costly shimming of plates for wall forms, or even greater costs of chipping concrete.

6 Wall Forms

Wall forms apply to many different types of construction. Figure 12 illustrates a typical series of steps in construction of a wall form. Relatively low forms

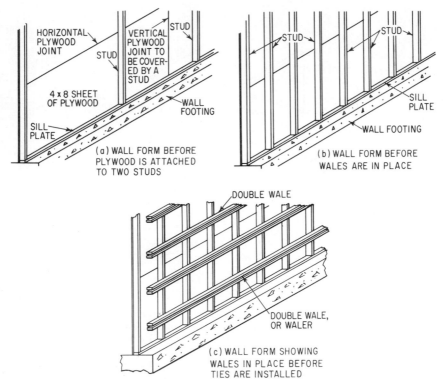

Fig. 12. Wall-form construction can be broken down into a series of simple steps.

for grade beams and foundation walls ranging from 18 in. to 6 ft in height are wall forms of simple design, for which 3,000-lb working-load ties 2 ft on center both ways should be adequate under most circumstances, as the lateral pressure exerted by concrete placing at the rate of 3 ft per hr at 60° temperature would not exceed 600 psf.

For walls of this type where there is little or no repetition, the forms are often built in place, that is, constructed right in the field. The single 2 × 4's, acting as both walers and studs, are held in place by means of special devices consisting

FIG. 13.   Combination snap ties and waler brackets used in connection with single waler forming system.   (*P.D.H. Photo Service.*)

FIG. 14.   Typical engineered-in-place forms for a building.   (*American Pecco Corporation.*)

of a combination snap-tie clamp and waler bracket (Fig. 13). If sheathing is used for the form facing with this type of application, sheathing ties can be used; their working load is less than 1,000 lb and so the ties are spaced 2 ft on center horizontally and 6 in. on center vertically. This type of tie does not break back within the wall.

Where there is repetition in this type of forming, prefabricated (either factory- or job-fabricated) wall forms should prove very economical. The forms are built in modules, and drilled or notched symmetrically to allow tie holes to match each other, so they may be stacked horizontally or vertically to meet the various heights and lengths of wall required.

Fig. 15. Multiple-use engineered forms for a large dam are sometimes cantilevered so the previously placed concrete supports the form and helps align it. (*U.S. Bureau of Reclamation.*)

Wall forms for structures of greater height, such as deep basement walls, high retaining walls, many stories up the face of a tall building, or multiple lifts on the face of a dam, must be engineered for each individual project. The complexity of structural forms may be seen in Fig. 14, and Figs. 15, 16, and 17 illustrate forms for dam construction. Typical form tieing devices are shown in Fig. 18.

Forms for these types of structures at one time were built all of wood. However, this practice has given way to some all-steel forms, some combination wood and steel, and those of reinforced-fiber-glass faces with wood or steel for support members. To justify these more durable and expensive forms, the designer must increase the reuse factor to the fullest, and maximize every laborsaving feature in connection with the form use.

FIG. 16. Methods of attaching the form to she-bolts in the previously placed concrete are clearly visible. (*California Department of Water Resources.*)

FIG. 17. Inclined tie rods, fastened to anchors in the surface of the concrete, are screwed into she-bolts in the form to hold it rigidly in place. When the form is stripped and moved up to the next lift, the cantilevers will be attached to the same she-bolt and tie-bar assembly. (*R. F. Adams.*)

Forming costs can best be reduced by eliminating completely, or reducing to an efficient minimum, the number of repetitive operations. An example of eliminating an operation is bolting rustication strips permanently to the form rather than reattaching them each time the form is set (erected in position). Form alignment is a procedure that is required each time a form is set. The cost of this operation can be substantially reduced by use of mechanically adjustable braces. Figure 19 shows braces attached to a column form. However, they are equally adaptable to wall forms.

The installation of form ties is required for most wall forms, a procedure that is normally accomplished by hand labor (Fig. 20). This costly operation can

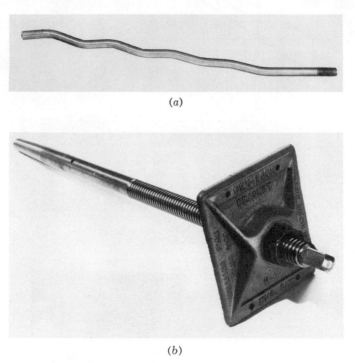

(a)

(b)

FIG. 18. Typical tieing devices for heavy forms: (a) a pigtail anchor that is embedded in the concrete; (b) a heavy duty she-bolt and locking clamp. (*Williams Form Engineering Corp.*)

be simplified in a number of ways: (1) be sure normal tie spacing is such that the ties are loaded to their full recommended working-load capacity, thus minimizing the number of ties required to install; (2) use ties of larger capacity than normally used, thus reducing the number of ties required; (3) design forms for the easy reception of form ties, cutting down on length of time needed to position each tie.

For example, design forms that can be tied at form joints wherever possible to eliminate the cost of fishing ties through holes in the forms. In cases where it is necessary to feed ties through the forms, holes should be made as large as practicable to facilitate tie installation. For jobs where the number of reuses does not readily justify the expense of building a crane-handled form, that is,

FIG. 19. Alignment of vertical or inclined forms is greatly aided by the use of adjustable braces.

FIG. 20. Coil ties are installed by hand after one wall of the form has been erected.

FIG. 21. A good example of gang-assembled factory-prefabricated forms. (*Universal Form Clamp Co.*)

a form large enough to be economically handled by a crane rather than hand labor, it has often proved economical to lock modular prefabricated factory-type forms together to make a crane-handled form (Fig. 21). It is possible to build the forms into a variety of shapes by using different combinations of sizes of form units. When patented forms are used for this type of application a heavy-duty tie is used to cut down on the number of ties required. A built-up form of this type can be lined with 4 × 8-ft plywood sheets, if a finish of this type is specified.

When designing wall forms, good judgment should be used in the selection and spacing of the members that will support the form-facing material. The form-facing material is chosen to satisfy the type of concrete finish required by the particular job plans and specifications, which should also stipulate the allowable deflection. If the specifications do not provide for other values, $\frac{1}{360}$ of the span is usually acceptable for particularly smooth surfaces and those near eye level, and for less particular sufaces or those not subject to close scrutiny $\frac{1}{270}$ of the span is allowable. See previous discussion under Deflection.

7   Column Forms

Concrete columns are molded into a variety of sizes and shapes depending upon their relationship to the structure they serve, the three commonest shapes being

FIG. 22. Square column forms are rapidly fabricated from ¾-in. plywood and Econ-o-clamps, a patented system for columns from 8 to 30 in. square. Note closer spacing of clamps toward the bottom of the form.

square, rectangular, and round. Column shapes are designed for strength, for decoration, or to serve both purposes. Selection of column forming material is governed by the finish and number of reuses required.

Square and rectangular column forms were for many years made almost exclusively with sheathing or planking as form-facing material, a practice that has given way to the use of plywood except in those cases in which boards produce a desired architectural effect. Labor reduction in both form fabrication and form maintenance as well as the quality of the finished surface is responsible for this change. The plywood-faced column form, supported by 2×4's either flat or on edge depending upon the particular design, is tighter and permits practically no grout leakage, a very desirable feature when compared with sheathing and planking forms that swell by absorbing moisture from the wet concrete, then warp and check upon drying.

Column forms are secured together by means of yokes, clamps, or strapping. Extremely large columns are through-tied when required; however, through tying is expensive and should be avoided if possible.

Symons adjustable column clamps are available in standard sizes to form square or rectangular columns from 14 to 72 in. on a side. The clamps, manufactured of high-carbon flat steel stock, consist of two pieces pinned (for hinging) together at one end to make one unit, which is half a clamp set. Two units, locked

FIG. 23. Bench-assembled column form, externally secured with she-bolts. (*Williams Form Engineering Corp.*)

together by a sliding bracket and a steel wedge that is driven into slotted holes already punched in the free ends of each unit, when correctly locked in place automatically square the column form. Clamp spacing is shown in Table 4.

Another type of column clamp for use with $\frac{3}{4}$-in. plywood is designed to eliminate the use of backup material. The Econ-o-clamp (Fig. 22) greatly reduces the cost of bench fabrication because its large bearing area fully supports the plywood. Rectangular and square columns can also be formed by using patented prefabricated forms.

For column forms that are completely fabricated horizontally on the bench (Fig. 23) and installed as a unit by slipping over the previously set column reinforcing steel, clamping is accomplished by outside bolting (as shown in the figure), column clamps, or banding. There are a number of ways column forms can be banded, one method being by use of a bull-winch clamp with permanently attached band strapping. The winch is attached to the column form, strapping is wrapped

Table 4. Recommended Spacing for Symons Column Clamps

SYMONS 36" EVERSQUARE COLUMN CLAMP

Spacing of column clamps in inches (depth from top, 1'–10')

SPAN OF COLUMN CLAMP	29"	25"	14"–20"
1'	30	30	24
2'	30	30	24
3'	22	24	24
4'	22	24	24
5'	18	18	24
6'	18	18	20
7'	18	16	20
8'	12	12	20
9'	12	12	20
10'	6	6	6

SYMONS 48" EVERSQUARE COLUMN CLAMP

Spacing of column clamps in inches (depth from top, 1'–10')

SPAN OF COLUMN CLAMP	40"	35"	30"
1'	17	21	30
2'	17	21	30
3'	11	21	22
4'	11	15	22
5'	9	15	18
6'	9	15	18
7'	7	11	18
8'	6	11	12
9'	6	9	12
10'	6	6	6

SYMONS 72" SPECIAL COLUMN CLAMP

Spacing of column clamps in inches (depth from top, 1'–10')

SPAN OF COLUMN CLAMP	60"	54"	48"
1'	18	18	22
2'	15	16	22
3'	11	14	22
4'	10	14	18
5'	8	12	18
6'	8	10	14
7'	6	9	14
8'	6	8	12
9'	6	8	10
10'	6	6	6

SYMONS 84" SPECIAL COLUMN CLAMP

Spacing of column clamps in inches (depth from top, 1'–10')

SPAN OF COLUMN CLAMP	72"	66"	60"
1'	16	15	18
2'	12	15	15
3'	10	14	11
4'	8	7	10
5'	6	7	8
6'	6	6	8
7'	*	6	6
8'		*	6
9'			6
10'			6

NOTE: (*) INDICATES MAXIMUM RATE OF PLACING FOR COLUMN DEFLECTION WITHIN $\frac{L}{270}$

around the form and attached to the winch, then the winch is tightened by use of a ratchet, and the form is secured. When it is time to strip, the ratchet winch is loosened and the whole unit is salvaged for future use. The other method of banding uses expendable band steel which is wrapped around the column form,

Cubic Content of Capitals and Shafts

Tabulated below is the volume of concrete, expressed in cubic feet, for various diameter capitals and shafts. The volume indicated for capitals includes the full capital and the portion of the shaft above a point 3 ft below the top of the capital. Also indicated is the volume of concrete per lineal foot of shaft.

To calculate the number of cubic feet required for any given column with a capital, deduct 3 ft from the net or overall height of the column and multiply the result by the volume per lineal foot figure for the diameter of the shaft in question. To this add the volume of the top 3 ft of the column found by selecting the correct capital diameter and reading down to the required shaft diameter.

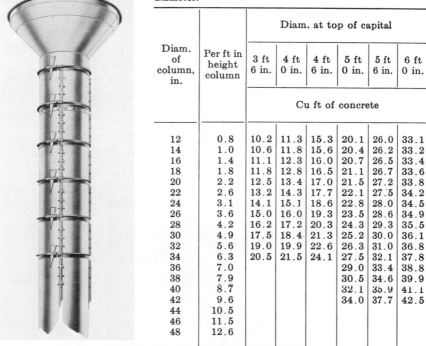

Diam. of column, in.	Per ft in height column	Diam. at top of capital					
		3 ft 6 in.	4 ft 0 in.	4 ft 6 in.	5 ft 0 in.	5 ft 6 in.	6 ft 0 in.
		Cu ft of concrete					
12	0.8	10.2	11.3	15.3	20.1	26.0	33.1
14	1.0	10.6	11.8	15.6	20.4	26.2	33.2
16	1.4	11.1	12.3	16.0	20.7	26.5	33.4
18	1.8	11.8	12.8	16.5	21.1	26.7	33.6
20	2.2	12.5	13.4	17.0	21.5	27.2	33.8
22	2.6	13.2	14.3	17.7	22.1	27.5	34.2
24	3.1	14.1	15.1	18.6	22.8	28.0	34.5
26	3.6	15.0	16.0	19.3	23.5	28.6	34.9
28	4.2	16.2	17.2	20.3	24.3	29.3	35.5
30	4.9	17.5	18.4	21.3	25.2	30.0	36.1
32	5.6	19.0	19.9	22.6	26.3	31.0	36.8
34	6.3	20.5	21.5	24.1	27.5	32.1	37.8
36	7.0				29.0	33.4	38.8
38	7.9				30.5	34.6	39.9
40	8.7				32.1	35.9	41.1
42	9.6				34.0	37.7	42.5
44	10.5						
46	11.5						
48	12.6						

FIG. 24. A sheet-metal column form showing wedge locking devices and a sheet-metal capital form. (*Deslauriers Column Mould Company.*)

tightened with a tightening device, and secured by clamps. The banding is cut and scrapped when forms are stripped.

Fabrication of banded column forms is substantially different from that used for column clamps, as the form must be designed so that the lateral pressure of the concrete is uniformly distributed to and supported by the banding material,

FIG. 25. Another type of sheet-steel column mold. *(Deslauriers Column Mould Company.)*

FIG. 26. A fiber-tube mold with sheet-metal capital form attached. *(Deslauriers Column Mould Company.)*

FIG. 27. Timber joists for formwork can be supported by wire beam hangers hung over the steel beam of the building.

FIG. 28. Overhang braces are commonly used on structures such as bridges when a narrow outboard band of deck is constructed. Note the shear connectors in the top of the concrete girders.

FIG. 29. Viewed from below, the joists and stringers supporting the deck form are visible. Overhang braces are attached to the precast-concrete girder.

accomplished by placing a vertical 2×4 or similar timber on edge near the center of the column side. If this is not done the column form will bulge and lose its correct shape.

The Quick-Strip column clamping system (Fig. 30) allows all the column clamps on any one column form to be locked or unlocked simultaneously. All clamps, being attached to a common latching rod, can be latched or unlatched by a single operation. Cam action of the latches assures secure closure of the form.

Circular columns can be formed by using ready-made sheet-metal, steel, reinforced-fiber-glass, or fiber-tube forms, some of which are illustrated in Figs. 24, 25, and 26. The type of forming material selected for this application depends on the quality of concrete finish required and the number of reuses. Fiber-tube

(a)   (b)

FIG. 30.   The Quick-Strip column form, showing (a) the form in the open condition, and (b) the form in the closed position.   (*Waco Scaffold & Shoring Co.*)

FIG. 31.   A maze of lightweight scaffold shoring supports formwork for this arch. (*Bil-Jax, Incorporated.*)

forms, available for either class A finish or for less rigid standard of finish, are discarded after one use.

Fiber-glass-reinforced plastic or steel forms are used where top-quality concrete finish is required for many reuses. Sheet-metal forms work very well where there are a large number of reuses and normal finish.

FIG. 32. Scaffold shoring towers support the forms for the T section of a heavy bridge pier. (*Superior Scaffold Company.*)

FIG. 33. Bridge-deck forms over sloping ground are supported on adjustable shores. (*Superior Scaffold Company.*)

FIG. 34. Heavy-duty scaffold shoring is combined with Jr. beam stringers and lightweight horizontal shores for this large deck form. Compare Fig. 13. (*American Pecco Corporation.*)

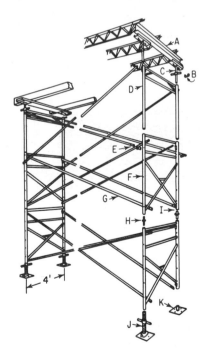

FIG. 35. Elements of a scaffold shore. (*A*) 4 × 8-in. steel Jr. beams serve as stringers. (*B*) Beam clamps secure the stringers to the jack head in such a way as to prevent rotation of the beam. (*C*) Screw jacks provide 18 in. of vertical adjustment. (*D*) Extension scaffold frames telescope into base frames to give height adjustments of 1, 2, 3, 4, or 5 ft. (*E*) Adapter pins fit into holes on legs of base frames to support the extension frames at the required height. (*F*) Base frame with X bracing. (*G*) Tubular steel cross bracing. (*H*) Coupling insert pins. (*I*) Special locking devices attach cross bracing to frame. (*J*) Swivel-base screw jacks compensate for uneven ground conditions and give 12 in. of height adjustment, fitting between legs of frame and base plate (*K*). (*Superior Scaffold Company.*)

The top of a round column is often flared the last 2 to 3 ft below the deck the column is supporting. This section of the column, commonly called the column capital, can be formed by using the Denform system, consisting of a series of small curved modular steel panels that can be assembled into a variety of capital sizes. Denform capitals have a seat for fitting on top of round fiber-tube forms. Column capitals are also formed by sheet-metal forms (Figs. 24 and 26), particularly applicable where beams run into the capital for support. When this is the case a section of the sheet-metal capital is cut out and fitted to the beam soffit and sides.

When it is necessary to design a form to fit unusual configuration and there is a good reuse factor, the form designer is wise to consider steel- or plastic-reinforced fiber glass. The form should be fabricated strong enough to be tied with the minimum of labor for the rate of concrete placement already established. A good example is illustrated in Fig. 36.

FIG. 36. A good example of pier and cap configuration which permits multi-use of forms. (*Symons Manufacturing Co.*)

8 Suspended Slabs and Beams

The principal concern of every form designer when designing forms for suspended structural concrete is to be sure that the form will carry the load until the concrete has attained sufficient strength to support itself. The type of form support depends upon how the suspended slab itself is to be supported.

When the permanent horizontal support for the suspended slab (commonly called a deck) is furnished by either structural steel or a precast-concrete member, some type of hanging device is usually used to support deck forms, as shown in Figs. 27, 28, and 29. However, if the suspended member is to be self-supporting or to be supported by integral beams, shoring will be required until the concrete is self-supporting. Various types of shoring methods and equipment are described in Chap. 14, and installations are pictured in Figs. 31, 32, 33, and 34. Elements of scaffold shoring are detailed in Figs. 35 and 37.

Loads (dead and live) imposed upon deck forms by concrete being placed on them must be correctly calculated to ensure against failure. The weight of concrete (pounds per cubic foot) times the concrete depth in feet (in decimals) equals the normal dead load (Table 5). An additional amount, allowed for the live load imposed by men and equipment used to place and finish the concrete, varies from 40 to 100 psf depending upon the particular job requirements. Most form designers allow 50 psf as the basic live-load allowance. It can be adjusted to

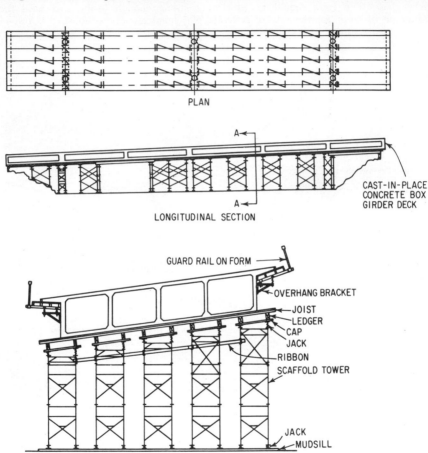

PLAN

CAST-IN-PLACE
CONCRETE BOX
GIRDER DECK

LONGITUDINAL SECTION

GUARD RAIL ON FORM

OVERHANG BRACKET
JOIST
LEDGER
CAP
JACK
RIBBON
SCAFFOLD TOWER

JACK
MUDSILL

SECTION A-A

Fig. 37. Scaffold framing and forming for a cast-in-place box-girder bridge on a curve. (*Superior Scaffold Company.*)

compensate for impact loading of concrete being dropped or for placing concrete with mechanized equipment such as power buggies.

The normal steps taken to design deck forms after the dead and live loads have been determined are:

1. Select the thickness of decking material. Minimum ⅝-in. plywood.
2. Calculate the size and maximum spacing of joists. Joists should be loaded to the maximum safe capacity for economy.
3. Calculate the size and maximum ledger or stringer spacing.

Table 5. Weight of Concrete, psf

Slab thickness, in.	Slab thickness, ft	Wt per sq ft, 150-pcf concrete	Wt per sq ft, 110-pcf concrete
3	0.25	37.5	27.5
4	0.33	49.5	36.3
5	0.42	63.0	46.2
6	0.50	75.0	55.0
7	0.58	87.0	63.8
8	0.67	100.5	73.7
9	0.75	112.5	82.5
10	0.83	124.5	91.3
11	0.92	138.0	101.2
12	1.00	150.0	110.0

4. Calculate the best spacing for shores, and select a shore that will safely support the loads anticipated.

5. Work out the best method for reshoring subsequent deck loads, where required.

The above method of deck forming design sequences applies equally well to flat slabs, flat slabs with drop heads, and beam-and-slab structures.

Designing deck forms for the waffle or grid-slab system, in which dome pans or fiberboard boxes are used to form a two-way joist system, is similar to the method for flat slabs. Table 6 will assist the form designer in calculating dead load for this deck system.

Pan joists use a joist spacing that has been determined by the structural engineer for the project at hand, the commonest center-to-center spacing being 30 in.: however, 20-in. spacing is also used. Standard metal pans for either spacing are available for rent or to be installed in place. The pan form creates a void in the slab which cuts down on the required concrete volume, at the same time creating a joist that acts as a small beam. Table 7 includes information for form design with this type of pan. The two types of joist beam are manufactured

Table 6. Equivalent Flat-slab Thickness for Dome Pans

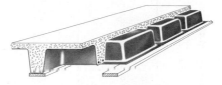

Depth of dome form, in.	Equivalent slab thickness, in.		
	Regular pan, 30 × 30 in.	Filler pan, 20 × 30 in.	Filler pan, 20 × 20 in.
8	3	3½	4
10	3½	4½	5
12	4½	5½	6
14	5½	6½	7

NOTES:
1. Add topping to above equivalents.
2. All joists in the above system are 6 in. wide.

Table 7. Equivalent Concrete Thickness for Pan Joists

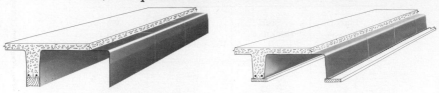

Depth of pan, in.	20-in. widths		30-in. widths	
	Width of joist, in.	Equivalent in. of concrete	Width of joist, in.	Equivalent in. of concrete
6	3½	1	5	1
6	4	1	5½	1
6	4½	1½	6	1½
6	5	1½	6½	1½
6	5½	1½	7	1½
6	6	1½	7½	1½
8	3½	1½	5	1½
8	4	1½	5½	1½
8	4½	2	6	1½
8	5	2	6½	2
8	5½	2	7	2
8	6	2½	7½	2
10	3½	2	5	2
10	4	2	5½	2
10	4½	2½	6	2
10	5	2½	6½	2
10	5½	2½	7	2½
10	6	3	7½	2½
12	3½	2	5	2½
12	4	2½	5½	2
12	4½	2½	6	2½
12	5	3	6½	2½
12	5½	3	7	3
12	6	3½	7½	3
12	7	3½		
14	3½	3	5	3
14	4	3	5½	2½
14	4½	3	6	3
14	5	3½	6½	3
14	5½	3½	7	3½
14	6	4	7½	3
14	7	4		

NOTES:
1. Add topping to slab equivalent shown above for overall thickness.
2. If tapered end spans are used, add ½ in. to equivalent inch of concrete.

for both center spacings of the joist and for five different joist depths. The flange type has a flange that supports the pan by resting on the wood-form joist. Pans of different depths are required for the different depths of joist. The second type of pan is an adjustable type that slips in between the joists and can be adjusted up or down to the required joist height and nailed to the edge of the joist.

Quite often structural-steel buildings are constructed with metal ribbed decking as a permanent deck form for the concrete floor. The decking is ribbed, giving it strength to work together with the concrete to support the loads imposed upon

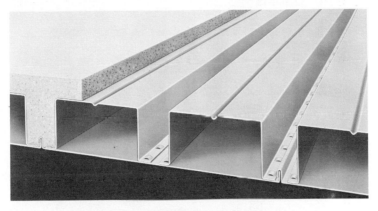

FIG. 38.   One type of steel corrugated deck and form unit.   (*Fenestra, Incorporated.*)

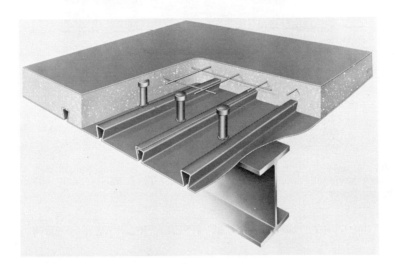

FIG. 39.   A corrugated deck and form, showing attachment to steel I beam and shear connectors for composite action.   (*Fenestra, Incorporated.*)

the deck. See Figs. 38 and 39 showing two different types of metal decking available.

9   Special Forming Problems

Forming for domes, vaults, folded plates, and other specialized construction follows the same general procedures described in this chapter. In Chaps. 25 and 26, covering placing of concrete, mention is made of tunnel and paving forms. A special form for suspended slabs in a building is shown in Figs. 40 and 41. Called a "flying shoring and deck," it is shown in Fig. 40 being pulled out from a completed deck unit of the building and being lifted in Fig. 41 to the second story above where concrete for another deck will be placed. Bridge-deck and guardrail forming is depicted in Figs. 42 and 43.

FIG. 40. A unit of "flying shoring and deck form" is being moved out, the concrete deck having developed sufficient strength. The complete unit will be lifted free by the tower crane and set in place above. (*Waco Equipment Company.*)

Forms for curb and gutter are quite simple forms, usually consisting of only timbers staked to line and grade (Fig. 44).

CARE OF FORMS

The most ingeniously designed, well-constructed form will not reflect the low cost of which it is capable unless good judgment is used in handling and care, as efficient handling of form materials can result in a substantial saving in forming cost. For example, mechanized movement and distribution of forms about the job reduce the amount of hand labor required.

A fork lift is a very useful piece of equipment for handling prefabricated forms, horizontal and post shores, lumber, and other material. Fork lifts, gasoline- or diesel-engine-powered in a wide range of sizes, can be fitted with a variety of special lifting or gripping devices to handle almost any construction material. Large, oversize pneumatic tires are required on rough terrain, while the small-tired units are suitable in smooth areas.

Fig. 41. The form unit shown in Fig. 40 is now being lowered into position preparatory to placing concrete for another increment of floor for the building. (*Waco Equipment Company.*)

Fig. 42. Details of a form for a bridge handrail. (*Illinois Tollway photograph.*)

Fig. 43. Forming for a composite bridge on prestressed-concrete girders.

Fig. 44. Curb and gutter forms are simple construction of timbers held in place with stakes. Normal practice is to strip the front form of the curb as soon as the concrete will stand, so the concrete can be finished.

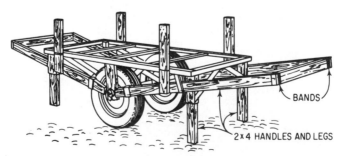

Fig. 45. Small handcarts are convenient for moving forms and other equipment about the job.

Horizontal transportation of material and equipment can be accomplished with a variety of carts (Fig. 45) and conveyors. Vertical movement is frequently done with hoist towers, but sometimes inefficiency in getting materials to and from the hoist is apt to be expensive. Tower cranes traveling on rails (Fig. 14) and climbing tower cranes (Fig. 1), besides providing vertical movement, have a large horizontal capability and can move materials, not only from one end of the building to the other, but also from locations on the ground some distance from the structure right into place on the structure. While these machines cannot be used on every structure, the possibility of their use should be explored.

Forms should be cleaned after every use, and form oil applied. In cleaning plywood, care should be exercised to avoid abrading or roughening the surface. It is better to scrape the surface with a small piece of wood rather than to use a wire brush or metal scraper. Avoid chipping the edges. Steel forms can be wire-brushed provided they are not brushed down to bright metal.

When stripping and cleaning forms, boxes, kegs, or buckets should be on hand to receive the many reusable small metal parts, especially when handling prefabricated forms. Careless loss of these small items can result in a substantial increase in cost. Solvents are available that will aid in cleaning these parts.

Forms when not in use should be carefully stacked to provide ventilation and prevent warping. Usual practice when hauling prefabricated units is to stack them face to face and back to back. They should not be carelessly dumped on the ground or left in untidy piles.

REFERENCES

1. ACI Committee Report, Recommended Practice for Concrete Formwork (ACI 347-68), report of Committee 347.
2. "Plywood for Today's Construction," American Plywood Association, Tacoma, Wash.
3. "Wood Structural Design Data," National Lumber Manufacturers Association, Washington, D.C.
4. "Manual of Steel Construction," American Institute of Steel Construction, New York.

Section 5

BATCHING, MIXING, AND TRANSPORTING

Chapter 16

PLANT LAYOUT

JOHN K. HUNT

A concrete plant consists of equipment and facilities for receiving, storing, handling, and proportioning the materials required to make concrete, and for delivering the proportioned materials to transport equipment, either before or after mixing the materials together to make concrete. For convenience, plants are classified by use as *mass concrete, paving, ready-mix,* or *concrete-products* plants; by function as *central-mix, transit-mix,* or *dry-batch* plants; and by ease of movement as *permanent, portable,* or *mobile plants.* Plants where materials flow continuously downward as they are proportioned, mixed, and placed in delivery units are called *gravity* or *tower* plants; plants where materials are elevated after proportioning are called *low-profile* plants. Thus a plant may be described as a *low-profile mobile central-mix paving* plant. Standards for concrete plants are published by the Concrete Plant Manufacturers Bureau, 900 Spring St., Silver Spring, Md. 20910.

Plant layout depends upon the method of concrete delivery from the plant, material flow through the plant, material delivery to the plant, and auxiliary equipment needed in particular installations. Typical characteristics of various types of plants are discussed below. Details of plant equipment and operation are discussed in Chaps. 17 through 23.

MASS CONCRETE PLANTS

1 Production Rate

Mass concrete plants are large installations with large concrete-making capacity. They are often manufactured and set up for a specific construction project such as a dam or navigation lock requiring a large amount of concrete in a relatively small area. The plants are usually portable, shipped in large sections by rail, and assembled by bolting the parts together. Six to eight weeks are required for erection. Project specifications of the contracting authority (in the United States typically the U.S. Army Corps of Engineers, the U.S. Bureau of Reclamation, a public resources board, or a public utility) control plant design to a large extent. Because such specifications are similar to each other in most respects, and because the purposes of the plants are similar, the resulting plants have many characteristics in common. Mass concrete plants must produce high-quality concrete of several mix designs at a steady rate for large placements at a minimum cost per cubic yard. They are usually as automatic in operation and as reliable as it is feasible to make them so that labor costs will be held to a minimum (Figs. 1 and 2).

The primary factor in mass-concrete-plant design is the nominal concrete production rate when making mass interior and mass exterior mixes as determined by specifications or by the number and size of concrete pours required to meet construction progress schedules. This production rate is limited by the number and size of mixers and typical mixing cycle times (Table 1). Where two mixing times are shown in Table 1 for the same size mixer, the longer times are generally typical of U.S. Army Corps of Engineers work. In some plants an extra mixer is supplied beyond that required by the production rate. Reduced production is expected because of reduced mixer capacity when mixing grout or high-slump concrete for small pours such as powerhouse walls. Most specifications include a provision for reducing mixing times as a result of mixer-efficiency tests, but the plant capacity should be established based on the initial mixing times. Further, the rated capacity of the mixers is usually used in establishing plant capacity,

Fig. 1. A large mass concrete plant of this type has a capacity of several thousand cubic yards per day. (*C. S. Johnson Division of Koehring Company.*)

though it is logical that the guaranteed capacity should be used. Multiple mixers are desirable to minimize the effect of interruptions. The smallest plants typically have two 2 cu yd tilting mixers; the largest have six or eight 4 cu yd mixers. Most tilting mixers used in the United States are either 2 or 4 cu yd; mixers larger than 4 cu yd are rarely used because specification mix times are not favorable to larger mixers. For example, a plant with four 6 cu yd mixers costs more than a plant with six 4 cu yd mixers per cu yd per hr of production capacity. If mixing times can be reduced and larger concrete buckets used more of the larger mixers will be used in the future.

Turbine mixers have been used in a few installations. Because turbine mixers are generally unsuitable for crushed aggregates larger than 2½ in. or gravel larger than 3 in., and because of higher wear and maintenance typical of turbine mixers, tilting mixers are preferred for mass concrete.

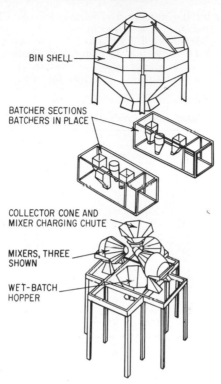

BIN SHELL

BATCHER SECTIONS
BATCHERS IN PLACE

COLLECTOR CONE AND
MIXER CHARGING CHUTE

MIXERS, THREE
SHOWN

WET-BATCH
HOPPER

FIG. 2. Blown-up schematic diagram of a large batching and mixing plant for a dam construction project.

2 Plant Location

The next most important factor is plant location to permit expeditious movement of the concrete to where it is used. Movement is usually in concrete buckets or rail cars or trucks to a position where the buckets can be picked up by a crane or cableway. The plant is located on a flat area, often excavated for the purpose, near the end of a crane trestle or cableway.

Mass concrete plants are usually of the gravity type with overhead storage for five sizes of aggregate with a maximum of 6 or 8 in., cement, and pozzolan. Overhead storage, like multiple mixers, is used to ensure continuity of operations in event of material-handling-equipment failure. Overhead storage is provided for 1 to 2 hr of operation at rated production capacity when producing concrete with the largest-size aggregate; thus the total cubic yards of aggregate storage in the bin is 1.25 to 2.5 times the rated production capacity in cubic yards per hour, and the cement and pozzolan storage capacity combined is usually about 4 cu ft per cu yd of aggregate stored in the overhead bin.

Aggregate materials for 3 days to 1 week of operation are stored over a reclaiming tunnel and moved by belt conveyors across rinsing screens, thence to the top of the overhead bin, where they are rescreened and delivered by chutes directly to the bin compartments. Cement and pozzolan for 3 days to 1 week of operation are normally stored in large steel silos and moved pneumatically or by screw conveyors and bucket elevators to compartments in the bin (see Chap. 19).

Production of chilled concrete is common in mass concrete plants, and the plant structure is modified by the method of cooling adopted. Crushed or flaked ice may be used as an ingredient. It is stored in insulated compartments and weighed as a granular material. Where chilled air is used to cool aggregates in the overhead bin, storage for about 2 hr operation is required in the bin to permit adequate cooling. Where inundation in chilled water or vacuum cooling is used at ground level, the bin capacity is kept smaller; sufficient for about 1 hr of operation (see Chap. 20).

Table 1. Tilting-mixer Production Rates—Mass Concrete

Nominal or rated mixer capacity, cu yd	Typical mixing time, min	Typical cycle time, min	Batches/hr	Cu yd/hr
2	1½	**2**	**30·0**	**60·0**
3	2	2½	24.0	72.0
3½	2¼	2¾	21.8	76.4
4	2½	3	20.0	80.0
4½	**2⅝**	**3⅛**	**19·2**	**86·4**
	2¾	3¼	**18·5**	**83·1**
5	2¾	3¼	18.5	92.3
	3	3½	17.1	85.7
5½	2⅞	3⅜	17.8	97.8
	3¼	3¾	16.0	88.0
6	3	3½	17.1	102.9
	3½	4	15.0	90.0
7	3¼	3¾	16.0	112.0
	4	4½	13.3	93.3
8	3½	**4**	**15·0**	**120·0**
	4½	**5**	**12·0**	**96·0**

The above production rates are for mixers with a batch equal to the nominal capacity shown. Mix times shown are typical for specification work in the absence of mixer-performance tests. Mixer sizes in boldface type are rated capacities of new (1970) Plant Mixer Manufacturers Division standard mixers. They are guaranteed to mix rated capacity when slump is between 1½ and 3 in. and aggregate size is not over 3 in. Before 1966, mixers were rated in cubic feet with 28 cu ft of rated capacity per cubic yard of nominal capacity, and guaranteed to mix a 10% overload. For mass concrete, a batch of the nominal capacity shown is usually mixed in these obsolete-size mixers, and batches of 4 or 8 cu yd are common in 4½ and 8½ cu yd mixers because of large aggregate and because of concrete bucket, crane, and cableway limitations.

Because tilting concrete mixers produce shocks that affect the operation of batching scales, either the mixers or the batchers are located on supports separate from the bin structure. The mixer supports must be designed for the added shock load. Vibrating screens located on the bin do not produce objectionable vibration of the batchers provided the bin is about 300 cu yd or more capacity and at least partially filled. Screen supports must be designed so their members will not vibrate excessively at the screen frequency.

3 Scales and Controls

Except on the smallest works each ingredient is weighed in a separate hopper with a separate scale, batch weights are recorded, and controls are automatic, interlocked, and have a means of rapid selection among 12 to 32 preset mixes. Sixteen to twenty different mix designs are commonly used on a project. Mechanical lever scales and full-reading springless dials are used. Some method for determining moisture in sand and sometimes ¾-in. aggregate, and for correcting the aggregate and water weights quickly is usually required (see Chap. 17).

An average to large mass concrete plant requires at least two men to control material flow, one of whom should be located at the top of the bin where he can observe the material flowing from the chutes and in the bin. One operator is located at the control stand where he can see and hear the batching equipment and see the mixers. It is desirable that this man be able to see into at least one mixer. A plant maintenance man and, on larger plants, a man to control discharge of concrete from the wet concrete hopper are required. The inspector is located on the floor with the operator and batching equipment where he can observe the material gradation, batching, recording, and mixing, and one or two technicians may be located near the wet-concrete hopper to operate a wet-concrete sampling device.

CENTRAL-MIX PAVING PLANTS

4 Portable and Mobile Plants

Contractors use central-mix paving plants of both portable and mobile types. The portable plants are movable by truck trailer in large sections, while the mobile

FIG. 3. Mobile central-mix paving plant. (*C. S. Johnson Division of Koehring Company.*)

plants are supplied with wheels and a kingpin for truck-tractor movement (Fig. 3). Most plants are low-profile to reduce erection time and keep erection lifting requirements within the height and weight capacities of cranes used by road contractors. Erection usually takes 1 or 2 days. Since fresh concrete is perishable and requires special hauling units, the plant is located in an area near the place where the concrete is used. The plant may be moved several times on one job to keep haul distances less than about 6 miles.

A large central-mix paving plant may easily use as much as 200 short tons of cement, 1,200 short tons of aggregate, and 175,000 lb of water per hr. It is obviously impractical to store enough material at the site to operate the plant for long periods, and arrival of materials must be steady and carefully controlled. Aggregate materials sufficient for 1 to 10 days of operation are usually delivered to the plant by truck and frequently stored in piles on the ground. Thence they are placed in the aggregate bins, which seldom hold enough aggregate for more than 5 to 20 min of plant operation, by clamshell bucket or by belt conveyors elevating material from low hoppers filled by large front-end loaders. Belt-conveyor charging of the bin is controlled automatically by bin-level signals. Cement is most commonly delivered in pneumatic semitrailers and pumped into the cement-storage compartment of the plant. Auxiliary storage may be provided by mobile units designed to be moved empty and holding about 200 short tons. Two to six cement hauling units may have to be unloading at the same time. Occasionally cement is delivered by rail or by bottom-dump hauling units and handled by screw conveyor and elevator. Water is often supplied by wells, pumped from nearby streams or quarries, or hauled by truck to the plant. Admixtures are delivered in tank trucks or drums and pumped into storage tanks at the plant. The expeditious scheduling and handling of materials in these quantities require skill and experience.

Mixers are either large tilting mixers or, in a few plants, turbine mixers. Reduction of mixing times for pavement concrete so that they are the same for all mixer sizes has greatly encouraged use of 8½ to 10 cu yd tilting mixers. These are usually of a special design with a forward trunnion or linkage to permit a high discharge height and low charging height consistent with the low-profile plant arrangement.

5 Production Rate

Plant capacity should be controlled either by the time required to batch and discharge one batch, or more commonly by the time required to charge, mix, and discharge the mixer. As a practical matter it is frequently controlled by the number of concrete-hauling units available. For any given plant there is a minimum mixing time below which the plant cycle time will limit production. Also mixer charging time varies from plant to plant in the range of 15 to 35 sec, not including water, which is usually allowed an additional 15 sec. Mixing time is started when all aggregate and cement are in the mixer. Figure 4 shows typical production-capacity ranges of central-mix paving mixers. Guaranteed mixer capacity is usually used where consistent with loads that can be hauled away from the mixer. Mixers most often discharge directly into trucks to eliminate added height required for a wet-concrete hopper.

In Fig. 4, the smaller batch sizes and shorter mix times are typical of turbine mixers, seldom used in sizes above 3½ cu yd, with typical mix times of 30 to 60 sec. Tilting mixers smaller than 4½ cu yd and mix times less than 40 to 50 sec are seldom used. The trend is to reduce mixing times for large tilting mixers from the 60- to 90-sec range to 40 to 60 sec when blended materials are fed into the mixer.

Plant production capacity is computed as follows:

$$\text{cu yd/hr} = \frac{3{,}600YN}{T}$$

where T = cycle time per batch, sec
   Y = cu yd per batch
   N = number of mixers

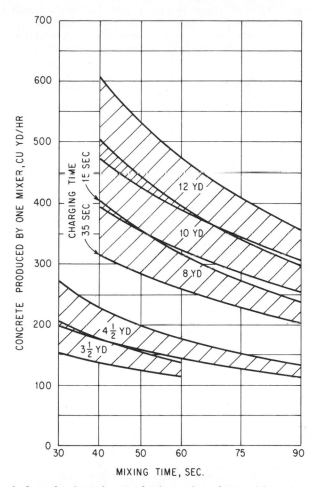

Fig. 4. Typical production of central-mix paving plants with various mixer sizes, mixing times, and charging times.

Cycle time *T* for a low-profile central-mix plant with one mixer is determined as in the following example, which is typical of a plant with a large tilting mixer and a cumulative aggregate batcher:

Operation	Mixer cycle, sec	Batcher cycle, sec
Discharge batchers...................	. . .	25
Travel time on charging belt.........	. . .	(6)
Charge mixer......................	25	
Mix...............................	45	
Batch.............................	. . .	45
Discharge mixer....................	10	
Prepare mixer for next batch........	6	
	86	70

Since the mixer cycle is longer, it will control, and 86 sec would be used for the value of T. Travel time on the charging belt is not counted because the batchers can begin to discharge as soon as the mixer has begun to prepare for the next batch and the batchers can begin to batch as soon as they are empty. The operational cycle for this example would be:

Time from start of cycle, sec	Operation
0	Mixer empty; starts to get ready for next batch
$6 - 6 = 0$	Batchers begin to discharge
$0 + 6 = 6$	Material starts into mixer
$0 + 25 = 25$	Batchers empty; begin to batch
$6 + 25 = 31$	All material in mixer except water. Start of mix time
$25 + 45 = 70$	Batchers ready to discharge
$31 + 45 = 76$	End of mix time. Mixer starts to discharge
$76 + 10 = 86$	Mixer empty; starts to get ready for next batch

Determination of cycle time is more difficult if there are two or more mixers. The following example is typical of a low-profile central-mix plant with two tilting mixers and a separate batcher for each material:

Operation	Mixer cycle, sec	Batcher cycle, sec
Discharge batchers............................	. . .	25
Travel time on charging belt.................	. . .	6
Charge mixer.............................	25	
Mix.......................................	45	
Shift charging mechanism to second mixer.....	. . .	10
Batch......................................	. . .	(14)
Discharge mixer............................	10	
Prepare mixer for next batch.................	6	
	2)86 43	41

Because the batcher discharge can be initiated as soon as the mixer has gone into an automatic cycle to prepare for the next batch (starting to close mixer discharge gate or right a tilting mixer), the mixer will complete this 6-sec operation at the same time the material reaches the mixer end of the charging belt. The charging mechanism cannot be shifted until the mixer-charging belt is clear. Clearing the belt and then shifting the charging mechanism requires 16 sec, which is longer than the 14-sec batching time; so the batchers will have finished batching before this operation is completed, and it will control. The batchers can produce a batch every 41 sec and the two mixers can make a batch every 43 sec; so the plant cycle time will be 43 sec. Time available for batching is $43 - 25 = 18$ sec. Using procedures similar to these examples it is possible to compute the capacity of a plant and to balance the various operations to obtain maximum capacity.

Except with both very short mixer-charging times (about 20 sec or less) and long mix times (about 75 sec or more), a three-material cumulative aggregate batcher (all three aggregates weigh in succession on one scale) will be too slow for use with more than one mixer. Single-material batchers (a separate batcher for each material and all materials weighed simultaneously) will permit maximum production with two mixers except when very long mixer-charging times (30 sec or more) are combined with short mix times (50 sec or less). Thus single-material batchers or dual cumulative batchers are usually necessary if there is more than one mixer in the plant.

6 Scales and Controls

Either beam scales or dial scale indicators with mechanical lever scales are almost universally used on paving plants in order to meet the requirements of accuracy, ruggedness, reliability, and insensitivity to temperature changes required. Rapid mix selection is not important because a plant usually makes the same batch formula all day. However, a method for determining sand moisture content and for correcting sand and water weights is essential. Typically the water content is changed to control slump as well, although this practice is questionable because often the slump change is caused by variations in concrete temperature and sand moisture and gradation. Mixer motor-load indicators are of assistance in controlling slump.

Controls are automatic and highly interlocked. The operator discharges the mixers when there is a truck in position, and all other operations except material delivery to the plant are often automated. The operator must be able to see the trucks and mixer discharge, so he is usually located on the ground or at mixer level, either in the open or in a small enclosure or trailer in which the controls are mounted. Graphic or digital recording of some ingredient quantities is often required.

DRY-BATCH PAVING PLANTS

7 Plant Types

Dry-batch paving plants are used to supply track-mounted mobile paving mixers with proportioned ingredients (except water and liquid admixtures) for making concrete. Where most plants were formerly portable with overhead bins, most modern plants are mobile (Fig. 5). The portable plants evolved into a great variety

Fig. 5. Mobile one-stop dry-batch paving plant. (*C. S. Johnson Division of Koehring Company.*)

of types, varying from a separate plant for each material and several batchers in each plant (so several batches could be weighed and discharged simultaneously) to single-stop plants with one or more sets of single-material batchers to permit rapid truck charging.

Normal batch sizes for 34-E paving mixers are 1.25, 1.385, and 1.50 cu yd, the latter sizes representing overloads of 10 and 20% where they are permitted by specifications. The dry-batched materials are carried in separate compartments on batch trucks with dump-truck bodies to the paving-mixer skip, and discharged into the mixer one batch at a time. Some specifications require the cement to be placed in cans or small compartments separated from the moist aggregates; others permit hauling both cement and aggregates in the same compartment in contact with each other. The commonest mixing time is 60 sec.

8   Production Rate

Dry-batch-plant capacity must be keyed to paving-mixer production, which frequently equals the theoretical rates shown in Table 2 for long periods. Plant

Table 2. 34-E Dual-drum Paving-mixer Production and Material Requirements

Mixing time, sec..............	50	60	75	90
Cycle, sec/batch..............	36.5	41.5	49.0	56.5
Batches/hr....................	98.6	86.7	73.5	63.7
Cu yd/hr.....................	136	120	101	88.2
8 in. × 24 ft pavement, lin ft/hr.	230	203	172	149
Cement, short tons/hr.........	38.0	33.4	28.3	24.6
Aggregate, short tons/hr.......	221	194	164	143
Water, lb/hr..................	36,482	32,079	27,195	23,569

Quantities are based on a 1.385 cu yd batch consisting of 771 lb cement, 4,474 lb aggregate, and 370 lb water. Comparable output for other pavers: 34-E triple drum 144%, 34-E single drum 57%, 16-E dual drum 47%, 27-E single drum 46%.

capacity is greatly influenced by the combinations of batchers and the number of compartments per batch truck, and plant capacity must be computed separately for each combination.

The average time required for a plant to charge a batch truck compartment if T is greater than f:

$$C = \frac{He + (H-1)f + T}{K}$$

or if T is less than f:

$$C = \frac{H(e+f)}{K}$$

where C = average cycle time per batch, sec
   K = number of batches per truck
   H = number of plant cycles to charge one truck
     = K/n to next larger whole number
   n = number of identical batchers in tandem in the plant or number of batches weighed simultaneously
   e = time to empty batchers, sec
   f = time to fill batchers, sec
   T = time from completion of discharge of last batch until the next truck is in position and discharge initiated, sec

For a 1.385 to 1.5 cu yd batch, $e = 7$ to 9 sec when cement discharges into the truck by gravity, 5 to 9 sec when aggregate discharges by gravity, and about

8 to 11 sec for a low-profile plant where aggregates are elevated to the truck by belt conveyor; f = 8 to 13 sec for cement batchers, 6 to 8.5 sec for single-material aggregate batchers, and 7 to 12 sec plus 3 sec per material for a cumulative aggregate batcher. (Note that T may be less than f in this instance.) T = 19 to 28 sec and averages about 22 sec with cement placed in separate cement cans on the batch truck, and T = 15 to 25 sec and averages about 19 sec for aggregate plants and for cement plants if cement and aggregate are mixed together in the truck. Hourly capacity is given by

$$\text{No. of batches/hr} - \frac{3,600}{C}$$

Where the batchers discharge into a hopper with a gate, the batchers can discharge one batch into the hopper and begin to rebatch before the truck is in position, which may increase plant capacity. The following formula is used if $2f + e$ is equal to or less than T, and K is greater than $2n$:

$$C = \frac{d + t + (R + 1)e + Rf + T}{K}$$

If n is less than K, and K is equal to or less than $2n$:

$$C = \frac{d + t + e + T}{K}$$

If n is equal to or greater than K:

$$C = \frac{d + T}{K}$$

and if $2f + e$ is greater than T:

$$C = \frac{d + t + (R + 2)(e + f)}{K}$$

where all factors are as in the paragraph above except d = time to empty dry-batch hopper, sec (about 6 sec); t = time for truck to shift for next discharge, sec (about 5 sec); and $R = (K - 2)/n$ to next larger whole number.

Table 3 shows typical production of a portable plant as influenced by the number

Table 3. Typical Portable Dry-batch Paving-plant Production, Batches per Hour

Cement batchers, gravity discharge, e = 9.0, f = 13.0, t = 24

Compartments per batch truck	No. of batches weighed simultaneously			
	1	2	3	4
2	131	218	218	218
50 % 2; 50 % 3	136	207	273	273
3	140	196	327	327
50 % 2; 50 % 4	138	240	240	327
50 % 3; 50 % 4	143	229	295	382
4	145	262	262	436
5	149	234	327	327
6	151	281	394	394
7	153	255	327	458

of batchers in tandem and by the number of compartments per batch truck. A typical low-profile mobile plant with one set of batchers where aggregate is elevated on a belt conveyor to the batch truck has a theoretical capacity of about 180 batches per hour with single-material aggregate batchers and about 133 batches per hour with a cumulative aggregate batcher. Comparable practical capacities when charging five-compartment batch trucks are about 160 and 125 batches per hour.

The total truck-loading time at the batch plant is influenced by the number of stops made to receive a full load. For each stop in a multiple-stop plant, the truck time can be approximated from the average cycle times at each stop C multiplied by the number of compartments in the truck. The truck delay is, of course, longer if more stops are made. For example, using a four-compartment truck and two sets of batchers, a single-stop plant causes a delay of about 30 sec, while a four-stop plant might result in a total delay of about 125 sec, but this will vary greatly with skill of the drivers.

Materials may be received at the batch plant by rail or truck or a combination of the two. Plant location is not so critical as for central-mix plants because dry batches may usually be retained in batch trucks for 45 min or longer if necessary. The plant is located where it will be most convenient for material delivery and still be within a reasonable distance from the pavement site, since this distance largely determines the number of batch trucks required (see Chap. 23).

To enhance their portability or mobility, dry-batch paving plants often have little material-storage capacity in the plant. Bins seldom hold enough aggregate for more than 10 to 30 min of operation, and may be even smaller. If continuously recharged by belt conveyors (often automatically controlled), bins may hold aggregate for only a few batches. A 1- to 10-day supply of aggregate is usually stored on the ground, and bulk cement storage of about 60 to 200 short tons is provided. Quantities of materials used are large, and careful scheduling of material deliveries as well as close supervision of unloading operations may be required. Water is delivered to storage tanks on the paving mixers by trucks.

9   Scales and Controls

Batcher scales for dry-batch plants are of the lever type, either with weighbeams and a beam balance indicator or with springless dial indicators. Weighbeams are usually more rugged and more sensitive than dials but cannot be used in some areas because of specifications or because of control attachments required that are incompatible with beam-scale mechanisms.

A dry-batch-plant operator must be located where he can see that a truck is in position to receive the batcher discharge. It is desirable that he be able to see the scale indicators or some other indication that the weighing operation is functioning properly. Some fully interlocked plants have been equipped to rebatch automatically and be discharged by the truck drivers. This is particularly useful where there are several stops because one operator can then watch over several of the stops.

COMMERCIAL READY-MIX PLANTS

10   Plant Types

The outstanding characteristic of commercial ready-mix plants is the need for a large variety of concrete formulations. While smaller plants may use only 3 aggregate sizes and 1 type of cement, larger plants require 4 to 12 aggregate sizes or types, 2 to 4 types of cement and pozzolan, and as many as 6 different admixtures. Plants are customarily of the portable type in that they are assembled by bolting, but larger plants are usually installed permanently, enclosed with siding, and perhaps roofed to keep out rain and snow and to facilitate aggregate heating (Fig. 6). Controls are designed to permit easy formula changes and to reduce operating costs.

FIG. 6. Large ready-mix concrete plant. (*C. S. Johnson Division of Koehring Company*.)

FIG. 7. Small mobile ready-mix concrete plant. (*C. S. Johnson Division of Koehring Company*.)

Mobile plants have become common in the small and medium plant sizes. They are often used in conjunction with a fixed plant by the same owner to reduce haul distances on large jobs, since after material costs, concrete delivery is the major cost item. Figure 7 shows a small mobile plant with a capacity of about 50 cu yd per hr. Larger mobile plants are similar in appearance to the dry-batch plant shown in Fig. 5.

In plants producing over 10,000 cu yd per yr, material costs are 53 to 63% and delivery costs are 18 to 23% of total costs. Plants costs, including plant labor, are about 6 to 10. Over half the material cost is cement. Material and delivery costs are not appreciably affected by production volume, but plant and other costs are reduced by increased production. Most plants produce between 10,000 and 60,000 cu yd per yr.

Commercial ready-mix plants may be *transit-mix*, where concrete is mixed in the hauling units, *central-mix*, where concrete is completely mixed before dumping it into hauling units, or *shrink-mix*, where concrete is partially mixed in a plant mixer and mixing is completed in a transit-mix hauling unit. Of these methods, transit-mix is by far the most common.

11 Production Rate

Batcher sizes are coordinated with truck capacities so that one or two, or occasionally three, batches will fill the largest truck used. The maximum truck capacity is usually established by legal load limits because it is economical to use large trucks to reduce delivery costs. Aggregate batchers usually weigh cumulatively in the smaller plants while two cumulative batchers or single-material batchers are used in large plants or in plants where several batcher cycles are used to charge one truck. Cement should be weighed in a separate batcher. Water is weighed in a separate batcher or measured by volume in a calibrated tank or water meter.

High-capacity plants usually have driveways through the plant. Lower-capacity plants may be designed so trucks back into the charging area, which some operators prefer because trucks are more easily kept clean. In central-mix plants mixers of the tilting type are preferred because they discharge rapidly and will mix low-slump concrete. Vertical-shaft (turbine) mixers are used where mix time must be held to a minimum and sometimes where little headroom is available. They require more cleaning and maintenance than other types. Nontilting revolving-drum mixers are used in some installations. They do not discharge low-slump concrete as well as other types, they require more cleaning than tilting mixers, and maintenance is intermediate between tilting and vertical-shaft mixers.

Plant capacity is usually given as though the plant cycled continuously with material loaded as rapidly as it can be discharged, or

$$P = \frac{3,600\,Y}{e + f}$$

in which P = maximum production capacity, cu yd per hr
Y = batcher capacity, cu yd
e = time required to empty batchers, sec
f = time required to fill batchers, sec

As a practical matter, transit-mix plants almost never reach their rated capacity.

Average time required to charge a transit-mix truck:

If T is greater than f:

$$C = He + (H - 1)f + T$$

or if T is less than f:

$$C = H(e + f)$$

where C = average cycle time per truck, sec
H = number of batcher cycles required to fill one truck
e = time to charge the truck or discharge the batchers, sec
f = time to fill the batchers, sec
T = time from completion of filling one truck until next truck is in position and discharge initiated, sec

Typically, e is 4 to 13 sec per cu yd because of the limited rate at which transit-mix trucks digest the material, with 7 to 9 sec per cu yd common; e is

5 to 6 sec per cu yd when discharging to a hopper or mixer; f for an automatic plant is 5 to 8 sec per cu yd plus 3 sec per aggregate material for cumulative aggregate batchers and 5 to 7 sec per cu yd plus 3 sec for single-material batchers, usually controlled by cement-batching time. Except for simple batches, manual filling requires about twice as much time as automatic filling. T is commonly about 25 sec for drive-through plants and about 30 sec if trucks back under the plant.

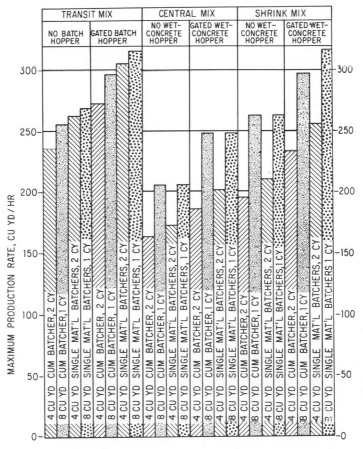

Fig. 8. Comparison of maximum ready-mix plant production rates. Central mix 60 sec; shrink mix 30 sec; $e = 7$ sec per cu yd for charging transit-mix truck; 5 sec per cu yd for discharging to hopper or mixer or for discharging mixer to truck; $f = 6$ sec per cu yd plus 3 sec per material; mixer discharge to hopper in 16 sec.

Plant capacity in cubic yards per hour is $3,600/C$ multiplied by the average truck size. To obtain usable production for a given installation, the average truck-load in cubic yards is substituted for the truck size. In commercial use and where some variety of truck capacities is available, the average load leaving the plant is about 77% of truck capacity. Where only large trucks are available this average load may be reduced to 65 to 70%.

Figure 8 shows a comparison of maximum ready-mix-plant production rates for a typical set of conditions for transit-mix, central-mix, and shrink-mix and for both one and two batcher cycles per load. In general, there is little reduction in capacity of a transit-mix plant if two batcher cycles are used. The maximum capacities shown in Fig. 8 are in excess of the requirements of most plants. For example, it would require about thirty 8 cu yd transit-mix trucks to utilize a plant capacity of 240 cu yd per hr if the average truck turnaround time were 1 hr. Capacity in excess of 240 cu yd per hr would serve only to reduce the delay on the last trucks charged at the start of the day. For example, while a capacity of 240 cu yd per hr would permit 30 trucks to be dispatched in 1 hr, 300 cu yd per hr would permit 30 trucks to be dispatched in 48 min. There is little further benefit from the excess capacity, since the number of trucks and the average turnaround time limit production for the remainder of the day to 242 cu yd per hr. Maximum production of the 240 cu yd per hr plant is 1,920 cu yd in 8 hr, compared with 1,980 cu yd in 8 hr for the 300 cu yd per hr plant.

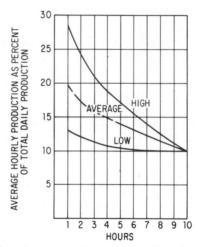

FIG. 9. Volume-duration curve for commercial ready-mix concrete plant.

Also in Fig. 8 if open-top hauling units are used for central-mix, maximum production would be as shown for a gated wet-concrete hopper even though no hopper were used. A plant with more than one cumulative aggregate batcher with not more than two materials batched in each batcher would have the same capacity as the plants with single-material batchers because cement-batching time would be longer than the aggregate-batching times. A plant with an adequate belt conveyor elevating materials to a transit mixer would operate at the same capacity as given above for transit-mix plants except that there may be a delay of 3 to 6 sec per truck because of travel time on the belt. A central-mix plant where materials are conveyed on a belt from the batchers to the mixer should be analyzed as was done for a central-mix paving plant earlier in this chapter.

Because of the nature of their business, commercial ready-mix plants usually produce only a fraction of their possible capacity. The average daily hourly production is about 55% of the peak daily production for 1 hr. Figure 9 is a volume-duration curve for a typical commercial plant based on a 10-hr day. If, for example, production is estimated at 500 cu yd per day, the peak hour would be $0.29 \times 500 = 145$ cu yd; the peak 3-hr period would be $0.21 \times 3 \times 500 = 315$ cu yd. On an average 500 cu yd day the average peak hour would be $0.20 \times 500 = 100$ cu yd; the average for a 5-hr period would be $0.14 \times 500 = 70$ cu yd per hr.

Ready-mix plants frequently need to operate at peak rates for only 1 or 2 hr a day, but the cost of delaying hauling units is such that plants must be able to operate at peak rates for at least several hours. Also an occasional job, such as a pavement or a massive foundation, may require capacity operation for an extended time. For these reasons, aggregate-handling equipment is usually designed to operate continuously at the peak rate rather than the average hourly rate.

Aggregates usually arrive at a ready-mix plant by rail or truck. Several days' supply is commonly stored over a tunnel conveyor and elevated to the bin as needed. Bin capacity varies from about 20 min operation in mobile plants to as much as 10 hr operation in elevated bins in cold areas or where storage on the ground is precluded by the location. Cement arrives by rail and is elevated to storage compartments in the bin or silos, or more commonly is received in pneumatic trucks that pump the cement directly into storage. Table 4 shows typical storage capacities. Water comes from private wells or municipal water supplies and is often arranged so some is stored in an overhead reservoir for

Table 4. Typical Storage Capacity for Commercial Ready-mix Plants

Peak concrete production, cu yd/hr	Aggregate storage, cu yd				Cement storage, bbl*		
	Mobile plant, bin	Fixed plant		Total storage	Mobile plant total storage	Fixed plant†	
		Bin with 3 to 5 compartments	Bin with 6 or more compartments			Overhead storage	Total storage
10	0–15	...	...	110	165‡	...	.....
25	0–30	30	40	275	165‡	70	350‡
50	35	50	65	550	165‡	130	650
100	40	100	125	1,100	350‡	260	1,300
150	45	225	270	1,650	525	390	1,950
200	50	400	480	2,200	700	520	2,600
300	60	650	780	3,300	1,050	780	3,900

* Cement-storage capacity is shown at 4 cu ft per bbl; cubic feet of storage space required is four times the figure shown. Because of fluffing in handling, actual capacity may be 70 to 85 % of the storage capacity shown.

† Overhead and total cement storage should be increased about 20 % if there are three or more types of cement.

‡ 165 bbl is the minimum storage for cement delivered in 125-bbl pneumatic trucks. 450 bbl is the minimum storage for cement delivered in 400-bbl rail hopper cars.

emergencies. A booster pump may be required to increase water pressure for high-speed batching. Admixtures are mixed or stored in tanks and pumped to dispensers as needed.

12　Scales and Controls

Manual controls have been replaced by sophisticated automatic controls on most larger ready-mix concrete plants. Scales are nearly always of the lever type, although some load-cell scales have been used. Small plants have dial indicators or multiple weighbeams and may be manual or automatic in operation, with the mix weights selected by the operator manually. Large plants typically have automatic dial scales with the mixes selected manually or electrically by the operator from among a large number of preset formulas or selected electrically from a formula card inserted by the operator. Electrical selection methods permit the operator to be located remotely from the batcher scales, often in an office near the plant and adjacent to the driveway where he can communicate easily with the truck drivers, receive orders, and dispatch trucks in addition to operating the plant. Most plants do not record batch weights, but the trend of many specifications is to require recording. Graphic and digital recording are most common, and there are a few photographic recording installations. Most of these produce a record maintained as a plant record. Some recent installations record digitally in multiple copies so tickets with batch weights can be dispatched with the concrete.

CONCRETE-PRODUCTS PLANTS

13   Plant Arrangement

Concrete-products plants are permanent installations arranged to make concrete block, pipe, precast or prestressed structural members, and similar products. Other similar plants are used to line or encase metal pipe for special applications. Plants are characterized by carefully organized material and product routing and are frequently highly automatic in operation. They differ from concrete plants discussed before in that concrete is often of special formulations, is made in small batches, and is mixed in concrete mixers of types not often used in other concrete plants. Mix-formula changes are infrequent. The quantity of concrete produced per day is often quite small.

Materials are delivered to the plants by rail or truck and stored or elevated into bins as they are in ready-mix plants. The number of materials is typically four aggregate sizes or types and one or two types of cement, but some products such as ornamental concrete may require a large variety of aggregate. Where materials are stored in bins at ground level or over a tunnel conveyor, aggregate handling is easily adapted to keep bins filled automatically. Since plants usually operate in cold weather, provision for heating materials is common. Plants, except in the warmest areas, are totally enclosed.

14   Scales and Controls

Concrete is batched in hoppers suspended from lever scales under storage bins and discharged by chutes or by trolleying hoppers into the mixers. Batcher hoppers may travel on rails with the scales stationary, the entire batcher including the scales may travel, or in some plants a trolley hopper separate from the batching hopper may convey the materials. Belt conveyors are also used to convey weighed batches to mixers. Cement may be weighed on the same scale as aggregates, but a separate scale is advisable to obtain more accuracy and economy in cement usage. Some materials, such as cinders or other light aggregate, may be batched by volume instead of weight in special applications where specific gravity of the material varies greatly. Water weigh batchers are seldom used. In some plants volumetric water tanks or water meters are used, but the plants using drier concrete add water based on judgment of the operator or have electrically controlled water addition based on conductance of the concrete.

In some plants controls are completely manual with the batcher and mixer operator setting weights on weighbeams or observing a dial indicator, and adding water from a volumetric tank or feeling the concrete and adding water until it is at the proper consistency. Automated plants frequently batch, mix, and deliver concrete to the product machines without attention other than setting the mix formula at the start of the production run and perhaps adjusting the formula occasionally (see Fig. 10). Long production runs with infrequent changes in mix formulas are typical, and only a few different basic formulas, determined by trial, are usual in a given plant. Batch quantities are seldom recorded for each batch.

Mixers are usually of the pug-mill type with a single rotating horizontal shaft to which blades are attached. Some pan-type or turbine mixers are used. These have a vertical shaft or shafts to which blades are attached, and the pan may

Fig. 10. Concrete-block-plant control center, showing the batching scales on the left of the three control panels. (*C. S. Johnson Division of Koehring Company.*)

Fig. 11. Machinery installation in a large concrete-block plant, showing batchers on the mezzanine and the mixer on the main floor. (*C. S. Johnson Division of Koehring Company.*)

rotate in addition to the blades in some mixers. Nontilting construction mixers are often used for cast-concrete products (see Chap. 18).

Concrete from the mixers is conveyed by gravity, belt conveyors, or skip hoists to concrete hoppers that feed machinery for machine-made products. Concrete for cast products is moved in concrete buckets or occasionally small rail cars to the point of use. Figure 11 shows a typical concrete-products-plant layout.

Concrete production capacity of these plants is tied to product capacity, usually determined by product machine capacity. Mix times are typically quite long, on the order of several minutes where a mixer is used for each machine. Production capacity of the batching equipment may be found using the formulas in the preceding section on commercial ready-mix plants.

POLLUTION

Pollution and pollution control are becoming more of a problem each day to concrete producers. Air and water pollution have received the most publicity and are certainly important. Another form of pollution which has been largely ignored but which now shares the limelight is noise. These three are tightly bound together. Some forms of wet dust collection could easily add to water pollution. The large and powerful blowers needed for air pollution control certainly add to the noise level, and the production of power to operate the systems adds to the air, water, and noise pollution. There is a point of diminishing return which must be established. The entire area of pollution will probably be very cloudy for quite some time.

Some of the types of dust collectors that might be used to combat air pollution are the electrostatic precipitator, dry centrifugal collector, fabric collector or bag house, or one of the several wet collectors.

The electrostatic precipitator utilizes a high voltage (60,000 to 100,000 volts) to impart a negative charge to the dust particles which causes them to gather on a grounded or positively charged plate. Efficiency is high regardless of particle size but cost is also usually high. The dust may be removed by vibrating the collecting plates.

Dry centrifugal collectors, such as the cyclone collector, use centrifugal force to throw the dust particles to the outside of a stream of air. These have relatively low cost and low maintenance. Since they do not normally remove the fine particles, they are sometimes used as a pre-cleaner, followed by a wet collector or a bag house.

Fabric collectors, or bag houses, pass the dust-laden air through tubular bags, which are usually cotton but may be wool, paper, glass fiber, or synthetic cloth. The dust is trapped as the low-velocity air passes through the fabric to the atmosphere. The dust builds up on the bags and helps improve the efficiency of the collection. The air flow is stopped and the bags are shaken or vibrated to remove the dust from the bags so it can be collected in hoppers at the bottom of the bag house. Efficiency is high. This is the most commonly used type of dust-collection system in batch plants.

Several designs of wet collectors are available. When and if the problem of disposing of or reusing the considerable quantity of water required is solved without adding to water pollution, these will probably gain more acceptance by concrete producers.

None of the above will reduce dust from the dirt haul roads, but frequent watering can help keep the dust down.

Chapter 17

BATCHING EQUIPMENT

JOHN K. HUNT

Volume and surface area of aggregates, volume of water, and weight of cement are the most important factors in proportioning concrete. Unfortunately concrete materials cannot be measured as accurately by volume as by weight. Aggregates "bulk" or change volume depending upon the method of handling and moisture content of fine aggregates, cement volume varies greatly with air content, and

Table 1. Typical Composition of Highway and Airport Paving Concrete

Values in lb per cu yd of resultant concrete based on saturated surface-dry aggregates

	Lowest observed value	Second centile	Median	Ninety-eighth centile	Highest observed value
Paving mixes, all mixes:					
Cement.........................	385	442	557	674	677
Water.........................	157	150	231	312	316
Sand..........................	975	955	1,185	1,415	1,580
Total coarse aggregate...........	1,660	1,790	2,030	2,300	2,380
Total aggregate.................	3,020	3,040	3,230	3,500	3,560
Wet concrete....................	3,906	3,896	4,018	3,140	4,142
Paving mixes, 2 aggregates:					
Sand..........................	1,066	1,043	1,180	1,359	1,342
Coarse........................	1,660	1,775	2,013	2,193	2,163
Paving mixes, 3 aggregates:					
Sand..........................	975	950	1,100	1,430	1,580
Coarse, smaller quantity..........	360	520	930	1,220	1,190
Coarse, larger quantity..........	956	910	1,160	1,410	1,740
Coarse, smaller aggregate........	360	520	1,000	1,300	1,274
Coarse, larger aggregate..........	683	750	1,100	1,410	1,740
Paving mixes, 4 aggregates:					
Sand..........................	986	...	...	...	1,311
Coarse, smallest quantity........	143	...	...	...	528
Coarse, middle quantity..........	733	...	...	...	992
Coarse, largest quantity..........	850	...	...	...	1,203

water volume varies with temperature. On the other hand, the specific gravity of most materials is relatively constant; so proportioning by weight with appropriate corrections is almost universally used. Exceptions are cinders and similar materials that have a wide range of specific gravities. Critical materials such as cement and water are usually weighed on a separate scale for each material. Each size group of aggregates may be weighed on a separate scale, or they may in some applications where speed and accuracy are less critical be weighed *cumulatively,* one after the other on the same scale.

Most batching specifications allow tolerances of plus or minus 1% in weight of cement and in weight or volume of water, 2% in weight of aggregates, and 3% in weight or volume of admixtures. Capabilities of modern batching equipment are well within these tolerances when operated slowly, and some allowance is

Table 2. Typical Composition of Commercial Ready-mix Concrete

Values in lb per cu yd of resultant concrete based on saturated surface-dry aggregates

	Lowest observed value	Second centile	Median	Ninety-eighth centile	Highest observed value
All mixes except pavement and lightweight mixes:					
Cement........................	188*	330	552	785	877
Pozzolan........................	0	...	...	...	75
Cement + pozzolan..............	230	330	522	785	877
Water.........................	210	209	269	372	383
Sand..........................	985	1,019	1,342	1,596	1,670
Total coarse aggregate...........	1,454	1,524	1,888	2,150	2,267
Total aggregate.................	2,755	2,990	3,233	3,410	3,442
Wet-concrete weight..............	3,809	3,845	4,011	4,148	4,180
1½ in. max, 2 aggregates:					
Cement........................	370	352	490	655	658
Water.........................	210	208	246	302	292
Sand..........................	1,018	950	1,310	1,550	1,550
Coarse aggregate................	1,780	1,750	1,950	2,150	2,143
Total aggregate.................	3,017	3,000	3,265	3,450	3,442
Wet-concrete weight..............	3,844	3,820	3,990	4,165	4,165
1½ in. max, 3 aggregates:					
Cement........................	380	426	533	640	750
Water.........................					
Sand..........................	985	960	1,200	1,440	1,410
Fine aggregate..................	750	660	1,010	1,400	1,362
Coarse aggregate................	607	650	1,010	1,250	1,155
Total coarse aggregate...........	1,736	1,690	2,010	2,300	2,267
Total aggregate.................	3,040	3,000	3,220	3,440	3,410
Wet-concrete weight..............					
1½ in. max, 4 aggregates:					
Cement........................	432	...	...	...	658
Water.........................	321	...	...	...	339
Sand..........................	1,122	...	...	...	1,311
Fine aggregate..................	201	...	...	...	202
Medium aggregate...............	1,167	...	...	...	1,170
Coarse aggregate................	644	...	...	...	646
Total coarse aggregate...........	2,012	...	...	...	2,018
Total aggregate.................	3,134	...	...	...	3,329
Wet-concrete weight..............	4,086	...	...	...	4,113
¾ to 1 in. max, 2 aggregates:					
Cement........................	188*	305	505	705	750
Water.........................	225	225	270	345	347
Sand..........................	1,040	1,080	1,360	1,630	1,670
Coarse aggregate................	1,660	1,670	1,860	2,140	2,170
Total aggregate.................	3,021	3,040	3,235	3,430	3,440
Wet-concrete weight..............	3,809	3,860	4,005	4,150	4,131

Table 2. Typical Composition of Commercial Ready-mix Concrete (*Continued*)

Values in lb per cu yd of resultant concrete based on saturated surface-dry aggregates

	Lowest observed value	Second centile	Median	Ninety-eighth centile	Highest observed value
¾ to ⅜ in. max, 2 aggregates:					
Cement..........................	393	360	585	910	877
Water..........................	267	250	325	400	383
Sand...........................	1,000	1,210	1,430	1,650	1,625
Coarse aggregate................	1,454	1,400	1,680	1,920	1,885
Total aggregate.................	2,755	2,710	3,160	3,320	3,300
Wet-concrete weight.............	3,960	3,970	4,020	4,065	4,180
Lightweight insulating or fill concrete:					
Cement..........................	318	300	530	770	700
Water..........................	401	360	460	560	509
Lightweight aggregate............	214	180	460	2,350	1,900
Wet-concrete weight..............	989	850	1,600	3,050	2,265
Lightweight structural concrete:					
Cement..........................	461	415	590	900	852
Water..........................	250	220	365	510	473
Sand...........................	250	100	1,200	1,410	1,380
Lightweight aggregate...........	710	640	850	1,680	1,530
Total aggregate.................	1,592	1,450	2,030	2,270	2,260
Wet-concrete weight..............	2,410	2,310	3,000	3,340	3,274

* Plus 45 lb pozzolan.

included for high-speed operation. Closer tolerances are not logically required because they are exceeded by the variability of the materials being batched.

After a desired concrete formula has been established, usually by trial mixes, the objective in batching concrete materials is to reproduce the predetermined formulation within the allowed tolerances. Thus repeatability is more essential

Table 3. Typical Concrete Mixes for Specific Uses

lb per cu yd

Use	Cement or cement plus pozzolan	Water	Sand	Total coarse aggregate	Total aggregate	Wet-concrete weight
Grout...............	590–2,360	318–730	316–2,700	. . .	316–2,700	3,406–3,645
Pumped concrete.....	500– 580	280	1,555	1,485	3,040	3,820–3,900
Nuclear shielding.....	564	300	900	3,860	4,760	5,624
Prestressed concrete..	658– 752	267–287	1,125–1,200	1,920	3,045–3,120	4,045–4,064
Lightweight pre-stressed concrete...	846	459	180	1,293	1,473	2,780
Regular concrete block.............	455	177	1,425–1,950	1,425–1,950	3,375	4,007
Lightweight concrete block.............	340	280–560	. . .	. . .	1,750–2,200	2,490–2,940
Concrete pipe........	640– 830	250–420	1,120–2,230	985–1,860	2,960–3,330	4,115–4,315
Porous concrete......	596	213	. . .	2,835	2,835	3,644

than absolute accuracy, and many factors affecting accurate weighing are ignored because they existed at the trial-mix stage and are thus largely canceled or because they are so small as to be unimportant. Among these are flotation by displaced air and ignoring scale and test weight-calibration errors in setting batch weights. Batching equipment is checked to establish that it is within certain weighing tolerances, using commercial test weights with tolerances of about one-fourth of

Table 4. Typical Composition of Mass Concrete

Weights in lb per cu yd of resultant concrete based on saturated surface-dry aggregates. W/C or W/(C + P) indicates water-cement ratio or water cement + pozzolan ratio by weight

	Lowest observed value	Second centile	Median	Ninety-eighth centile	Highest observed value
Grout mixes:					
Cement	590	...	810	...	1,020
Pozzolan	0	:..	...	...	260
Cement + pozzolan	590	...	...	...	1,020
Water	318	...	325	...	550
W/(C + P), %	32	...	39	...	78
Sand	2,065	...	2,350	...	2,700
Wet weight	3,465	...	3,600	...	3,645
¾ in. max mixes:					
Cement	318	410	610	740	711
Pozzolan	0	...	...	...	105
Cement + pozzolan	423	410	610	740	711
Water	225	270	307	355	342
W/(C + P), %	45	44	51	63	59
Sand	991	970	1,255	1,445	1,450
No. 4 to ¾-in. aggregate	1,355	1,685	1,775	1,835	2,080
Total aggregate	2,801	2,765	3,025	3,155	3,280
Wet-concrete weight	3,794	3,790	3,920	4,090	4,043
1½ in. max mixes:					
Cement	282	375	590	805	765
Pozzolan	0	...	...	...	100
Cement + pozzolan	374	375	590	885	852
Water	163	160	270	310	310
W/(C + P), %	25	20	45	64	60
Sand	825	795	990	1,180	1,180
No. 4 to ¾-in. aggregate	830	810	1,050	1,210	1,576
¾- to 1½-in. aggregate	444	990	1,110	1,330	1,372
Total coarse aggregate	1,844	1,960	2,180	2,330	2,300
Total aggregate	2,996	2,990	3,145	3,300	3,397
Wet-concrete weight	3,823	3,820	3,990	4,190	4.277
3 in. max interior mixes:					
Cement	188	160	280	550	500
Pozzolan	0	...	...	...	89
Cement + pozzolan	255	240	320	550	500
Water	152	140	185	290	252
W/(C + P), %	50	49	57	79	74
Sand	748	680	875	1,050	997
No. 4 to ¾-in. aggregate	642	590	800	1,010	938
¾- to 1½-in. aggregate	745	685	830	980	927
1½- to 3-in. aggregate	750	840	1,010	1,160	1,115
Total coarse aggregate	2,490	2,430	2,660	2,870	2,769
Total aggregate	3,330	3,250	3,540	3,850	3,766
Wet concrete	3,980	3,930	4,090	4,250	4,193

Table 4. Typical Composition of Mass Concrete (*Continued*)

Weights in lb per cu yd of resultant concrete based on saturated surface-dry aggregates. W/C or W/(C + P) indicates water-cement ratio or water cement + pozzolan ratio by weight

	Lowest observed value	Second centile	Median	Ninety-eighth centile	Highest observed value
3 in. max exterior mixes:					
Cement........................	245	250	400	550	625
Pozzolan.......................	0	...	...	...	82
Cement + pozzolan..............	286	275	425	575	625
Water.........................	160	145	210	280	200
W/(C + P), %..................	42	41	49	64	61
Sand..........................	685	650	845	1,140	1,061
No. 4 to ¾-in. aggregate.........	395	300	760	980	910
¾- to 1½-in. aggregate...........	675	650	840	980	927
1½- to 3-in. aggregate...........	750	700	1,030	1,450	1,370
Total coarse aggregate...........	2,445	2,420	2,610	2,920	2,884
Total aggregate.................	3,210	3,140	3,470	3,840	3,750
Wet-concrete weight..............	3,980	3,940	4,100	4,260	4,235
6 to 8 in. max interior mixes:					
Cement........................	148	150	205	400	400
Pozzolan.......................	0	0	56	110	118
Cement + pozzolan..............	198	200	255	410	400
Water.........................	94	108	165	305	309
W/(C + P), %..................	36	40	64	85	86
Sand..........................	714	705	825	1,210	1,170
No. 4 to ¾-in. aggregate.........	410	415	565	725	721
¾- to 1½-in. aggregate...........	333	380	580	770	913
1½- to 3-in. aggregate...........	675	650	815	1,140	1,096
3- to 6- or 8-in. aggregate........	433	700	950	1,150	1,150
Total coarse aggregate...........	2,560	2,540	2,900	3,260	3,211
Total aggregate.................	3,542	3,520	3,750	3,980	3,925
Wet-concrete weight..............	3,914	3,910	4,150	4,390	4,306
6 to 8 in. max exterior mixes:					
Cement........................	247	247	307	441	441
Pozzolan.......................	0	0	50	105	106
Cement + pozzolan..............	247	247	368	441	441
Water.........................	130	132	175	232	226
W/(C + P), %..................	41	41	49	56	56
Sand..........................	710	680	790	1,030	1,022
No. 4 to ¾-in. aggregate.........	280	260	500	650	679
¾- to 1½-in. aggregate...........	201	460	575	670	675
1½- to 3-in. aggregate...........	635	620	790	1,060	1,069
3- to 6- or 8-in. aggregate........	764	760	950	1,290	1,230
Total coarse aggregate...........	2,500	2,420	2,885	3,090	3,055
Total aggregate.................	3,440	3,410	3,630	3,970	3,913
Wet-concrete weight..............	4,036	4,040	4,185	4,310	4,290

the scale tolerance, and the errors introduced by these tolerances are then ignored. Since these errors are relatively constant, the repeatability required of the batching function is not impaired.

Typical ranges of material quantities useful for establishing batching-equipment capacities are shown in Tables 1 through 4. These are based on a large number of batches from many locations. Two formulas out of one hundred taken at random can be expected to be less than the second centile value, and two can be expected to be more than the ninety-eighth centile value. Half of a large number of formulas can be expected to be above and half below the median. Slag aggregates have been excluded but would be about 90% of the aggregate weights shown.

Concrete materials are usually weighed in *hoppers* with a *fill valve* or *gate* fastened to a bin above the hopper to control flow of material into the hopper and a *discharge valve* or *gate* to control flow of material from the hopper. The hopper is suspended from or supported on scales to indicate the weight of material in the hopper after the fixed weight of the hopper is balanced or *tared* off the scale. Modifications of this arrangement are sometimes used. The hopper may be filled by belt conveyors or front-end loaders; or the hopper may be filled, tared, and then indicate the weight of material discharged. Aggregates may be weighed continuously while on a moving conveyor by integrating devices or by timed flow with materials fed at a controlled rate onto a constant-speed weighing conveyor.

HOPPERS AND GATES

1   Aggregate Hoppers and Gates

Aggregate-hopper capacity should equal or exceed about 1.2 cu ft per 100 lb of aggregate to be batched unless the weight per cubic foot of material is known. The hopper must be of such size that it can collect material from the necessary fill gates, and it should have net valley slopes of about 40° or more to encourage complete discharge. Hoppers may be conical, but rectangular hoppers are far more common. The hopper is usually constructed of relatively light steel, stiffened where necessary, and lined with heat-treated steel plates at points of most wear. Three-sixteenths to three-eighths liners are typical. Liners are usually tack-welded into place. Liner bars 4 to 6 in. wide in sloping corners help the discharge. Fill gates are nearly always of the clam type with one clam leaf or two opposing leaves meeting at the center of the gate. Provision is often made to prevent a fill gate for large aggregate from grabbing a large particle as it closes, permitting smaller particles to dribble out, by either hinged weights or a resilient material against which the gate closes. Discharge gates may also be single- or double-clam type but must open sufficiently so the hopper can discharge completely. Provision to remove overload is seldom provided, since with reliable modern equipment this is required infrequently and can be accomplished from the hopper top or discharge gate in most installations if absolutely necessary. Typical nominal gate size is about 18 by 18 in., with about 10 by 17 in. net material opening, and a capacity of about 2.2 cfs for coarse gravel, about 3.6 cfs for free-flowing fine material. Larger gates or multiple gates are used for high-speed operation and for aggregates larger than 3 in. Gate capacity varies approximately as the cube of the net linear gate dimension or the $\frac{3}{2}$ power of the net gate area and is practically independent of the depth of aggregate over the gate. Thus a gate with an opening 15 by 17 in. will pass about $(1.5)^{\frac{3}{2}} = 1.8$ times as much aggregate per second as a gate 10 by 17 in. If the flow rate is too high for accurate weighing, the area of the opening can be reduced by adjusting the gate stroke or, on a long narrow gate, by welding an angle across the center of the gate to reduce the area of the opening. Fill gates are often opened and closed rapidly (jogged) when nearing final weight to reduce the flow rate for the final 100 lb or so.

2   Cement and Pozzolan Hoppers and Gates

Cement and pozzolan can be batched with the same type of equipment, and in some plants are weighed cumulatively on a single scale. Hoppers may be conical but are often rectangular with 50 to 60° net bottom slopes. Capacity should be at least 9 cu ft per cu yd plus 3 cu ft for regular concrete. Material is usually 10- or 12-gage steel. Lining to resist wear is not necessary, but special low-friction linings of epoxy or plastic are sometimes used. Hoppers must be enclosed and connected by cloth or rubberized cloth shrouds to fill gates to prevent excessive dust. A 4- to 6-in. vent to *atmospheric* pressure is essential to accurate weighing since even small pressure or vacuum in the hopper can cause a weighing error equal to the pressure or vacuum times the area of the fill opening in the top of the hopper. An inadequate or partly clogged vent will result in the indicated weight's drifting after the fill gate is closed. Small vibrators assist in discharging the last few pounds of cement or pozzolan. Cement must flow readily when it reaches the hopper, but if held in the hopper for more than a few minutes it may lose air and be difficult to discharge unless aeration is provided in the lower part of the hopper. The cement fill gate is usually a closely fitting rotary plug valve, a motor-driven vane feeder, a screw conveyor, or a short pneumatic gravity conveyor with a damper for positive shutoff. In addition to the above types, the discharge valve may be a conical drop valve with a resilient seal or a flexible tube with means of squeezing it closed. Flow of aerated cement or pozzolan through a typical 3- by 8-in. opening is about 200 to 300 lb per sec, but this varies widely with the air content. Cement may be so sluggish it will not flow or so wild it spouts through a small hole like water, or anywhere in between. Gate openings or air pressure often must be adjusted to obtain satisfactory flow rates. Openings in cement hoppers are often provided for removing overloads but are rarely used.

3 Liquid Hoppers and Valves

Water and liquid admixtures are weighed in hoppers of light-gage corrosion-resisting alloy or galvanized steel. Any quick-acting valve may be used. Steel pipe and iron valves are satisfactory for calcium chloride solution, the most corrosive liquid usually handled. Different metals, such as steel and brass or aluminum, in contact with each other should be avoided to reduce electrolytic corrosion. Admixtures, like some air-entraining agents, may be viscous or sticky, and require higher valve operating forces and provision for cleaning. Liquid-hopper shape is not critical unless a viscous admixture is used, and then should be steep to reduce discharge time. Since liquid-hopper tops are open, no provision is required for removal of overloads. Liquid-valve capacity depends upon pressure and area of the valve opening. An orifice with sharp edges in a water-hopper bottom discharges

$$Q = 0.18a(h)^{\frac{1}{2}}$$

where Q = discharge, U.S. gal per sec
 a = orifice area, sq in.
 h = head, or depth of water over the orifice, ft

The coefficient can vary from 0.15 to 0.29 depending on valve construction. Published flow tables are available for most commercial valves. For a water hopper consisting of a vertical cylinder with a sharp-edged orifice in a nearly flat bottom, the time required to empty the hopper is about

$$t = \frac{8.0d(G)^{\frac{1}{2}}}{a}$$

where t = time, sec
 d = hopper diameter, in.
 G = water quantity, U.S. gal
 a = orifice area, sq in.

The coefficient can vary from 4.9 to 9.5 depending on valve construction. Pumps may be used to speed up water discharge.

4 Volumetric Hoppers

Water and admixtures are often measured in volumetric hoppers, tanks with graduated gages or automatic liquid-level controls. Correction may be required for temperature when hot water is measured volumetrically. The weight of 1 U.S. gal (231 cu in.) of water at various temperatures is shown in Table 5. Volumetric correction is important when a mix is designed for a final temperature of 60°F with 24 U.S. gal of water at 60°F, for example. If 140°F water is used, $(24)(8.337)/8.206 = 24.38$ U.S. gal required. Neglecting the correction results in an error of 1.6%. This correction need not be made when water is weighed. Volumetric aggregate hoppers are sometimes used in concrete-products plants for special materials and where batch quantities are repeated time after time. They are similar to other aggregate hoppers except that they must have an adjustable side or top and be filled full to the fill gate each batch. The fill gate is then closed, leaving a fixed volume of material in the hopper.

Table 5. Weight of 1 U.S. gal of Water

Temp, °F	Weight, lb
32	8.344
40	8.345
60	8.337
80	8.317
100	8.288
120	8.250
140	8.206
160	8.155
180	8.099
200	8.039

5 Water Meters

Either a nutating piston or a propeller drives an indicator showing the *volume* of water that has passed through a water meter. Nutating-piston (or disk) meters are generally accurate within 1% in the ranges shown in Table 6 and are commonly

Table 6. Typical Water-meter Capacities

Three capacity figures are given: minimum, maximum continuous, and maximum intermittent flow. Figures in boldface type are for propeller meters; other figures are for nutating-piston meters. Manufacturer's data should be consulted for capacities of specific meters

Nominal size, in.	Cold water, U.S. gal/min, 32–100°F			Hot water, U.S. gal/min					
				100–150°F			150–200°F		
1½	5	66	100	9	36	50	9	18	36
2	8	110	160	13	52	95	12	30	52
	18	**90**	**170**	**16**	**100**	**160**	**16**	**100**	**160**
2½	10	150	225	20	80	112	19	45	80
3	16	210	330	23	100	140	23	55	100
	30	**185**	**350**	**30**	**150**	**350**	**30**	**150**	**350**
4	40	330	500	50	200	350	45	110	200
	50	**300**	**600**	**50**	**225**	**600**	**50**	**225**	**600**
6	50	660	1,000	100	400	500	90	220	400
	85	**625**	**1,400**	**90**	**500**	**1,400**	**90**	**500**	**1,400**

more accurate over a wider range of flows than are propeller meters. Propeller meters are accurate within 1 to 1.5% from about 40 to 140% of rated continuous capacity, and within 2 to 3% for the balance of their operating range. They have higher maximum capacities and lower head loss than nutating-piston meters. Meters read low at very low flows, slightly high at low flows, and slightly low at flow rates near the maximum. In Table 6, three capacity figures are given: minimum, maximum continuous, and maximum intermittent flow. Batching meters are operated intermittently between the latter two values. When used with nearly constant water pressure, repeatability is better than accuracy. Cold-water meters cannot usually be used with hot water, but hot-water meters can be used with cold water at cold-water flow rates with some sacrifice in accuracy. A strainer (about 50 mesh) is used ahead of the meter to prevent damage from solid particles. Some meters can be installed vertically; most must be horizontal. Meters must be protected from freezing and from pressure surges (water hammer) in water lines. Electrical or mechanical controls to stop flow at a preset amount and automatic, remotely controlled meters are available.

ADMIXTURE DISPENSERS

A wide variety of additives or admixtures are used in the production of concrete, and the quantities used may be from 1 oz to 2 qt per 100 lb of cement. Since many admixtures become somewhat gummy when in contact with air it is advisable to arrange the piping so air will not contact the solenoid valves used with admixtures because the solenoid valves would not open or close completely, would overheat, and eventually would fail completely. If the solenoid valves are exposed to the air they should be washed out thoroughly with water at the end of each day.

Some admixtures are not compatible, and if two or more are put into the water batcher or water piping together they may react with each other and nullify the effect of all. It may be necessary to add one admixture to the water batcher, one to the sand, and one or more directly into the mixer, mixer truck, or dry-batch truck.

6   Timed Flow Dispensers

One of the simplest methods of dispensing admixtures is by adjusting a valve under a storage tank to a predetermined flow rate and then allowing the admixture to flow for the proper length of time to produce the desired amount. Usually an electric timer is used which can be preset for the correct time interval and actuated by a start button. A solenoid valve in the admixture line is opened to allow flow as long as the timer is timing but is closed as soon as the time runs out. If the solenoid valve is gravity-fed then it is important to maintain a relatively constant head in the storage tank to maintain a constant flow rate. In some instances a pump is used instead of an overhead storage tank.

There is no practical way to provide a positive electrical interlock to be certain the correct amount of admixture is furnished. Quite frequently the admixture goes into a calibrated sight glass so the operator can make a visual check on the quantity dispensed. If the operator is not alert the storage tank may be empty, the piping or the solenoid may be partially or completely clogged in such a way that the flow rate would be changed, or the solenoid valve may fail to function, thus causing an error in the amount dispensed.

7   Displacement Dispensers

The volume of a closed chamber is calibrated and adjusted to the amount of admixture needed. Admixture is then drawn into the chamber on the back stroke of a connecting air cylinder and is forced out on the forward stroke. No air is allowed in the admixture chamber at any time to give a false quantity.

8   Meters

Metering the admixture eliminates some of the problems encountered with a timing system. If no admixture goes through the meter it gives no indication and the operator is alerted to the problem. In an automatic system a meter with an impulse head may be used to produce one impulse for each ½ oz, 1 oz, 2 oz, on up through 1 gal if necessary depending on the quantity used. A counter is connected to the meter to count the impulses, and when the preset number is reached the counter causes the flow to stop. An electrical interlock can be provided to ensure that the counter has reset before the system proceeds to the next operation.

9   Weight Dispensers

Measuring admixture by weight is the most positive, most nearly foolproof, and most accurate method, but it is also the most expensive because it requires a complete scale system including either a dial or beams and also a full set of controls. It is not feasible to dispense all admixtures with one scale system because the quantity required may vary considerably. An admixture scale can be used down to about 10% of the scale capacity and still stay within the normal delivery tolerance.

SCALES

Practically all scales used for batching materials for concrete employ a set of lever scales with either a weighbeam and balance indicator or a dial indicator.

Weighbeam-type scales are simpler, more rugged, more easily maintained, more easily adjusted, have smaller weight divisions when used for cumulative batching, and always indicate balance at the same point after the required weight is set on the weighbeam.

Dial-indicator-type scales allow quick changes of material quantities without changing poise weights, are much easier to adapt to graphic or digital recording, and are much easier to use with electronic controls.

Electronic scales and other types of scales normally simplify the mechanical portion of the system by eliminating part or all of the levers, but they add complications due to the complexity of the entire sensing systems and the highly specialized knowledge necessary to service them.

10   Lever Scales

Most batcher scales use the principle of the unequal-arm beam so that a heavy weight can be balanced by a relatively small known weight. Normally the weigh hopper is suspended by four hanger rods from two main levers, sometimes known as rocker arms. Each rocker arm consists of two simple levers connected by a torque tube that transmits the load from the short lever in the back to the longer lever in the front. The load from the weigh hopper is reduced by the ratio in the rocker arms, collected and applied to the second lever in the system. The load is reduced again by the ratio in the second lever and applied to the third lever. More levers may be used if necessary to reduce the total load of the material in the weigh hopper to the correct load to pull into the indicating unit.

The ratio of a scale lever is the ratio of the weight on the load pivot or knife-edge and the weight necessary to balance the lever when placed on the power pivot or knife-edge. The ratio may be determined by measuring the distance between the load knife-edge and the fulcrum knife-edge, and the distance between the power knife-edge and the fulcrum, and calculating the ratio of the two distances. Normally the ratio is expressed as 2:1, 3:1, 5:1, or 6⅔:1, which indicates that the distance between the load knife-edge and the fulcrum is 2, 3, 5, or 6⅔ times the distance between the power knife-edge and the fulcrum. Since the primary function of a scale system is to enable a small known weight to balance a considerably larger unknown weight, the ratio is almost always more than 1:1.

The ratio of a lever system is the product of the ratio of the individual levers in the system. The two rocker arms are considered a single lever for this purpose because their load is combined and transferred to a single lever. If the ratio of the rocker arms is 5:1, the ratio of the second lever is 5:1, and the ratio of the third lever is 4:1, then the ratio of this system is $5 \times 5 \times 4$ or 100:1. If the scale is balanced by the tare weight and a load of 500 lb is placed in the weigh hopper, a 5-lb weight at the power knife-edge of the third lever is required to balance the system again.

It is virtually impossible to manufacture levers that are perfect and to install knife-edges that are in exactly the right place. If the distance from one knife-edge to the fulcrum is 2 in., an error of 0.001 in. would cause a ratio error of 0.05%, which would put the scale system at the extreme limit of some tolerances. This is not so critical as it might seem because each scale system has provision for adjusting the ratio to compensate for the small inaccuracies that occur in the construction and assembly.

The most common means of adjusting scale ratio is a nose iron. The seats for the knife-edges on the rocker arms at the point where the load is combined are bolted to the rocker arms. A small amount of play in the boltholes allows the seats to be shifted slightly to change the ratio of either rocker arm.

Most knife-edges are keyed to the levers, but if one is free to rotate in its hole it can be used to adjust the scale ratio. The barrel of the knife-edge has a flat side on the top or bottom. Two setscrews, one on each side of the centerline of the knife-edge, can be used to rotate the knife-edge in either direction, which will move the centerline or bearing edge and will change the ratio of the lever. There may be adjustable knife-edges in several levers in a scale system.

The bolts in the nose iron or the setscrews in an adjustable knife-edge must be tightened after each adjustment and before a load is put on the scale to prevent them from moving unintentionally.

11 Beam Scales and Weighbeams

A beam scale consists of a *tare beam,* one or more *weighbeams,* and a *balance indicator* (Fig. 1). The pull from a set of lever scales is connected to a tare beam which is equipped with an adjustable sliding weight or roller weight to balance the scale system and the empty hopper. A balance indicator is connected to the tare beam to show whether the scale is in a balanced condition. The balance indicator usually must have an overweight and underweight range of 4 to 5% of the scale capacity or a minimum of 200 lb so the operator will know when the scale is approaching a balanced condition. If this overweight and underweight range is calibrated to be read directly in pounds it is an *over and under indicator* instead of a balance indicator. This is not normally required or furnished for concrete-batching scales.

A weighbeam is connected to the tare beam so that it can be quickly lifted off the scale to check the balance condition when the hopper is empty. Notches are cut into the top or side of the weighbeam to subdivide the beam into the major increments of its capacity. A major inlay is used to identify the notches, but sometimes the weight is marked directly on the side of the beam. A major poise weight equipped with a pawl to fit firmly in the notches can be moved along the beam to set the weight, and a separate fractional bar or a minor inlay and minor poise weight on the main beam can be used to set the smaller increments of weight. Normally the minor inlay is equal to the amount between notches for the major inlay to eliminate confusion. The total weight set on the weighbeam equals the sum of the amounts set by the major and the minor poise weights.

The scale is balanced with the weigh hopper empty and the weighbeam locked off the scale. The desired weight of material is now set on the weigh beam by the major poise weights and the weighbeam is dropped on the scale. The balance indicator should show an underweight condition. The fill gate is opened and material put in the weigh hopper until the balance indicator shows the scale is in balance.

If several materials are weighed cumulatively in the weigh hopper there should be one weighbeam for each material to be weighed. After the first material is weighed, the second weighbeam (which has been set for the net amount desired for the second material) is dropped on the scale. The balance indicator should

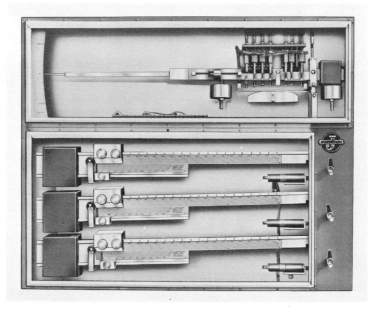

Fig. 1. Beam box with three weighbeams, transmission beam, and balance indicator. (*C. S. Johnson Division of Koehring Company.*)

show an underweight condition. The second fill gate is opened and material admitted until the scale balances, continuing the process until all the materials have been weighed. Note that before the weighing cycle begins all the weighbeams should be locked off the scale. They should be dropped on the scale one at a time as the materials are weighed. Once the weighbeams have dropped on the scale they should be left on until the weighing cycle is complete.

After all materials have been weighed all the weighbeams should be locked off the scale and the discharge gate opened. The discharge gate should remain open until the balance indicator shows a balance condition, which means that all the material has been emptied out of the weigh hopper.

12   Dial Indicators

A dial indicator consists of a pendulum weight, gear segment, pinion, rack, indicator, and dial face. The pull from a set of lever scales is transmitted to a draft band that partially wraps around a hub on the shaft that supports the pendulum. As the pull from the scales increases, the pendulum shaft must rotate further to counteract the pull. A gear segment is operated from this same shaft and turns the pinion which is mounted on the indicator or pointer shaft. This provides continuous indication of weight from zero to full-scale capacity.

The scale levers should be approximately level at half dial capacity (or half the maximum intended use) to decrease the angular movement of the levers. With a beam scale the levers should be approximately level for all loads.

Calibration for full-scale reading is accomplished by raising or lowering the pendulum weight. Adjusting a small weight on the indicator shaft will remove errors at one-fourth and three-fourths capacity.

13 Electronic Scales

An electronic scale is one in which part of all of the lever system is replaced by an electronic device to determine weight. The resistance of an electrical conductor increases when the conductor is placed in tension and decreases when the conductor is placed in compression. A strain gage is generally made by forming an electrical conductor into a small grid which is then attached to the member in which the strain is to be measured. Several strain gages are attached to a small steel column to make a load cell.

Usually the weigh hopper is placed on or suspended from one or more load cells and the resistance changes can be compared with calibrated resistors by using a Wheatstone bridge. The readout device used with strain-gage load cells can be visual indication or a digital printing system.

14 Other Scales

Nuclear weighing can be applied to bulk material moving on a belt conveyor or falling through the air without the use of a conventional scale system.

Pneumatic and hydraulic scales in several configurations could be used to weigh materials for concrete.

None of these weighing systems has achieved widespread acceptance.

15 Scale Accuracy and Sensitivity

The accuracy of a scale is normally checked by comparing the actual observed weight indication and the value of standard test weights placed on or suspended from the scale. Since no scale will be perfect some tolerance must be allowed in checking accuracy. The most restrictive tolerance is essentially ±0.05% of the scale capacity. The most common tolerance is ±0.1% of the scale capacity. Some specifications allow ±0.4 or 0.5% of the net load on the scales. Other specifications allow up to ±0.5% of the scale capacity.

If a tolerance is specified as some percentage of the net load on the scales there must also be some minimum tolerance. For example, in checking a 25,000-lb scale with 25-lb graduations to a tolerance of ±0.5% of the net load it would be unrealistic to place one 50-lb test weight on the scale and expect it to be accurate within ±0.25 lb (0.5% of 50 lb is 0.25 lb) because this is only $\frac{1}{100}$ of one graduation. The minimum tolerance should be a function of the capacity of the scale or the size of the minimum graduation. Frequently these tolerances are applied only within the normal operating range of the scale.

The sensitivity of a beam scale with a balance indicator is defined in at least two ways by various specifications. Some specifications require that the indicator show movement when a specific weight (usually 0.1% of the scale capacity) is placed in the hopper. Other specifications require that the indicator must move a definite distance (usually $\frac{1}{8}$ in.) when a specific weight (usually twice the value of the minimum graduation interval on the weighbeam) is placed in the hopper. There is normally no specific sensitivity requirement for a dial scale, but if there is enough friction or drag in the scale system to allow the dial pointer or indicator to be held off more than the tolerance of the scale it is impossible to determine whether the scale meets the accuracy requirements. Under these circumstances the scale is not sensitive enough.

The scale should be adjusted so that it consistently returns to zero with no weight in the hopper before the accuracy test is started. Any hooks, bars, platforms, or other material used to facilitate the application of the test weights should be balanced out by adjusting the tare weight or adding more tare weight before checking the zero reading of the scale. If this is not feasible, then enough additional weight should be placed on the hopper to bring the indicator to some convenient reference point.

Enough standard test weights should be available to check the scale to full capacity or to the maximum reading expected to be used. For large-capacity scales this is sometimes not possible. In this case the scale should be checked in the normal way until all test weights are on the scale. Extra small weights should be added to produce a readily reproducible scale reading, such as the next higher graduation on a dial scale or exact balance on a beam scale. If a beam scale is indicating too high it may be necessary to move the minor poise to obtain an exact balance for reference. Neither the tare weight nor anything used to balance the scale at zero should be disturbed at any time during the scale check because this would make it impossible to determine if the scale returns to zero balance after the check is completed. The test weights should then be removed carefully and enough material added to the platform or in the hopper to bring the scale back to the same position as before with the test weights. The extra small weights should be removed or the minor poise moved back to zero to restore the scale to the same condition as existed when the test weights were first placed on the scale. The test weights should be placed on the scale again and the check resumed. Extreme care must be exercised in using other material to replace test weights since any error will be multiplied by the number of times the other material is used and could easily result in an error larger than the allowable tolerance.

Since almost any scale that is reasonably free from friction can be calibrated to read correctly at zero and at any other one point, it is essential that a scale be checked at several points. A dial scale should be checked at least at one-fourth, one-half, three-fourths, and full capacity. A beam scale should be checked at least at one-fourth, one-half, three-fourths, and full capacity and, if there are notches for the major poise, at any notches that look worn or otherwise deteriorated.

Standard test weights for checking scales should have an allowable error of no more than 25% of the minimum tolerance of the scale. If a scale is to be checked to $\pm 0.05\%$ of full scale the 50-lb test weights should be within ± 0.00625 lb or about ± 2.84 g. If a scale is to be checked to $\pm 0.1\%$ of full scale the 50-lb test weights should be within ± 0.0125 lb or about 5.68 g.

Some means of damping oscillations should be provided so the scale will come to a stop in a reasonably short period of time, but this must be so constructed that it will not seriously impair the accuracy or sensitivity of the scale. Normally a dashpot is connected near the dial or beam box for this purpose. A dashpot is a cylinder filled with fluid in which a piston with a head moves up and down with the scale movement. If the scale moves rapidly a considerable force is applied to slow it down but if the scale moves slowly there is very little force applied. Some dashpots have adjustable holes or slots in the piston head to control the amount of damping action while others require a change to a fluid with a different viscosity for this adjustment. If the viscosity of the fluid changes radically with a temperature change then the damping action will also change with temperature.

The scale accuracy and sensitivity are determined by static-load tests before actual batching begins. A scale that meets the applicable tolerances for accuracy is usually considered perfect, and any errors in the scale are ignored when applying the control tolerances or delivery tolerances during the actual batching process. These will be discussed later.

MOISTURE COMPENSATION

The final batch weights of water and aggregate should be adjusted to compensate for the moisture in the aggregate. Significant errors can be introduced if this process is not done consistently and correctly.

Normally the moisture content is based on the saturated surface-dry (SSD) weight of sand, but it can also be based on the wet weight of sand or the ovendry weight of sand, as follows:

$$\% \text{ surface moisture, SSD basis} = \frac{A - B}{B} \times 100 \tag{1}$$

$$\% \text{ surface moisture, wet basis} = \frac{A - B}{A} \times 100 \tag{2}$$

$$\% \text{ total moisture, ovendry basis} = \frac{A - C}{C} \times 100 \tag{3}$$

where A = wet weight of sand sample
B = SSD weight of sand sample
C = ovendry weight of sand sample

For example, in a sand sample of 500 g, if $A = 500.0$ g, $B = 471.7$ g, and

$$C = 467.0 \text{ g}$$

then

$$\% \text{ surface moisture, SSD basis} = \frac{500.0 - 471.7}{471.7} \times 100 = 6.0\%$$

$$\% \text{ surface moisture, wet basis} = \frac{500.0 - 471.7}{500.0} \times 100 = 5.66\%$$

$$\% \text{ total moisture, ovendry basis} = \frac{500.0 - 467.0}{467.0} \times 100 = 7.1\%$$

To illustrate the problems involved, assume a 4 cu yd batch with a saturated surface-dry sand weight of 4,717 lb and a total water weight of 883 lb. The correct amount of moisture to be compensated is 283 lb, the final sand-batch weight is 5,000 lb, and the final water-batch weight is 600 lb.

If the % moisture was calculated by Eq. (1), this is 6% of 4,717 lb.

If the % moisture was calculated by Eq. (2), this is

$$\begin{aligned}
5.66\% \text{ of } 4,717.0 \text{ lb} &= 267.0 \text{ lb} \\
5.66\% \text{ of } 267.0 \text{ lb} &= 15.1 \text{ lb} \\
5.66\% \text{ of } 15.1 \text{ lb} &= \underline{0.9 \text{ lb}} \\
&\ 283.0 \text{ lb}
\end{aligned}$$

Since the % moisture in Eq. (2) was based on the total wet-sand weight, all the sand weight must be used to calculate for the amount of compensation.

If the % moisture was calculated by Eq. (3), the absorption must be subtracted to get the % moisture based on the saturated surface-dry weight.

A change in moisture or water of ¼ to ⅓% of the sand weight will change the slump about ½ in. The following errors are not unusual.

Error	Error, lb of water	Error as % of total water	Error as % of batch water
1% of water-batch weight..................	6.0	0.68	1.00
2% of sand-batch weight of 5,000 lb..........	5.7	0.63	0.95
½% in moisture content [if the actual moisture content by Eq. (1) is 5.5% the correction should be 259.4 lb].......................	23.6	2.67	3.93
½% in moisture content [if the actual moisture content by Eq. (2) is 5.16% the correction should be 256.7 lb].......................	26.3	2.98	4.38
Using 267 lb instead of 283 lb as moisture correction [Eq. (2) applied incorrectly]......	16.0	1.81	2.67
Using 334.9 lb instead of 283 lb as moisture correction [Eq. (3) applied incorrectly]......	51.9	5.88	8.65

One way to reduce these errors is to maintain a fairly constant moisture content in aggregate material. If this is not or cannot be done, then close attention should be given to proper moisture compensation.

16   Moisture Measurement

The drying method as described in ASTM C 566 is widely used to determine the moisture content of fine aggregate. This is satisfactory if the moisture content is relatively constant but, because of the time necessary to perform the drying and weighing, is obviously not adequate if the moisture content varies from one batch to the next. Errors may result from inaccurate or insensitive scales, absorption of moisture from the air while the aggregate is cooling before the final weighing, or improper sample selection. The errors normally encountered probably are between 0.2 and 1.0% of the aggregate weight.

Moisture meters that give an instantaneous reading have been available for many years. These operate on the principle that the amount of electric current that will flow through moist fine aggregate is a function of the amount of water and the amount and type of minerals and chemical present in the sand and water. Errors may result from polarization (if direct current is used), temperature change (the effective resistance increases with an increase in temperature), improper sampling (due to improper design or positioning of the probe), incorrect or too infrequent calibration of the meter, change in gradation, changes in minerals or chemicals, or power-line fluctuations. Errors may range from 0.2 to 1.5% of the aggregate weight.

A quick check of moisture content may be made using a given quantity of fine aggregate mixed in a closed chamber with a small amount of a calcium carbide additive. The gas formed varies with the amount of moisture in the fine aggregate, and the pressure can be read directly on a gage calibrated in percent moisture. Errors from 0.5 to 1.0% of the fine-aggregate weight are not unusual.

17   Correcting Material Weights for Moisture Content

In many plants mixes are designed for an average sand and fine-aggregate moisture content and no adjustments are made unless there is a radical change in the moisture content. In other plants, charts are prepared that show the weight of each material for different amounts of moisture so the operator does not have to do the actual calculation each time he batches. Some aggregate weighbeams have inlays with different moisture contents marked so the weight can be set directly without calculations.

No calculations or weight-setting changes are necessary in an aggregate scale system that utilizes a moisture-compensating lever. This lever has a movable knife-edge which changes the scale ratio and is calibrated in percent moisture. The operator determines the moisture content, sets the lever to the proper point, and the compensation is made without further attention. This is applicable to manual, semiautomatic, or automatic equipment, but it can normally be used only with systems which have a separate scale for each aggregate. If a water scale is used the same moisture-compensating lever can be connected to the water scale so one moisture setting changes both scales. This is done by transferring part of the scale pull from the aggregate-weigh-hopper scales to the water-indicating scale unit.

Electronic controllers usually have either a selector switch or a potentiometer calibrated in % moisture to change the preset voltage or resistance in the aggregate and water-measuring circuits to produce the correct compensation. It is relatively easy to change the aggregate weight because the change is a direct percentage of the aggregate preset weight. To compensate in the water circuit it is necessary to have the system calculate the number of pounds of water to be subtracted from the water preset weight, convert that figure to the number of volts or ohms of resistance it represents in the water circuit, and then make the subtraction. This is necessary because the ratio of sand weight to water weight varies considerably with different mix designs, different types of aggregate, different slump requirements, and other factors. This problem is further complicated because the water correction is much more critical than the aggregate correction due to the fact that the change in the aggregate will normally be from 2 to 12% of the aggregate preset weight but the change in the water may be as much as 60 to 80% of the water preset weight.

CONTROLS

Control systems are available in all degrees of complexity from simple manual hand levers to the highly sophisticated electronic controls that allow the operator to select a preset or prepunched mix design and quantity from a remote location, push one button to batch all materials, verify that the batch is correct, produce a printed record, and punch cards or tape for automatic invoicing and inventory control. More maintenance on a higher level will be needed with the more elaborate controls, and troubleshooting problems in the equipment will require an electronic technician instead of a mechanic.

18 Manual Controls

Manual controls consist of one hand lever for each fill gate and discharge gate. A well-muscled operator with very good coordination is required to obtain both speed and accuracy. Operator fatigue is a critical problem. The desired weights are set by moving the poise weights on beam scales or by observing the pointer position on dial scales. The operator opens the proper fill gate with the hand lever, then closes the gate when the scale indicates the correct weight.

Manual controls permit some variations not possible with even the most sophisticated automatic systems. The gate opening and the flow rate can be readily changed as needed for different amounts of material or to compensate for changing characteristics of a material. A very small quantity can easily be weighed in a large batcher by just "cracking" the fill gate. The rate of discharge can be changed to adapt to fast-charging trucks or slow-charging trucks. The sequence of material discharge can be changed if different trucks require it. Most of the water can be discharged first with some water held back for washing down the last of the material. Most of these functions can be performed with an automatic control system but not to the extent or with the infinite variations possible with manual controls.

However, the efficiency of the operator declines during the day and operator fatigue becomes an important factor in the proper functioning of the plant. If the primary operator is unable to work, his understudy must be equally capable or the accuracy, production, and correct operation will suffer. Since there can be no interlock except the operator's eye to indicate that the correct amount of material goes into the batchers, an inefficient, inexperienced, or inept operator can be very expensive in lost production, poor quality of concrete, and wasted materials.

19   Manual Controls with Power Assist

The operator selects the desired weight in the same way as with manual controls. Instead of using hand levers he pushes a button which causes an air cylinder to open a fill gate to feed material into the weigh hopper. When the scale indicator nears the desired weight he releases the button and the fill valve closes. He may have to open the fill gate for a short time by pushing the button again if the weight was not near enough to the desired amount. The operator's fatigue is not so critical as with manual controls, but his reflexes may be slower near the end of the day and the batches will probably vary more than when he is fresh in the morning. The operator cannot slow down the filling as he nears final weight except by alternately opening and closing the fill gate. This increases batching time.

20   Semiautomatic Controls

The desired weights are set by moving the poise weights on beam scales or adjusting a cutoff point on dial scales and the fill gate is opened manually. The fill gate closes automatically when the desired weight is in the hopper. The accuracy of the system is determined by the cutoff and the material flow instead of operator reflexes. If the rate of flow of the material changes appreciably (if sand or fine-aggregate moisture content changes, if gradation of any material changes, or if the aeration in the cement changes) or if the speed of operation of the fill gate changes, then an adjustment must be made in the cutoff point or a sizable error may result.

21   Automatic Controls

Automatic controls normally provide more consistent and more accurate results by reducing the number of operator actions and decisions. The operator sets the desired weight of material and actuates a Start button. The control system starts the flow of material and then stops the flow when the proper amount of material has been batched. The operator then actuates a Discharge button and the material is emptied from the batcher. Because much of the responsibility for a correct batch has been delegated to the control system there should be some form of electrical interlocks to notify the operator if something is wrong. The batching cycle should not begin unless:

1. The scales are in balance at zero weight within the required tolerance.
2. The discharge gate is closed.

The discharge cycle should not begin unless:

1. The scales are at the desired weight within the required tolerance.
2. The fill gate is closed.
3. The mixer is empty.
4. The mixer is in the correct condition to receive a new batch.
5. The two-way chute or swivel chute is positioned to charge a mixer, which is ready to receive a new batch.

The mixer should not discharge unless the proper mixing time has elapsed.

For practical purposes, if the fill gate is not closed more material will flow into the weigh hopper and the scales will register overweight and prevent discharge; therefore, the necessity for this interlock is questionable.

Many automatic control systems have some type of preset mix selection so more than one weight for the same material may be set in advance of the day's operation and the proper weight may be selected by the operator as required.

Just after the fill gate first opens there is a column of material in the air which has not yet reached the weigh hopper and therefore is not registered on the scales. In some cases where only a small batch is being weighed there may be more material in the air than is desired in the total batch. The same problem exists with large batches because when the cutoff mechanism senses that there is enough material on the scale the column of material is added to it and then there is an overweight condition. Some amount of time is required for the gate to close, and the extra material which comes through as the gate is closing increases the overweight condition.

At least two approaches to this problem are currently being employed. In some systems, particularly with electronic controls, the sensing device is actuated shortly before the final cutoff point is reached. Unfortunately, the amount of this compensation is difficult to calculate in advance; so the adjustment must be made on a trial-and-error basis for each different weight of each material and for each different combination of materials. If a different amount of any one material is to be batched the column of material will be longer or shorter. If the gate-closing time or the rate of flow changes, either more or less material will be added after the gate receives the signal to close. If several different materials are being weighed cumulatively in the same weigh hopper each material will probably have a different rate of flow, possibly a different gate-closing time, and certainly a different amount of material in the air since the first material will necessarily be near the bottom of the weigh hopper and the last material may end quite near the top of the weigh hopper; so each material should have a separate adjustment for compensation of the material column in the air.

Another approach to this problem is the use of two cutoff points. During the initial part of the weighing cycle the fill gate is opened wide to allow material to flow rapidly to get most of the material in the weigh hopper in a short time. When the first cutoff point is reached the rate of flow is reduced so the amount of material in the air is relatively small. This may be done by restricting the fill-gate opening or by alternately opening and closing the fill gate a small amount. When the final cutoff point is reached the fill gate closes completely.

22 Electronic Controls

Electronic controls (Fig. 2) consist of three basic parts, a weight-setting device, a transducer, and a balance detector. The weight-setting device is used to preset an electrical quantity, usually voltage or resistance. Typical examples are a punch card, push card, thumbwheel switches, and potentiometer with dial or digital knob. The transducer converts the scale reading to an electrical quantity so it may be compared with the preset amount. Typical examples are potentiometers, linear variable differential transformers, special inductive devices, and pulse generators. The balance detector compares the preset quantity with the output of the transducer and should be capable of distinguishing between an underweight condition where the scale weight is less than the preset amount, an overweight condition where the scale is greater than the preset amount, and a balance condition where the scale weight is equal to the preset weight within plus or minus the applicable tolerance. Typical examples are vacuum-tube or transistor amplifiers, magnetic amplifiers, and sensitive meters.

When punch cards or push cards are used to set the desired weight any number of mix designs can be preset by having a separate card prepunched for each mix design. Some systems using potentiometer presets have a plug-in unit so any number of presets may be kept on hand ready to plug in. If potentiometer or thumbwheel presets are built into the control panel there normally are several sets so that several mixes may be preset as required for a particular day's operation and one set is usually left open for use with odd mixes which are used infrequently during the day.

A proportioning control or cubic yard selector is a very useful item frequently supplied with electronic controls. This permits selection of a range of batch sizes with any of the preset mix designs. For example, if a 4 cu yd batcher is used the selector may be set from 1 to 4 cu yd in ¼ cu yd increments. This may be arranged for automatically batching two or more batches if the truck capacity exceeds the batcher capacity.

A "slump control" is normally merely a means of increasing or decreasing the amount of water included in the batch without changing the preset mixes.

A "harshness" control is used to replace some of the fine aggregate with coarse aggregate or vice versa to increase or decrease the harshness of the mix.

Since many electronic control systems are located remote from the batch plant complete interlocking is even more important than in the conventional automatic-

Fig. 2. Electronic controller with digital printer. (*C. S. Johnson Division of Koehring Company.*)

control systems. Remote location of the controls protects the system from dust and reduces operator fatigue from the harsh sounds of gates opening and closing and aggregate hitting the steel hoppers and chutes. Remote operation also permits operators of smaller plants to perform other functions more easily, such as taking telephone orders or expediting truck traffic via two-way radio. However, if the operator cannot see the batchers and scales he must have some substitute method of determining what is actually taking place. Remote indicators or slave dials, meters which indicate the relative amount of imbalance of the scales, progress or sequence-indicating lights, overweight and underweight lights, and lights to indicate which gates are open are frequently supplied to give the operator the necessary information.

23   Delivery Tolerances

Scale accuracy and sensitivity are determined by a static test but the delivery or weighing accuracy is necessarily determined during actual batching operation. Normally the scale error is ignored in the check of delivery tolerance.

Delivery tolerances are usually ±0.5 to ±1% for cement and water, ±0.5 to

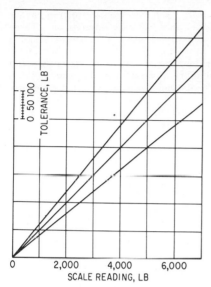

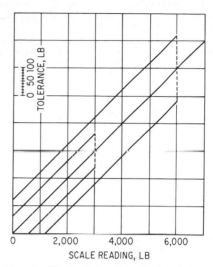

Fig. 3. Method 2. Cumulative batching tolerance. (This shows a delivery tolerance of ±2% of the desired cumulative scale reading for any number of materials and any amount per material to be batched in any sequence.)

Fig. 4. Method 3. Cumulative batching tolerance. (This shows a delivery tolerance of ±2% of the total batch weight for 1 and 2 cu yd batches for any number of materials and any amount per material to be batched in any sequence.)

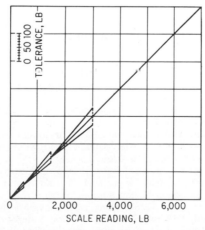

Fig. 5. Method 1. Cumulative batching tolerance. (This shows a delivery tolerance of ±2% of the net weight of each material for a 1 cu yd batch with net batch weights of 500, 1,000, and 1,500 lb to be batched in that sequence.)

±3% for aggregate, and ±3% for admixtures. The commonest tolerances are ±1% for cement and water, ±2% for aggregate, and ±3% for admixtures. These tolerances apply to the net weight of each material when a separate scale is used for each material.

Delivery tolerances for cumulative aggregate systems are very deceptive and often misunderstood and misinterpreted. There are at least three different methods of specifying cumulative tolerances. The tightest tolerance (method 1) is a percentage of the net weight of each individual material. The next (method 2) is a percentage of the cumulative desired scale reading for each material. The most liberal (method 3) is a percentage of the total aggregate weight as a tolerance for each individual material. The commonest tolerance is ±2%.

The maximum possible error at any one cutoff point occurs if one material is at the extreme upper or lower limit of its tolerance and the next material

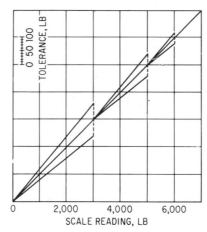

FIG. 6. Method 1. Cumulative batching tolerance. (This shows a delivery tolerance of ±2% of the net weight of each material for a 2 cu yd batch with net weights of 3,000, 2,000, and 1,000 lb to be batched in that sequence.)

is at the opposite extreme limit. As a practical matter this would rarely occur with most control systems but it is possible, particularly if a very small batch is being weighed in a large batcher.

Many examples would be necessary to demonstrate fully the complexity of this concept but only two will be used here. A tolerance of ±2% will be used for both examples.

In the following examples the maximum possible error is expressed as a percentage of the net weight of the material being batched and is not necessarily the largest error in terms of actual weight.

Assume a 1 cu yd batch with three materials, each requiring a net weight of 1,000 lb (Figs. 3 and 4 and Table 7).

The maximum possible error would be ±4% for method 1, ±10% for method 2, and ±12% for method 3. The final batch weight would be within ±⅔% for method 1 and within ±2% for methods 2 and 3.

For the second example assume a 1 cu yd batch with three materials with net weights of 500, 1,000, and 1,500 lb (Figs. 3, 4, 5, and 6 and Table 8). For this example the sequence of weighing becomes important.

The maximum possible error varies from ±3⅓ to ±8% for method 1, ±5 to ±22% for method 2, and ±12 to ±24% for method 3 depending on the weighing sequence. The final batch weight could vary from ±⅓ to ±1% for method 1 and would be within ±2% for methods 2 and 3.

Note that for both examples and for all three methods the final batch weight is within ±2% of the total desired amount of aggregate. Under- or overweighing of one material (within the applicable tolerance) is automatically compensated by the next material since the tolerance must be applied at the desired scale reading. The effect of these errors can be reduced if the weighing sequence is according to size of material (e.g., ¾ to 1½ in., No. 4 to ¾ in., sand; or sand, No. 4 to ¾ in., ¾ to 1½ in.); so the error of one material is compensated by the next size of material. Quite often the tolerances on gradation of aggregate are larger than the maximum possible errors of any of the three methods described above.

Table 7. Comparison of the Three Methods of Establishing Tolerances for Cumulative Aggregate Batching Using Three Materials, Each Requiring a Net Weight of 1,000 lb

Desired net wt of each individual material, lb	Desired cumulative scale reading, lb	± delivery tolerance, lb	Min and max possible scale reading	Max possible error, lb	Min and max possible net wt of each individual material, lb	Max possible error as ±% of desired net wt of each individual material	Delivery tolerance as ±% of desired cumulative scale reading
Method 1:							
1,000	1,000	20	980–1,020	20	980–1,020	2	2
1,000	2,000	20	1,980–2,020	40	960–1,040	4	1
1,000	3,000	20	2,980–3,020	40	960–1,040	4	⅔
Method 2:							
1,000	1,000	20	980–1,020	20	980–1,020	2	2
1,000	2,000	40	1,960–2,040	60	940–1,060	6	2
1,000	3,000	60	2,940–3,060	100	900–1,100	10	2
Method 3:							
1,000	1,000	60	940–1,060	60	940–1,060	6	6
1,000	2,000	60	1,940–2,060	120	880–1,120	12	3
1,000	3,000	60	2,940–3,060	120	880–1,120	12	2

Since there are many possible variations of these methods the specifications for any particular job should be checked carefully to determine exactly what delivery tolerances are applicable. Some specifications do not clearly spell out the delivery tolerances applicable to cumulative aggregate batching.

Many factors influence the delivery accuracy for automatic and semiautomatic control systems. Some of the most common factors are:

1. False loading of the scale by the cutoff or sensing device. Limit switches and reed switches are often used as cutoff devices. Even the most sensitive limit switches must have some preoperating force applied; so this applies an external load to the scale. Reed switches are operated by a magnet. When the reed switch is near the magnet the magnetic pull applies an external load to the scale.

2. Friction added in the scale by the cutoff or sensing device. When a potentiometer or other sensing device is mounted on the dial pointer shaft friction is added to the scale. The effect of this is sometimes reduced by increasing the force available to move the dial pointer.

3. Accuracy or linearity of the cutoff or sensing device. The most commonly used potentiometers or linear variable differential transformers have a linearity of from ±0.05 to 0.5% of full range. This is ±½ to 5 graduations on most standard dial-scale charts.

4. Change in the rate of flow of material. Unless a slow feed of some type is used when nearing the desired weight, a change in the gate-closing time or the rate of flow of the material can cause a considerable error. The rate of flow of aggregate changes with a change in moisture content, impurities in the

Table 8. Comparison of the Three Methods of Establishing Tolerances for Cumulative Aggregate Batching Using Three Materials with Net Weights of 500, 1,000, and 1,500 lb. All Possible Sequences of Batching Are Shown

Desired net wt of each individual material, lb	Desired cumulative scale reading, lb	± delivery tolerance, lb	Min and max possible scale reading	Max possible error, lb	Min and max possible net wt of each individual material, lb	Max possible error as ±% of desired net wt of each individual material	Delivery tolerance as ±% of desired cumulative scale reading
Method 1:							
500	500	10	490–510	10	490–510	2	2
1,000	1,500	20	1,480–1,520	30	970–1,030	3	1⅓
1,500	3,000	30	2,970–3,030	50	1,450–1,550	3⅓	1
Method 2:							
500	500	10	490–510	10	490–510	2	2
1,000	1,500	30	1,470–1,530	40	960–1,040	4	2
1,500	3,000	60	2,940–3,060	90	1,410–1,590	6	2
Method 3:							
500	500	60	440–560	60	440–560	12	12
1,000	1,500	60	1,440–1,560	120	880–1,120	12	4
1,500	3,000	60	2,940–3,060	120	1,380–1,620	8	2
Method 1:							
500	500	10	490–510	10	490–510	2	2
1,500	2,000	30	1,970–2,030	40	1,460–1,540	2⅔	1½
1,000	3,000	20	2,980–3,020	50	950–1,050	5	⅔
Method 2:							
500	500	10	490–510	10	490–510	2	2
1,500	2,000	40	1,960–2,040	50	1,450–1,550	3⅓	2
1,000	3,000	60	2,940–3,060	100	900–1,100	10	2
Method 3:							
500	500	60	440–560	60	440–560	12	12
1,500	2,000	60	1,940–2,060	120	1,380–1,620	8	3
1,000	3,000	60	2,940–3,060	120	880–1,120	12	2
Method 1:							
1,000	1,000	20	980–1,020	20	980–1,020	2	2
500	1,500	10	1,490–1,510	30	470–530	6	⅔
1,500	3,000	30	2,970–3,030	40	1,460–1,540	2⅔	1
Method 2:							
1,000	1,000	20	980–1,020	20	980–1,020	2	2
500	1,500	30	1,470–1,530	50	450–550	10	2
1,500	3,000	60	2,940–3,060	90	1,410–1,590	6	2
Method 3:							
1,000	1,000	60	940–1,060	60	940–1,060	6	6
500	1,500	60	1,440–1,560	120	380–620	24	4
1,500	3,000	60	2,840–3,060	120	1,380–1,620	8	2
Method 1:							
1,000	1,000	20	980–1,020	20	980–1,020	2	2
1,500	2,500	30	2,470–2,530	50	1,450–1,550	3⅓	1⅕
500	3,000	10	2,990–3,010	40	460–540	8	⅓
Method 2:							
1,000	1,000	20	980–1,020	20	980–1,020	2	2
1,500	2,500	50	2,450–2,550	70	1,430–1,570	4⅔	2
500	3,000	60	2,940–3,060	110	390–610	22	2
Method 3:							
1,000	1,000	60	940–1,060	60	940–1,060	6	6
1,500	2,500	60	2,440–2,560	120	1,380–1,620	8	2⅖
500	3,000	60	2,940–3,060	120	380–620	24	2

Table 8. Comparison of the Three Methods of Establishing Tolerances for Cumulative Aggregate Batching Using Three Materials with Net Weights of 500, 1,000, and 1,500 lb. All Possible Sequences of Batching Are Shown (*Continued*)

Desired net wt of each individual material, lb	Desired cumulative scale reading, lb	± delivery tolerance, lb	Min and max possible scale reading	Max possible error, lb	Min and max possible net wt of each individual material, lb	Max possible error as ±% of desired net wt of each individual material	Delivery tolerance as ±% of desired cumulative scale reading
Method 1:							
1,500	1,500	30	1,470–1,530	30	1,470–1,530	2	2
500	2,000	10	1,990–2,010	40	460–540	8	½
1,000	3,000	20	2,980–3,020	30	970–1,030	3	⅔
Method 2:							
1,500	1,500	30	1,470–1,530	30	1,470–1,530	2	2
500	2,000	40	1,960–2,040	70	430–570	14	2
1,000	3,000	60	2,940–3,060	100	900–1,100	10	2
Method 3:							
1,500	1,500	60	1,440–1,560	60	1,440–1,560	4	4
500	2,000	60	1,940–2,060	120	380–620	24	3
1,000	3,000	60	2,940–3,060	120	880–1,120	12	2
Method 1:							
1,500	1,500	30	1,470–1,530	30	1,470 1,530	2	2
1,000	2,500	20	2,480–2,520	50	950–1,050	5	⅘
500	3,000	10	2,990–3,010	30	470–530	6	⅓
Method 2:							
1,500	1,500	30	1,470–1,530	30	1,470–1,530	2	2
1,000	2,500	50	2,450–2,550	80	920–1,080	8	2
500	3,000	60	2,940–3,060	110	390–610	22	2
Method 3:							
1,500	1,500	60	1,440–1,560	60	1,440–1,560	4	4
1,000	2,500	60	2,440–2,560	120	880–1,120	12	2⅖
500	3,000	60	2,940–3,060	120	380–620	24	2

material, or gradation. The rate of flow of cement changes with the temperature and air content.

RECORDERS

Recorders serve many purposes. When used with a control system that has adequate interlocks (i.e., overweight, underweight, return to zero, etc.) the primary function should be to enable anyone to check later to determine that the proper mix design was used for a particular series of batches, that each material was actually included in the batch, and that the proper number of batches was produced. Extreme accuracy and readability are not needed for this. If the control system does not have adequate interlocks or if they are not functioning properly then the recorder must essentially replace the interlocks as a check on the accuracy and repeatability of the batching system. This would necessitate accuracy and readability sufficient to determine if the material weights are within the delivery tolerances. Sometimes the record is used as a delivery ticket, and there is some difference of opinion whether it should indicate the exact weights of material as batched or the desired weights with the stipulation on the ticket that all materials have been batched within the required tolerance.

24   Mechanical Graphic Recorders

A continuous-line record of the scale movement from zero weight to the final batch weight and back to zero weight is made on a lined chart. The pen, coupled directly to the scale, records every movement of the scale on the chart. This makes it very difficult for an operator to hold the scale at the correct reading without having this deception detected. Accuracy is near 0.1% of scale capacity, but if the chart width is too narrow readability will not be the same as the accuracy. A quick check on consistent operation is available for repeat batches. A permanent record is produced immediately without any delay in the batching operation. The record for all material in a batch can be placed on a single chart in addition to time, date, serial number, mix identification number, and notations by an inspector. Duplicate records are difficult to obtain.

25   Electrically Coupled Graphic Recorders

A record is produced by driving a pen electrically from the scale movement and is similar to the mechanical graphic recorder. The accuracy is frequently about 0.5% of scale capacity. There is necessarily some time lag in producing the recording so the accuracy will be affected in high-speed batching unless enough time delay is introduced to allow the recording pen to catch up with the scale. It is difficult to obtain this type of recorder commercially to record all batch weights on one strip of paper unless the recordings for various materials are superimposed. The recording equipment is more complicated and therefore more difficult to adjust or repair than mechanical graphic recorders. Circular charts are generally unsuitable for high-speed batching because of chart size and closeness of recorded weights. It is also difficult to apply batch numbers or other identification to circular charts.

26   Digital Recorders

A printed digital record may be made on tickets or on a continuous tape. Accuracy and readability are usually about 0.1% of scale capacity but may be near 0.5% in some cases. An immediately available permanent record is produced with all batch weights, time, date, serial number, and mix-identification number on the same ticket or tape. Up to five copies are feasible. It is usually more difficult and more expensive to provide the time, date, and serial number on a tape record than on a ticket. The tickets can be preprinted with serial numbers and are convenient to use as customer delivery tickets.

Comparison of a series of identical batches is time-consuming and subject to human error in correlating digital data. A continuous history of the scale operation is not shown, and it may be relatively easy for the operator to hold the scale at the "correct" reading to obtain a "correct" record. Zero scale weights are not always recorded. Weighing operations usually must be delayed to allow time for the print cycle. Most digital printing systems require the use of dial scales rather than beam scales. Some specifications require that all scale linkages be enclosed when a digital printer is used.

27   Photographic Recorders

A camera installed at some convenient location makes a photographic record of all dial readings, time, date, and serial number on the same film, except that, in the case of a cumulative batcher, a separate film must be used for each material since several records must be made of the same dial. Accuracy can be quite high but care must be taken to reduce parallax errors. The record may not be available for several days and there is no assurance that a usable record is being made or that it will not be damaged or lost in processing. Weighing operations must be delayed to operate the recorder. A continuous history of the scale operation is not shown, and it may be relatively easy for the operator to hold the scale at the "correct" reading to obtain a "correct" record. Zero scale weights are not always recorded.

THE CONCRETE-MOBILE

The Concrete-Mobile is a concrete manufacturing plant that can be truck-chassis mounted, truck semi-trailer mounted, or in-plant stationary mounted. It is a self-contained plant to handle stone and sand aggregate, water, and cement in separate bins and tanks (Fig. 7). Concrete materials, mechanically dispatched at a specified engine speed through calibrated aggregate bin openings and constant cement discharge, are discharged into a rapid mix-conveyor of proprietary design. No scales are required. A predetermined continuous flow of water is introduced into the mix-conveyor through a calibrated flow-control valve.

The mixing system consists of a U-trough with a tough, flexible rubber boot (bottom). Through it, longitudinally, runs a hydraulically driven shaft that rotates at a predetermined speed during the mixing operation. During mixing, the mix-conveyor system is operated at an angle from the horizontal, with the receiving

FIG. 7. The Concrete-Mobile shown mixing and distributing concrete for a small slab job. Note the discharge chute attached to the end of the mix-conveyor. Units with capacities up to 10 cu yd are available.

end lower than the discharge end. A combination of auger flights and angled paddles is attached to the lower or receiving end of the shaft. From this point in the mix-conveyor system, auger flighting moves the mixed concrete to the discharge point, where placing or distributing chutes can be attached. Mixing action is completed in about 20 sec. The mix-conveyor is so constructed that the operator and inspector can look into the mixer and observe the continuous mixing and discharge operation.

Before concrete-mix design formulas can be calculated, it is necessary to know the physical characteristics of the aggregates, including gradations, specific gravities, and dry rodded unit weights, as well as the type of cement to be used. Separate mix-setting charts are designed for each machine for each set of materials and mixes, including admixtures. Any mix design, whether stated by weight or by volume, can be converted into a mix-setting chart from which the operator sets each aggregate dial and the water control valve. The cement meter register indicates the cement being used.

Advantages claimed are that transit time is eliminated because dry ingredients are mixed in any volume required on the job. This eliminates the danger of partial set of the concrete in transit and in waiting prior to placement. Close control of workability is possible, as the operator and inspector can see the concrete being mixed and can adjust water as required to maintain slump.

Chapter 18

CONCRETE MIXERS

JOHN K. HUNT

A concrete mixer is a machine for combining portland cement, water, aggregate, and other ingredients to make concrete. They are broadly classified as *plant mixers* that mix in a stationary location (but may be on wheels for moving between locations), mobile *paving mixers* mounted on wheels or tracks so that they can be moved while mixing, and *truck* (or transit) *mixers* mounted on a truck or trailer for mixing while en route to the point of concrete delivery. In the United States, mixer capacity ratings and, to some extent, construction details have been standardized by various agencies. Nearly all specifications in North America require mixers to mix and discharge one batch at a time; continuous mixers are seldom used for concrete but are used for cement base-course materials.

Table 1. Standard Sizes and Types of Concrete-plant Mixers

These sizes are established by standards of the Plant Mixer Manufacturers Division, Concrete Plant Manufacturers Bureau, 900 Spring St., Silver Spring, Md. 20910. Common sizes are shown in boldface type

Mixer type	*Size and rated mixing capacity, cu yd*
Single-compartment two-opening nontilting type..............	1, **2**, 3½
One-opening tilting-type mixing angle of drum 15° with horizontal...	2, 4½, 6, **8**, 9, **10**, **12**, 15
Two-opening front- or rear-charge and front-discharge tilting type..	2, 4½, 6, **8**, **9**, **10**, 12, 15
Vertical-shaft type.....................................	1, 1½, **2**, 3½, 4½

Mixer design is empirical and based on the experience of manufacturers. Minimum drum volumes for mixers of each type (and maximum drum volumes for some types) are established by standards of industry associations. The location, shape, and angle of blades, shape of the mixing chamber, speed of rotation, and horsepower are determined by manufacturers to meet the wide range of types of concrete usually encountered. For a given concrete composition and slump, there are probably one or more combinations of the above factors that will result in mixing concrete in the shortest time, but the changes necessary to accomplish this are not easily understood, observed, or effected. Some factors are not critical, but minor changes in others (for example, blade shape or angle) may result in unsatisfactory operation or reduced mixer capacity. The recommendations of the manufacturer regarding blades and speed of rotation should be followed.

Table 2. Capacity of Truck Mixers

These sizes are established by standards of the Truck Mixer Manufacturers Bureau, 900 Spring St., Silver Spring, Md. 20910.

Mixing capacity	Agitating capacity*
6	7¾
7	9¼
7½	9¾
8	10½
8½	11¼
9	12
10	13¼
11	14¾
12	16
13	17½
14	19
15	20¼

* Unless otherwise restricted on the manufacturer's data plate on the mixer. Some mixers will agitate this amount only when the drum opening is closed.

Table 3. Comparison of Concrete-plant Mixer Types

Excluding mixers seldom used in North America

Typical characteristics	Single-compartment two-opening nontilting type	One-opening tilting type, mixing angle of drum 15° with horizontal	Two-opening rear-charge and front-discharge horizontal-drum tilting type	Vertical-shaft type	Horizontal-shaft type
Slump range, in........	2–5	0–3[a]	1½–3[a]	0–5	0–5
Max aggregate size, in.[b]	2½	6–10	6–8	1½–2½	1½
Mixing time, sec[c]......	60–90	50–180[c]	50–180[c]	20–60	180–300
Peripheral drum or blade speed.........	270–300	280–310	260–370	550–750	270–290
Revolutions per minute	10–17	8–13½	9–15½	18–25	20
hp/cu yd.............	10–18	8.6–10	8.9–15	28–40	20–35
kwhr/cu yd..........	0.21–0.42	0.13–0.54	0.14–0.71	0.33–0.87	0.79–2.12
Charging time, sec[d]....	15	10–15	10–15	15–30	15
Discharge time, sec.....	30–40	10[e]	15[e]	20–30	15–25
Cycle time, sec........	105–145	85–215	90–220	65–120	220–340
Liner and blade wear[f]..	Moderate	Low	Low	Very high	High
Hydraulic discharge, hp/cu yd...........	[g]	3–6	2–5		
Air discharge, cu ft free air/cu yd..........	1–2	4–7	3–10	9–14	1–2
Concrete temp rise, °F per min of mixing....	0.4	0.3	0.3	1.0	0.9

[a] Higher slumps may usually be mixed at reduced capacity.

[b] Lower maximum particle size applies to smaller mixers.

[c] Shorter mixing times are usual for concrete with high cement factors. Long mixing times for tilting mixers are typical of large mixers with low-slump low-cement-factor concrete on specification work.

[d] Charging time may be longer because of factors external to mixer.

[e] Additional 5 to 10 sec required to right mixer after discharge is complete.

[f] Varies greatly with abrasiveness of coarse aggregate.

[g] When used usually operates from mixer drive.

Concrete production capacity may be computed from the rated mixing capacity (Table 1 for plant mixers or Table 2 for truck mixers) and the mixer cycle time (Table 1, Chap. 16, for mass concrete mixers; Table 2, Chap. 16, for paving mixers; and Table 3 for plant mixers). Concrete production is often controlled by factors external to the mixer, as discussed in Chap. 16.

In the United States and Canada, standard mixers are rated in cubic yards of mixed concrete. Some other mixers are rated in cubic feet of loose material charged into the mixer. About 40 cu ft of loose solid material is required to make a cubic yard of concrete. In countries using metric measurements, ratings are usually in liters or cubic meters of mixed concrete but may be in liters or cubic meters of loose material charged into the mixer.

PLANT MIXERS

Standard types and sizes of concrete-plant mixers are shown in Table 1. Table 3 is a comparison of plant-mixer types in common use. Some of these mixers were formerly covered by standards of the Mixer Manufacturers Bureau, the Associated General Contractors of America. Typical capacity ratings were in cubic feet of mixed concrete, such as 210S, with 10% overload capacity guaranteed. Two types of *tilting* mixers and three types of *nontilting* mixers are commonly used in concrete plants.

1 Tilting Mixers

Tilting mixers are rotating-drum mixers that discharge by tilting the drum about a fixed or movable horizontal axis at right angles to the drum axis. The drum axis may be horizontal or inclined upward at the discharge end while charging and mixing, and is rotated downward until it is 50 or 60° below the horizontal during discharge. Tilting is usually done by hydraulic rams, occasionally by air rams. Power requirements for tilting are high, but several mixers may be operated in succession from a common hydraulic pump or air compressor. Figure 1 shows a typical hydraulic tilting circuit. Mixers are usually weld-coated or lined with abrasion-resisting steel-plate liners welded or bolted in place.

One-opening tilting mixers with the mixing angle of the drum inclined 15° with the horizontal are used for a variety of concrete types including base-course material, soil cement, pavement concrete, commercial ready-mix concrete, and low-slump concrete with large aggregate for dams (Fig. 2). The single opening is typically large for fast charging through a retracting charging chute or from the discharge trajectory of a high-speed belt conveyor. Mixers discharge by tilting and dumping concrete rapidly into a concrete hopper, guide chute, or directly into open-top hauling units. Three or four blades lift and circulate the concrete to the rear of the drum, where it is sheared by the following blades and tumbled over the concrete mass.

Two-opening front charge and discharge tilting mixers with the mixing angle of the drum inclined 5° with the horizontal are no longer commonly used. They are similar to the horizontal-drum type in characteristics except that they are charged through the discharge opening.

Two-opening rear-charge and front-discharge tilting mixers with the mixing axis of the drum horizontal are used for a variety of concrete types including pavement concrete, commercial ready-mix concrete, and low-slump concrete (Fig. 3). The rear-charging opening is normally large and equipped with a retracting chute for fast charging. Discharge is by tilting and dumping through a smaller front opening into a concrete hopper, guide chute, or directly into open-top hauling units. Various types of blades lift and circulate concrete longitudinally in the drum and over the concrete mass.

Charging tilting mixers for minimum mixing time is done by blending materials, either by metering materials onto a belt conveyor or by adjusting the sequence and discharge speed of batchers. Where the materials pass through a chute, water and the largest aggregate should start first, followed after a few seconds by the other materials. The largest one or two sizes complete discharge early, the discharge of the finer materials spreads over the charging time, and a medium-sized aggregate may be lagged to scour the chute. Water started as early as possible helps keep the mixer clean, and all water and admixtures should be in the mixer with the

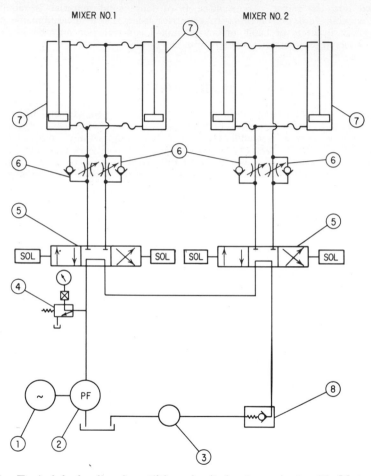

FIG. 1. Typical hydraulic-mixer tilting circuit for two mixers. (1) Motor. (2) Pump. (3) Filter. (4) Relief valve. (5) Four-way solenoid valve. (6) Flow-control valve. (7) Hydraulic cylinder. (8) Check valve.

last of the solid materials, but in some instances where the plant cycle is not adversely affected, the water may finish discharge 15 sec after the solid materials.

2   Nontilting Mixers

Nontilting mixers discharge through gates in the mixing compartment or by chutes or blades that direct concrete from a rotating drum.

Single-compartment two-opening nontilting mixers have a revolving cylindrical

FIG. 2.   One-opening tilting-type mixer, mixing angle 15° with horizontal.   (*C. S. Johnson Division of Koehring Company.*)

FIG. 3.   Two-opening tilting-type mixer, horizontal drum.   (*Rex Chainbelt, Inc.*)

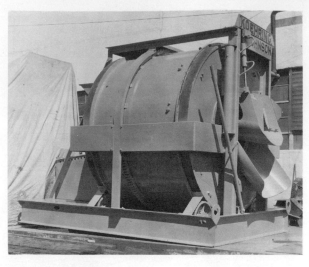

Fig. 4. Single-compartment two-opening nontilting mixer. (*C. S. Johnson Division of Koehring Company.*)

drum (Fig. 4). The charging chute is usually fixed and projects into the rear opening. A hinged chute projects into the front opening to discharge concrete dropped by pickup blades. This chute is usually operated by compressed air or an integral hydraulic system in larger mixers. In addition to pickup blades that lift and drop concrete, other blades may direct concrete longitudinally, usually toward the front. Mixing action is primarily by picking up and spilling concrete over the concrete in the bottom of the drum. Mixers are usually lined with abrasion-resisting steel plate. These mixers are used mostly for ready-mix concrete or for higher-slump concrete for cast-concrete products. Smaller sizes may be obtained with wheels and charging skips. Because of cleaning and discharge problems, this mixer is not well suited to low-slump concrete or concrete with large aggregates.

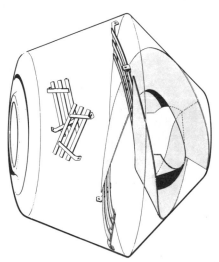

Fig. 5. Reversing-drum mixer. (*Winget, Ltd.*)

A variety of two-opening nontilting mixer seldom found in North America is the *reversing-drum mixer* (Fig. 5). This mixer has conical ends on a horizontal drum and is charged through a rear opening and discharged by screwing the concrete from a front opening when the drum is reversed, much as does a truck mixer. The front cone of the mixer has helical blades and the rear cone of the mixer has tilted blades to direct concrete toward the front cone. It is suitable for medium- to high-slump concrete with small maximum aggregate size.

Vertical-shaft mixers (sometimes called *turbine, pan,* or *compulsory* mixers) have a mixing compartment with an essentially level floor and a cylindrical outer wall (Fig. 6). They may have an annular mixing chamber because of a cylindrical center well, where the motor or gears may be mounted. There are one or more rotating vertical shafts to which blades or paddles are attached, and the mixing chamber may be stationary or rotate about a vertical axis during mixing. Blade speeds and power requirements are high and mixing is rapid. Liner and blade or paddle wear is higher than in other types of mixers, excepting possibly high-speed horizontal-shaft mixers seldom seen in North America. Liners and blades are often made of nickel cast iron; some mixers have liners of abrasion-resisting steel plate. Vertical-shaft mixers are suitable for a wide range of concrete slumps but power requirements are higher for concrete of very low slump, often resulting in reduced mixer capacity. Vertical-shaft mixers will not ordinarily start easily under load. They are most often used in concrete-products plants, but the larger sizes find some application in ready-mix concrete plants, central-mix paving plants, and in heavy construction where maximum aggregate size does not exceed 2½ to 3 in. Dis-

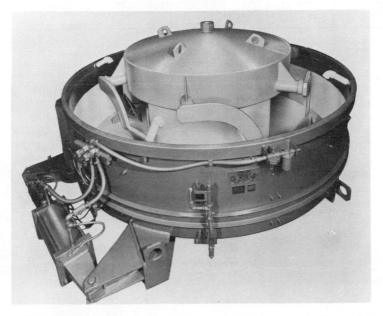

FIG. 6. Vertical-shaft mixer. (*C. S. Johnson Division of Koehring Company.*)

charge is through one or more doors located in the floor of the mixer, permitting direction of the concrete to several points depending upon which door is opened.

A motor-load indicator (usually an ammeter) is valuable in operation of a vertical-shaft mixer, particularly when trying different mixes or different charging procedures, and will indicate overloads much more rapidly than motor-protective circuits. The following are typical readings, but the manufacturer's recommendations for the particular mixer should be followed.

The load indicator may read over 150% of full motor load momentarily when the mixer starts, and it may read 125% for 2 or 3 sec while charging, but it should not exceed 125% at any other time and it should not exceed 100% for more than 10 or 15 sec. Twenty to thirty-five percent is normal when the mixer is running empty. With a dry mix the load will rise as the mixer is charged and remain high, and the quantity mixed may be limited by the power available. With about 1½ in. or more slump the motor load rises to perhaps 115% as the mixer is charged, drops rapidly to about 85%, then drops more slowly to about 75% and remains steady. At this point the concrete is usually well mixed. The above percentages are typical for a capacity mix but will vary with size and type of aggregate, with cement content, and with slump.

The vertical-shaft mixer is sensitive to methods of charging and to changes in blade size, shape, and angle when mixing large batches. It mixes by shearing material in contact with the blades against material in contact with the walls and floor of the mixing chamber. If the friction of the concrete on the blades exceeds the friction on the mixing chamber, the mass may rotate without mixing. When mixing properly, an undulating or weaving motion of the concrete is visible, and power requirements are higher than when the mixer is not mixing properly. After the mixing charging procedure and capacity have been established, mixing should continue satisfactorily until the mixer becomes worn or until some change is made, perhaps inadvertently, in the procedure.

FIG. 7. Horizontal-shaft mixer. (*Besser Company.*)

Ideally aggregate and cement should be blended before reaching the mixer or ribboned into the mixer concurrently, but the mixing time is reduced less by blending than it is in rotating-drum mixers. The cement and at least one aggregate, preferably sand, should be put into the mixer together, followed by the coarse aggregate. Cement should not enter the mixer before the aggregate and water, and all cement should be in before the last of the aggregate. Materials can usually be introduced at any point around the periphery of the mixer.

Improper mixing may result if water is not introduced rapidly and well into the mixer at several points around the mixer, because water dribbling down the mixer wall may lubricate it. Improper mixing may also result if water is too hot. Water should start into the mixer with the first of the other materials. If a batch tank is used its discharge valve should be at least 6 ft above the mixer top and the pipe between them should be large. One 5-in. or two 3-in. pipes are typical. If the pipe is connected to the tank, a vent is necessary so that water can clear the pipe rapidly after the tank discharge valve closes. If a water meter is used, it should be sized to discharge the maximum batch quantity in not more than 30 sec at the pressure available. The power requirements for a given mix determine the point in the mixing cycle at which the water must finish discharging. For example, crushed rock requires more horsepower than gravel, and the water must be in the mixer sooner to prevent overload. Often 15 to 20% of the water can be put in after all other materials. All materials can be introduced into the mixer in 15 to 30 sec. The slower rate may be necessary for batches approaching the mixer capacity to permit some shrinking of the mix during charging.

Overmixing is particularly objectionable because of unnecessary wear and heat developed in high-speed mixing. Doubling the mixing time nearly doubles the blade and linear wear. About 700 Btu per cu yd per min is added in mixing, raising the concrete temperature about 1°F per min. The effect on the concrete appears to be more pronounced than this, probably because of the slow rate of absorption of heat by the coarse aggregate, and the mortar fraction of the concrete may be heated more nearly 1.5 to 1.7°F per min. This is further aggravated by an increase in fines which decreases the slump.

Horizontal-shaft mixers have a stationary horizontal cylindrical mixing compartment and one or more rotating horizontal shafts to which blades or paddles are attached (Fig. 7). They are best suited for mixing very dry concrete or base-course materials and find use mostly in concrete-products plants. Blades are usually helical and positioned to work the materials back and forth longitudinally as well as radially over materials in the bottom of the mixer. Paddles of various shapes and sizes may be used instead of blades. Blades and liners are usually nickel cast iron. Materials are introduced through an opening at the top and discharged through a door in the lower part of the cylinder. Power requirements and liner and blade wear are high. Horizontal-shaft mixers are often rated in cubic feet of loose material per batch rather than in cubic yards of resulting concrete.

Charging procedure for these mixers depends on the product and materials. For base-course materials or dry concrete partially blended material may be fed in batches (or continuously) into the mixer and water added from a spray bar (a pipe with nozzles or holes) extending lengthwise over the mixer. Some lightweight aggregates require prewetting and are introduced and mixed with water for 30 sec to several minutes before cement is introduced. Open-grid mixer covers may be used for safety, or closed covers may be used where dust is objectionable.

PAVING MIXERS

Paving mixers are track-mounted machines that can move at a rate of about 1 mph while mixing (Fig. 8). Modern machines have two or three drums built

Fig. 8. Tri-batch paver. (*Koehring Division of Koehring Company.*)

into a single drum shell. Cement and aggregate are charged into the first drum by a skip from trucks that carry several batches. Water and admixtures are introduced from volumetric batchers built into the machine. After mixing for a time in the first drum, the concrete is transferred by a pivoted chute to the second drum and another batch is fed into the first drum when the transfer is complete. Concrete is discharged from the last drum to a bucket supported by a boom on the machine and discharged at some distance from the mixer, usually on the ground. Mixing time is usually measured from the time the skip reaches the top of its travel until the concrete starts to discharge to the bucket, and the time used to transfer concrete between drums is included as mixing time. Mixing time is 50 to 90 sec, with 60 sec common, controlled by a mechanical or electric batchmeter that ensures mixing for the set time. Batch size varies from 1.26 to 1.51 cu yd, with 1.38 cu yd commonly permitted on grades up to 6%. A drum speed of about 15 rpm results in peripheral speeds of about 300 fpm. Paving-mixer standards are published by the Mixer Manufacturers Bureau, the Associated General Contractors of America, 1957 E St. N.W., Washington, D.C. 20006.

TRUCK MIXERS

Truck mixers designed to mix or agitate concrete while mounted on a moving truck or trailer are of three types: *inclined-axis revolving drum, horizontal-axis*

Fig. 9. Truck mixer. (*Cook Bros. Equipment Company.*)

revolving drum, and *open-top revolving blade or paddle.* The inclined-axis type (Fig. 9) is commonest because it has slightly larger capacity with equal drum volume and discharges high so concrete can be chuted into place more easily with chutes built onto the mixer. Some inclined-axis truck mixers speed up discharge by tilting the drum until the axis is nearly horizontal. Inclined-axis truck mixers have variable-speed drum drives and drum revolution counters, and are discharged by reversing the drum so that spiral blades screw the concrete to the drum opening. Truck mixers may be used to mix concrete (*truck* or *transit mixing*), complete mixing concrete partially mixed in a plant mixer (*shrink mixing*), or agitate and deliver concrete completely mixed in a plant mixer (*central mixing*). Standard mixing and agitating capacities are shown in Table 2. Truck mixers will mix about 10% more shrink mix than truck mix, and will usually charge mixed or partially mixed concrete more rapidly than loose ingredients.

Mixing speed is from 4 rpm to the speed resulting in a maximum peripheral velocity of 225 fpm (about 9 to 12 rpm depending upon diameter) for revolving-drum mixers, or to a maximum blade speed of 16 rpm for open-top mixers. *Agitat-*

ing speed is 2 to 6 rpm. Drums are often operated at speeds up to 20 rpm while charging or discharging to save time, and each mixer design may have different optimum charging and discharging speeds. At rated capacity concrete should be mixed 70 to 100 revolutions at mixing speed after all ingredients are in the drum. Truck mixers filled to not more than ½ cu yd under their rated capacities, and trucks used for shrink mixing, should mix 50 to 100 revolutions at mixing speed. Any operation in excess of 100 revolutions should be at agitating speed unless additional water is added later, requiring an additional 30 revolutions at mixing speed.

Water is added to the mix with the other ingredients at the batch plant or metered from a water tank on the truck to an accuracy within 1% of the net amount of water needed.

When mixing at rated capacity, truck mixers are relatively inefficient mixers. Mixing time is typically 8 to 15 min and concrete temperature is typically increased about 5°F during mixing, perhaps much more if agitated for some time after mixing. Solar radiation adds further heat. This temperature rise can result in troublesome loss of slump in hot weather. Solar-heat pickup can be reduced by painting drums gloss white. Cool materials and corrected batch proportions can also compensate for heat rise. About ½ in. of slump is lost if concrete temperature rises from 75 to 85°F. The change is slightly more at lower temperatures, and less at higher temperatures. A typical 3-in.-slump 1½-in.-aggregate concrete requires about 0.64 lb of added water per degree Fahrenheit to maintain slump constant. This added water requirement varies from about 3 to 5% of the total water per inch change in slump for various medium-slump concrete mixes.

Truck mixers are not generally suited for concrete of less than 2 in. slump or with aggregate over 2½ in. Charging may require as much as 13 sec per cu yd when materials are fed into a rear drum opening, but time is much shorter for batch charging. Discharge rates are typically 1 to 1½ min per cu yd for 2-in.-slump concrete, about 30 sec per cu yd for 3-in.-slump concrete. Some newer models and special mixers made for paving concrete charge and discharge much more rapidly; charging time may be as low as 4 sec per cu yd and discharge time about 4 sec per cy yd under ideal conditions.

CONSISTENCY

Consistency is a loosely used term that describes the fluidity of concrete. The slump test is used in the absence of other easy tests as a measure of consistency but is a static test and does not measure consistency or workability under the more important dynamic conditions. *There is no wholly satisfactory method of measuring consistency (or slump) of concrete while it is in the mixer.* Mixer-consistency meters do not measure slump; they measure a similarity or difference in the effect of different batches of concrete on the mixer. Properly understood and applied, mixer-consistency meters can be of assistance to an operator in detecting a difference between supposedly identical batches.

3 Center-of-gravity Meters

These meters measure the reaction of the concrete on some part of the mixer frame, usually the tilting rams of a tilting mixer, by either electromechanical or hydraulic means. They were used for some time but have fallen into disuse because they are difficult to keep operating properly and work well only with certain shapes of mixer drums.

4 Ammeter Instruments

A current transformer reduces the load current used by the mixer motor, usually on only one phase line, to the proper scale so an ammeter can present a visual indication of the change in motor current. These meters are reasonably useful with some mixers and certain types of concrete for comparing batches of identical size and composition if line-voltage fluctuations are less than a few percent. They have been used on nontilting horizontal-drum, tilting, and vertical-shaft mixers for commercial ready-mix and pavement concrete with generally, but not always, useful results. Sometimes a modification in the mixer blades is necessary before the change in current is large enough to be useful.

5 Wattmeter Instruments

Both the voltage and current are utilized with these instruments to indicate the power drawn by the mixer motor, usually on only one phase line. Two- and three-phase meters are available, but the value of the added cost is questionable. Wattmeters are used the same as ammeters except that they compensate for line-voltage changes.

6 Vacuum Instruments

These instruments have been used on internal-combustion engines. Their use is similar to that of ammeter instruments.

7 Factors That Affect Consistency

While water content has the most marked effect on consistency, consistency meters do not invariably indicate a need for more or less water in the mix. Consistency is also affected by changes in fine-aggregate gradation, concrete temperature, batch weights, specific gravity of materials, aggregate porosity, and melting time of ice when used in the mix. In most plants satisfactory control of consistency is best achieved by controlling segregation of aggregate, controlling batch weights, and by measuring moisture in sand and fine aggregate and controlling mix water to compensate for aggregate moisture and concrete temperature. To control slump within ½ in. it is necessary to compensate for aggregate moisture within about ¼ to ⅓% of the sand weight. Concrete temperature can be equally important. A 10°F change results in about ½-in. change in slump, the change being greater at lower temperatures.

With experience an operator can estimate consistency of familiar mixes about as accurately as they are measured by consistency meters if he can see the concrete in the mixer, discharging from the mixer, or flowing from chutes.

MIXER TESTS

Various specifying agencies perform mixer-efficiency tests. While there is similarity in some of the items measured, there is a large variation in the selection of items tested, methods of sampling, and interpretation of test results. This results in a variety of mixing times for a given mixer depending upon the agency performing the test. Several or all of the following items may be included in tests of the concrete: slump, ball penetration, air content, unit weight of concrete, coarse aggregate retained on the No. 4 sieve (larger aggregate may be removed by hand in some tests), distribution of aggregate larger than 3 in., unit weight of air-free mortar, compressive strength, flexural strength, cement content by extraction, and water content of mortar. Usual acceptance criteria are based on variation of test results of two or three samples taken from various parts of the mixer discharge. Other criteria are based on tests of variability of consecutive batches. Individual specifications must be consulted for specific work.

OMNI MIXER

The Omni Mixer uses an inclined-axis wobble plate attached to a deformable rubber bowl (Figs. 10 and 11). The rotation of the drive shaft causes the wobble plate to move the particles near the plate in many different directions with varying acceleration. This produces a nearly random motion of the particles and greatly enhances the mixing action. Neither the slump of the concrete (within realistic

Fig. 10. Omni Mixer, 1 cu yd. (*Gar-Bro Manufacturing Company.*)

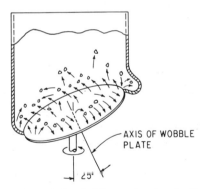

Fig. 11. Schematic drawing of Omni Mixer showing inclination of wobble plate. (*Gar-Bro Manufacturing Company.*)

limits) nor the order of charging the mixer has an appreciable effect on mixing time, which is between 15 and 30 sec. The mixing time is affected by the angle and diameter of the wobble plate and the height of material above the plate. Generally, the height of material should be less than or equal to the diameter of the mixer bowl for most efficient mixing.

A low profile cone may be attached to the center of the wobble plate for mixing under particularly difficult conditions. This projection reduces the possibility that material will gather at the center where motion is at a minimum.

The mixer motor load as indicated by an ammeter is a useful indication of the consistency of the mix. When the mixer motor is first started, with all material in the mixer, the ammeter may read 175% of full motor load. It will remain high for 2 to 3 sec, then drop to about 125% of full motor load in 5 to 10 sec. The mixing is essentially complete at this point.

Zero slump concrete with 2½ in. maximum size aggregate has been mixed satisfactorily in 15 sec in the 1 cu yd mixer, which is the largest size that has been built. The cycle time should be from 45 to 60 sec, with a production rate of 60 to 80 cu yd per hr.

Chapter 19

MATERIALS HANDLING AND STORAGE

JOHN K. HUNT

The extreme difference in characteristics of the materials used for concrete necessitates different methods of handling and storage. Aggregate is usually elevated by clamshell bucket, belt conveyor, or bucket elevator; cement by bucket elevator or air blower; and water and admixtures by pumps. The methods used to transport materials to the plant site are a major factor in determining the handling system used.

Aggregate conveyor belts and bucket elevators should be rated in tons per hour, cement bucket elevators and other cement-conveying equipment should be rated in barrels per hour, aggregate bin storage should be stated in terms of cubic yards of aggregate, and cement bin or silo storage should be stated in terms of barrels of cement. Capacities are defined by Concrete Plant Standards published by Concrete Plant Manufacturers Bureau, 900 Spring St., Silver Spring, Md. 20910.

AGGREGATE HANDLING

Aggregate stockpiles should have a hard, clean base with provisions for keeping the sizes separated to reduce contamination. Clams should not swing over piles of other sizes, particularly sand. Excessive handling after material is sized will result in segregation and degradation. An accumulation of undersize material in bin compartments due to breakage and discharging characteristics of the bin can be eliminated by completely emptying the bin compartments once each day. Sand drainage for about 24 hr is required to reach a reasonably stable moisture content. Three to six percent moisture is common in natural sand after draining. Manufactured sand typically has a higher moisture content.

1 Unloading and Stockpiling

Delivery of aggregate to the plant site may be by rail, truck, or barge. Railroad cars may have bottom discharge to dump into a hopper which feeds a belt conveyor. Clamshell buckets may be used to unload the cars and stockpile directly.

Bottom-dump or end-dump trucks may discharge into a hopper which feeds a belt conveyor. End-dump trucks may be driven on the stockpile to deposit the material in layers but care must be taken to reduce contamination. The material may be dumped on the ground and stockpiled by clamshell buckets or front-end loaders. If the batch plant or stockpile is near a cliff the trucks may be driven onto an open grille and discharged directly into the bin or stockpile.

Barges are normally unloaded with clamshell buckets.

2 Clamshell Buckets

Rehandling buckets are most suitable for handling aggregate materials. Sand may be piled in a cone. Damp sand tends to segregate less in handling and storage than very dry sand. Coarse-aggregate stockpiles should be made by placing each bucketful in a consistent pattern and building up layer by layer instead of indiscriminately creating many small stockpiles and filling in between them. The material should be laid in place, not thrown at the pile. Throwing the material causes breakage and degradation and also causes the larger particles to run down the sides of the pile, which creates segregation.

3 Belt Conveyors

A radial stacking conveyor (Fig. 1) stockpiles material on an arc and a parallel stacking conveyor stockpiles material in a straight line. The material can be

FIG. 1. Radial stacker with raising and lowering boom. (*Barber-Greene Company.*)

reclaimed by a tunnel conveyor or clamshell buckets. The stacker should be equipped with a vertical adjustment at the head end or stone ladders should be used to reduce the vertical drop of the material. Low-capacity low-profile plants may be fed with one stacking conveyor.

A shuttle conveyor (Fig. 2) distributes material in a straight line. It may be used for stockpiling or on top of a bin. Reclaiming is usually done with a tunnel conveyor. Stone ladders are almost a necessity. There is less dead storage than with a radial stacking conveyor if a tunnel conveyor is used.

Material may be stockpiled in storage bins, in piles separated by walls, or in open piles over a tunnel conveyor (Fig. 3) which is fed by a series of gates under the stockpiles. The tunnel conveyor delivers the material to a rinsing screen or to an inclined conveyor which delivers to the bin. Material should be discharged onto the centerline of the belt with a double-clam gate or in the direction of the belt travel using a single-clam gate to avoid lateral force on the belt which would cause it to run off center. If material is to be blended on the belt the large material should be nearest the bin to avoid fouling the other gates. The number of gates per material depends on the size of the stockpiles.

Inclined conveyors (Fig. 2) carry material to the top of the aggregate bin or to intermediate points. They require trusses, A frames, or towers for support. Normally one belt is used if the bin is high and the bin has a large storage capacity. It may feed a shuttle conveyor, pivoted distributor, flop chute, or screen. One belt

FIG. 2. Inclined belt conveyor with reversible-shuttle belt conveyor. (*Atlas Conveyor Company*.)

FIG. 3. Tunnel belt conveyor with tunnel gate. (*Barber-Greene Company*.)

for each material is sometimes used for low-profile low-storage-capacity and high-production plants. In this case each belt is fed from a hopper and quite often the belts are kept fully loaded and are controlled by a high bin signal to start and stop automatically to keep the bin full.

4   Bucket Elevators

Bucket elevators may be either the open inclined type or vertical enclosed type. The open inclined type usually costs less but material can spill from the buckets and create a nuisance or safety hazard.

Aggregate bucket elevators are almost always the continuous type which has the buckets closely spaced so the material will discharge at the top of the elevator by flowing across the back of the preceding bucket and into the discharge chute. Material is fed directly into the buckets instead of going into the bottom of the elevator to be picked up by the buckets.

Either chain- or belt-type elevators can be supplied. If the pitch of the sprocket and chain are very much different the chain will be forced to slip on the sprocket and cause rapid wear of the sprocket. Belt-type elevators run slightly faster and smoother than the chain type. Belt-type elevators should be equipped with a gravity take-up instead of a screw take-up to compensate for the expansion and contraction of the belt due to temperature changes.

Bucket elevators cost less and require less ground space than belt conveyors but have lower capacity and more maintenance. Over a long term belt conveyors are more economical and should be used wherever it is feasible.

5   Pivoted Distributors and Flop Chutes

A pivoted distributor is placed at the top of a bin to receive material from a belt conveyor or bucket elevator and direct it to the proper bin compartment. A flop chute or two-way chute can be used if the material is to be directed to one of only two bin compartments. The pivoted distributor or flop chute can be manually operated by a cable or the pivoted distributor can be motor-driven and the flop chute can be operated by an air cylinder.

High- and low-level signals can be placed in each bin compartment to indicate by a light or buzzer whether a particular compartment is almost full, almost empty, or in between. These high- and low-level signals can be electrically connected to the pivoted distributor or flop chute, the inclined belt conveyor or bucket elevator, and the tunnel conveyor, to provide a fully automatic continuous bin-filling system. When the material in one compartment is low enough to actuate the low level, the pivoted distributor or flop chute moves to the proper position to charge the compartment, the inclined belt conveyor or bucket elevator is started, the tunnel conveyor is started, and the tunnel gate under the proper stockpile is opened. When the material in this compartment reaches the high-level signal the tunnel gate closes immediately. A time delay is built into the system to allow the tunnel conveyor, the inclined conveyor or bucket elevator, and the pivoted distributor or flop chute to clear all material that was on the way. If another compartment has reached the low signal the entire procedure repeats. If no other compartment is low the charging equipment will shut down until it is needed again. If the plant operation is such that some material is feeding most of the time it will probably be advantageous to let the charging equipment run all day instead of starting and stopping it many times.

6   Stone Ladders

Stone ladders are used to reduce segregation and degradation of coarse aggregate by shortening the vertical drop. These are arranged so material flows from one step to another. Stone ladders are often used on 1½- to 3-in. or larger material. The steps should be not more than 30 in. apart.

7   Bins

Aggregate storage bins can be obtained with almost any amount of storage from 25 to 3,000 cu yd or more. They can be square, rectangular, hexagonal, octagonal, or round. Bins can be for aggregate only with a separate bin for cement, with the aggregate surrounding the cement, with the aggregate on one end and the cement on the other end, with the aggregate on one side and the cement on the other side, with the aggregate on both ends and the cement in the middle, or almost any other conceivable arrangement.

Wind load, particularly on a high bin, must be carefully analyzed and the foundation loads given careful consideration. Seismic loading must be included where it is required. Vibration in the bin due to mixers or screens needs special attention. The belt tension on long conveyor belts attached to the bin must be considered. Special provisions must be made if heavier than normal aggregate is to be handled to prevent overloading the bin.

RESCREENING OF AGGREGATE

After aggregate has been screened to establish the proper gradation for each size and stored in stockpiles it is often desirable and sometimes necessary to rescreen the aggregate to reduce the effects of degradation and contamination due to handling.

8 Purpose and Methods

Since there is usually only one screen deck for each size material, rescreening can only reject undersize material passing through a screen and cannot and is not intended to rearrange the gradation. Usually the material is blended in the approximate proportions to be used by feeding onto a belt conveyor simultaneously from several stockpiles with the belt conveyor feeding the screens. Sending one material at a time to the screens would overload one screen and would result in wasting the oversize and undersize material. Material passing the No. 4 screen is normally the only material wasted.

Since the specifications for gradation usually apply to the material as batched, it is more desirable to place the screens on top of the batch plant with the material going directly into the proper compartments instead of screening on the ground and then elevating the material, because screening on the batch plant eliminates one handling operation after screening and reduces the possibility of more contamination and degradation.

Rescreening structures on top of a batch plant must be carefully designed to keep vibration at a minimum. If very much vibration is transmitted to the automatic batching controls, scales, and recording, the operation may be seriously affected.

9 Screen-size Selection

The most important factors affecting screen capacity are:
1. Screen size
2. Amount of oversize
3. Amount of undersize
4. Density of material
5. Whether crushed or gravel material
6. Amount of moisture
7. Whether screen is top, second, or third deck on one machine
8. Shape of screen opening
9. Open-area ratio of screen
10. Desired efficiency

On a typical installation for mass concrete, the coarse aggregate is divided into four sizes, 6 to 3 in., 3 to 1½ in., 1½ to ¾ in., and ¾ in. to No. 4.

For the larger installations either two double-deck screens one above the other or two parallel sets of two double-deck screens (four double-deck screens) are used. For small installations one triple-deck screen can be used with the bottom deck split for two materials.

Required capacity is computed based on the average mix design, the maximum production, or the expected hourly production, and considering use of crushed stone. For the three top sizes 90% efficiency is normally adequate because most gradation specifications allow 15 to 20% of slightly undersize material. The efficiency of the No. 4 screen could be as low as 50% since in general removal of one half the undersize material below No. 4 mesh will bring the material within the usual gradation specifications. This usually permits carrying a depth of material on the No. 4 screen from six to ten (or more) times the screen opening.

The highest loaded screen usually determines the size of all the screens, although in some installations the upper set of screens may be one size larger than the lower set.

When the same conveyor is used for coarse aggregate and sand a diverting gate is usually used to allow the sand to bypass the coarse-aggregate screens and go directly to the sand-bin compartment. Approximately one-third of the total loading time will be needed for sand; thus only two-thirds of the total loading time is available for rescreening of the coarse aggregate. Higher-capacity plants use a separate conveyor for the sand, which increases the available screening time by 50%.

Table 1. Typical Screen Sizes Used in Plants with One Conveyor Handling Both Coarse Aggregate and Sand

Mixers	Production, cu yd/hr	Screen
2–2 cu yd	80–120	One 5 × 14 or 5 × 16 triple deck
2–4 cu yd	130–160	Two 5 × 12 double deck or one 5 × 16, 5 × 20, or 6 × 20 triple deck
3–4 cu yd	200–240	Two 5 × 14 or 5 × 16 double deck
4–4 cu yd	250–320	Two 6 × 16 or 7 × 16 double deck or two parallel sets of two 5 × 12 double deck
6–4 cu yd	400–480	Two 6 × 16 or 7 × 16 double deck or two parallel sets of two 5 × 12 double deck (with a separate conveyor for sand)

Typical screen sizes used in plants with one conveyor handling both the coarse aggregate and the sand are shown in Table 1.

CEMENT AND POZZOLAN HANDLING

Since the handling problems of cement and pozzolan are similar the following comments apply generally to both. It is safer to use separate structures and handling facilities for cement and pozzolan to avoid contamination, but if the same handling equipment is used for both materials the valves that determine to which compartment each is fed should be watched carefully and inspected regularly. A double wall or extra-strong partition should be placed between two adjacent compartments.

Cement must be protected from water during handling and storage. Problems often arise on handling and storage capacities because of the fluffing of cement when it is well aerated. At least two ratings should always be given for cement. The maximum capacity should be based on the assumption that a barrel of cement occupies 4.0 cu ft and the minimum capacity should be based on the assumption that a barrel of cement occupies 4.8 cu ft. Handling cement is complicated by the fact that when cement is "dead" it will hardly flow at all and when it is hot and well aerated it will flow wildly in all directions.

10 Pneumatic Conveyors

Pneumatic conveying of cement has been used for many years. The delivery of cement to concrete-batching plants is now being done more and more by trucks equipped with air blowers for the discharge of cement to bins or cement silos. Cement companies have set up transfer plants close to delivery points so that trucks can often deliver cement within a few hours after it is ordered.

This method reduces the amount of rehandling equipment required at the plant site. Air-blowing systems are also used to transfer cement from auxiliary storage to the batching bin.

11   Screw Conveyors

Screw conveyors are normally used to transport cement horizontally but may be inclined upward with some loss of capacity and increase of power. By using a half-pitch screw flight and covers shaped to conform to the top of the screw, it is possible to recover part of this loss of capacity. A screw conveyor with a full-pitch screw operated at 20° from horizontal would deliver approximately 57% as much as a horizontal unit, and a screw conveyor with a half-pitch screw and shaped covers operated at 20° from horizontal would deliver approximately 84% as much as a horizontal unit.

The most common sizes for screw conveyors are 10 and 12 in. but they may be either smaller or larger for special applications.

Special consideration must be given to the motor required if it is to be started under load because of the high torque required.

Two-speed operation is often used in order to get more accurate weighing when a screw conveyor feeds a weigh hopper directly.

12   Bucket Elevators

The centrifugal-discharge type of bucket elevator is commonly used for cement. There is a considerable space between buckets. The relationship of speed and the diameter of the head pulley is carefully selected so the material will be thrown from the bucket to the discharge chute. Increasing or decreasing the speed of the elevator will reduce the capacity because much of the material will miss the discharge chute and fall to the bottom of the elevator.

Cement is fed into the boot or bottom of the elevator slightly more slowly than it can be carried away. The buckets scoop up the cement as they pass around the tail pulley. If the elevator is fed by a screw conveyor they can be driven by the same motor. If a separate drive is used the motor starters should be electrically interlocked so the screw conveyor cannot be started until the elevator is running, and if the elevator motor stops for any reason the screw conveyor motor should automatically stop. This is to prevent plugging the boot of the elevator and eventually plugging the screw conveyor.

13   Bins and Silos

All joints, manhole covers, doors, or gates in a cement bin or silo should have a watertight seal to prevent the formation of chunks which would later plug up or jam the weigh hopper or cement-handling equipment.

A manhole and inside ladder should be provided to permit periodic checking, cleaning, or other maintenance.

Since it is quite often difficult to get cement to flow properly, a low-pressure aeration system should be provided for furnishing diffused air near the discharge of the bin or silo. High pressure or direct air tends to cut a hole through the cement and does not aerate it.

An adequate vent should be used to prevent air pressure from building up inside a cement bin or silo.

Chapter 20

HEATING AND COOLING

JOHN K. HUNT

This chapter is concerned with all aspects of heating and cooling concrete in the plant. Chapter 30 covers hot- and cold-weather operations, and Chap. 46 is concerned with cooling mass concrete both during manufacturing and after placing.

Concrete must often be heated in cold weather to ensure that it can be placed and protected before it freezes. Concrete is also cooled, principally in mass-concrete work, to permit lower water and cement contents for a given strength and slump, and to offset temperature rise in mass concrete caused by heat of hydration of the cement. Drying shrinkage of concrete is reduced by the lower cement content and more particularly by the lower water content of cool concrete. Plastic or drying cracking of slabs is reduced by slower moisture loss at reduced temperature. Depending upon the type of cement, hydration releases 80 to 180 Btu per lb of cement in the first 3 days and about twice this amount in 28 days. In mass concrete this heat develops faster than it can be radiated by the concrete and results in temperature gains of about 25 to 30°F in 28 days for concrete with 376 lb of cement per cu yd. To offset these effects, concrete is often placed at 40 to 50°F so that it will be substantially cooler than the average ambient temperature.

1 General

Heat is energy that raises the temperature of a substance as the amount of stored heat increases. Heat flows from a substance at a higher temperature to a substance at a lower temperature by *radiation*, by *conduction* where they are in contact, and by *convection* when heat is carried by a moving gas or liquid. Heat flow requires time.

Time required for heating or cooling depends upon a variety of factors. Neglecting surface resistance, the minimum time T in minutes required to change the initial temperature t_0 of an aggregate particle to an average temperature t_a when its surface temperature is held at t_s is given by

$$T = \frac{CD^2}{\alpha} \tag{1}$$

where C is a value from Table 1 depending upon the percentage of heating or cooling remaining to make t_a equal to t_s. Either centigrade or Fahrenheit temperatures may be used if used consistently. Typical values of thermal diffusivity α in square feet per hour are given in Table 2; and $\alpha = k/c\rho$ where k is conductivity in Btu/(hr)(sq ft)(°F/ft), c is specific heat in Btu/(lb)(°F), and ρ is density in pounds per cubic foot. This equation for α is only approximately correct for porous

Table 1. Values of C for Eq. (1)

% of heating or cooling remaining $\dfrac{t_a - t_s}{t_0 - t_s}$	C
0.001	0.0678
0.005	0.0510
0.01	0.0436
0.02	0.0361
0.03	0.0318
0.05	0.0264
0.10	0.0191
0.20	0.0118
0.30	0.0077
0.40	0.0051
0.50	0.0032
0.60	0.0019
0.70	0.0009
0.80	0.0004
0.90	0.0001

materials. D is particle diameter in inches. Root-mean-square values of D^2 for common mixtures of particle sizes are:

Diam range, in.	D^2
3–8	32.33
3–6	21.00
1½–3	5.25
¾–1½	1.31
No. 4 to ¾	0.27

and for typical blended aggregate for mass interior dam concrete are:

Diam range	D^2
No. 4 to 8 in.	12.34
No. 4 to 6 in.	8.64
No. 4 to 3 in.	2.50

These values of D^2 may be used with Eq. (1) to give minimum times for a mixture of particle sizes to reach an average temperature t_a.

Since time is proportional to D^2, note the effect of large aggregate sizes on cooling and heating times. If oversize cobble is permitted, minimum cooling and heating times are substantially increased. Inclusion of 10% of 9-in. "plums" in 3- to 6-in. aggregate increases the time for this material by 29%, for example. Cooling cobble graded 3 to 7 in. requires about 25% longer than cobble graded 3 to 6 in.

The assumption is often made that particles of the same material of any shape other than a sphere will cool more rapidly than a sphere. This is true only if the particle has the same volume as the sphere. Aggregate, however, is screened by linear dimension, and it is probable that cooling time is not greatly different from that for spheres. For example, cooling a 6 × 6 × 6-in. cube requires about 20% more time than a 6-in. sphere. Cooling a 6 × 6 × 12-in. "brick" requires about 65% more time than a 6-in. sphere. Cooling a 4 × 6 × 9-in. "brick" requires very nearly the same time as a 6-in. sphere.

In practice, heating and cooling times for concrete aggregates are always longer than the minimum times given by Eq. (1). The surface of the particles being heated or cooled cannot be maintained at the temperature of the surrounding medium. Often part of the aggregate surface will be in contact with the surface of other particles, exposing less than the full surface to the cooling medium. Also in most heating or cooling processes some time is required to change the temperature

Table 2. Thermal Characteristics of Concrete Materials

These are typical values. In some instances, as with aggregate, the values may vary considerably depending upon the sample

Material	Specific heat c		Conductivity k, Btu/(sq ft)(°F/ft)	Density ρ, pcf	Thermal diffusivity α, sq ft/hr
	Range of reported values	Typical value for computations			
Cement, portland............	0.12–0.22	0.22	0.17	94	0.008
Pozzolan....................	....	0.22			
Water......................	....	1.00	0.32	62.4	0.0052
Ice at 32°F.................	0.49–0.50	...	1.28	57	0.045
At 14°F..................	0.49	0.50	1.35		
At −4°F..................	0.47	...	1.41		
Sand, saturated surface-dry..	0.16–0.22	0.20	0.19	95	0.011
Sand, 4% moisture..........	....	0.20	0.40	100	0.020
Crushed rock, 4% moisture..	....	0.20			
Granite...................		...	0.62	110	0.030
Traprock.................		...	0.50	110	
Gravel, 4% moisture........	....	0.20	0.75	110	0.036
Aggregate particles, saturated surface-dry.............	....	0.20			
Basalt....................	0.18–0.24				
Cinders..................	0.18				
Dolomite.................	0.19–0.22	...	1.0	167	0.03
Gneiss...................	0.18–0.20				
Granite...................	0.18–0.20	...	1.6	168	0.050
Limestone................	0.18–0.22	...	1.2	160	0.030
Magnetite or hematite.....	0.16–0.17				
Marble...................	0.21–0.22	...	1.3	168	0.037
Rhyolite..................	0.18				
Quartzite.................	0.16				
Sandstone................	0.21–0.25	...	1.5	144	0.043
Serpentine...............	0.25–0.26				
Traprock.................	....	...	......	....	0.029
Concrete, hardened........	0.15–0.25	0.22	0.54	144	0.019
Concrete, wet, fresh........	0.22–0.28	0.25			
Grout, wet, fresh..........	0.27–0.38	0.32			
Steel.....................	0.11–0.12	0.12	26	490	0.48
Steam, constant pressure, 15 psia..................	....	0.48			
150 psia................	....	0.61			
Air at 14.7 psia............	....	0.24	0.014	0.081	0.72
Calcium chloride "standard solution" (29.12%) at 60°F.................	....	0.66	......	80.1	

of the surrounding medium to t_s. In addition, the cooling medium must be able to remove heat as rapidly as it can be given up by the aggregates.

Figure 1 shows typical practical cooling times for various processes. Of the cooling methods shown only inundation, vacuum, and air flow through aggregates in deep beds are presently used in large concrete plants in North America. The curves for tumbling and air flow through shallow beds are from tests by J. Poulter of the Koehring Company using high-velocity air flow. Vacuum-cooling tests are by E. Goldfarb, and tests of air cooling of deep aggregate beds are by T. B. Appel and G. L. Marks for the C. S. Johnson Division, Koehring Company. Time required to fill and empty tanks is excluded, but time to lower temperature of coolant in contact with the aggregate is included in Fig. 1. Minimum curves shown for 1½- to 3- and for 3- to 6-in. material are for α equal to 0.03.

Temperature (or dry-bulb temperature) is a measure of the property of a substance that determines whether it will gain or lose heat in proximity with another substance. Temperature is usually determined by a thermometer based on expansion or vapor pressure of a liquid. It may also be measured by expansion or pressure of a gas or by its electrical effect. While the 0 and 212°F points

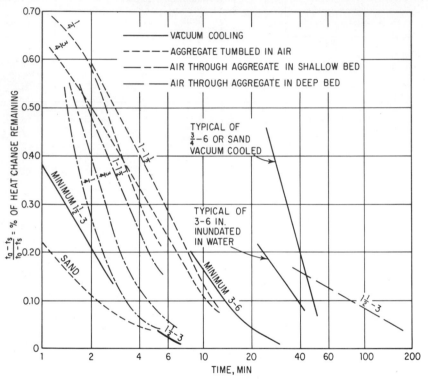

FIG. 1. Typical samples for vacuum cooling, inundation, and air cooling in deep beds are obtained with initial high surrounding temperatures (surface temperatures), which are then gradually reduced (depending on size of cooling units) to the final surface temperatures t_s on which the shown percentages are based. t_a = initial uniform aggregate temperature. t_0 = average aggregate temperature. t_s = temperature of cooling medium. Minimums shown are for ideal conditions with thermal diffusivity = 0.030.

can be determined accurately, intermediate points are somewhat arbitrary and may not agree exactly on different thermometers even when they are properly calibrated. Also thermometers can get out of calibration by several tenths of a degree as a result of alternate heating and cooling and must be calibrated against a standard for accurate work.

Temperature of water, flake or slush ice, sand, cement, pozzolan, and wet concrete can be taken by inserting a thermometer well into the material. In fresh concrete the thermometer bulb should not be against large aggregate because its temperature may be higher or lower than the other ingredients until sufficient time has passed for the temperature of the concrete to stabilize. If coarse aggregate is warmer or cooler than the mortar, temperature of the mortar may increase or decrease several degrees after concrete leaves the mixer. Temperatures of small and medium aggregates may be taken with a thermometer inserted in a sample in an insulated bag or container. Temperatures of block ice and aggregates over about 1 in. are determined in an insulated container with a hole in the lid for a thermometer (called a *calorimeter*) by mixing a known weight of ice or large aggregate with a known weight of water at a known temperature. For example, if a sample of aggregate weighing m lb and with a specific heat of c (see below) is placed in a calorimeter containing w lb of water at a temperature t, and the mixture is allowed to reach a stable temperature t_2, the initial temperature of the aggregate t_1 is computed from

$$t_1 = \frac{(w + K)(t_2 - t)}{cm} + t_2 \qquad (2)$$

K is the *water equivalent* of the calorimeter and is the amount of water that requires the same amount of heat to raise or lower its temperature 1°F as was required to raise or lower the calorimeter temperature 1°F. K is determined as follows: Fill the calorimeter with water at a temperature near that used in making tests (say 60°F). After about 10 min measure the temperature t_3, empty the calorimeter, and fill it with a known weight W of water at a higher temperature t_4. After an additional 10 min read the temperature t_2 of the water in the calorimeter:

$$K = \frac{W(t_4 - t_2)}{t_2 - t_3} \qquad (3)$$

K should approximate the weight of the calorimeter lining multiplied by its specific heat; K equals about 12% of the weight of the lining if it is steel.

A British thermal unit (Btu) is a unit of heat. One Btu will raise 1 lb of water 1°F, and 180 Btu are required to heat 1 lb of water from 32 to 212°F.

Specific heat is the number of Btu required to raise the temperature of 1 lb of a substance 1°F. The specific heat of water is 1. Table 2 gives specific heats of various materials. Specific heat of most solids and liquids varies only slightly with temperature insofar as concrete computations are concerned. Specific heat of cement is assumed to be 0.22, and that of aggregate is 0.20. These values are generally used in computations. Specific heat of aggregate is determined in a calorimeter when the weights and temperatures of aggregate and water are known.

Heat of fusion of water is heat absorbed by ice in melting or given off by water in freezing, 144 Btu per lb.

Heat of vaporization is heat absorbed in changing a liquid to vapor or given off by vapor in condensing to liquid, for water 970.2 Btu per lb at normal atmospheric pressure (14.7 psi and 212°F). Heat of vaporization of water is higher at lower temperatures, about 1,070 Btu per lb at 38°F. Because of its high specific heat and high heats of fusion and vaporization, water can store or release large amounts of heat when it changes temperature, as shown in Fig. 2.

Vapor pressure of water is the pressure exerted by water in a gaseous state (Table 3). It is also called *elastic pressure, gaseous pressure,* and *vapor tension.* This pressure exists at all temperatures when water is present. When the vapor pressure equals the pressure at the water surface, water vaporizes freely (boils).

Wet-bulb temperature is the minimum temperature to which water can be cooled by evaporation under existing conditions of pressure, dry-bulb temperature, and humidity. It is obtained with a *psychrometer,* which consists of a dry-bulb thermometer and a wet-bulb thermometer that has a wick saturated with distilled water. When fanned or whirled vigorously, the wet-bulb thermometer will show a wet-bulb temperature lower than the dry-bulb temperature. Relative humidity and actual vapor pressure can be obtained from tables such as the "Smithsonian Meteorological Tables" published by the Smithsonian Institution, Washington, D.C., if these temperatures and the barometric pressure are known.

Relative humidity is the actual vapor pressure divided by the saturated vapor pressure (Table 3) at the dry-bulb air temperature.

Dew point is the temperature at which the water vapor in the air is just sufficient to saturate the air (increase the relative humidity to 100%). Moisture condenses (changes from vapor to water) on objects with temperatures below the dew point in still air and condenses if the air temperature is reduced to the dew point, releasing the heat of vaporization, about 1,000 Btu per lb of water condensed.

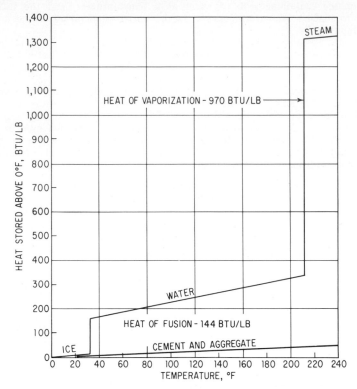

FIG. 2.   Heat content of concrete materials under a pressure of 14.7 psi.

Boiler horsepower is a unit of steam boiler capacity equal to 33,472 Btu per hr.

Steam quality is a measure of the liquid water carried in steam. The liquid water has lower heat-carrying capacity than water vapor. Quality is the dry steam (vapor) weight per pound of wet steam.

A ton of refrigeration is the cooling necessary to make 2,000 lb of ice from 32°F water in 24 hr. It is equivalent to 12,000 Btu per hr.

2   Typical Material Temperatures

Cement is usually discharged from the grinding mills at 200 to 300°F and may be cooled to 120 to 180°F in cement coolers before being placed in storage. Thermal conductivity of dry cement is fairly low, about 2 Btu/(hr)(sq ft)(°F/in.), and it loses heat very slowly in storage. A typical loss is 3 to 10°F per month in large concrete silos, slightly more in large steel silos. A further reduction of about 10°F may occur as cement is handled for shipment. Shipping temperatures are usually 120 to 180°F, with the higher temperatures probable in late summer

Table 3. Vapor Pressure of Water*

Air temp t, °F	Vapor pressure V, in. Hg	t	V	t	V
30	0.164	60	0.517	90	1.408
31	0.172	61	0.536	91	1.453
32	0.180	62	0.555	92	1.499
33	0.187	63	0.575	93	1.546
34	0.195	64	0.595	94	1.595
35	0.203	65	0.616	95	1.645
36	0.211	66	0.638	96	1.696
37	0.219	67	0.661	97	1.749
38	0.228	68	0.684	98	1.803
39	0.237	69	0.707	99	1.859
40	0.247	70	0.732	100	1.916
41	0.256	71	0.757	101	1.975
42	0.266	72	0.783	102	2.035
43	0.277	73	0.810	103	2.097
44	0.287	74	0.838	104	2.160
45	0.298	75	0.866	105	2.225
46	0.310	76	0.896	106	2.292
47	0.322	77	0.926	107	2.360
48	0.334	78	0.957	108	2.431
49	0.347	79	0.989	109	2.503
50	0.360	80	1.022	110	2.576
51	0.373	81	1.056	111	2.652
52	0.387	82	1.091	112	2.730
53	0.402	83	1.127	113	2.810
54	0.417	84	1.163	114	2.891
55	0.432	85	1.201	115	2.975
56	0.448	86	1.241	116	3.061
57	0.465	87	1.281	117	3.148
58	0.482	88	1.322	118	3.239
59	0.499	89	1.364	119	3.331

* U.S. Weather Bureau.

or when cement stocks are low and shipments are made from recently ground supplies. Shipping temperatures as high as 300°F are not unknown but are not common because cement is nearly always cooled after grinding. There is often little temperature loss in shipment (but it may be 20°F or more), and unloading (by other than compressed air) may reduce temperatures by a few degrees. Thus cement usually arrives at a concrete plant at about 90 to 150°F, and occasionally as high as 170°F.

Water from wells typically has a temperature of about the mean annual air temperature, but there is a large variation. For example, a study of about 700 wells by the Illinois State Water Survey shows an average temperature of 56°F at the pump discharge, but a range from 47 to 77°F. Water from surface streams commonly approximates the mean monthly air temperature but may be 5 to 15°F lower in spring and summer and 5 to 15°F higher in fall and winter.

Aggregate is usually assumed to be at approximately the mean monthly temperature of air in the shade as reported by the Weather Bureau. It may be 10 to 25°F warmer than this if dry and stored or handled in the sun, and

2 to 10°F cooler if sprayed with water or stored moist in shade. Aggregate that is processed by washing and placed in large storage piles without delay tends to remain at the wash-water temperature for several days.

3 Heat Balance

Formulas for concrete temperature are complicated because of the heat of fusion of water and the variety of materials. They can be derived as needed by remembering that the number of Btu above the final temperature of the mixture must equal the Btu below the final temperature, or computed as in the following example:

Material	lb/ cu yd	Specific heat above 32°F	Btu/°F above 32°F	Initial temp, °F	Btu/cu yd above or below 32°F (ice melted)
Cement............	516	0.22	113.5	80	113.5(80 − 32) = 5,448
Sand.............	1,340	0.20	268.0	18	268(18 − 32) = −3,752
Moisture (ice) in sand, 6%	80.4	1.00	80.4	18	80.4[0.50*(18 − 32) − 144*] = −12,140
Coarse............	1,885	0.20	377.0	20	377(20 − 32) = −4,524
Moisture (ice) in coarse, 1%.......	18.9	1.00	18.9	20	18.9[0.50*(20 − 32) − 144*] = −2,835
Added water (270 − 80.4 − 18.9).....	170.7	1.00	170.7	160	170.7(160 − 32) = 21,850
	4,011		4,011)1,028.5		1,028.5)4,047
Specific heat of wet concrete = 0.256					3.9 +32.0
				Concrete temp =	35.9°F

* Specific heat of ice is 0.50; heat of fusion is 144 Btu per lb. Notice the important effect of ice in the aggregate, that hot water is an important source of heat, and that it is not easy to make warm concrete with frozen moisture in the aggregate. Computations such as this are accurate only if temperatures, specific heats, and material weights are accurate. For this example of typical ready-mix concrete:

An error of	*Changes concrete temp, °F, by*
1°F in water temp...................................	0.17
1°F in sand temp....................................	0.30
1°F in coarse-aggregate temp.........................	0.38
1°F in cement temp.................................	0.11
1% in sand moisture................................	1.83
1% in coarse-aggregate moisture......................	2.66
1% in water weight.................................	0.21
1% in sand weight.................................	0.17
1% in coarse-aggregate weight........................	0.09
1% in cement weight...............................	0.05
0.01 in specific heat of sand........................	0.23
0.01 in specific heat of coarse aggregate...............	0.29
0.01 in specific heat of cement.......................	0.22

Two quarts of standard calcium chloride solution at 40°F would lower the concrete temperature about 0.5°F. These changes apply only to this set of computations but indicate the magnitude of changes. When the temperature of any material is near the final concrete temperature, the effect of an error in specific heat becomes less important.

HEATING CONCRETE

4 Heating Water

Figure 3 shows mix-water temperatures required to produce 60°F concrete with aggregate at various temperatures and moisture contents for a typical commercial mix. Moisture percentage for all aggregate is usually near half the moisture percentage for the sand. Unless sand is quite dry, hot water alone will not produce 60°F concrete from frozen aggregate. It is customary to use heated water if air temperatures are below 45°F.

Flash set results when cement contacts water that is too hot. This may occur at relatively low temperatures if cement and water are combined directly, and can be avoided by mixing aggregate and water together for about 15 to 20 sec before adding cement if either the water or aggregate is hotter than 100°F. This increases mixing time, both because of the above delay and because the cement and aggregate are not blended before reaching the mixer. Water as hot as 150°F is mixed with blended cement and cold aggregate in some plants, and water as hot as 200°F can be used by mixing the hot water with cold aggregate for about 30 sec before adding cement.

Heat required depends upon temperatures and plant capacity. Commercial concrete is usually delivered at 55 to 75°F when air temperatures are near freezing, and 60 to 80°F when temperatures are substantially below freezing. Mass concrete is usually produced at 40 to 60°F. A plant heating 60°F water to 180°F, using 200 lb per cu yd of added water, and producing 100 cu yd per hr needs 100 × 200(180 − 60) = 2,400,000 Btu per hr. Allowing 15% for losses in piping, this requires 2,400,000 × 1.15 ÷ 33,472 = 82.5 bhp minimum, or about

$$2{,}400{,}000 \times 1.15 \div 1{,}015.6 = 2{,}718 \text{ lb}$$

per hr of steam at 15 psig at sea level if the steam temperature is reduced to 180°F in heating the water and steam quality is 100%. The value 1,015.6 is obtained from pressure tables for saturated (100% quality) steam, and is the difference in total heat of 1 lb of steam at 29.7 psia and the total heat of water at 180°F. If steam quality is 90%, 2,996 lb of steam is required per hour.

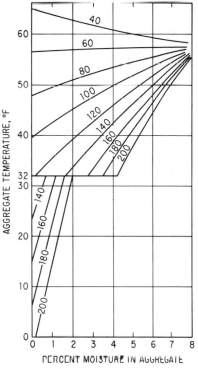

Fig. 3. Required temperature of added water to produce concrete with a temperature of 60°F. Batch weights (per cubic yard) are cement 516 lb at 80°F, aggregate (SSD) 3,225 lb at temperatures shown, and water 270 lb total.

Sparger nozzles in a water reservoir will heat water rapidly by dispersing steam in the water, where it condenses and remains. A 2-in. pipe nozzle with 32 holes ¼ in. in diameter, with deflectors to mix steam and water, will heat 24 gpm through an 80°F rise at 5 psig steam pressure at the nozzle and will heat 36 gpm at 15 psig. A steam valve controlled by a thermostat regulates water temperature. If boiler feedwater does not require treatment this is the most economical method of heating large quantities of water rapidly, but when feedwater treatment is required to prevent boiler scale or sludge, the large quantities of steam used make the water treatment too expensive. About 1 lb of steam is used for each 8 lb of batch water at 180°F. Heating with live steam cannot be done in a batcher because water weight is increased by the added steam. The reservoir where the water is heated should have a vent and overflow pipe.

Steam coils in a water reservoir heat the water without loss of boiler water because the condensed steam is returned to the boiler. Feedwater treatment is required to make up losses due to leaks only. The required coil surface area in square feet is given by

$$\frac{W}{U} \ln \frac{t_s - t_1}{t_s - t_2} \tag{4}$$

where W is pounds of water heated per hour, t_s is steam temperature at the pressure supplied, t_1 is initial water temperature, and t_2 is final water temperature, all in degrees Fahrenheit. U is the coefficient of heat transmission in Btu per square foot of coil surface area per hour. From condensing steam to stationary water in a tank, U is usually about 125 for thin tubes and about 100 for pipe, perhaps as low as 50 for old steel pipe. U values for hot water to water would be about half those for steam to water. In the example above for 100 cu yd per hr production and 200 lb per cu yd added water heated from 60 to 180°F using 15 psig steam, the pipe coil outside surface area required would be about

$$\frac{20,000 \times 1}{100} \ln \frac{249.75 - 60}{249.75 - 180} = 200 \text{ sq ft}$$

and that is about 266 ft of 2½-in. pipe. Smaller pipe would cause excessive pressure loss. Pressure losses can be reduced by using a system of coils in parallel.

In-line water heaters have steam coils built into an enlarged section of water pipe. They are installed on the inlet to a water reservoir and heat the water as it flows through the pipe automatically with thermostatic control. Heat transfer is improved because both water and steam are moving, so much less heating surface is required than with tank coils. Some auxiliary heat in the reservoir may be required for start-up and to make up reservoir heat losses during erratic production.

Package water heaters with water coils, either light-oil- or gas-fired, with automatic start-up and thermostatic control are convenient for heating water where little other heating is required. They start fast, recover rapidly, and are lower in cost and occupy less space than boilers. Efficiency is about 70% of fuel input. Water treatment is less essential than with boilers or steam generators because fewer minerals are precipitated at normal heating temperatures. Typical units are in the 175,000 to 1,000,000 Btu per hr heating range (5 to 30 bhp).

Steam generators of the flash type produce steam in a few minutes from a cold start. They are usually gas-fired. They are compact and low in cost but of limited capacity and durability. Feedwater treatment consisting of at least softening is needed in most areas. Typical capacities are 5 to 30 bhp.

Package or **scotch boilers** are multiple-pass fire-tube units, automatic in operation and typically available from 50 to 300 bhp with insulation, burners, and controls installed. Larger units from 400 to 700 bhp are also available. Usual fuels are natural gas regulated to 1,000 Btu per cu ft or fuel oils in grades 1 to 3 (light, medium, and heavy domestic) or grades 4 to 6 (light, medium, and heavy industrial). Grades 5 and 6 are also called "Bunker B" and "Bunker C." Typical caloric value is 18,500 to 20,000 Btu per lb (140,000 to 150,000 Btu per U.S. gal). Industrial-grade oils usually require heating before use. Coal, now seldom used in concrete plants, has caloric values of 7,000 to 14,000 Btu per lb. Boiler fuel-conversion efficiency is about 60 to 75% for stoker-fired coal, 70 to 80% for oil or gas. These efficiencies are for steady load, and fuel consumption is higher in practice because of warm-up and operation at part load. Large amounts of fresh air are required for combustion, about 10 cu ft per cu ft of gas, for example. While package boilers can be used to heat water or oil, they normally make steam. Whether high- (above 15 psig) or low-pressure boilers are used depends somewhat upon local regulations, amount of pipe, and steam quality required. Operators are frequently required for boilers operating over a fixed pressure or capacity. Some steam systems operate with more efficiency slightly above or below atmospheric pressure by using vacuum pumps on condensate return lines. Feed-water treatment is nearly always required to protect the investment in and efficiency of a good boiler.

Fixed boilers are high-capacity built-in units most suitable for permanent plants where large quantities of steam are required. Other comments on package boilers apply to these as well.

Storage tanks are useful in reducing the size of heating equipment required by allowing it to operate over longer periods. Tanks of 1,000 to 1,500 gal capacity are used in smaller plants, tanks up to 20,000 gal in large plants. The tanks and piping should be insulated. Large tanks are sometimes buried 3 ft or more

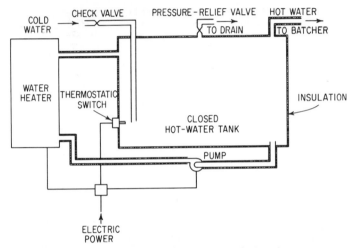

Fig. 4. Connections for a closed-tank, circulating hot-water system.

below ground, which provides effective insulation, and heated at night when heating equipment is operating at less than capacity. With small tanks temperature of mix water is regulated by the heater thermostat. Large tanks are better kept at a temperature as high as the maximum used and the batch water temperature regulated by a thermostatic mixing valve, which blends hot and cold water to produce the set temperature. Heating of a closed tank (kept full of water under pressure) is by steam coils or by circulating the water through a water heater with a pump typically 30 to 50 gpm at 80- to 120-ft head (Fig. 4). The pump should move about the quantity that can be heated through a 40°F rise at the pressure needed to overcome friction in the heater and piping. Buried tanks must be kept nearly full or well anchored to prevent their "floating" out of the ground. Small open tanks (not under pressure) are usually located above the batchers and filled through a water heater, by circulating water through a heater, or with steam coils.

Storage-tank size is determined by demand and heating capacity. For example, a ready-mix plant makes the following amounts of concrete for 10 successive hours of a peak 500 cu yd winter day: 94, 70, 65.5, 46.5, 59, 44.5, 46, 49.5, 20, 5. Water in the tank is heated to 175°F at the start of the day and is to be kept above 165°F minimum. The water heater or steam valve will cut in at 174°F and cut out at 176°F. An average cubic yard uses 230 lb of added water, and supply water is at 60°F. A heater that will provide something over half the load for the peak hour is desirable: $\frac{1}{2} \times 94 \times 230(170 - 60) = 1{,}189{,}100$ Btu per hr heating required. Say 1,500,000 Btu per hr heater capacity is provided. This will make 165°F water faster than it is used if production is less than

$$\frac{1{,}500{,}000}{230(165 - 60)} = 62.1 \text{ cu yd/hr}$$

so for a period of 3 hr the demand will exceed the supply. During this period approximately $(94 + 70 + 65.5)(230)(170 - 60) = 5{,}806{,}350$ Btu will be used and $1{,}500{,}000 \times 3 = 4{,}500{,}000$ Btu heated; so 1,306,350 Btu will have to come from storage without reducing tank temperature more than 10°F. Thus a minimum of $1{,}306{,}350 \div 10 = 130{,}635$ lb of water or 16,100 U.S. gal storage will be needed (water weighs 8.13 lb per gal at 170°F).

By limiting the amount of water fed into the reservoir to

$$1{,}500{,}000/(170 - 60 = 13{,}636 \text{ lb per hr}$$

(good for 59.3 cu yd per hr) and using the stored hot water when production exceeds 59.3 cu yd per hr, the minimum reservoir size for the above application (using a 1,500,000 Btu per hr heater) is

$$(94 + 70 + 65.5)(230) - (3 \times 13{,}636) = 11{,}877 \text{ lb}$$

or 1,461 U.S. gal storage needed.

If the heating unit can be operated 24 hr per day another arrangement which could possibly be the most economical would be to size the heater for the total heating required over 24 hr, $500 \times 230 \times (170 - 60) \div 24 = 527{,}083$ Btu per hr, say a 600,000 Btu per hr heater, and limit the water rate into the reservoir to $600{,}000/(170 - 60) = 5{,}455$ lb per hr (good for 23.7 cu yd per hr). The minimum storage required is then

$$(94 + 70 + 65.5 + 46.5 + 44.5 + 59 + 46 + 49.5)(230) - (8 \times 5{,}455)$$
$$= 65{,}610 \text{ lb or } 8{,}070 \text{ U.S. gal}$$

In the first example given above, it is assumed that the storage tank will be kept full of water at all times at a temperature between 165 and 175°F. This would ensure availability of more than a 10-hr supply of hot water if the heating system had to be shut down for any reason. In the second example the storage tank is full at the beginning of the day but would be almost empty at the end of the first 3 hr. After the third hour water would be entering the tank faster than it is used and at the end of the day the tank would be full. In the third example the storage tank is full at the beginning of the day but would be almost empty at the end of 8 hr. At the end of the day there would be 5,160 lb or about 63.5 U.S. gal in the tank. About 11 hr would be required to finish filling the tank for the next day.

Heat stored by a steel tank is negligible. If a 20,000-gal tank 12 ft in diameter by 24 ft long is used it has a surface of 1,131 sq ft. Two inches of polystyrene-foam insulation with a k value of 0.32 (see Table 4; k is increased by 0.05 because of higher temperature) and a 30°F air temperature will have a heat loss of about

$$0.32 \times 1{,}131 \frac{175 - 30}{2} = 26{,}240 \text{ Btu/hr}$$

and 20,000 gal $\times 8.13 \times 1 = 163{,}000$ Btu per °F, so tank temperature will drop about $26{,}240 \div 163{,}000 = 0.16°$F per hr if no water is used. If thick insulation were used, the mean area of the insulation would be used in the above computation in place of the tank surface area.

Estimation of heat loss from buried water tanks is complicated. Burial under 3 ft of earth is roughly equivalent to $1\frac{1}{2}$ to 3 in. of insulation, under 8 ft of earth equivalent to 2 to 4 in. of insulation, for 5,000- to 30,000-gal tanks. The higher equivalent insulation thicknesses apply to larger tanks with the smaller diameter range for each size.

5 Heating Aggregate

Heating of frozen aggregate to a temperature above 32°F is usually required to make warm concrete. Aggregate should not be heated above about 100°F because of the danger of flash set, and if all aggregate were heated to this temperature,

Table 4. Thermal Conductivity k of Various Materials

For most insulating materials k increases with density and also increases about 0.05 per 100°F. Most insulating materials become less effective as temperatures increase. Values given are for about 70°F and for typically low densities for insulating materials

Material	Approx k, $Btu/(hr)$ $(sq\,ft)(°F/in.)$ thickness
Urethane foam	0.15
Polystyrene foam	0.27
Commercial insulating blanket materials of vegetable fiber, hair, glass fiber, or rock fiber	0.28
Corkboard	0.30
Board of pressed corn, cane, or wood fiber	0.33
Glass foam	0.37
Sawdust, wood shavings	0.40
Wood across grain: White pine, fir, redwood	0.80
Cypress	0.90
Yellow pine	1.0
Oak	1.1
Air space greater than $\frac{1}{2}$ in.	1.2
Sand or gravel, dry	2.3
Sand or gravel, moist	5–15
Soil, wet	5–13
Concrete, regular, dry	6–10

the concrete would be too hot. If aggregate can be kept above 32°F in storage, additional heating can be kept to a minimum. Heat required to thaw and warm frozen aggregate is given by

$$\text{Btu} = A[0.20(t_2 - t_1) + p(128 - 0.5t_1 + t_2)] \tag{5}$$

where A is the weight of saturated surface-dry aggregate, t_1 and t_2 are initial and final temperatures in degrees Fahrenheit, and p is surface moisture expressed as a decimal. If the aggregate weight available is total weight including moisture, substitute $A/(1 + p)$ for A. Heating 3,200 lb of aggregate plus 3% average surface moisture at 30 to 60°F for a typical cubic yard of concrete requires about 35,800 Btu or about 36 lb of steam. Total boiler capacity for aggregate and water heating combined is in the range of 1 to $2\frac{1}{2}$ bhp per cu yd per hr of concrete depending upon severity of winter temperatures. Heating aggregate in covered stockpiles may be necessary to permit handling in cold weather, to permit moisture content to stabilize, or to permit rescreening of the aggregate before it is placed in the bin. Stockpile heating also permits round-the-clock heating for more efficient use of heating capacity and may be essential where small bins are used in high-production plants because of the short retention time in the bin.

Live steam jets at pressures of 10 to 25 psi in storage bins will heat aggregate rapidly because steam disperses through the material and condenses on the aggregate particles, giving up the heat of vaporization. Time required for heating is about five times that given by Eq. (1). The condensed steam usually raises the moisture content of the aggregate to saturation and water drains from the material. Total water in concrete is in the range of 6 to 12% of the aggregate weight. When moisture reaches this level, no added batch water is needed and aggregate must then be heated to an average temperature near the final concrete temperature without having the water in the aggregate exceed the requirement for total water. If live-steam heating is continued long enough, temperature of aggregate near the nozzles will approach 212°F, while aggregates farther away are cooler, resulting in uneven heating. Boiler feedwater treatment may be expensive because of the large amount of makeup water required. Despite these objections, steam jets are a useful method, but control of the heating process and control of water in the mix become difficult. Two or more ½-in. pipe fittings with ½-in. globe valves on each bin compartment with a 2-in. steam header are usually adequate. Drain channels can be built into a bin to remove much of the water drainage before it reaches the fill gates, but some water will nearly always drip from the gates. Live-steam heating of storage piles where some remixing and drainage take place before material is delivered to an overhead bin is more easily managed.

Steam, hot-water, or hot-oil coils in the aggregate heat the material slowly because sand and gravel are relatively poor heat conductors. This method tends to dry the aggregate some and stabilize the moisture content and so has few of the objectionable features of live-steam heating. Also the aggregate must be heated only sufficiently to melt any ice because generous quantities of hot batch water can be used. Where usage is moderate and concrete mixes dry, as in products plants, coil heating may be the only practical method. Coils should be located near the center of a bin or pile to minimize heat loss and near the main material flow. When used in stockpiles to prevent freezing slowness of heating is not so objectionable, but frozen lumps may still be drawn down through the piles without thawing unless coils are located to intercept them. The amount of coil surface area in square feet is given by

$$\frac{0.20A}{U} \ln \frac{t_s - t_1}{t_s - t_2} \tag{6}$$

where A is pounds of aggregate heated per hour, t_s is steam, water, or oil temperature, t_1 is initial aggregate temperature, and t_2 is final aggregate temperature, all in degrees Fahrenheit. U is heat transmitted in Btu per square foot of coil surface area per hour, and is probably in the range of 2 to 10. Oil heated in a boiler similar to a steam boiler or heater similar to a water heater has an advantage in that corrosion, need for water treatment, and danger of coil freezing are eliminated. The specific heat of oil is about half that of water and viscosity is higher than water requiring more pump capacity or larger pipe. Temperatures up to 600°F are possible but temperatures over 212°F will have the undesirable effect of removing absorbed water from the aggregate. A pump is required for circulation of oil or hot water. The possibility of oil leakage into the aggregate is the main objection to the use of oil.

Electric resistance-heating elements attached to a bin bottom and covered by insulation can effectively prevent freezing in the lower part of the bin and are easily controlled by a thermostat. Each 1,000 watts of heating element supplies 3,413 Btu per hr, of which about 90% may reach the aggregate.

Radiant heaters or space heaters below a bin bottom provide an effective means of keeping aggregate temperature above freezing but usually do not produce enough heat to melt ice at any appreciable distance from the bin bottom. If a bin is enclosed and the enclosure is spaced out from the sides of the bin, warm air from the batcher space below the bin will envelop the bin, provide some heat, and effectively prevent freezing if aggregate can be delivered from a stockpile at temperatures above freezing.

Inundation heating by immersing aggregates in hot water is seldom used. Dewatering of the aggregate is difficult.

Vacuum heating wherein air is partially exhausted from a tank of aggregate and replaced by steam is very effective but too expensive to be used unless the same equipment is provided for cooling. If a tank containing 400,000 lb of aggregate at 20°F and 1% surface moisture is reduced to a pressure of 5 psia and the air removed replaced by 2,900 lb of steam at 144 psia the steam condenses on the aggregate and raises the temperature to about 50°F. Air is then introduced until the tank returns to atmospheric pressure. Moisture is increased from 1 to about 1.8%, and some of the increase in moisture can be removed by a barometric drain or pump at the bottom of the tank. Heating is fairly uniform because steam is drawn into the mass.

Warm-air heating is feasible for coarse aggregate but quite expensive and seldom used.

COOLING CONCRETE

Commercial concrete usually can be produced at reasonable temperatures in hot weather without use of refrigeration by using the coldest water available (from wells, for example), by shading and sprinkling aggregate piles, and by painting storage, handling, and mixing equipment white if it is exposed to the sun. While most common surfaces absorb solar radiation nearly as well as they radiate heat at lower temperatures, and some metals absorb solar heat more efficiently than they radiate, white paint radiates heat at 100°F several times as efficiently as it absorbs heat from the sun, and white-painted surfaces will remain cooler than other surfaces in sunlight.

When concrete must be cooler than about 80°F in hot weather, refrigeration is usually required. Cooling by refrigeration is more expensive than heating and requires more expensive equipment, so equipment must be accurately sized. It is even less economical to provide cooling for the maximum possible temperatures than to provide heating for the minimum possible temperatures. Some calculated risk must be taken after study of reported climatologic data for the area and the cost of providing excess cooling capacity. Even then it must be assumed that combinations of climatic conditions can occur that will make operation at the design temperature impossible.

The following heat balance for a typical 6-in.-maximum exterior dam will be used for illustration, but actual mix designs should be used when available. Ma-

Material	Lb/cu yd	Specific heat, Btu/°F	Btu/cu yd/°F	Initial temp	Btu/cu yd above 48°F
Cement......................	310	0.22	68.2	120	4,910
Pozzolan.....................	58	0.22	12.8	120	922
Sand........................	780	0.20	156.0	90	6,552
Water in sand 8 % maximum...	62.4	1.00	62.4	90	2,621
Aggregate:					
3–6 in.....................	964	0.20	192.8	90	8,097
1½–3 in....................	800	0.20	160.0	90	6,720
¾–1½ in...................	584	0.20	116.8	90	4,906
No. 4 to ¾ in..............	506	0.20	101.2	90	4,250
Water in aggregate, ½ % ¾ –6..	11.7	1.00	11.7	90	491
2 % No. 4 to ¾.............	10.1	1.00	10.1	90	424
Mix water (180 − 62.4 − 21.8)	95.8	1.00	95.8	70	2,108
Concrete weight	4,182		)987.8		42,001
Specific heat of concrete			0.236		
Heat generated in mixing 0.70 × 10 hp/cu yd × 42.4 Btu/hp min × 2.5 =					742
(assumes 70 % of rated motor power goes into concrete)					
Heat gain in screening and handling materials.........................					1,000
				987.8)	43,743
					44.3
					48.0
Concrete temperature out of mixer without cooling.....................					92.3°F

terial temperatures and moisture content of aggregates will also vary depending upon the method of processing and cooling and upon the local climate. Concrete has a maximum placing temperature of 50°F in this example. Allowing for 2°F temperature in transporting to the forms, concrete must come out of the mixers at 48°F.

Cooling required per 100 cu yd per hr of concrete is $100 \times 43,743/12,000 = 365$ tons of refrigeration. About 15% of this can be accomplished by sprinkling coarse aggregate; so a minimum of 310 tons of mechanical refrigeration will be required per 100 cu yd per hr.

The effect of errors in estimating temperature and specific heat of coarse aggregate is most important. For this typical mass-concrete mix:

An error of	Changes the cooling required, %, by
1 min in mixing time	0.68
1°F in water temp	0.22
1°F in sand temp	0.50
1°F in coarse-aggregate temp	1.35
1°F in cement temp	0.16
1 % in sand moisture	0.55
1 % in No. 4 to ¾ moisture	0.38
1 % in water weight	0.05
1 % in sand weight	0.21
1 % in weight of one of coarse aggregates	0.11–0.19
1 % in cement weight	0.11
0.01 in specific heat of sand	0.75
0.01 in specific heat of coarse aggregate	2.74
0.01 in specific heat of cement	0.51

At least 43,700 Btu per cu yd must be removed by cooling to make 48°F concrete. There are several ways of removing heat; in approximate order of increasing cost:

	Approx. practical concrete temp. reduction, °F
1. Sprinkle coarse-aggregate stockpiles with water (10°F reduction)	6
2. Chill mix water to 34 or 35°F	3
3. Substitute ice for 80 % of chilled mix water	12
4. Cool coarse aggregate by vacuum to 35 or 38°F	31
5. Cool ¾- to 6-in. coarse aggregate by air to 40°F	25
6. Cool coarse aggregate by inundation to 40°F	30
7. Cool sand to 34 to 80°F	2–12
8. Cool cement to 80°F	3

Sand can be cooled to about 50 to 70°F by contact coolers. The higher-temperature reduction is for vacuum cooling of sand in bulk. Ice is limited to about 80% of the added mix water because some water must be used to dilute admixtures.

Some combination of these methods adding up to more than $92.3 - 48 = 44.3$°F will be required, and the combinations cannot include various methods of cooling the same materials. For example, vacuum cooling of coarse aggregate and sand plus chilled mix water gives $31 + 12 + 3 = 46$. Cooling coarse aggregate by inundation, chilled mix water, and substitution of ice for 80% of mix water gives $30 + 3 + 12 = 45$. As a practical matter in this example coarse aggregate must be cooled to make 48°F concrete.

6 Plant Start-up

The time of plant start-up is important in making cool concrete regardless of the method of cooling used. It is best to schedule the start of pours for early morning hours when air and material temperatures are lowest. After starting with proper concrete temperatures, the plant will take higher ambient temperatures up to the design temperature in stride. If a plant must be started late in a hot day, all materials must be as cold as possible and equipment in contact with the materials or concrete such as batchers, mixers, hoppers, and buckets must be washed thoroughly with cold water.

7 Sprinkling Coarse Aggregate

Sprinkling coarse aggregate with clean water will, under most conditions, perform a substantial part of the coarse-aggregate cooling, reducing refrigeration requirements. Sprinkling may be intermittent on a short time cycle but must be sufficient to keep the aggregate continuously wet, which requires more than enough water to replace moisture evaporated. Where the active zone (Fig. 5) of a large-aggregate

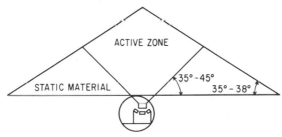

FIG. 5.   Active zone of aggregate storage pile.

storage pile holds more than is used in 1 day, the temperature of the pile can be approximated by balancing the heat gained and lost from material additions, radiation, convection, and evaporation. With all temperatures above 32°F, a formula for approximate pile temperature t in °F is

$$t = \frac{t_x A(s+p) + t_x'A'(s+p') + t_w W + t_w'W' + R' + R + t_a C + t_a'C' - HE - H'E'}{A(s+p) + A'(s+p') + W + W' + C + C'}$$

(7)

where t_x = daytime temperature of aggregate added to pile
   t_w = daytime temperature of water sprinkled on pile
   t_a = daytime temperature of air
   A = lb aggregate added during day
   W = lb water sprinkled during day
   R = Btu radiation gain (or loss) during day
   C = convective-heat gain (or loss) during day, Btu per °F
   s = specific heat of aggregate
   p = percentage of moisture in aggregate as received at the pile during the day, expressed as a decimal
   H = heat of vaporization of water at temperature t_w, Btu per lb
   E = lb water evaporated during the day

Prime symbols are the same except that they are for night conditions. In some applications it is necessary to consider day and night conditions separately as above, but where $t_x = t_x'$, $A = A'$, $p = p'$, $t_w = t_w'$, $W = W'$, and where operation is continuous with day and night of equal length, this formula reduces to

$$t = \frac{2t_x A(s+p) + 2t_w W + R + R' + t_a C + t_a'C' - HE - H'E'}{2A(s+p) + 2W + C + C'}$$

(8)

with A, W, E, and E' in pounds per hour, R and R' in Btu per hour, and C and C' in Btu per hour per °F.

The difficulty in applying the above formulas is in evaluating evaporation, radiation, and convection, as discussed below.

Evaporation occurs when the dew-point temperature of water vapor in the air is less than the water temperature. Approximate evaporation for large areas, adapted from the formula developed be R. E. Horton (*Engineering News-Record,* Apr. 26, 1917), is

$$E_i = 0.0167(WV_w - V_a) \tag{9}$$

where E_i is evaporation in inches per hour, V_w is vapor pressure in inches of mercury at the water temperature t_w (Table 3), V_a is vapor pressure in inches of mercury at the air temperature t_a, and W is a factor from Table 5 to correct for wind velocity and convective vapor removal at low wind velocities. Actual evaporation over small areas can reach a value of $E_i = 0.0167(WV_w - hV_a)$, where h is relative humidity, when the vapor is removed rapidly as on the windward side of a pile; otherwise the accumulation of vapor near the surface increases h to 1.00. Evaporation always occurs if $WV_w > V_a$. When $V_a > WV_w > hV_a$, some evaporation occurs on windward areas but cannot be relied upon for cooling aggregate piles. When $hV_a > V_w$, condensation occurs in still air. Evaporation with a 10-mph wind, 70% relative humidity, 80°F air temperature, and 70°F sprinkler water is

$$E_i = 0.0167[(1.86 \times 0.732) - (0.70 \times 1.022)] = 0.11 \text{ in./hr}$$

maximum over small areas, and a reasonable figure for large piles is

$$E_i = 0.0167[(1.86 \times 0.732) - 1.022] = 0.0057 \text{ in./hr}$$

With an active pile area of 30,000 sq ft, this is 890 lb per hr of water; and at 1,053 Btu per lb, 940,000 Btu per hr is removed by evaporation. Evaporation increases rapidly at higher water temperatures or lower air temperatures.

Table 5. Evaporation from Large Areas.   Values of Factor W

Wind velocity, mph	0	1	2	3	5	10	15	30 or more
$t_w \leq t_a$	1.00	1.18	1.34	1.45	1.64	1.86	1.95	2.00
$t_w - t_a = 1$	1.18	1.34	1.45	1.64	1.64	1.86	1.95	2.00
$t_w - t_a = 5$	1.38	1.49	1.58	1.64	1.64	1.86	1.95	2.00
$t_w - t_a = 10$	1.48	1.57	1.64	1.64	1.64	1.86	1.95	2.00
$t_w - t_a = 20$	1.64	1.64	1.64	1.64	1.64	1.86	1.95	2.00

t_w = temperature of water.
t_a = temperature of air.

Radiation of heat from the sun and sky generally increases aggregate temperatures during the day, and aggregate piles may lose heat by radiation at night. Radiation gain varies with season, time of day, altitude, latitude, and weather. Maximum radiation at the earth's surface at altitudes up to a few thousand feet is about 250 to occasionally 300 Btu per (sq ft)(hr) and varies but little with latitude. In mid-latitudes there is little variation in the maximum on a normal surface summer and winter. Average daytime radiation is about 58% of the peak values. Average total radiation considering weather at mid-latitudes is about 750 Btu per (sq ft)(day) in winter and about 2,000 Btu per (sq ft)(day) in summer. The average exposed normal area of a typical wedge-shaped aggregate pile three times as long as its width during a day is 67 to 71% of the horizontal area, depending upon pile orientation. Resultant radiation at the surface of the pile is about 2,000 Btu per (sq ft)(day) or about 145 Btu per (sq ft)(hr) average during a summer day.

The ability of a material to absorb (or emit) thermal radiation is its *emissivity* B expressed as a percent of the radiation absorbed by a perfect absorber. B

values for aggregates with an effective increase because of roughness vary from about 0.40 to 0.90. Assuming a typical effective B of 0.70, about 100 Btu per hr is absorbed by the aggregate per square foot of horizontal pile area.

Heat from the pile is lost by radiation, mostly to the sky but some to other surroundings, if their temperature is lower than that of the aggregate-pile surface. Net heat loss is on the order of $17.3 \times 10^{-10}B(t_1{}^4 - t_2{}^4)$, where B is emissivity of the aggregate and t_1 and t_2 are absolute Fahrenheit temperatures of the aggregate surface and of the surroundings, respectively ($t = 459 + °F$). The daytime effective sky temperature is a few degrees above air temperature. A cloudy night sky has an effective temperature a degree or two lower than air temperature; a clear night sky may have an effective temperature as low as $-50°F$. Heat loss on a cloudy night with aggregate temperature of 90° and air temperature of 70°F is about $17.3 \times 10^{-10} \times 0.70(549^4 - 529^4) = 15$ Btu per (sq ft)(hr). This is usually neglected.

Convection heat gain or loss per degree Fahrenheit difference between aggregate and air temperature is about as follows:

Air velocity, mph....	2	5	10	15	20	30	40
Btu/(sq ft)(hr)(°F)..	1.7	2.7	4.3	5.8	7.3	10.0	12.5

While this applies to the surface area of the pile, horizontal area can be used to correct approximately for reduced velocity at the material surface.

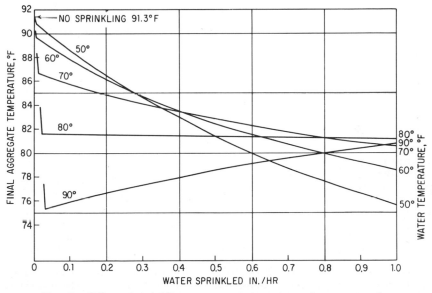

Fig. 6. Effect of sprinkling water on temperature of aggregate pile.

Figure 6 shows computed aggregate-pile temperatures assuming 913,000 lb per hr of aggregate is added day and night at 90°F to a pile 100 by 300 ft, with air temperatures of 80°F day and 60°F night and wind velocities of 5 mph day and 2 mph night. The results indicated for sprinkling with high-temperature water are not attainable in practice because the cooling of water drops and films by evaporation reduces their temperature and thus reduces evaporation. The opposite is true if cold water is used because aggregate and air warm the water. Thus aggregate temperatures shown for sprinkling with small quantities of water would be lower than shown for cold water and higher than shown for warm water, with the temperatures more nearly correct as the quantity sprinkled increases. As a practical matter, ⅛ to ½ in. per hr must be sprinkled to keep piles wet, and it is better to use the smallest quantity that will keep the pile wet if the water is warm. If water is cool, larger quantities are beneficial. The spray must be coarse because water evaporating from a fine spray will increase the relative humidity near the pile and decrease the water temperature, both of which decrease evaporation. Sprinkler heads of the type used on golf courses can put out as much as 150 gpm at 100 psi on a 250-ft diameter, with quantities up to ½ in. per hr. Above 3 mhp wind velocity, the diameter decreases about 2% for each added mile per hour of wind velocity. Drains are needed to remove excess water. Sand and some fine aggregates cannot be cooled in practice by sprinkling because they accumulate too much water and are difficult to dewater.

8   Refrigeration Processes

Vapor-compression refrigeration (Fig. 7a) depends upon the expansion and vaporization of a liquid refrigerant, commonly ammonia or refrigerant 12 or 22, to absorb heat. The gas is compressed and then cooled in a *condenser* by air or water until it becomes a liquid, and then introduced through an *expansion valve* or nozzle into an *evaporator*, kept at lower pressure by the compressor intake. In the evaporator, the liquid becomes a gas, absorbing its heat of vaporization, and then returns to the compressor. Air or water in contact with the evaporator is cooled by conduction and may be used to cool other materials. More complex combinations are also used (Fig. 7b). Vapor-compression refrigeration is convenient and economical for small installations; electrically driven package units are available, and condenser water requirements are lower than for vacuum systems. Power required is about 1 hp per ton of refrigeration.

Absorption refrigeration (Fig. 7c) is similar to vapor compression but more complicated in that two fluids are used, a *refrigerant* and an *absorbent*. The refrigerant passes from a *condenser* where it has heat removed and is converted to a liquid, thence through an *expansion valve* to the *evaporator* where it picks up heat and becomes a vapor, thence to an *absorber* where it is dissolved or absorbed in the absorbent, loses heat, and becomes liquid. It is then pumped with the absorbent to the generator, where it is heated. The refrigerant becomes a vapor, separates from the absorbent, and goes to the condenser. The absorbent circulates from the generator to the absorber and is then pumped to the generator with the refrigerant. The absorbent must absorb or dissolve the refrigerant vapor easily. Both the generator and evaporator absorb heat from air or water in contact with them, and heat must be removed from the condenser and absorber. Very little mechanical work is used in pumping the liquid absorbent and refrigerant. Ammonia-water and water–lithium bromide are the only common fluid combinations. The former requires equipment between the generator and condenser to keep water vapor out of the condenser. The latter cannot use an air-cooled absorber and operates at low vapor pressures, requiring large pipe. Absorption refrigeration operates efficiently at reduced load.

Vacuum refrigeration (Figs. 7d and 8) as applied to concrete is a vapor-compression system using water as the refrigerant. As the air is exhausted from a closed tank, any water in it vaporizes rapidly (boils) at the temperature shown in Table 3 corresponding to the pressure. For example, at a pressure of 0.247 in. of mercury water boils if the temperature is 40°F or higher. In vaporizing, the water absorbs about 1,070 Btu per lb of vapor. If this vapor is removed any water present will continue to boil until its temperature is reduced to 40°F. The water that vaporizes is the refrigerant, and it cools other water or aggregate with which it is in contact because it must draw the heat required for vaporization from them. Practically, because of the large amounts of vapor that must be removed, high-pressure steam jets are used to pump the vapor, and the mixture of steam and vapor is condensed. Surface or evaporative condensers can be used, but if condensing temperatures are below 100°F, barometric contact condensers are most economical and efficient. These mix sheets of water and vapor, with the water and condensed vapor moving down a tailpipe about 32 ft long, with its lower end submerged in an open water tank or well. The vacuum tank is equivalent to an evaporator, the steam-jet pump to a compressor, and the condenser to

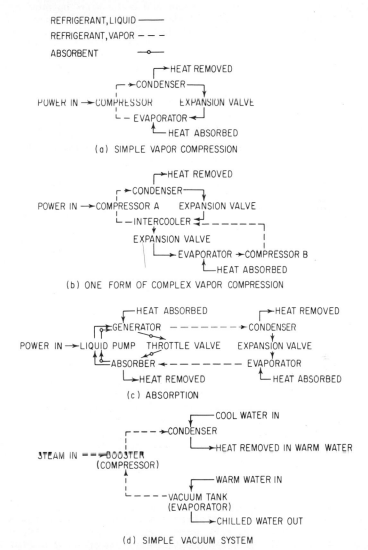

REFRIGERANT, LIQUID ——
REFRIGERANT, VAPOR – – –
ABSORBENT ——o—

(a) SIMPLE VAPOR COMPRESSION

(b) ONE FORM OF COMPLEX VAPOR COMPRESSION

(c) ABSORPTION

(d) SIMPLE VACUUM SYSTEM

Fig. 7. Schematic diagrams of typical refrigeration systems.

the condenser in a simple vapor-compression system. Since refrigerant (water) is usually available in the evaporator (vacuum tank), it is not always necessary to recirculate the refrigerant, but recirculation may be necessary to maintain a moisture level sufficient for satisfactory cooling or to permit more efficient operation. Dirty water can be used in barometric condensers if the condenser output is not used for boiler water. More commonly the condenser water is cooled by evaporation in a *cooling tower* and reused. Air pumps (*primers*) are required to reduce the pressure in the vacuum tank initially and other air pumps (usually two-stage *ejectors* with a small condenser between them) are used to remove air present in the system through small leaks or dissolved in water because air cannot be removed by condensation. *Boosters* perform the compressing operation at low pressure. Primers, ejectors, and boosters are operated by steam. Condensate reclaimed from primers and ejectors is hotter and cleaner than that from the large condensers and is reused for boiler water. Boosters cannot be operated at part load by throttling the steam because they operate at supersonic velocities near 4,000 fps. Steam must be of high quality. Vacuum booster-condenser combinations are closely sized for a minimum steam pressure in the 100-

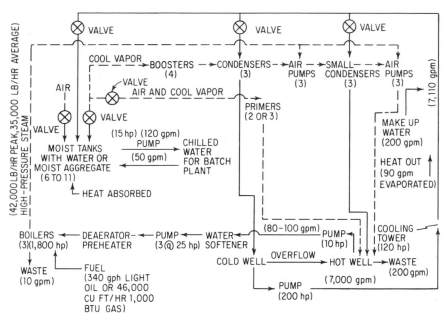

Fig. 8.   Typical vacuum-cooling system for 320 cu yd per hr concrete production.

to 130-psig range and a maximum condenser-water temperature. When the condenser-water temperature is lower than the maximum, the booster steam pressure can be reduced, saving steam. Steam and condenser-water consumption vary greatly with water temperature and steam pressure. Typical values are about 6 gpm condenser water at 80°F and 25 lb per hr steam at 100 to 130 psig per ton of refrigeration.

The triple point of water, where it exists as solid, liquid, and vapor at the same temperature and pressure, is about 32.01°F at 4.6 mm (0.180 in.) of mercury. If pressure is reduced below this value water freezes and continues to lose temperature slowly. It is possible to freeze the refrigerant (water) in a vacuum system unless the pressure is carefully controlled.

Advantages of the vacuum system are that heat exchangers are not required between the refrigeration system and water or aggregate being cooled, moving parts consist only of valves and pumps, no special refrigerant is required, and for large installations first cost and maintenance are generally lower than with other methods of refrigeration. Vacuum systems require more condenser water than other systems and require clean, soft boiler water and gas or oil for fuel.

9 Cooling Water

Chilled water is useful mostly as an adjunct to other cooling or for cooling other materials. Water can be chilled economically by vacuum or by vapor-compression refrigeration units and heat exchangers. Most heat-balance computations assume near maximum moisture in aggregates, resulting in minimum batch-water requirements. Excess chilled batch-water capacity is needed for times when aggregate moisture content is low.

Vacuum water chillers used in concrete plants usually have two or three cooling chambers. Water is partly chilled in the first chamber, then goes to the second and then to the third chamber, in each of which it is further chilled. From the last chamber it is pumped through insulated pipe to an insulated batcher reservoir or other point of use. Overflow or return water is fed to the intake of the chiller to reduce the cooling load. Water level is automatically controlled by a float valve and temperature automatically controlled by sensors in the unit. Safe temperature is about 34°F with temperatures of 32.5 to 33°F possible with careful adjustment and a constant rate of water use. About 0.2 connected electric horsepower and 0.75 bhp are required per ton of refrigeration, including condensed-water pumps and cooling-tower fans.

Vapor-compression water chillers are available as packaged units in capacities from 12,000 to 3,000,000 Btu per hr (1 to 250 tons) with about 1.0 to 1.2 electric motor horsepower per ton of refrigeration. When a water-cooled condenser is used it is usually built into the unit, but air-cooled or evaporative condensers are often separate units. Where the chilled water is used in heat exchangers about 2% makeup water is required and the unit remains reasonably clean. Where water is used as mix water or for inundation of aggregates, makeup or return water may require treatment to prevent fouling the chiller. Automatic control for varying rates of water use is by compressor unloading or starting and stopping multiple compressors depending upon water temperature or suction pressure. Temperature of chilled water leaving the unit cannot be less than about 35°F without danger of freezing the water. Condenser water is required at 3 to 5 gpm per ton of refrigeration for water-cooled units. Chillers will not function properly if condenser water or air temperature (for air-cooled units) is too low.

10 Cooling with Ice

Ice is very effective in cooling concrete, but the quantity that can be used is limited. Ice can be delivered to the plant in blocks and ground up for use in batching. It is then elevated in bucket elevators to an insulated daily use tank or bin (commonly of wood) holding several tons of ice at 60 cu ft per ton with ice depth not more than 8 or 9 ft, and moved by a screw conveyor to an ice weigh batcher similar to an aggregate batcher. In smaller plants ice may be added to the water batcher, but it tends to jam in conventional water batchers. Ice does not flow readily in bins and hoppers. Ice usually enters the bin below 32°F, frequently about 10°F. If the insulated bin can be kept free of water and moist air, handling is easier. Some refrigeration of the bin is required for long-time storage. Ice melts slowly in contact with cold aggregates. Flake ice usually melts during about 2 min of mixing if only coarse aggregate is cold; more time is required if all other ingredients are cold or ice particles larger.

Automatic icemakers are available as packaged, electric-driven units with capacities to 30 U.S. tons of ice per day. They usually make ice at 0 to 10°F to keep the ice from sticking together, but this subcooling is not relied upon when computing concrete temperatures because much of it is lost in handling. A battery of ice machines is commonly required for a high-production concrete plant. Machines that make flakes instead of cubes or chunks are used to reduce melting time in the mixers. Several methods are used but basically water is frozen on a stationary or revolving cylinder or belt and mechanically broken off into a storage bin as white ice. A few machines thaw ice loose from tubes or plates and crush it. Input water temperatures are 40 to 110°F, with more efficiency at the lower temperature. Vapor-compression refrigeration is used. The condenser may be air-cooled, but water cooling is used for larger units. Power required is about 2 to 3 hp per U.S. ton of ice per day, not including condenser-water pumps or ice-transporting equipment. About 5,000 gal of condenser water at 70°F is required per ton of ice.

An advantage of ice is that it can be stored, and icemakers can operate nearly continuously. A shutdown for 2 hr each day is allowed for maintenance.

11 Cooling Coarse Aggregate

Aggregate can be cooled at ground level before being placed in a bin over the batchers or after it is placed in the bin. When cooled on the ground it may be cooled in separate size groups and conveyed to the bin as sized material or blended and separated into size groups by screens on the bin. It may also be blended before cooling and separated into sizes as it is fed into bin compartments, and this method reduces the size of cooling, conveying, and rescreening equipment but requires more careful process control. When aggregates are cooled before conveying them to the bin they pick up about 2°F from radiation, conduction, convection, and condensation of atmospheric moisture as they are moved to the bin. Handling equipment should be rinsed with cold water before start-up, shielded from wind, and painted white on surfaces exposed to the sun. Coarse aggregates can also be cooled after they are placed in separate size groups in the overhead bin structure. Aggregate temperatures at the batchers depend upon the method of cooling used and where the cooling is done. In order of increasing cost for large installations, cooling may be by vacuum, cold air, or inundation in cold water. Aggregates have been cooled at rates of 240 U.S. tons per hr by 36 to 40°F water and 42°F air while moving slowly on belts in long tunnels, in India. Exposure time is 20 to 25 min. Cooling by spraying with large quantities of cold water and by spraying while forcing cold air through the material is also feasible but is not currently used in large concrete plants.

Cold aggregate gains about 1°F per hr in insulated storage bins at 80 to 90°F air temperature. Regardless of the cooling method used, the storage bin (other than a vacuum tank) should be enclosed on the sides and bottom and covered with about 2 in. of commercial insulation or 1 in. of urethane foam, which may be sprayed in place. The bin should be roofed. Cooling coils may be built into the bin bottom for cold-water circulation, to reduce heat gain primarily while the plant is shut down temporarily and chilled water is not in use elsewhere.

Vacuum cooling of coarse aggregates (Fig. 8) from No. 4 to the largest size used is done by placing a batch of moist aggregate in one of a group of vacuum tanks and removing the air and vapor to maintain a pressure corresponding to the desired temperature (Table 3). Typical tanks hold 150 to 300 tons. Three tanks are typical of ground installations, eight of an overhead installation. The amount of water removed as vapor is about 1% of the aggregate weight, and this can be replaced by sprays inside the tank. Even so, for satisfactory operation the aggregate should be sprinkled in storage piles so that it is saturated and wet on the surface before being placed in the vacuum tank to ensure even moisture distribution and to provide a reserve for recooling should it be needed. Dry aggregate may absorb 1 to 3% moisture. Dewatering is not a problem because aggregates can be drained through a pump or barometric drain for at least an hour while in the vacuum tank. When aggregates are held in a tank under vacuum or recooled after a shutdown, water vapor continues to be extracted. Once a tank is cool it can be sealed under vacuum and the temperature held for several hours because heat gain is quite slow.

While ducts are arranged to permit reasonably free egress of vapor, all parts of the tank are not at the same pressure during the cooling process and some margin must be left to prevent freezing of material near the duct entrance. Practical average temperature is about 34°F. When aggregate is accidently frozen it thaws very slowly; so steam jets built into the tank are used to break it up.

The process sequence is complicated by filling and cooling tanks in rotation, and by interchange of vacuum between warm and cold tanks, but basically a sealed tank is *primed* for about 5 min to remove air and reduce the pressure to about 0.7 in. Hg. Then a *booster* is connected until the pressure in the tank is reduced to about 0.19 in. Hg. The pressure may be held at this level for a short time for added cooling effect. About 40 to 50 min is required to cool a tank, depending on the aggregate temperature. Cooling time (Fig. 1) is controlled more by the time required for the process than by the time required for heat flow from the aggregate. Sensitive vacuum gages are used to determine pressure and thus temperature. After the desired pressure is reached the main duct is shut off and the tank is held under vacuum until the material is required. Vacuum is broken by an air-inlet valve or partially broken by connecting the cold tank to a warm tank rapidly, and then by an air inlet to bring the tank to atmospheric pressure before opening the vacuum seals and discharging material. About 0.18 connected electric horsepower and 0.75 bhp are required per ton of refrigeration, including power for condenser-water pumps and cooling-tower fans.

Cold-air cooling of sized coarse aggregates from ¾ in. to the largest size used to about 40°F is usually done in overhead bins by circulating about 38°F air through them. Aggregates smaller than ¾ in. have such high resistance to air flow that they are not cooled by this method economically. Cold air is forced by a fan through shielded openings in the bin bottom, travels up through the aggregate, and is reclaimed from the top of the bin and rechilled. Some cooling is accomplished by evaporation, especially initially in the smaller sizes. Vapor-compression refrigeration is used, with air passing over the evaporator to be chilled. Air temperatures below about 35°F at the evaporator may result in stoppages caused by icing of the evaporator. Package air-cooling units may be mounted on the bin or large units may be on the ground. Cooling time is about 2 hr because of the poor heat transfer from air to aggregate, and bins should be sized to hold material for this time and should be kept well filled. About 1.9 electric horsepower is required per ton of refrigeration.

Inundation cooling of coarse aggregates from No. 4 to the largest size has been used on several projects. It is usually done on the ground in groups of 5 to 10 tanks, each of about 120 to 180 tons aggregate capacity. Vapor-compression refrigeration has been used to supply chilled water, but vacuum water chilling could be used economically. A typical cycle starts with a tank about one-third full of 35°F water to which the aggregate is added during about 10 min. Water at 35°F is then pumped into the bottom of the tank, flows through the aggregate and over weirs near the top, and is piped to a reclaiming (surge) tank, whence it is rechilled and reused. The water has a tendency to accumulate fines that may coat aggregate particles, necessitating filtering or changing the water. After about 40 min, the tank is drained and the material is held in the tank to drain until it is needed. It may be conveyed over vibrating dewatering screens (sometimes with air suction or blowers) before delivery to the overhead bin. Cycle time is about 2 hr for each tank. About 1.2 electric horsepower is required per ton of refrigeration.

Table 6 shows advantages and disadvantages of the various methods of aggregate cooling in use in North America in 1967.

12 Cooling Sand

Sand is cooled most efficiently by the *vacuum process*. Because it impedes vapor flow, sand must be exposed to the vacuum in thin layers or small increments to obtain uniformly low temperature. One method uses a two-compartment vacuum tank. Sand from the upper chamber is fed to the lower chamber in the sealed tank, is cooled quickly as it is exposed to the vacuum, and accumulates in the lower chamber. With delicate control sand can be frozen into granules like popcorn by this method, but temperatures of 33 or 34°F are more easily controlled. Another type of sand vacuum tank has a built-in bucket elevator to circulate sand from the bottom to the top. Most cooling is performed as sand enters the elevator buckets, in the buckets, and at the elevator discharge. Sand can be frozen into chunks small enough to pass through material gates, but again 33 to 34°F sand is more easily controlled. Freezing the moisture in sand can remove about an additional 7,000 Btu per cu yd of concrete. This is useful in emergencies but should not be relied upon for routine operation. About

Table 6. Methods of Cooling Aggregates

Cooling method	Typical aggregate temp at batchers, °F	Advantages	Disadvantages
Vacuum cooling of sized aggregate overhead	35	Low operating cost. All sizes including sand can be cooled. Aggregates can be recooled after shutdown. High initial aggregate temperature requires relatively small added cooling time. Easily used for heating. Electric-power requirement is low	Moderately high equipment cost. About 1 hr may be required for plant start-up. Boilers require 1 hr additional. Requires fuel supply for boilers. Fine particles accumulate in water pumps
Vacuum cooling of blended aggregate on ground	38	Low equipment and operating cost. All sizes including sand can be cooled. High initial aggregate temperature requires relatively small added cooling time. Easily used for heating. Electric-power requirement is low	Aggregate left in bin over long shutdown must be removed. About 1 hr may be required for start-up. Boilers require 1 hr additional. Requires fuel supply for boilers. Fine particles accumulate in water pumps
Cold-air cooling of sized aggregate in overhead bin	40	Low equipment cost if extreme low temperatures not required. Aggregate can be held at low temperature for long periods. Continuous process. Can be all electric power. Can be used for heating with addition of heat supply	Requires large overhead bin capacity. No. 4 to ¾ in. cannot be cooled economically. Sand cannot be cooled. Cooling time is extended if initial aggregate temperature high. Efficiency lowered by air leakage. Dust accumulates in cooling system. Power requirements high because of large fans
Inundation cooling of sized aggregate on ground	40	Water chillers available in package units. Can be all electric power. Can be used for heating with addition of heat supply	Aggregate left in bin over long shutdown must be removed. High equipment cost. Finer aggregate difficult to dewater. Sand cannot be cooled. Chilled water picks up fine material and may coat aggregate if not filtered or changed frequently

1½% sand moisture is removed when cooling sand from 90 to 34°F, 1% when cooling from 70 to 34°F.

Sand can also be cooled in *screw-conveyor coolers* with jackets and hollow flights through which 35 to 40°F water or 20 to 30°F brine circulates. Because the sand is handled in thin layers while exposed to air, cooling below the dew point results in condensation on the sand, and 40 to 50°F is about the lower limit for efficient cooling with brine, 50°F with chilled water.

13 Cooling Cement

Cement and pozzolan are cooled when necessary by contact with metal jackets chilled by 35 to 40°F water. One type of cooler is like the screw-conveyor cooler described for sand. In another type thin layers of cement move upward in contact with the inside of a vertical metal cylinder with cold water running down the outside of the cylinder. Capacities to 265 bbl per hr with a 100°F temperature reduction, or higher capacity with less reduction, are available. Cement cooling is relatively expensive because cement loses heat slowly. Cement should not be cooled below the dew point of air in contact with it or it will pick up moisture from the air. This limits cement temperature to 60 to 70°F. About 20 hp and 25 gpm of 35°F water are required per 100 bbl per hr cooled.

14 Practical Cooling Combinations

As a practical matter certain combinations of cooling methods show more economy than randomly selected methods. In many combinations the same equipment (such as boilers or water chillers) can be used for several materials. For example, if coarse aggregate, sand, and water are cooled by a vacuum installation, additional chilled water can be made economically for cooling cement if further cooling is required, and addition of icemakers and an ice batcher is not economical. If coarse aggregate is cooled by air, ice is usually required. If inundation cooling is used for coarse aggregate, either ice or chilled water and sand cooling are usually needed, and cement cooling may be required. In any event, the combination must remove enough heat units to reduce concrete temperature to the desired level, and specialists in concrete-material cooling should be consulted for economical design of other than the most simple systems.

Chapter 21

PLANT OPERATION

JOHN K. HUNT

For maximum economy consistent with good practice it is necessary that the complete operation of the plant from the arrival of the basic materials to the final delivery of the concrete be carefully planned and checked periodically. Preventive maintenance is essential to avoid very costly shutdown time. Properly operated and maintained equipment is necessary for personnel safety.

1 Materials Handling

Planning and scheduling material delivery to coincide with production needs is difficult but important. Many costly errors can be prevented by checking and carefully recording material deliveries.

2 Batching

To produce consistently good concrete an operator should know as much as possible about what is happening throughout the plant. He should know any specifications he must meet in the batching operation. If final concrete temperature is critical he must have some knowledge of the temperature of the various materials. Moisture in the aggregate must be checked and compensations made. If the operator is remote from the plant, closed-circuit television or a good system of indicating lights or remote dials may be employed to give him information about the batching process or the proper location of a truck under the plant. Some means of manual operation is essential in an automatic plant to prevent a complete shutdown in the event of a malfunction of the automatic controls.

3 Mixing

Many tests have indicated that proper sequencing or blending of materials into a mixer is very important for proper mixing. Some type of slump indication is useful. A timer interlocked with the mixer discharge should be provided to ensure a minimum mixing time. A maximum mixing time should also be established. It is desirable to be able to see into a mixer to check the mixing action.

4 Dispatching

Before a mixer truck is brought to the batch plant it should be washed out and the water tank filled so it is ready to receive concrete as soon as it arrives. The use of two-way radio communication between all phases of a job can help to coordinate all operations. If trouble occurs on a paving spread the batch-plant operator can be notified immediately to stop sending concrete. A mixer-truck driver can notify the dispatcher that he is on the way in and that the job will need more concrete or that the job is finished and the trucks being held in reserve are free to be used for another customer.

5 Plant Maintenance and Repair

Manufacturers' recommended practices should be followed for cleaning, lubricating, and general maintenance of all equipment. A maintenance check sheet and repair record for each piece of equipment is very useful in determining what spare parts are necessary and in predicting and preparing for breakdown of equipment.

Scales should be inspected frequently to ensure that there are no rubbing parts or chipped knife-edges. A regular scale inspection and testing procedure should be set up and carried out to ensure that any potential scale problems are found and remedied before they become expensive problems.

Minor adjustments on a scale system can be made by the operator, but any major overhaul or repair should be done by a specialist.

Bins and hoppers should be inspected regularly and cleaned and painted when necessary. The bottoms and sides of bins and hoppers should be checked for wear and liner plates replaced as needed.

Elevators, screw conveyors, and belt conveyors should be kept cleaned, painted, checked for wear, and lubricated as needed.

Water meters should be calibrated regularly. They should be drained in cold weather to prevent damage due to freezing.

Mixers should be kept cleaned, lubricated, and properly aligned. Blade wear should be checked periodically and worn blades and liners repaired or replaced.

Recorders should be kept clean at all times.

Controls are particularly susceptible to dust and moisture. Damaged or malfunctioning parts should be repaired or replaced.

Many of the sophisticated and highly complex control systems must be serviced by experts. Troubleshooting some of this equipment requires specialized tools or testing equipment. Any changes in the control systems should be recorded on the wiring diagrams so they will be current at all times.

Gates should be cleaned and lubricated.

Hydraulic equipment must be kept free from contamination.

Air equipment must be kept free from moisture and well lubricated.

Chapter 22

PLANT INSPECTION AND SAFETY

JOHN K. HUNT

INSPECTION AND TESTING

Inspection and testing procedures are discussed in detail in Sec. 3 of this handbook, and need not be repeated here. However, there are several points that need to be emphasized. Inspection and testing at the batch plant are intended to ensure close control over all phases of the operation with the objective of uniform quality of concrete. Since concrete manufacturing is not an exact science, it requires some exercise of judgment.

1 Materials Handling

Materials must be received, stored, and used without excess handling and in a way to reduce contamination, segregation, and degradation. Adequate storage must be provided. All material-handling vehicles must be emptied as completely as possible and full and empty weights checked to reduce unintentional shorting of material.

Samples of aggregate should be taken at the batcher discharge for checking gradation.

2 Batching Equipment

Hoppers and valves should be inspected for faulty operation. Scales should be inspected for drag or jerky operation and to be certain they are operating within the applicable delivery tolerances. Scales should be given a static-load test with standard weights periodically. Graphic recorders should be checked frequently during the day to be sure the pens are marking properly and accurately. Digital recorders should be checked frequently during the day to be sure all digits are printing correctly.

Water and admixture meters should be calibrated as often as necessary. Admixture timers and the calibration of the flow controls should be checked frequently to be sure they are as they should be. If calibrated sight glasses are used this can be done by simply watching the sight glasses for a few batches.

Moisture meters should be calibrated at least daily.

Any unusual condition should be noted and checked immediately.

3 Mixers

Mixers should be cleaned daily to prevent buildup of material, and blade and liner wear should be checked as needed. Mixer-drum speed should be checked against the manufacturer's recommended speed. Any changes in material-charging sequence to the mixer should be noted and the mixing action and results checked. Mixing time should be spot-checked.

4 Dispatching

Each batch of concrete should be identified by type and the time of dispatching recorded. A record should be kept of wasted batches and the reasons for them.

SAFETY AND ACCIDENT PREVENTION

5 Responsibility

The matter of safety and accident prevention is primarily a responsibility of the owner and operator of the plant. However, construction is inherently a hazardous occupation, and it is the responsibility of every person employed on the job to be alert to dangerous conditions and to take the necessary precautions for his own safety as well as that of others.

Construction is a highly competitive industry in which deadlines impose the need for speedy construction regardless of delays occasioned by unfavorable weather. Much of the work is done outdoors where the workmen are exposed to heat and cold, rain, wind, and all other vagaries of the weather. Because of these hazardous conditions, accident prevention requires even closer attention to be effective.

Industrial commissions and similar political bodies of the states provide rules and regulations governing nearly all phases of construction and machinery operation, and most of them employ safety inspectors to enforce these laws. The National Safety Council and insurance companies supply posters and safety literature to their clients—material that can be posted about the jobsite, especially in the vicinity of dangerous conditions. Many large contractors employ a safety engineer whose duties are to educate the workmen by use of posters, literature, motion pictures, and similar aids, as well as to supervise installation and maintenance of safety equipment, first-aid stations, fire-fighting facilities, machinery guards, and similar safeguards. Smaller companies usually designate one employee as a safety supervisor in addition to his other duties.

6 Methods

Accident prevention is primarily a matter of education and organization. Accident prevention and safety, to be successful, must be publicized and must be presented to the workmen so they are constantly aware of the program and voluntarily cooperate in observing safety rules. A new employee should be introduced to the company's safety program at the time of his employment by being given a letter briefly explaining the accident-prevention features.

One method of maintaining safety awareness is to establish a system of awards in which the department, section, or gang with the best safety record receives publicity for its record, and perhaps a small cash bonus.

Most persons are apt to measure an accident by the spectacular results of the accident, the injury or death of workmen, or property damage. Actually, even though there was no personal injury or property damage, any unexpected or unintentional interruption of the orderly progress of the work is an accident and should be analyzed as such to determine why it happened and how it can be prevented in the future.

Accidents should not be hidden. It is wrong to "cover up" for any reason whatever. Openly investigating and reporting an accident makes it possible for the workmen to benefit thereby, so they can avoid a similar situation in the future.

Specific recommendations and warnings in and about concrete construction include those relating to formwork, shoring, hoists and cranes, conveyors, bucket elevators, pneumatic and electric machinery and tools, motor trucks, railroad cars, pumps, handling and storing aggregates and cement, ladders, scaffolding, and runways. Special regulations govern the use of boilers for supplying hot water or steam for heating aggregates and curing concrete. Quarry and aggregate excavation operations have special dangers in drilling, blasting, and excavating.

Protective clothing is necessary for some occupations. Welders and sandblasters require special helmets, face guards, and goggles. Gloves should be worn only when necessary for protection. Loose-fitting clothing should be avoided. Every new employee should be issued a safety helmet ("hard hat") with instructions to wear it at all times on the job. Protective shoes ("hard toes") are recommended.

These are but a few of the precautions. The safety engineer or supervisor should become familiar with all potential hazards on the job. One way to do this is to study the literature that is available from manufacturers, state industrial safety commissions, insurance companies, the Associated General Contractors, and similar organizations.

Chapter 23

TRANSPORTING CONCRETE

JOHN K. HUNT

Concrete should be transported so that segregation of ingredients, loss of ingredients, and loss of slump are at a minimum. Because of its plastic consistency, low-slump air-entrained concrete is more easily handled without detriment than the runnier mixes used in the past. Some segregation of the larger aggregate is usually not objectionable if the aggregate is recombined with the fine materials or ends up on top of the concrete when it is placed in its final location, because vibration will consolidate the materials. Deleterious segregation is avoided by dumping, dropping, or chuting concrete only when necessary, and then so that materials are recombined as they are discharged or so that the segregated larger aggregate, if any, ends up on top of the concrete. Concrete lowered vertically in full tubes or trunks segregates very little. Concrete turned by baffles as it is dropped or chuted is usually recombined satisfactorily, but best location for the baffles may have to be determined by trial. Stirring of the concrete while it is being moved reduces segregation and permits more time to elapse between mixing and placing. Excessive jolting or vibration without stirring tends to segregate concrete.

Loss of ingredients, usually grout, is avoided by using gates that do not leak grout, by coating mixing and handling equipment with grout at the start of a pour (often by adding cement, sand, and water to the initial batch), by preventing excessive buildup of grout on equipment in contact with the wet concrete, and by not dropping concrete through water. Belt conveyors need scrapers to keep fine materials from carrying around head pulleys.

Loss of slump is caused by increase in concrete temperature and by drying as well as by loss of grout. It is best avoided by handling concrete expeditiously, protecting concrete from sun and wind, and by painting containers exposed to the sun gloss white if concrete is held in them more than a short time.

CONCRETE HOPPERS

1 Normal-slump Hoppers

Mixers do not discharge vertically into concrete hoppers, and there is a tendency for larger materials to roll to the far side of the hopper. Materials are best recombined if the hopper discharges vertically on the centerline of a symmetrical hopper. Thus hopper gates are best located centrally with double-clam gates for vertical discharge from the hopper. Rubber gate wipes may be required to reduce loss of grout. The gates must be large for efficient handling of low-slump concrete. Operating forces required for gates are typically 3 to 4 psi of gate area. Hopper shape may be conical or nearly conical with six or more sides. Hoppers with four sides require steep side slopes because the net valley angle should be not less than 60° from the horizontal for held concrete, and the sharper corners tend to accumulate the smaller ingredients. A guide chute or hopper through which concrete moves without stopping should have net valley angles of at least 50°. Baffles may be used to reduce segregation in hoppers, but their optimum location must be found by trial. Regardless of planned capacity, concrete hoppers should be designed for a full load of concrete of the maximum weight shown in Tables 1, 2, 3, or 4, Chap. 17.

2 Low-slump Hoppers

Hoppers for very dry concrete, as used in base courses and concrete products, present a different problem. Shape is less critical because the material is usually a granular mixture of coated particles and does not tend to segregate as much as regular concrete. However, it is tacky and tends to build up in corners unless valley slopes are steep and the hopper is vibrated each time it is nearly empty.

CONCRETE BUCKETS

Concrete buckets are available in a variety of types and sizes for specific purposes. Typical sizes are shown in Table 1.

Table 1. Sizes and Typical Weights of Concrete Buckets
Sizes in boldface type are most commonly used

Size, cu yd	Typical bucket weights, lb			
	Lightweight buckets†	General-purpose buckets	Lay-down buckets	Low-slump buckets
⅓	300–400	330–450		
½	**330–460**	430–530	500–550	
¾	**440–590**	450–1,030	500–630	
1	**500–700**	**640–1,170**	**620–1,580**	**970–1,910**
1½	**560–940**	**730–1,560**	**860–2,040**	1,050–2,080
2	610–1,080	**1,100–1,680**	**1,080–2,480**	**1,700–3,600**
3	. . .	1,550–2,600	1,460–3,500	2,500–4,230
4	. . .	1,800–3,800	. . .	**3,650–5,180**
5	. . .	. . .	. . .	4,400–5,200
6	. . .	. . .	. . .	4,900–7,200
7	. . .	. . .	. . .	5,800–7,000
8	. . .	. . .	. . .	6,600–8,300‡
8*	. . .	. . .	. . .	8,600–9,800*
12	. . .	. . .	. . .	9,500–12,700‡

* These weights are for an 8 cu yd bucket with two 4 cu yd compartments.
† Aluminum buckets weigh about three-fourths of the minimum weights shown.
‡ Some 8 cu yd buckets weigh as much as 13,000 lb. Some special 12 cu yd buckets weigh as much as 14,500 lb.

3 Low-slump or Mass-concrete Buckets

Some specifications limit the amount of concrete that can be placed in one pile to 4 cu yd; so larger buckets are sometimes compartmented with separate gates on each compartment. Low-slump buckets (Fig. 1) have bottom slopes of 60 to 70° and large gate areas for handling 1- to 3-in.-slump concrete with aggregate up to 8 in. The smaller sizes may be operated manually, but compressed-air (or sometimes hydraulic) operation is used on larger sizes because of the large gate forces required. Air-operated buckets are operated by air connections from hoses at the placing site, either by making or breaking the air connection or by a manual valve, or occasionally by air lines connected to the bucket from the crane and manual valves, or by built-in rechargeable high-pressure air receivers

Fig. 1. Concrete bucket for low-slump concrete. (*Blaw-Knox Company.*)

and manual valves. About 10 cu ft of free air at 125 psi should be allowed for each gate cycle, and about two cycles per bucket load of concrete. Gates should close automatically if the valve is released or if the air line fails. Hydraulic buckets have devices built into them to use the bucket weight to provide energy to operate the gates several times.

4 Lightweight or General-purpose Buckets

These buckets, as shown in Fig. 2, are used for handling 2- to 6-in.-slump concrete with about 2½- to 3-in. maximum aggregate. They are usually manually operated and often have rubber spout attachments to aid in placing walls and columns. Gate areas are smaller than in low-slump buckets and hopper slopes are 50 to 60°.

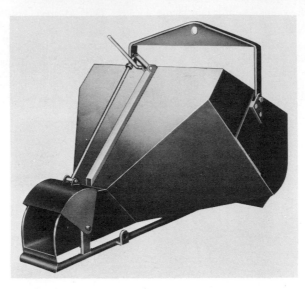

FIG. 2. One type of lightweight lay-down concrete bucket for general usage. (*Gar-Bro Manufacturing Company.*)

5 Lay-down Buckets

Normally lay-down buckets (Fig. 3) are used in building construction where they are filled from transit mixers and low filling height is required. They usually have small manual gates of the single-clam type.

FIG. 3. A lay-down bucket frequently used in building construction. (*Gar-Bro Manufacturing Company.*)

6   Special Buckets

Special buckets are made for placing concrete underwater, lowering concrete in caissons, and other special applications. *Underwater* or *tremie buckets* are frequently covered to prevent loss of grout while the bucket is being lowered and operate upon contact with the bottom or with a gate line from above (see Chap. 26 also).

While some buckets for special applications must compromise on desirable features, to minimize segregation concrete buckets should have hopper slopes steep enough to discharge thoroughly and have a symmetrical vertical discharge. An asymmetrical discharge opening causes a suspended bucket to kick sideways during discharge. The minimum dimension of the discharge opening should be at least five times the nominal size of the largest aggregate. While low-slump buckets may open full and close upon release of the valve, buckets for building construction require regulated gate openings for feeding concrete into forms and do not always close automatically.

In building construction the concrete bucket is often filled from a transit mixer and handled by a crane. Mass concrete is usually hauled from the mixing plant in buckets with two or more buckets on a rail car or truck and lifted and moved to final position by a trestle crane or cableway. An empty bucket is often placed on the car or truck before a full bucket is lifted; so there is usually space for one bucket more than the number of full buckets carried on the hauling unit.

AGITATING TRUCK BODIES

7   Truck Mixers

When used as agitating hauling units, truck-mixer capacity is shown in Table 2, Chap. 18, unless otherwise limited by truck chassis capacity or legal load limits. Drums (or paddles) are revolved at 2 to 6 rpm during agitation. Discharge times are the same as for mixers and can be quite long for low-slump concrete. Typical discharge times are 30 to 90 sec per cu yd, but mixers built especially for low-slump

FIG. 4.   Agitating open-top end-dump hauling unit. (*T. L. Smith Company.*)

concrete may discharge in 5 to 10 sec per cu yd. Standards for agitators are published by the National Ready Mixed Concrete Association, 900 Spring St., Silver Spring, Md.

8   Open-top Hauling Units

Units of this type have specially shaped bodies with rotating blades (Fig. 4). Concrete discharges down a chute at the rear by tilting the body. Typical capacities are 4, 6, 8, and 12 cu yd. Discharge times are comparable with those of transit-mix trucks except that low-slump concrete can be discharged more readily. Typical discharge time is 30 to 40 sec.

NONAGITATING TRUCK BODIES

9   Side-dump and End-dump Units

These nonagitating hauling units are available in capacities of 4, 6, and 8 cu yd. Truck-trailer side-dump combinations increase capacity to 16 cu yd per unit. Nonagitating units work well with plastic air-entrained concrete with slump up to about 3 in. Typical discharge time is 30 to 40 sec. In addition, ordinary dump bodies may be converted to haul low-slump concrete by welding fillets in the corners where necessary to prevent concrete buildup, and by making sure the tail gate seals well enough to prevent loss of grout. Dump bodies discharge rapidly, and the rate is difficult to control. The tail gate may be chained to provide some control of discharge rate. On hauls up to 8 or 10 miles covers are not usually used on hauling units, but they may be required if drying due to exposure is troublesome. A vibrator or frequent scraping may be required to avoid accumulation of concrete. Nonagitating units are also suitable for hauling very dry concrete like that used for base courses.

NUMBER OF HAULING UNITS REQUIRED

10   Estimate of Hauling Units Required

A shortage of concrete hauling units (or batch trucks) results in inefficiency in concrete operations. The number of hauling units required may be estimated from the formula

$$N = \frac{Y}{60y} \left(\frac{60L}{S} + T \right) + C \tag{1}$$

where N is the number of hauling units required; Y is the concrete-plant batching capacity, mixing capacity, or concrete placing and finishing capacity, whichever is less, in cubic yards per hour; y is cubic yards hauled per truck; L is the round-trip distance in miles; S is the average road speed in mph (about 25 to 40); T is the total road-trip delay time in starting, stopping, loading, dumping, washing, etc., in minutes (usually about 4 to 8); and C is an allowance for truck bunching and other interruptions to the truck cycle, such as refueling. C should be not less than 1, and a value between 2 and 3 may be required. Figure 5 will give a graphical solution for the approximate number of hauling units required. When a job is operating, a more accurate estimate may be obtained by timing several round trips and using the formula

$$N = \frac{Y}{60y} \text{ (avg round-trip time, min)} + C \tag{2}$$

but it is better to find the values of S and T so the number of trucks can be adjusted for conditions on other days.

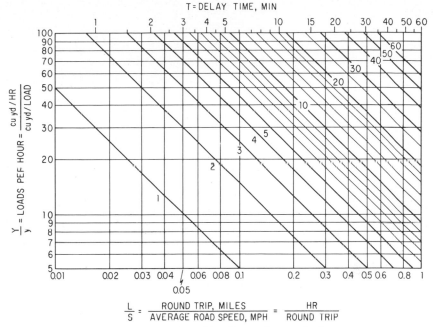

FIG. 5.   Number of hauling units required.

When the number of trucks has been established for a known haul, each added mile of round-trip haul requires Y/yS additional trucks. In paving, it is common for the required number of trucks to increase or decrease by three or four during a day because of the changed haul distance.

11   Effect of Delay on Hauling Units Required

When a central-mix plant is supposedly operating at capacity, add the delays for 1 hr from the time a mixer is ready to discharge until a truck is in position. Each 36 sec per hr of delay indicates a production loss of 1%. If 19 trucks are operating, for example, a total delay of 5% or 180 sec per hr indicates an additional hauling unit is required to maintain concrete production. More specifically,

$$\text{No. of trucks required} = \frac{3,600 \times \text{No. of trucks operating}}{3,600 - \text{total delay, sec/hr}}$$

Whether additional hauling units should be added is then an economic decision depending upon the cost of the delays and the ability of the placing and finishing equipment to use additional concrete. The reverse of this procedure is deceptive: if there are sufficient trucks to maintain plant capacity, timing truck delays while trucks are waiting for the mixer to complete mixing should produce an excess because of the tendency of trucks to bunch.

When used for batch trucks hauling to mobile paving mixers, it is convenient to substitute batches required per hour for Y and average number of batches hauled per truck for y in the above formulas.

The above formulas can also be used to estimate the number of transit-mix trucks needed to operate a plant at capacity if appropriate factors are used. For transit mixers the average road speed S is commonly lower and the delay time T is much greater than given above.

Section 6

PLACING CONCRETE

Placing concrete consists of moving or transferring the concrete from its delivery point at the jobsite, placing it in forms, and consolidating it to provide a concrete structure having adequate structural integrity, durability, serviceability, and appearance in accordance with the design and specifications. Placing concrete to provide such a finished structure requires knowledge of a few fundamental facts about the behavior of concrete when handled under different conditions, selection of equipment suitable for the job, and close attention to details that make the difference between a satisfactory job and a mediocre job. This has resulted in a large background of "do's and don'ts" and "correct and incorrect practices and methods" which are discussed in this section.

Chapter 24

METHODS AND EQUIPMENT

ROBERT F. ADAMS

DISTRIBUTING CONCRETE ON THE SITE

1 Equipment for Moving or Handling Concrete

Equipment should be carefully selected to move concrete from its delivery point into place in the finished structure. The slump, sand content, maximum size of aggregate, or concrete mix should not be governed by the equipment; rather the equipment should be capable of expeditiously handling, moving, and discharging concrete of such slump, sand content, maximum aggregate size, or mix proportions considered otherwise suitable, and which can be placed by vibration or other suitable placing methods. Description of some of the basic equipment used for moving and handling concrete after it is received at the jobsite follows. (Equipment for transporting concrete to the job is covered in Chap. 23.)

Chutes (Fig. 1) are a simple and expeditious way of transferring or moving concrete to a lower elevation. The commonest example is the chute used to discharge concrete from a mixer into a bucket or other equipment, or directly into

FIG. 1. Chute delivering concrete from transit-mix truck into hopper of drop chute in bridge pier. (*California Department of Water Resources photograph.*)

the forms. Chutes must have sufficient slope so that concrete will readily move down them by gravity.

Wheelbarrows (Fig. 2) are frequently used to move small quantities of concrete horizontally a short distance (suggested maximum 200 ft). A runway or relatively smooth path is required. Wheelbarrows have capacities of 2½ to 3½ cu ft, and one

FIG. 2. Wheelbarrows transport concrete short distances. Hopper used to transfer concrete releasing truck mixer sooner. (*Gar-Bro Mfg. Co. photograph.*)

FIG. 3. Carts or buggies transport concrete short distances. Carts having rubber tires are best. (*California Department of Water Resources photograph.*)

man and wheelbarrow can place up to 1 to 1½ cu yd concrete per hr. Wheelbarrows with pneumatic tires are far superior to those with steel wheels.

Concrete Carts. Hand-operated carts or buggies (Fig. 3) having capacities of 6 to 8 cu ft and power-operated carts (Fig. 4) with capacities of up to ½ cu yd are suitable for moving concrete horizontally for relatively short distances. Suggested maximum distances are 200 ft for the handcart and 1,000 ft for the rider-type power-driven cart. Hand-pushed carts can move from 3 to 5 cu yd concrete per hr each under favorable conditions. Power carts can move 15 to 20 cu yd concrete per hr each on a 600-ft haul. Runways or relatively level smooth paths are required. Pneumatic tires are more suitable than steel tires for the hand-operated carts, and are universally furnished on power buggies.

Buckets. The commonest method of handling concrete is in buckets. Buckets are transported or handled by crane, derrick, hoist, monorail, truck, railway car, cableway, or helicopter. Two types of buckets, the conventional type, either cylindrical or square with bottom discharge (Fig. 5), and the lay-down type (Fig. 6), are available. The conventional cylindrical or square bucket is available in many sizes from ⅓ to 12 cu yd and in many designs having various slopes to their sides and sizes of gate openings. Some, having steep side slopes and large gate openings,

FIG. 4. Power-operated cart transports concrete for bridge deck. Note runway supported by forms, not reinforcing steel. (*Gar-Bro Mfg. Co. photograph.*)

are suitable for low-slump and mass concrete. The lay-down bucket is for use where the headroom for charging the bucket is more limited. These are available in capacities up to 5 cu yd. Discharge of the buckets may be either by hand-operated gates for smaller buckets or by air-actuated gates. Air-actuated gates are operated by air supplied and connected at the placing site or by an air tank carried on the bucket. The gates for buckets should be grout-tight, particularly if transported by truck or railway car for some distance. Lightweight buckets made of magnesium are available and advantageous where weight is a factor, such as transporting them by helicopter. See also Chap. 23.

Conveyors. Belt conveyors (Fig. 7) are used to transfer concrete horizontally and modest distances vertically. Conveyors are relatively inexpensive and may eliminate the need for other more expensive auxiliary equipment, such as cranes. Conveyors are particularly useful in areas of limited room, such as in tunnels for transferring concrete from the delivery car or truck to the hopper on a concrete pump or tunnel-invert lining machine, but are widely used on large areas such as floor slabs and bridge decks. A short drop chute should be used at the discharge end of a conveyor to help control segregation. A side-discharge feature on a conveyor (Fig. 8) provides more flexibility in discharging concrete at the desired location on bridge decks or slabs.

Fig. 5. Concrete bucket used to place concrete in bridge foundation. Bucket is transported from mixing plant on shore by helicopter. Use of lightweight bucket reduces deadweight and increases payload. (*Sikorsky Aircraft photograph.*)

Fig. 6. Lay-down-type concrete bucket, used when headroom is limited, discharging concrete in wall form. (*California Department of Water Resources photograph.*)

Concrete Pumps. The principal advantage of concrete pumps is that they can transfer or move concrete through or into congested areas. Pumps are particularly advantageous for tunnel lining, although they are used for many other applications, particularly in buildings in congested areas. The subject of concrete pumps is discussed in more detail in Chap. 36.

Fig. 7. System of belt conveyors used to transport concrete on a large bridge deck. Note short sections of drop chute at end of each belt. (*Morgen Mfg. Co. photograph.*)

Fig. 8. Belt conveyor placing concrete on a bridge deck. Side-discharge feature is movable and permits discharge of concrete at any point across span of conveyor, facilitating coverage of area. (*Morgen Mfg. Co. photograph.*)

Receiving hoppers (Fig. 9, also Figs. 2, 4, and 5) are used in combination with other equipment to provide temporary storage or surge capacity or to facilitate transfer of concrete from one type of handling equipment to another. Hoppers with center discharge and with side slopes steep enough to facilitate discharge of the concrete are preferred.

FIG. 9. Receiving hoppers provide temporary storage or surge capacity, or facilitate transfer of concrete. (*Gar-Bro Mfg. Co. photograph.*)

Drop chutes or elephant trunks (Fig. 10) are for the purpose of delivering concrete to a lower elevation without segregation or for the purpose of correcting segregation that might otherwise occur because of concrete's hitting reinforcing steel or other obstructions. They also keep the reinforcing steel and forms from becoming coated with mortar, where this may be considered objectionable. (Where concrete placing is rapid enough that this mortar coating does not dry out, this may not be objectionable.) These are made of rubber tubing, plastic tubing, sheet metal, or short sections of steel tubing fastened together so they are flexible

FIG. 10. Drop chutes used in dropping concrete to a lower elevation. (*California Department of Water Resources photograph.*)

and can be readily shortened. Hoppers for drop chutes should be large and steep enough so they readily discharge the concrete.

PLACING CONCRETE IN THE FORMS

2 Correct and Incorrect Methods of Handling and Placing Concrete

Figures 11 through 15 (from Ref. 1) show some of the correct and incorrect ways of handling and placing concrete. A study of these figures will give a better understanding of the principles involved and will greatly assist in selecting proper equipment for placing concrete, using equipment properly, and analyzing and correcting placing difficulties.

A basic rule is that concrete should be deposited as nearly as possible in its final location. The production of uniform concrete which materially assists the placing operations is discussed in Chaps. 16 through 21. Concrete consists of coarse aggregate and mortar and will separate if conditions and opportunity exist. To

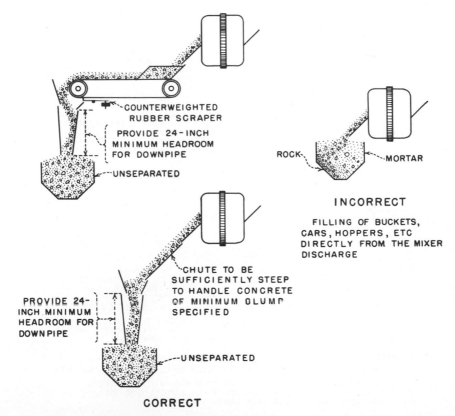

Fig. 11. Unless discharge of concrete from mixers is correctly controlled, uniformity resulting from effective mixing is destroyed by separation. (*By permission of American Concrete Institute.*[1])

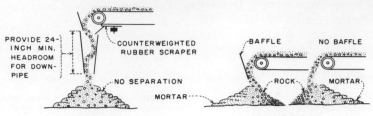

PROVIDE 24-INCH MIN. HEADROOM FOR DOWN-PIPE

COUNTERWEIGHTED RUBBER SCRAPER

NO SEPARATION

MORTAR

BAFFLE   NO BAFFLE

ROCK   MORTAR

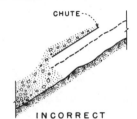

CORRECT

THE ABOVE ARRANGEMENT PRE-VENTS SEPARATION OF CONCRETE WHETHER IT IS BEING DISCHARGED INTO HOPPERS, BUCKETS, CARS, TRUCKS, OR FORMS.

INCORRECT

IMPROPER OR COMPLETE LACK OF CONTROL AT END OF BELT.

USUALLY A BAFFLE OR SHALLOW HOPPER MERELY CHANGES THE DIRECTION OF SEPARATION.

CONTROL OF SEPARATION OF CONCRETE AT THE END OF CONVEYOR BELT

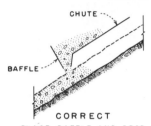

CHUTE

BAFFLE

CHUTE

CORRECT

PLACE BAFFLE AND DROP AT END OF CHUTE SO THAT SEPARATION IS AVOIDED AND CONCRETE REMAINS ON SLOPE.

INCORRECT

TO DISCHARGE CONCRETE FROM A FREE END CHUTE ON A SLOPE TO BE PAVED. ROCK IS SEPARATED AND GOES TO BOTTOM OF SLOPE.

VELOCITY TENDS TO CARRY CON-CRETE DOWN SLOPE.

PLACING CONCRETE ON A SLOPING SURFACE

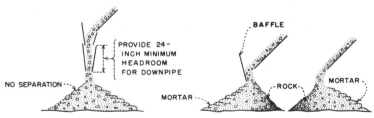

NO SEPARATION

PROVIDE 24-INCH MINIMUM HEADROOM FOR DOWNPIPE

MORTAR

BAFFLE

ROCK   MORTAR

CORRECT

THE ABOVE ARRANGEMENT PRE-VENTS SEPARATION, NO MATTER HOW SHORT THE CHUTE, WHETHER CONCRETE IS BEING DISCHARGED INTO HOPPERS, BUCKETS, CARS, TRUCKS, OR FORMS.

INCORRECT

IMPROPER OR LACK OF CONTROL AT END OF ANY CONCRETE CHUTE, NO MATTER HOW SHORT.

USUALLY A BAFFLE MERELY CHANGES DIRECTION OF SEPARATION.

CONTROL OF SEPARATION AT THE END OF CONCRETE CHUTES

THIS APPLIES TO SLOPING DISCHARGES FROM MIXERS, TRUCK MIXERS, ETC AS WELL AS TO LONGER CHUTES, BUT NOT WHEN CONCRETE IS DISCHARGED INTO ANOTHER CHUTE OR ONTO A CONVEYOR BELT.

FIG. 12. Control of separation of concrete. (*By permission of American Concrete Institute.*[1])

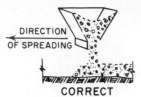

DIRECTION OF SPREADING

DIRECTION OF SPREADING

ROCK POCKETS FORM AT BOTTOM OF SLAB

CORRECT
TURN BUCKET SO THAT SEPARATED ROCK FALLS ON CONCRETE WHERE IT MAY BE READILY WORKED INTO MASS.

INCORRECT
DUMPING SO THAT FREE ROCK ROLLS OUT ON FORMS OR SUBGRADE.

DISCHARGING CONCRETE

CORRECT
DROPPING CONCRETE DIRECTLY OVER GATE OPENING.

INCORRECT
DROPPING CONCRETE ON SLOPING SIDES OF HOPPER.

FILLING CONCRETE HOPPERS OR BUCKETS

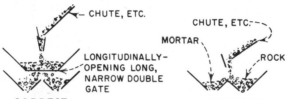

CHUTE, ETC.

LONGITUDINALLY-OPENING LONG, NARROW DOUBLE GATE

CHUTE, ETC.

MORTAR

ROCK

CORRECT
THE ABOVE ARRANGEMENT SHOWS A FEASIBLE METHOD IF A DIVIDED HOPPER MUST BE USED. (SINGLE DISCHARGE HOPPERS SHOULD BE USED WHENEVER POSSIBLE.)

INCORRECT
FILLING DIVIDED HOPPER AS ABOVE INVARIABLY RESULTS IN SEPARATION AND LACK OF UNI-FORMITY IN CONCRETE DELIVERED FROM EITHER GATE.

DIVIDED CONCRETE HOPPERS

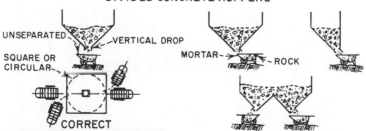

UNSEPARATED

VERTICAL DROP

SQUARE OR CIRCULAR

MORTAR

ROCK

CORRECT
DISCHARGE FROM CENTER OPEN-ING PERMITTING VERTICAL DROP INTO CENTER OF BUGGY. ALTERNATE APPROACH FROM OPPOSITE SIDES PERMITS AS RAPID LOADING AS MAY BE OBTAINED WITH OBJECTIONABLE DIVIDED HOPPERS HAVING TWO DIS-CHARGE GATES.

INCORRECT
SLOPING HOPPER GATES WHICH ARE IN EFFECT CHUTES WITHOUT END CONTROL CAUSING OBJECTION-ABLE SEPARATION IN FILLING THE BUGGIES

DISCHARGE OF HOPPERS FOR LOADING CONCRETE BUGGIES

FIG. 13. Correct and incorrect methods for loading and discharging concrete buckets, hoppers, and buggies. Correct procedure minimizes separation of coarse aggregate from mortar. (*By permission of American Concrete Institute.*[1])

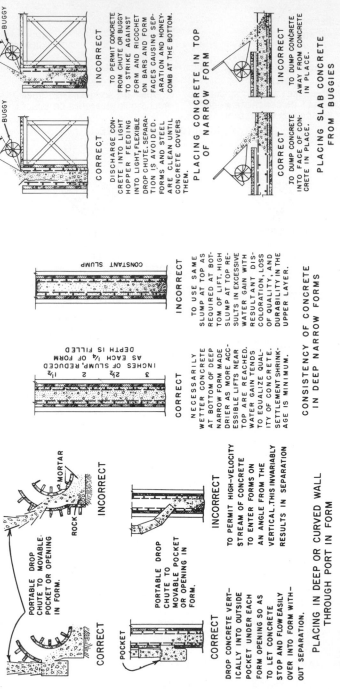

Fig. 14. Placing concrete in curved and narrow form and in slabs. (*By permission of American Concrete Institute.*[1])

minimize segregation or separation, it is desirable to drop concrete vertically, rather than at an angle, whenever possible (Figs. 11 to 14). There should be a short vertical drop chute at the end of sloping chutes or the end of belt conveyors. Baffles should generally be avoided as they may only cause separation to occur on the opposite side. Baffles that can be moved or tilted can sometimes be used satisfactorily provided they·are properly positioned to direct the stream of concrete into previously placed concrete. If baffles are used, they must be changed as

CORRECT

START PLACING AT BOTTOM OF
SLOPE SO THAT COMPACTION
IS INCREASED BY WEIGHT OF
NEWLY ADDED CONCRETE.
VIBRATION CONSOLIDATES.

INCORRECT

TO BEGIN PLACING AT TOP OF
SLOPE. UPPER CONCRETE TENDS
TO PULL APART, ESPECIALLY
WHEN VIBRATED BELOW, AS VIBRA-
TION STARTS FLOW AND REMOVES
SUPPORT FROM CONCRETE ABOVE.

WHEN CONCRETE MUST BE PLACED IN
A SLOPING LIFT

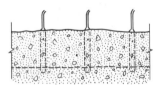

CORRECT

VERTICAL PENETRATION OF
VIBRATOR A FEW INCHES INTO
PREVIOUS LIFT (WHICH SHOULD
NOT YET BE RIGID) AT SYSTEM-
ATIC REGULAR INTERVALS
FOUND TO GIVE ADEQUATE
CONSOLIDATION.

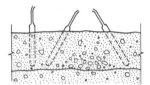

INCORRECT

HAPHAZARD RANDOM PENETRA-
TION OF THE VIBRATOR AT ALL
ANGLES AND SPACINGS WITHOUT
SUFFICIENT DEPTH TO ASSURE
MONOLITHIC COMBINATION OF
THE TWO LAYERS.

SYSTEMATIC VIBRATION OF EACH NEW LIFT

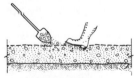

CORRECT

SHOVEL ROCKS FROM ROCK
POCKET ONTO SOFTER, AMPLY
SANDED AREA AND TRAMP OR
VIBRATE.

INCORRECT

ATTEMPTING TO CORRECT
ROCK POCKET BY SHOVELING
MORTAR AND SOFT CONCRETE
ON IT.

TREATMENT OF ROCK POCKET WHEN PLACING
CONCRETE

Fig. 15. Proper use of vibrator and treatment of rock pocket. (*By permission of American Concrete Institute.*[1])

the height of the concrete or other conditions change to be used effectively and intelligently. Conveyor belts should have a scraper or wiper to remove mortar adhering to the belt and return it to the concrete (Fig. 12).

A hopper should be filled by dropping concrete into the center and should discharge vertically from a center opening; hoppers with side discharge or sloping gates should be avoided as they may cause separation (Fig. 13).

Drop chutes or elephant trunks should be avoided when the concrete can be satisfactorily placed without them. A crane bucket moving along the top of and discharging directly into a wall form is entirely satisfactory provided segregation does not occur. Although some specifications may limit free fall of the concrete to 5 or 6 ft, much greater free fall can give entirely satisfactory results provided the fall is vertical and segregation does not occur (see Chap. 26). When used, drop chutes should be arranged so they can be quickly moved and shortened, or a sufficient number should be supplied to cover the placing area without moving. If the use of drop chutes or elephant trunks slows concrete placing, it only adds to the problems as concrete loses slump rapidly. The ends of elephant trunks should remain vertical; to push them to the side results in a sloping condition and causes separation. When concrete is placed in a deep or curved wall form through a port in the side of the form, the concrete should drop vertically into an outside pocket, then overflow into the form without separation (Fig. 14). Sometimes such a drop chute and pocket can be built into the form, between studs.

Concrete should be discharged from equipment and placed rapidly enough so that "stacking" does not occur. When concrete dribbles slowly from a bucket, chute, drop chute, or elephant trunk, it piles up and coarse aggregate rolls down the sides, separating. By opening the bucket faster or speeding up the flow of concrete it does not stack and separate and its impact force materially assists in its placing and consolidation.

Concrete being placed in a slab should be dumped into the face of the concrete in place, not away from it (Fig. 14).

If concrete has separated, the coarse aggregate should be shoveled onto concrete and incorporated into the concrete. Concrete or mortar should not be shoveled onto separated coarse aggregate (Fig. 15).

In placing concrete on a slope, placing should start at the bottom of the slope and move up the slope (Figs. 12 and 15).

Because concrete usually bleeds, that is, the solids settle and water moves to the top, concrete in a wall placement should be placed with a lower slump (lower water-cement ratio) in the uppermost layer, thus compensating for water gain, providing concrete of uniform appearance, and ensuring more durable concrete at the top where freezing and thawing conditions are usually more severe (Fig. 14).

Concrete should be deposited as nearly as possible in its final location, as mentioned previously. It should not be moved long distances by vibration. Rather, it should be deposited as close intervals and each pile consolidated by vibrating it in its final location, particular attention being given to vibrating the junction between piles.

Considerable thought should be given to the job conditions and in selection of equipment to place the concrete. For example, placing concrete in a floor slab or foundation walls directly from a ready-mixed concrete truck discharging from only a limited number of places results in having to move the concrete a considerable distance with a vibrator or gravity, resulting in demands for wet, lower-quality concrete to accomplish this movement readily, or the use of a long chute with a flat slope which would require wet concrete to flow down the chute. The wet concrete would have less strength and durability than concrete of moderate slump, and is more subject to cracking, dusting, abrasion, leakage or dampness, and other undesirable qualities. The desirable alternative to the above is to provide more access to the placing site so that the truck can discharge the concrete at as many locations as needed to cover the area properly, or to use other additional placing equipment to move the concrete from the truck mixer to the location

Table 1. Range of Characteristics, Performance, and Applications of Internal Vibrators*

(1)	(2)	(3)	(4)	(5)	(6)	(7)	(8)	(9)
			Suggested values of			Approximate values of		
Group	Diameter of head, in. (cm)	Recommended frequency, vibrations per min (Hz)	Eccentric moment, in.-lb (cm-kg)	Average amplitude, in. (cm)	Centrifugal force, lb (kgf)	Radius of action, in. (cm)	Rate of concrete placement, cu yd per hr per vibrator (m³/hr)	Application
1	¾–1½ (2–4)	10,000–15,000 (170–250)	0.03–0.10 (0.035–0.12)	0.015–0.03 (0.04–0.08)	100–400 (45–180)	3–6 (8–15)	1–5 (0.8–4)	Plastic and flowing concrete in very thin members and confined places. May be used to supplement larger vibrators, especially in prestressed work where cables and ducts cause congestion in forms. Also used for fabricating laboratory test specimens.
2	1¼–2½ (3–6)	9,000–13,500 (150–225)	0.08–0.25 (0.09–0.29)	0.02–0.04 (0.05–0.10)	300–900 (140–400)	5–10 (13–25)	3–10 (2.3–8)	Plastic concrete in thin walls, columns, beams, precast piles, thin slabs, and along construction joints. May be used to supplement larger vibrators in confined areas.
3	2–3½ (5–9)	8,000–12,000 (130–200)	0.20–0.70 (0.23–0.81)	0.025–0.05 (0.06–0.13)	700–2,000 (320–900)	7–14 (18–36)	6–20 (4.6–15)	Stiff plastic concrete [less than 3-in. (8 cm) slump] in general construction such as walls, columns, beams, prestressed piles, and heavy slabs. Auxiliary vibration adjacent to forms of mass concrete and pavements. May be gang mounted to provide full width internal vibration of pavement slabs.
4	3–6 (8–15)	7,000–10,500 (120–180)	0.70–2.5 (0.81–2.9)	0.03–0.06 (0.08–0.15)	1,500–4,000 (680–1,800)	12–20 (30–51)	(15–40) (11–31)	Mass and structural concrete of 0 to 2-in. (5 cm) slump deposited in quantities up to 4 cu yd (3 m³) in relatively open forms of heavy construction (powerhouses, heavy bridge piers and foundations). Also auxiliary vibration in dam construction near forms and around embedded items and reinforcing steel.
5	5–7 (13–18)	5,500–8,500 (90–140)	2.25–3.50 (2.6–4.0)	0.04–0.08 (0.10–0.20)	2,500–6,000 (1,100–2,700)	16–24 (40–61)	25–50 (19–38)	Mass concrete in gravity dams, large piers, massive walls, etc. Two or more vibrators will be required to operate simultaneously to melt down and consolidate quantities of concrete of 4 cu yd (3 m³) or more deposited at one time in the form.

Notes:
Column 3—While vibrator is operating in concrete.
Column 5—Peak amplitude (half the peak-to-peak value), operating in air.
Column 6—Using frequency of vibrator while operating in concrete.
Column 7—Distance over which concrete is fully consolidated.
Column 8—Assumes insertion spacing is 1½ times the radius of action, and that vibrator operates two-thirds of time concrete is being placed.
Columns 7 and 8—These ranges reflect not only the capability of the vibrator but also differences in workability of the mix, degree of deaeration desired, and other conditions experienced in construction.
* From ACI Standard 609-72 by permission of American Concrete Institute.

where it is deposited. This equipment, discussed previously, might include wheel-barrows, concrete carts, crane buckets, conveyors, or pumps.

Concrete in walls or in sections having considerable thickness (more than 18 in.) should be placed systematically in layers having a thickness of not more than 15 to 20 in. Each layer should be systematically vibrated, care being taken to ensure that the junction between deposits is adequately vibrated (Fig. 15).

Large areas having several layers such as mass concrete should be placed systematically in a stair-step manner. In placing a large area or large section, the placing should be arranged to proceed systematically and have the least area of concrete exposed, thus helping to avoid cold joints. More details on placing procedures will be covered in Chap. 26 for different kinds of construction.

CONSOLIDATION

Concrete after being deposited in forms must be consolidated into a homogeneous solid mass without such defects as rock pockets, voids, or sand streak. Until the advent of mechanical equipment that consolidates concrete by vibration, concrete was consolidated by laborers using spades, shovels, hand tampers, or their feet. It should again be pointed out that the purpose of vibration is to consolidate concrete, not move or place it. Almost all concrete is now compacted or consolidated with vibrators of many designs and characteristics. High-frequency vibration transforms low-slump concrete into a semiliquid mass so that gravity causes it to flow together into a compact mass. ACI Standard 609-72[4] lists the different types of internal vibrators and their characteristics and gives considerably more detail on consolidation than is covered here. Table 1 showing the performance of internal vibrators is reproduced from this report.

3 Equipment

Vibrators used on construction jobs for mass, structural, paving, or similar concrete are usually high-frequency low-amplitude vibrators. Low-frequency high-amplitude vibrators are sometimes used, particularly in fixed installations in plants producing precast elements or in concrete-block machines. One power float for heavy-duty concrete floors imparts a tamping action to the concrete which amounts to low-frequency vibration, effectively tamping the surface of earth-moist concrete.

Most vibrators are powered by either electricity or air. Small gasoline engines are also used, either direct-connected or driving a generator to supply electricity. For electric operation, adequate electric service is required. For air-operated vibrators, an adequate air supply is likewise required. Difficulties might occur with air vibrators slowing or stopping during cold weather because of freezing of moisture in the air supply. Sometimes the use of alcohol-type antifreeze agents or air driers may alleviate this difficulty; ethylene glycol should not be used since it "gums up" the vibrator. All vibrators, regardless of type, should be kept in good repair in order that they will operate satisfactorily. A marked loss in frequency will seriously reduce the effectiveness of a vibrator. The vibration frequency of a vibrator can be determined by a small inexpensive tachometer of the vibrating-reed type called a Vibra-Tak (Fig. 16). The frequency should be determined with the vibrator operating in concrete. A marked difference in the frequency when operating in and out of concrete or a frequency lower than normal or previously obtained for the same vibrator may indicate repairs are necessary, or an inadequate power or air supply, and corrective action should be taken. Pressure gages can be used to measure the air line pressure near the vibrator (Fig. 17). Some vibrators are variable-frequency to accommodate concrete of different consistencies.

Vibrators are likely to fail frequently and for this reason require considerable maintenance. Standby vibrators must be readily available for replacement should any in use fail to operate properly. One standby vibrator is a bare minimum where only one or two vibrators are used on a job. Larger jobs require several standby vibrators.

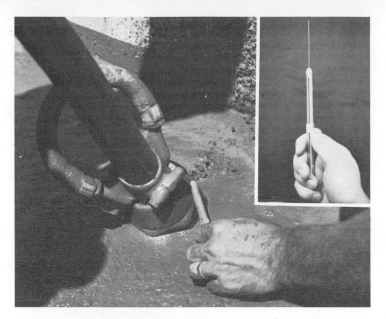

FIG. 16. Checking vibration frequency of a concrete vibrator with a Vibra-Tak. Measurement is made while vibrator is operating in concrete and measures vibrations per minute. (*California Department of Water Resources photograph.*)

FIG. 17. Pressure gage inserted in air line near vibrator to determine adequacy of air pressure to vibrators. Insertion is by hypodermic needle or line couplings.

Vibrators may be grouped into three classes:

Internal vibrators, or immersion vibrators (Figs. 18 through 24), are used internally in the concrete. These vary considerably, ranging from vibrators with small vibrating heads on flexible shafts suitable for thin walls or small sections congested with reinforcing to large vibrators suitable for mass concrete. In some, the motor is in the head; in others, it is separate with the vibrating element

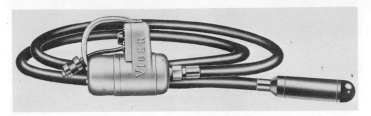

FIG. 18. Electric internal vibrator. (*Viber Co. photograph.*)

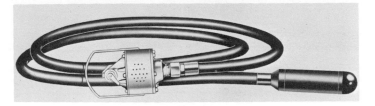

FIG. 19. Pneumatic internal vibrator. (*Viber Co. photograph.*)

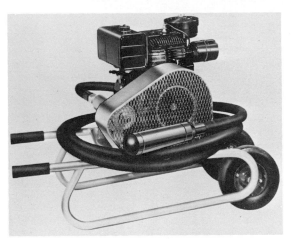

FIG. 20. Gasoline-engine-driven internal vibrator. (*Viber Co. photograph.*)

on a flexible shaft. Some have a saberlike extension on the vibrating element to use in congested areas such as around prestressing steel (Fig. 22). Several vibrators may be mounted together such as the multiple-tube or gang vibrators used on paving or canal-lining machines (Fig. 23) or gang-mounted on tractors as sometimes used for mass concrete, the latter application being more common in other countries (Fig. 24).

External vibrators (Fig. 25) are clamped or attached to the forms and vibrate the concrete from vibration of the form. External or form vibrators are usually used for thin or congested sections, and for locations such as in small tunnels or the arch of larger tunnels, where internal vibration is not possible or practical. They are also particularly useful for tunnel lining, either alone or with internal vibrators, and in plants for concrete pipe and other precast-concrete products.

Surface vibrators consist of vibrating-pan or screed vibrators, which vibrate the concrete from the surface—usually at the time the concrete is struck off or screeded. Surface vibrators are usually used as screeds for slabs or pavements. They are considered effective for a depth of about 8 in. If more than 8-in. depth is to be vibrated, the ability to vibrate the greater depth should be determined or internal vibrators should be used in conjunction with a vibrating screed. A vibrating screed is particularly useful as a screed for horizontal surfaces. (See Chap. 27, Figs. 27, 28, and 32.)

FIG. 21. Malan 6-in. one-man pneumatic vibrator for mass concrete. (*Malan Vibrator Co. photograph.*)

4 Consolidating Concrete with Vibration

Concrete should be consolidated to its maximum practical density so it is free of rock pockets and entrapped air and closes snugly around all form surfaces and embedded materials. Vibration should be applied systematically to cover all areas immediately after the concrete is deposited, as shown in Fig. 15. The vibrator should penetrate the layer of concrete vertically and into the underlying layer previously vibrated when possible. Immersion vibrators are sometimes used in slabs and pavements where they cannot be used vertically. In this usage, the vibrator should be kept immersed. The contact between adjacent piles or loads of concrete should be vibrated thoroughly. Vibrators should not be used to move concrete in the form appreciable distances.

Vibration should be continued until the concrete flattens and takes on a glistening appearance, the rise of entrapped air ceases, the coarse aggregate blends into the surface but does not completely disappear, and the vibrator, after an initial slow-down when first inserted in the concrete, resumes its speed. The vibrator should then be slowly withdrawn to ensure closing the hole resulting from insertion of the vibrator. Satisfactory mixes, properly placed, seldom require more than 5 to 15 or 20 sec of vibration at any one location to accomplish the above.

Fig. 22. Saber vibrator for vibrating thin sections and sections containing closely spaced reinforcing steel. Vibrator is homemade using a pneumatic external vibrator mounted on a piece of ¼- by 1½-in. steel flat bar. (*Model made by John Gebetsberger of Lathrop Construction Co.*)

Fig. 23. Gang-mounted internal vibrators vibrating 15-in. airfield parking apron. (*Viber Co. photograph.*)

FIG. 24. Internal vibrators mounted on a tractor vibrating mass concrete on a dam in Japan. Note bulldozer used to spread and level concrete prior to vibration.

FIG. 25. External or form pneumatic vibrators on forms for concrete pipe. (*Viber Co. photograph.*)

Overvibration can and does occur, although it is unlikely in well-proportioned concrete with normal-weight aggregate having the correct slump. Overvibrated concrete will have excess mortar on the top and a frothy appearance. The correction is to reduce the slump and also to reduce the sand content if the mix is oversanded. Overvibration can occur in lightweight-aggregate concrete, resulting in "floating" the coarse aggregate causing finishing difficulties, particularly if the concrete is too wet. Overvibration can occur in heavy concrete as there is a greater tendency for the heavy aggregate to settle in the fresh concrete, particularly if the concrete is too wet.

It is difficult to vibrate concrete on a sloping surface or where a grade must be held on a lower surface, such as a sloping wall surface or a step, unless positive means are taken to hold the concrete at lower elevations. Such situations frequently result in inadequate compaction of upper concrete as the amount of vibration

Fig. 26. Blemish on concrete surface caused by vibrator hitting plywood form surface and damaging it.

is frequently controlled by the difficulties caused by bulging of concrete at lower unformed surfaces, rather than by the needs of the concrete being vibrated above. Such a condition should and usually can be eliminated by proper selection of equipment or construction practices, such as use of temporary holding forms or heavy sliding forms. Examples of these are given in Chap. 26, in discussions of placing mass, canal-lining, or slope-paving concrete. Stub walls, which frequently are not satisfactorily vibrated, can and usually should be eliminated.

Revibration of previously vibrated concrete is generally beneficial when done at a time the vibrator will still penetrate the concrete with its own weight and make it plastic again. Revibration will improve strength, improve bond to horizontal reinforcing steel by eliminating the water pocket under the steel, remove air and water pockets, and eliminate settlement cracks such as occur at the junction of floor slabs and columns in the same placement or over window or door openings or other blockouts. Revibration into a layer of concrete partially hardened to the extent that it cannot be made fluid with vibration may cause a wavy line on the surface of walls, which is undesirable on exposed concrete.

Vibrators should not be allowed to strike the forms as the forms may be damaged, resulting in an unsightly blemish on the surface of the concrete, as well as damage to the forms (Fig. 26).

It is sometimes useful to vibrate the reinforcement in congested or inaccessible areas, although it is not advocated that the reinforcement system be regularly used for this purpose. This is best done with form vibrators. Fear that vibration transmitted to the reinforcing steel is detrimental is unfounded, even when part of the steel is embedded in partially hardened concrete. Contact between a vibrator and reinforcing steel will not cause damage (unless the steel is displaced—and it should be adequately tied and supported to prevent displacement).

In some instances, especially in architectural concrete, vibration may be supplemented by rodding or spading along the form, especially at corners or angles. Such spading as well as additional vibration will sometimes minimize or eliminate "bug holes." It is virtually impossible to eliminate bug holes from inward-sloping forms. Fear that ample or extra vibration necessary to reduce or eliminate bug holes will adversely affect the durability of air-entrained concrete because of removal

Fig. 27. Concrete slabs removed from a structure. Surface finish was excellent. Concrete below surface was inadequately compacted and full of rock pockets, which may have contributed to its failure.

of part of the air is unfounded, provided the concrete originally contained the amount of entrained air recommended for durability in ACI Standard 211.1-70[5] and in Chap. 11. Research has shown that the remaining air-void system is unimpaired in its ability to improve durability in freezing and thawing. Less than the recommended initial air content may result in not enough air remaining after vibration to provide adequate durability.

5 Consolidating Concrete without Vibration

Modern concrete technology and construction practice assume the use of vibration in placing concrete. Even for small jobs, vibrators are relatively low in cost or can be rented, and their use is encouraged. Where vibration is not used, concrete can be adequately placed but with more physical effort. The concrete will need to be more workable for hand-placing methods with more slump and perhaps a higher sand content, although excessive high slumps and sand contents are not necessary and are undesirable as they result in lower-quality concrete. After the concrete is placed in the forms, it must be worked with spades, tampers, shovels, or with the feet to consolidate it, eliminating voids. Simply dumping concrete into a form from a ready-mixed-concrete truck without some means being taken to consolidate it, even with high-slump concrete, cannot achieve a satisfactory job. This is seen in observing the results of such practices, particularly on small jobs such as foundations for houses or small buildings or sidewalks, slabs, or other flat work. Slabs such as those in Fig. 27 had an excellent surface finish but were inadequately compacted—and full of "rock pockets."

REFERENCES

1. ACI Standard 614-59, Recommended Practice for Measuring, Mixing, and Placing Concrete.
2. "Concrete Manual," 7th ed., Chap. VI, U.S. Bureau of Reclamation, 1963.
3. Waddell, Joseph J.: "Practical Quality Control for Concrete," McGraw-Hill Book Company, New York, 1962.
4. ACI Standard 309-72, Recommended Practice for Consolidation of Concrete.
5. ACI Standard 211.1-70, Recommended Practice for Selecting Proportions for Normal Weight Concrete.

Chapter 25

PREPARATION FOR PLACING CONCRETE

ROBERT F. ADAMS

Much preparatory work needs to be done before concrete is placed; foundations must be excavated, cleaned, and prepared; forms must be built; reinforcing steel must be placed; construction joints against previously placed concrete must be cleaned; embedded items such as bolts for anchoring machinery or equipment, pipe, conduit, castings for manholes or catch basins, water stops, and blockouts for gate guides, machinery bases, doors, windows, drains, and sumps must be prepared and placed.

All the above must be supervised and inspected by the contractor's representative and by the owner's engineer during progress of the work and given a final inspection before concrete is ordered for placing. The supervision and inspection are for the purpose of building the structure required by the owner as shown in his plans and in accordance with the specifications.

FOUNDATIONS AND JOINTS

1 Foundation Preparation

Foundation surfaces against which concrete is to be placed should conform to the specified location, size, and shape and should have adequate bearing capacity to carry the anticipated loads, both during and after construction.

As foundation excavation and preparation proceed, construction forces should be alert to changed or unforeseen conditions that would affect design of the foundation. Such conditions should be reported to the engineer for review and investigation.

Steel shells for cast-in-place piles, drilled pile openings, shafts, or caissons may be inspected by lowering a light in them. They should be clean and free of water when placing concrete starts.

Excavation. Rock should be excavated to sound material and be completely exposed, and rock surfaces should be normal to the direction of load. (Sloping rock surfaces have caused failure due to sliding of the foundation or footing on the rock.) Blasting should be done so that the foundation rock which remains is not damaged and excessive overbreak does not result. Controlled blasting, supervised by specialists in this field, can minimize rock problems caused by careless blasting and result in much better construction. Much blasting is carelessly done and results in considerable damage to remaining rock, or overbreak that must be removed and replaced with concrete, increasing the cost for concrete and for the job. Good practices, such as closer and better spacing of drill holes, controlled powder charges, and delays, can result in closer control of rock excavation to required line and grade and less damage to the rock (Fig. 1).

Surfaces of rock subject to air slaking or raveling need special attention prior to placing concrete on them. A covering of shotcrete, bituminous material, or other sealer is used to protect such rock from air slaking until concrete can be placed over it (Fig. 2). Covering rock with shotcrete or other sealer usually improves working conditions, particularly if wet conditions are encountered. The workmen do not work in mud; tools and equipment, forms, and reinforcing steel are kept cleaner; and cleanup prior to placing concrete is minimized.

Loose or unsound rock or debris should first be removed and open fissures cleaned to a suitable depth and to firm rock on both sides. Rock should be cleaned by use of brooms, picks, sandblast, water jets, air-water jets, or other effective means followed by thorough washing and blowing out the remaining water by air jets (Fig. 3). The rock surface should be dry and free of any

FIG. 1.   Close spacing of drill holes and careful blasting result in rock excavation in accordance with the plans. (*California Department of Water Resources photograph.*)

free surface water (indicated by a shiny surface) that will prevent bond of concrete to the rock.

Earth foundations should be generally excavated to original undisturbed material if practical. If the original undisturbed material is unsatisfactory for the foundation, it should be removed and backfilled. Backfill should be suitably compacted to the specified density. Subgrades or bases for pavements or slabs should be filled with specified material and suitably compacted to the required density in accordance with the plans and specifications. Soft spots in the foundation are eliminated by removal of unsuitable material and replacement with suitable material adequately compacted (Fig. 4) or lean backfill concrete. Overexcavation may be corrected by backfill with suitable earth material adequately compacted, or lean backfill concrete. Backfill concrete has sometimes been used with cement contents as low as 1 sack per cu yd. Earth subgrades should be damp or moist but not wet when concrete is placed on them.

FIG. 2. Excavation in shale which slakes covered with shotcrete to eliminate slaking and mud and improve working conditions. In contrast, note muddy area on right side outside shotcrete area and structure limits. (*Delta Pumping Plant of California Water Project. California Department of Water Resources photograph.*)

FIG. 3. Cleaning rock surface with air-water jet and removing loose rock preparatory to placing concrete. (*California Department of Water Resources photograph.*)

Drainage. All foundations should be free of debris, frost, ice, snow, mud, or water. Water which enters a foundation area should be drained to a sump and pumped out or drained to the outside of the foundation area. Sometimes permanent concrete or clay pipe or porous gravel underdrains are placed to provide drainage for water entering the area and to relieve uplift pressures on the concrete (Fig. 5). Such drains are usually designed as part of the job and are provided for in the plans and specifications. Drains may be constructed by placing the concrete or clay pipe on a clean gravel or crushed-stone blanket in a trench

FIG. 4. Backfilling and compacting an overexcavated foundation for a concrete culvert. (*California Department of Water Resources photograph.*)

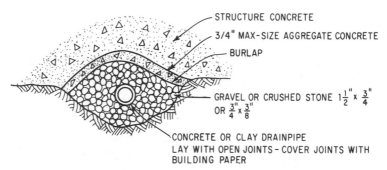

FIG. 5. Underdrain for temporary or permanent use.

or depression in the foundation, backfilling around it with the same clean gravel or crushed stone (¾ by ⅜ in. or 1½ by ¾ in.). Usually the aggregate is covered with burlap to prevent mortar from infiltrating the gravel when concrete is placed over it, and it may be covered with a thin layer (1 to 2 in.) of ¾-in. maximum-size-aggregate concrete to protect it from the hazards of construction if some time will elapse before the concrete is placed over it. The concrete or clay pipe is laid with open joints so that water drains into the pipe. Sometimes the concrete or clay pipe is omitted, and gravel alone serves as the drain. Some drains may be temporary and may be grouted later.

Sometimes the entrance of water into a foundation excavation can be prevented by installation of wells or well points and pumps outside the excavation, or it can be controlled by diverting the flow by constructing dikes of sandbags or sacked concrete.

Concrete should not be deposited in water, particularly in running water, unless approved by the plans and specifications of the engineer and suitable underwater concrete-placing procedures such as tremies or special concrete buckets are used. When concrete is placed underwater and it is impossible to unwater the excavation for cleanup, cleanup can be accomplished by suction pumps to remove mud and other debris. Underwater concrete-placing procedures are discussed in more detail in Chapter 26. In some cases, inspection of the underwater area by divers may be required.

Construction of forms and other related work is facilitated where it is difficult to control water in an excavation if a foundation seal is placed. This is a slab of concrete placed in the bottom of the excavation usually within a sheetpile cofferdam or caisson to seal off the flow of water. If such a seal is placed underwater, it should be placed by one of the underwater concrete-placing methods. Concrete for the seal is placed so the top is approximately at the elevation of the bottom of the foundation.

2 Construction-joint Preparation

When fresh concrete is placed upon or adjacent to previously placed concrete, whether old or new concrete, and where a bond between the two surfaces is

Fig. 6. Evidence of leakage of water through a poorly prepared construction joint. In many cases, concrete deteriorates at this point because of freezing and thawing.

required, it is essential that the surface of the previously placed concrete be clean and properly prepared. This is particularly important where watertight and durable joints are required (Fig. 6). The surface should be free of laitance, carbonation, scum, dirt, oil, grease, paint, curing compound, and loose or disintegrated concrete and should be slightly roughened. Several methods of cleanup are available, depending on the size of areas to be cleaned, age of the concrete, skill of the workmen, and availability of equipment.

Whichever method of surface cleanup is used, it is necessary only to remove undesirable surface dirt, laitance, or unsound mortar. It is necessary only to expose the sand in sound surface mortar (Fig. 7). Removal of sound mortar or concrete to expose coarse aggregate or to create roughness is not considered essential. Deep cutting, resulting in removal or undercutting of coarse aggregate, is also considered unnecessary and may be undesirable if the coarse aggregate is loosened.

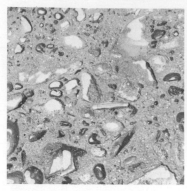
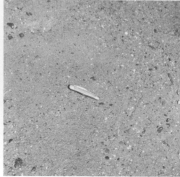

FIG. 7. Sandblasted construction joints. Concrete on left has been sandblasted more than really necessary, exposing a considerable amount of coarse aggregate. Additional cutting which would have undercut aggregate would have been undesirable. Concrete on right has been sandblasted only enough to remove laitance and soft mortar to expose sound mortar. The appearance is clean and bright.

Wet sandblast and high pressure water blasting are considered to be the most satisfactory methods of joint preparation (Fig. 8). If there is any question about the adequacy of other joint-preparation methods, their results should be compared with that obtained with wet sandblast. A properly prepared joint will have a clean, clear, sharp appearance, similar to that of a fresh break in sound concrete.

FIG. 8. Left, wet sandblast of a large area of mass concrete. Smaller sandblast units for smaller jobs are available. Right, high-pressure water jet cleanup. (Pipe grid shown on concrete surface is cooling pipe for next concrete placement.)

Many times construction-joint cleanup, particularly on horizontal surfaces, can be facilitated by foresight and care in seeing that the surface of the concrete placement is left relatively smooth and free of depressions, foot tracks, or other surface irregularities that will impede cleanup which is to be performed later. Excessive job traffic and overworking of the surface of fresh concrete should be particularly avoided. On mass concrete, finishing beyond that left by the vibration of the concrete is not necessary. On other concrete, finishing beyond a screed strike-off or float finish is not required. Workmen who find it necessary to walk on the surface of fresh mass concrete to place form anchors for future placements or for other reasons should wear "snowshoes" (oval wooden boards strapped to the feet) to reduce footprint depressions that make cleanup more difficult (Fig. 9). Planks can be laid on the surface for other foot traffic—or traffic should be detoured.

Hand tools such as wire brushes, wire brooms, hand picks, or bushhammers, used to remove dirt, laitance, and soft mortar, may produce satisfactory results but their use is time-consuming and considered practical only for small areas. If bushhammers or chippers are used, particularly those powered by pneumatic or electric hammers, care must be exercised to see that their use does not result in broken or shattered aggregate at the surface, leaving an undesirable surface for bonding new concrete, and consequently reduced bond.

Etching with acid is considered practical only for small areas. The concrete should be thoroughly wet with water prior to application of the dilute acid so the acid does not soak into the concrete. Commercial hydrochloric acid or muriatic acid diluted with water in the proportions of about 1 part acid to 3 or 4 parts water should be satisfactory for initial trials. The acid should be applied by spray, a bristle brush, or broom and vigorously scrubbed onto the concrete. After application of the acid and scrubbing, the surface should be thoroughly washed with water. The acid water should be disposed of where it will not create problems. Although the danger connected with using hydrochloric acid is not great, workmen should be cautioned regarding the danger and should be protected with rubber gloves, rubber boots, goggles, and rubber or plastic outerwear to the extent necessary to remove such danger.

FIG. 9. "Snowshoes" made from ¾-in. plywood permit workmen to work on top of concrete without leaving depressions which interfere with construction-joint cleanup.

Green cutting or initial cleanup method consists of washing the surface of freshly placed concrete with high-pressure water or air-water jets at the proper time in the hardening process (when the concrete has hardened sufficiently so it will not ravel) to remove surface laitance. This is satisfactory only for horizontal surfaces and is usually used only on large jobs. Considerable skill is required to get a satisfactory job with this method because the cutting must be done at the proper time, not too early so as to avoid too much cutting, and not too late in order to obtain sufficient removal of the surface laitance. The proper time is dependent upon the hardening characteristics of the concrete, which in turn is affected by many factors, such as temperature, characteristics of cement, use of admixtures, and water-cement ratio. (This technique is similar to that used to produce exposed-aggregate concrete on horizontal surfaces, although the removal of mortar to expose aggregate is more than is considered necessary for construction-joint cleanup.) Water and a wire broom may be found effective for small areas instead of air-water jets.

Use of retarder consists of treatment of the surface with a retarder and subsequent removal of the unhardened surface mortar with water and brushes or brooms, high-pressure water, or air-water jets. Retarders suitable for use on both horizontal and vertical surfaces are commercially available. The retarder is most conveniently applied to a horizontal concrete surface by a sprayer after the final finishing operation and before the concrete has set. For vertical surfaces, the retarder is usually applied to the forms. The manufacturer's instructions for application and coverage rate should be followed. The removal of the retarded surface layer by wire brushes or brooms and water or air-water jets may be done the following day or two. The longer it is delayed, the less retarded surface layer will be removed. This method is subject to some of the same disadvantages as the green-cutting method but to a lesser extent. It is dependent upon the hardening characteristics of the surface, which are dependent upon the same factors as mentioned previously plus the effect of the retarder. (This technique is also used for exposed-aggregate concrete surfaces. See Chap. 27.)

Wet sandblast is considered the most satisfactory method of joint preparation, as mentioned before. This is done with water in order to eliminate the dust problem (Fig. 8). Sandblasting is considered practical for both large and small construction projects and is suitable for either horizontal or vertical surfaces. Suitable equipment is readily available for both large and small jobs.

Sandblasting should be delayed as long as practical, preferably until just before the forms are erected for the next placement. If done after forms are erected, the forms must be protected from the sandblasting. Embedded items such as conduit or water stop must not be damaged by the sandblasting. As mentioned previously, only laitance and soft mortar should be removed. Overcutting resulting in exposure and undercutting of coarse aggregate is not necessary and is considered undesirable (Fig. 7). The advantage of wet sandblast over green cutting or the use of surface retarders (which must be done at the right time) is that sandblasting can be delayed, thereby reducing the chance that the surface will become unsatisfactory because of a scum forming from curing water, second rise of laitance, carbonation, or dirt from construction activity, and thus it can be done at a more convenient time.

Sand for sandblasting should be hard, not readily broken down, and sufficiently dry that it feeds through the equipment satisfactorily. Specially prepared dried sand from which the particles larger than No. 4 or No. 8 mesh and smaller than No. 16 or No. 30 mesh are removed is most satisfactory. Sand can be prepared on the job by using simple hand-screening methods, or on large jobs in mechanical screening equipment. Prepared sandblast sand is also commercially available in most areas.

Water blasting using a high-pressure water jet has become popular recently because of the development of suitable equipment and is considered by many to be equal to and an acceptable alternate to wet sandblast for construction joint cleanup (Fig. 8). This method is different from "green cutting" in that much higher water pressure is used, up to 10,000 psi, and no air. Water blasting is also done at a later time than possible in green cutting. The time at which it is done is not so critical as in green cutting; although the longer it is delayed, the less the production will be, as with sandblasting. One advantage of water blasting over wet sandblast is elimination of the sand problem (supply, scatter, and disposal). Some of the same comments above regarding wet sandblast apply equally to water blasting. Equipment for water blasting is available from several manufacturers.

Protection of construction joints is required after construction-joint cleanup by any of the above methods. All require a final thorough washing of the surface to remove the debris resulting from the cleanup operations. Once a joint is properly prepared, it should be protected to prevent its becoming unsatisfactory again. Protection includes some of the curing procedures such as damp sand, burlap, or cotton mat covers, and waterproof paper or plastic covering. Should a prepared surface become unsatisfactory because of a second rise of laitance, carbonation, or dirt, the cleanup should be redone.

The Mortar Layer. A rock surface or a construction joint in previously placed concrete should be properly prepared to receive the concrete placement, as discussed previously. Research by the Corps of Engineers[1,2,3] indicates that it may not make much difference whether or not a layer of mortar is broomed onto a properly prepared surface of either structural or mass concrete prior to placing concrete upon the surface. This research does indicate that a thin mortar either flowed or broomed onto the surface is superior to a thick mortar. In some cases superior joints were obtained without mortar. Also dry joints appeared to give superior results to wet joints.

Tests by the California Department of Water Resources showed that satisfactory construction joints were obtained in mass concrete including lean mass concrete (cementing materials equal to 188 lb per cu yd in the Oroville Dam core block) without the use of a layer of mortar on rock or a previously placed concrete lift. Inspection of cores taken across and along the joints and tests confirms this (Fig. 10). In some cores, it has been difficult to locate the construction joint. Successful elimination of mortar in structural concrete with smaller maximum-size aggregate is also being satisfactorily accomplished.

The advantage of eliminating mortar is that some of the problems involved in obtaining and using the mortar are eliminated. The quantities of mortar may be rather small. Only large equipment is oftentimes available for its delivery. The use of mortar sometimes results in coating forms, reinforcing steel, and equipment with mortar which sometimes dries before concrete is placed over it and must be cleaned, or is otherwise "messy."

Regardless of whether or not mortar is used on rock or concrete construction joints, good construction-joint cleanup or rock cleanup is also necessary and the surface should be saturated, surface-dry, or free of water which would be indicated by shininess. More than usual care must be taken to ensure that the first layer on the construction joint is adequately vibrated to secure good bond.

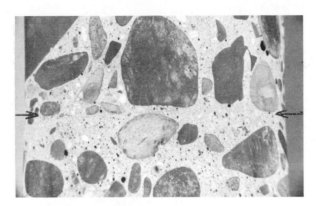

FIG. 10. Polished concrete core showing excellent construction joint between two lifts of mass concrete. Joint was wet sandblasted. Upper layer of concrete was placed on joint without use of mortar coating. (Arrows locate construction joint.) In some cores it has been difficult to locate the construction joint.

Mortar, if used on rock or concrete construction joints, should be made with the same materials and the same mix as in the concrete except that the coarse aggregate is omitted. On some jobs where a small amount of mortar is needed, materials and a small mixer are kept on the job to supply the mortar as needed. As mentioned above, thin mortar flowed on the surface gives as good or better results than a thick mortar broomed in. Mortar will readily spread itself on a large area if dumped from a concrete bucket held at a height of several feet, although the resulting splatter may be objectionable. The surface upon which mortar is placed should be free of puddles of water.

In structural concrete, such as wall sections where mortar is not used, a more workable bottom layer of concrete is sometimes used. A mix with more sand, more slump, or with part of the coarse aggregate omitted may be used. This provides not only a more suitable mix for bonding but a softer concrete in the bottom of the form for the next layer to work into.

FORMWORK

3   Form Preparation

The design and construction of forms are discussed in Sec. 4 and in Refs. 4, 5, and 6.

Details of form construction are covered in Sec. 4, but certain details are important to those concerned with preparation for and placing concrete. Particular care must be taken with forms for concrete that will be permanently exposed to view or in which special architectural effects are required. Since deflections in formwork cause waviness in finished surfaces that show under certain light conditions and may be objectionable, forms for architectural concrete must be designed carefully, and deflections may govern rather than design loads. For example, waviness in panels caused by deflection of plywood or sheathing between the studs would not be objectionable in a location not subject to view, but such waviness would be objectionable where good appearance is desirable, and heavier plywood or closer stud spacing would be required. Many times it is desirable to construct test sections or panels to determine that proposed material or techniques will achieve

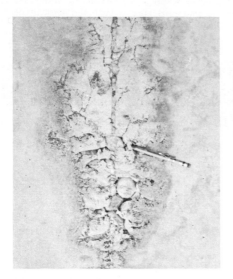

Fig. 11. Sand streaks and rock pockets left in concrete because of loss of mortar at gap between form panels during vibration of concrete.

Fig. 12. Rock pocket left because of leakage of mortar at opening for conduit during vibration of concrete.

the desired architectural effect, and to acquaint the workmen with techniques to be used. Sometimes this test section can be incorporated in a portion of the structure constructed at the start of construction and which will not be exposed.

Forms should be mortar-tight. Vibration will liquefy the mortar in the concrete, allowing it to leak from any openings in the forms, leaving voids, sand streaks, or rock pockets (Figs. 11 and 12). Sometimes wood-board forms will need to be soaked with water before concrete placing starts, to close the joints and prevent leakage of grout or mortar.

Forms should be tied together in an approved manner and braced so that movement does not take place (Figs. 13 and 20). Formwork should be anchored and braced to adjacent sections and to shores below so that upward or lateral movement of any part of the formwork system will be prevented during concrete placing. Tie rods which act as spacers for forms are available and are preferred. Wood spreaders, if used, should not remain in the concrete, and means should be provided so that they can be easily removed. Sometimes they can be tied to wires and lifted out as concrete placing proceeds. It is the responsibility of the inspector and

FIG. 13. Well-braced form for downstream sloping face of a spillway. Upper row of ties used to tie forms for lift above this one. Struts are removed after form is partially filled with concrete. Grade strip is on form at top of oiled area.

placing foreman to see that they are removed. Wire form ties may be used in unimportant work if both sides of the concrete will be covered with backfill or not exposed to view. Wire ties should be cut off flush with the concrete surface. For concrete surfaces exposed to view, metal ties or anchorages should terminate not less than 1 to 1½ in. from the surface of the concrete. Such ties should be made so that the projecting portion can be easily removed without causing spalling of the face of the concrete. Shebolts and tie rods or other similar form-tie devices should have sufficient threads engaged to develop full strength of the tie without thread stripping. Ties which when removed will leave an opening through the concrete should not be permitted. Directions for filling cone holes and other holes left by form ties are given in Chap. 48.

FIG. 14. Form watcher or carpenter checking alignment of form during placement of concrete by checking clearance of form from string stretched between ends of form. "Pigtails" are placed in concrete to serve as anchors for forms for next lift.

Line and grade on forms should be checked before and sometimes during placing (Fig. 14). Sometimes telltale devices are desirable so that any undesirable movement or excessive deflection of the forms can be detected. These can consist of lines or wires stretched between reference points with intermediate measurements, plumb bobs, plumb lines, check with a surveyor's transit or level, check with carpenter's level, or other methods.

When placing successive lifts of concrete on previously hardened concrete (construction joints), rustication strips (Fig. 15) or grade strips (Figs. 15 and 16) can be used to obtain a straight and neat horizontal joint. The grade strip (Fig.

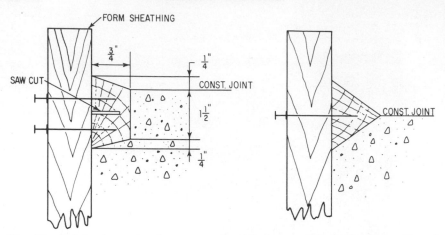

FORM SHEATHING

SAW CUT

CONST. JOINT

CONST. JOINT

FIG. 15. Rustication strips for construction joints or architectural effects. Use of double-head form nails facilitates removal of forms without removing strip. Saw cut facilitates removal of strip.

16) should be set as shown with its bottom edge about ½ in. below the finished elevation for the placement. Rustication strips are frequently required at locations other than construction joints to provide a pattern for architectural effect and at control or expansion joints. The grade and particularly the rustication strip should be planned beforehand and should be straight and continuous across the structure so as to give a pleasing appearance. Form anchorage should be provided about 4 in. below the top of a construction joint to anchor the forms for the placement above (Fig. 16). When the form is set for the succeeding lift, the sheathing should overlap the previous concrete not more than 1 in., as shown, and the forms should be drawn tightly to the existing concrete by means of the anchorage below and by ties close to the bottom of forms for the upper lift. This assures a neat joint and will prevent the unsightly offset which many times occurs (Fig. 17). (More than 1-in. lap prevents pulling the forms tight against the concrete.)

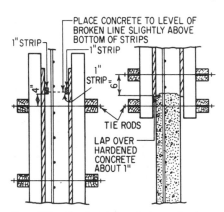

PLACE CONCRETE TO LEVEL OF BROKEN LINE SLIGHTLY ABOVE BOTTOM OF STRIPS

1" STRIP

1" STRIP

1" STRIP

TIE RODS

LAP OVER HARDENED CONCRETE ABOUT 1"

FIG. 16. Grade strip at top of placement ensures neat straight line at construction joint. Minimum overlap of form permits tightening form to prevent unsightly joint (see **Fig. 17**). (*Reproduced by permission of American Concrete Institute.*)

All elements of form ties, such as nut washers (sometimes called cat-heads), clamps, wedge clamps for snap ties, and wooden wedges used for alignment, should be secured to the forms with nails or other means so that vibration of the concrete does not loosen them, causing loss of form alignment or more serious difficulties (Fig. 18). Likewise, form shoring and bracing should be well secured to prevent similar difficulties.

Side openings should be provided and strategically located in the side of tall forms to facilitate placing of concrete, vibration, or inspection, when their use will facilitate such operations or where required. They should be made so they can be conveniently and quickly closed by the form carpenter when necessary and will not cause a blemish on the surface of the concrete where appearance is important. Runways should rest on the forms, not on the reinforcing.

Falsework, shoring, or centering should be adequately designed, supported, and braced to prevent undue sagging or movement and provide adequate support of the forms. Particular attention should be paid to bracing forms, shoring, and falsework against lateral movement. Provision should be made to leave all or at least part of falsework or shoring in place under beams, slabs, or arches after concrete has hardened until the concrete has obtained sufficient strength to permit safe removal. Forms should be designed so they can be removed and the shoring left in place when required. Vertical shores for multifloor forms must be set plumb and in alignment with lower tiers so that loads from the upper tiers are transferred directly to the lower tiers. Lateral loads must be transferred to the ground or to completed construction of

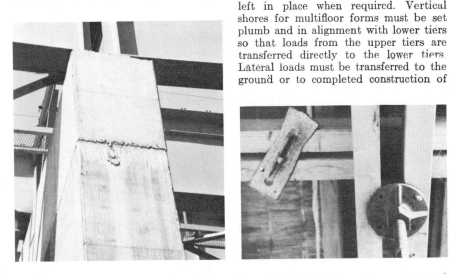

FIG. 17. Unsightly construction joint caused by form not fitting tightly against previous placement.

FIG. 18. Nailed form-tie wedges and nut washers prevent their loosening when concrete is vibrated during placing.

adequate strength. All joints in falsework and shoring members should be butt joints when practical, not lap splices. Adequate spread footings, mudsills, or other supports should be provided. Mudsills should not rest upon frozen ground—a thaw would be disastrous. Sometimes jacks or wedges can be used in falsework or shoring to adjust the forms or take up settlement before or during concrete placement. Wedges should be nailed in place to prevent their coming loose.

Tolerances for forms and concrete suggested in ACI Recommended Practice for Concrete Formwork,[4] in the absence of specification requirements, are as follows:

Where tolerances are not stated in the specifications or drawings for any individual structure or feature thereof, permissible deviations from established lines, grades, and dimensions are suggested below. The contractor is expected to set and maintain concrete forms so as to insure completed work within the tolerance limits.

Tolerances for reinforced concrete buildings*:

   1. Variation from the plumb:
      (*a*) In the lines and surfaces of columns, piers, walls, and in arrises ¼ in. per 10 ft, but not more than 1 in.
      (*b*) For exposed corner columns control-joint grooves, and other conspicuous lines:
         In any bay or 20 ft maximum.................................... ¼ in.
         In 40 ft or more... ½ in.

* Variations from plumb and linear building lines on upper stories of high rise structures (above 100 ft high) are special cases which may require special tolerances.

2. *Variation from the level or from the grades indicated on the drawings:*
   (a) In slab soffits,* ceilings, beam soffits, and in arrises:
   In 10 ft... ¼ in.
   In any bay or 20 ft maximum................................. ⅜ in.
   In 40 ft or more.. ¾ in.
   (b) For exposed lintels, sills, parapets, horizontal grooves, and other conspicuous lines:
   In any bay or 20 ft maximum................................. ¼ in.
   In 40 ft or more.. ½ in.

3. *Variation of the linear building lines from established position in plan and related position of columns, wall, and partitions:*
   In any bay or 20 ft maximum................................. ½ in.
   In 40 ft or more.. 1 in.

4. *Variation in the sizes and locations of sleeves, floor openings, and wall openings.* . ¼ in.

5. *Variation in cross-sectional dimensions of columns and beams and in the thickness of slabs and walls:*
   Minus.. ¼ in.
   Plus... ½ in.

6. *Footings:*
   (a) Variation in dimensions in plan:
   Minus.. ½ in.
   Plus... 2 in.†
   (b) Misplacement or eccentricity:
   2% of the footing width in the direction of misplacement but not
   more than... 2 in.†
   (c) Reduction in thickness:
   Minus.................................... 5% of specified thickness

7. *Variation in steps:*
   (a) In a flight of stairs:
   Rise... ⅛ in.
   Tread.. ¼ in.
   (b) In consecutive steps:
   Rise... 1⁄16 in.
   Tread.. ⅛ in.

Tolerances for special structures:

1. *Concrete canal lining:*
   (a) Departure from established alignment:
   2 in. on tangents
   4 in. on curves
   (b) Departure from established profile grade.......................... 1 in.
   (c) Reduction in thickness of lining.................. 10% of specified thickness: *provided*, that average thickness is maintained as determined by daily batch volumes
   (d) Variation from specified width of section at any height.... ¼ of 1% plus 1 in.
   (e) Variation from established height of lining.............. ½ of 1% plus 1 in.
   (f) Variations in surfaces:
   Invert ¼ in. in 10 ft
   Side slopes ½ in. in 10 ft

* Variations in slab soffits are to be measured *before* removal of supporting shores; the contractor is not responsible for variations due to deflection, except when the latter are corroboratory evidence of inferior concrete quality or curing, in which case only the *net* variation due to deflection can be considered.
† Applies to concrete only, not to reinforcing bars or dowels.

2. *Monolithic siphons and culverts:*
   (a) Departure from established alignment............................ 1 in.
   (b) Departure from established profile grade......................... 1 in.
   (c) Variation in thickness:
      At any point: minus 2½ % or ¼ in., whichever is greater
      At any point: plus 5 % or ½ in., whichever is greater
   (d) Variation from inside dimensions............................... ½ of 1 %
   (e) Variations in surfaces:
      Inverts ¼ in. in 10 ft
      Side slopes ½ in. in 10 ft

3. *Bridges, checks, overchutes, drops, turnouts, inlets, chutes, and similar structures:*
   (a) Departure from established alignment............................ 1 in.
   (b) Departure from established grades............................... 1 in.
   (c) Variation from the plumb or the specified batter in the lines and surfaces of columns, piers, walls, and in arrises:
      Exposed, in 10 ft.. ½ in.
      Backfilled, in 10 ft.. 1 in.
   (d) Variation from the level or from the grades indicated on the drawings in slabs, beams, horizontal grooves, and railing offsets:
      Exposed, in 10 ft.. ½ in.
      Backfilled, in 10 ft.. 1 in.
   (e) Variation in cross-sectional dimensions of columns, piers, slabs, walls, beams, and similar parts:
      Minus... ¼ in.
      Plus.. ½ in.
   (f) Variation in thickness of bridge slabs:
      Minus... ⅛ in.
      Plus.. ¼ in.
   (g) Footings: Same as for footings for buildings
   (h) Variation in the sizes and locations of slab and wall openings.......... ½ in.
   (i) Sills and side walls for radial gates and similar watertight joints. Variation from the plumb or level....................... Not greater than ⅛ in. in 10 ft.

4. *Tolerances in mass concrete structures:*
   (a) All structures:
      (1) Variation of the constructed linear outline from established position in plan:
         In 20 ft.. ½ in.
         In 40 ft.. ¾ in.
      (2) Variations of dimensions to individual structure features from established positions:
         In 80 ft or more...................................... 1¼ in.
         In buried construction.................... Twice the above amounts
      (3) Variation from the plumb, from the specified batter, or from the curved surfaces of all structures, including the lines and surfaces of columns, walls, piers, buttresses, arch sections, vertical joint grooves, and visible arrises:
         In 10 ft.. ½ in.
         In 20 ft.. ¾ in.
         In 40 ft or more...................................... 1¼ in.
         In buried construction.............. Twice the above amounts
      (4) Variation from the level or from the grades indicated on the drawings in slabs, beams, soffits, horizontal joint grooves, and visible arrises:
         In 10 ft.. ¼ in.
         In 30 ft or more...................................... ½ in.
         In buried construction.................... Twice the above amounts
      (5) Variation in cross-sectional dimensions of columns, beams, buttresses, piers, and similar members:
         Minus.. ¼ in.
         Plus... ½ in.
      (6) Variation in the thickness of slabs, walls, arch sections, and similar members:
         Minus.. ¼ in.
         Plus... ½ in.
   (b) Footings for columns, piers, walls, buttresses, and similar members:
      (1) Variation of dimensions in plan:
         Minus.. ½ in.
         Plus... 2 in.*

* Applies to concrete only, not to reinforcing bars or dowels.

    (2) Misplacement or eccentricity:
        2% of footing width in the direction of misplacement but not more than.. 2 in.*

    (3) Reduction in thickness....................... 5% of specified thickness

  (c) Sills and side walls for radial gates and similar watertight joints:
    (1) Variation from plumb and level.......... Not greater than ⅛ in. in 10 ft

5. Tolerances for concrete tunnel lining and cast-in-place conduits:

  (a) Departure from established alignment or from established grade:
    Free-flow tunnels and conduits............................... 1 in.
    High velocity tunnels and conduits........................... ½ in.
    Railroad tunnels... 1 in.

  (b) Variation in thickness at any point:
    Tunnel lining.. minus 0
    Conduits.................... minus 2½% or ¼ in, whichever is greater
    Conduits...................... plus 5% or ½ in., whichever is greater

  (c) Variations from inside dimensions............................. ½ of 1%

4 Use of Forms

Safety Provision. Provision should be made, particularly on wall forms, for safe working conditions for the placing crew. Ladders, working platforms, hand railing, kick boards, and other safety necessities should be provided, as required by all safety codes (Figs. 19 and 20).

Form Coatings. Forms must be treated with a form oil or other coating material to prevent adhesion to the concrete. Many form oils and parting compounds, such as plastic, lacquer, or shellac, are available. Major oil companies formulate or recommend form oil for either metal or wood forms. Sometimes the same oil will work for both. The form coating should be formulated for the particular usage and material to which it is to be applied and should protect the form from water, strip readily from the concrete, not interfere with subsequent curing, painting, or other surface treatments, or stain or cause softening of the surface. Form oil or coatings should preferably be applied before the forms are erected. Care should be taken to avoid getting form oil on the reinforcing steel or other embedded items or on construction joints.

Cleaning within Forms. All dirt, sawdust, shavings, loose nails, and other debris should be removed from within the forms before concrete placing is started. Provision should be made in wall or similar forms, particularly those for thin sections, for washing or blowing out all such debris, by leaving a panel or the bottom board loose or otherwise providing an opening at the bottom or end of the form for washout. Sometimes nails and wire clippings can be picked up by a magnet on a pole. The ends of reinforcing-steel tie wires should not rest against the form where they will cause unsightly rust stains.

Concrete Placing. The placing inspector and contractor's form watcher should carefully watch the forms during concrete placing for any signs of difficulties. Should any develop, placing should be immediately stopped or slowed as conditions warrant so that the conditions can be inspected and corrected.

Form Removal. Factors to be considered in determining when forms are to be removed are the effect of the form-removal operations in damaging the concrete, the structural strength or deflection of the concrete, curing and protection, finishing requirements, and requirements for reuse of the forms. Forms may be removed from a few minutes to days after the concrete is placed.

Forms should be removed using tools and equipment that will not damage either the concrete or forms if the forms are to be reused or salvaged. Wooden wedges, not steel tools, should be used to separate the forms from the concrete. Sometimes air or water pressure can be used to remove forms such as for larger slabs or pans for waffle slab construction, in which case fittings are provided for attaching the air or water lines. Forms should be designed and made so that they can be easily removed without damage to either the concrete or forms, particularly those which are to be reused.

---

* Applies to concrete only, not to reinforcing bars or dowels.

FIG. 19. Safety provisions on a column for a freeway viaduct. Note ladder and safety cage, working platform and railing, and barricade around excavation.

FIG. 20. Form insulated with 2½-in. insulation placed between studs. Form is well braced by ties from top of other structures located outside picture. Note working platform and hand railing.

Strength of the concrete and structural requirements determine the form-stripping time for arches, beams, girders, and similar load-carrying structural members. Strength of the concrete for this purpose is best determined by strength tests of field-cured concrete cylinders subjected to the same curing conditions as the structure. Some requirements specify a safety factor of 2 or more for form removal for such structures.

Form removal at 1 day is common on walls, columns, sides of beams, girders, and slabs, and other places where strength is not a problem. Forms of the soffits of arches, beams, and girders and floor slabs are commonly left in place 7 days, 14 days, or more depending upon strength and deflection considerations and strength gain of the concrete.

In summer concreting, forms are not considered curing and forms should be removed as soon as practical and specified curing started immediately, as discussed in Chap. 29. Sometimes forms are loosened without being removed and curing water is allowed to run down between the forms and concrete. Only under this condition should forms be left on as long as practical. However, it should be noted that leaving the forms in place is better than removal if it is known the concrete will not be properly cured, a not uncommon practice despite specification requirements for curing.

In winter concreting when insulated forms are used, they should be left as long as practical to protect the concrete from cold, thermal shock, or freezing conditions (Fig. 20). Reference 7 gives further information on the insulated forms. See also Chap. 30.

Early form removal is sometimes desirable to permit finishing the concrete to eliminate bug holes such as in inverts, on spillway surfaces, or for other reasons, or to facilitate repairs to the concrete while it is still "green" and bond of the repairs is improved. Early form stripping to permit earlier reuse of the forms may be facilitated by acceleration of the strength gain of the concrete. This can be accomplished by such means as warmer concrete, steam curing, or heated enclosures (with care taken to prevent drying), more cement (lower water-cement ratio), lower slump (lower water-cement ratio), use of high-early-strength cement, use of accelerators (when permitted by specifications), vacuum-processed concrete, and insulation of forms. Form removal in a few hours to permit reuse of the forms on a 24-hr cycle or sooner is common in tunnel lining and in precast-concrete-products plants.

Reuse of Forms. As mentioned previously, forms should be designed to permit easy removal and reuse. Forms should be carefully removed, cleaned, repaired, handled, and stored so they are not damaged. Cleaning may be done by a wooden scraper and stiff fiber brush on wood forms or steel scrapers and wire brush on steel forms. Mechanical wire brushes are used for cleaning large numbers of form panels on large projects. Damaged places are repaired as required. The holes are patched with metal plates, corks, or plastic inserts. When coating with some coating materials or oils, a drying period is required before stacking or storing forms.

Reusable form hardware such as form ties, she-bolts, nut washers, and wedges should be sorted, cleaned if necessary, and stored for reuse.

REINFORCEMENT

Reinforcing steel, commonly called "re-steel" or "re-bars," is placed in concrete to reinforce the concrete structure adequately so that it will support expected loads. Concrete is weak in tension, and most reinforcing is used to provide tensile reinforcement. Steel is also used to provide compressive reinforcement, in columns and arches, for example, and to resist shrinkage or temperature stresses. Reinforcing should be accurately placed. Inaccuracies can reduce the strength of the structural unit containing the steel or, if the steel is placed in the wrong place (the wrong side of a cantilever beam, for example), might lead to failure of the structure. It is particularly important that reinforcing steel be properly placed in areas subject to earthquakes, tornadoes, and hurricanes. Investigation of structural failures following such disasters shows many violations of specifications and good practice in placing the reinforcing steel and making adequate connections between structural elements. This caused or contributed to the failure. Also indicated is a lack of competent inspection. Plans and specifications give the grade of steel, its location in the structure, and other requirements. Placing plans, bar lists, and bar schedules give details regarding the dimension and location of each bar in the structure. Prior to placing concrete, inspection by both the contractor and owner's representatives should be made to see that the reinforcing steel is properly placed. Particular attention should be given to grade of steel, size, number, location, and spacing of bars, correctness of bends, clearance from forms or required cover, cleanness of steel, supports and ties, and splices.

Those concerned with reinforcing steel will find Chap. 3 and Refs. 8 through 13 helpful. The ACI Building Code, ACI Standard 318-71 is quite detailed.

The various grades of reinforcing steel, their properties, sizes and weights of bars available, and markings used to designate the producer, size, and grade are discussed in Chap. 3. Bars received on the job are tagged or marked for identification. Bent bars are identified with tags giving order number, number of pieces in the bundle, size, length, and the bar mark.

5 Bends

Typical Bar Bends. Types of bar bends are standardized as shown in Fig. 21. This permits bar lists to refer to the standard types. Type 3, for example, immediately describes a trussed bar hooked at each end. When no value is given for any dimension, distance G for Type 3 bar, for example, the bar does not have a hook on that end. An example of a bar list which is a bill of materials for a structure or a part of a structure is shown in Fig. 22. This bar list does not refer to the standard bar bends but shows the shape and location of each bar in the structure. This type of bar list gives complete information for both fabricating the reinforcing steel and for placing it, making the list more useful. Bar lists are usually prepared by the reinforcing-steel fabricator.

Tolerances in Fabrication. Standard tolerances accepted in cutting and bending bars are given in Refs. 9 and 10 and shown in Fig. 23. Rolling mills are allowed permissible variations in weight of plus or minus $3\frac{1}{2}\%$ for a lot of deformed bars and of minus 6% for an individual bar.

Hooks and Bends. The ACI Building Code, ACI Standard 318-71,[8] defines a hook as either:

1. A semicircular turn plus an extension of at least four bar diameters but not less than $2\frac{1}{2}$ in. at the free end of the bar

2. A 90° turn plus an extension of at least 12 bar diameters at the free end of the bar

3. For stirrup and tie anchorage only, either a 90 or a 135° turn plus an extension of at least six bar diameters but not less that $2\frac{1}{2}$ in. at the free end of the bar

The radii of bend measured on the inside of the bar for standard hooks should not be less than the values of Table 1, except that for sizes No. 6 to No. 11,

Table 1. Minimum Diameter of Bend*

Bar size	Min diameter, inside of bar
No. 3–8	6
9, 10, or 11	8
14 or 18	10

*From ACI 318-71, permission of American Concrete Institute. For No. 3 through 11 grade 40 bars having 180° hooks, minimum diameter is 5 bar diameters.

inclusive, in structural and intermediate grades of bars only, the minimum radius should be $2\frac{1}{2}$ bar diameters. For stirrup and tie hooks and bends other than standard hooks,

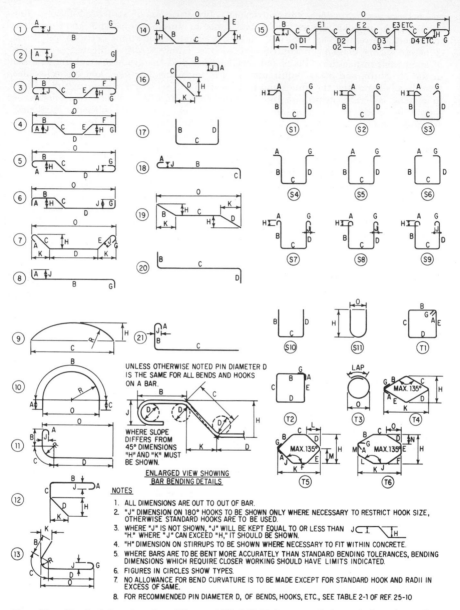

Fig. 21. Typical bar bends. (*From ACI 315-65 by permission of American Concrete Institute.*)

MEMBER	NO. PER MEMBER	TOTAL NO.	SIZE	LENGTH	SHAPE	LOCATION	MARK

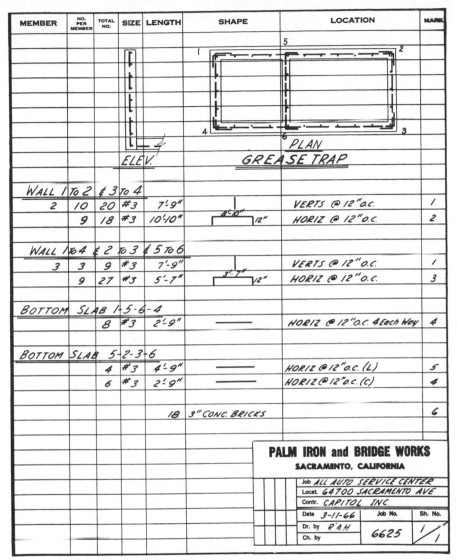

ELEV.   PLAN   GREASE TRAP

MEMBER	NO. PER MEMBER	TOTAL NO.	SIZE	LENGTH	SHAPE	LOCATION	MARK
WALL 1 To 2 & 3 To 4							
	2	10	20	#3	7'-9"		VERTS @ 12" o.c.	1
		9	18	#3	10'-10"	8"-10" 12"	HORIZ @ 12" o.c.	2
WALL 1 To 4 & 2 To 3 & 5 To 6							
	3	3	9	#3	7'-9"		VERTS @ 12" o.c.	1
		9	27	#3	5'-7"	3'-7" 12"	HORIZ @ 12" o.c.	3
BOTTOM SLAB 1-5-6-4							
			8	#3	2'-9"	——	HORIZ @ 12" o.c. 4 Each Way	4
BOTTOM SLAB 5-2-3-6							
			4	#3	4'-9"	——	HORIZ @ 12" o.c. (L)	5
			6	#3	2'-9"	——	HORIZ @ 12" o.c. (c)	4
			18			3" CONC. BRICKS		6

PALM IRON and BRIDGE WORKS
SACRAMENTO, CALIFORNIA

Job	ALL AUTO SERVICE CENTER
Locat.	64700 SACRAMENTO AVE
Contr.	CAPITOL INC.

Date 3-11-66	Job No.	Sh. No.
Dr. by R'A.H.	6625	1/1
Ch. by		

FIG. 22. Bar list for a simple structure. Such a bar list, which shows identification mark, shape, and dimension and location of bars in structure, provides complete information for the reinforcing-steel fabricator, bar setter, and inspector. (*Palm Iron Works bar list.*)

1. Inside diameter of bends for stirrups and ties shall not be less than 1½ in. for No. 3, 2 in. for No. 4, and 2½ in. for No. 5 bars.

2. Bends for all other bars should have radii on the inside of the bar not less than the values of Table 1.

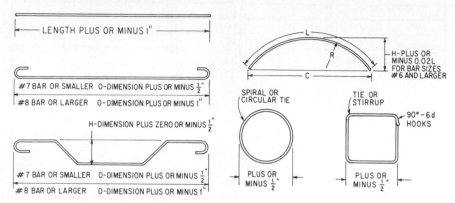

Fig. 23. Standard tolerances for cutting and bending steel reinforcing bars. (*From CRSI "Recommended Practice for Placing Reinforcing Bars."*)

All bars should be bent cold, unless otherwise permitted by the engineer. Bars partially embedded in concrete should not be field-bent except as shown on the plans or specifically permitted by the engineer.

Radial Bending. Reinforcing bent to a radius for circular tanks, silos, chimneys, tunnels, and similar uses can be furnished as straight bars and sprung into place, provided the radius is great enough. Radii less than that shown in Table 2 require the bars to be prebent to the specified radii.

Table 2. Maximum Radius Requiring Prebending of Reinforcing Bars*

Bar size, No.	Max radius, ft
2	8
3	10
4	15
5	25
6	40
7	60
8	80
9	110
10	130
11	150
14S	Prebend all
18S	Prebend all

* By permission of Concrete Reinforcing Steel Institute.

6　Placing Reinforcement

Tolerances in Placing. The ACI Building Code requires that reinforcement, prestressing steel, and prestressing steel ducts should be placed within the following tolerances:

1. For clear concrete protection and for depth d in flexural members, walls, and compression members, where d is

8 in. or less...............	± ¼ in.
8 to 24 in...............	± ⅜ in.
24 in. or more..........	± ½ in.

but the cover should not be less than one-third the specified cover.

2. For the longitudinal location of bends and ends of bars: ±2 in., except that for discontinuous ends of members tolerance should be ±½ in.

Spacing of Bars. The Building Code further requires that: (1) The clear distance between parallel bars (except in columns and between multiple layers of bars in beams) should be not less than the nominal diameter of the bars, nor 1 in. (2) Where reinforcement in beams or girders is placed in two or more layers, the clear distance between layers shall be not less than 1 in., and the bars in the upper layers should be placed directly above those in the bottom layer. (3) In walls and slabs other than concrete-joist construction, the principal reinforcement should be centered not farther apart than three times the wall or slab thickness nor more than 18 in. (4) In spirally reinforced and in tied columns, the clear distance between longitudinal bars should be not less than 1½ times the bar diameter, nor 1½ in. (Fig. 24). (5) The clear distance between bars should also

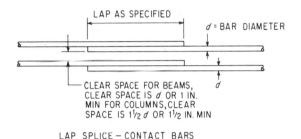

LAP SPLICE — CONTACT BARS

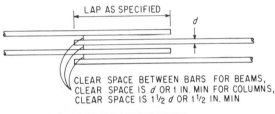

LAP SPLICE — SPACED BARS

Fig. 24. Typical minimum spacing in reinforcing bars. (*From CRSI "Recommended Practice for Placing Reinforcing Bars."*)

apply to the clear distance between a contact splice and adjacent splices or bars. (6) Groups of parallel reinforcing bars bundled in contact to act as a unit must be deformed bars with not over four in any one bundle and should be used only when stirrups or ties enclose the bundle. Bars larger than No. 11 should not be bundled in beams or girders. Bars in a bundle should terminate at different points with at least 40 bar diameters stagger unless all the bars end in a support. Where spacing limitations are based on bar size, a unit of bundled bars should be treated as a single bar of equivalent area.

Usually the exact spacing between bars is not critical. For example, in placing bars in flat slabs, if bar spacing was specified at 6½-in. centers, and bar supports with 1-in. corrugations were used, bars spaced alternately at 6 and 7 in. would provide the required steel and should be satisfactory otherwise. The required number of bars is important, and they should be fairly evenly spaced.

Sometimes some modification can be made to the reinforcement location to accommodate embedded materials such as inserts, conduit, or blockouts, to facilitate concrete placing. The effect of this on the structural integrity of the structure should be considered and the designer should be consulted. Indiscriminate change, removal, or cutting of the reinforcement should not be permitted. Sometimes bars can be either temporarily shifted or removed to facilitate concrete placing, being replaced when concrete placing reaches the point where the bars are to be covered, at which time they are replaced.

Splices. Splices in reinforcing steel are commonly used and cannot be avoided because of manufacturing, fabrication, or transportation limitations. Therefore it is necessary to provide for properly designed splices. The ACI Building Code gives detailed requirements for splices and insists that they be made only as required or permitted on the design drawings, in the specifications, or as authorized by the engineer. Splices should preferably be made at locations where the stresses or forces on the bars are at their minimum.

Splices may be made by lapping the bars, by welding, or by mechanical connectors. In most cases, lapped splices are more economical; however, in the case of larger bars, other types might be more economical. Because of close spacing of bars, it is not always possible to use lap splices.

Stress in a tension bar can be transmitted through the concrete and into another adjoining bar by a lapped splice. The amount of lap depends upon the bar size, grade of steel, and concrete strength, and it is computed according to the ACI Building Code or is given in the specifications. Tension lap splices in bars larger than No. 11 are not permitted. Compression lap splices in bars larger than No. 11 are permitted in some cases.

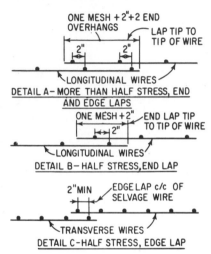

ONE MESH + 2″+2 END
OVERHANGS
LAP TIP TO
TIP OF WIRE
2″   2″
LONGITUDINAL WIRES
DETAIL A— MORE THAN HALF STRESS, END
AND EDGE LAPS

ONE MESH + 2″
END LAP TIP
TO TIP OF WIRE
2″
LONGITUDINAL WIRES
DETAIL B— HALF STRESS, END LAP

2″MIN   EDGE LAP c/c OF
SELVAGE WIRE
TRANSVERSE WIRES
DETAIL C—HALF STRESS, EDGE LAP

FIG. 25. Lap requirement for welded-wire fabric. (*From ACI 315-65*,[10] *by permission of American Concrete Institute.*)

Bars carrying compression only may be butted together and held in concentric alignment by a mechanical connector (see Chap. 3). The ends of such bars must usually be cut square, within 1½°. Such splices may be used only in members containing closed ties, closed stirrups, or spirals. Such connections are required to develop 125% of the specified yield strength of the bar.

Welded splices may also be used. They are also required to develop 125% of the specified yield strength of the bar. All welding should conform to AWS D12.1, Recommended Practices for Welding Reinforcing Steel, Metal Inserts and Connections in Reinforced Concrete Construction.[11,12] Chapter 3 discusses various types of welded splices and welding procedures in detail.

Welded-wire fabric used as reinforcement in structural slabs should be spliced in accordance with the following provisions:

1. Lapped splices of wires in regions of maximum stress (where they are carrying more than one-half of the permissible stress) should be avoided wherever possible; such splices where used should be so made that the overlap measured between outermost cross wires of each fabric sheet is not less than the spacing of the cross wires plus 2 in.

2. Splices of wires stressed at not more than one-half the permissible stress should be so made that the overlap measured between outermost cross wires is not less than 2 in.

These requirements are shown in Fig. 25.

Cover. Concrete protects reinforcing steel from corrosion and serves as fire-proofing. Adequate cover or embedment of the steel in concrete to provide this protection is specified in the ACI Building Code.

The following minimum concrete cover shall be provided for reinforcing bars, prestressing tendons, or ducts. For bar bundles, the minimum cover shall equal the equivalent diameter of the bundle, but need not be more than 2 in. or the tabulated minimum, whichever is greater.

	Minimum cover, in.
1. Cast-in-place concrete (nonprestressed):	
Cast against and permanently exposed to earth.....................	3
Exposed to earth or weather:	
No. 6 through No. 18 bars....................................	2
No. 5 bars, ⅝ in. wire, and smaller...........................	1½
Not exposed to weather or in contact with the ground:	
Slabs, walls, joists:	
No. 14 and No. 18 bars....................................	1½
No. 11 and smaller.......................................	¾
Beams, girders, columns:	
Principal reinforcement, ties, stirrups, or spirals.................	1½
Shells and folded plate members:	
No. 6 bars and larger.....................................	¾
No. 5 bars, ⅝ in. wire, and smaller...........................	½
2. Precast concrete (manufactured under plant control conditions):	
Exposed to earth or weather:	
Wall panels:	
No. 14 and No. 18 bars....................................	1½
No. 11 and smaller.......................................	¾
Other members:	
No. 14 and No. 18 bars....................................	2
No. 6 through No. 11......................................	1½
No. 5 bars, ⅝ in. wire, and smaller...........................	1¼
Not exposed to weather or in contact with the ground:	
Slabs, walls, joists:	
No. 14 and No. 18 bars....................................	1¼
No. 11 and smaller.......................................	⅝
Beams, girders, columns:	
Principal reinforcement....................................	d_b but not less than ⅝ and need not exceed 1½
Ties, stirrups, or spirals..................................	⅜
Shells and folded plate members:	
No. 6 bars and larger.....................................	⅝
No. 5 bars, ⅝ in. wire, and smaller...........................	⅜
3. Prestressed concrete members—prestressed and nonprestressed reinforcement, ducts, and end fittings:	
Cast against and permanently exposed to earth.....................	3
Exposed to earth or weather:	
Wall panels, slabs, and joists..............................	1
Other members..	1½
Not exposed to weather or in contact with the ground:	
Slabs, walls, joists.......................................	¾
Beams, girders, columns:	
Principal reinforcement....................................	1½
Ties, stirrups, or spirals..................................	1
Shells and folded plate members:	
Reinforcement ⅝ in. and smaller............................	⅜
Other reinforcement.......................................	d_b but not less than ¾

The cover for nonprestressed reinforcement in prestressed concrete members under plant control may be that given for precast members.

Cover specified in type 3 above is for prestressed members with stresses less than or equal to the limits of Section 18.4.2.2 of the ACI Building Code. When tensile stresses exceed this value for members exposed to weather, earth, or corrosive environment, cover shall be increased 50%.

In corrosive atmospheres or severe exposure conditions, the amount of concrete protection shall be suitably increased, and the denseness and nonporosity of the protecting concrete shall be considered, or other protection shall be provided.

Exposed reinforcing bars, inserts, and plates intended for bonding with future extensions shall be protected from corrosion.

When a fire-protective covering greater than the concrete protection specified in this section is required, such greater thicknesses shall be used.

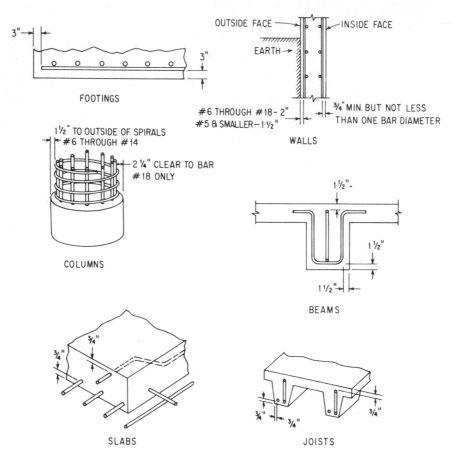

Fig. 26. Concrete cover required for concrete protection. In no case shall reinforcement be within one bar diameter of the surface of the concrete. (*From Ref. 9 by permission of Concrete Reinforcing Steel Institute.*)

Figure 26 shows concrete-cover requirements.

Lateral Reinforcement. The ACI Building Code requires spiral compression member reinforcement to consist of evenly spaced continuous spirals held firmly in place and true to line by vertical spacers. At least two spacers should be used for spirals 20 in. or less in diameter, three for spirals 20 to 30 in. in diameter, and four for spirals more than 30 in. in diameter. When spiral rods are 5/8 in.

or larger, three spacers should be used for spirals 24 in. or less in diameter and four for spirals more than 24 in. in diameter. The spirals should be of such size and so assembled as to permit handling and placing without being distorted from the designed dimensions. For cast-in-place construction, the material used in spirals should have a minimum diameter of ⅜ in. Anchorage of spiral reinforcement should be provided by 1½ extra turns of spiral rod or wire at each end of the spiral unit. Splices when necessary in spiral rods or wires should be tension lap splices of 48 diameters minimum but not less than 12 in., or welds. The clear spacing between spirals should not exceed 3 in. or be less than 1 in. The reinforcing spiral should extend from the floor level in any story or from the top of the footing to the level of the lowest horizontal reinforcement in the slab, drop panel, or beam above. In a column with a capital, the spiral should extend to a plane at which the diameter or width of the capital is twice that of the column. Where brackets or beams are not present on all sides of a column, ties should extend above the termination of the spiral to the bottom of the slab or drop panel.

Nonprestressed bars for tied columns should be enclosed by lateral ties, at least No. 3 in size for longitudinal bars No. 10 or smaller, and at least No. 4 in size for No. 11, 14, and 18 and bundled longitudinal bars. The spacing of the ties should not exceed 16 longitudinal bar diameters, 48 tie bar diameters, or the least dimension of the column. The ties should be so arranged that every corner and alternate longitudinal bar should have lateral support provided by the corner of a tie having an included angle of not more than 135°, and no bar should be more than 6 in. away on either side of such a laterally supported bar. Ties should be located vertically not more than half a tie spacing above the floor or footing; they should be spaced to not more than half a tie spacing below the lowest horizontal reinforcement in the slab or drop panel above, except that where beams or brackets provide closure on all sides of the column, the ties may be terminated not more than 3 in. below the lowest reinforcement in these beams or brackets. Welded wire fabric of equivalent area may be used. Where the bars are located around the periphery of a circle, a complete circular tie may be used.

Compression reinforcement in beams or girders should be enclosed by ties or stirrups satisfying the Building Code or by welded wire fabric of equivalent area. Such stirrups and ties should be used throughout the length where the compression reinforcement is required.

Lateral reinforcement for flexural framing members subject to stress reversals or torsion at supports should consist of closed ties, closed stirrups, or spirals extending around main reinforcement.

Closed ties and stirrups may be formed in one piece by overlapping standard stirrup or tie end hooks around a longitudinal bar, or in one or two pieces spliced or anchored in accord with the Building Code.

Typical column ties meeting the above requirements are shown in Fig. 27, and typical arrangement of column reinforcing in Fig. 28.

Corner and Wall Intersection Bars and Connection Details. Horizontal bars at corners and at all intersections are required to hook around the corner or into the intersecting wall to provide strength at the corner or wall intersection. This is usually accomplished by either putting a hook on a long bar or by using a short corner or elbow bar. Suitable corner and wall intersection details are shown in Fig. 29.

It is important that good connections between various structural elements be made in order that the structure be tied together and perform satisfactorily. A number of connection details are shown in Fig. 30.

7 Bar Supports, Spacers, and Ties

Bars should be supported, anchored, and tied to hold them in place before and during concrete placing (Fig. 31). Supports, anchors, or ties should not permit subsequent leakage of water into the hardened concrete, which would cause corrosion.

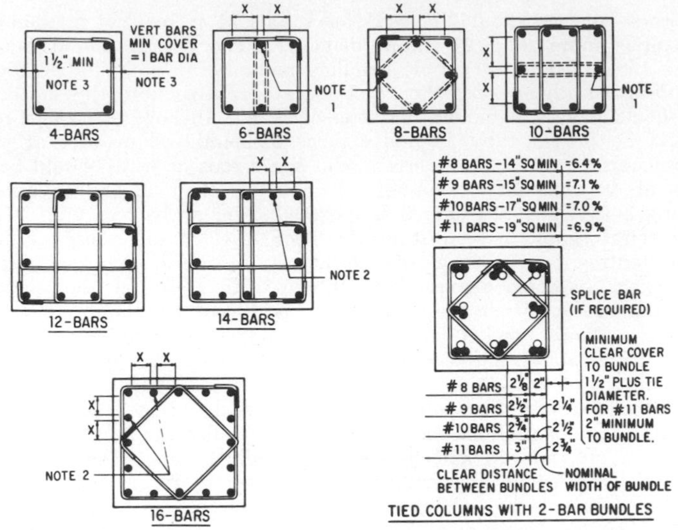
VERT BARS MIN COVER =1 BAR DIA
1½" MIN
NOTE 3
NOTE 3
4-BARS
X X
NOTE 1
6-BARS
8-BARS
NOTE 1
10-BARS
12-BARS
NOTE 2
14-BARS
NOTE 2
16-BARS
#8 BARS-14"SQ MIN =6.4%
#9 BARS-15"SQ MIN =7.1%
#10 BARS-17"SQ MIN =7.0%
#11 BARS-19"SQ MIN =6.9%
SPLICE BAR (IF REQUIRED)
MINIMUM CLEAR COVER TO BUNDLE 1½" PLUS TIE DIAMETER. FOR #11 BARS 2" MINIMUM TO BUNDLE.
#8 BARS 2⅛" 2"
#9 BARS 2½" 2¼"
#10 BARS 2¾" 2½"
#11 BARS 3" 2¾"
CLEAR DISTANCE BETWEEN BUNDLES
NOMINAL WIDTH OF BUNDLE
TIED COLUMNS WITH 2-BAR BUNDLES

BUNDLED BARS AS COLUMN VERTICALS*
DESIGN AND DETAIL DATA

BUNDLE	EFFECTIVE NUMBER OF BARS	BAR SIZE	TOTAL AREA, IN.^{2}	EQUIV. DIA., IN.	EFFECTIVE PERIMETER OF BUNDLE, IN. — AT A SPLICE BAR: SPLICE BAR	AT A SPLICE BAR: REMAINDER OF BUNDLE	WITHOUT SPLICE BAR	MINIMUM CLEAR DISTANCE, IN. — BETWEEN BUNDLES	BUNDLE TO EDGE ‡
SPLICE BAR (IF USED)**	2	#8	1.58	1.42	2.61	5.24	6.28	2⅛	1½
		#9	2.00	1.60	2.94	5.91	7.09	2½	1⅝
		#10	2.54	1.80	3.33	6.65	7.98	2¾	1¾
		#11	3.12	2.00	3.67	7.39	8.86	3	2
SPLICE BAR (IF USED)**	3	#8	2.37	1.75	2.35	7.07	7.85	2⅝	1¾
		#9	3.00	1.95	2.65	7.97	8.85	3	2
		#10	3.81	2.20	3.00	8.97	9.97	3⅜	2¼
		#11	4.68	2.44	3.33	9.96	11.06	3¾	2½
**	4	#8	3.16	2.01	–	–	9.42	3	2
		#9	4.00	2.26	–	–	10.63	3½	2¼
		#10	5.08	2.55	–	–	11.97	3¾	2½
		#11	6.24	2.82	–	–	13.29	4¼	2¾

* BARS IN A BUNDLE SHALL TERMINATE WITH AT LEAST 40 BAR DIAMETERS STAGGER EXCEPT WHERE THE BUNDLE TERMINATES.

** SPLICE BARS, WELDING, OR POSITIVE CONNECTION MUST BE PROVIDED FOR SPLICES REQUIRED TO CARRY FULL TENSION OR TENSION IN EXCESS OF THE CAPACITY OF THE UNSPLICED PORTION OF THE BUNDLE. COMPRESSION MAY BE TRANSMITTED BY END BEARING OF SQUARE-CUT ENDS.

‡ THESE MINIMUM DISTANCES APPLY TO BUNDLES ONLY. WHERE TIES OR SPIRALS ARE PRESENT, THE 1½ IN. MINIMUM COVER TO THEM WILL CONTROL IN SOME CASES.

NOTES: – 1. THESE BARS MUST BE TIED AS SHOWN BY DASHED LINES WHEN X DISTANCE IS OVER 6 IN.
2. THESE BARS NEED NOT BE TIED WHEN X DISTANCE EQUALS 6 IN. OR LESS.
3. APPLICABLE TO ALL TIED COLUMNS.

A DIFFERENT PATTERN OF TIES MAY BE SUBSTITUTED PROVIDED THAT DETAILS OF THE REQUIREMENTS ARE SHOWN ON THE CONTRACT DRAWINGS.

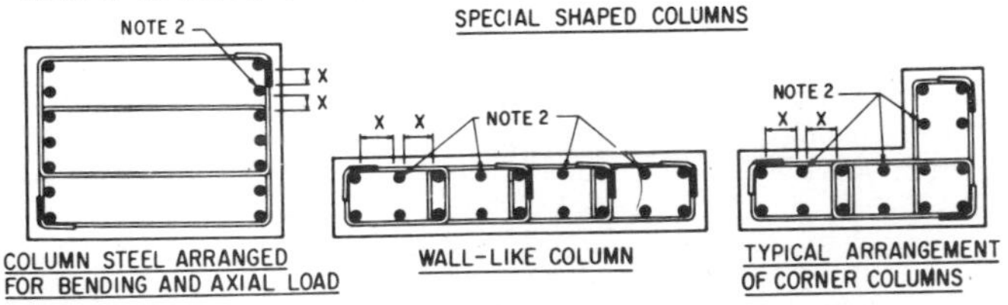
SPECIAL SHAPED COLUMNS
NOTE 2
X
X
NOTE 2
X X
NOTE 2
X X
COLUMN STEEL ARRANGED FOR BENDING AND AXIAL LOAD
WALL-LIKE COLUMN
TYPICAL ARRANGEMENT OF CORNER COLUMNS

Fig. 27. Column ties and bundled bars. *(From ACI 318-65,[10] by permission of American Concrete Institute.)*

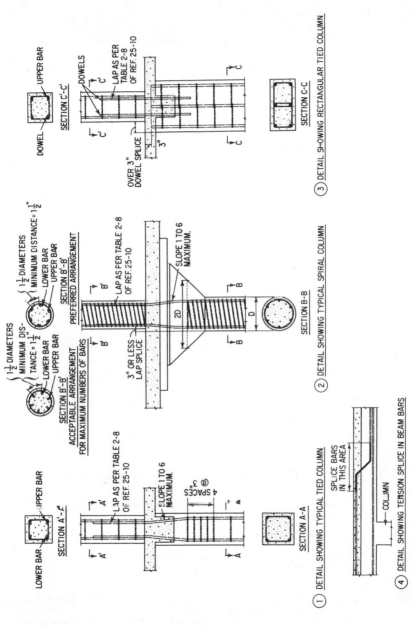

Fig. 28. Typical arrangement of column reinforcing and splices. (*From ACI 315–65*,[10] *by permission of American Concrete Institute.*)

In footings, mats of bars can be supported by precast concrete blocks made for this purpose, or chairs. Chairs made for this purpose have a large bearing pad of sheet metal (called a "sand pad") to prevent their sinking into the soil. Mats should not be placed on the ground with the expectation they will be pulled up into place after concrete is placed. Such a procedure is uncertain at best—it is far better to support the mat adequately beforehand or to place concrete to

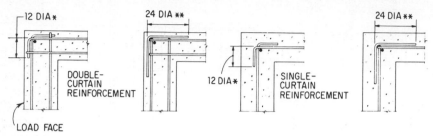

TYPICAL CORNER DETAILS

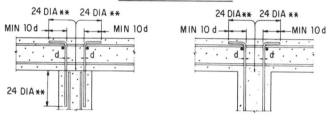

TYPICAL INTERSECTION DETAILS FOR DOUBLE-CURTAIN REINFORCEMENT

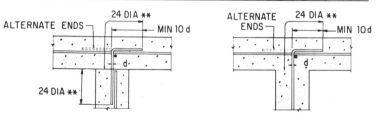

TYPICAL INTERSECTION DETAILS FOR SINGLE-CURTAIN REINFORCEMENT

NOTES:
ALL 90° BENDS AS SHOWN UNLESS OTHERWISE INDICATED ON DESIGN·DRAWINGS.
VERTICAL BARS SHOWN AT HOOKS ONLY.
∗ OR NOT LESS THAN ½ MINIMUM LAP SPECIFIED (12″ MIN).
∗∗ OR NOT LESS THAN MINIMUM LAP SPECIFIED (12″ MIN).

FIG. 29. Corner and wall intersection details. (*From ACI 315-65,*[10] *by permission of American Concrete Institute.*)

the mat elevation, place the mat on it, then resume placing. This latter practice is commonly used in paving reinforced with mats.

For joists, slabs, beams, and girders, various wire bar supports shown in Chap. 3, Fig. 11 are available.

Precast mortar or concrete blocks should not be used to support or block steel where they will be exposed and detract from the appearance of the structure. (Precast blocks are many times made of poor-quality mortar or concrete. The quality of the blocks should be considered.)

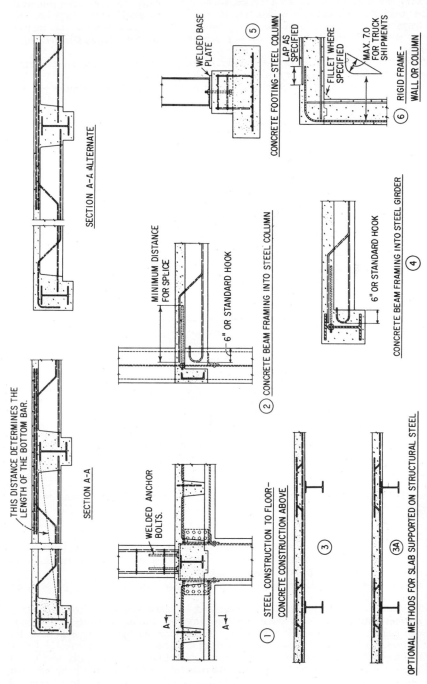

FIG. 30. Connecting details for structural elements. (*From ACI 315-65*,[10] *by permission of American Concrete Institute.*)

Wire ties are commonly used to assemble and support steel in mats and walls. Wire used for this purpose is No. 16, 15, or 14 gage, black, soft iron wire. No. 14 gage wire should be used for heavy work. No. 18 gage wire is too light to do a good job usually. Types of ties are shown in Fig. 32 and described as follows:

Simple or snap tie A, used particularly on horizontal mats where least strength is required. Double-strand or snap tie E provides more strength.

Wall tie B, similar to A but with a turn around a vertical bar, is more effective on light vertical mats of steel.

FIG. 31. Footing showing reinforcing neatly and accurately placed and well supported. (*California Department of Water Resources photograph.*)

Saddle tie C does not exert the twisting effect on the bars such as that caused by A, B, and E. The saddle tie with a twist D is used for vertical mats or preassembled mats to be lifted in place by a crane.

Cross or figure-eight tie F does not cause a twist in the bars and has more strength than the simple ties.

Slab bars or mats that are being assembled in place should be tied sufficiently to prevent shifting, at least three times in any bar length, at every intersection around edge, about every 5 to 6 ft for No. 5 bars and smaller, every 8 to 10 ft for No. 6 to 9 bars, and every 10 to 12 ft for No. 10 to 11 bars.

Wall bars or mats being assembled in place should be tied sufficiently to prevent shifting even under the movement of concrete, at least three times in any bar length, every intersection around edge, every third intersection, about every 3 to 4 ft for No. 5 bars and smaller, every 4 to 5 ft for No. 6 to 8 bars, every 6 to 8 ft for No. 10 to 11 bars. Preassembled mats should have at least the above ties and more, if necessary for proper handling. Templates may be used for accurately preassembling bars.

Wall reinforcing should be adequately supported to prevent its being pushed toward or against the form when concrete is placed. This may be done by tying it to the form tie rods, tying two curtains of steel together to prevent their spreading, by concrete blocks between the steel and form where such is permissible, or by nails driven into the form at the proper locations so bars can rest against the nailheads, and be tied to the nails, thus properly spacing the steel from the form (this used only where the nails would be unobjectionable. Nails might rust, causing unsightly staining on exposed concrete).

8   Cleaning Reinforcement

Reinforcement at the time of placing concrete should be free of mud, oil, or other foreign matter that will reduce or destroy bond. Mill scale or a light coating of rust tightly bonded to the steel is not objectionable. Loose, flaky, scaly rust,

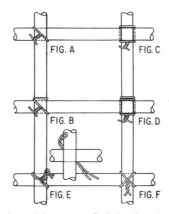

FIG. 32. Common wire ties. (*Concrete Reinforcing Steel Institute photograph.*)

which would affect the bond, should be removed by scraping, shock treatment (dropping, hammering, vibration), wire brushing, or possibly sandblasting. It is not necessary to have a bare metal surface, only to remove loose, flaky rust. Bars rusted to the extent that their size is reduced below the ASTM requirement for new bars should not be used. This can be determined by removing the rust and weighing a measured length of bar. Bars should be kept free of mud and mud should be washed off before using. Loose mortar should be removed. Mortar so tightly bonded to the steel that it is not removed by vigorous wire brushing can remain. Grease or oil should be removed by a propane torch, using care not to overheat the bars, or washed off with kerosene or gasoline with adequate safety precautions for using such materials. When oiling forms, oil may get onto the bars. If the form oil evaporates or is a very thin coating, it will do no harm. If it is a thick coating, it should be removed.

9   Check List for Reinforcing Steel

The Concrete Reinforcing Steel Institute recommends the following check list for inspection of reinforcing steel placing:

Preliminary Arrangements

Latest approved placing plans are being used and followed.

Reinforcing materials are the kind, type, and quality specified.

All test reports required by the contract documents are received promptly and that materials qualify.

Materials are on hand in advance of job requirements.

Reinforcing steel is well spread out, in convenient spots, readily identified, and not in mud.

Fabricated bars are correctly cut and bent (verified by visual check and occasional measurement).

Bar setters are briefed on just what is to be done and on critical details that require special attention.

During Steel Setting

Every footing, wall, slab, beam, girder, and column, before concreting, has the right number of the right type, size, and style of reinforcing bars, bar supports, and welded wire fabric securely fastened in the proper place, especially as regards:

Total depth of member as established by formwork and screeds. Proper amount of cover or fire-proofing to the near surface of the concrete. This should provide the effective depth from center of bar to compression surface of concrete.

Proper setting of dowels for number, size, location and projection, as well as determination of all points needing dowels.

Type and location of bar supports to hold bars firmly in place until concrete sets.

Correct sizes of bars (because sometimes a bar of the same shape but of a different size than specified would fit into the space equally well).

Required number of bars (because only close attention by bar setter and inspector can prevent omission of bars with consequent decrease in safety).

Correct length of bars (because a different length bar might go into the space but would not carry out the engineer's intentions).

Proper location of bars within cross section of the member, i.e., top bars to top and bottom bars in bottom; bottom layer on the bottom and superimposed layer just above by the amount called for; outside bars on the outside and inside bars on the inside; stirrups, ties,

FIG. 33. Concrete placement for small pumping plant showing several embedded items (conduit, anchor bolts, sumps) and blockouts (circular blockout in foreground is damaged and will need to be replaced). (*California Department of Water Resources photograph.*)

and spirals enclosing the prescribed longitudinal bars; bars hooked around others as specified; bars hooked in concrete with hooks in proper position at the correct end; all bars securely wired or otherwise held in place, with tops of all bars the proper amounts below finished concrete, to allow passage of screed and finishing machine.

Proper placement of bars lengthwise of a member, i.e., hooked end of a bar at the end specified, properly embedded in concrete; end of each bar within 2 in. of the location indicated on the placing plans; truss bars, stirrups, and other bent bars right side up as called for (foundation mats and similar members that carry upward pressure require bent bars exactly reversed from slabs and beams carrying downward loads); bend-down points and bend-up points within 2 in. longitudinally of where they work out on placing plans; stirrups and ties spaced as scheduled from the exact starting points specified within ±1 in.; specified overlap in lapped splices and designated welding in welded splices.

Spacing between bars, between layers of bars, and between bars and formwork. For architectural concrete and exposed work, it is usual to provide 1½ in. of cover over bars, with some positive way of maintaining it.

EMBEDDED ITEMS AND BLOCKOUTS

10   Embedded Items

Most concrete has various items embedded in it such as catch basins, manholes, sumps, traps, conduit, pipe, grounding cable; bolts for anchoring machinery, equipment, or fixtures; inserts for supporting or attaching wall material or panels; water stops, drains, and form hardware, as ·shown in Fig. 33. The plans and specifications should be carefully checked to determine the presence of and location of such embedded items.

Usually embedded items are installed in the forms prior to start of placing concrete and held firmly in place by attaching to the forms or reinforcing steel, or by templates. Bolts for anchoring machinery or equipment must be accurately placed. Commercially available bolt anchors (Fig. 34) or job-made anchors (made from steel plate, pipe, and bolts) provide a greater length of bolt to take strain and permit some shifting of the bolts when the machinery is installed to allow for inaccuracies in the machine base or in locating the bolts. Embedded items that are to be installed during concrete placing should have their location determined and referenced on the form and they should be available for embedment during placing. Any sizable embedded items such as heating ducts must be securely anchored to prevent floating due to uplift of the concrete. Pipes, conduits, and other such items having voids should be capped or plugged to prevent concrete from getting into them.

Some embedded items are to bond to the concrete. These should be kept free of mud, oil, grease, form oil, or other foreign material that would affect bond. Others should not bond to concrete and should be coated with asphalt, tar, grease, or similar material to prevent bond.

11   Blockouts

Blockouts are openings made in concrete to provide places where gate guides, handrail posts, seals, doors, or windows can be placed later (Fig. 33). Blockouts may be formed of wood, Styrofoam, or metal. Wood blockout forms should be made so as to facilitate removal later. This may be done by making them in several sections, splitting sections, making tapered pieces, and making angle cuts. Wood blockouts should be soaked in water ahead of concrete placing so they will not expand and crack the concrete if located close to an edge or corner. Blockouts made of Styrofoam can be easily removed. Blockouts formed with steel (cans) should be removed before the final installation; otherwise the steel may rust, causing an unsightly rust stain on the concrete surface. Rusting can also cause the concrete to spall if close to an edge or surface. The use of aluminum cans can also cause similar problems if the cans are left in place.

Blockout openings are filled later, after installation of the item to be placed, using concrete, mortar, or grout as appropriate. Sometimes the use of aluminum powder in the concrete, mortar, or grout or the use of nonshrink or fast-setting cements may be considered desirable.

Shifting or cutting reinforcing steel to permit installation of embedded items or blockouts should be permitted only if shown on the plans or specifications

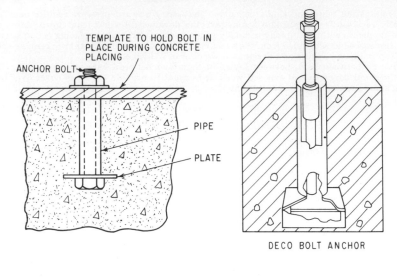

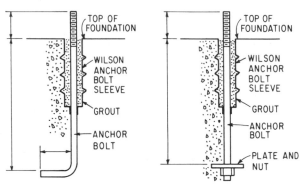

Fig. 34. Anchor-bolt details. Pipe-sleeve device as shown or commercially available bolt anchors provide more flexibility for fit of bolt to machine base and longer length of bolt to take strain. (*Photographs courtesy of Wilson Anchor Sleeve, Inc., and Decatur Engineering Co.*)

or if approved by the engineer after consideration of the effect of such shifting or cutting. The effect of the location of items on the strength of structural members should be considered and approval of the engineer obtained if there is any question.

FINAL INSPECTION

During all phases of construction activity prior to start of placing concrete, continuing inspection of the preparations by both the contractor and the inspector representing the owner should ensure that the work is progressing satisfactorily. The progress of the work is the responsibility of the contractor. The inspector should be available for consultation and interpretation of the plans and specifications. The inspector should refrain from directing or running the job for the contractor. To do so assumes responsibility for the work.

DEPARTMENT OF WATER RESOURCES

FEATHER RIVER FISH HATCHERY

CHECK OUT SHEET FOR CONCRETE PLACEMENT

Monolith No. _5-9_   Lift Elev. _Foundation Rock_   to Elev. _143.0_

Placement No. _4_

CHECK ITEM	APPROVED			
	Frazier-Davis		Dept. Water Resources	
	Date	Initials	Date	Initials
Foundation Adequacy	7-17-62	A.E.M.	7-18-62	D.H.
Rock Cross Sections Taken	—	—	7-18-62	R.C.B.
Formwork - Line & Grade	7-19-62	H.L.S.	7-19-62	E.R.G.
Formwork - Construction	7-19-62	H.L.S.	7-19-62	E.R.G.
Embedded Items-Waterstop	7-19-62	H.L.S.	7-19-62	E.R.G.
~~Resteel~~	—	—		
~~Mechanical~~	—	—		
~~Piping~~	—	—		
~~Misc. Metalwork~~	—	—		
~~Electrical~~	—	—		
Wet Sandblast	—	—	ok w/o	D.H.
Formwork - Oiled	7-19-62	A.E.M.	7-19-62	CHE
Final Concrete Cleanup	7-19-62	A.E.M.		CHE
Concrete Placing Equipment	7-19-62	A.E.M.		CHE
Concrete Curing Equipment	7-19-62	A.E.M.		CHE
OK for Concrete	7-19-62	A.E.M.		CHE

Remarks _Close washout openings in form before placing starts. 9/19/62 RFA_

Fig. 35. Check-out sheet for concrete placement. (*California Department of Water Resources form.*)

On large jobs, it is desirable to have an inspection card such as that shown in Fig. 35 at the location of each placement being prepared. As each item is completed it is inspected by both the foremen and inspector assigned to inspect that particular item and their signatures are entered on the card after completion and approval. After all applicable preparations are completed and approved, the placing inspector makes a final inspection to see that all applicable items have been previously inspected and are still satisfactory. The inspector should review details of the concrete to be used such as slump, cement content or water-cement ratio, maximum size of aggregate, air content, temperature, concrete-delivery schedule, and placing procedure. The placing inspector determines that suitable equipment for delivering and handling the concrete; vibrators including standbys; sufficient workmen; preparations for protection of the concrete in either hot, cold, or rainy weather; equipment for finishing and equipment and material available for curing are all ready or arranged for. If concrete placing or finishing is to continue after dark, adequate lighting should be ready.

Quite obviously many of the details concerning the concrete placing should be discussed and agreed upon in discussions or meetings with the contractor prior to the final inspection. If previous discussions, preparations, and inspections are adequately done, the final inspection takes only a few minutes.

The inspector authorizes concrete to be ordered only after he assures himself that everything is or will be ready.

REFERENCES

1. Investigation of Methods of Preparing Horizontal Construction Joints in Concrete, U.S. Army Engineer Waterways Experiment Station, *Tech. Mem.* 6-518, July, 1959.
2. Investigation of Methods of Preparing Horizontal Construction Joints in Concrete, U.S. Army Engineer Waterways Experiment Station, *Tech. Mem.* 6-518, Report 2, February, 1963.
3. Wuerpel, Charles E.: Tests of the Potential Durability of Horizontal Construction Joints, *ACI Proc.*, vol. 35, p. 181, June, 1939.
4. ACI Standard 347-68, Recommended Practice for Concrete Formwork.
5. Hurd, M. K.: Formwork for Concrete, *ACI Spec. Publ.* SP-4, 1963.
6. "Forms for Architectural Concrete," Portland Cement Association, Skokie, Ill., 1949.
7. Tuthill, L. H., R. E. Glover, C. H. Spencer, and W. B. Bierce: Insulation for Protection of New Concrete in Winter, *ACI Proc.*, vol. 48, p. 253, November, 1951.
8. ACI Standard 318-71, Building Code Requirements for Reinforced Concrete.
9. "CRSI Recommended Practice for Placing Reinforcing Bars," Concrete Reinforcing Steel Institute, Chicago, 1968.
10. ACI Standard 315-65, Manual of Standard Practice for Detailing Reinforced Concrete Structures, *ACI Separate Publ.*, 1965.
11. AWS Standard D12.1-61, Recommended Practices for Welding Reinforcing Steel, Metal Inserts and Connections in Reinforced Concrete Construction, American Welding Society, 1961.
12. "The Welding of Concrete Reinforcing Bars," Booklet 527-B, Bethlehem Steel Co., 1960.
13. "Manual of Standard Practice for Reinforced Concrete Construction," Western Concrete Reinforcing Steel Institute, 1964.
14. ACI Standard 311-64, Recommended Practice for Concrete Inspection.
15. ACI Standard 614-72, Recommended Practice for Measuring, Mixing, Transporting, and Placing Concrete.
16. Manual of Concrete Inspection, *ACI Spec. Publ.* SP-2, 1967.
17. Waddell, J. J.: "Practical Quality Control for Concrete," McGraw-Hill Book Company, New York, 1962.
18. "Concrete Manual," 7th ed., U.S. Bureau of Reclamation, Denver, 1963.

Chapter 26

PLACING CONCRETE

ROBERT F. ADAMS

STRUCTURES AND BUILDINGS

Structures and buildings are made up of various elements such as foundations, floors, walls, columns, and beams (Fig. 1). The general principles of placing concrete in these structures are covered in Chap. 24; preparations are covered in Chap. 25.

Equipment should be carefully selected, as discussed in Chap. 24, in order to handle satisfactorily and expeditiously the concrete specified and otherwise considered satisfactory.

Concrete mixes are usually specified or may be selected from the principles

FIG. 1. Typical high-rise concrete building frame consisting of cast-in-place and precast elements. Columns, floors, and central service core containing elevators, plumbing, and other services are cast in place. Floor beams are precast. Precast panels will be placed later in exterior walls. Movable crane is used for setting precast elements and for placing concrete as well as handling other materials. Note shores in place supporting floors.

outlined in Chap. 11, based upon strength, durability, workability, and other considerations.

1 Foundations

Foundations for structures are of course important in that they support the structure. Foundations vary widely depending upon subsurface conditions and the loads to be supported. Their design for important structures must be carefully considered by design engineers and by experts in foundation engineering. Construction personnel should be alert to foundation conditions differing from those assumed in the design or expected from foundation explorations which might affect the structural integrity of the structure.

Foundations vary from piling of various types to caissons to spread footings to mat or raft foundations. With such a wide variety of possibilities, procedures for placing concrete can vary considerably depending upon the job situation, but they should follow the principles outlined in Chaps. 24 and 25 and in other parts of this book. Some foundations are massive, and the principles of placing mass concrete discussed later in this chapter apply.

Problems caused by water lead to many difficulties in foundation concrete, and water should be controlled where possible. In some cases it may be necessary to place concrete underwater—which is covered later in this chapter.

Concrete in cast-in-place piles should be carefully placed to be sure that voids are not left. The shells or drilled openings should be carefully inspected immediately prior to placing the concrete to determine that they are open and free of water.

2 Buildings

Construction of buildings by such techniques as slipform, tilt-up, precast, prestressed, or lift-slab is covered in Chaps. 33 and 34. Reference 1 gives considerable information for building construction.

Slabs in buildings are used for either floors or roofs. They might be cast on the ground, in which case they may or may not be reinforced; if cast on forms they are reinforced. Construction of floors is discussed in Ref. 2, and heavy-duty floors are discussed in Chap. 27. Floors on grade may be placed in sections with control joints as discussed in Chap. 28. For a large building some of the paving equipment discussed later in this chapter can be used to place the floor slabs on grade.

Walls, columns, and beams provide the framework of the building. For economy, the design should be such that maximum simplicity and reuse of forms are achieved. It is likewise important that forms be tight and have a satisfactory finish for appearance sake.

Construction joints against which fresh concrete is to be placed should be cleaned by one of the methods outlined in Chap. 25. The use of drop chutes will depend upon the placement conditions. If used, they should be of a size and shape that can readily handle the concrete considered otherwise suitable. Drop chutes should not be a "bottleneck." If necessary to move drop chutes, they should be so arranged that they can be quickly and easily moved or shortened as necessary and not hold up placing concrete. If concrete leaving the drop chute "stacks" and causes segregation, this can be alleviated by keeping a vibrator working in the pile to move concrete away from it, lowering the pile and thus eliminating the segregation. It is good practice for the first few inches of concrete in a wall or column to be more workable, i.e., have more sand (or less coarse aggregate) to assist in starting. Concrete should be systematically placed in uniform layers of not more than 15 to 18 in. and be followed by systematic consolidation effort, vibration preferably. It is also good practice to decrease the consistency of the concrete as the concrete level rises to compensate for bleeding, providing more uniform and durable concrete. This should be done particularly in the top 2 ft of walls (see Fig. 14, Chap. 24). In placing concrete around blockouts or door, window, or ventilation openings, care should be taken to determine that concrete fills the underside of the opening (Fig. 2). This can be assured by vibrating concrete placed on one side of the opening until it fills under and emerges at the opposite side to the level of the bottom of the opening, then placing it on the other side.

Where walls and floor slabs or columns and beams and floor slabs are placed integrally in one placement, cracking can occur at the junction of the two elements because of shrinkage of concrete in the plastic state and restraint of the forms. This can be prevented by placing the concrete to the top of the wall or column and

FIG. 2. Void on underside of window opening due to failure to place concrete properly.

waiting until the bleeding of the concrete ceases and the concrete has stiffened considerably before placing the beam or floor concrete. Vibration of the concrete at the junction of the wall or column and beam or floor slab after bleeding has ceased will accomplish the same end. A similar condition exists at the top of blockouts such as door or window openings, and similar preventive measures should be taken.

It is also good practice to cross-slope the tops of walls which will later be permanently exposed to the elements so that water will drain off rapidly.

Arches, Domes, Paraboloids, and Folded Plates. Structures of these types are characterized by thinner concrete sections, complicated formwork with curves, slopes, intricate reinforcing, and difficult concrete placing and finishing conditions,[3] and with considerable hand placing and finishing. Steep slopes require top forms to hold the concrete in place during vibration. These should be made in small panels and be readily removable for finishing. Concrete mixes for these types of structures need to be proportioned to provide the required strength and to be readily placeable. Sometimes the maximum aggregate size is limited to ½, ¾, or 1 in. for thinner sections.

Domes are sometimes built without form support by constructing the dome on the original earth which has been trimmed to the required shape or upon fill trimmed to the desired shape. The earth is later excavated. In some cases, the completed dome is lifted into place.

Whatever the type of structure, concrete placing should be consistent with the general principles outlined in Chaps. 24 and 25, and every effort should be made to obtain good-quality, dense concrete.

3   Heavy Piers and Retaining Walls

The construction of heavy piers and retaining walls varies from that of structural to mass concrete depending upon the design and size of the structure. Adequate foundations are most important, as discussed earlier in this chapter and in Chap.

25. Construction of the portions above the foundations should follow the principles outlined in Chaps. 24 and 25.

4　Bridges and Viaducts

As with most other structures, foundations are most important and should be given the consideration discussed earlier in this chapter and in Chap. 25. The portions of the structures above the foundations should be constructed utilizing

Fig. 3.　Freeway viaduct. Hollow-box girder is about 330 ft long; columns are about 110-ft centers.

Fig. 4.　Columns for freeway viaduct, a hollow-box girder, in various stages of construction.

the general principles outlined in Chaps. 24 and 25 (Figs. 3 to 6). Particular care should be taken in constructing the bridge deck, particularly bridges for high-speed highway traffic, to obtain a smooth surface for safe travel. A number of finishers are available for finishing the surface (Figs. 7 to 9). Care should

FIG. 5. Shoring and forms for hollow-box-girder freeway viaduct. Forms are designed and constructed to permit frequent reuse, the shoring remaining in place until concrete has attained sufficient strength to permit removal.

FIG. 6. Underside of a simple concrete bridge across a flood channel showing excellent concrete placing. Foundations are cast-in-place piles, columns cast in fiber-tube forms; deck is a simple flat slab. (*California Department of Water Resources photograph.*)

be taken in supporting and aligning the forms and the surfaces the finishers operate from to assure a smooth surface. The forms should be carefully watched for excessive deflections during concrete placings (Ref. 3). In some cases, designers require the bridge deck to be placed in sections following a planned sequence or in alternate sections to provide uniform loading. This is true particularly of the deck in a suspension bridge and may be required in continuous-beam and arch bridges.

Fig. 7. Gomaco finisher finishes concrete deck of bridge. (*Gomaco photograph.*)

Fig. 8. Clary screed finishing a bridge deck. (*Clary Manufacturing Co. photograph.*)

Bridges around water in climates subject to freezing and thawing conditions are particularly vulnerable to deterioration from these conditions. Bridge decks are most vulnerable when ice-removal salts are used. When these conditions are to be encountered, every effort should be made to use concrete having a sufficiently low water-cement ratio and adequate air entrainment to assure durability. Reinforcing steel should have adequate cover. Only competent and conscientious construction and inspection can assure that a structure with the ability to withstand the deteriorating influences of freezing and thawing is achieved.

Fig. 9. Capital finisher finishes lightweight concrete in a deck of a suspension bridge. Deck was placed in sections following the sequence required by the designers to balance loading. (*California Department of Water Resources photograph.*)

MASS CONCRETE

Mass concrete may be defined as any large volume of concrete cast in place generally as a monolithic structure incorporating a high proportion of large coarse aggregate and a low cement content, and intended to resist loads by virtue of its mass. Mass concrete is generally thought of in connection with gravity dams but is used in arch dams, spillways, large massive foundations, and in large structures such as power and pumping plants or bridge piers. Considerable information on mass concrete is given in Refs. 4 through 11.

Mass concrete is basically no different from any other concrete except that it generally uses large-sized aggregate. The larger aggregate size and grading permit a lower sand, water, and cement content. Such concrete, because of its lower cement content, undergoes a lower temperature rise which minimizes the buildup of heat and consequent temperature problems in the structure.

Mass concrete is placed in blocks, or monoliths, from 30 to 70 ft wide and 10 to 200 ft long. With modern temperature control of mass concrete there is no limit to the length of a block. Lifts are usually 5 or 7½ ft high in this country (Figs. 10 and 11).

5 Mixes

Concrete for mass concrete usually contains aggregate with a maximum size of 6 or 7 in. but smaller maximum sizes, such as 4½, 4, or 3 in., may be used. There is no advantage to using aggregate larger than about 6 in. Concrete with aggregate larger than 6 in. shows little gain in any benefits, is more difficult to use, and is harder on the equipment. Cement contents vary from about 4 to as low as 2 sacks per cu yd. It is not uncommon to use mixes having different cement contents in the same placement. For example, a mix with 2½ or 3 sacks of cement per cu yd might have sufficient strength and would be used for the interior concrete. For the exterior concrete exposed to freezing and thawing, sulfate, or other deteriorating conditions where durability is a problem, a mix having additional cement such as 3½ or 4 sacks might be used. Mass-concrete mixes have from

Fig. 10. Mass concrete being placed in Thermalito Dam of California Water Project. Twelve cubic yard buckets of concrete are transferred from rail-mounted concrete-delivery car from mixing plant by cableway. Forms are steel. Concrete is placed in blocks and 7½-ft lifts. Concrete placed in five 18-in. layers in each lift, in "stairstep" arrangement. Note forms for gallery (extreme right side of photograph). (*California Department of Water Resources photograph.*)

Fig. 11. Another view of dam shown in Fig. 10. Note numerous blocks of concrete at various stages of construction. Note burlap shading for recently placed concrete and water curing. (*California Department of Water Resources photograph.*)

about 19 to 24% sand by absolute or solid volume of total aggregate. Type II, moderate heat- and moderate sulfate-resisting portland cement, is largely used for mass concrete now and has largely replaced Type IV, low-heat cement. Modern temperature-control methods make it practical to use Type II cement. Type V cement may be used where more severe sulfate conditions are encountered.

Natural sand and gravel aggregates are considered best because of the better workability of concrete made with them. However, crushed aggregates are satisfactorily used. Crushed aggregates should be prepared in equipment which produces good particle shape, with minimum "flats and slivers."

The coarse aggregate may contain as much as about 35% of the 6- to 3-in. fraction with successively smaller percentages of smaller sizes. Gap gradings are sometimes used in other countries. A study of the aggregate sources and trial mixes should be made for large jobs to utilize the aggregate deposit and available concrete materials most efficiently. For example, in the materials investigations for Glen Canyon Dam,[9] the aggregate deposit contained 10% 6 × 3-in. size. Computations of aggregate requirements for the various classes of concrete showed that

Fig. 12.   Four cubic yard bucket of mass concrete just dumped, ready for vibration. (*California Department of Water Resources photograph.*)

the coarse aggregate for the mass concrete could contain 13% of the 6 × 3-in. fraction and utilize the deposit most efficiently. Trial mixes showed that this amount of cobbles (6 × 3-in. fraction) could be satisfactorily used. A larger amount of cobbles would have required overexcavation and considerably more waste. Similar studies can sometimes justify the use of smaller maximum aggregate size or sand grading different from that usually specified.

Admixtures are widely used in mass concrete. Air entrainment is particularly beneficial in improving workability as well as durability for the surface concrete. The amount of air entrained is usually in the range of 3 to 3½%. Pozzolans can be used as a part of the cementing material, usually about 15 to 35%. Pozzolans can have many effects such as improving workability, reducing the temperature rise of the concrete, and sometimes reducing cost. Water-reducing and -retarding admixtures are beneficial in that they reduce the water required by concrete, thus improving the quality of the concrete. They can also permit a small cement reduction with less temperature-rise potential with the same strength potential. Retardation may be beneficial in placing the concrete. The strength of mass concrete is usually judged on the basis of longer age tests than the standard 28-day strength for structural concrete. Common ages are 3 to 6 months or 1 year.

A mass-concrete mix looks harsh to those unfamiliar with this type of work because of the large aggregate (Fig. 12). The consistency of mass concrete should be judged by the consistency of the mortar, not by the looks of the total concrete. A properly designed mix with 1- to 2-in. slump responds well to vibration and is easily placed (Fig. 13). The slump test and sometimes air-content test are made on the minus 1½-in. fraction of the concrete obtained by removing the larger aggregate by either hand picking or wet screening. The unit weight of

FIG. 13. Vibrating mass concrete shown in Fig. 12 with three one-man 6-in. pneu-matic vibrators. Note how readily concrete responds to vibration. (*California Department of Water Resources photograph.*)

the full mass mix is sometimes made in a 1 cu ft measure; however, some prefer a larger container, 3 to 5 cu ft.

6   Preparation for Placing

Forms for mass concrete can be of wood or steel. They are usually made in large sections and handled by crane or cableway or by other lifting equipment. Forms are usually anchored to the rock foundation by drilling and setting anchors, or to previously placed concrete by embedded pigtails, hairpins, or other form-anchor accessories, and are braced by struts (Figs. 10, 11, 18, and 19). For the downstream-sloping surface of a dam forms may be hinged at about half height to permit depositing concrete in the lower half of the lift closer to the form. After concrete is placed to hinge level, the top half is folded in place and anchored.

Forms are designed for maximum reuse and economy. Forms for curved surfaces in either a horizontal or vertical direction or double curvature are designed and built with the required curvature or with adjustable curvature if curvature of the structure varies. Cantilever forms are sometimes used. These eliminate the internal bracing or ties that sometimes interfere with concrete placing. Deflection rather than strength usually governs in the design of cantilever forms. Starter forms must be specially constructed to conform to the irregular rock surface.

Cleanup and joint treatment prior to placing concrete are accomplished by the methods outlined in Chap. 25. The joint may be grouted or not as specified and as discusssed in Chap. 25. All pools of water should be removed from the rock or concrete surface prior to grouting or placing concrete on them. Grout, if used, may be spread by dumping it from the bucket from a height allowing it to spread itself or by brooms, pushing it across the block as placing proceeds. Too much surface should not be grouted ahead of placing and allowed to dry out. If grout is not used, the surface of the rock or concrete should be at a saturated surface-dry condition. The surface can be kept at approximately this condition by blowing off excess water with air, and by preventing excess drying by periodic fogging with an air-water jet (Fig. 14).

7   Placing and Finishing

Placing mass concrete starts at one side of the form. Buckets are dumped systematically across the width of the placement. If a richer face or exterior concrete is used, the concrete is delivered to provide the number of buckets of exterior concrete required to cover the width of the placement. Buckets may

FIG. 14. Periodic fogging of construction joint with air-water fog nozzle keeps joint at saturated surface-dry condition. Surface or edge of concrete may also be fogged to prevent excessive drying during hot or windy weather.

FIG. 15. Shoveling cobbles which rolled down edge of pile onto concrete to correct segregation.

be flagged to indicate the kind of concrete being delivered. Each bucket of concrete is vibrated after it is dumped (Figs. 12 and 13). Usually 6- and 4-in. vibrators are used. If grout is not used, particular care is taken to vibrate the bottom layer thoroughly to secure good bond to the preceding lift. Particular care is taken to vibrate thoroughly the place where adjacent piles meet. Sometimes this is done with the smaller vibrator. If large-sized coarse aggregate rolls to the edge of the pile, this is shoveled onto the concrete to eliminate the segregation which has taken place (Fig. 15). For 5-ft lifts, the concrete may be placed in four 15-in. layers or three 20-in. layers. For a 7½-ft lift, five 18-in. layers can be used.

Placing proceeds across the placement, "stairstepping" the layers to keep a minimum amount of concrete exposed (Fig. 10). In hot weather, drying of the surface of the concrete can be prevented by frequent fogging with the air-water jet mentioned previously (Fig. 14), care being taken only to keep the concrete damp, replacing only the water that has evaporated. Also, the use of air-water foggers during hot weather can keep the air temperature in the vicinity of the placing several degrees cooler (Fig. 16 and Ref. 10).

The top layer in a lift should be vibrated sufficiently to embed all the coarse aggregate and provide a relatively smooth surface which will aid cleanup for the next lift. Further finishing of the surface is not necessary unless the lift being placed is the top lift in the block, in which case the required finish should be provided as specified and discussed in Chap. 27. Once the top layer in a lift is completed, workmen should be kept off the completed surface except as necessary to place pigtails, hairpins, or other embedded items. Workmen doing this should wear "snowshoes" to keep from leaving footprints (Fig. 9, Chap. 25).

Placing mass concrete in sections having flat slopes such as in spillway buckets, spillway crests, or spillway slopes having little slope imposes problems. It is difficult to screed concrete having large aggregate satisfactorily. In addition it is sometimes desirable to finish the concrete, particularly where water flows over it, to eliminate bug holes and provide a satisfactory surface; therefore, any forms used must

FIG. 16. Air-water foggers keep air temperature several degrees cooler during hot weather.[10] (*California Department of Water Resources photograph.*)

FIG. 17. Weighted slipform screeding mass concrete on a sloping spillway floor. Note finishing behind slipform which is supported on pipe screeds and is slowly pulled up the slope by hand-operated winches (not shown). (*California Department of Water Resources photograph.*)

be removed in time to permit finishing. There are two ways in which placing of such sloping surfaces can be accomplished. One is to use a heavy slipform similar to that used in slope paving (Fig. 17). The slipform is pulled up to the slope slowly as placing proceeds, the concrete ahead of the form being vibrated. Finishers follow the trailing edge of the form. The other method is to provide temporary holding forms, which can be easily and rapidly placed and removed

FIG. 18. Removable top panels used to hold concrete in place in bucket section of Thermalito Dam spillway. Ribs hold panels in place. Panels are quickly set in place and removed. After concrete has stiffened, panels are removed and concrete finished by hand. Mortar which squeezes into gaps in panels provides mortar for finishing. Ribs are removed also to permit finishing. (*California Department of Water Resources photograph.*)

as placing proceeds. Forms at the lower part of the placement can be removed at the proper time to permit finishing operations. Placements of this type in a spillway bucket and a spillway-crest placement are shown in Figs. 18 to 20.

8 Temperature Control

Control of temperature in mass concrete can be achieved in various ways. Temperature control and cooling of concrete are discussed in detail in Chap. 46. For large structures, it is common practice in the design stages to make a temperature study of the structure and concrete to provide information to determine the most economical and satisfactory way to provide temperature control.[4,5]

Temperature control can be achieved by several means. Pipe cooling, that is,

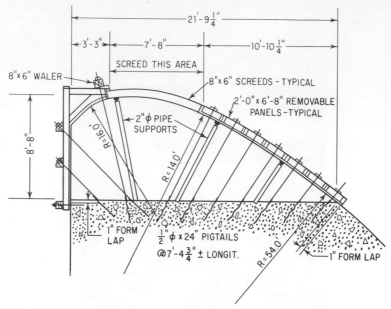

FIG. 19. Sketch for removable-panel arrangement for Oroville Dam spillway crest. Note ribs, removable panels, form ties, and supports.

FIG. 20. Spillway crest for Oroville Dam (see Fig. 19).

embedment of pipe coils in the structure, can be used (Fig. 8, Chap. 25). The coils are usually placed at construction joints and either river water or refrigerated water or refrigerant is circulated through the coils to remove the heat generated by hydration of the cement. The concrete may be cooled initially to provide a lower starting temperature and consequently a lower final temperature in the structure, resulting in less temperature drop to the stable temperature range of the structure. Initial concrete temperatures as low as 40, 45, and 50°F are sometimes specified. Initial cooling can be accomplished by various methods or combinations thereof. The temperature of the concrete can be computed by means of the equations given in Chaps. 30 and 46.

Aggregate can be cooled to a limited extent by shading the aggregate storage bins and piles, by water sprinklers, and by circulating refrigerated air or water through storage bins. Evaporative cooling of aggregate in storage bins resulting from a vacuum created by steam jets is sometimes used also.

Cement can be cooled by a longer storage period after manufacture or by cement coolers at the cement mill. Cooling cement is less effective than cooling aggregate or water, since the amount of cement is small in relation to the aggregate and the heat capacity of water is five times that of cement. Water can be cooled by refrigeration, or flake, tube, or crushed ice can be substituted for a portion or all of the mixing water. Limiting the minimum period for placing a lift upon a previous lift (such as 72 hr) and limiting the height difference between adjacent lifts (such as 20 ft) are sometimes specified for temperature-control purposes. Subsequent to placing concrete, shading the concrete can assist in keeping the surface cooler during hot weather (Figs. 11, 18, and 20 and Ref. 10).

Mass concrete is usually water-cured using sprinklers (Chap. 29).

PAVEMENTS AND CANAL LININGS

9 Street, Highway, or Airport Paving

Paving streets, highways, airport runways, or other large areas is usually accomplished by a "paving train" consisting of several pieces of equipment designed to spread, consolidate, screed, float, and finish the concrete. Two types of equipment are now in general use, the type that utilizes and rides on side forms, and the slipform type that does not use side forms. The slipform type of paver has become popular in recent years because it eliminates the need for and considerable labor involved in placing, removing, and cleaning the side forms as well as the investment in them. Slipform pavers are capable of providing paving with excellent riding qualities. Lighter-weight paving equipment of the type which utilizes side forms is also available for small jobs such as city streets. Considerable helpful information is contained in Refs. 12 through 20 and in Chap. 27.

Foundation. The subgrade, subbase, and base for the paving should be adequately compacted in accordance with the plans and specifications and should extend a foot or more beyond the form line to support properly the forms and equipment moving on it. Greater width is required where slipform paving is used, to support the slipform equipment adequately. When compacted, the final grade should be slightly higher than the finish grade to permit fine grading to the finish grade, providing firm foundation for the forms and equipment which must operate on the subgrade. The subgrade material should be of a type suitable for the purpose; if not suitable, it should be replaced.

Many types of trimmers are available to accomplish the final grading between the forms, varying from a grader to specially built trimmers which operate from the side forms or which operate from preerected grade lines set on either one or both sides of the paving. Such equipment has electrical, electronic, and hydraulic controls that follow the grade line to trim the base accurately. Accurate trimming of the base outside the paving is desirable for slipform paving because of its effect upon concrete quantity, slab thickness and uniformity, and resulting paving smoothness and riding quality. Fine grading to close tolerances sometimes restricts the size of the aggregate in the base material to about ¾ in., particularly for slipform paving. When cement-treated bases are used, their construction and fine grading must be completed in a limited period of time. A longer time period can be allowed for construction of untreated bases.

Grade lines, where used, should be set accurately and supported at 50-ft intervals on tangents and 25-ft intervals on vertical and horizontal curves. The same set of level control stakes can and should be used for the subgrade, subbase, and base, for setting the side forms, or for the slipform guidelines to reduce discrepancies and inaccuracies in surveying. The same set of stakes and brackets can be used to support guidelines for different operations. The line may be removed and replaced. Guidelines are usually piano wire or nylon cord (fishing line is good).

Concrete for paving can be mixed and delivered by several methods as discussed in Chap. 23. Mixing can be done in paving mixers at the placing site supplied with materials by batch trucks (Figs. 21 and 22). The current trend for large jobs is to use large central-mix plants having 8 or 10 cu yd mixers, delivering

Fig. 21. Side-form-type paving train placing 24-ft-width pavement in Denver-Boulder Turnpike. Concrete batches delivered in batch trucks are mixed in two 34E pavers and discharge concrete on subgrade. Concrete is spread, consolidated, screeded, floated, and finished by equipment following as described in text. (*Koehring Co. photograph.*)

Fig. 22. Slipform paver paving 36-ft-width California freeway. Batches delivered in batch trucks to three 34E pavers which mix concrete and discharge it into a spreader box on front of paver (see Fig. 24). Paver moves slowly forward following guide wires along side of pavement, consolidating and finishing concrete.

the concrete in trucks (end-dump, Agitor or Dumpcrete, or transit mixers used as agitators). Central-mix plants may be relocated several times on a paving job to provide a shorter haul of concrete to the placing site. Uniform concrete is essential for a smooth paving operation, particularly for slipform paving. Concrete can be made with 1½- or 2-in. maximum-size aggregate for paving having 8- or 9-in. thickness and up to 3-in. aggregate for greater thickness as in airport runways. Mix proportioning should be in accordance with methods outlined in Chap. 11 and in Recommended Practice for Selecting Proportions for Concrete, ACI Standard 211.1-70. Slumps should be in the lower range, 2 to 2½ in. or less, and sand contents should be suitable for paving concrete. Cement content should be adequate to provide required strength and with water-cement ratio for required durability. Air entrainment should be used for durability, and adequate inspection should ensure by test that the required amount of air is obtained and that poor construction practices such as finishing too soon or overfinishing of overwet concrete do not result in loss of entrained air in the top surface where durability is most important.

Side-form Paving. Forms must be set accurately to line and grade and be supported uniformly by a firm foundation. The smooth-riding quality of the paving depends upon the care with which the forms are set and maintained since the finishing equipment rides on the forms. Forms should be maintained in good condition. After final form alignment with instrument, a final check should be made by sighting along the tops of the forms—any gross errors will be readily apparent.

The paving train capable of paving widths up to 36 ft may consist of either one or two spreaders (two are sometimes used if reinforcing mats are used), a finishing machine, longitudinal float, mechanical belt, and burlap or pipe drag (Fig. 21).

Concrete should be dumped on the subgrade as nearly as practical in its final location in accordance with the good practices for handling concrete outlined in Chap. 24. The mechanical spreader distributes the concrete between the forms. If reinforcing is used, the first spreader strikes off the concrete at the correct depth and the reinforcing steel is placed. The second layer of concrete is placed and spread by the same spreader making a second pass, or by a second spreader. The final spreader should leave a slight excess of concrete for the finisher.

Consolidating the concrete is usually done by vibrating pans or screeds, or immersion spud or tube vibrators attached to the spreader or finisher or other independent equipment. Vibrating pans or screeds are suitable for paving thicknesses of 8 to 12 in. Immersion vibrators are required for greater thicknesses. Concrete must be thoroughly vibrated throughout its full depth with just enough mortar brought to the surface to finish satisfactorily. Excess soupy fines or mortar indicates overwet concrete, too high a sand content, or overfinishing.

The **slipform paver** consists of equipment mounted on crawler tracks having moving side forms and incorporating the spreading, consolidation, finishing, and floating operation usually all in one piece of equipment (Figs. 22 and 23). As the machine moves forward, concrete at the trailing edge of the siding has stiffened sufficiently so that it does not slump as the side-form support is removed. The automatic machine is guided in both line and grade by sensors that follow guide wires set accurately in line and grade outside one or both edges of the slab. The nonautomatic machine is controlled in grade by the base upon which it rides. Slipform pavers commonly pave 24-, 36-, or even 48-ft widths.

Slipform pavers incorporate most of the same features of side-form-type pavers. Concrete is delivered to the paver and spread laterally. A spreader operating ahead of the slipform is desirable to spread the concrete to a uniform thickness to produce more uniform conditions, helping the slipform paver to produce a smooth surface. Sometimes a spreader box is incorporated in the paver (Fig. 24), or it may be separate. When separate, the paving mixer or delivery trucks discharge into the spreader box and pull it along, metering the concrete onto the base. Augers or screeds spread the concrete laterally. Discharge of concrete should be done so as to eliminate or minimize the effects of segregation. A strike-off or a conforming screed molds the concrete to the required cross section (Fig. 23). Concrete is usually vibrated ahead of the screed both by spud vibrators and by a vibrating tube set just ahead of the leading edge of the screed to prevent surface tearing. The proper location of both types of vibrators is critical in order to prevent tearing of the concrete by the leading edge of the screed

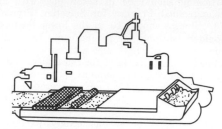

Fig. 23. Artist's schematic sketch of conforming screed slipform paver showing spreader box, vibrators, side form, conforming screed, rotating (Clary type) screed, and pan float. (*California Division of Highways photograph.*)

and to prevent the surge or boiling of concrete behind the screed. Variable-frequency vibrators are sometimes used so the operator can compensate for variations in the consistency of the concrete, reducing the surge. Following the screed, the concrete is floated, utilizing a pressure pan, vibrating or oscillating screeds, vibrating pan floats, or a rotating screed plus a pan float. Final finishing consists of pipe floating, dragging a 6- or 7-in. aluminum pipe forward and backward over the

Fig. 24. Front of slipform paver showing spreader box. Also shown is dowel inserter. Workman loops a wire around each end of dowel; wires pass over drum and are pulled into concrete as paver moves forward.

surface (Fig. 25). The pipe is placed diagonally across the slab. The pipe can be moved either manually for narrow widths or by machine for wider paving. Straightedging may next be done so that irregularities can be detected and removed.

When reinforcing is required and slipform paving is used, the two-lift method of placing the reinforcing is usually followed. A stripped-down paver or other spreading equipment places the first lift to the required thickness for the reinforcing steel. The width is slightly narrower than the paving width. The mat of steel is placed and anchored and the second paver follows, placing the top layer of concrete. Internal spud vibrators, which would interfere with the reinforcing steel, cannot be used with this paver; so vibration must be obtained by vibrating screeds or pans or with tube vibrators.

There is normally a slight amount of edge slump with slipform pavers. This can be corrected by the edging operation or by addition of a small amount of mortar during the edging.

The final operation is to texture the concrete with a burlap drag as mentioned for side-form pavers (Fig. 25).

Fig. 25. Pipe and burlap drag for 48-ft pavement placed by slipform paver. Aluminum tube is at an angle to the centerline. Pipe may be manually pulled for narrower-width paving. (*Photograph by J. J. Waddell.*)

Good control of concrete deliveries and uniform consistency are essential to a smooth paving operation with slipform pavers. Variation in consistency will cause differences in surge of concrete behind the screed and consequent finishing difficulties. Overwet concrete causes edge slump that causes irregular edge surfaces. Sometimes concrete that is too wet or has a bleeding tendency will have "slip-outs" at the edge some time after the paver has passed. (Air entrainment will reduce bleeding and reduces this problem.) If concrete deliveries slow or cease for a short period, it is better to slow the slipform paver until deliveries resume rather than stop it completely.

Pavers for Small Jobs. Paving equipment discussed previously is suitable for large jobs. Some small jobs, such as city-street paving, are done by lightweight equipment such as that pictured in Fig. 26. Such equipment is also sometimes used for large floor slabs.

Dowels, tie bars, and reinforcing should be placed as required by the plans and specifications. Dowels should be held in place and in alignment so they will function as intended. Tie bars are conveniently inserted in the concrete by equipment designed for this purpose. They may be inserted by hand or by a hydraulic inserter following the side-form paver, or they may be carried into the concrete by wires as the slipform paver moves ahead (Fig. 24).

Joints should be made as required by the plans and specifications. Contraction joints are for the purpose of controlling the cracking of concrete due to shrinkage.

These may be either preformed or sawed. Preformed joints may be made by a strip of sheet metal set and left in place during paving operations. Another type consists of inserting a wood or metal strip into a slot in the surface formed by vibrating a bar into the concrete. The final finishing produces a smooth surface over the strip which is removed the next day and the crack is later filled with joint sealant. Care should be taken that the strips do not cause irregularities in the concrete.

Joints may be sawed in the concrete by either carborundum or diamond saws. The time at which sawing is done is critical. It must be done late enough that the sawing does not ravel the concrete surface and before the concrete cracks. This may be from about 4 to 24 hr after the concrete is placed. Sometimes such saw cuts are made diagonally at a slight angle across the slab to break the rhythmic effect on traffic. Sometimes it is desirable to saw every other joint initially when

Fig. 26. Light-duty paver for street paving. (*Curbmaster of America, Inc., photograph.*)

cracking is imminent, sawing the intermediate joints later. Joints in one lane should line up with joints in previously placed lanes.

Longitudinal joints (hinge or warping joints) can be sawed, formed with a strip of sheet metal left in place, or formed by a strip of polyethylene inserted in the concrete by an attachment on the slipform paver (Fig. 27).

Construction joints are installed at the end of a day's run. They should be placed to correspond to a regular contraction joint when possible. A transverse construction joint not intended as a contraction joint should have tie bars and keyways. Longitudinal construction joints should likewise have tie bars and keyways.

Expansion joints are not usually required except where the paving joins a bridge or other structure, curb, sidewalks, or at intersections. Preformed expansion-joint material such as cork, sponge rubber, or fiber treated with bitumen should be placed perpendicular to the concrete surface and held in place while concrete is placed against it. The expansion-joint filler should cover the entire surface of the joints; otherwise spalling will occur with expansion. The joint filler can be held to previously placed concrete with bitumen. See also Chap. 28, where joints are discussed in detail.

Curing paving can be accomplished by methods outlined in Chap. 29. Water curing by ponding, sprinklers, wet earth, sand, or wet cotton or burlap mats is considered best. Covering with impermeable paper, plastic sheets, wet straw, or white-pigmented curing compound is also used. Curing should be started as soon as practical after the concrete is finished—as soon as it will not injure the surface—and should be continuous for the required period. Curing compound should be applied after the concrete surface has lost its shininess but before it has dried. Curing compounds should be thoroughly stirred before and during spraying, should not be diluted, and should be applied at the specified coverage to form a continuous uniform covering. Damaged curing compound should be resprayed after thoroughly wetting with water. The edges of slabs as well as the surface should be cured.

Fig. 27. Trailing edge of slipform paver showing inserter for polyethylene plane of weakness tape (about 2 in. wide). Tape is carried on rolls, not visible, behind inserter. Note also electronic control riding guide wire to control elevation of trailing edge of slipform.

Traffic should be kept off the paving until it has attained its design strength, which is usually determined by flexural strength of beams cured alongside the slab.

Every effort should be made to construct smooth paving having satisfactory riding qualities as correction of roughness is expensive and not too satisfactory at best. Where satisfactory smoothness is not obtained as determined by measurements of the completed paving by a profilograph or other equipment, the bumps can be removed with a "bump cutter." One bump cutter is about 2 ft wide, consisting of about 100 diamond-saw blades mounted on a shaft. Another is a cylinder mounted on a shaft with diamonds set into the surface in a spiral pattern.

10 Canal Lining and Slope Paving

Concrete lining of canals, drainage channels, reservoir slopes, and other slopes such as around abutments on bridges and fills can be done in several ways. The method selected will depend primarily on the size of the job. The excavation work for the canal lining can likewise be done in several ways. Rough excavation is done by any of the excavation methods considered suitable. Fine grading is done with trimming machines, some of which are made particularly for the purpose. During excavation, the possibility of slides and unsuitable material such as expansive clays should be watched for. If overexcavation is required or occurs because of careless excavation, backfill should be with suitable material adequately compacted.

Shotcrete is suitable for small canals or areas. This method is discussed in Chap. 36.[21]

Subgrade guided slipforms (Fig. 28) can be used for small canals. The canal section is excavated and trimmed to the required section. The form rides on sledlike runners, which are a part of the form and which space it away from the excavation to give the required lining thickness. The slipform is built so that

Fig. 28. Subgrade guided slipform placing concrete in a small canal. Canal section is excavated. Conforming screed slipform mounted on sledlike runners is pulled along as concrete deposited in its hopper is molded to the canal section. (*Portland Cement Association photograph.*)

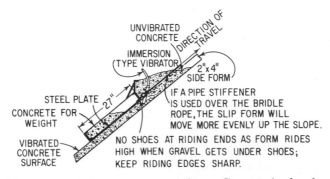

Fig. 29. Slipform for placing concrete on a slope. Concrete is placed and vibrated as slipform is slowly pulled up the slope. For placing unformed concrete on slopes slipform screed should be steel-faced, weighted, and unvibrated. Concrete should be vibrated ahead of slipform. (*By permission of American Concrete Institute.[7]*)

it can be pulled along the canal with a winch or other suitable means. Concrete is delivered into the hopper of the form. As the slipform is pulled ahead slowly, the concrete in the hopper is spaded or vibrated, and fills into the space between the excavation and the pan or form which shapes the concrete. Vibration must be done with internal vibrators in the concrete in the hopper, not to the form.

Slope slipforms in which the concrete is placed in sections up the slope can be used on small jobs or jobs where other equipment is not suitable. A suitable slipform and its essential features are shown in Figs. 29 and 30. The slipform

Fig. 30. Simple slope slipform in use. Concrete is placed in alternate panels. Larger and more elaborate ones are used.

is supported on side forms or previously placed sections of concrete. The leading edge must be sharp, not rounded, where it rests on forms or concrete to keep concrete from getting between it and the form or concrete it rides on, thereby raising it. The form must be heavy—sufficient to resist the hydrostatic uplift of the concrete. As the form is slowly and uniformly moved up the slope by winch, air tuggers, or other means, the concrete in the "trough" at the leading edge is vibrated with internal vibrators. The vibrators must not be run under the form except at the start of a section before the slipform moves. The slipform must not be vibrated, as this would cause concrete to "boil" behind it, causing bulging and a wavy surface. The form must be long enough that concrete does not boil out below the form. When slipforms of this type are used, concrete is usually placed in alternate sections, the first sections being placed using side forms. The second sections are placed between the first sections. Inverts are placed using conventional flat-slab paving equipment. The slipform used for the slope may sometimes be used, or another slipform built particularly for the invert section might be used.

Canal-lining paving trains (Figs. 31 to 34) used for paving larger and longer sections, where the size of the job justifies the large equipment expenditure, consist of the trimmer that fine-grades the section to required line and grade (Fig. 31); the lining machine, a longitudinal-operating slipform that places the concrete; the finishing jumbo where finishers use hand tools to obtain a better finish, where required, and to tool joints, place joint strips, and other handwork; and a curing jumbo or other equipment used for applying curing compound where membrane curing is permitted (Figs. 32 and 34).

The canal-lining machines may place the canal in one pass (Fig. 32) or in sections (Figs. 33 and 34) and are supported and guided either by rails set to line and grade (this method is now obsolete) or by crawler tracks following guidelines set to line and grade. The equipment can also be controlled by sensors riding on the subgrade, which provides the most accurate control of the thickness of the lining.

Canal-lining machines consist of a concrete-distribution system utilizing chutes, drop chutes, conveyor belts, or drags. The concrete is vibrated, usually by vibrating-tube internal vibrators in the hopper section ahead of the slipform portion forming the concrete to the required section as the machine moves slowly forward. Pressure plates at the rear of the machine impart a troweling to the concrete, eliminating most of the hand troweling. Equipment of this type was used in lining the California Aqueduct of the state water project (Figs. 33 and 34).

Fig. 31. Trimmer for fine-grading California Aqueduct described in Figs. 33 and 34.

Control joints may be required at intervals as small as 12½-ft centers in canal lining. These can be conventional planes of weakness that are usually made by grooving the concrete. Such grooves are filled later with various types of joint-sealing compound. Longitudinal grooves can be made by the canal-lining machine with a projection on the underside of the pan that cuts the groove. These grooves may require hand tooling from the finishing jumbo. Longitudinal grooves can also be conveniently formed by a plastic strip that is intruded into the concrete through a tube from the front of the machine. The plastic strips are carried on reels on the front of the machine (Fig. 34).

Transverse grooves can be cut into the concrete by a cutter bar working off an eccentric which is actuated at the required groove interval. These grooves are tooled by hand from the finishing jumbo. In some cases plastic joint former is placed in the groove to hold it to the required dimension until the concrete stiffens enough that the plastic can be removed, or the plastic is removed after the concrete sets.

Water stops are sometimes placed in the concrete. One design of water stop includes vertical wings that serve as a plane of weakness to control the crack (Fig. 35). Installation of this type of water stop, and water stop at construction joints, is satisfactorily accomplished longitudinally by intruding it into the concrete through a tube or other device from the front of the machine (Fig. 34) and transversely with special equipment carried on the finishing jumbo. Efforts to

Fig. 02. Front and rear views of a canal paving train placing concrete in South Bay Aqueduct of California Water Project. Canal has a bottom width and depth of about 6 ft and concrete is 3½ in. thick. Train consists of a paving mixer, lining machine which places full section of concrete, a finishing jumbo, and spray equipment for applying curing compound. (*California Department of Water Resources photograph.*)

seal the intersections satisfactorily have not been too successful at the time of this writing. Some consider the leakage at intersections inconsequential.

Concrete mixes for canal lining are conventional mixes with additional concern for workability and finishing properties. Concrete with excessive slump is not used, as such concrete will not hold on the slope. For 4-in. lining on a 1½:1 slope, 3-in. slump is about the maximum that can be used. Canal lining 4 in thick with 1½-in. maximum aggregate size can be satisfactorily placed. One such mix satisfactorily used in the aqueduct of the California Water Project contains 40% sand and a cement content of 4.4 sacks per cu yd, of which 70 lb is a pozzolan. The average 28-day strength is in excess of 3,500 psi. A water-reducing retarding admixture is used. Entrained air is between 2 and 3%. It is common practice

FIG. 33. Lining one-half of canal of California Aqueduct with canal-lining train. See Fig. 34 for closeup. (*California Department of Water Resources photograph.*)

in canal-lining concrete to split the ¾ in. × No. 4 size coarse aggregate into two sizes on the ⅜-in. sieve. This allows the ⅜ in. × No. 4 size to be kept at a small amount since an excess of this size sometimes causes finishing problems. The coarse-aggregate grading being successfully used on two jobs is:

Coarse-aggregate size	Job A	Job B
1½ × ¾ in.	57 %	45 %
¾ × ⅜ in.	33 %	35 %
⅜ in. × No. 4	10 %	20 %

Sand-grading specifications sometimes require that the No. 50 to 100-mesh fraction be 15 to 20% instead of the 12 to 20% used for other concrete, to provide sufficient fines and added workability. Lining as thin as 3 in. has been placed using 1½-in. maximum-size aggregate. Table 23 of the "Concrete Manual"[8] gives details of mixes for a number of canal-lining jobs.

Fig. 34. Closeup of canal paving train shown in Fig. 33 used to line California Aqueduct in two sections. Concrete is 4 in. thick. Canal prism has a 40-ft bottom width, 32.6-ft depth, 1½:1 slopes, and 137.8-ft top width. Concrete mixed in 8 cu yd batches in 10 cu yd mixers in a central-mix plant is hauled to job in 8 cu yd Agitor dump trucks. Paver places one slope and half the invert. Reels on front of paver carry plastic groove former or water stop inserted in concrete through tubes through front of machine. Note transverse groove cut into concrete to rear of paver and white plastic former being placed in groove to hold its shape. Finishers work from jumbos when inserting groove former and finishing concrete. Joints are at 12½-ft centers. Curing jumbo applies white-pigmented curing compound. (*California Department of Water Resources photograph.*) [Similar canal of San Luis Project[22] has 110-ft bottom width, 32.8-ft depth, 2:1 slopes, top width of 250 ft, and 4½-in. concrete thickness. Section was placed in four passes: one side slope and 10 ft of invert in one pass, and 90 ft of invert in two passes (45-ft width each). Same lining machine, with adjustments, was used for both side slopes and invert.]

Subsequent to placing the concrete, curing should be started as soon as possible. On large jobs, curing with white-pigmented curing compound is almost universally used.

11 Curbs, Gutters, Sidewalks, Etc.

Construction of curbs, gutters, sidewalks, drainage ditches, and median barriers is auxiliary to highway or street and residential construction. Small amounts are usually done by hand methods. Forms are placed after excavation and fine grading are completed; concrete is placed in the forms and compacted, then screeded, and hand-finished following conventional methods. On larger jobs, slipform or small paving equipment, which will accomplish such work, is available. Equipment such as that shown in Figs. 36 and 37 is capable of placing concrete in curbs; curbs and gutters; sidewalks; combined sidewalks, roll curbs, and gutters; drainage ditches; and median barriers. Attachments are available, or the machines can be adjusted to give a variety of cross sections.

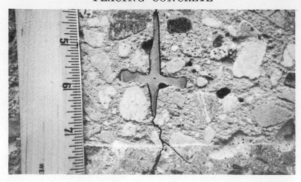

FIG. 35.   Section of canal lining showing embedded polyvinyl water stop and plane-of-weakness former inserted to form and seal contraction joints.

FIG. 36.   Form-guided slipform placing concrete in drainage ditch. Note control for varying vibrator power as concrete varies. This versatile machine can be converted to place concrete in sidewalks; sidewalks, roll curbs, and gutters; curbs and gutters, median strips, or other similar placements. (*Curbmaster of America, Inc., photograph.*)

Concrete work should be done with as much care as in any other work in order that the concrete will perform satisfactorily. Concrete with high slump (more than 4 in.) should be avoided. Air entrainment should be used in climates where durability is a problem. Concrete should be carefully finished and properly cured.

TUNNELS AND CONDUITS

12   Tunnels and Subways

Tunnels for railroads, highways, sewers, water supply, and other utilities are sometimes lined with concrete. This may be for the purpose of supporting the opening, improving flow conditions for water, containing water under pressure, sealing off the inflow of water, or usually a combination of these factors. Lining tunnels with concrete is one of the most difficult kinds of concrete work. Limited working room; difficult placing conditions due to confined spaces, reinforcing steel, steel or timber support, steel or timber lagging and blocking; water problems; and difficult cleanup conditions make unpleasant and discouraging conditions where it is difficult to get a good concrete job. For this reason, more than usual planning should be done to select the best procedures and equipment to accomplish the job.[8,23,24,25,26,27]

FIG. 37. Slipform placing curb and gutter. Same machine, with adjustments or other attachments, can place other sections. (*Gomaco photograph.*)

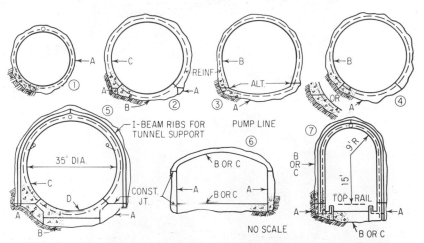

FIG. 38. Several tunnel sections showing sequence of placing concrete.

Concrete is usually placed in tunnels using concrete pumps of either the piston or "air-gun" types, described in Chap. 36. Concrete is also sometimes placed using the preplaced aggregate method decribed in Chap. 38.[25]

Sequence of Tunnel Lining. The size of the tunnel, length, section, and sometimes available equipment determines the sequence of tunnel-lining operations. The usual method of lining is to place the concrete in sections between bulkheads. However, tunnel lining can be placed continuously (particularly applicable to long tunnels) by having sufficient length of forms so designed that they can be telescoped and leapfrogged ahead after concrete has hardened sufficiently that the last form section can be removed to obtain continuous placing.

Lining may be placed in a complete section, applicable particularly for small and circular tunnels, or placed in several placements in the section.

Several sections of tunnels are shown in Fig. 38. Sequences which have been or might be used are described as follows:

Tunnel section 1, a small circular tunnel. This can be placed in a single placement in sections 40 or 50 ft long or more or in continuous placing (Figs. 39 through 43).

Tunnel sections 2, 3, 4, and 5 show circular or horseshoe tunnel sections placed in more than one placement. In these either the curb section or invert is placed first. The purpose of a curb section is to provide support for rails or for the invert screed or for supporting and anchoring forms. The curb or invert section is cast with appropriate embedded anchors for attaching rails if required, and for anchoring the arch or sidewall forms.

Tunnel section 5 is that for two diversion tunnels for Oroville Dam (Figs. 44 through 49). This was placed in four placements. Curbs *A* were placed first (Fig. 44). Next the subinvert *B* was placed, which was used to facilitate the remainder of the construction in providing a smooth surface for support and travel of the form carriers, concrete-delivery trucks, and the concrete pumps. The sidewalls and arch *C* were placed in one operation in a section 48 ft long (Fig. 45). Finally the invert was placed using a slipform mounted on rails attached to the lower edge of placement *C* for one tunnel (Figs. 46 and 47). For the other tunnel, the invert section was formed and concrete was pumped into the section to fill it (Fig. 48). (This did not result in a satisfactory placement. The slipform

Fig. 39. Rail-mounted form in place ready to receive concrete for lining a 6-ft-diameter circular tunnel. Note bulkhead lumber at end of form held to angle bolted to steel form at one end and cut to fit rock and braced from rock at other end. Note also doors in side of form, slick line from concrete pump entering form at top, and rail-mounted carrier for slick line which is backed out as placing proceeds and connection for air line near bend in slick line for "slugging" concrete. See Figs. 40 through 43 for subsequent concreting operations. (*California Department of Water Resources photograph.*)

FIG. 40. Form vibrators fastened to form shown in Fig. 39. Internal vibrators also used through door openings. Note hinge in form. (*California Department of Water Resources photograph.*)

FIG. 41. Small-diameter special 6 cu yd concrete car for delivering concrete from mix plant outside into 6-ft-diameter tunnel. (*California Department of Water Resources photograph.*)

Fig. 42. Single-piston Pumpcrete machine pumping concrete into forms for 6-ft-diameter tunnel. Concrete is transferred from the delivery car to the pump hopper by belt conveyor. (*California Department of Water Resources photograph.*)

Fig. 43. Finishing invert concrete after removing the invert form when concrete has stiffened sufficiently that it can be done.

Fig. 44. Curb form being placed for lining one of the 35-ft-diameter circular diversion tunnels for Oroville Dam. Top is tied and braced to rock. Bottom is anchored to rock. Note adjusting screws at bottom for elevation. Space between bottom of form and rock was filled with wood to hold concrete (water shown removed before concrete placed). Figures 45 through 49 show subsequent concreting operations. (*California Department of Water Resources photograph.*)

Fig. 45. Front form is in place following completion of one 48-ft-long placement, and rear form has been moved ahead by telescoping through front form and is being set. This permitted a 48-ft section to be placed each day. (Forms were required to be left on for 32 hr.) Forms are hinged at two places on upper part of circle and are supported by curb section (Fig. 38, section 5 and Fig. 44). Note also working platforms on side of form, pump line located below upper platform to place concrete into openings to fill sidewalls. (*California Department of Water Resources photograph.*)

26–33

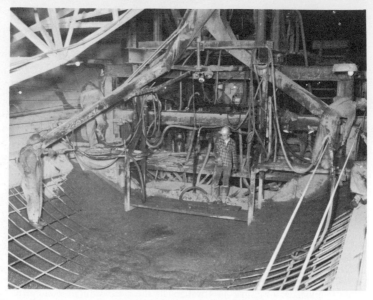

Fig. 46. Slipform placing invert concrete in a diversion tunnel. Slipform supported on rails fastened to previous placement and slowly pulled ahead by air-operated tuggers or winches. Concrete delivered to top of chutes shown by conveyor receiving concrete from a delivery truck. Chutes shown delivered concrete to each side of invert. (*California Department of Water Resources photograph.*)

Fig. 47. Finishing invert concrete being placed in Fig. 46 from finishing platform at rear of slipform. (*California Department of Water Resources photograph.*)

FIG. 48. Form for invert concrete in one diversion tunnel. Form held in place by anchors in previous placement. Concrete was pumped under form. Placing was not nearly so satisfactory as with slipform in Fig. 46. (*California Department of Water Resources photograph.*)

FIG. 49. Concrete transfer for one diversion tunnel. Concrete transferred horizontally on belt conveyor to vertical pipe where it dropped into hopper shown, then transferred into Agitor truck and hauled to delivery point. (*California Department of Water Resources photograph.*)

produced much more satisfactory results.) In these diversion tunnels concrete with a maximum placing temperature of 50°F was specified to achieve a lower maximum temperature and result in relatively crack-free concrete. This resulted in strength gain in the concrete that would not permit removal of sidewall and arch forms and reuse on a 24-hr cycle. Specifications required forms to be left in place for 32 hr, and they were reused in a 48-hr cycle. Two sets of forms were used and concrete was placed each day. The forms were designed so one would telescope inside the other and could be leapfrogged ahead.

Fig. 50. Form for sidewalls and arch of railroad tunnel carried on truck. Note openings in form for access for vibration and for inspection. (*California Department of Water Resources photograph.*)

Tunnel section 6 is similar to a large double-track railroad or highway tunnel carrying two or more lanes. In the section shown, the sidewalls were placed first. The next placement is usually the arch, although the invert may next be placed to provide better working conditions. However, the invert is usually placed last.

Tunnel section 7 is for a single-track railroad relocation around the reservoir for Oroville Dam and was common for five tunnels (Figs. 50 and 51). The same forms were used for all tunnels (with modification for the fifth tunnel, which was on a curve). The curb sections were placed first. In these, the forms were anchored to and supported by steel tunnel support and rock. In the case of two tunnels, the tunnel-drilling jumbo and mucking equipment were rail-mounted. These same rails were used for the form carrier and concrete equipment. The sidewalls and arch were placed in the second placement (Fig. 51). These were placed in 50-ft sections. The invert was the third placement. In the other three tunnels, the tunneling equipment was rubber-tired wheel, not rail-mounted. The invert was placed following the curb placement to provide a smooth working surface for the form carrier and placing equipment. The invert was placed with paving equipment supported and operating off the curbs.

The curved tunnel utilized the same forms. Permission was granted to place the tunnel in straight tangents about 40 ft long. The sidewall and arch forms were modified by insertion of a wedge-shaped section at one end and widened to provide required clearance by insertion of a section (dutchman) of the required width at the crown.

Concrete in the sidewalls and arches of all the above railroad tunnels was placed on a 24-hr cycle. Concrete had attained sufficient strength for the forms to be removed in 10 to 14 hr.

Equipment Used in Tunnel Lining. Forms for tunnel lining vary considerably, depending upon the section involved. Forms for curbs or invert sections can be either wood or steel. Forms for the curb section, *A* placement in section

Fig. 51. Raiload-tunnel form in place. Note hinges at two points, working platforms at three levels corresponding to locations of access doors, and braces across forms at two levels. Form is supported by anchor bolts in curb section previously placed. (*California Department of Water Resources photograph.*)

7, were steel and were reused many times in five railroad tunnels. Such forms usually position inserts used for aligning and supporting the sidewall and arch forms. The curb forms for this placement were supported and aligned from the steel tunnel-support ribs.

Forms for sidewalls and arch are usually specially designed for the job and made of steel (Figs. 40, 45, and 50). Circular pressure tunnels with a steel liner use the liner as a form which remains in place. In such tunnels, the reinforcing steel may be placed on the liner outside the tunnel and moved into place in the tunnel. For small tunnels (such as section 1) the form may have a removable invert, which permits finishing the concrete at the time it can be satisfactorily finished where a smooth invert is desired (Fig. 43). Wall and arch forms are usually hinged at one or more places to facilitate placing and removal (Figs.

40, 45, and 50). Forms are usually carried and moved by a traveler or carrier mounted on rubber-tired wheels (Fig. 50) or on rails and are supported, positioned, aligned, braced, and removed by use of screw jacks, hydraulic jacks, rams and cylinders, and steamboat jacks. Sidewall and arch forms have openings in the sides and arch to permit access for vibration and inspection (Figs. 40 and 50). In some cases concrete can be placed through such openings (Fig. 45). Form openings must be capable of being opened and closed easily and be tight. Openings at the crown should not be on the centerline but should be placed alternately on each side of the centerline to eliminate interference with the concrete pump line, which is usually placed on the centerline. Working platforms are built into the larger forms (Figs. 45 and 51).

Bulkhead forms in tunnels in rock may be built as shown in Fig. 39. Each tapered piece of wood is individually cut to fit the rock. Tunnels which are drilled with a mole may have reusable bulkheads which fit the outer more regular surface of the tunnel sufficiently closely so that individual fitting is not required.

Concrete-placing equipment for curb and invert varies considerably from conveyors, chutes, concrete pumps, or other means depending upon the job (Figs. 42 and 46). Placing concrete in the sidewalls and arch is almost universally done with either piston or pneumatic ("air-gun") concrete pumps (Chap. 36). These may be located outside the tunnel for short tunnels where pumping distances are short, or inside the tunnel. Concrete is usually consolidated by vibrators, internal type in invert and curb sections, and either internal or external (form) vibrators, or usually both, in the sidewalls and arch placement.

Concrete is mixed either outside in a central-mix plant or inside the tunnel in a variety of mixers including transit and paving mixers or special mixers, materials being delivered in batch trucks or cars. Concrete is delivered in a variety of equipment ranging from special trucks or rail cars to conventional concrete conveyance equipment such as transit mixers mounted on rail cars or trucks, or tilting-type concrete trucks. On small tunnels special small-diameter cars or concrete mixers are frequently employed (Fig. 41). On long tunnels it is desirable to mix the concrete near the placing site to eliminate time delays, reduce the amount of concrete in transit, and minimize difficulties caused by slump loss, equipment delays, and breakdown. In long tunnels concrete or concrete batches may be delivered from the portals or concrete may be mixed above the tunnel and either pumped into the tunnel or dropped through a vertical shaft or a drilled hole into the tunnel. Concrete for one diversion-tunnel job was mixed in a central-mix plant above ground, moved horizontally about 100 ft on a conveyor belt, and dropped vertically about 150 ft through a drop chute placed in a hole drilled in the rock, thence into a collecting hopper. The concrete in the hopper discharged into Agitor trucks that transported the concrete to the placing location (Fig. 49). Concrete for the adjacent diversion tunnel came from the same central-mix plant but was hauled in transit trucks into the tunnel, requiring a considerably longer time.

Some thought should be given to expediting concrete deliveries by providing passing tracks for trains or arrangements for trucks passing, or turnaround. For one of the diversion tunnels the width between curbs was insufficient for trucks to pass. A portable ramp and turntable were constructed to a elevate the trucks to provide additional clearance and to permit trucks to turn around near the placing site, which eliminated the necessity for trucks to back in slowly a considerable distance. In some of the railroad tunnels a portable ramp was constructed so that trucks were elevated sufficiently to clear the curb and pass. (These trucks were not able to turn around and had to back into the tunnel, however. The passing ramp permitted two trucks to be inside the tunnel at one time.)

Concrete Mixes for Tunnel Lining. Concrete pumps having 8-in. lines are capable of placing concrete having maximum aggregate size up to 2½ or 3 in. Pumps with 6-in. lines usually limit the aggregate to about 1½-in. maximum size. The mix used must be a workable cohesive mix without excessive bleeding suitable for pumping. Slumps as low as 3 in. are pumpable, and air-entrained concrete is pumpable. Concrete with 2½-in. maximum-size aggregate (natural sand and gravel) used in the railroad tunnel (Fig. 38), section 7, contained 5 sacks of cement per cu yd, 33% sand, 5- to 7-in. slump, 3% entrained air, and had an average 28-day compressive strength of 4,000 psi. (This higher slump was required by the placing conditions to be discussed later, not by the concrete pumps.) In the lining of the diversion tunnels, using 8-in. pumps, 1½-in. maximum-size aggregate (natural sand and gravel) concrete was used because the smaller aggregate was more efficient for the higher-strength concrete required. The cement content was 5.5 sacks per cu yd (including 70 lb pozzolan), 35% sand, 3% entrained air, 5-in. slump, and average 28-day compressive strength was 4,200 psi. (Required 1-year strength of 5,500 psi was exceeded.) Concrete suitable for similar placements elsewhere from a strength, durability, and workability standpoint is suitable for curbs and inverts for tunnels. There are no special requirements otherwise.

Control of Seepage and Dripping Water. Control and effective removal of seepage and dripping water are some of the most difficult and ineffective parts of tunnel-lining work. Such water can cause damage to the concrete and should be controlled. Water entering the sidewalls and arch can be guided down to the invert and into drains by means of corrugated sheet metal placed so it fits closely against the rock. This is called "panning." If large amounts of water are entering, these may be sealed off at the time of driving the tunnel or later by grouting. Gravel or tile drains can be placed in the invert to carry water to a sump where it is pumped out, or drained outside the tunnel. Branches may be placed to pick up water from springs or that coming down the sides behind the pans.

In some cases holes are drilled into the rock and pipe is grouted into place with quick-setting cement to collect and drain away water.

Drains temporarily placed for controlling water during construction and not a part of the design are usually grouted later. Their location should be accurately detailed so they can be effectively grouted.

Cleanup. Cleanup of rock should be done by washing with water or air-water jets. In hard-rock tunnels, loose rock should be removed. In soft rock, cleanup should be consistent with the specified required results. Soft, loose material should be removed by pumping. Sandbags can sometimes help keep water out of the invert placing area.

Forms and Reinforcing Steel. Reinforcing steel is placed in accordance with the plans and specifications. Ties with No. 14 tie wire instead of No. 16 are recommended to avoid reinforcing steel displacement by the concrete-lining operations. Forms should be placed and set accurately in alignment and grade. Particular care should be taken to see that the forms are securely anchored and braced. Embedded items such as form anchors, inserts, drains, conduit, and grout pipe must be securely anchored so they will not become displaced.

Placing Concrete. Inverts can be placed with conventional methods. These include a heavy slipform or screed pulled by winches longitudinally along the tunnel on rails or other supports provided for support as concrete is placed and vibrated ahead of it. For smaller tunnels a transverse screed operated longitudinally may shape the surface. Wide curved inverts can be shaped by setting transverse screed guides and screeding with a straightedge parallel to the centerline. Concrete is delivered usually by transit mixers and chutes, conveyors, pumps, or other means.

Concrete must not be placed in water. If pools of water are still present and it is impractical to remove them or keep them dry by pumping, suction pumps, or blowing with air, concrete should be placed on previously placed concrete, and the toe of the concrete slope vibrated ahead, pushing water ahead of it, and subsequently out of the form or placing area.

Concrete in sidewalls and arches is placed behind the forms, using concrete pumps which are mentioned previously in this chapter and described more fully in Chap. 36. Construction joints against previously placed concrete should be clean prior to erecting the forms and recleaned if necessary prior to placing concrete. Small tunnels can be placed with a single pump line in the crown. Large tunnels are best placed with three pump lines, two located at or above the spring line through which concrete is delivered to both sides of the tunnel simultaneously. The pump line in the crown is used later to fill the upper or arch portion. For tunnels placed continuously, placing is done through a crown line only. The pump is moved as placing continues. The diversion tunnels, section 5 (Fig. 38), were placed with three pump lines as shown. The lines on the side were located outside the form and concrete was chuted through the form openings (Fig. 45). Pneumatically operated gates at each side opening permitted concrete to be delivered through an opening (openings were at 8-ft centers). Location of steel rib support and reinforcing steel prevented placing the pump line through the bulkhead and behind the form as is sometimes done. When the line can be placed behind the form, concrete is placed first in the far end of the form and either the pump line is shortened or the pumps are backed away from the form as concrete rises to the level of the pump line. Concrete is vibrated with internal vibrators by workmen working behind the forms or working through form openings. The level of concrete must be balanced on both sides of the form. Intermittent operation of form vibrators may also assist the placing. When the sidewalls are filled to the level of the side pump lines, pumping concrete is started through the crown line. This likewise starts at the far end of the form. As the concrete builds up and flows down the side its movement and consolidation can be assisted by form vibrators operated at proper locations. As the concrete level rises, it is more difficult to use internal vibrators, and form vibrators must be depended upon to move the concrete into the sidewalls and arch below the line. As the concrete builds up in the arch, the end of the line should become buried 5 to 10 ft in concrete. When the pumping resistance builds up, a section of line should be removed or the pump backed away, shortening the buried length. Keeping the line buried allows concrete to be forced up higher than the line, thus filling the space above the line. During this portion of the work, form vibrators near the area being filled should not be operated as this will pull the concrete down, losing the fill which has been achieved. The end of the pump line should be positioned between steel supports so they do not block the flow of concrete. With piston pumps, the air slugger should be used sparingly. It should be used in accordance with the manufacturer's recommendations that "Its sole purpose is to push concrete away from the discharge end of the pipe in order to fill the farthermost recesses of the form first—but it should be used sparingly if at all after the pipe is buried." Air slugging is used to move concrete away from the line, not scatter it. A skillful operation can use the air slugger to obtain this result.

Placing concrete with an air gun is likewise carelessly done many times, and only skilled and careful operation can achieve good results. The gun should be operated so that a small amount of concrete remains in the reservoir. Uncontrolled blasting of all concrete out of the gun into the forms results in dangerous conditions for workmen and much segregation of concrete. Air guns may create fog behind the forms, which makes it difficult to see. Uncontrolled use of air guns or air sluggers may cause loosening, dislodging, or moving of reinforcing steel or other embedded items.

Railroad tunnels, section 7, are best placed similarly to the diversion tunnels using three pump lines as described above. However, some were placed fairly satisfactorily with only one crown line as shown. The end of the pump line was moved laterally to move concrete into one side or the other. The concrete was flowed from one end of the form to the other with vibrators (violating the principles of good practice outlined in Chap. 24; but as mentioned at the start of this section, tunnel lining is sometimes not so good as one likes). The considerable amount of steel ribs, lagging and blocking, and reinforcing steel required 5- to 6-in.-slump concrete to flow and fill satisfactorily. Experience showed the necessity of operating form vibrators near the end of the line in the crown to eliminate a premature buildup and hardening of concrete on the form at this point, which results in poor consolidation—evident when the forms were removed.

Sometimes construction joints in the first placement in a tunnel, conduit, or dam having sloping sides are detailed as shown in Fig. 52 with a short slope of concrete normal to the surface. Such sections are easily detailed but more difficult to place and secure adequate compaction, particularly if reinforced. Adequate vibration results in losing such a slope. It is usually much simpler to keep the entire construction joint horizontal as shown. The placement above can be adequately placed with careful placing, vibration, and inspection.

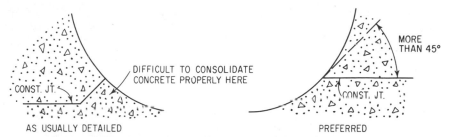

AS USUALLY DETAILED   PREFERRED

Fig. 52. Sloping construction joint normal to surface in tunnel, conduit, or other construction is easily detailed but difficult to consolidate satisfactorily. Horizontal joint as shown in right sketch is satisfactory with good concrete-placing procedures.

Keyways on horizontal construction joints are another detail which should be avoided if possible. Their use makes it more difficult to clean the construction joints and creates a depression that is a place for dirt to lodge and mortar used for joint preparation to fill. With good construction practices to obtain a good construction joint, many consider keyways unnecessary.

Curing. Curing of tunnels is best done by water sprinklers. Excessive drying conditions caused by movement of air through the tunnel can be prevented by closing one end where this is practical and can be done. Curing compounds, if used, must be of a type that will not be toxic to workmen in confined spaces.

13   Culverts and Cast-in-place Pipe and Conduits

Structures such as single- or multiple-box culverts and cast-in-place pipe and conduits are usually constructed for water supply, sewers, or vehicular or pedestrian traffic. The principles outlined in Chaps. 24 and 25 apply to these types of structures.

Culverts are usually placed in two or more placements (Fig. 53). The floor slab is placed as the first placement. The walls are placed next followed by the roof slab (Fig. 54), or the walls and roof may be placed in one placement. If the latter is done, the placing of the roof should be delayed until the wall concrete has settled, or the concrete at the junction of the wall and roof slab should be revibrated later to eliminate cracks which may form because of settlement of the wall concrete and restraint of the forms. Wing walls, if used, can be placed integrally with the walls or as a separate placement. Sloping surfaces on the wing walls should be covered with forms to retain the concrete during vibration and permit adequate consolidation. These forms are usually made in sections so they can be easily placed in position as concrete placing proceeds and to permit their removal shortly thereafter to permit finishing.

Cast-in-place pipe and conduit usually have a horseshoe or circular cross section. These may be cast in one placement, or usually the invert section is placed first followed by the sidewalls and arch in the second placement. Forms should be constructed so that concrete can be easily placed in the forms, vibrated, and inspected (Fig. 55). These structures may be heavily reinforced, which makes concrete placing difficult. The use of a drop chute outside the form, which spills concrete through the side of the form as shown in Fig. 14, Chap. 24, may be satisfactorily used.

The upper portion of the outside form which slopes over the wall may be

Fig. 53. Multiple-box culvert in various stages of construction. (*California Department of Water Resources photograph.*)

Fig. 54. Placing concrete in wall of a multiple-box culvert, without use of drop chutes. Note "trough" which assists in getting concrete into narrow form without spillage. Sometimes a short trough is made and is moved along the top of the wall to funnel concrete into a narrow form. (*California Department of Water Resources photograph.*)

hinged to facilitate placing concrete in the sidewalls to the height of the hinge, the hinged section is then lowered and fastened in place, and placing continues (Fig. 55). Inspection openings may need to be placed in the side of the form to

Fig. 55. Forms for conduits. In left-hand picture heavily reinforced section and overhanging portion of outer form made placing of concrete, vibration, and inspection very difficult. Overhanging portion of form in right-hand picture is hinged, which permits concrete to drop into the lower portion of the form more easily and allows better access for vibration and inspection. When concrete is placed to height of hinge, hinged portion is quickly lowered and fastened in place. (*California Department of Water Resources photograph.*)

permit vibration access, observation, and inspection. Such openings should be located to provide convenient access and should be quickly and easily closed when concrete placing reaches their elevation.

Cast-in-place unreinforced-concrete pipe can be placed continuously by a slipform process (Refs. 28). Such pipe is used for drainage or irrigation-water supply. Its strength depends upon arch action. In one method (Fig. 56) a ditch having a semicircular bottom is excavated with an excavator specially built for this purpose.

Fig. 56. Placing 72-in. cast-in-place unreinforced pipe with a slipform. Trench is excavated to a semicircular cross section and slipform places concrete as it is pulled forward by a winch.

The lining machine consists of a steel sled or slipform that closely fits the sides and semicircular bottom of the ditch. A gasoline engine mounted on the forward end of the form operates a winch that moves the slipform forward by a cable attached to a deadman ahead. A generator provides power to operate a variable-frequency vibrator attached to the bottom of the inside form. Concrete placed in the hopper from a ready-mixed concrete truck alongside is vibrated and mechanically spaded into place. Collapsible aluminum-alloy semicircular forms about 4 ft long are fed into the front of the hopper assembly, hooked together, and support the top and sides of the pipe as the traveling form moves forward. The collapsible forms are held in place by wood struts or spreader bars or frames. These support

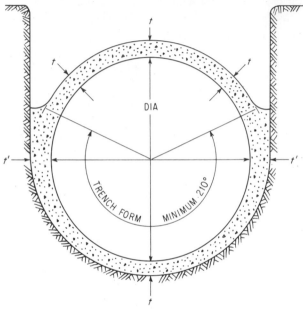

Fig. 57. Typical section of 12- to 96-in. cast-in-place nonreinforced concrete pipe. Minimum thickness t is $\frac{1}{12}$ diameter plus $\frac{1}{2}$ in. Thickness of sides t' is governed by clearance between pipe machine and trench wall. Thickness on horizontal diameter normally varies from t plus $\frac{3}{4}$ in. to t plus 2 in. (t plus 2 to 5 cm) depending on size of pipe. (*From American Concrete Institute.*)[28]

forms are usually removed the following day. The invert of the pipe is finished by hand by a finisher working inside the slipform. This process produces pipe having inside diameters of from 24 to 84 in. (Fig. 57). Concrete mixes suitable for structural concrete with slumps of 1½ to 3 in. are suitable for this type of concrete. The maximum size of aggregate should not exceed one-third to one-fourth the wall thickness. In another process, an inflated inner form of rubber and fabric is used to form the inner pipe circle.

PLACING CONCRETE UNDERWATER

Placing concrete underwater should be avoided whenever practical to do so. However, with care and attention to important details concrete can be placed underwater satisfactorily by four methods: underwater bucket, tremie, preplaced aggregate, and concrete pump, depending on the nature of the placement. References 29 through 33 describe several methods of concrete-placing operations in bridge foundations, graving docks, drydocks, wharfs, or caissons. Concrete should not be placed underwater in water colder than 35°F. Underwater concrete-placing methods require a very workable concrete with slumps as high as about 7 in. and cement contents of up to 650 lb per cu yd. Large placements of this rich concrete have a considerable internal temperature rise.

Placing underwater concrete is usually within caissons, cofferdams, or forms. Foundation cleanup imposes problems but can be done by hydraulic jets and pumps. Concrete must not be placed in running water or allowed to fall through water. To do so would wash the cement from the concrete.

14 Underwater Bucket

Underwater-bucket placing consists of lowering a special bucket containing concrete to the bottom of the foundation and opening it slowly to allow the concrete to flow out gently and without causing turbulence or mixing with the water. Subsequent buckets are landed on the previously placed concrete, and the gate should sink into contact with the previously placed concrete to prevent fall of concrete through water. Buckets should have drop-bottom or roller-gate openings. The gates obviously should have means for being opened from above water. If air is used to open the bucket, the air should discharge through a line to the surface to prevent disturbance. The top of the bucket must be covered in some way to prevent water from washing the surface of the concrete. One way is to cover with canvas or plastic sheets. Special buckets are made for underwater placing with a sloping top having a small opening to minimize water surge (Fig. 58).

Fig. 58. Special concrete bucket for placing concrete underwater. Note top, which covers concrete, preventing water surge from washing cement from concrete. (*Gar-Bro. Manufacturing Co.*)

Concrete for this type of concrete placing will have 6 or 7 sacks of cement per cu yd, 5- or 6-in. slump, 1½- to 2-in. maximum-size aggregate, and sand contents higher than normal, 40% or more. Concrete for foundations for the San Francisco–Oakland Bay Bridge was placed in water as deep as 240 ft with buckets.[32]

Bucket placement is a good way to start a placement by the tremie method; the latter is probably faster and possibly superior once it is well started without washing the first concrete placed. After a depth of at least 2 ft, preferably 3 ft, of concrete has been placed by bucket at the point the tremie pipe is to be operated, the empty watertight tremie pipe with its lower end covered with the pressure-seal plate shown in Fig. 59 is lowered and thrust deeply into the concrete already in place. Concrete is then placed in the tremie pipe. When

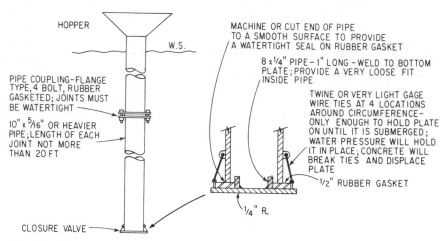

FIG. 59. Tremie and closure-valve seal details for underwater concrete placement.

it reaches a level somewhat above that sufficient to balance the water and concrete pressure, it will push off the lightly tied seal plate, and concrete will then extrude into and expand the mass already in place.

15 Tremie

Tremie placing consists of placing concrete through a vertical pipe, allowing the concrete to flow from the bottom (Fig. 59). The bottom of the tremie is kept submerged in the concrete at all times, and the concrete flows into the mass of previously placed concrete. References 30, 31, and 33 describe concrete placing with tremie.

A tremie consists of a pipe usually 10 or 12 in. in diameter with a hopper or funnel at the upper end. The equipment must be adequately supported and arranged so it can be raised and lengths of pipe can be removed if necessary as the level of concrete rises in the forms. The spacing of tremies should not be more than 12 to 16 ft apart or 8 ft from the sides or ends of the enclosure, although greater spacing has been successfully used in placements of large vertical dimensions. Concrete is delivered to the hopper by bucket, truck mixer or transport, pump, or conveyor. This delivery must be at a good rate and without interruption. Delays for any reason impair the free-flowing mobility of the mass into which concrete is being placed and may endanger the success of the placement. Plugged pipe and cold joints can result from delays.

Starting a tremie placement is very important. Unless great care and proper methods are used, much of the first concrete placed will become badly diluted and washed with water. Initial discharges should be controlled to extrude very slowly into the water and should consist of cohesive starter mixes of high cement (and possibly pozzolan) content.

There are two basic methods of starting a tremie placement which might be called the "wet-pipe" and the "dry-pipe" method. The wet-pipe method is practical in water depths under 100 ft and for large area work. The dry-pipe method is preferable for deeper work and for confined areas where nearly total freedom from any impairment of the first concrete on the foundation is important.

The wet-pipe method consists of placing a plug or a "go-devil" in the tremie pipe at about where the water level will be, with the pipe standing on the bottom in the water ready to start. This will be some distance below the concrete hopper on the upper end of the pipe, so there will be a sufficient weight of concrete in the pipe above the plug to overcome its friction and start it toward the bottom. These plugs may consist of a roll of burlap or cloth sacks, a cylinder of closed-cell PVC, or an inflated ball tight enough to stay well in place until the concrete pushes it down, forcing out the water in the pipe below it. The pipe is not lifted off the bottom until it is evident from the amount of concrete in the pipe and rising in the hopper that the plug is down to the bottom. The pipe is then slowly raised just enough to permit the plug to escape. This will be evident by a sinking of the concrete in the center of the hopper. The pipe is then quickly dropped to the bottom to shut off rapid flow of concrete into water; then it is repeatedly raised and lowered by small amounts as necessary to let concrete extrude rather than jet from the lower end of the pipe, until soundings show that the end of the pipe is embedded in concrete. Then the rate of flow can be increased because the concrete will be flowing into concrete and not into water.

The dry-pipe method requires a tremie pipe that is watertight, including gasketed, four-bolt flanges at the couplings. The pressure-seal plate shown in Fig. 59 is attached to the bottom of the tremie pipe in such a manner that the water pressure will make it completely watertight and the interior of the pipe will be empty and free of any water. With the pipe wall heavy enough to prevent flotation, the pipe is rested on the bottom to start, then filled with concrete, preferably a very rich, plastic, cohesive starter mix. When the pipe is lifted slightly, the weight of the column of concrete will break the strings holding the seal plate, and the concrete will extrude outward. With this arrangement, as in Fig. 59, it is evident that the pipe does not have to be lifted as high as for the plug to escape in the wet-pipe method. Accordingly, with careful manipulation of raising and lowering small distances, the concrete extrudes with a low velocity that causes a minimum of washing and dilution until sufficient concrete has been released to embed the end of the pipe; thereafter concrete will flow into concrete and not into water.

With either method of starting, once the discharge is well embedded in fresh concrete, greater concrete velocities and the resulting kinetic energy developed (which is achieved by a greater rate of concrete delivery) produce better results. Continuous delivery of concrete is desirable. Better results are obtained and there is less likelihood of a tremie plugging if concrete is discharged into the hopper in a manner to avoid segregation. If a tremie plugs, it can sometimes be freed by a quick lift of the pipe a few inches; quick dropping may plug it tighter. The end of the tremie must be kept buried in concrete at all times; from 3 to 5 ft is considered satisfactory. The pipe should not be lifted more than 6 to 12 in. at any time. Soundings should be taken during placing, and the elevation of the concrete surface and position of the lower end of the pipe should be frequently noted from markings on the pipe and plotted to show continually their relation to each other. Placing should start at a tremie in the center of a large placement, and others should be started from the center outward.

Air-lift pumps at the outside edge of the placement can remove mud pushed to the outer edge by the advancing concrete.

Very workable concrete is required for tremie placement. Seven-inch-slump, 7-sack concrete is considered standard. Maximum-size aggregate of 1½ to 2 in. and sand contents of up to 45% are used. Anything that will improve the workability of the concrete will improve tremie placing. These include such things as use of air entrainment, pozzolans, natural gravel instead of crushed stone, reduction in amount of crushed material in coarse and fine aggregate, and increase in fines in sand if a coarse sand is being used.

16   Preplaced Aggregate

The preplaced aggregate method briefly consists of placing aggregate within the designated area and injecting grout into the mass. This method is described in Chap. 38.

17   Pumps

Concrete pumps have placed concrete satisfactorily underwater. Reference 29 describes placing concrete in a pier caisson of a wharf with a concrete pump. The caissons were 4½-ft-diameter steel cylinders driven to bearing about 90 ft below pier top and 67 ft below water. Muck was removed from the caisson. The end of the concrete pump line was plugged and lowered to the bottom of the caisson. The line was filled with concrete, forcing the plug out, the concrete at the bottom end sealing it. Concrete pumping continued and the pump line was periodically raised as resistance increased. Sections of pipeline were removed above water as work progressed. The caisson was filled and allowed to overflow to remove laitance and excess water in the top concrete.

FARM AND HOME CONCRETE

18   Concrete Construction around the Farm and Home

Satisfactory work can be done by those unfamiliar with concrete work by observing the good practices pointed out in other chapters of this book. Much helpful information and literature can sometimes be obtained from service engineers of the Portland Cement Association, cement companies, and local ready-mixed concrete producers. Some helpful hints for this type of work are:

1. Plan the job.
2. Prepare the forms and other details required before placing the concrete.
3. Visit jobs where professionals are at work.
4. Secure sufficient help. Divide a large job into smaller placements. Concrete work is hard work physically.
5. Use ready-mixed concrete if a quantity of concrete is involved.
6. Use air-entrained concrete for its superior durability and helpful effect on workability.
7. Use concrete with a consistency such that a little effort is required to place it—and see that such effort is used. The concrete should have a plastic, fatty, not soupy, appearance. Avoid overwet, soupy, runny concrete. Concrete which can be "poured" is "poor" concrete. Concrete should be placed, not poured. If vibrators are used, use a drier consistency. If vibrators are not available, work the concrete into place with spades, shovels, tampers, or the feet.
8. Use concrete with sufficient cement. The cost of an extra sack of cement per cubic yard is good insurance against some of the deficiencies of this type of work.
9. After placing and screeding the concrete, avoid further finishing operations, floating, and troweling until the concrete is ready. In other words, avoid overworking fresh concrete. Finish as outlined in Chap. 27.

10. After finishing, cure the concrete with frequent sprinkling with water for a few days (Chap. 29). Cover the concrete with burlap, old carpeting, sand, waterproof paper, or plastic to retain moisture. Concrete should be cured with water, not "dried." (Water is required for concrete to gain strength.)

PROBLEMS IN CONCRETE PLACING

19   Placing under Adverse Weather Conditions

Concrete placing during either cold or hot weather is covered in Chap. 30 and Refs. 34 and 35, and the recommendations in those chapters and references should be followed.

Concrete placing during rain is, of course, not considered advisable. If rain is threatening at the time concrete placing is to start, its effect on the concrete and the advisability of canceling placing should be considered. If placing is allowed to proceed, plans should be made to take necessary precautions if rain starts. If rain starts after concrete placing starts, it is sometimes better to continue placing than to stop. The following precautions should be taken:

1. Cover the working area with tarpaulins or other protection to keep rain off the concrete, if possible.

2. Use a lower-slump concrete.

3. Dry up puddles of water collected on the foundation or on the construction joint before new concrete is placed.

4. Keep the surface of the concrete on a slight slope so water will run off, and provide a place for water to drain off.

5. Avoid working the surface of the fresh concrete.

6. If rain is so heavy that puddles cannot be removed and the surface of fresh concrete is being washed, it may be advisable to suspend concrete placing. If this is done, bulkheads, dowels, headers, or steps should be placed so that the placement is left in the best condition possible for continuing at a later time. The construction joint so formed because of discontinuing placing will, of course, be treated as a construction joint with suitable cleanup preparatory to placing the remainder of the concrete later.

7. During thunderstorms of short duration, cease placing concrete until the storm passes and cover the concrete with tarpaulins, plastic sheets, or other protection during the storm.

20   Avoidance of Cold Joints

Cold joints occur in a concrete placement when a layer of previously placed concrete hardens or sets to the extent that a newly placed layer does not bond to it. Cold joints are more likely to occur during hot or windy weather because concrete either dries or sets faster because of high temperatures and loss of moisture. In addition to lack of bond, the cold joint is visible and sometimes unsightly because of a lack of consolidation, or there is an offset due to the form's bulging from the pressure of the overlying concrete. Cold joints can be eliminated by planning placing procedures so that the concrete is not exposed too long before it is covered by subsequent layers. This can sometimes be accomplished by "stair-stepping" the layers or by speeding up placing. This speedup is achieved by more or different equipment, more workmen, or elimination of bottlenecks that cause slow placing.

Conditions that cause slower setting of the concrete such as the use of colder concrete, placing concrete at night during hot weather, shading the work area, maintaining high humidity at the concrete surface by fog sprays, or use of a retarder in the concrete are sometimes beneficial.

If the underlying concrete has hardened to the extent that a cold joint is likely to occur, the subsequent layer of concrete should be preceded by a layer of mortar, or the same procedure for starting concrete on a construction joint as discussed previously in this chapter should be followed. If cold joints cannot be prevented, the concrete placements should be reduced to a size that can be handled without cold joints.

21 Slump-loss Problems

As stated previously, concrete stiffens or loses consistency or fluidity rapidly, commonly called slump loss. This is most likely to be troublesome during hot, dry, windy weather, especially when the concrete is hauled long distances or transported on long chutes or conveyors or otherwise exposed to drying from wind and heat. The following can be helpful in reducing slump loss or its effects:

1. Organize and coordinate the job to transport and place concrete faster, and avoid delays. Select and use equipment that will expedite the work, not slow it. Schedule concrete delivery so that trucks can be unloaded and concrete used as soon as they arrive—do not have several trucks waiting to unload. Provide communications with the concrete plant so that deliveries can be changed to fit the progress at the placing site.

2. If using dry absorptive aggregate in concrete, dampen it before use.

3. If subgrade is dry, dampen it to lessen absorption.

4. Reduce temperature of concrete by

   a. Precooling aggregate with cold water, cold-air jets in the batcher bins, or vacuum evaporative cooling.

   b. Cooling cement.

   c. Cooling mix water by using chipped ice in mix water.

   d. Shading aggregate storage, water tanks, and lines.

   e. Painting truck mixers and water tanks and lines white.

5. Avoid overmixing.

6. Work at night.

7. Shade work and use windbreaks.

8. Use fog (not water) sprays in the concrete placing area to keep the humidity high and the air temperature low.

9. Investigate the possibility of false set or premature stiffening in the cement, and avoid cement having it.

Usually, the faster concrete can be handled and placed, the less difficulty will be encountered with slump loss and handling and placing problems resulting from such slump loss.

REFERENCES

1. ACI Standard 301-72, Specifications for Structural Concrete for Buildings.
2. ACI Standard 302-69, Recommended Practice for Concrete Floor and Slab Construction.
3. Hurd, M. K.: Formwork for Concrete, *ACI Spec. Publ.* SP-4, 1963.
4. Townsend, Charles L.: "Control of Temperature Cracking in Mass Concrete," paper presented at 62d annual meeting of American Concrete Institute, 1966.
5. Townsend, Charles L.: "Control of Cracking in Mass Concrete Structures," U.S. Bureau of Reclamation Engineering Monograph 34, 1965.
6. Symposium on Mass Concrete, *ACI Spec. Publ.* SP-6, 1963.
7. ACI Standard 614-72, Recommended Practice for Measuring, Mixing, Transporting, and Placing Concrete.
8. "Concrete Manual," 7th ed., U.S. Bureau of Reclamation, Denver, 1963.
9. Price, W. H., L. P. Witte, and L. C. Porter: Concrete and Concrete Materials for Glen Canyon Dam, *Proc. ACI,* vol. 57, p. 629, December, 1960.
10. Stodola, P. R., J. E. O'Rourke, and H. G. Schoon: Concrete Core Block for Oroville Dam, *Proc. ACI,* vol. 62, p. 617, June, 1965.
11. Mass Concrete for Dams and Other Massive Structures, ACI Committee 207 Report, *Proc. ACI,* vol. 67, p. 273, April, 1970.

12. ACI Standard 325-58, Recommended Practice for Design of Concrete Pavements.
13. ACI Standard 617-58, Specifications for Concrete Pavements and Concrete Bases.
14. "Concrete Pavement Inspector's Manual," Portland Cement Association, Skokie, Ill., 1959.
15. "Sawed Joints in Concrete Pavement," Portland Cement Association, Skokie, Ill., 1957.
16. Waddell, J. J.: "Practical Quality Control for Concrete," McGraw-Hill Book Company, New York, 1962.
17. Symposium for Slipform Paving, *Highway Res. Board Record* 98.
18. Gillis, L. R., and L. S. Spickelmire: Slipform Paving, *Calif. Highways Pub. Works,* vol. 44, nos. 1–2, 3–4, pp. 63, 68, January–February, March–April, 1965.
19. Gillis, L. R., and L. S. Spickelmire: Slip-form Paving in the United States, 1965, prepared for "Concrete Yearbook," 1965–1966, Bundesverbund der deutschen Zementindustrie, E. V., Germany.
20. Ray, Gordon K., and Harold J. Holm: Fifteen Years of Slipform Paving, *Proc. ACI,* vol. 62, p. 145, February, 1965.
21. ACI Standard 506-66, Recommended Practice for Shotcreting.
22. Johnson, Max R.: Slipform Lining of the San Luis Canal, *Proc. ACI,* vol. 62, p. 1313, October, 1965.
23. "Tunnel Lining with the Rex Pumpcrete," Chain Belt Co., 1945.
24. Wilson, R. J.: Lining of the Alva B. Adams Tunnel, *Proc. ACI,* vol. 43, p. 209, November, 1946.
25. Davis, R. E., Jr., G. D. Johnson, and G. E. Wendell: Kemano Penstock Tunnel Liner Backfilled with Prepacked Concrete, *Proc. ACI,* vol. 52, p. 287, November, 1955.
26. Placing Concrete by Pumping Methods, ACI Committee 304 Report, *Proc. ACI,* vol. 68, p. 327, May, 1971.
27. Tuthill, L. H.: Tunnel Lining with Pumped Concrete, *Proc. ACI,* vol. 68, p. 252, April, 1971.
28. ACI Standard 346-70, Specifications for Cast-In-Place Nonreinforced Concrete Pipe, Recommendations for Cast-In-Place Nonreinforced Concrete Pipe, ACI Committee 346 Report, *Proc. ACI,* vol. 66, p. 252, April, 1969.
29. Messina, R. F.: Pumpcrete Conveys Tremic Concrete to 90 Ft. Loading Wharf Caisson, *Concrete,* December, 1943, p. 2.
30. Halloran, P. J., and K. H. Talbot: The Properties and Behavior Underwater of Plastic Concrete, *Proc. ACI,* vol. 39, p. 461, June, 1943.
31. Angas, W. M., E. M. Shanley, and J. A. Erickson: Problems in the Construction of Graving Docks by the Tremie Method, *Proc. ACI,* vol. 40, p. 249, February, 1944.
32. Stanton, T. E., Jr.: Cement and Concrete Control—San Francisco–Oakland Bay Bridge, *Proc. ACI,* vol. 32, p. 1, September–October, 1935.
33. Gerwick, Ben C., Jr.: Placement of Tremie Concrete, Symposium on Concrete Construction in Aqueous Environments, *ACI Spec. Publ. 8,* 1964.
34. ACI Standard 300-66, Recommended Practice for Cold Weather Concreting.
35. ACI Standard 305-72, Recommended Practice for Hot Weather Concreting.

Section 7

FINISHING AND CURING

Chapter 27

FINISHING

CHARLES E. PROUDLEY*

Concrete surfaces are classified according to the appearance as called for or described in the plans and specifications. The finish is the result of the characteristics of the form or form liners against which the concrete is cast and any subsequent manipulation or treatment of the hardened concrete after form removal, or the result of working and treatment of unformed surfaces.

The plans and specifications for the job will describe procedures to be followed to secure the desired finish for concrete surfaces resulting from the use of forms or molds. For precast units the finish will have been applied at the manufacturing plant and there will be little that can be done at the construction site except for the arrangement of the units as they are placed, by the application of coatings to the finished structure, or by rejection of defective units.

Formed surfaces are for vertical structures such as walls and columns, and the finished product will show the imprint of the forms or form liners except as this may be modified by rubbing or grinding or the application of a coating of mortar or coloring. The texture of the surface depends on the manner of rubbing, grinding, sandblasting, or bushhammering, or the architectural finishes of limitless variation in sculptured or plain geometric designs. Depending upon the type of form material used, the formed surface without further treatment may range from very rough to glassy smooth. Forms are discussed in detail in Sec. 4 and in Refs. 1 and 2.

Unformed surfaces are finished as specified to present either a rough or smooth surface depending on the purpose to which the concrete is to be put. A rough finish is used for a base to be topped out with a cement mortar or surface course of other materials as for roadways, walks, floors, roofs, and similar construction where the concrete is stiff enough when placed to retain its shape without forms. In most cases the final finish on an unformed surface is secured before the concrete has hardened or attained any appreciable strength, through the processes of screeding, floating, troweling, and the application of colors, special aggregates, or other surface treatments.

Tilt-up construction permits architectural designs and effects on vertical surfaces, while taking advantage of the simplified formwork permitted by casting the units in a horizontal position. Either the design may be on the bottom of the mold with the panel backing on the top of the slab, or the architectural effect may be obtained by the application of specially prepared aggregates or colored mortars to the upper surface with the panel backing on the bottom of the slab. These will be described later.

For either formed or unformed surfaces it is common practice among many specifying agencies and organizations to designate finishes by a numerical system that defines the texture of the surface and the maximum deviation from a plane of specified dimensions, such as a fraction of an inch from a straightedge 5 ft

* Deceased. Chapter revised by Joseph J. Waddell.

long held in any direction on the surface of the slab, that any projection or hollow may have.

1 Classification of Finishes

The amount of finishing and dressing to be applied to a concrete surface, either formed or unformed, depends upon the type of surface and whether it is to be exposed to view. The following classifications, based on practice of the U.S. Bureau of Reclamation and the Concrete Industry Board of New York City, should be considered when preparing specifications for concrete work. In these classifications, the F finishes apply to formed surfaces, and the U finishes apply to unformed surfaces.

Finish F-1. Finish F-1 is applied to surfaces where roughness is not objectionable, such as surfaces to be backfilled, permanently submerged, or otherwise concealed from view. The only surface treatment after removal of forms is filling of tie-rod holes on walls under 12 in. thick, or on all surfaces to be waterproofed, and repair of defective concrete. Forms may be quite rough, as long as they do not leak mortar.

Finish F-2. This is used for all permanently exposed surfaces where a higher-quality surface is not specified, such as external portions of bridges (except grade separations); hydraulic structures such as tunnels, canals, culverts, siphons, spillways, and dams; retaining walls not subject to close public view; docks and wharves; and certain rough buildings. Forms must be carefully and accurately built, without conspicuous offsets or bulges. Sheathing may be plywood, shiplap, or steel (not thin sheet steel). Surface treatment includes removal of mortar fins, filling of tie-rod holes, and dressing of offsets greater than $1\frac{1}{4}$ in. and bulges greater than $\frac{1}{2}$ in. in 5 ft.

Finish F-3. Finish F-3 is used for all surfaces subject to close public view such as inside and outside of buildings (except those surfaces with special architectural treatment), parapets, grade-separation structures, and powerhouses. Forms must be carefully and accurately built by skilled workmen, without visible offsets or bulges. Sheathing should be plywood or tongue-and-groove boards. Special care is required at construction joints. Uniform color and texture are necessary, which may necessitate sack rubbing, grout cleaning, or other treatment, as specified. Materials should be from the same source throughout the work, except for surfaces to be painted.

Finish F-4. Finish F-4 provides a special smooth surface for conduits carrying high-velocity water where cavitation might occur. Forms must be especially tight, strong, and smooth, without perceptible offsets or bulges. Surface treatment consists in removal of all rough spots, offsets, and bubble holes by filling or grinding.

Finish F-5 This is a special rough surface for bonding to plaster or stucco. Concrete is cast against rough, unoiled form boards. Surface treatment includes removal of fins and projections, and repair of defective areas.

Finish U-1. Finish U-1 is screeded only, and it is used for surfaces to be backfilled, construction joints, base for two-course floor, and rough exposed slabs. Also, it is the first stage for further finishing.

Finish U-2. Screeded and floated, finish U-2 is used on all outdoor concrete, unless another finish is specified, such as tops of bridge piers, outside decks at industrial buildings, hydraulic structures such as canals and tunnel inverts, and reservoirs.

Finish U-3. This finish is screeded floated, and troweled and is used for inside floors; at tops of walls and parapets subject to close public view, and for roof slabs, sidewalks, and pools.

Finish U-4. This is a special simplified trowel or slipform finish for canal linings. It is equivalent to bull-float or darby slab finish.

Finish U-5. Various special finishes designated in the specifications, requiring certain finishing procedures in each case, come under this category. It is also used for architectural finishes.

FORMED CONCRETE

2   Rubbed Surfaces

Rubbed or stoned surfaces are commonly applied to concrete cast at the jobsite. Rubbing is performed immediately upon removal of the forms or, if this is impracticable, a covering of wet burlap is used to keep the surface from drying too much before rubbing. Forms are removed within 12 to 24 hr, and all patching and pointing are done immediately with a 1:1½ mortar of sand that passes a No. 16 sieve and only enough water to give a dry, adhesive consistency. When

Fɪɢ, 1.   Wet rubbing with power grinder.   (*Courtesy of Stow Mfg. Co.*)

repairs have hardened enough to withstand it, the entire concrete surface to be rubbed is cleaned with a strong jet of water to remove loose surface materials. While still wet, the concrete is hand-rubbed using a No. 60 grit carborundum stone or a wood float, rubber float, or burlap sacking, depending on the texture or surface desired. The rubbing should result in a coating of mortar that will fill all small voids and airholes, leaving a smooth surface. Stoning with water may raise sufficient mortar but a mortar of creamy consistency must be used when rubbing is done with wood or rubber floats. A 1:1½ mortar of fine sand passing the No. 30 sieve, and cement to which is added white portland cement to give a light color matching the concrete, is used for the preliminary rubbing (see Figs. 1 and 2).

Mortar used for rubbing will usually be darker than that produced by grinding the formed concrete surface; therefore, it is recommended that some white cement be added to the mortars for rubbing, especially if the structure is subject to critical observation by the public. The amount of white cement is usually in the neighborhood of 20% but should be adjusted as found necessary by experiment on a small inconspicuous area at the beginning of the job.

Final rubbing with a burlap-covered or sponge-rubber float results in the coarsest

FIG. 2.　Equipment for rubbing or grinding ceiling.　(*Courtesy of Stow Mfg. Co.*)

texture, a wood float with a neat cement slurry of about 2 lb cement per gal gives a smoother finish, and a stoned surface over a sack-rubbed finish using only the mortar rubbed up by the No. 50 or No. 60 grit stone produces the smoothest of the rubbed finishes. A soft-bristle brushing of the freshly rubbed, moist surface adds further to the fine texture.

Power-operated grinding stones are often used, especially in the preliminary stages of rubbing. After the final rubbing operation the surface should be moist-cured for 14 days to give time for the mortar produced by rubbing to gain strength for permanence and resistance to dusting.

The color of the finished surface is affected by a number of factors, some of which must be given attention during mixing and placement of the concrete. The sources of cement, fine aggregate, and coarse aggregate must be continued throughout wherever changes in color of the structure are of concern. The same consistency must be maintained from batch to batch and, of course, the form oil or other substances with which the concrete will come in contact must be the same throughout.

Grout Cleaning.　After all concreting operations have been finished and curing completed, a simple method for the final cleaning of a formed surface consists of rubbing the surface lightly with a coarse carborundum stone or cork float when the forms are removed 24 to 72 hr after placing the concrete. Exposing the concrete by stripping forms should be kept to a minimum in advance of the initial rubbing operations. Fill all air bubbles' and water holes and other imperfections immediately with a mortar of the same proportions as was used in the concrete. Then rub the entire area only enough to break the surface and fill all voids. The last contact should be such that no trace of the circular rubbing action is left on the surface. When a uniform color and texture are obtained rubbing is discontinued and curing is resumed, keeping the surface wet for at least 7 days and preferably longer. After the initial rubbing has been performed satisfactorily, the surface is sprayed with water until it is thoroughly saturated; then, starting at the top and working down, a grout coat is applied with a whitewash brush. The grout consists of 1½ parts of light-colored clean sand that will pass a No. 50 sieve and 1 part of a mixture of white and gray cement with enough water to make a creamy paste. A dark color of gray will result unless some white cement is used. No more water is added to the surface to be cleaned. After a uniform color and texture are obtained by brushing on the mortar, it is allowed to stand until the grout becomes tacky. Excess grout is then removed by rubbing with a folded burlap bag, being careful that a layer of grout does not remain on the wall, as this would scale off later. See also Chap. 45 for a discussion of sack rubbing.

Grinding. Carborundum stone is used extensively for grinding concrete and can be handled manually or with a power grinder, as shown in Fig. 3. The stone comes in various grit numbers; the higher the number the finer the grit and the texture it will produce. Grinding can be done for the sole purpose of removing fins without affecting the finish otherwise, and for this purpose a coarse grit is preferable. Continued rubbing results in a smoother surface and, with simultaneous applications of a soft mortar, will fill airholes and surface honeycomb. When it is proposed to remove the surface skin by machine grinding,

FIG. 3. Wet grinding to remove form marks and fill surface voids. (*Courtesy of Stow Mfg. Co.*)

the surface is kept wet to reduce the health hazard from silica dust, or it can be ground dry with the operator wearing a dust mask. A variety of shapes of grinding stones are available to adapt to curved surfaces, coves, and interior angles. Power grinders are driven by electric, gasoline, or compressed-air motors attached to the grinding disk by a flexible cable. For large horizontal surfaces such as floors and pavements the machine used for troweling can be equipped, in the case of most machines, with carborundum blocks instead of trowels (Fig. 4).

Sandblasting. A hardened concrete surface can be given a sandblast treatment for artistic effect or to provide a rough nonskid surface. Sandblasting can give an exposure as light as etching, revealing only the particles of sand; or heavy sandblasting can result in a depth of reveal of as much as ¾ in. Light sandblasting is very effective for cleaning-up operations to remove stains, grout accumulations, sealing compound, and other substances stuck to the surface. For ornamental sandblasting, the coarse aggregate used in the base concrete can vary from those normally available to selected materials such as granite or quartz when special

FIG. 4. Grinding a horizontal surface using grinding blocks instead of blades in a troweling machine. (*Courtesy of Stow Mfg. Co.*)

color effects are required. Often local gravel will serve as exposed aggregate, or it may be necessary to use special aggregate to impart the color tone and texture desired by the architect. Such aggregates will usually be more expensive than local material.

It is important that forms be sufficiently tight at the joints to prevent leakage. Form joints can be sealed by using one of the special pressure-sensitive tapes made for that purpose. The imprint of the tape on the concrete will be removed by the sandblasting. Plaster of paris, calking compound, or a mixture of beef tallow and portland cement can also be used for joint sealing. Unsealed joints leave lines on the surface of the concrete that cannot be removed by the blasting operation.

Abrasives include hard angular sand, silica sand, and special blasting grits. Sand should be finer than No. 30. A ⅜-in.-diameter venturi-type nozzle is normally large enough, although larger sizes are occasionally used. Each nozzle requires about 300 cfm of air at a minimum pressure of 90 psi. Blasting is begun at the top, and the nozzle is directed from side to side somewhat as in spray painting, keeping the nozzle about 24 in. from the surface of the concrete. A final washdown with water completes the cleaning.

When sandblasting is done for ornamental effect, outlines and designs can be applied to the concrete by means of a mask of rubber cement of heavy consistency that can be peeled off readily after sandblasting to the depth desired.

Normally, the sandblasting should be done as early after stripping the forms as economically feasible, even within 24 hr. Scheduling problems may make it necessary to delay sandblasting, even though it becomes more difficult to blast as the concrete gains strength. For the sake of surface uniformity of architectural surfaces, all areas must be blasted at approximately the same age.

Acid etching of a concrete surface to change the texture can be done with a muriatic (hydrochloric) acid solution of one part commercial acid to three or four parts water. Etching is difficult to apply to cast-in-place concrete, its use usually being confined to treating precast elements. Acid etching produces a rather fine texture, without the reveal that is obtained by other methods. Carbonate aggregates, such as limestone, dolomite, and marble, should not be used in concrete to be etched.

Etching is accomplished by immersing the concrete casting in a tank of acid. Agitating and warming the acid will hasten its action. After etching, the casting is thoroughly washed and scrubbed with brushes and clean water. In some plants, the casting is neutralized in an alkaline bath after etching and before washing.

The acid is highly corrosive; hence workmen must wear protective clothing and take care to avoid splashing the acid onto surfaces or materials other than the one being treated.

Bushhammering can be done with hand hammers or with pneumatic equipment, as shown in Fig. 5. Its purpose is to give a relatively uniform surface finish in which the form marks are removed, joints between lifts of concrete in the forms are obliterated, aggregate particles on the surface are chipped, and mortar is roughened. Obviously, the color of the coarse aggregate in the concrete will be revealed by bushhammering, and this may be the desired effect. If

FIG. 5. Using the bushhammer to texture the surface.

so, the concrete mix should be of low slump and have a minimum of mortar consistent with satisfactory placement in order to have the maximum amount of coarse aggregate at the surface to be hammered.

Hand hammers are good for small jobs, at corners, and for restricted areas where a power tool cannot reach. The hammer has a face on which are a number of points spaced about $\frac{3}{8}$ in. apart, all in the same plane. Hammerheads for power tools are much the same as the hand hammers. Power bushhammers are usually air-operated, although other types are available.

Sharp corners on the concrete should be avoided, as they will surely spall off if struck with a bushhammer. For this reason chamfered corners should be formed when it is intended to bushhammer the concrete. Concrete should have attained a compressive strength of at least 3,500 psi before being hammered.

3   Form Liners

Architectural effects of almost unlimited variety can be produced by selection of sheathing or use of liners attached to the inside of the form that will transfer relief and designs to the concrete cast against the lining.[3] Forms may be vertical when cast, or horizontal as in the tilt-up method of construction. The latter is simpler to cast and results in fewer flaws, especially for intricate patterns. Examples are shown in Figs. 6 and 7.

A sample of the desired finish or finishes, either by means of a sample panel or by reference to a completed structure, will be of more help to the contractor than an attempt to describe the procedure by which the result is to be secured, since each contractor is most familiar with his own methods for reaching an end result. Contractor and architect should work together and agree on the methods in advance of actual concrete operations.

Very smooth, glasslike surfaces, not obtainable with ordinary forms, are sometimes called for, requiring the use of a plastic form liner. After the forms are erected joints between the liners are filled with a caulking compound to give a joint-free surface. Caulking compounds are available at most building supply dealers, or the joints may be filled with patching plastic, cold-water putty, or a mixture of equal parts of beef tallow and portland cement. Light sanding with No. 0 sandpaper will make the joint smooth and practically invisible. Some form-coating materials such as epoxy-resin plastic, when properly applied to plywood, give almost the same effect as plastic liners. Imperfections on glossy concrete surfaces are very noticeable and not easily repaired. Special putties and sticks of resin compound are available in various colors for repairing glossy concrete surfaces, but their permanence is uncertain. Horizontal casting with the design on the bottom of the form reduces the number of air bubbles at the surface, especially if internal vibration is used. Exterior vibration of forms should not be used as this seems to draw air bubbles to the vibrated face.

Fig. 6. Strips attached to forms to give paneling or rustication effect. Striated plywood panels are placed alternately for architectural effect. (*Courtesy of Portland Cement Association.*)

Board marks and simulated wood grain are easily produced to give rustic or natural effects. Liners of plywood that have been striated, wire-brushed, or sandblasted to accentuate the grain, as shown in Fig. 8, may be prepared by the contractor or purchased from a mill that handles form materials for concrete work. Unfinished sheathing is often used without lining. By selecting boards having the grain effect desired, varying the width of boards and spacing between them, many attractive and striking results are possible. Spraying the forms with ammonia raises and emphasizes the grain. Minor defects in the concrete surface are not noticeable, but major imperfections such as severe honeycomb are very difficult to cover up in wood-grain surface finishes.

Whenever possible, wooden strips and inserts used for rustication, paneling, and geometric designs should be fastened to the forms by nailing through the form from the back, but if nailing must be done from the inside, finishing nails should be used. This permits the form to be removed but leaves the inserts in the concrete, to be removed later when the probability of damaging edges of the designs is much less. See Fig. 6 and Fig. 15, Chap. 28.

Wooden forms and form liners require some type of parting agent or form oil to prevent bonding to the concrete. A nonstaining parting agent or oil will prevent discoloration of the concrete. It should be carefully and uniformly applied and any excess wiped off.

Tempered fiberboard, or hardboard, is frequently used as form lining, attached to the sheathing with either the smooth side or the screen side to the concrete depending on whether a smooth or slightly textured surface is desired.

Hardboard is manufactured in two thicknesses, $\frac{3}{16}$ and $\frac{1}{4}$ in., and in sheets 4 ft wide by 8 or 12 ft long. It is treated by the manufacturer by an oil-tempering process to minimize absorption, and only the tempered board of concrete-form-board grade should be used.

The board should be preconditioned at least 12 hr before use by wetting and stacking the panels back to back. When applying sheets of lining material it is important to start attaching each sheet at its center and work toward the edges to prevent buckling, using at least one 3-penny flathead blue shingle nail in every square foot and at least one every 8 in. along the edges. In lieu of nailing, the board can be attached to the sheathing with a special waterproof adhesive. Edges of abutting sheets must be nailed to the same backing board, and care must be exercised to butt the edges tightly together to avoid revealing the joints. If edges do not come together tightly, the crack should be filled with patching plaster or similar material. Form oil or a suitable parting agent must then be applied.

Hardboard can be worked with ordinary woodworking tools, but for straight cuts a power saw having an 88-tooth, 10-gage, 14-in.-diameter blade, or an 8-13-8 gage for miter-ground saws will give cleanest cuts. If there are any burrs on the edges of the hardboard liners after they are attached to the sheathing, they should be removed by means of a block plane. When boring holes a worm-center bit will avoid tearing the fiberboard if drilled from the face side of the liner. Slight burrs can be easily smoothed with fine sandpaper.

Plastic Form Liners. Plastic liners can provide a finish ranging from satin smooth to any texture desired at relatively low cost. Straight plastic sheet can

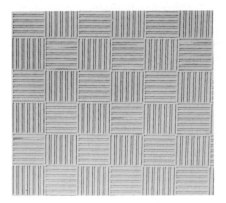

FIG. 7. The effect of using sheet-rubber floor covering as a form liner. (*Courtesy of Portland Cement Association.*)

FIG. 8. Plywood form or form liner treated to accentuate grain gives a natural or rustic effect. (*Courtesy of Portland Cement Association.*)

be glued and welded to intricate patterns and can be laminated to lumber using a water-soluble adhesive. The sheets can be formed by bending, but because of their flexibility, overbending by about 15° is necessary to secure a desired shape. Thermoplastic sheets are used frequently, premolded to a pattern by a vacuum-forming process at a temperature of 325 to 375°F. Fiber-glass-reinforced plastics, or plastics formed into shape by heat and pressure, have great durability for many reuses. Since plastic is quite rigid and its surface smooth, concrete cast against it is smooth and uniform in color. Furthermore, because plastics do not bond to concrete no parting agent is necessary.

Thermoplastics may be high-impact styrene, linear polyethylene, acrylonitrile butadiene styrene, or other similar materials. Sheet size is 4×7 or 4×8 ft, and this must be remembered when laying out a design for large panels. Overly ornate and complicated patterns should be avoided and the plastic sheet must be sufficiently rigid that it will not deform under the weight of the concrete. Tilt-up slabs, cast downward, can normally support only 3 or 4 in. of concrete without backing. For vertical walls a lift of 3 or 4 ft can be placed against it, regardless of thickness, without appreciable distortion.

Before the advent of thermoplastic materials, rubber form liners were used widely. Sheet rubber is available in many designs as in floor coverings, runners, and matting. It is not generally necessary to fasten the rubber sheeting to a horizontal surface because friction between the supporting surface and the matting is enough to prevent movement during casting. Furthermore, it is not necessary to use a parting agent, although they must be cleaned well after each use, brushing with a soft brush and a jet of water if needed. When a parting agent is used it should be a vegetable oil such as castor oil or lanolin. Of course, the pattern in the design must have no undercutting. Forms should not be stripped in less than 5 to 7 days or until the concrete has attained strength to give a clean break and avoid "plucking." A 6° angle of draft, with all edges and corners rounded, is generally recommended because the resulting surface will not have recesses in which water can accumulate. Information about the rubber material should be obtained from the supplier to secure the best results in durability and service because many of the synthetic rubbers behave differently from the natural rubber.

4 Architectural Finishes

Metal Molds. Metal forms and molds such as steel embossed with a uniform design are used to some extent for special designs and intricate details. Some stock patterns are available and if economy suggests it for repeated use of the forms, a sheet-metal-working plant will prepare special designs. They should be made of black iron since galvanized iron has a tendency to stick to the concrete more than black iron. The forms must, of course, be oiled before use and wire-brushed and washed between uses. Plain sheets are generally unsatisfactory where the formed surface will be subject to observation by the public because they are difficult to keep satisfactorily flat and the joints between sheets are almost impossible to conceal. Corrugated-metal sheets are used frequently for such purposes as fluting in pilasters, piers, and spandrels. Corrugated sheets should be cut exactly at the center of a corrugation to make a tight joint between the metal and the backing to which it is nailed. Cutting of metal for forms should be done in a sheet-metal shop equipped with shears to give straight, smooth edges. If corrugated sheets are not long enough to cover the entire form they may be butted or lapped. When lapped, the upper sheet is lapped over the lower one so that the offset, when stripped, will not cast a shadow and the joint will be less conspicuous.

Wood Molds. Wood molds are best adapted to ornament consisting of simple moldings, combinations of moldings, or shapes that can be made with a band saw. Detail involving carving or undercuts should be formed with plaster waste molds. Wood molds are easier to erect and strip than plaster molds, less work is required to prepare them, and less care in handling is necessary. Joints in wood molds should be tight enough to prevent leakage of mortar that would create objectionable fins. Wherever possible, at corners and elsewhere in a wood mold, the joints should be made by overlapping the pieces instead of butting them, as shown in Fig. 9. In this way slight movement during construction will not open the joints. Square-edged butt joints are acceptable for moldings applied to a solid backing. When members of the form must be joined where there is no return or reveal, use tongued-and-grooved or shiplap lumber, or spline the joints.

If several pieces are required to make a complete mold, the joints in the different members should be staggered to secure maximum rigidity. Much time can often be saved in erecting and stripping forms for a detail involving many pieces of run moldings if brackets are made in the mill to a template to fit the general profile of the detail. This is illustrated in Fig. 9.

Plaster Waste Molds. Plaster waste molds are for casting intricate ornamentation as shown in Fig. 10, where the mold must be broken to remove it. In most cases the molds are built by the ornamental plasterer working closely with the architect and the contractor. A model is first made of wood, plaster, clay, or a combination of these and similar materials for a "positive" from which the "negative" can be cast. The negative is the mold from which the finished concrete is cast.

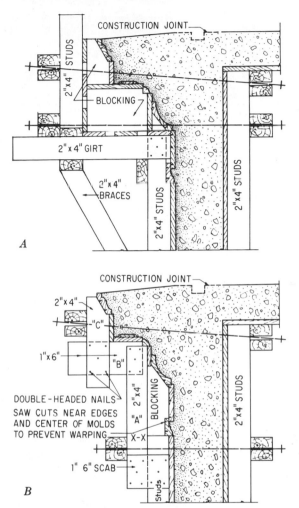

FIG. 9. Wood mold for intricate architectural effect. *A* is original construction of mold. *B* is the same mold reassembled for reuse. (*Courtesy of Portland Cement Association.*)

When the pattern, which is the positive, has hardened it is coated with a parting compound and the waste mold is cast against it. First, a ¼-in. layer of colored plaster (laundry bluing makes a good color) is applied to the pattern and then a mixture of plaster is mixed with jute fibers to give a thickness of at least 1 in. at all points of the mold. After the plaster is thoroughly set and hard the mold is removed by cutting it into sections along lines that will permit easiest removal from the pattern with a minimum damage to the details of the mold. The colored plaster at the face of the mold aids in detecting joints in need of repair. Repairs are made with fresh plaster and finished to the texture of the form sheathing or whatever texture is desired.

The contact side of the mold is waterproofed with shellac or some other compound to prevent the mold from absorbing water from the concrete placed against it,

Fig. 10. Plaster waste molds are used for complicated detail and must be broken away for removal. These panels were cast against plaster molds. (*Courtesy of Portland Cement Association.*)

which would weaken the concrete and, also, the mold. The sections of the mold are wired or cemented together firmly and fastened securely to the forms in the required position. This may require application of additional reinforcing plaster, wire mesh, reinforcing rods, or other mechanical devices to prevent dislocation of the mold during the concreting operation. None of the sections of the mold should weigh more than what one or two laborers can handle conveniently when setting in the form (150 lb or less). The colored plaster will help to avoid damage to the concrete upon removal of the waste mold since it can be readily distinguished from the concrete product.

The shape of the back of the mold depends upon the depth of the ornamentation in the design. If relatively shallow and less than 3 in. thick at any point the backing is more readily made flat or roughly so, with a flat surface at the edges of the back of the mold to bear against the form sheathing or a framework of studs, as will be seen in Fig. 11. The two strips are attached to the back to facilitate handling but are removed when the form is set in place. The molds are nailed to the supporting formwork with common nails, countersunk, and the holes are patched with plaster. Number 14 gage wire passed through two holes drilled through the mold and concealed with plaster will aid fastening to the back. Molds with very irregular shape or deep relief are made about 1½ to 2 in. thick to conform to the front of the pattern, and wads of plaster-soaked jute or reinforcing fiber are placed on the back to give points of contact with the form framing, as seen in Figs. 11 and 12.

The edges of the mold require special attention. If possible, the joints between the mold and the form sheathing should be hidden at reveals or returns. The edges of each piece should be rabbeted as illustrated in Fig. 11 so they will fit closely with the form sheathing to make a tight joint. All joints between pieces of the mold and mold to sheathing should be pointed with plaster.

Before concrete is placed the waste molds must be greased lightly to aid in the removal and chipping of the plaster when the concrete has hardened.

Slope of Form Designs. Form liners of any kind of material except waste molds require draft or sloping sides so that they can be easily removed after the concrete has hardened. Sheet-metal forms need only a slight slope; wooden forms need a slightly greater draft of 12 to 15%; plaster forms should have still more. The deeper the pattern the greater the amount of slope that should be allowed in order that there will be minimum damage to the concrete or the forms. If strips of wood are attached to the liner or the sheathing to form geometric

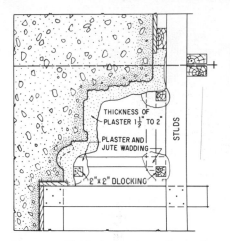

Fig. 11. Typical cross section of a plaster waste mold. (*Courtesy of Portland Cement Association.*)

Fig. 12. Back of a plaster waste mold that has deep relief showing supports in the hollow part of the mold. The mold is in position in the form and is supported by studs and braces. (*Courtesy of Portland Cement Association.*)

designs or the appearance of joints between panels, especially if the material is relatively thick and wide, one or more saw cuts in the back of the strips or boards will help to prevent warping and allow enough "play" or spring so the mold will not wedge too tightly because of swelling. Stripping will be easier and danger of spalling of the concrete lessened.

Exposed aggregate panels make an excellent architectural finish that can be varied by using aggregates of different sizes and colors. The size of coarse aggregate depends on esthetic values, including what might be called the "viewing distance." The viewing distance is the distance from the observer to the concrete surface under normal conditions and should be as shown in Table 1. Color is obtained

Table 1. Viewing Distances for Exposed Aggregate Panels

Distance, ft	Size of aggregate, in.
20 to 30	¼ to ½
30 to 75	½ to 1
75 to 125	1 to 2
125 to 175	2 to 3

by selecting coarse aggregate of different mineralogical compositions, or even by using crushed ceramic and glass materials.

Several methods have been used for making the panels, the easiest being to cast in a horizontal position, as in tilt-up construction (Fig. 13). For one method, the coarse aggregate is spread on the bottom of the form in a single layer with particles placed as close together as possible. Dry, fine sand is sprinkled uniformly over the aggregate and brushed with a soft-bristle brush through the openings until about one-third of the depth of the layer of aggregate is filled. An air hose to blow the sand into place is sometimes used. Finally the aggregate is moistened with a fine spray of water to remove dust from the aggregate and to firm the sand in place (Fig. 14). If the panel is to be white the sand bed should be made with white sand. The next step is to place the mortar over

Fig. 13. Hand-placing aggregate in the form for exposed-aggregate tilt-up panel.

the aggregate. A typical mortar mix is 1 part cement to 2½ parts of well-graded, clean sand finer than a No. 8 sieve, and enough water to make a creamy mix. If the panel is not being made on a vibrating table the mortar can be tamped gently to penetrate between the coarse-aggregate particles, using a narrow wooden tamping board but being careful not to disturb the coarse aggregate and sand bed. The reinforcing, with bolts and inserts attached, is then set in position and the backup of standard structural concrete is placed in the forms. The backup concrete can be vibrated with a surface vibrator or screed. A finished panel is shown in Fig. 15.

Fig. 14. A fine water spray removes dust from the aggregate and firms the sand bed. (*Courtesy of Portland Cement Association.*)

Fig. 15. A finished panel after the backup concrete has hardened and the sand bed has been washed away. (*Courtesy of Portland Cement Association.*)

A similar method that will accomplish much the same effect is to place a layer of fine sand on the bottom of the form to a depth of about one-third of the diameter of the coarse aggregate to be used. Particles of aggregate are pushed into the sand to about one-third their diameter and as close together as possible to obtain a dense coverage, but only one particle deep. For large areas and for large aggregate sizes the aggregates may be tamped into place. A fine spray of water is used to settle the sand around the aggregates so each particle is held firmly in place. As in the previously described procedure, mortar can be spread and worked into and around the bed of aggregates, or if the mortar at the face and that in the backup concrete are the same color, the reinforcement, bolts, and inserts can be placed on plastic or galvanized chairs and the concrete placed and vibrated, being careful not to disturb the bottom layer of aggregate, especially if a spud vibrator is used. Where the thickness of the panel will permit, it is advisable to place the concrete in two layers, inserting reinforcement, etc., on the first layer. Placing the facing mortar and the subsequent lifts of concrete should follow each other promptly to avoid cold joints.

The panels should be cured at least 5 days or until they have gained sufficient strength to be lifted and handled, after which they are raised and any sand clinging to the exposed-aggregate surface is removed by brushing, air blast, or washing with a stream of water.

In a third method, used especially where the slab is to remain horizontal as for floors, walks, patios, etc., the form is filled with standard concrete, after which the one-size coarse-aggregate particles are broadcast on the surface, tamped into the concrete until well embedded, and the excess mortar is broomed, brushed, and flushed away when the concrete has set just enough that the coarse-aggregate particles will not be dislodged. Proper curing methods for 7 days or longer should follow.

The one-size aggregate should be clean, washed aggregate free of fines that would interfere with a finished surface of uniform appearance. The size selected may vary from standard No. 6 (¾ to ⅜ in.), preferred for horizontal slab work such as sidewalks, to a massive effect given by rocks of 6- to 8-in. size. Standard sizes No. 2 (1½ to 2½ in.) and No. 3 (1 to 2 in.) are more generally used. Should it be desired to use flat pieces of stone, they should be not less than 1½ to 2 in. thick, placed on the bottom of the form, and fine dry sand placed in the space between the particles to a depth of one-third to one-half of the thickness of the stone. The rest of the procedure is the same as the first method described above.

Aggregate transfer can be accomplished with vertical forms by making form liners to which the surface aggregate particles have been glued. Such liners are prepared in a horizontal position, of course, usually on ¼-in. form plywood although sheet metal, cardboard, or heavy waterproofed paper can be used under some conditions, as for curved surfaces. After the liners have been cut to size and fitted, where necessary, the working surface is oiled lightly with a No. 10 motor oil or a good grade of form oil, after which some of the water-resistant adhesive is thinned with about an equal amount of an appropriate thinner and is brushed on the liners and allowed to dry 24 hr to provide a protective treatment for the liner. When dry the liner is coated with special adhesive and the aggregates are applied immediately so as not to allow the adhesive to form a skin and reduce adhesion. The adhesive must be water-resistant so that it will not be softened by wet concrete or by rain. It should be strong enough so aggregate particles will not be dislodged when the liners are handled or the concrete is placed, but it should not be so strong that it will damage the liners when they are stripped from the concrete, thus preventing their reuse. The most successful adhesive consists of nitrocellulose, dammer gum, and acetate. Thickness of the adhesive will vary depending on size of the aggregate being used and the type of finish desired.

The aggregates should be dry, and of as uniform size as it is possible to obtain, which may require the producer to use special screening methods at the plant. Thin, flat, and elongated pieces will not work satisfactorily. It is very important also that the aggregate particles be structurally strong so that they will pull away from the adhesive and remain embedded in the concrete when the liner is removed.

After the aggregate has been placed, the adhesive should be allowed to dry for at least 24 hr (longer in cool, damp weather) before the liner is used. Once prepared, the liners may be stored indefinitely while awaiting use and will improve with age. Should they, upon inspection, show areas that are not satisfactory they may be repaired by pressing aggregate into a fast-setting adhesive piece by piece.

When the liners are positioned they are fastened to the sheathing with ⅝-in. wire brads spaced about 6 in. on centers near the edges and on 16-in. centers elsewhere, being careful not to loosen the aggregate. Needless to say, placing of reinforcement in the forms must be done with great care to avoid damage to the liners.

Before concrete is placed, the forms, and any concrete previously placed and hardened, must be flushed with water with a moderate spray; then excess water in the forms is allowed to escape. The concrete is placed in approximately 12-in. lifts, vibrated sufficiently to assure that there will be no honeycomb, being very careful, however, not to let the vibrator spud touch the form liner at any time. The concrete should be placed through a chute or tremie to avoid abrasion of the liner aggregate.

When the concrete has hardened sufficiently the forms are removed, but the liners are left in place as long as operations will permit, but not less than about 5 days. Left in place, the liners aid in curing the concrete and will strip more readily. Stripping is started at a corner of the liner, prying it off with a beveled 2 × 4, not with hard, sharp tools that will mar the concrete surface. Further curing of the concrete should be provided until at least 14 days of proper curing have been obtained.

UNFORMED SURFACES

Horizontal surfaces require only side forms when on grade or other foundation, or both side and bottom forms for suspended floor slabs, roofs, or panels for tilt-up construction. Special attention to the immediate finishing of the exposed surface is essential if a satisfactory job is to be obtained. Several steps enter into this concrete work, and the success of each depends on the care exercised in the performance of the preceding operations.[4]

For all concrete slab work, the concrete should be placed and spread in a manner that prevents segregation. A properly designed mix is essential, and the consistency of every batch as it is placed must be the same or it will not be possible to secure a uniform quality and appearance of the finished slab. In general, a slump of about 2 in. will give best results with regard to strength and durability and, in the case of floor finishing, will avoid excessive waiting for evaporation of surface water before troweling. If the concrete is vibrated, a drier consistency, as low as ½-in. slump, can and should be used. Some aggregates or working conditions may make wetter consistencies desirable, but in any event slump should never exceed 4 in. for hand placement or 2 in. for vibrated.

5   Setting Forms

For highway or airport work accurate elevations and alignment require the use of surveyor's instruments, and even for interior floors of considerable area a wye level should be used. On small jobs a carpenter's level will be adequate. In this chapter only the special jobs of comparatively small size will be described.

6   Tools for Concrete Work

Workmen soon become familiar with methods that, for each individual, will be the most effective and laborsaving for handling concrete from time of placing to final curing of the finished product. Certain basic tools and machines are common to practically all, however, and include most of those described below, depending on the type of construction at hand.

FIG. 16. Hand tamper for compacting earth base for slab. (*Courtesy of Portland Cement Association.*)

FIG. 17. Using the straightedge to bring the concrete to the elevation of the grade points. (*Courtesy of Portland Cement Association.*)

Compactors are used for consolidating the soil, sand, crushed stone, gravel, cinders, or other base on which the concrete is to be placed. Small areas can be compacted with an 8- or 10-lb hand tamper that consists of a handle, about waist high, attached to a wooden or metal face about 8 in. square (Fig. 16). For large areas, mechanical tampers speed production and give better consolidation. One type of tamper consists essentially of a vertical ram that is operated by

FIG. 18. One type of job-made strike-off or screed.

compressed air. Compaction is accomplished by the tamping action of the ram. Another machine, usually driven by a gasoline engine, accomplishes compaction by high-intensity vibration achieved by several thousand blows per minute delivered through a sole plate with an area of 1 or 2 sq ft. The vibratory machine is especially suitable for noncohesive granular soils.

Straightedge or strike-off rod is usually a straight piece of 2 × 4-in. or 1 × 4-in. lumber with a ½ × 2-in. metal shoe strip attached to the bottom. It can be made entirely of metal, but in any case it must be rigid enough to produce a plane surface in the concrete being worked. Its length should be 1 ft or so more than the widest distance between the edge forms. It is the first tool used after placing concrete in the form and, by a sawing motion, tamping, and manipulation, brings the concrete surface to the proper grade, as shown in Figs. 17 and 18.

Hand tamper, or, as it is sometimes called, the "jitterbug," is used to compact concrete and remove entrapped air, especially in very low slump concrete. It

can be made of a metal grill about 6½ in. wide by 3 or 4 ft long attached to a sturdy U-shaped handle long enough so the operator can stand upright when tamping. This tool must be used sparingly and then only to force large particles of aggregate below the surface enough to allow smoothing of the surface with the darby or bull float that follows immediately. The jitterbug should not be used for bringing a thick layer of soupy mortar to the surface (Fig. 19).

Darby is a long, flat, rectangular piece of wood, aluminum, or magnesium 2½ to 6½ ft long and 3 or 4 in. wide, with a narrower handle attached to the top as seen in Fig. 20. It is used to eliminate high and low spots and irregularities left by the straightedge, thus preparing the surface for floating and troweling.

Bullfloat consists of a wood, aluminum, or magnesium blade 8 in. wide and 3½ to 5 ft long attached to a handle up to 16 ft long that permits coverage

Fig. 19. Hand tamper or "jitterbug" for compacting low-slump concrete.

of the slab without the operator's getting on the concrete. The bullfloat serves the same purpose as the darby but is especially useful on large slabs in open areas where there are no walls or posts to interfere with the handle, as shown in Fig. 21.

Edgers are used to produce a radius at the edge of a slab to improve appearance and to reduce danger of spalling. They are about 6 in. long, varying from 1½ to

Fig. 20.   A darby for smoothing concrete.

4 in. in width where they bear on the top surface of the slab, with a side lip of ⅛ to ⅝ in. The radius for the rounded edge can be anything from ⅛ to 1½ in. depending on the design of the tool.

Jointers or groovers are similar to edgers, but in addition to rounding the edges of adjacent slabs a jointer is used to cut a joint partly through fresh concrete to form a control joint or contraction joint. The cutting edge or bit may be from 3⁄16 to ¾ in. in depth.

Hand floats, used to prepare the surface for troweling, are made of wood, aluminum, or magnesium. Wood floats are 12, 15, or 18 in. long and 3½ or 4½ in. wide. Aluminum and magnesium floats are 12 or 16 in. long by 3½ in. wide.

Power floats, driven by electric motors or gasoline engines, have a rotating disk about 2 ft in diameter. The concrete must be stiff enough to support the

Fig. 21. Using the bullfloat for smoothing the concrete.

Fig. 22. A wooden hand float.

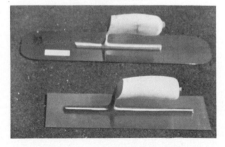

Fig. 23. Two types of steel finishing trowels.

weight of a man before the power float is used to compact the concrete. Hand floats supplement the power float in corners and wherever the power float does not reach.

Steel hand trowels come in sizes ranging from 10 to 20 in. long and 3 to 4¾ in. wide. The blades give better performance after they have been used enough to break them in by removing the sharp edges. For the first troweling of flatwork a long, wide trowel is generally used. For troweling the last few times, most cement masons prefer a "fanning" trowel 14 to 16 in. long and 3 to 4 in. wide. By the skillful use of the steel trowel the surface of the slab is given a dense, smooth finish.

Power trowel, driven by electric motor, gasoline engine, or air motor, consists of three or four steel blades attached to a rotating vertical shaft. Slope of the

FIG. 24.   A groover and an edger.

FIG. 25a.   A power float. (*Courtesy of Portland Cement Association.*)

FIG. 25b.   A power-operated troweling or finishing machine.

FIG. 26.   An electric saw fitted with abrasive blade for cutting joints. (*Courtesy of Portland Cement Association.*)

through a flexible-shaft drive from a gasoline engine or air motor. A shatterproof blade with diamond or other abrasive embedded in it produces a groove about ⅛ in. wide that minimizes spalling.

These hand tools are shown in Figs. 22, 23, and 24, and power float and trowel in Fig. 25. Figure 26 illustrates a joint cutter.

7   Placing and Striking Off

When the concrete is deposited in the area being concreted, it should be spread by means of hand shovels to a depth slightly more than called for by the plans, perhaps 1 or 2 in. It is then struck off by means of a heavy, rigid, strike-off board or template that has the contour of the desired finished surface. A sawing motion of the strike-off is used as it rests on the side forms. Plow handles attached

FIG. 27. Hand-operated vibrating strike-off board or screed sliding on side forms. (*Courtesy of Stow Mfg. Co.*)

FIG. 28. Vibrating screed riding on pipe rails above bridge deck. (*Courtesy of Stow Mfg. Co.*)

to each end of the strike-off board will make this job easier for two men, especially when the distance between the side forms is more than about 6 ft. If there are voids on the surface after the first passage of the strike-off, they are filled with more concrete and the striking off is repeated until a comparatively smooth and uniform surface results. Should the concrete mix be too dry, contain coarse-aggregate particles so large that they tend to tear the surface, or otherwise be difficult to work, the strike-off board can be used as a tamper to bring mortar to the top for later finishing. However, it is preferable to make adjustments in the proportions of the concrete so that excessive tamping will not be necessary. Avoid oversanded mixes that will reduce the strength of the concrete.

Strike-off boards, commonly referred to as screeds, can be operated by hand or by means of vibrators mounted on them to make vibrating screeds, as shown in Figs. 27 and 28. Mechanical vibration has the ability to consolidate relatively dry mixes rapidly and with much less effort over longer spans between side forms than could be done by hand. For still larger jobs the strike-off can be fitted with rollers to ride the forms, and if sufficient area makes it economically feasible, self-propelling motors can be attached to the rollers or wheels. Concrete should not be overworked, as this brings an excessive and unnecessary amount of mortar and water to the surface, resulting in poor resistance to abrasion and otherwise inferior quality at the surface where durability and abrasion resistance are especially needed.

8 Floating and Troweling

The next operation is to remove irregularities remaining after striking off, accomplished with a wooden float about 8 in. wide and 4 to 5 ft long, attached to a handle long enough so the workmen can reach easily to all points without

FIG. 29. Troweling with a steel or magnesium trowel when the concrete is almost dry and no free water is brought to the surface. (*Courtesy of Portland Cement Association.*)

stepping on the concrete. The float is attached at an angle that permits forcing down the high spots with the forward movement or cutting them down on the backward pull. Low spots are filled with concrete from the high spots or by adding a shovelful of fresh concrete and working it in with the float. Filling low spots entirely with soupy mortar is bad practice, resulting in a soft, nonuniform surface.

Instead of the float just described, a wooden float about 2 ft long by 4 or 5 in. wide with a handhold near one end, called a darby (Fig. 20), is used in places where it would be difficult to operate a long-handled float. The operator kneels on a wide board on the untroweled concrete while using a darby.

Troweling of the floated surface gives a smoother finish and should be done after all excess water has evaporated, thereby giving a harder and more abrasion-resistant surface. Troweling may be necessary to reach points not finished satisfactorily by the previous floating. If a float finish is acceptable a wooden trowel should be used to match the texture produced by the long-handled float. Troweling with a steel or magnesium float when the concrete is almost dry gives a very smooth finish (Fig. 29).

9 Textured Surfaces

A nonskid sandy texture is obtained by brushing the surface with a wide brush. Heavy brushing with stiffer bristles, or working softer concrete, produces increasingly rougher finishes. Deep brushing is usually required where there is danger that vehicles or pedestrians may slip on sloping roofs, walkways, revetments, and ramps. If an ornamental effect is desired, the brush marks can be drawn in alternating directions, zigzagged, or in such patterns as the architect or finisher may devise. The nature of the patterns will depend on the stiffness of the bristles, time elapsed after placing the concrete, consistency of the concrete when brushed, and the effort used in applying the brush.

Other texture variations can be produced by floating the surface. A smooth matte finish is attained by holding the steel trowel flat and finishing with swirling motions. Medium textures are produced with aluminum or magnesium floats, and coarse textures with wood, cork, or carpet floats.

In a climate where it seldom or never freezes, walkways, patios, hallways, and floors imitating travertine can be produced as follows: A finish coat is made by mixing 1 bag of white portland cement and 2 cu ft of sand with about ¼ lb of mineral oxide color pigment, usually yellow. Enough water is used to give a soupy mix of the consistency of thick paint. Proportions for every batch must be exactly the same. After the freshly prepared concrete slab has been broomed with a stiff-bristle broom to ensure bond, the mortar is thrown vigorously on the slab, using a dash brush to make an uneven surface with depressions and ridges about ¼ to ½ in. high. The mortar is allowed to harden enough to support a cement mason on it with a kneeboard, then is steel-troweled to flatten the ridges, leaving the slab surface rough or coarse-grained in the low spots. Depending on the amount of mortar thrown on the slab and the extent of troweling on the mortar coat, many attractive texture variations can be produced. After texturing, the slab may be scored into regular or random geometric designs, and then cured. A similar texture can be produced by scattering coarse rock salt over the smooth troweled fresh concrete surface and then pressing the salt into the concrete with a trowel. After the concrete has completely hardened, the salt is dissolved away by flooding with water, leaving pits and holes in the surface. Salt can be ordinary rock salt or water-softener salt. All the salt particles should pass through a ⅜-in. screen and at least 85% should be retained on the No. 8 screen. Salt should be spread at a rate of about 5 lb per 100 sq ft and embedded so the tops of the grains are barely exposed. For the geometric designs previously mentioned, there are numerous tools on the market that, when pressed into the soft mortar on the surface of a slab, will give a ribbed effect, a fan texture, brick and tile design, or other textures representing various flooring materials. Leaves and ferns, pressed into the freshly troweled surface and subsequently removed, provide interesting patterns.

10 Retarded Surface

An attractive finish results from treatment of a freshly finished concrete surface with a cement retarder (see Chap. 4). These retarders are marketed by a number of firms supplying concrete admixtures and related compounds. The manufacturer's directions must be followed exactly. For unformed surfaces the compound is sprayed or painted on the soft, freshly placed concrete at a uniform rate, and when the concrete below the surface has hardened sufficiently to resist dislodging the coarse-aggregate particles at the surface, the retarded concrete is brushed and washed away until the desired effect is attained (Figs. 30 and 31). Too much time should not elapse before the mortar is removed, since its hardening is only retarded and complete setting will take place eventually. The proper time can be determined by probing with a knife blade from time to time. Subsequent cleaning with muriatic acid removes the dull cement coating and gives a pleasing result. Special effects of texture and color can be obtained by selecting the color and size of particles that will give the desired appearance.

11 Exposed Aggregate

An exposed-aggregate effect can be obtained on horizontal surfaces by scattering selected, one-size aggregate on the fresh concrete that has been struck off and floated. The aggregate particles are tamped into the concrete with a float until

Fig. 30. Brushing and flushing away the mortar surface that has been treated with a set retardant to reveal the coarse aggregate. (*Courtesy of Portland Cement Association.*)

Fig. 31. Exposed-aggregate surface obtained by use of a surface retardant. Surface mortar has been brushed and washed away and the surface cleaned with acid. (*Courtesy of Portland Cement Association.*)

they are embedded to a depth that will assure their being held securely during later treatment. If a small-sized aggregate such as pea gravel is used it can be completely submerged. When the concrete has hardened enough to hold the embedded particles firmly in place, the mortar can be brushed and flushed away gently until the desired appearance is secured.

12 Colored Concrete

There are five methods of producing colored concrete finishes: (1) use of stains and dyes, (2) paints, (3) one-course concrete method, (4) two-course method, and (5) dry-shake procedure. The one-course method, the two-course method, and the dry-shake procedure give the best and most economical results.

Stains and Dyes. Stains are used normally to color an existing slab, either new or old, usually in interior locations where the surface can be protected by waxing. When using stains or dyes the surface must be clean and free of all foreign material such as hardeners, waterproofing, curing compounds, oil, or any substance that will prevent penetration. The concrete should be at least 6 weeks old, properly cured and allowed to dry thoroughly. Instructions furnished by the manufacturer must be followed in detail in order to assure a uniform color without streaks or spots.

Paint. A good appearance can be achieved with paint where the slab is not subjected to much traffic. A wide range of colors and types is available including oils, plastics, resins, gums, and cement. The condition of the concrete to be painted is very important, for the paint will not repair a dusting, spalling, or cracking concrete. The surface must be entirely free of dirt, water, oil, grease, and any substance to which the paint will not adhere. The concrete must be thoroughly cured prior to painting, and if an oil-base paint is used the alkalies in the concrete should be neutralized with a wash consisting of 2 to 3 lb of zinc sulfate crystals per gal of water, or a solution containing 2 lb of magnesium fluosilicate per gal of water. After 48 hr or more for the treatment to dry all protruding crystals should be removed by brushing. A preferred procedure is to allow the concrete to dry for several weeks after curing, then apply generously a solution of 3 oz of zinc chloride and 5 oz of orthophosphoric acid (85% phosphoric acid) per gal of water. After drying for 24 to 48 hr, the surface is brushed to remove dust but should not be rewetted before the paint is applied. If the surface is very smooth a better bond will result by etching it with a 10% solution of muriatic acid followed by a thorough washing and drying. Three coats of paint are necessary for even light traffic, and repainting will be required from time to time.

One-course Color. By adding pigment to the concrete in the mixer, the color is distributed throughout the entire slab thickness. To maintain a uniform color, proportioning of all materials in the mix must be carefully controlled by weight. Mixing time should be longer than usual to assure uniform distribution of the pigment. It is preferable to mix the dry cement and pigment thoroughly before adding to the aggregates in the mixer.

The pigment must be a pure mineral oxide especially prepared for use in concrete as described in Chap. 4. Both synthetic and natural pigments are satisfactory if they are insoluble in water, free of soluble salts and acids, and fadeproof to sunlight. They should be inert to alkali and weak acids, should have only limited amounts of calcium sulfate, and should have at least 90% passing the 325-mesh sieve. Generally 3 lb of color per 100 lb of cement will produce light shades. Medium shades require 5 lb per 100 lb and deep shades 7 lb. Extra-dark shades will take a maximum of 10 lb per 100 lb of cement. It is advisable to make test mixtures before starting work so as to use the least pigment for the color desired since pigment is not only expensive but reduces strength of the concrete to some extent.

Bleeding can cause variations in pigment concentration and spotty coloring; hence the least water should be used for needed workability. The use of air-entrained concrete will minimize bleeding.

White portland cement will produce cleaner, brighter colors and should be used in preference to normal gray portland cement, except for shades of gray secured by adding black pigment.

Two-course Color. Because of the economies it offers, a two-course system is frequently used, in which the base course is mixed and placed in the normal way and allowed to stiffen. After the surface water has evaporated a topping of ½ to 1 in. of colored concrete is placed using the appropriate size of coarse aggregate.

Dry Shake. Also referred to as the dust-on method, a dry shake is more economical than either the one-course or two-course method, since only the top ⅛ to ¼ in. of the slab is colored. It usually produces more brilliant colors because there is no dilution by surface water. The dry material, usually purchased ready to dust on, is composed of properly proportioned mineral oxide pigment, white portland cement, and specially graded silica sand.

After the concrete has been struck off and darbied, and free water and excess moisture have evaporated, the surface is floated by either a power or hand float. If by hand, a magnesium or aluminum float should be used. Preliminary floating should be done before the dry-shake material is applied so as to bring enough moisture to the surface for combining with the dry material. Immediately after floating, the dry material is shaken evenly by hand over the surface. If too much color is applied in one spot the color will not be even and there is danger of peeling. About two-thirds of the amount of dry material specified by the manufacturer is sifted through the fingers while shaking low over the surface.

When the material has absorbed enough water from the slab to be thoroughly damp, a wooden float is used to float the color into the surface. All tooled edges and joints should be run before and after each dust-on. Avoid overworking. Use of a power float is preferred. After the first application of dry-shake material has been floated into the surface, the rest of the material is distributed as before, taking care to distribute it so as to produce an even color and texture, and immediately floated into the surface. Never use additional water.

As the concrete begins to stiffen it should be steel-troweled, either by hand or, preferably, by machine until the characteristic ring of the steel trowel is heard and the desired finish secured.

Curing should be begun immediately. A procedure that has been found to be especially satisfactory is the application by roller coater of a waxing-curing compound that has been formulated especially for this purpose. After thorough curing and drying, it is recommended that the concrete be washed and polished with wax prepared especially for the purpose to bring out the color.

FLOORS

A floor finish can be applied to the top of a slab, in which case it is called a monolithic floor, or a special topping mix can be applied to the base concrete, either while the latter is still plastic or after it has hardened.[5]

13　Preparation of Base or Subgrade

A floor may be placed on grade where the base is a soil free of organic matter, sticky clay, muck, or other spongy or highly expansive materials, or it can be placed on an insulating layer of free-draining clean sand, stone screenings, crushed stone, graded gravel, or cinders 6 in. thick. Frequently, a porous base is covered with saturated felt or polyethylene film, known as a vapor barrier, to prevent movement of water from the subgrade through the slab to the floor surface. Occasionally, a new concrete floor is placed over or in place of an old floor that has become unserviceable.

A firm and stable base or subgrade is necessary for a concrete slab on ground if it is to give long and useful service without excessive maintenance. This requires removal of substances of unsuitable characteristics, scarifying to a depth of at least 6 in., and rolling or tamping the acceptable soil until it is firm. When the soil is adequately compacted, footprints and wheelbarrow tracks will not make appreciable indentations. Tamping of the insulating layer must be done with

FIG. 32.　Spreading and compacting concrete. Concrete is discharged from the mixer on the grade between forms and spread with shovels to elevation slightly above finished floor. A hand-operated power screed compacts the concrete to the required elevation. (*Courtesy of R. J. Anderson & Co.*)

the same care as required for soil. However, it is usually necessary to dampen the base to lend stability to clean granular aggregates while tamping or rolling. If the area will be subject to percolation of water at any time, regardless of whether it is an interior floor or exterior pavement, underdrains should be provided to carry the water away. Farm drain tiles are effective where heavy loads will not be imposed on the slab; otherwise it is advisable to use perforated pipe installed in sloped ditches backfilled with porous material such as coarse sand or screenings, pea gravel, and cinders.

If the base dries out before concrete is placed on it, it should be saturated with water well ahead of the time that concrete placing starts. The base should be damp but free of puddles and mud when concrete is placed on it.

Grade is established while the subgrade is being prepared. Side forms are set for the final elevation of the floor. For areas wider than a strike-off board can span, intermediate stakes or screeds are placed. The grade is checked against the side forms by means of a template that hangs from the forms or screeds. As the template slides along, it shows the high and low spots that then can be cut down with a shovel or filled in and retamped. An alternative method, for a flat surface, is to stretch a string line between the forms and measure down from the string to establish grade. Just before placing the concrete the base is sprinkled with a hose only sufficient to prevent the base from drawing water out of the concrete and not so much that puddles or muddy spots are formed.

Figure 32 is a view of heavy-duty floor construction.

Floor forms above grade, as in multiple-storied buildings, should be prepared according to instructions given by the plans and following recommendations in Sec. 4. Reinforcing must be held in place rigidly, all trash removed from the forms, the forms oiled and, if called for, sprayed with water before concreting commences. However, any accumulation of water in the forms must be removed.

Old floors allowed to stay in place can be resurfaced with a thin overlay of new concrete. Best results are obtained with an overlay at least 2 in. thick, although overlays as thin as 1 in. have been successfully completed. The old surface must be thoroughly cleaned so that a bond will be secured to it by the new surface. Grease and oil can be removed by scrubbing with a detergent, paint must be chipped, ground, or sandblasted, and all loose or unsound areas removed by brooming, chipping, or other effective means that will result in a sound surface to which the new concrete will adhere. If the new surface must have the same elevation as the old, the old concrete must be chipped away to a depth of at least 1 in. Repair of scaled concrete is discussed in Sec. 14.

14   Placing Concrete

Proportions for the ingredients of suitable mixtures for floors and surfaces should include only enough water to make the concrete workable. For machine floating, the water should be kept at not more than 33 lb per 100 lb of cement, and for hand floating, not more than 5 gal (42 lb).

Concrete must be proportioned in accordance with the procedures described in Chap. 11. When correctly proportioned, concrete for the base course should have a 28-day compressive strength in excess of 3,500 psi if it was properly mixed, placed, finished, and cured. A surfacing concrete will have a strength in excess of 4,000 psi.

Spreading and compacting the base course follows usual practice for placing concrete except that the base is struck off at least 1 in. below the finish grade. The grade must be protected from damage when depositing the concrete by board runways for wheelbarrows or buggies. Spreading should be done with shovels, although for low-slump concrete it is permissible to use a hoe or the back of a rake for leveling or moving the concrete short distances. Concrete is spread to slightly above required elevation for the base and then leveled with a straight-edge. Compaction is obtained by vibrating screed, hand tamper, or if care is exercised to avoid disturbing the base or overvibrating, an internal spud vibrator can be used. The internal vibrator is not recommended for depths of less than 8 in.

When the base course has stiffened enough to bear the weight of a man without distortion, it is brushed with a stiff-bristled broom to remove scum and laitance and to assure a rough surface to provide mechanical bond. The surfacing concrete can be placed at this time while the base concrete is plastic enough to bond readily with the topping, or the base can be wet-cured for at least 5 days before proceeding with the surfacing course. The latter procedure is preferred in building construction so as to protect a carefully finished final surface from damage during other construction operations.

A **bonded floor** will result when the topping is completely bonded to the base. Even though the base has hardened before the topping is applied a good bond can be had by scrubbing the base with plenty of clean water and a stiff brush and removing all foreign particles. If the base has dried out since it was placed, it must be wetted for several hours, preferably overnight; then free water is removed before application of a slush coat of cement and water mixed to the consistency of thick paint. Best results are obtained when the slush coat is applied by vigorous scrubbing, followed by brushing to level it to uniform thickness. It should not become dry before the topping concrete is placed on it.

Old concrete floors that are used as a base should be cleaned as previously described and saturated with water overnight. Wire reinforcement, weighing not less than 30 lb per 100 sq ft, should be used to control cracking. The reinforcing is laid on the old floor as flat as possible and lifted very slightly after the topping concrete has been placed on it, being careful that none of the reinforcing wires is closer than 1 in. from the finished surface.

15   Placing and Compacting the Topping

The freshly mixed topping concrete is deposited so as to require a minimum of movement when spreading over the base. Shovels or rakes are used for spreading to a depth slightly above the required elevation, about ½ in. for a 2-in. thickness. Tampers or rollers or both are used to consolidate the concrete to slightly above the required level. For small areas a straightedge is used to bring the concrete to finish grade; for larger surfaces a vibrating screed is more efficient. Special care and attention must be given to corners, edges against walls, and around columns or other obstacles not easily compacted with mechanical equipment.

A level or flat surface on narrow walkways or floors of less than 8-ft width is quite simple to refer to form boards at the sides, but for larger areas it is necessary to use reference points spaced 8 to 10 ft apart in two directions. The reference points can be steel pins driven to elevation using a transit or level for accuracy, or concrete blocks precast to useful dimensions about 2 in. square set in the mortar with tops at the finished floor level. Having established the reference points, the straightedge is laid over two of the reference points and sawed downward into the fresh concrete so as to show a line of compacted mortar that becomes a reference line. Additional reference lines are made closer together using the previously established straightedge marks so that the entire area is marked off into rectangles with sides 4 or 5 ft long. Steel pins and concrete blocks must be removed and the holes tamped full of concrete; pins or concrete blocks should not be left in place.

Excess concrete above the elevation of the guidelines is removed by means of a scraper made about 5 ft long from a ¾ × 5½-in. board, beveled on the bottom and faced with a strip of steel to provide a cutting edge.

Floating, whether by hand or by machine, is for the purpose of compacting the concrete and bringing enough mortar to the surface to permit troweling, at the same time smoothing out irregularities. The concrete for hand floating is a wetter mixture than that for mechanical floating, but in either case the operation is continued only long enough to fill the surface with mortar and provide enough excess above the coarse aggregate to give a smooth texture.

Troweling with either hand or mechanical trowels must be done at the proper time to avoid troweling a soupy surface that will surely result in a finished floor that has poor resistance to abrasion and will be dusty and generally unsatisfactory. When all free water and water sheen have disappeared and the concrete has begun to stiffen, troweling should begin. If, however, more water and cement are brought to the surface the operation should be further delayed.

The concrete mix designed for mechanical floating will usually be dry enough that troweling can be started immediately after floating. The hand-floated mix is wetter and will require a waiting period. Under no circumstances should dry cement or cement and sand mixture be dusted on the too-wet surface for the purpose of speeding the drying, nor is it permissible to sprinkle water on a floated surface that appears to be too dry for easy troweling; either addition will reduce the strength of the final surface and cause scaling, crazing, and poor performance as well as poor appearance.

The first troweling is for the purpose of removing any marks, swirls, or ridges left by the floating operation. For the first troweling the blade should be held as flat against the surface as possible to avoid creating washboarding or chatter marks. As the surface stiffens or hardens, smaller trowels are used to assist the cement mason to exert more pressure as needed. Final troweling is done after the concrete is hard enough that the steel trowel remains clean and makes a ringing sound as it is swept over the surface. If the final surface is too smooth and dangerously slippery for its intended use, it can be given a satin, nonslip finish by drawing a soft-bristled push broom over it after troweling. Coarser textures for steeper slopes or walks, corridors, ramps, etc., are obtained by using brooms with stiffer bristles.

Air-entrained concrete topping does not bleed so much as non-air-entrained concrete, and the floating and troweling operations must be performed earlier before the surface becomes too dry or tacky. An aluminum or magnesium float and not a wood float is essential because of the somewhat sticky nature of air-entrained concrete. Power floating is not affected by the air entrainment other than that it can be started earlier.

Curing must follow final troweling as soon as possible to avoid drying of the concrete surface that causes a dusting surface subject to rapid wear. A number of curing methods are discussed in Chap. 29. By whatever means, the concrete must be kept from drying during the first 3 days as a minimum. Interior floors will not dry so quickly as those outdoors, and it will often be found that frequent wetting with the fine spray of a hose not only will be adequate but will avoid possible discoloration or scarring of the finish that might occur if cover materials such as burlap, paper, plastic film, and sand are used as moist covers. Liquid membrane-forming compounds are good curing for many kinds of concrete work, but for interior floors they create a problem of removal since they are, generally, not intended to be a permanent coating. If color treatment or other applications are to follow (including application of adhesive for a floor covering), the compound must be completely removed, requiring wire brushing or grinding that may still not leave the surface in an acceptable condition. Polyethylene film of any thickness from ½ to 4 mils, tightly sealed at the edges of the slab with sand, boards, or similar weights, will seal in the moisture but should not be applied until the floor has been thoroughly wet by at least one sprinkling from a hose. Waterproof paper blankets serve the same purpose as plastic-sheet material. When burlap is used, it must be washed free of all filler or contamination to prevent permanent disfigurement and must be kept wet by sprinkling to prevent the burlap from becoming dry. Continuous curing for 10 days or longer will assure attainment of maximum strength. The same comments apply to outdoor curing, except that closer attention will be necessary to prevent drying.

Temperature conditions play an important part in curing, and whether inside or outside, protection should be provided to maintain a temperature between 60 and 80°F in the concrete. Enclosed spaces can be heated with space heaters that do not create carbon dioxide, or if they do, the gases must be vented to the outside. Carbon dioxide has been found to combine with the alkali in fresh concrete, resulting in carbonates in the surface that have no wear resistance and will cause a variety of unsatisfactory conditions. Flambeaux, salamanders, and heaters that rest on the floor usually make hot spots of uneven temperatures that are likely to be hotter than the concrete can tolerate safely, and should be avoided. If such heaters must be used they should be insulated from the floor or placed where they cannot do any damage.

Outdoor work in cold weather should be covered with an insulating layer of straw or similar porous material with a waterproof sheet material over it to hold the original heat of the concrete plus the heat of hydration of the cement in the slab. If temperatures drop very low, artificial heat from radiant heaters may be needed. Curing under cold-weather conditions should be carried on for longer periods to accommodate the slower gain in strength and quality at the lower temperatures. In hot weather the outdoor slab will be protected from excessive heat if a white or light-colored reflective covering is used for the curing sheet material or membrane. See Chap. 30 for precautions for hot or cold weather.

Jointing and edging have not been mentioned as a part of the floor-finishing operations, but there are many occasions when they are required. Use of the jointer or edger is similar in that these tools are used immediately before and after floating, being careful not to force the tools so deeply into the concrete that the face of the tool is below the finished floor surface. It is wise to use a board as a straightedge to guide the jointing or grooving tool, but the edger will follow the side form. If less obvious joints are desired, a concrete saw with a narrow blade will usually prove satisfactory. Time for sawing depends on the type and hardness of the aggregate in the concrete and on the strength that has developed in the concrete at the time of sawing. Experimenting with the time for cutting will show when a clean cut can be secured—usually from 4 to 12 hr after the concrete has hardened.

Resilient Floor Coverings. Frequently floors are to be covered with asphalt or vinyl tile, linoleum, carpet, or other decorative and utilitarian material. Most of these materials are cemented to the concrete with an adhesive of some kind. When floor coverings of this nature are to be installed, it is important that no material is used on the concrete that might interfere with adhesion of the covering. Curing compounds and tilt-up parting compounds should not be of a greasy or oily nature that interferes with the adhesive bond to the concrete, nor should they saponify in the presence of the alkalies that exist in concrete. Low-volatility solvents that might still be present when the floor covering is installed can be a source of trouble. Any compound that remains on the concrete surface at the time of application of the covering either must be easily removed or should remain as a harmless film that will not affect the adhesive or the covering.

16   Heavy-duty Surfacing

The preceding instructions will result in a floor that will give good service under heavy traffic; however, when it is known that exceptionally heavy and abrasive loads and service conditions will be encountered, there are further considerations that will assure even better service.

The coarse aggregate has much to do with wear resistance of concrete subjected to steel-wheeled warehouse trucks, abrasion of materials dragged over the surface, and similar conditions. Laboratory tests using the Los Angeles abrasion machine (ASTM Methods C 131 and C 535) give a good indication of quality for wear resistance, a percent of wear of less than 35 being preferred. Generally, siliceous aggregates such as granites are less subject to wear than calcareous aggregates such as limestones, dolomites, and marble. Tough sandstones and quartzites are likely to have superior wearing qualities. Siliceous sands of low mica content should be used in preference to fine aggregate composed of soft, friable, and weak grains.

Strength and other desirable properties of the concrete are related to the water-cement ratio; hence any action that helps to minimize total water requirement, such as good aggregate grading, will serve to improve these desirable properties.

Floor topping ¾ to 1 in. thick, applied the same day as the bottom or base portion of the slab, makes possible the selection of special materials with less expense. The base is struck off about ¾ or 1 in. below the finish elevation and allowed to harden without troweling. If the topping is not applied while the base is still green enough to assure a good bond, care must be exercised to see that the base is scrubbed and moist when the topping is spread on it. This can be several days or longer.

Mix proportions for the topping are 1 part cement, 1 part fine aggregate, and 1½ to 2 parts of coarse aggregate by volume. Batch proportioning should be by weight. Only sufficient water should be used to provide a very stiff consistency that can just be worked with a straightedge, certainly not more than 33 lb of water per 100 lb of cement. The topping mix should form a ball when squeezed in the hand, with about 0- to ½-in. slump.

When the base has been properly prepared, a thin neat grout having been broomed into the area to be topped if the base is more than 1 day old, the topping is placed, straightedged, tamped, or rolled, taking care that there are no cold joints where successive batches of topping come together. Floating, preferably with a power float, is commenced as soon as the topping can be worked without bringing free water to the surface. A straightedge is used occasionally to detect high or low spots which are removed by scraping with a metal straightedge or adding a small amount of fresh topping to a low spot. The difference in temperature between the base slab and the topping should not be more than 10°F. Any joints that have been constructed in the base must be continued through the top. After the desired finish has been secured, the slab must be properly cured for at least 7 days.

Dry Shake. A dry shake or dust coat is applied to a one-course floor to improve its resistance to abrasion and impact.

Abrasive material for this usage consists of especially hard and durable particles, such as aluminum oxide, silicon carbide, or malleable-iron particles, the size of particles depending upon traffic conditions and type of finish desired. Particles are seldom larger than those passing a No. 8 screen or smaller than those retained on a 60-mesh screen. Because of rusting of the iron exposed on the surface, iron aggregate should not be used outdoors, or where the floor is exposed to moist conditions.

In using a dry shake, recommendations of the manufacturer should be followed. Lacking such recommendations, a mixture is made consisting of 2 parts dry abrasive and 1 part cement. The amount of shake to be used depends upon anticipated traffic conditions but ranges between ½ and 1½ lb of aggregate per sq ft of floor. About two-thirds of the required quantity of shake should be broadcast over the surface immediately before power floating. Considerable care is necessary to spread the material evenly over the slab. After the material has been blended into the surface with the power float, the remaining material is spread, taking care to make the complete coating as uniform as possible; then the surface is power-floated again, followed by machine troweling, then hand troweling. No water should be used during any of these finishing operations.

The cement and special aggregate mixture must not be placed on the concrete base too soon, as this results in some of the aggregate's being worked beneath the surface of the concrete, with a thin layer of cement and water on top. This surface cement water paste will scale or peel later. Overworking of the surface material might also contribute to scaling.

Curing must be performed in the usual manner for a good, durable floor.

Armored floors are those which have embedded in them some material of greater wear resistance than is possible with portland-cement concrete alone. Factory floors, unloading platforms, floors subjected to heavily loaded steel-wheeled carts, trucks, etc., sometimes have surface reinforcement of steel grids, assemblies of steel strips bolted together or welded in various designs so that they remain firm in the concrete under impact, or specially prepared and graded iron aggregate that is applied in accordance with the manufacturer's specifications. This last is usually mixed with portland cement and dusted on the freshly floated floor and troweled in the usual manner.

Liquid hardeners will improve the durability of a concrete floor surface but cannot be depended on to correct poor quality resulting from improper construction and curing. Some of the solutions that are applied to harden the finished surface contain magnesium fluosilicate, zinc fluosilicate, sodium silicate, aluminum sulfate, chinawood oil, linseed oil, and various gums, resins, and paraffins. The floor must be clean and free of plaster, paint, oil, or other foreign substances, and should be as dry as feasible to permit good penetration. For preliminary estimating purposes it may be assumed that 1 gal of the liquid will treat 150 to 200 sq ft for each application.

Fluosilicates of zinc and magnesium seem to be preferred, and either will give good results. A mixture of 20% zinc fluosilicate and 80% magnesium fluosilicate has been found to perform better than either of the salts alone. The solution is made with ½ lb of the fluosilicate per gal of water for the first application and 2 lb per gal for subsequent applications. It is applied with a mop or sprinkling can and spread evenly to wet the entire surface. Each coat is allowed to dry before the next is applied. Two or more coats should be used depending on the porosity of the concrete. After the last coat has dried the floor is mopped with water to dissolve and remove any crystals that would cause white stains if allowed to remain.

Sodium silicate (commercial water glass) comes in about 40% solution which is too viscous for good penetration; therefore, it is diluted with water at a rate of about 1 gal of silicate to 3 gal of water. Two or three coats are scrubbed in with stiff fiber brushes or floor-scrubbing machines, allowing each coat to dry before the next is applied.

Aluminum sulfate is difficult to put in solution. About 2½ lb of the powdered sulfate per gal of water is mixed in a nonmetallic container. The solution is acidulated by adding 1 teaspoonful ($\frac{1}{16}$ fl oz) of commercial sulfuric acid for each gallon of water. Stirring occasionally for 2 or 3 days may be required to put the sulfate completely into solution. The first treatment should be diluted with 2 parts water to 1 part solution. Later applications should be full strength at intervals of at least 24 hr.

Zinc sulfate treatment will darken the color of the floor somewhat. The solution is made with 1½ lb of zinc sulfate and 1 teaspoonful of commercial sulfuric acid per gal of water, in a nonmetallic container. Two coats are applied, full strength, the second about 4 hr after the first. Between coats the surface should be scrubbed with hot water and mopped dry.

Oil treatments such as those with chinawood, linseed, or soybean oil consist of about equal parts of the oil cut with gasoline, Varsol, naphtha, or turpentine, applied with a mop or large brush. It is necessary to renew the treatment about every 6 months, depending on traffic, taking care not to apply too much on the later applications that would not penetrate but create a disagreeable slippery and tracking condition for several days. Either raw linseed oil or the more rapid drying boiled oil can be used.

17   Terrazzo Floors

Terrazzo is a method of floor construction and finish in which specially selected coarse aggregate is embedded in colored mortar. After the concrete has hardened, the surface is ground and polished. Designs are obtained by using metal or plastic strips to separate one color from another.

Selection of special tough and hard natural aggregate and the use of mortar of low water-cement ratio assure a floor that will have exceptionally good wear resistance.

It is generally recognized that terrazzo floors are the highest type of concrete floor if the combination of decorative effect, durability, and ease of maintenance are the criteria. A terrazzo floor offers greater opportunity of color and artistic design than any other concrete floor finish. As in all floor construction, however, quality of workmanship is the primary factor governing the quality of the floor.

Stairs, ramps, coves, bases, and wainscots are sometimes made in terrazzo. However, these are more readily precast in a shop equipped to handle the forming,

Table 2. Nomenclature °

Acetylene carbon black: Used as the vehicle to conduct electricity through a floor system. The commonest agent used in hospital floors today.

Admixes: Materials used and mixed with the terrazzo topping, i.e., epoxy, latex, etc.   *Note.* Normal terrazzo does not contain any admixes, and the use of such materials is not covered in this text.

Aggregate: A granule used in the topping other than marble, i.e., abrasives.   For the purpose of this text, marble shall be referred to as chips.

Art marble: Same as artificial marble, precast terrazzo.   Terrazzo that is fabricated in a shop (see Precast).

Bonding agents: Materials generally applied to "thin-set" terrazzo, i.e., latex, epoxy, Thiokol, etc.   Used to aid or assist adherence of the terrazzo mix to an existing base slab.

Broken marble: Fractured slabs of marble (not crushed by machine into chips).

Byzantine mosaics: The art of mosaic in vitreous material.

Chips: Marble granules screened to various sizes.

Cleaners: A neutral soap used to remove any accumulated surface dirt.

Color pigments: Matter used in the terrazzo mix to vary the color.   A powdered substance which, when blended with a liquid vehicle, gives the cement its coloring.

Divider strips: Divider strips used in terrazzo floors are made of half-hard brass, white alloy zinc (99 % zinc), and plastic in various colors.   The thickness of the strips is specified using Brown & Sharpe gages or fractions of an inch.   Strips less than $\frac{1}{8}$ in. thick are made of uniform thickness for their entire depth.   Strips $\frac{1}{8}$ in. and thicker are made of the "heavy-top" type with the top member having a minimum wearing depth of $\frac{1}{4}$ in. and a thin bottom member.

Flats: Along with flakes, this refers to marble chips.   A shale of marble produced when crushing.

Grouting: A matrix applied to the floor to fill the voids and pits after rough grinding.

NTMA: National Terrazzo & Mosaic Association.

Oxalic acid: An acid sometimes used to give ornamental and precast terrazzo a highly honed finish.

Oxychloride terrazzo: A flooring resembling portland-cement terrazzo but which incorporates oxychloride cement.

Panels: The spaces formed by the division strips.

Plate numbers: Numbers appearing adjacent to the various terrazzo illustrations in the NTMA's color catalog.   Often used in lieu of formulas in specifications.

Precast: Terrazzo fabricated in a shop or factory by a compression and vibratory method in watertight molds.

Sand: Small noncoherent rock particles.

Screed: The act of pulling the mortar level.

Sealer: A protective coating or treatment applied to preclude absorption of any foreign liquid or matter by closing off the pores of the grout.

Setting bed: Related to vertical surfaces.   Consists of mortar and used as a bond, base, and level for the terrazzo topping.

Surfacing: The grinding, grouting, and finishing operations done to the terrazzo topping.

Tesserae: Thin slices of marble, colorful stone, or a glasslike highly colored vitreous enamel material cut into squares or other shapes of any size.   Used in mosaic work.

Topping: The wearing surface of the terrazzo floor (see Types of Terrazzo).

Underbed: Related to flooring   Consists of mortar and is placed over the slab as a base, bond, and level for the terrazzo topping, as well as a bed for the division strips.

Venetian mosaic: The same as Byzantine, but because it flourished in Venice it took its name to distinguish it from the mosaics that are formed by tesserae of marble.

   Note.   The definitions listed here are used to clarify further the terms used in the terrazzo trade, and are not necessarily interchangeable with similar terms of other trades.

---

   * Adapted from "Specifications and Technical Data," National Terrazzo & Mosaic Association, Inc.[6]

---

curing, grinding, polishing, and other treatments that are required. Production of these specialty products requires close contact with the field operations for the job so that colors, textures, finishes, and dimensions will match.

   There are several terms, listed in Table 2, that have a special meaning when used in the terrazzo trade.[6]

Types of Terrazzo. Several hundred years of experience have led to the development of a number of variations in the methods for making a terrazzo floor, some of which are considerably more economical than others. Some of the types are described briefly as follows:

1. *Sand-cushion terrazzo.* The oldest form, and still considered the best, consists of a sand cushion about ¼ to 1 in. deep placed on the structural base, followed by a mortar bed approximately 2 in. thick with a terrazzo mix ½ to ¾ in. thick on top. Saturated felt is spread over the level sand bed, the mortar is spread uniformly to the required depth on the paper, and while the mortar is still plastic, dividing strips are set in place according to plans, and the terrazzo mix is placed in the spaces formed by the dividing strips. Additional terrazzo chips are spread and worked in with rollers, after which floating and troweling proceed, followed by curing and grinding.

2. *Bonded terrazzo* uses no sand cushion. Instead, a thin grout of neat cement is broomed thoroughly over the cleaned structural base to aid in obtaining a complete bond with the mortar underbed, which is then applied to a thickness of about 1 in. The terrazzo mix is placed on the plastic underbed and completed as indicated in the preceding paragraph.

3. *Monolithic terrazzo* consists of a ⅝-in. layer of terrazzo bonded directly to the supporting concrete slab, using a slush coat of cement and water to obtain bond. This method is the most economical, but it has the disadvantage of providing little opportunity to control random cracking.

4. *Chemical adhesive* used, instead of portland cement, to secure a bond between the concrete substrate and the terrazzo topping.

5. *Terrazzo topping* bonded to substrate with a flexible adhesive such as neoprene polysulfide and epoxy-polysulfide adhesives.

6. *Synthetic matrices* for the terrazzo topping instead of portland cement. These synthetic resins or polymers may be epoxy or neoprene. They have the advantage of a very high tensile strength, permitting the topping to be feathered to thin edges as thin as ⅛ in. They are inherently skid-resistant.

7. *Special terrazzos* include *Venetian* terrazzo, which is a standard terrazzo with large aggregate chips seeded into the surface. Usually thicker toppings and deeper dividers are required because of the aggregate size. *Palladiana* or *Berliner* contains large pieces of marble, up to 140 sq in. in area with standard terrazzo dividers. *Conductive* terrazzo, used to eliminate the danger of sparks in explosive atmospheres such as in operating rooms, contains a paste of acetylene black and isopropyl alcohol which is added to the underbed and topping mixes.

Terrazzo Construction. Terrazzo floor construction is considered to be a special art within the portland-cement concrete industry. The tools, equipment, labor, and expert judgment in the successive steps and the selection of materials usually lead the builder to call on the terrazzo flooring contractor for a guarantee of a satisfactory job. Where the expert is not available, however, success can be obtained by following instructions, specifications, and general knowledge of the behavior of concrete.

After portland-cement terrazzo floors are finished they are often given a treatment of hardener, delayed 30 days or more to ensure against any chemical reaction between the cement and some types of treatment. Waxing, sometimes with a special colored wax, while usually considered to be of little value, does have advantages for certain situations such as dance floors, lobbies, locker rooms, and areas subject to travel by the public and needing frequent cleaning. Several coats of paste wax are usually needed, after which paste or liquid wax can be used.

Sand-cushion Terrazzo. Beginning with the structural slab as a base, a uniform layer of fine sand is spread, varying in thickness depending on the condition and irregularities of the surface of the structural slab. The sand layer is about ¼ in. thick when the base is smooth and level, increased to as much as 1¼ in. for structural floors having cracks, ridges, and depressions of considerable magnitude. A heavy grade of saturated building felt is laid over the sand carefully and lapped at all edges at least 2 in.

The mortar bed consists of a mixture of 4 parts of clean, coarse concrete sand and 1 part portland cement by weight with only enough water to cause the mixture to cling together loosely when squeezed in the hand. This mixture is spread on the paper to a depth sufficient to bring the underbed to within $\frac{1}{2}$ to $\frac{3}{4}$ in. below the finished surface elevation. The mortar bed should be reinforced with 4×4 No. 14 wire mesh.

Division strips are set in the underbed for crack control or for ornamental effect (Fig. 33). Strips are of brass, stainless steel, nickel silver, zinc, or plastic, of 12 to 18 gage and from 1 to $1\frac{1}{8}$ in. wide. Wider strips of the "heavy-top" type are available where the joints are to be more than $\frac{1}{8}$ in. wide. The strips are set in the mortar bed according to the design, preferably not more than 4 ft apart, to keep any shrinkage openings that might occur adjacent to the strips to a width small enough so they will not be observable and will not accumulate foreign matter. Strips are set to a height of $\frac{1}{32}$ in. above the level of the finished floor.

After the bed has cured for at least 1 day, it is saturated and washed with water, the excess water is removed, and a neat cement slurry is broomed into the surface.

The terrazzo topping mix is composed of 1 sack portland cement, 200 lb of the terrazzo aggregate, and color pigment in the amount required to give the desired shade. Use only pure mineral pigments that are nonfading and lime-resistant, and weigh the amount of pigment for every batch. Where clear colors are important, white portland cement will be necessary. Terrazzo aggregate is generally marble chips graded according to sizes adopted by the National Terrazzo and Mosaic Association as shown in Table 3. Chips should be crushed uniformly so that all dimensions are reasonably close to the recommended size and with a minimum of flat, flaky, or elongated chips.

Table 3. Terrazzo Aggregate Sizes

Chip size No.	Passes through screen, in.	Retained on screen, in.
0	$\frac{1}{8}$	$\frac{1}{16}$
1	$\frac{1}{4}$	$\frac{1}{8}$
2	$\frac{3}{8}$	$\frac{1}{4}$
3	$\frac{1}{2}$	$\frac{3}{8}$
4	$\frac{5}{8}$	$\frac{1}{2}$
5	$\frac{3}{4}$	$\frac{5}{8}$
6	$\frac{7}{8}$	$\frac{3}{4}$
7	1	$\frac{7}{8}$
8	$1\frac{1}{8}$	1
For Venetian terrazzo the larger chips should be used in the following groups:		
No. 4 plus No. 5	$\frac{3}{4}$	$\frac{1}{2}$
No. 6, No. 7, and No. 8	$1\frac{1}{8}$	$\frac{3}{4}$

The cement and pigment are mixed dry, preferably in a paddle-type (plaster) mixer, to a uniform color before adding the aggregate and finally the water. Water must be held to not more than 35 lb per 100 lb of cement, and the mixture must be of the driest consistency possible to be placed with a sawing motion of a straightedge. If the mix is too wet it is dried by increasing the amount of coarse aggregate; if too dry, the amount of coarse aggregate is reduced to give better workability. Immediately after striking off the topping, it is rolled by means of heavy rollers, adding aggregate if necessary so that the finished surface shows at least 70% aggregate. The surface is then floated and troweled in the usual manner, using mechanical floating and troweling whenever possible.

The concrete should be protected from damage by barricading against any kind of traffic and the maintenance of a temperature between 60 and 80°F under a continuously moist condition for at least 7 days for normal cement or 3 days when high-early-strength (Type III) cement is used. A 1-in. layer of wet sand is usually used for curing terrazzo, although other curing procedures are acceptable if they provide adequate curing and will not stain the surface.

After the concrete has been cured, the floor is machine-rubbed, using a No. 24 grit abrasive stone for the initial rough grind and a No. 80 grit abrasive stone for the second rubbing, keeping the floor wet and removing the ground-off material by flushing with water and squeegeeing during the process, as shown in Fig. 34. A grout of portland cement, water, and pigment of the same kind and color as was used in the matrix is now applied to the surface by brushing and squeegeeing to fill all voids.

The objective of applying the grout is to fill all pinholes and voids that were exposed by the grinding. Grout should be a fairly thick paste, troweled on the surface and worked into the voids. Excess grout is scraped off. After the grout starts to set, it is hand-rubbed with pads of burlap or excelsior, or on large areas a grinding machine fitted with wood blocks can be used.

The grout is moist-cured for at least 72 hr, after which the surface is given a final light polish using a machine with stones no coarser than No. 80 grit. The finished terrazzo surface is cleaned by scrubbing with warm water and detergent and then is mopped dry. The use of a penetrating sealer such as has been developed through research of the National Terrazzo and Mosaic Association is recommended as the best means for preventing absorption of liquids that are liable to stain or cause deterioration of the floor. Surface coatings such as waxes are not recommended for terrazzo floors because they increase cleaning problems without adding to the appearance or durability of the floor.

A wood floor can serve as a base for terrazzo provided the floor is sound and rigid and can carry the additional dead load without deflection. A membrane

Fig. 33. Dividing strips for terrazzo set in place to separate the topping into panels.

Fig. 34. Initial grinding of the terrazzo topping.

of saturated felt is spread over the wood floor, followed by the reinforced mortar bed and topping as described for sand-cushion terrazzo.

Bonded terrazzo differs from sand-cushion terrazzo, described in detail above, only in that no sand cushion is placed on the thoroughly cleaned structural base which will be only about 1¾ in. below required finish elevation. After brooming a slush coat into the base, the underbred mix, consisting of a 1:4 cement-sand mortar of stiff consistency, is spread and struck off at least ½ in. and not more than ¾ in. below the final surface level. Division strips are set and the topping mix, prepared as described for the sand-cushion method, is rolled to a compact surface and cured. Floating, troweling, curing, rubbing, grouting, and final polishing are done as described for sand-cushion terrazzo. One disadvantage of bonded terrazzo is the possible occurrence of cracks, usually at division strips, and sometimes elsewhere in the slab.

Monolithic terrazzo uses neither sand cushion nor underbed but is bonded directly to the structural floor. Unlike sand-cushion terrazzo, any crack that occurs in the structural floor will persist in the terrazzo topping because the topping is monolithically united to the structural slab. However, by anticipating where structural cracks are most likely to occur, construction joints or sawed joints cut partially through the base floor can be used for installing the division strips, thereby keeping random cracks under some degree of control. These joints are placed at such locations as at doors, over beams and columns, exterior corners, and wherever movement and stress variations would be expected to cause a crack.

The terrazzo topping mix is approximately ⅝ in. thick and can be placed on an old concrete floor provided the floor is sound and reasonably level. Large variations in thickness of the topping will tend to create random cracks; therefore, the levelness of the old floor will govern to a considerable extent the levelness of the terrazzo. The topping is bonded to the base with a portland-cement slush coat or slurry as described for bonded terrazzo. The successive steps in the process of obtaining the finished terrazzo surface are as previously described.

Plastic adhesive membrane can be substituted for the portland-cement slurry. The substrate must be trowel-finished, however, to receive the adhesive. There are two approaches to the bond; one provides a rigid bond of high tensile strength; the other is sufficiently flexible to permit differential movement between the topping and the base. In either case the dividing strips are cemented to the base using the adhesive selected for the bond. Epoxy and vinyl adhesives give a rigid bond; neoprene polysulfide and epoxy polysulfide adhesives have flexibility. The clean substrate must be thoroughly dry when the bonding material is applied. The adhesive is applied by means of brush, broom, roller, or spray, depending on the consistency of the liquid and the area to be covered. If the adhesive is a water emulsion it must be allowed to become tacky before fresh terrazzo topping is placed on it. Under any circumstances, directions of the manufacturer of the adhesive must be followed in every detail of temperature, time, thickness of the adhesive coating, curing period, and sometimes precautions to be used to avoid harm to workmen while handling the materials.

Synthetic-matrix toppings using synthetic resins or polymers instead of portland cement in the topping mixture have extremely high tensile strength. They can be placed in thicknesses varying from ⅛ to ⅝ in., depending on the size of terrazzo aggregate particles used. Larger particles of aggregate can be seeded on these thin toppings if desired. The neoprene matrix is usually placed ⅜ to ⅝ in. thick, and dividing strips are used as mentioned above. The epoxy toppings must be handled quickly because of their rapid rate of set. A soupy mix is easier to place and work, seeding the aggregate withheld from the mix, on the surface after it has been spread. Any water present or temperatures below 50°F will prevent proper setting. Experiment and experience are required to learn the best procedures to follow. Instructions accompanying the synthetic materials must be followed with great care. Some of the manufacturers will furnish expert supervision initially on jobs of major proportions.

18   Mosaic

Mosaic is similar to terrazzo, in that a colorful surfacing is applied to a mortar underbed. The design, made up of small thin slices of marble, other colorful stone, or ceramic (the tesserae), is laid out by the artist and glued to heavy paper.

The concrete slab should be screeded off 2 in. below finished-floor elevation. A 1:3 mortar underbed is then placed and screeded off 1 in. below the floor line; then the setting bed is placed. Mortar for the setting· bed consists of 1 part cement and 2 parts sand with sufficient water to make a plastic, workable mix. Usually a small amount of lime is included for additional plasticity. The paper-backed tesserae should then be embedded in the setting-bed mortar, tamped, and pressed into place. The paper backing is removed and the joints between the tesserae grouted. After the grout has hardened the floor is rubbed and the joints grouted again. Finally, after the floor has cured, it is thoroughly cleaned and polished.

19 Skidproof Floors and Walks

Portland-cement concrete floors that have been troweled to a smooth surface are likely to be slippery, especially when wet. Where danger of accidents by slipping of vehicles or pedestrians is a consideration, as on steps, ramps, or walks having an appreciable slope, some effort is advisable to eliminate the hazard.

Brushing or brooming a freshly troweled surface will roughen it, and depending on the location, a light brushing to dull the surface or heavier brooming with stiffer bristles will provide for friction with tires that should prevent skidding of normal traffic. Such scoring of the surface may eventually wear away under traffic and other treatment might be needed, such as epoxy mortar coatings.

Siliceous coarse aggregates such as granites, quartzites, sandstone, gneiss, and schist will provide a good degree of friction whereas limestone, marble, dolomite, and similar calcareous and argillaceous aggregates have a tendency to wear smooth and slick with a low coefficient of friction.

Crushed aggregate of hard, durable characteristics can be mixed with the concrete placed on the surface, or sprinkled on the wearing course just prior to finishing if the cost of the special aggregate is a consideration. About ¾ to 1 lb of nonslip aggregate is needed per sq ft of floor when mixed in the top ⅝ in. of concrete, but only ¼ to ½ lb is necessary if scattered on the surface while finishing. After the floor has hardened so the particles of aggregate will not be loosened, the film of cement should be scrubbed or ground away using a scrubbing machine and steel-wool pads.

Artificial abrasive aggregates or grit such as aluminum oxide or corundum should be used in floors or areas that are to be subjected to extreme abrasion and must be at the same time as slipproof as possible.

Stair treads are precast of concrete containing coarse grit to maintain good traction and safety, or they can be made skid-resistant by applying the abrasive grit in a matrix of synthetic adhesive spread on the treads ⅛ to ¼ in. thick.

PAVEMENTS

20 Spreaders

Concrete should be deposited on the base only after the base has been brought to the required elevation, density, and contour and has been wetted to minimize absorption of water from the concrete. To the extent that it is practical, concrete should be distributed on the grade by the mixer in amounts and to a depth that will result in a moderate excess to be moved forward by the spreader. Piling a large excess of concrete on the grade necessitates moving it greater distances with consequent likelihood of segregation. Pavement construction is discussed in detail in Chap. 26.[7]

On small jobs concrete is spread by means of hand shovels after being deposited. On large jobs mechanical spreaders give much greater efficiency (see Figs. 21, 22, and 24, Chap. 26).

There are two general types of mechanical spreaders, both of which can be operated to move the concrete toward the sides or toward the center of the slab. The butterfly type has a large blade that moves from side to side between the forms with the blade swinging to an angle that pushes the concrete forward and sideways at the same time. The blade is adjustable in height above the grade so that it does not move concrete that is already in place.

The screw type of spreader has two helical screws operating from the center of the slab and reversible so that concrete can be moved in either direction as required. Spreaders are moved forward or backward by motors attached to wheels riding on the forms. Following the spreading device is a screed or strike-off that can be adjusted to give the required depth of concrete. The screed should be adjusted so that it leaves about ¼ in. or more of concrete above the finish elevation, or enough surplus to provide a roll of concrete 4 to 8 in. high in front of the finishing-machine screed. If the concrete is consolidated by vibration, the vibrating units are sometimes attached to the spreader immediately ahead of the strike off screed.

21 Finishing

Finishing Machines. Mechanically operated finishing machines, in general, have two reciprocating screeds oscillating transversely while the machine, traveling on the forms, moves forward 2 to 4 in. with each transverse oscillation of about 6 in. These motions can be changed by the operator by shifting gears.

Screeds are adjustable for the roadway crown. The first screed has a horizontal surface 12 to 15 in. wide, tilted a fraction of an inch upward on the leading edge to compress the concrete, and a vertical face of about 18 in. When in operation, a surplus of concrete is rolled ahead of the first screed to provide extra concrete required to prevent areas of honeycomb on the surface of the slab where the spreader did not completely fill the forms. The roll of concrete should be between 4 and 8 in. high for most effective compaction with the least effort.

The second screed has a reciprocating motion also that trowels the surface to approximately final elevation. This screed is about 8 in. wide with only a few inches of vertical face that pushes a small amount of mortar ahead of it.

The transverse finishing machine should be kept as close to the spreader as reasonable in order to take advantage of the softer consistency and easier workability of the fresh concrete prior to early stiffening that takes place during the first 15 min or so.

Some contractors prefer to attach vibrating screeds, or a number of vibrating spuds, to the finishing machine ahead of the first screed. Regardless of where the vibrating devices are used, they must be stopped when the forward motion of the spreader or finishing machine is stopped, or if the direction of the equipment is reversed.

Should it be found that the surface of the pavement is not uniformly consolidated, or forms are not filled to the required elevation, the finishing machine is moved back, with screeds raised, and successive passes made until the desired appearance is secured. Special attention must be given to consolidation of the concrete against the forms, and even when vibration is not required by the contract, it is necessary to have hand vibrators at each side and the concrete vibrated against the form either by hand or by means of vibrators supported in brackets designed for the purpose of holding the vibrators in position while the finishing machine is moving forward.

Longitudinal Float. To assure better riding characteristics of the pavement, a mechanical longitudinal float is used. Its use lessens the amount of work required of the straightedgers who follow. The float is at least 10 ft long and about 5 or 6 in. wide, and is moved with a sawing motion parallel to the centerline of the pavement as the entire machine moves forward under its own power. With each such lateral pass, the machine moves forward half the length of the longitudinal float. The screed rides very lightly and should have only enough mortar in front of it to show that it is performing its function. When the float reaches the pavement edge it is automatically lifted over the ridge of mortar before it begins the return trip. Adjustment of the motion and elevation of the longitudinal float require experience and care.

The longitudinal float is used as soon after the transverse finishing screeds as possible, but time must be allowed for most of the bleeding that might occur to take place. The longitudinal float will aid in removing the excess water that might otherwise contribute to scaling and crazing of the pavement surface.

A manually operated longitudinal float calls for a special bridge rolling on the forms with transverse walks about 2 ft farther apart than the length of the float, for the two operators who will handle it. The longitudinal float is 12 to 16 ft long with plow handles attached to each end. The width should be 6 in. and the vertical stiffener 8 in. high. The operators saw the float to and fro as they walk across the bridge, making as many passes as necessary, which is usually two. They then move the bridge forward about half the length of the float and repeat the floating operation.

Hand Finishing with Straightedge. The straightedge is used while the concrete is still soft enough to be manipulated without damage to the pavement. Its purpose is to check the surface for irregularities and, if necessary, to scrape down high spots and fill in low areas. The straightedge is generally 10 ft in length and is attached to a handle long enough to reach easily the center of the slab. It is made of wood or metal, and designed so as to maintain a true, rigid straightedge. The edge is used for cutting or scraping high areas. It is made of aluminum or steel for best results.

In use, the straightedge is pushed toward the center of the slab with the handle held low so that it slides easily without digging in. It is lifted over the ridge of mortar and pulled back with the handle raised to shoulder height until the blade is brought over the side of the form and any water or laitance is discarded outside the form. This is repeated until all points of the concrete have the correct alignment, following which the straightedge is advanced along the pavement about half its length and the process is repeated. High spots are scraped, and the handle is moved to various angles to cut off the excess concrete as seems best to accomplish the leveling procedure, raising the handle when pushing and lowering the handle when pulling, so that the forward edge of the blade cuts into the concrete.

Low spots may necessitate bringing fresh concrete from ahead of the spreader to fill in. The straightedge is not very effective for consolidating such repairs, and a long-handled float should be used. This is made of steel or wood, is about 5 to 8 in. wide and at least 4 ft long. The handle, which should be as long as for the straightedge, is attached and braced to the float so that by raising or lowering the handle the operator can do effective floating, as he does with a hand trowel. Sometimes the long-handled float is used for cutting and smoothing high spots.

After working with the straightedge and long-handled float, and when it is thought that the surface meets the requirements for smoothness (usually not more than $\frac{1}{8}$-in. variation in any 10-ft length), a straightedge that has been checked for accuracy against a master straightedge is used for a final check by resting it lightly on the concrete at a number of places across the slab, then raising it and moving along the pavement about half the length of the straightedge each time. This should be done before the concrete has set to the point that the pavement surface will be damaged if corrections are needed. Sometimes, when the concrete has taken its initial set, it is necessary to scrape the surface with the edge of a shovel or similar sharp cutting edge and patch with a little fresh mortar and a hand float, being careful not to disturb the surrounding concrete. Although this results in an unsightly remedy, specifications often call for grinding off the high spots after the concrete has been cured, an operation that is quite expensive as well as unsightly.

Surface Texturing. There are three popular methods for providing a relatively nonskid surface: belting, burlap drag, and brooming. Belt finishing is usually done by two laborers who hold the ends of a two-ply canvas or light rubber belt, 8 in. wide and about 2 ft longer than the width of the pavement, and with a sawing motion as the belt is advanced along the pavement produce a herring-bone texture that improves the traction of rubber-tired vehicles.

The burlap drag produces a sandy texture. A strip of wet burlap at least 3 ft wide and several feet longer than the width of the pavement is pulled by two laborers, so that 18 to 24 in. of the burlap is continuously in contact with the concrete. Sometimes the drag is mounted on a bridge which travels on the side forms or is attached to the final finishing machine (Fig. 25, Chap. 26). It is dragged without any sawing motion. It must be kept wet and free of accumulations of hardened concrete or mortar. This must be done at the right time, as too soon will tear the surface, and too late will do no good. This final drag operation is the only time that water may be used in finishing.

Broom finishing can be made to give different textures, depending on the condition of the concrete when broomed, weight of the broom head, and stiffness of the bristles. Special brooms with a single row of steel-wire bristles are now available, and rattan stable brooms have been used successfully. The broom is lifted to the center of the pavement and pulled by means of a long handle straight to the side form. This is repeated, lapping the broom marks about ¼ in. at the edge of each pass.

It is obvious that the timing of the texturing operation is of utmost importance. If it is done too soon, water might be brought to the surface, flattening and melting the texture. If it is done too late, the dragging or brooming will fail to make an impression, or the impression will not be uniform. Also, it is important that the consistency of the concrete be uniform from batch to batch so that the surface to be textured will have reached the desired consistency or workability as the texturing work proceeds along the pavement.

Edging. The final hand-finishing operation is edging of the slab against the forms and along the construction and other joints. The edger usually gives a ¼-in. radius on pavement work. It is important that this is not exceeded when edging transverse joints because of the objectionable shock to vehicles and to the pavement caused by unnecessarily wide spaces at joints. The edging tool is worked back and forth with light hand pressure until the edge is well troweled and free from honeycomb. A pointing trowel run along the form and each side of joints is an aid to starting the edging tool without tearing the concrete.

22 Combinations and Slipforms

Greater efficiency in paving operations has been achieved by combining several of the machines into one, or by the use of mechanical straightedging instead of by handwork. There are advantages provided preliminary planning has been carried out so that the concrete is of uniform quality, thus making it possible to continue all operations without delays at any point that might be caused by inadequate supply of concrete, alternate wet and dry batches, and other items that affect the quality of the pavement and the characteristics of the concrete the workmen must contend with. Another consideration is the resulting additional weight of mechanical equipment on the forms and the need for greater stability on the subgrade or base as well as the support of the side forms on which the equipment travels.

Slipform paving has had considerable success, with an important saving in labor. The slipform combines spreader, finishing machine, and screeds into one unit that travels on crawler treads directly on the prepared base. No side forms are required. It is essential that the concrete be uniform in consistency and have been formulated so there will be a minimum slump or slipping of the slab when the side forms that move with the machine have passed beyond the finished slab. Concrete that bleeds cannot be used successfully. Slipform pavement is reviewed in Chap. 26.

REFERENCES

1. Formwork for Concrete, *ACI Spec. Publ.* 4, Detroit, March, 1963.
2. "Forms for Architectural Concrete," Portland Cement Association, Skokie, Ill.
3. "Exploring Color and Texture in Concrete," Portland Cement Association.
4. "Cement Mason's Manual for Residential Construction," Portland Cement Association.
5. Recommended Practice for Concrete Floor and Slab Construction, (ACI 302-69).
6. "Specifications and Technical Data," National Terrazzo & Mosaic Association, Inc., Arlington, Va., November, 1963.
7. Woods, K. B.: "Highway Engineering Handbook," McGraw-Hill Book Company, New York, 1960.

Chapter 28

JOINTS IN CONCRETE

CHARLES E. PROUDLEY *

It is rare that a concrete structure can be built so monolithically that one or more joints of some type are not required. *Construction joints* occur wherever concreting is stopped or delayed so that plastic concrete is subsequently placed against hardened concrete. *Contraction joints* are placed where shrinkage of the concrete would cause a crack. *Isolation joints* are placed where movement of a part of a structure, due to temperature or moisture variations, settlement, or any other cause, results in harmful displacement of adjoining structural components. Other names are given to joints, such as dummy, expansion, hinge, and weakened plane, and each can be designed in a number of ways to accomplish special purposes or accommodate unusual conditions.

1 Construction Joints

A construction joint can be either bonded or unbonded, depending on design requirements. Usually a bonded joint is specified which, when properly made, results in a monolithic structure. Hardened concrete to which the new concrete is to be joined is treated to obtain a complete bond. The exposed line of contact either can be made as inconspicuous as possible or can be accentuated to become a part of the architectural treatment by attaching a rustication strip to the form. Horizontal construction joints will occur at the levels between lifts, usually designated by the plans and specifications. Vertical joints occur where the structure is of such length that it is not feasible to place the entire length in one continuous operation.

Water gain, or bleeding, is always objectionable, especially on a horizontal construction joint where a strong, tight bond is needed. It is common practice to begin with a satisfactory workable mix but omitting about half of the coarse aggregate in the first batch to assure the absence of honeycomb at the bottom of the form. Mixing water is reduced to a workable minimum to avoid water gain and, if it is seen that water is accumulating, further reduce the mixing water. An accumulation of soupy concrete, if left in the forms, will surely cause a poor joint that will deteriorate with age and exposure. The joint in Fig. 1 is an example.

A vertical construction joint in a thin section is made by installing a vertical header consisting of short, horizontal boards fitting tightly between the inner and outer forms and located opposite a pair of studs. Reinforcing bars are extended through the header at least 30 bar diameters. If a water stop is required, a crimped copper strip or a rubber or composition stop is installed on the inside of the bulkhead with one edge protruding into the concrete and the other to be straightened out when the bulkhead is removed so that it will be embedded in the concrete to be placed later. This will result in a watertight vertical construction joint if given proper attention, even though horizontal movement of the

* Deceased. Chapter revised by Joseph J. Waddell.

wall opens the joint slightly. In general, the business of making a vertical construction joint acceptable in performance and appearance is the same as for horizontal joints. A continuous and complete seal is necessary because even the smallest leak will make the seal ineffective.

Making of construction joints, especially those for mass concrete, is covered in detail in Chap. 25.

In many structures and pavements dowels, tie bars, or keys are installed to keep the joint from shifting out of alignment when forms are removed or under load conditions in service.[1] Plain bars are effective for dowels provided they are extended at least 30 bar diameters into each side of the joint. They must be kept straight and not displaced during placing and consolidating the concrete. Dowels are used for load transfer and to maintain alignment when the joint is not bonded. If the joint is to be bonded, deformed tie bars are used. Bars can be held in place by putting them through holes in the bulkhead, the holes being snug enough to keep concrete from running out when vibrated, but loose enough to permit easy removal of the bulkhead form without loosening the bars. For horizontal headers in floor slabs and pavements it is permissible to shove

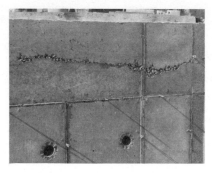

Fig. 1. Deficiencies of this joint are: (1) it is located too near the top of the wall, (2) the surface is uneven, (3) the old surface was not properly cleaned up, and (4) the subsequent lift of concrete was not adequately consolidated by vibration.

the bars through the holes after the concrete has been placed, being careful to insert them before the concrete stiffens.

Dowel bars can be any grade of steel, are plain without burrs at the ends, one-eighth the thickness of the slab in diameter, and are placed at regular intervals of 12 in. exactly parallel to the centerline of the slab and perpendicular to the joint. They are lubricated to prevent bond to the concrete by a coating of heavy oil or light grease. Many devices are available to hold the dowels in position while the concrete is being deposited around them, but the success of the installation depends upon the care exercised in placing them. If the dowels are not perpendicular to the joint and parallel to each other, major cracks are likely to occur over the dowels that are out of line.

If holes in the form are objectionable, a two-piece tie bar can be used, securing the female end of the bar against the bulkhead. Although sometimes permitted, bending the bar at right angles at the middle and stapling it to the inside of the header or bulkhead, to be straightened when the form has been removed, are not recommended. The bar installed in this manner is subject to breaking when straightened, or the concrete around the bar is loosened and damaged, thereby making the bar ineffective. Tie bars are not intended to be for load transfer, although they do so to some extent. Their chief function is to keep the construction joint from opening or separating noticeably.

Tongue-and-groove shear keys are sometimes cast in vertical construction joints, the groove being made by attaching a strip to the bulkhead or header. The forming strip should be trapezoidal in cross section so as to provide draft to permit easy withdrawal. It is usually about 1¾ in. thick, and its width at the widest dimension should be approximately one-half of the wall thickness, to cause the shear stress to be distributed equally on both sides of the key. If there are no reinforcing bars to be carried through the joint, tie bars can be used to add stability and prevent excessive opening of the joint.

Tie bars are either ½ or ⅝ in. in diameter, at least 24 in. long, and spaced 3 to 4 ft apart. They should be free of excessive rust and scale, oil, grease, paint, mud, or any material that will weaken bond with the concrete. Smooth dowels are greased or otherwise treated to prevent bond, so the joint can work as designed.

Chemical bonding agents are available and are used where conditions call for early use or rapid replacement of a structure, as described in Sec. 14. Bonding agents are composed of epoxy resins, liquid latex, or similar adhesive compounds.[2] The manufacturer's instructions must be followed in detail with respect to timing, temperature, quantities, method of application, curing, and in some instances precautions against health hazards. Epoxy adhesive, consisting of two components one of which is the synthetic resin and the other the activator, is used where high strength is needed. The components are mixed thoroughly in the proportions called for by the manufacturer, immediately prior to application, and are applied to the old concrete surface with a stiff bristle brush to obtain a continuous film about 0.01 in. thick. At temperatures in the range of 60 to 80°F the film will remain tacky for not more than an hour; so it is essential that the fresh concrete be applied within this time. The lowest slump consistent with good concrete practice should be used in order to secure the maximum benefit from the high-strength adhesive. Protective goggles, waterproof gloves, and protective clothing or creams should be used by workmen who come in contact with epoxy materials.

Liquid-latex adhesives are usually ready to apply without any additions, although some of them are mixed with water. The latex is brushed on the contact surface and the new concrete is placed immediately, before the latex dries. For additional protection against percolation of water, the liquid latex can be used in the concrete mix as a part of the mixing water. Dilution of the material, or other departures from the instructions of the manufacturer, must not be made.

2 Contraction and Control Joints

The expansion and contraction of concrete that occur with changes in moisture and temperature can set up stresses in concrete that sometimes exceed its strength and result in cracks and spalls unless provision is made to relieve the condition, as shown in Fig. 2. Control joints can be formed by attaching a strip of wood, plastic, or metal to the form, for vertical surfaces (Fig. 3) or by pressing the strip into the fresh concrete on a slab. A dummy or weakened plane joint (Fig. 4) is made by scoring the slab to a depth of about one-fifth the slab thickness, while the concrete is still plastic. Contraction joints can also be made by sawing the joint after the concrete is several hours old. This practice is especially common in pavement construction (see Chap. 26). Joints in large industrial floors are frequently sawed. Saw cuts should be flushed with a jet of water immediately after sawing to remove all sludge that, if allowed to remain, will set up and require sawing again. The joint thus formed by any of these methods is later filled with a poured sealing compound.

Preformed joint materials have been used extensively and are acceptable if installed with care to have them in a vertical position and correctly aligned. Special equipment is available to insert the filler into the fresh concrete on a slab. Usually a cap is used which aids in keeping the filler aligned in place and the correct distance below the slab surface. After the concrete has hardened, the cap is removed and the space is filled with a sealing compound.

The metal-insert type of contraction joint is similar in many respects to preformed joints. They are installed with a "planter" that vibrates the folded metal strip

FIG. 2. Failure to provide a control joint at the reentrant corner caused this driveway slab to crack.

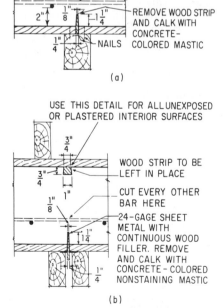

FIG. 3. (a) A wood strip attached to inside of sheathing forms a control joint. (b) Control joint can be formed by a removable metal strip. (*Courtesy of Portland Cement Association.*)

into place, where it remains until the job is completed. It is then collapsed to permit easy withdrawal and the clean groove is sealed with either a cold-poured or hot-poured elastic compound. Preformed neoprene or synthetic-rubber strip

Fig. 4. A dummy joint of adequate thickness serves to control cracking. This joint in fresh concrete has been finished on one side. Subsequently the other side was finished in a like manner, and the sidewalk was edged.

material of dimensions that fit tightly into the groove has been used successfully when cemented into place with an epoxy adhesive or similar material.

3   Isolation Joints

Isolation joints are necessary wherever a concrete floor or pavement abuts a fixed structure such as a wall or column. Isolation joints permit movement of the slab horizontally and vertically relative to the structure without damage to either the structure or the slab. When an interior floor, for example, a basement floor against foundation walls, is installed, an asphalt-impregnated strip ⅛ to ¼ in. thick with a width equal to the depth of the concrete floor is attached to the wall with asphaltic cement or masonry nails prior to placing of the slab concrete. The top of the strip should be at the elevation of the floor. Around columns or other structural features that must be independent of the floor, a circular or square wooden form is placed to the grade and elevation of the floor. When the floor has been finished except for these areas around columns, etc., the forms are removed, and a strip of the isolation-joint material is put in place against the finished floor and around the column. The small area is then concreted and finished to match the rest of the floor. When square forms are used to isolate a column, the sides of the square are set at an angle of 45° to the floor joints.

Isolation joints for structures outdoors must allow for greater movement because of greater temperature ranges and other differences. Roads, runways, driveways, and other areas of concrete having a length that will cause a thrust against a stationary, anchored structure or an intersecting roadway require isolation joints that are also expansion joints. A portion of the compressive force that can be exerted in the slab and the structure will be absorbed by elastic deformation of the concrete and by subgrade friction, but when this is exceeded the pavement will rupture or the structure will be displaced. To prevent this, the isolation joint is made thick enough to permit movement without damage. Under most conditions an isolation joint ¾ to 1 in. thick will suffice, but if there is doubt or a special situation, a second expansion joint placed 20 to 30 ft from the structure gives added protection. A preformed joint filler the full width of the slab and 1 in. less in depth than the slab thickness is installed with load-transfer dowels, if required, passing through holes in the filler. The filler can be a clear cypress board, an asphalt-impregnated fiberboard, nonextruding sponge rubber, cork, or cork rubber, or other materials manufactured for this purpose. The adjacent slab should be edged during the finishing. After the concrete has been cured and has dried out, the space above the filler is sealed with an adhesive sealant of the elastic type.

REFERENCES

1. "Joint Design for Concrete Pavements," Portland Cement Association, Skokie, Ill.
2. Thiokol LP-3/Epoxy Resin Adhesives for Concrete, Thiokol Chemical Corp. *Tech. Bull.*, Trenton, N.J.

Chapter 29

CURING

CHARLES E. PROUDLEY*

1 Purpose of Curing

Curing is the procedure used to assure that there is enough water present in the concrete to provide for continuous hydration of the cement.[1] It is obvious that if the supply of water, usually present in ample quantity in the finished concrete product as mixing water, disappears as the result of evaporation, hydration of the cement will stop and there will be no further gain in strength and durability of the concrete.

As concrete dries it shrinks, and if drying occurs when the concrete has little if any strength, cracks are sure to result. Furthermore, since drying occurs first on the surface, the cement will not be hydrated there but will be present as a dust coating having no strength to hold the aggregate particles together.

Temperature of the concrete during the chemical reactions of hydration has an important effect on the rate of strength gain. Temperatures near the freezing point of water retard the setting or hardening of cement to almost zero. Should the temperature fall below freezing, the free water in the concrete will turn into ice crystals, and since ice has much greater volume than the same water in a liquid state, the concrete is disrupted and upon thawing will have no strength. On the other hand, should the temperature be above 90°F there is great danger that the water will evaporate quickly and having once "boiled" off can be replaced only with considerable difficulty and much longer curing time for complete hydration. It is generally recognized that the problems of curing concrete during hot weather are much greater than during cold weather (see Chap. 30). Cracks in the surface of the concrete are likely to occur on a hot summer day because the setting reactions take place faster and the internal heat due to hydration causes expansion of the interior greater than at the surface.

In general, for the usual curing processes, special effort should be made to keep the concrete wet as evidenced by the continuous presence of moisture on the surface, and within a temperature of 50 to 80°F. Alternate wetting and drying can do more damage than no curing at all.

There are two principal systems of curing: (1) application of water directly or through some material that holds a reservoir of water in contact with the surface, and (2) a seal to prevent or retard the escape of moisture from the concrete.

Temporary protection of flat slabs placed during hot and windy weather can be accomplished by the application of a sprayed-on monomolecular film (Confilm) that reduces rapid evaporation of water from freshly placed concrete. Available from some admixture producers, this material is sprayed on the concrete surface

---

* Deceased. Chapter revised by Joseph J. Waddell.

immediately after screeding. It is not a curing compound and is no substitute for proper curing, its purpose being to reduce evaporation only while the concrete is in a plastic state. This material should not be used on concrete that bleeds excessively or under conditions of high humidity.

2 Water Curing

One method of direct application of water is inundation of the concrete, as when horizontal slabs are kept flooded by means of an earth dam around the edge that retains about an inch, more or less, of water over the entire surface. The

Fig. 1. Large areas of mass concrete are sometimes cured by means of water from spray pipes attached to the forms. Note that one block in the illustration is dry, the water having been shut off while the form is moved. (*U.S. Bureau of Reclamation photograph.*)

same effect is obtained by using a fine spray or mist from a number of nozzles arranged to cover the whole area. In either case, consideration should be given to water that runs off and whether it is objectionable in its effect on soil under or around the slab.

On dams and high piers it is sometimes the practice to attach spray pipes to the lower edge of the forms. As the forms are raised for successive lifts (Fig. 1), water curing is applied to the freshly exposed concrete. However, water running over the lower portions of the structure for a long period of time is apt to stain the concrete.

Burlap is used extensively as an absorbent cover for curing concrete. It should be closely woven and weigh at least 12 oz per 10 sq ft. Before it is used, burlap for curing concrete should be washed to remove soluble substances that are injurious to concrete or will mark the surface. Burlap is saturated with water before being placed over the concrete and successive strips are overlapped. The concrete must have set sufficiently so the wet burlap will not mar the surface; this will be before the concrete surface shows signs of drying. Later, additional layers of wet burlap can be added so that wetting will be required less often. Burlap must be kept continuously wet. Burlap curing is frequently followed after 12 hr or more by some other method, such as membrane curing, that does not require further applications of water.

Sand, sawdust, or earth curing is used successfully if kept saturated and if there are no salts or substances present that will damage or stain the concrete. Granular materials must be watched to see that they do not dry to the point that they are absorbing moisture from the concrete instead of supplying it. Oak sawdust must never be used because of the tannic acid it contains.

Straw is acceptable as a curing medium, spread to a depth of 6 in. or more, thoroughly saturated with water. It must be watched for fire hazard should it become dry between wetting, and also there is the possibility that the straw will be blown away unless it is held down by some other covering such as burlap or paper. The special advantage of straw is its insulating value against heat and cold.

Cotton mats or blankets, made of low-grade cotton bats quilted between two layers of 6½-oz burlap or osnaburg (cotton), and weighing a total of 25 oz per sq yd when dry, are excellent for curing because of their high absorbency and thermal-insulating properties. Their weight makes it necessary to use a lighter burlap cover or a water mist until the concrete has set enough to withstand marring by the weight of the wet mats.

3 Impervious-sheet Materials

Sealing the water in by means of waterproof, vaportight coverings has the advantage that it requires less attention subsequent to application of the curing, and also less water is needed, which can be a serious economic consideration where water is not readily available. One of the first waterproof or impervious-sheet curing materials was a paper blanket made by cementing sheets of kraft paper together by means of a bituminous adhesive, with a wide-mesh fiber fabric in between to give strength and durability. Sometimes the top or exposed surface is white to reflect the sun's heat. The kraft paper is treated to minimize dimensional changes as the paper comes in contact with water. The paper blankets come in widths up to 12 ft and are in rolls of 75-ft length. For wider blankets, sheets are cemented together to obtain a cover that can be sealed at the edges, usually by means of earth windrows on the paper, to prevent escape of moisture, as shown in Fig. 2.

Repeated use of impervious-sheet curing materials is permissible but the number of repetitions will depend on the condition of the sheets after removal from the concrete. Rips and tears, obviously, permit rapid loss of moisture and must be repaired to reseal the sheet material. Pinholes caused by walking or rolling equipment on the paper or impervious sheets are not so obvious but will permit the loss of moisture nonetheless. Pinholes are more difficult, if not impossible, to repair. It is necessary, therefore, to make a careful examination of the impervious-sheet curing covers between each use to determine if they will be effective. Sometimes an extra use or two may be made of slightly damaged or questionable covers by using them in double thickness. The careful workman may obtain as many as 20 reuses of a waterproof paper blanket for curing concrete, although under average conditions five or six uses are the limit for good curing.

The development of plastic-sheet materials has brought the use of polyethylene covering into prominence. It is available either clear or white-pigmented in rolls of varying widths up to 12 ft and in thicknesses from ½ to 4 mils. Clear polyethylene is generally used because it is less expensive than the white; however,

Fig. 2. Applying waterproof curing paper. Ends and edges are held in place by small windrows of earth. (*Courtesy of Portland Cement Association.*)

in hot climates the white reflects heat from the sun and provides some thermal regulation. Black polyethylene should not be used for curing except where heat absorption is not a problem. The ½-mil sheet is extremely light and fragile, making it difficult to handle and keep in place. If used, it is generally discarded after one use. The 4-mil thickness is comparatively stiff but very durable. For most purposes the 2-mil polyethylene film is preferred.

There are also available combinations of polyethylene fused to burlap or containing glass reinforcing fibers and several other innovations that are designed to give better service or economy.

Impermeable sheets must be well lapped and weighted at joints and edges. Pieces of lumber or earth can be used for weights. Be sure corners and edges are well covered.

4   Liquid Membrane-forming Curing Compounds

Liquid compounds to be sprayed on the freshly finished concrete that will dry quickly to an impervious membrane or film have been used extensively for vertical as well as horizontal surfaces. This material, variously known as sealing, curing, or membrane-forming compound, is a paint-like liquid that, when sprayed on concrete, forms an impervious membrane over the surface. The formulation contains resins, waxes, gums, solvents, and other ingredients such as a fugitive dye or a pigment.

Compounds can be classified as white-pigmented, gray-pigmented, and clear, the latter containing a fugitive dye to aid in observing the coverage. The dye should render the film distinctly visible for at least 1 hr after application and should fade completely in not more than 1 week. All compounds should be of a consistency suitable for spraying, should be relatively nontoxic, should adhere to a vertical or horizontal damp concrete surface when applied at the specified time and coverage, and should not react harmfully with the concrete. The clear compound should not darken the natural color of the concrete. Sometimes a thixotropic compound is added to hold the pigment in suspension, since the vehicle is usually so thin that pigments settle out rapidly, requiring special devices and constant care to keep the curing compound stirred and agitated for uniform application. Compound is formulated with either a wax-and-resin base or an all-resin base, some users specifying the all-resin type. One advantage claimed for the resin type is that it does not tend to separate as the wax-and-resin type does; hence there is less need for agitation.

Curing compounds are sprayed on at the rate of 1 gal per 200 sq ft for horizontal surfaces or 1 gal per 150 sq ft for vertical faces. Rough concrete surfaces require more compound than smooth surfaces. Recommendations of the manufacturer should be followed. For best results on all surfaces the time of application should be when all surface water has disappeared, the surface is still dark with moisture but not shiny, and there are no dry areas.

Compound should be applied to unformed surfaces as soon as the water shine disappears, but while the surface is still moist. If application of the compound is delayed, the surface should be kept wet with water until the membrane can be applied. Do not apply compound to areas on which bleeding water is standing.

Compound should be applied to formed surfaces immediately after removal of forms, unless the surface is to be rubbed, in which case a moist covering should be used to protect the concrete until the surface has been rubbed. If the concrete surface is dry, it should be moistened and kept wet until no more water will be absorbed. As soon as the surface moisture film disappears, but while the concrete is still damp, the compound is applied, taking special care to cover edges and corners. Patching of the concrete is done after the compound has been applied. Compound should be sprayed, not brushed, on patched areas.

When used solely for curing it is necessary that the life of the membrane be only about 30 days or less, after which it will be either worn, washed, or blown away, or removed readily with a stiff bristle brush. Water retention of the membrane is very good for the first 3 days, still effective for another 4 days or more, but of little value for curing after 14 days. It is available in a variety of colors for aesthetic or practical purposes but if the concrete is to be rubbed, as on handrails, bridge piers, or architectural concrete, it should be clear without pigment or dye in order to avoid a mottled appearance.

Pressure-tank equipment, as shown in Figs. 3 and 4, should be used for spraying the compound, and the compound should be agitated continually during application. Apply in one coat, consisting of two passes of the spray nozzle at right angles to each other.

Compound should never be thinned. However, during cold weather, the compound can be heated if it becomes too viscous for application. Heating is done in a hot-water bath, never over an open flame, and the temperature should never exceed 100°F. When being heated, the container should be vented and there should be room in the container for expansion of the compound. The compound should be agitated during heating.

Fig. 3. Automatic self-propelled spray machines are available for applying curing compound on large projects. (*Courtesy of Portland Cement Association.*)

Fɪɢ. 4. Even on small jobs, manual pressure-spray equipment should be used. (*Courtesy of Portland Cement Association.*)

5 High-temperature Curing

Special methods of curing that accelerate the curing process have been developed for concrete-products plants, permitting less storage space for finished units and quicker delivery on orders.[2,3] The presence of adequate water for hydration is always of primary importance. More rapid hydration of the cement in the product is accomplished by elevated temperature. Systematic control is essential, care being taken to raise the temperature slowly so the units do not develop critical internal stresses through differential thermal expansion.

The normal schedule for atmospheric steam curing is about as follows:

1. Preset. Also referred to as "delay," "presteaming," or "holding." A delay period between fabrication of the product and the start of steaming. Hydration of the cement starts during this period, stabilizing the product. Products are stored in the steaming chambers at normal temperatures and prevented from drying out, if necessary, by the application of a light fog. During cold weather a small amount of heat might be necessary to keep the temperature above 60°F.

2. Period of rising temperature. During this time the products are enclosed in the curing chamber (sometimes called a "kiln") and heat is applied to raise the temperature to the maximum curing temperature. The temperature rise is controlled at a rate of 20 to 60°F per hr, depending on the size and character of the products.

3. Maximum-temperature period. The temperature within the chamber is held at a constant temperature for whatever period of hours is necessary for the product to develop the necessary strength. In some plants this period approaches zero. See the following soaking period.

4. Soaking. After the product reaches maximum temperature the steam is shut off and the product "soaks" in the chamber as the temperature slowly drops.

5. Cooling. The product is permitted to cool to atmospheric temperature at a controlled rate, the rate again dependent on size and shape of product.

In a typical plant, after a preliminary curing at atmospheric temperature for 2 to 4 hr, the temperature of the curing chamber or enclosure is raised at a rate of 40° per hr to a maximum of 140 to 175°F, holding at this temperature as a soaking period for 8 to 18 hr or longer if tests show an advantage in a longer period, and reducing the temperature at not more than 1° per min. Units having thin walls or small dimensions can be heated and cooled at a faster rate. Temperature recording or automatic temperature controls in the curing chamber are a necessity if uniformity of product is to be assured.

Some products plants supply moist curing at atmospheric temperatures by means of wet steam that maintains temperatures at about 135°F. Steam from the boiler is released through numerous holes in pipes laid along the floor between rows or racks of the units to be cured. If the temperature exceeds 135° the steam is shut off or reduced to prevent units from becoming too hot, with consequent possibility of dehydration and undercuring. The units should always show a film of moisture while curing with low-pressure or atmospheric steam. Not more than 2 days of atmospheric steam is customary, provided tests show that adequate strength has been attained. When the chamber is opened there should be a period of several hours for the concrete to adjust to the temperature of the outside air; otherwise cracks may appear, especially in cold weather. Under high-temperature conditions both heat and moisture may be lost to the outside through the material used to cover the concrete; hence the importance of providing tight covers and kilns, insulated if possible. More moisture is required to maintain 100% relative

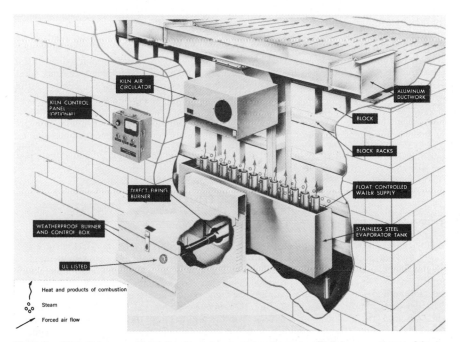

FIG. 5. The Johnson direct-fired curing system, here applied to a masonry block-curing kiln, makes use of heat, steam, and carbon dioxide for curing the block. See Chap. 42. (*The Johnson Gas Appliance Company.*)

humidity at higher temperatures than at low temperatures. At 70°F, 1.6 lb of water is required for 100 lb of air for 100% humidity, but 15 lb is required at 140°F and 66 lb at 180°F.

Under steam-curing conditions, the heat of hydration serves to raise the concrete temperature in the same way that it does under normal curing. Therefore, there comes a time when the concrete temperature will reach the ambient temperature within the enclosure, then rise above if control measures are not taken. After the concrete reaches the ambient temperature, the amount of heat should be reduced gradually in order to minimize the temperature differential between the concrete and the atmosphere within the enclosure and to permit a gradual cooling of the concrete.

When the concrete temperature exceeds the enclosure temperature, moisture will be lost from the concrete, unless wet steam is used as a source of heat. If heat is provided by other means than wet steam, there must be a source of additional moisture, such as hot-water sprays.

High-pressure steam curing can be used for any concrete product but is usually applied to masonry units in which, through choice of materials, there will be some reaction with the siliceous sand in the mix. An autoclave is necessary for high-pressure steam curing because the cycle of curing includes a period in dry steam at about 360°F. The cycle can be varied, but a typical sequence starts with a preset period at room temperature for 2 to 4 hr; then the autoclave chamber charged with units is sealed and the temperature and pressure are raised over a period of about 3 hr by means of saturated steam to about 360°F and 140 to 150 psig. Pressure is released during the final 20 to 30 min and cooling is rapid. Higher temperatures and pressures with shorter timing are often used, and also lower temperature and pressure with longer timing, depending on the manufacturer's facilities or choice based on materials and the quality or appearance of the concrete units resulting from the schedule.

Other curing methods involve the use of various means for applying heat while retaining moisture for hydration. Dry heat is applied where the concrete unit is sealed from loss of moisture that occurs when there is a vapor-pressure differential resulting from a difference between the temperature of the concrete and that of the surrounding atmosphere. The heat can be supplied from a heated casting bed, a removable enclosure in which there are heating units to heat the air as is the case when concrete is cured in the mold, or whatever arrangement is most convenient for the characteristics, dimensions, or design of the concrete product to be cured.

See Chap. 42 for a detailed discussion of high-temperature curing, including a direct-fired system that utilizes carbon dioxide during part of the cycle (Fig. 5).

SELECTION OF METHOD TO USE

6 Pavement, Roofs, Bridge Decks, Exterior Floors

Horizontal slabs such as pavement, sidewalks, parking areas, bridge decks, runways, roofs, and all concrete exposed to climatic conditions should be cured in a manner to provide for temperature control insofar as possible. The first step under any condition of exposure is to prevent drying of the surface.[4] Before the concrete has set enough to withstand marring, a lightweight sheet material such as polyethylene film should be used, sealing it along the edges to retard loss of moisture and to hold it in place by keeping wind from getting under it and ballooning the film. A fine fog to provide a mist that will envelop the area is excellent in hot weather if economically feasible since it will usually hold the temperature of the concrete uniformly within the recommended range while providing the required water for curing. The labor and facilities needed for water spray make this method more adaptable to curing large precast units at the manufacturing plant. Preliminary curing with a single layer of burlap is the most popular procedure for concrete pavement where construction moves along rapidly.

The most extensively used method of curing for surfaces that are outdoors is clear membrane-forming liquid compound since it can be applied almost immediately after the concrete is finished and no further attention is required. However, this is true only when temperature conditions are between 40 and 80°F. When the sun's rays and air temperature are likely to cause concrete temperatures to exceed 100°F, white-pigmented compounds should be used. When air temperatures are below 40°F and there is the possibility of subfreezing weather during the first 72 hr, further temperature control is necessary. This is readily provided by placing a layer of straw about 12 in. thick on the slab as soon as possible without damage to the surface, and covering the straw with a blanket material such as polyethylene, waterproof paper, or burlap to hold it in place. The straw insulation can be used without the liquid membrane-forming compound if straw curing alone is desired. In such cases only enough water should be applied to the straw to assure that the concrete surface will not become dry during the curing period since excessive use of water may cause freezing in spots subject to greatest heat loss such as at edges and corners.

Sand, sawdust, soil, and similar granular materials are satisfactory for curing and provide considerable protection against high temperatures and a limited protection when temperatures dip below the freezing mark for short intervals during each 24 hr. The sand, sawdust, etc., must be kept wet by sprinkling several times a day with a hose or a spray bar on a tank truck operating outside the paved area.

Bridge decks must be protected from below during cold weather unless the temperatures are not likely to be lower than about 15 or 20°F during the night, under which circumstances wood forms furnish ample insulation. To provide more protection and to accelerate curing, heaters placed at a safe distance below the formwork of the bridge floor will be helpful. The same recommendations apply to any slab above grade.

7   Interior Floors, Walls

In hot weather it is necessary only to prevent drying of the concrete inside buildings, and this can be done with liquid membrane-forming compounds, impervious sheet materials such as paper or plastic, with a fine spray from a hose if the room is closed to maintain high humidity, or any of the absorbent covers such as burlap, cotton mats, sand, or sawdust. If workmen will be moving about on the concrete during the curing period, wet sand or sawdust is preferable to sheet materials or membrane curing because of the protection against abrasion; however, sand or sawdust must be kept wet and deep enough to be a protection until the curing period is over.

In cold weather the room or building should be closed and heated in addition to the application of curing materials as described above. Clear plastic sheets are used to cover windows and doors, and space heaters are employed to keep the temperature up to 60°F or better. Combustion gases from space heaters must be vented to the outside to avoid carbonation of the concrete surface and subsequent dusting (see Chap. 30). If the concrete slab is not protected by a building, a temporary enclosure of plywood, or lightweight frames or trusses covered with plastic or canvas, and heated by atmospheric steam, electric radiant heaters, or similar devices, can be constructed for protection against almost any cold-weather severity during concreting, finishing, and curing.

8   Exterior Walls, Columns, Bridge Piers

Vertical surfaces to be cured after removal of forms can be protected from moisture loss, heat, and cold by enclosures, as has been described for horizontal slabs. Liquid membrane-forming compound sprayed or brushed on in one or two applications as may be necessary to obtain the required thickness of film is the least trouble for the degree of resulting moisture retention. If early rubbing is required, however, the removal of the film is troublesome, and unless it is completely removed the rubbed finish will not have a uniform appearance. If, however, the surface is rubbed immediately after the removal of forms, the sprayed-on curing can be applied after the rubbing, either clear without fugitive dye, white, or light gray, and the final appearance will be acceptable.

Where there is an abundance of fresh water available, good curing is obtainable by draping burlap against the piers or walls and keeping it saturated by a continuous trickle of water from pipes or hose at the top. Unless the water is free from minerals such as sulfides and iron, permanent stains may occur that cannot be removed satisfactorily by rubbing. Flowing water over a vertical surface under a covering of polyethylene film so attached as to be relatively airtight will cure satisfactorily but will not give much protection against cold or solar heat. In cold weather a framework around the walls or piers covered with impervious sheets of plastic or waterproof paper will give considerable thermal protection

FIG. 6. A polyethylene enclosure, with heat supplied inside and straw on the deck. provides curing protection for this structure during cold weather.

as long as the water flowing over the concrete does not freeze. Sometimes cotton mats are wrapped around columns or piers to preserve heat of hydration and prevent freezing (see Fig. 6).

9 Architectural Concrete

If the architectural concrete is cast in place in the structure and the forms must be removed before curing time has been completed, further curing is obtainable as described for vertical walls. Where the concrete is an overhang, strips of wood are attached by whatever means is most convenient to hold the blanket or sheet material as close to the concrete as feasible.

Architectural shapes, units, designs, or figures that are made in the concrete-products plant or shop will have been properly cured before being delivered to the job (Fig. 7). Job curing and plant curing must be coordinated to result in the same color and appearance of the finished concrete if it is desired to have them match in the structure. If curing results in differences that cannot be corrected otherwise, rubbing with a carefully chosen mortar mix sometimes improves the appearance.

FIG. 7. Precast units in the casting yard are covered with heavy waterproof canvas for curing.

10 Masonry Units, Precast Members, Concrete Pipe

Precast-concrete products will arrive on the job ready-cured for immediate use. The method of curing used by the manufacturer will depend upon his analysis of the economics of his operations, and these vary geographically and with the demands of the marketing area. The majority of plants use atmospheric steam curing since this can be controlled for changing weather conditions and does not require much financial investment. Water-fogging nozzles instead of steam are good but require supplementary heat during cold weather.

Large units such as culvert and sewer pipe that must be cast rather than machine-made are molded over a steam line that vents atmospheric steam to the interior of the mold. After casting and finishing the top end of the pipe, a jacket of dimensions to allow several inches of space around the outside and top of the mold is dropped over the entire assembly, leaving openings at the top and bottom so the warm, steam-laden air can circulate entirely around the mold. After 24 hr the mold is removed and the cover replaced for more steam curing that can be continued for whatever curing period strength tests of the concrete in the pipe indicate is required.

Masonry units, sand-lime brick, concrete brick, cement-asbestos pipe, and other mass-production shapes are autoclaved with high-pressure steam when it is found by experience that the rapid production and shipment of units make it profitable, or when the characteristics of the autoclaved unit show marked improvement over any other means of production.

11 Tilt-up and Lift Slab

For curing the concrete slab on which tilt-up or similar slabs are to be subsequently cast, there are available liquid curing compounds that act as a sealer to the concrete and as a parting compound to prevent the later concrete from bonding to the floor slab. The first coat is applied to the fresh concrete in the usual manner to serve as a curing compound. A second coat to serve as a bond breaker is applied after the forms have been placed but before the reinforcement and inserts are placed.

REFERENCES

1. ACI Standard 308-71, Recommended Practice for Curing Concrete.
2. ACI Standard 517-70, Recommended Practice for Atmospheric Pressure Steam Curing of Concrete.
3. ACI Committee 516 Report, High Pressure Steam Curing: Modern Practice, and Properties of Autoclaved Products. *J. ACI*, August, 1965, pp. 953–986.
4. "Current Road Problems, 1-2R, Curing Concrete Pavements," Highway Research Board, National Academy of Sciences, Washington, D.C., May, 1963.

Section 8

SPECIAL CONCRETE AND TECHNIQUES

Chapter 30

ADVERSE-WEATHER CONCRETING
AND ENVIRONMENT

PERRY H. PETERSEN

Concreting performed under *standard* conditions as set forth by reference to ACI or ASTM standards and recommendations properly involves a temperature of 70 to 75°F and relative humidities of 50 and 100%. These standard conditions of humidity and/or temperature are stipulated as being all-inclusive as well as being applicable to the concrete molds, work areas, moist closets, water-cure tanks, and testing environment. Furthermore, strength determinations made to ascertain conformance of concrete to established building-code requirements involve cylinders tested at 7 and 28 days of age, wet- or moist-cured at 70 to 75°F in conformance to those same standards until tested.

Theoretically, therefore, conditions of environment outside the above might be adjudged technically as being nonstandard and thus might be interpreted as being adverse to the attainment of quality concrete. However, this supposition is not always true inasmuch as *cool* concreting can be very beneficial to quality whereas *warm* or *hot* concreting usually is detrimental, at least to some extent. For instance, the amount of water required for proper mixing and for optimum workability increases markedly with increase in temperature of the concrete; Fig. 1 based on Bureau of Reclamation data[1] illustrates that with constant slump, concrete at 100°F requires up to 40 lb more water per cu yd than does the same concrete at a temperature of 40°F. Paul Klieger[2] of the PCA indicates that this value of 40 lb may be as low as 20 to 25 lb per cu yd of concrete. This greater water content with increase of concrete temperature is detrimental at least in the fact that it causes a reduction in strength and an increase in shrinkage. Furthermore, strength of concrete is also very much dependent upon the temperature at which the concrete is placed and cured; Fig. 2 based on Bureau of Reclamation data[3] indicates that cool-placed concrete which is kept cool at least for its first 7 days of age is eventually stronger than is warm concrete placed and cured at higher temperatures but of the same water-cement ratio.

As to moisture during curing, enough water must be retained in concrete so that hydration of cement may proceed to some required or desirable point. If concrete is dried at any particular age (1 day, 3 days, 1 year, etc.), strength development is essentially arrested; rewetting at a later age may result in very little additional strength gain. Figure 3 indicates that later strength is much dependent on noninterruption of moist curing; in this reported study by the National Ready Mixed Concrete Association,[4] a loss in 28-day strength of 27% was observed solely because moist curing was not begun until the concrete was 21 days old. These conditions simulate all too well a 24-hr removal of forms by a formwork contractor followed by a much later and likely too late realization by the prime contractor that this bared concrete should have been cured. Therefore, it is imperative that someone be designated specifically to see that all concrete is protected against early drying.

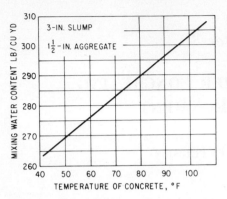

FIG. 1. Relationship of mix-water content vs. temperature of concrete at constant slump.

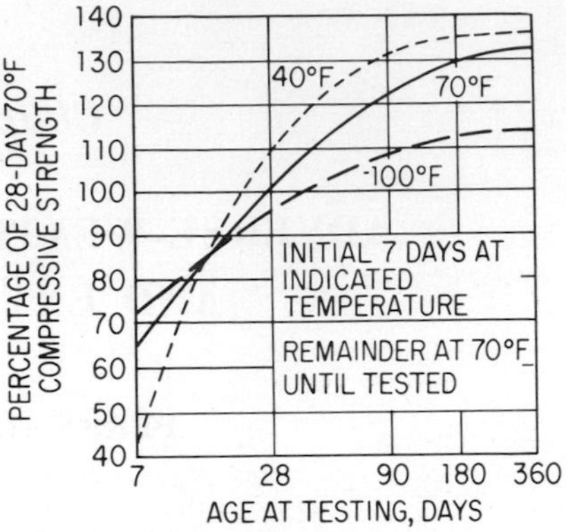

FIG. 2. Effect of temperature during initial 7-day curing on concrete subsequently stored at 70°F.

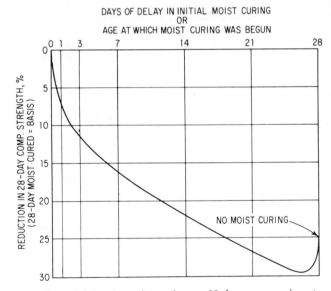

FIG. 3. Effect of delay in moist curing on 28-day compressive strength.

HOT-WEATHER CONCRETING

As indicated above, high temperature of itself will result in lowered strength of concrete. Also under such conditions, concrete will stiffen more rapidly, and any inherent false set, or flash set, will be aggravated. Time of placement of the concrete must be reduced, or otherwise too much water will have to be added to maintain constant and required slump. Energy exerted in mixing concrete supplements the heat of the day and that contributed by sunlight impinging on the mixer. With increase in concrete temperature, there is an accompanying increase in the rate of hydration of the cement, which in turn results in further generation of more heat; this heating in effect is self-energizing or spontaneous in its nature since high temperature means faster hydration, which results in even higher temperatures, which means still faster hydration, and so forth. Thus under such circumstances, concrete has been known to set up in the mixer in a short time.

Positive steps against the occurrence of hot-concreting difficulties with regard to mixing, placing, finishing, and curing are given in Ref. 5. Additional measures are described in Chap. 20. Briefly, these include:

1   Cooling of Aggregate and Cement

Aggregates may be wetted down some few hours before they are batched, enough so that interim evaporation of moisture will cause their temperature to be appreciably reduced by the time they are used; this evaporative cooling is especially marked and beneficial in an environment of low humidity. Cement can be cooled only by storing it in shaded structures or bins, and in some instances limitations are invoked as to its maximum temperature on delivery or when used.

2   Use of Cool Mixing Water

Cool well water or relatively cool deeply piped city water will help to reduce the temperature of concrete at time of mixing, particularly because it has a high heat capacity as compared with the other ingredients. Otherwise, flaked or crushed ice is substituted in whole or in part (by weight) in place of the mix water. When ice is used, mixing must be continued until all of it has melted and before the concrete is placed. The use of liquid nitrogen further contributes to cooling without any effect on water content.

3   Shades, Spray Cooling, and Heat-reflectant Paint

Light- or white-colored and glossy-finished mixers absorb appreciably less heat under direct sunlight than do those dark or black in color. Similarly, the shade of a tree, a lean-to, or even a separate light-reflecting panel positioned and permanently installed over the mixing drum will do much to eliminate heating of the mixer by sunlight. Spray cooling of the mixer drum and consequential evaporation of this moisture are beneficial in keeping the drum cooler than it otherwise would be.

4   Use of Short Mixing Time

For long hauls, the coarse aggregate, the sand, and then the cement (in that order) can be dry-batched through a manhole in the side of the mixer; when concrete is needed at the jobsite, all the water (as cool as possible) can then be added and the whole immediately mixed and placed. The use of mobile on-site batching and mixing equipment (ASTM C-685) might also be helpful in hot weather in eliminating slump loss normally occurring during long hauls of mixed concrete.

5   Use of Admixtures

Approved water-reducer retarders can do much toward retarding the rate of hydration and the time of set under high-temperature conditions so that the concrete can be hauled farther or else held in the mixer longer. Steps should still be taken, however, to keep the concrete as cool as possible. The retarder or the use of cooled concrete will help eliminate undesirable cold joints or layered separations in large or continuous placements involving several loads or more of concrete. Time for placing and finishing of concrete in floor slabs and flat work can likewise thus be extended so that these operations are not unnecessarily hurried. Also, job addition of water to a concrete containing a water-reducer retarder has been determined not to be so detrimental[6] to quality as the retempering of concrete without the admixture; the practice of retempering partially set concrete, however, should be avoided in either case.

Air Entrainment. As to air-entrained concrete, the required level of air is attainable even at high temperatures although the required amount of air-entraining admixture per unit of concrete might have to be increased appreciably over that used in cooler concrete.

Accelerators. Set-accelerating admixtures normally are not employed in hot-weather concreting. Rate of hydration of cement is increased through their use, thereby causing an early rapid rise in concrete temperature; if such concrete is then subjected to cooling, and especially within the first 24-hr period such as may occur during the first night, detrimental crazing and shrinkage cracking may result. For these reasons also, the use of high-early-strength cement is often avoided in hot weather.

6 Care in Placement

Chutes, conveyors, pump lines, buckets, buggies, or wheelbarrows which have become hot in the sun should be cooled before using; cooling by evaporation of sprayed water alone may be enough. Forms, reinforcement, as well as tools for placing, distributing, vibrating, and finishing the concrete should also be cooled, care being taken that puddles of water do not remain in spots where concrete is to be placed. White polyethylene sheeting or wetted burlap laid over freshly placed concrete will do much toward keeping the concrete cool, meanwhile maintaining a uniform moisture content throughout the concrete; this is much to be desired in slab concrete in order that hydration may progress uniformly through the thickness of the slab. The work done in the next finishing operation is usually sufficient to erase any marring of the surface caused by dragging these coverings back and forth. It is most disconcerting to have the top surface of a slab harden and crust over a spongelike interior mass of concrete; crazing cracks are likely to develop almost immediately and it is impossible to obtain a satisfactory hard steel trowel finish. In addition to the above-mentioned use of plastic sheeting or wetted burlap, an evaporation-retardant spray-on liquid[7] has been used successfully to prevent surfaces from drying out from the time the concrete has been placed, and extending to and possibly through the flat-troweling operation. This spray-on application is best made immediately after the concrete has been screeded. In deep beams, concreting should progress in stages in both depth and length, the bottom layer projecting a little ahead of the next above it; in this way, one is always placing new concrete against fresh plastic concrete, thereby avoiding cold joints and unbonded layering.

7 Curing

Curing should start the moment the concrete is placed; concrete should not dry out at all until it has attained the desired strength and durability (see Fig. 3). Often, curing is said not to be practicable, until the concrete has set; even if it is not practicable it should be done. Formed concrete such as walls and columns can be cured almost immediately after the concrete has set simply by supplying water at the top of the forms; when the forms are removed, wetted burlap or polyethylene sheeting will provide continued curing if cover of the concrete is complete. Spraying of bared concrete with water for 3 to 7 days is often impracticable since it is often done sporadically, and curing is not certain of accomplishment. The use of a curing compound on floors is recommended, particularly if the compound is applied immediately after the last finishing operation is complete and not before the finish will be marred by the work involved in the application. Consideration should be given to the type of curing compound and to the method of application if the floor surface is later to serve as a base for resilient floor tile or other decorative treatment such as a paint or other coating. Some manufacturers or installers of resilient floor tile insist that many curing compounds are not compatible with the adhesives they employ; in particular this ban includes the wax-base curing compounds. Walls, columns, and other vertical surfaces can likewise be sprayed or roller-coated with curing compound if the intended concrete finish will not be detrimentally affected by the use and presence of the compound. White-pigmented curing compounds are recommended for concrete exposed to sunlight, although often the white coloration may be objectionable.

COLD-WEATHER CONCRETING

As previously mentioned, "cool-weather" concreting can mean a better concrete. Normally, however, the term cold-weather concreting involves placement, finishing, and curing at freezing and below-freezing temperatures. It has been established that there is nothing wrong with letting concrete freeze provided of course that the concrete has hardened sufficiently to attain certain minimum strength[8] and assuming it has had the necessary durability and freeze-thaw-resistant attributes built into it. However, no newly placed and finished portland-cement concrete can withstand freezing and thawing cycles without some deterioration, at least until it has had an opportunity to harden properly. There is still no known concrete admixture that will prevent freezing of concrete or freezing damage of concrete freshly placed and exposed to "30°F weather with temperature falling." The admixture coming closest is the accelerator; it has merit only because it increases the hydration of 40 to 60°F concrete so much that it can resist the effects of an environment of 30°F or lower temperatures in 1 day instead of 3, or in 3 days instead of 7. Hydration of cement or consequent hardening of concrete normally progresses very slowly at temperatures just above freezing; hydration that occurs in 1 day at 70°F may take 7 days at 40°F. Thus to attain a desired strength (structural or otherwise) in a reasonable length of time, heated concrete and heated or heat-insulated environments are resorted to in practice along with, or in lieu of, admixtures and accelerators.

Recommendations for cold-weather concrete[9] include the following.

8   Heating of Concrete

In using heated concrete (Fig. 4) care must be taken that it is not overheated; see the hot-weather concreting difficulties above. Freshly mixed concrete in cold weather should never exceed 70°F in temperature; the temperature of the concrete should be in the range of 40 to 60°F. The approximate temperature of any fresh concrete can be obtained by use of the formula

$$T = \frac{S_a T_a W_a + S_b T_b W_b + 0.22 T_c W_c + T_w W_w}{S_a W_a + S_b W_b + 0.22 W_c + W_w}$$

where a, b, c, w = coarse aggregate, fine aggregate, cement, and added water, respectively

T = temperature of each, respectively, °F

W = dry-batch weights of coarse W_a and fine aggregate W_b, weight of cement W_c, and weight of added water W_w

S_a = 0.22 plus decimal moisture content of coarse aggregate

S_b = 0.22 plus decimal moisture content of fine aggregate

In the above formula, the decimal 0.22 is an approximation for the specific heat of dry cement and dry aggregate. By adding the decimal moisture content (for sand of 6% total moisture, use 0.06) to 0.22 in computing S_a and S_b for the

Fig. 4. Heated concrete and enclosures (in background) are used to facilitate concreting during cold weather.

coarse and the fine aggregate, respectively, the heat capacity of the aggregate and that of the moisture in the aggregate are both taken into account. Thus with aggregate moisture already included, the terms T_w and W_w pertain only to the added water.

9   Insulation vs. Environmental Heating

After being placed, concrete should be immediately protected from cooling or from losing heat in cold weather; this can be done by employing heat-insulating forms and formwork or by artificially heating the environment. The ACI 306 Recommended Practice for Cold-weather Concreting[9] includes in tabular form information as to the amount of protective heat insulation recommended for the placement and finishing of cold-weather concrete having a temperature of only 50°F; the environmental air temperature is coupled directly with the amount of thickness of commercial blanket or bat insulation generally employed in such operations as well as the amount of cement used in the concrete.   Data are reproduced here covering concrete walls and suspended slabs up to 2 ft in thickness (Table 1), concrete slabs up to 1.5 ft in thickness on 35°F subgrade (Table 2), and comparative heat insulation of various materials (Table 3).

It must be realized that the indicated thicknesses of insulation in Table 1 are applicable to both sides of a wall and to the top and bottom of a suspended slab unless, of course, one of the surfaces acts as the interior wall, floor, or ceiling of a heated enclosure.   In Table 2, the thickness of insulation pertains only to the top surface of the slab, but in this case, the concrete should be afforded protection equal to that shown in the table, beginning with the screeding and finishing operations; the use of a temporary or portable enclosure during placing and finishing is one way in which this can be done.   Table 3 indicates equivalent thicknesses of various heat-insulation materials normally available to the contractor. By it, one will find that for heat-insulation purposes, 3 in. of lumber (table value equal to 0.333) is thus required as the equivalent of 1 in. of commercial blanket or bat insulation.   Also, damp sand is indicated as having little heat-insulative value; its contained moisture may have merit as a reservoir of heat, especially in its conversion to ice.

Table 1. Insulation Requirements for Concrete Walls and Floor Slabs above Ground; Concrete Placed at 50°F

Wall or floor thickness, ft	Min air temp, °F			
	Thickness of commercial blanket or bat insulation*			
	0.5 in.	1.0 in.	1.5 in.	2.0 in.
Cement content at 300 lb/cu yd				
0.5	47	41	33	28
1.0	41	29	17	5
1.5	35	19	0	−17
2.0	34	14	− 9	−29
Cement content at 400 lb/cu yd				
0.5	46	38	28	21
1.0	38	22	6	−11
1.5	31	8	−16	−39
2.0	28	2	−26	−53
Cement content at 500 lb/cu yd				
0.5	45	35	22	14
1.0	35	15	− 5	−26
1.5	27	− 3	−33	−65
2.0	23	−10	−50	
Cement content at 600 lb/cu yd				
0.5	44	32	16	6
1.0	32	8	−16	−41
1.5	21	−14	−50	−89
2.0	18	−22		

* See Table 3.

10   Slab Cold-weather Concreting

In placing slab concrete which of necessity is darbied, floated, and several times troweled, much moisture as well as heat can be lost to the environment. Figure 5 indicates in graphical form the approximate relative magnitude of moisture loss through surface evaporation which at a constant air velocity of 10 mph appears dependent upon the temperature of the concrete and the dew point of the air. Thus 100°F concrete exposed to moving air having a dew point of 10°F will dry out five times as much as does 70°F concrete in laboratory air (70°F, 50% relative humidity, 50°F dew point) and having the same velocity; the drying-tendency value of the 100°F concrete at a 10°F dew point is 0.47 whereas that for the 70°F concrete at a 50°F dew point is only 0.09. The greater the difference in these temperatures (concrete high, dew point low), the greater the evaporation.

Table 2. Insulation Requirements for Concrete Slabs Placed on the Ground Concrete at 50°F, Ground at 35°F

Slab thickness, ft	Min air temp, °F			
	Thickness of commercial blanket or bat insulation*			
	0.5 in.	1.0 in.	1.5 in.	2.0 in.
	Cement content at 300 lb/cu yd			
0.333†				
0.667†				
1.0	47	42	35	29
1.5	37	19	−1	−21
	Cement content at 400 lb/cu yd			
0.333†				
0.667	50	49	47	46
1.0	42	30	17	5
1.5	29	1	−27	−56
	Cement content at 500 lb/cu yd			
0.333†				
0.667	47	42	35	30
1.0	37	19	0	−19
1.5	21	−16	−54	−92
	Cement content at 600 lb/cu yd			
0.333†				
0.667	43	34	24	14
1.0	31	7	−18	−42
1.5	13	−33	−80	−127

* See Table 3.
† Heat insulation alone will not offer sufficient protection at these temperatures and for the indicated thicknesses.

Table 3. Insulation Equivalents of Various Materials

Insulating material	Equivalent thickness,* in.
1 in. commercial blanket or bat insulation	1.000
1 in. loose-fill insulation, fibrous type	1.000
1 in. insulating board	0.758
1 in. sawdust	0.610
1 in. (nominal) lumber	0.333
1 in. dead-air space (vertical)	0.234
1 in. damp sand	0.023

* Based on insulation of blanket type having assumed conductivity of 0.25 Btu/(hr)(sq ft)(°F)(in. of thickness). Values are based on still-air conditions or where windproof cover offers protection against air flow or infiltration into insulation.

Thus, if a contractor elects to place 90°F concrete in an environment such as that of hot desert air at 100°F and 8% relative humidity, or in a cold climate at 30°F and 85% relative humidity, both having a dew point of 26°F, the rate of evaporation will be the same in both instances and at a relatively high level. In this same instance, if the temperature in the indicated cold-climate atmosphere and outside a temporary shelter is 30°F or less, the dew point of the air within the shelter is likely to be at or only slightly above this same assumed figure of 26°F no matter what the temperature is inside the shelter; the moisture content and the dew point of the air in the enclosure will be in close agreement with

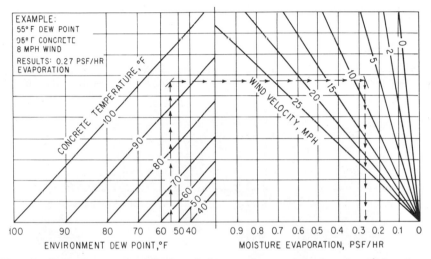

FIG. 5. Indication of evaporative drying tendency of fresh concrete based on concrete temperature and atmospheric dew point. (*From W. Lerch, Plastic Shrinkage, Proc. ACI, vol.* 53, *p.* 800, *February,* 1957.)

those of the outside air. Thus, even in extreme cold weather desiccation and dehydration of the top surface of a concrete slab will occur; therefore, the same precautions must be taken in covering the slab between operations in cold weather as are recommended for use in dry and hot desert climates (see also Table 1, Chap. 44).

11. Admixtures

Set acceleration is certainly desirable in cold-weather concreting especially if preferred "cool" concrete is employed, and particularly so in floor slabs. Concrete at 40 to 50°F temperature will exhibit a time of set double that of 70 to 75°F concrete, and thus the finishing time of cold concrete in floors or slabs is unnecessarily and often detrimentally long. If the concrete mason attempts to hurry his cold-concreting operations or to time them on the basis of using 70 to 75°F concrete, the slab surface can be ruined by overworking. Floating and troweling can be done properly only after hydration of the cement has progressed to a certain point, and only after the bleed water, if any, has either disappeared or been removed. Thus, calcium chloride or "set-acceleration admixtures" conforming to ASTM C 494, classes C and E, are employed in cold-weather slab concrete. High-early-strength portland cements contribute very little toward an early set, although these are often employed to enhance 1- or 3-day strengths in spite of their having a setting time quite similar to that of ordinary Type

I portland cements. Even here, however, the same admixtures mentioned above plus those of the class A (ASTM C494) water-reducer type contribute markedly to high early strength as well as to early set. Of course, one may use warm concrete rather than the optimum "cool" concrete in the attainment of early set and of high early strength, but this will be accomplished only with an accompanying reduction in ultimate strength.

Naturally, no calcium chloride or other chloride compound should be used in accelerating the set or strength gain of concrete used in prestressed pretensioned members. In this instance, steam curing of the prestressed members is usually employed along with nonchloride water-reducing accelerators and/or high-early-strength cements; at times, the cement factor of such concrete is also increased since the increase in heat of hydration of the extra cement contributes some to high early strength and somewhat to an earlier set of concrete.

12 Placement and Protection

A heated concrete placed in contact with ice in the bottom of a form cannot be expected to melt and displace the ice.[10] Thus it is imperative that all ice

Fig. 6. Note that the salamanders are raised up off the slab to permit air circulation underneath, and they are vented so the fresh concrete is not exposed to carbon dioxide.

be melted and that all free water be removed from the forms before concrete is placed. Formwork, subgrade, and reinforcing should preferably be at the same temperature as the concrete being placed. Further, the temperature of the concrete should not be permitted to drop below 50°F until the concrete has set and hardened sufficiently. With normal Type I portland cement, this interval of time should be at least 5 days, whereas the use of accelerators, Type III cements, or water reducers may shorten this period to 3 days depending on their efficiency. Attainment of strength for removal of formwork or shoring may require this protection to be continued much beyond the 5- and 3-day periods, this decision being often dependent upon indication of strength of field-cured specimens. In any case and aside from consideration of strength, the schedule of form stripping and removal should be such that gradual cooling of the concrete is accomplished; the drop in concrete temperature should not exceed 40°F in the first 24-hr period for structural concrete or 20°F for massive sections. Again, care must be taken that desiccation of the bared and comparatively warm concrete will not occur and be detrimental to further gain in strength of the concrete newly stripped of its protective forms.

While new concrete is resistant to freezing damage even at low strength (1,000 psi and possibly lower), water-cured new concrete is not, and thus water-saturated concrete must be protected against freezing for an even longer period. Membrane curing and polyethylene sheeting as well as insulation are recommended for the protection of newly placed cold-weather concrete.

As to jobsite enclosures, the greatest problem in connection with these involves the use of unvented heaters (Fig. 6), the comparatively large quantities of carbon dioxide that are given off by them, and the aggravated drying out of freshly placed slab concrete subjected to the direct blast or stream of air. Adverse carbonation[11] of the top surface before setting and hardening will occur if the heaters are not vented; these surfaces become soft, powdery, and without strength. If the carbon dioxide content of the air in contact with the fresh slab concrete is high, the job should be shut down regardless of the fact that all other work is to be done in conformance with good-practice recommendations otherwise necessary to get a good floor; through carbonation, the top surface will be so poor that even a chemical floor hardener will have little or no remedial benefits.

OTHER ADVERSE EXPOSURES

Concreting must sometimes be done not only in hot or cold weather but also under windy conditions, in rain or snow, or when subjected to heat (as around furnaces) or vibration (as in the repair of bridge decks or next to equipment and machinery in constant operation).

13 Wind

Wind, particularly at low humidity, causes desiccation and crusting of the top surface, meanwhile carrying sand, earth, paper, leaves, and other debris along with it. Prevention of drying out and surface crusting is discussed above under Hot-weather Concreting. Temporary coverings are recommended, being manipulated over the concrete surface and removed piecemeal only from those areas being immediately troweled and finished; the use of spray-on evaporation retardant is also of merit. Fences or vertical shields have similarly been advocated for the protection of flat-slab projects, but it so happens that many of these function much as a snow fence does; all the snow, or all the debris in the case of freshly placed concrete, is blown over the fence or wall, dropping then onto the freshly placed concrete because of the loss in velocity of the wind on the lee side. Therefore, the vertical shield might be more of a hindrance than a help.

14 Rain and Snow

Rain or snow are as detrimental as ice to the placing of concrete, especially if any accumulation of either has occurred in the forms beforehand. Conversely, the amount of water added by rain or snow during concreting (other than a downpour) might be insignificant in its effect on the quality of formed concrete which has considerable bulk and little surface area. Small amounts of moisture will, however, have a marked effect on freshly placed surface layers, especially if the excess water is not removed prior to darbying, floating, and troweling. To show the importance of rainwater, melted snow, or bucket and brush for that matter, one can readily ascertain that the top ¼-in. layer of concrete contains about 5 cu in. of water per sq ft of area. If laid out as one layer of water, it would be 0.035 in. deep, and thus the equivalent of three or four drizzles (a trace of rain is less than 0.01 in.) will double the existing water-cement ratio of that top ¼-in. layer of concrete. Therefore, there is no doubt that rain or snow can have an adverse effect on slab finishes, whose strength and durability are directly influenced by the water-cement ratio.

15   Exposure to High Temperature

The placing of concrete next to a furnace in constant operation, as in the replacement and repair of any adjacent floor slabs, calls for many of the precautions advocated for hot-weather concreting. Gloss-finished aluminum, stainless-steel, or galvanized-metal screens do much good in reflecting the heat back to the furnace and away from the work area; double sheets of metal with a 2-in. air space between them are a further improvement. Prior to placing concrete, the subgrade should be wetted; flaked ice can be spread and raked about so that it has all melted just prior to placing the coolest concrete available. Placing of the grout bond coat and of the concrete can then be carried out in the usual manner, but curing should be initiated immediately after final finishing. In this instance, wetted burlap should be employed, and it should be kept in a wet or moist condition for at least 24 hr and preferably for several days. Even after that time, an excellent grade of curing compound should be applied in a two-coat operation so that pinholes which might occur in the first coat will be closed by the second application.

16   Vibration

Vibration of freshly placed concrete by cars and trucks traveling on adjacent bridge lanes, or by adjacent machines, compressors, and the like in a plant or factory, can ruin the concrete during its initial and final set. Concrete roadways laid next to an operating and active railroad line or truck roadway are likely to be cracked before the concrete is set and hardened unless proper precautions are taken. Scheduling the placement of such concrete for off periods (midnight, 3 A.M., weekends, holidays) is often resorted to in such circumstances. Alternatively, the use of set accelerators and/or the revibration of the freshly placed concrete up to its vibration limit are second-best alternates but usually only a poor second. One point of great importance here is that all reinforcing steel should be tied down or otherwise fastened and dampened against vibration. Whereas the concrete may lie like a dead and vibration-absorbing mass until it attains some degree of strength, the reinforcing steel will tend to transmit vibrations long distances from the source of vibration and thus destroy bond of steel to concrete.

REFERENCES

1. "Concrete Manual," 7th ed., U.S. Bureau of Reclamation, Denver, Colo., 1963.
2. Klieger, Paul: Effect of Mixing and Curing Temperature on Concrete Strength, *Proc. ACI,* Vol. 54, pp. 1063–1083, June, 1958.
3. Effect of Initial Curing Temperatures on the Compressive Strength and Durability of Concrete, U.S. Bureau of Reclamation, *Rept.* C-625, Denver, Colo., July, 29, 1952.
4. Effect of Curing Condition on Compressive Strength of Concrete Test Specimens, National Ready Mixed Concrete Association *Publ.* 107, August, 1962.
5. Recommended Practice for Hot-weather Concreting, ACI 305, American Concrete Institute, Committee 305, 1972.
6. Field Experience Using Water-reducers in Ready-mixed Concrete, discussion by W. L. Cordon, STP 266, *ASTM Spec. Tech. Publ.,* 1960, p. 148.
7. Cordon, W. A., and D. Thorpe: Control of Rapid Drying of Fresh Concrete by Evaporation Control, *Proc. ACI,* Vol. 62, p. 977, July, 1965.
8. Powers, T. C.: Prevention of Frost Damage to Green Concrete, *PCA Res. Dept. Bull.* 148, 1962.
9. Recommended Practice for Cold-weather Concreting, ACI 306, American Concrete Institute, Committee 306, 1966.
10. Recommended Practice for Measuring, Mixing, Transporting, and Placing Concrete, ACI 614-72, American Concrete Institute.
11. Verbeck, G. J.: Carbonation of Hydrated Portland Cement, *PCA Res. Develop. Lab. Bull.* 87, February, 1958.

Chapter 31

LIGHTWEIGHT CONCRETE

PERRY H. PETERSEN

Lightweight concretes include those made with (1) lightweight aggregates; (2) foamed cementitious matrix, the end product usually known as cellular concrete; (3) gap-graded natural or lightweight aggregates wherein the sand fraction is usually omitted; and (4) combinations or variations of the above. The range in lightness in weight extends from a high of 120 pcf to a low of about 10 pcf. Structural lightweight concrete is generally in the range of 115 to 95 pcf. Semistructural, insulative, and/or fill concrete usually weighs from 90 to 50 pcf, whereas the least dense concretes (50 pcf and less) are employed solely as heat insulation. Some lightweight concretes made with cementitious binders other than portland cement require autoclaving or high-pressure steam curing in order to cause hydration to take place. Thus this one type is limited solely to precast shapes, slabs, planks, and the like, and these have a structural capacity sometimes greater than that indicated by the density of the concrete. However, lightweight concretes are mixed, placed, handled, and cured in a manner and by methods conventional to the concrete industry, and with little modification. Furthermore, lightweight concrete is often used as a complete and suitable substitute for normal-weight concrete even though it is light in weight. Its cost per cubic yard may be greater, but it reduces the dead load as well as contributes to fire resistance. The decreased dead load results in reduction of the size of footings, the number of piles required to support the structure, and the size of the foundation walls, columns, beams, and floor thickness; this reduction in mass of concrete (bulk as well as density) will result in savings that may far offset the increased cost of the concrete. Furthermore, the heat-insulation value of the lightweight concrete may be sufficient in itself, thereby eliminating need for additional insulative material.

1 Structural Lightweight Aggregates

This category encompasses aggregates that can be used in the making of concretes having a strength greater than 2,500 psi and a density of less than 120 pcf but nominally greater than 85 pcf. Lightweight aggregates commonly meeting these requirements and those of ASTM C 330, Standard Specifications for Lightweight Aggregates for Structural Concrete, include in general the following:

Expanded shale, clay, and slate
  Rotary-kiln process
  Sintered process
Expanded slag
Scoria
Cinders
Sintered fly ash

Compressive strengths of 6,000 psi and higher are attained with some of these aggregates, whereas others may just make the 2,500 psi minimum applicable to this class of concrete. Incidentally, the strength of lightweight-aggregate concrete has little if any relationship to density; density of concrete is more indicative of the density of the aggregate used in it, and concretes using the heavier lightweight aggregates are not necessarily the strongest. Actually, the highest concrete strengths are obtained with the rotary-kiln-processed expanded shales, clay, or slate aggregate. In the manufacture of this particular aggregate, the raw material must be of a type that bloats upon heating to fusion or vitrification. The raw clay, etc., is introduced into the upper end of a rotary kiln (Fig. 1), and in its travel down the inside of the sloping cylinder, virtually all is subjected to high heat with essentially all being brought to fusion. Therefore, the probability that underburned or unburned particles will appear in the final product is remote.

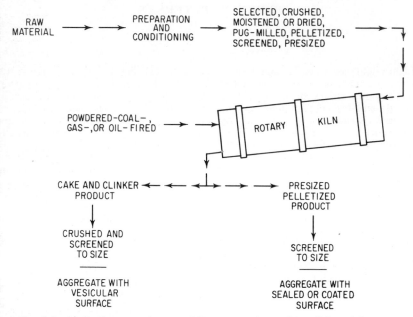

FIG. 1.   Schematic diagram of rotary-kiln process for making lightweight aggregate.

In preparing the raw clay, shale, or slate for firing, it may be mulled, mixed, or shredded before introduction into the kiln, but often it is pelletized into nuggets of predetermined size such that the subsequently fired or calcined particles conform to a prescribed gradation and maximum size. Usually, this firing of a pelletized particle results in a rounded or near-round particle having what appears to be a sealed surface enclosing a highly vesicular mass, the whole particle often having a density less than water.

In the sintering process (Fig. 2) as contrasted to the rotary-kiln process, the raw material is calcined on a fixed or moving grate, and it is therefore likely that even under perfect conditions, portions of the layered raw material will not reach the desired temperature. Thus the sintered product is normally rough-screened upon discharge from the grate, the fine material (comprising the unburned portion) being returned to the raw-material stockpile. The acceptable clinkerlike material is then crushed to size as is also done to that production of the rotary kiln not employing the pelletizing step in its operation. This crushing and sizing after firing causes the aggregate particles to be harsh in appearance, open-textured, and showing outwardly their highly vesicular structure. Absorption of water by the crushed particle is much more rapid than that exhibited by the pelletized-to-size production. However, in a properly designed lightweight-aggregate concrete mix, the workability of that with the harsh-appearing crushed aggregate is not much different from that made with the pelletized aggregate; workability and ease of finishing are too much dependent as well on other factors such as proportioning, water content, air content, and slump.

The term expanded slag could be applied to air-cooled slags which are normally somewhat porous in nature although usually not enough so that they can be classified as lightweight aggregate. Thus, in order to make slag lighter in weight, either the molten material obtained directly from a blast furnace is poured into a vat of water, or else the stream of molten slag and a stream of water are merged at a point, this combination then being spattered by mechanical means so that the slag becomes highly vesicular and porous. Although some of the expanded slag finds use in structural lightweight concrete, most of it is used in concrete block manufacture.

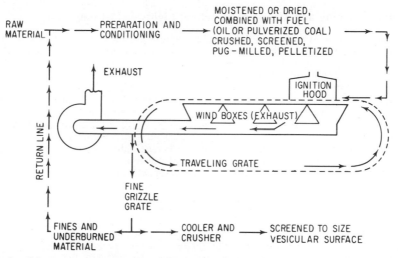

Fig. 2. Schematic diagram of sintering process for making lightweight aggregate.

Sintered-fly-ash lightweight aggregate is now being produced by pelletizing moistened raw fly ash, partially drying the pellets, and then subjecting them to sintering much the same as that described above for sintered clay, shale, or slate. Since fly ash is a waste product of power plants employing pulverized coal as fuel, this sintering of fly ash may well be an economical way of disposing of this waste product, especially when the bulk of such material crops up in the larger metropolitan areas where disposal of anything is difficult and expensive. Cinders, on the other hand, appear to be diminishing in supply because of modern advances in combustion equipment; cinders are used for limited structural purposes and primarily in the manufacture of concrete block, although in this field the volume of cinder block is small compared with that of block made with expanded shale, clay, slate, and slag. Scoria and some grades of pumice are essentially the only naturally occurring lightweight aggregates found suitable for concrete; their use is still quite limited inasmuch as they are economically available in only a few areas.

2 Availability of Structural Lightweight-aggregate Concretes

Few areas in the United States or in the world for that matter do not have locally available raw materials from which a lightweight aggregate may be made having a soundness, composition, and structure suitable for structural concrete. Also, whereas suitable clay, shale, or slate may be on hand in certain localities, normal-weight aggregates are often either lacking or in short supply there, or else they are deficient in some respect such as gradation, purity, and freedom from organics, freeze/thaw resistance, and the like. Of course, fuel to fire and calcine the raw material in the processing of lightweight aggregate must also be readily available and at economical cost. Even so, the production and use of structural lightweight aggregate is progressing rapidly, especially since it is now possible to make an aggregate from almost anything and yet design it to have certain attributes. These attributes might eventually be tailored so that certain required or desirable properties will be realized in concretes in which the aggregates are used. The quality of the aggregate itself certainly has an influence on thermal conductivity, compressive strength, flexural strength, bond strength, free/thaw resistance, sulfate resistance, modulus of elasticity, shrinkage and volume stability, creep properties, and the like.

3   Proportioning and Mixing of Structural Lightweight Concrete

Proportioning and mixing of lightweight concrete are discussed in detail in Chap. 11, but it might be mentioned here that properly proportioned and mixed lightweight-aggregate concrete can be handled by means conventionally employed with normal-weight-aggregate concrete. However, since it is light in weight, its slump for a workability equal to that of comparable normal-weight concrete is usually fixed at but one-half to two-thirds that of the normal-weight concrete. Whereas a 6-in. slump may be used when sand-and-gravel or crushed-rock aggregates are employed in a particular placement of concrete, 3- to 4-in. slump is used with lightweight-aggregate concrete. As to entrained air, normal-weight concrete might employ up to 4% for optimum workability, whereas lightweight-aggregate concrete would employ up to 6 or 8% dependent upon the aggregate texture and its vesicular structure. The harsher the aggregate, the greater will be the required air content. The use of a water reducer or water-reducer retarder (ASTM C 494, types A and D, respectively) is highly recommended here to effect a further reduction in water, to increase plasticity at a given slump, and to increase workability and ease of finishing.

Because of high absorption, lightweight aggregate is usually wetted before batching; or else it is introduced into the mixer with an appreciable amount of mix water in order that the absorption may in part be satisfied before the cement is introduced into the mix. Pumping of lightweight aggregate concrete appears to be greatly facilitated if the aggregate is saturated beforehand during manufacture through the use of a "hydrothermal" process or by a "vacuum" process;[1] high absorption is attained by controlling the aggregate's rate of cooling from its initial high temperature (2000°F or 1100°C) down to 300 to 400°F (150 to 230°C), whereupon it is sprayed with water and also inundated until it is relatively cool. In the vacuum process, the stocked aggregate is transferred to a tank and subjected to vacuum, whereupon water is introduced in copious amounts, thus causing the aggregate to be completely saturated. The technician or customer might benefit from suggestions of the producer of the aggregate or the local ready-mix supplier familiar with the particular lightweight aggregate being used.[2] More attention is given in the mix design of lightweight concrete to cement content, slump, and yield than is given to the water-cement ratio.

Normally, the design of a lightweight-aggregate concrete mix[3] involves the making of trial batches using a range of three cement factors rather than a range of three water-cement ratios normally associated with the design of normal-weight-aggregate concrete. Proportioning of lightweight fines and lightweight coarse material by weight might seem unnatural to the newcomer in the concrete industry. The end result, however, is that the volume of the fines will amount to about 50% of the total aggregate. Approximately 32 cu ft of lightweight aggregate ((16 fine, 16 coarse) will be required to make 1 cu yd of concrete. Since the fines have a specific gravity greater than that of the coarse material, the batch-weight ratios are much unbalanced as compared with those for normal-weight concrete; the fines in a batch will weigh about 50% more than the coarse aggregate. This unbalance, or rather this revised balance as applied to lightweight aggregate, is similarly influenced when one begins to introduce into the mix a natural sand as a partial or whole substitute for lightweight fines. The substitution of natural sand for lightweight fines is often resorted to, although it is always and directly accompanied by an increase in dead weight of the lightweight-aggregate concrete and by a decrease in its heat-insulation value. The cost of shipping lightweight fines to some point quite distant from the aggregate plant might be uneconomical. However, certain benefits may be realized by the use of normal-weight fines of good quality with regard to strength, increased modulus of elasticity, and decreased volume change; these attributes might well be worth the accompanying increase in weight of the concrete.

4 Placing and Curing of Structural Lightweight-aggregate Concrete

As mentioned above, the slump of lightweight-aggregate concrete should be about one-half that recommended for normal-weight-aggregate concrete in any particular application. The use of entrained air and of lowest optimum water content is essential in obtaining assurance that these lightweight concretes are of a required workability for proper placing and finishing, especially those made with crushed, angular, and highly vesicular aggregates; bleeding, segregation, and undesirable floating to the surface of the larger less dense aggregate particles are each thereby reduced to a minimum. Vibration of lightweight-aggregate concrete should be done with extreme care so as to avoid segregation and consequent separation of aggregates into layers of variable density. As to the finishing of slabs or flat work, a jitterbug[4] consisting of a flat coarse wire mesh is often used to depress any large coarse aggregate down just below the finish elevation of the slab at a time when the concrete will hold the particle in that position. With regard to curing, the same methods and timing apply to lightweight concretes as are set forth for normal-weight concretes. While one should not condone any reduction in prescribed curing, it may be that a greater factor of safety in this regard is inherent in lightweight-aggregate concrete because it contains much more absorbed moisture than does normal-weight aggregate; since this moisture requires more time to evaporate, curing of the interior mass of concrete should be better. Conversely, this reservoir of moisture within the concrete is not of itself beneficial in the curing of surfaces like floors unless, of course, a curing compound is applied soon after they are finished.

5 Testing and Control of Structural Lightweight-aggregate Concrete

Whereas control of slump, air content, and strength may be sufficient for normal-weight-aggregate concrete, this is not true for lightweight-aggregate concrete. A change in the density of lightweight aggregate may have a marked effect on the inherent strength of the aggregate particle and thus on concrete made with it. Rather than wait until 28-day strengths are obtained, one should check the unit weight of the fresh concrete as well as slump and air content. Uniformity of unit weight within a 2-pcf range coupled with close adherence to mix-established limits for air content and slump will give better assurance that the density and quality of the aggregate and of the concrete are what they were purported to be. In making an air-content determination of lightweight-aggregate concrete, preference is always given to the volumetric method (ASTM C 173), as described in Chap. 12, wherein the concrete is placed in a bowl, covered, and then carefully topped with water to a prescribed level. After the water and concrete are physically mixed by rolling the container on its side, the entrained air released from the concrete is measured as percent of entrained air simply by ascertaining the drop in the level of water. The pressure air meter (ASTM C 231) is not generally used with lightweight-aggregate concrete inasmuch as it does not differentiate between entrained air and any entrapped air or voids in the lightweight-aggregate particles. This entrapped air in the aggregate voids has no bearing on the quality of the cement paste and mortar fraction of the concrete, and thus there is no merit in measuring it. At times, it may be feasible to use the pressure air meter if readings with it have been correlated with those of the volumetric meter beforehand, and if the moisture content of the aggregate in the freshly mixed concrete is known to be reasonably constant from test to test.

OTHER LIGHTWEIGHT CONCRETES

6   Heat-insulation Concretes

These lightweight concretes of a density of 50 pcf or less and a compressive strength of less than 500 psi usually employ expanded perlite, exfoliated vermiculite, or a natural pumice as aggregate. Cellular or foamed concretes are also included in this category along with gap-graded or popcorn concretes, which in turn contain no fine aggregate. Heat-insulation concretes often make use of air contents of 20 to 30% or higher. They are mixed, handled, and placed by conventional methods except that overmixing must be avoided for fear of breaking down and pulverizing the rather fragile aggregate. Cellular concretes, on the other hand, may be made in any one of three ways. One method involves the beating of a cement slurry or of a cement-sand slurry into a highly foamed mixture in which foaming and/or air-stabilizing admixtures are used. A second method employs aluminum powder or other fine-sized alkali-reactive material which when incorporated in a cement slurry causes a gas to be generated in the form of small discrete bubbles; this foaming can be timed to take place during mixing but some operators rely on its taking place in the forms, rising up and filling the space much as bread does in the oven. The third, like the second, also employs a normal cement slurry except that, in this instance, a preformed and stable soaplike foam similar to that produced by foam fire extinguishers is folded and mixed with the slurry to make a foamed concrete.

One deviation from normal concreting is encountered here; i.e., in the making of precast cellular concrete panels or shapes which can be autoclaved or high-pressure steam-cured, little or no portland cement need be used if the foamed slurry consists of ingredients such as free lime and silica which react favorably under such curing to form cementitious compounds similar to those obtained by the hydration of portland cement. In the control and testing of any of these heat-insulation concretes, most importance is given to criteria covering the two properties: dry strength and weight. Here, however, the concrete often need only be strong enough to support its own weight, although some is used as roofing planks.

7   Insulation and/or Fill Concrete

This category covers in general the gap between structural lightweight concrete and the heat-insulation concretes. With mediocre heat-insulative properties, this concrete is usually called upon to carry light loads or to serve as roof decking and as an excellent base for built-up roofing. With a weight range of 85 to 50 pcf, these concretes may incorporate structural lightweight aggregates as well as aggregates having less strength. They may also include cellular concretes and especially those using a concrete sand (lightweight or normal weight) in order to conserve on cement. Those exhibiting strengths in the range of 500 to 2,000 psi[5] may well be employed particularly in one-story structures and dwellings because the stresses in even a load-bearing wall in such applications are quite low, on the order of 10 to 15 psi.

8 Nailing Concrete

Lightweight concretes in general incorporate an aggregate which will be deformed or crushed by the penetration of a nail so that the broken material will bind around the nail and make withdrawal difficult. Structural lightweight aggregates usually exhibit greater merit in this regard than do normal-weight aggregates. Natural-aggregate, crushed-stone, or sand-and-gravel concrete tends to fracture and break or split so that the nail may be easily withdrawn. Perlite and vermiculite heat-insulative concretes can be readily nailed, although their holding power is usually in direct proportion to their low compressive strength. Concrete made with sawdust[6] or with wood and other fibers as described below has evidenced good holding power. The Bureau of Reclamation finds that good nailing concrete is made by mixing equal parts by volume of portland cement, sand, and a pine sawdust with sufficient water to give a slump of 1 to 2 in. The proportion of sawdust is adjusted to result in the type of concrete suited for the purpose in hand, especially if some other source of sawdust is used (pine, fir, hickory, oak, birch, or cedar).

9 Fiber Concrete

In the past fiber concretes incorporated fibers or chips made from wood or other plant life, these aggregates being mixed in a dry no-slump concrete which was compacted and cured so as to exhibit special properties. In general such concretes have exhibited appreciable tensile strength, especially when tamped into the formwork. Recent findings indicate that the use of steel[6] synthetic, or glass fibers, filaments, or short fine wires is very beneficial in increasing the tensile strength of conventionally placed concrete much above that normally associated with portland-cement concrete and mortars. Under the process name of Wirand,[7] the Battelle Development Corporation shows that the addition to concrete of up to 4% steel fibers (by volume of concrete) results in appreciably higher tensile and flexural strengths, greater extensibility, and a beneficial change in cracking. The size and the amount of coarse aggregate in such concrete call for special consideration inasmuch as the beneficial aspects relate to mortar or to the mortar fraction of whole concrete; thinner sections of concrete can be employed.

REFERENCES

1. Reilly, W. E.: Hydrothermal and Vacuum Saturated Lightweight Aggregate for Pumped Structural Concrete, *J. ACI*, vol. 69, pp. 428–432, July, 1972.
2. Expanded Shale, Clay and Slate Institute, Information Sheet No. 3, Mix Design of Structural Lightweight Concrete.
3. Recommended Practice for Selecting Proportions for Structural Lightweight Concrete, ACI 211.2-69, American Concrete Institute.
4. Expanded Shale, Clay and Slate Institute, Information Sheet No. 7, Lightweight Concrete Floor Finishing.
5. Design Data for Some Reinforced Lightweight Aggregate Concretes, Housing Research Paper 26, Housing and Home Finance Agency, October, 1953.
6. Romualdi, J. P., and J. A. Mandel: Tensile Strength of Concrete Affected by Uniformly Distributed and Closely Spaced Short Lengths of Wire Reinforcement, *Proc. ACI*, vol. 61, p. 657, June, 1964.
7. Lankard, D. R.: "Steel Fibres in Mortar and Concrete," *Composites,* March, 1972. (Reprints available from Battelle Institute, 505 King Ave., Columbus, Ohio 43201.)

Chapter 32

HEAVY CONCRETE

PERRY H. PETERSEN

Heavy concrete is invariably more costly per pound of weight than comparable normal-weight concrete with a density of about 150 pcf, because more than normal care must be exercised in selecting an aggregate heavy enough and of a quality suitable for the purpose in hand, in mining or quarrying the material, in crushing and grading it, and in mixing it into a concrete mix as well as in placing and finishing it. Normal-weight aggregate is always obtained from a pit or quarry quite close to the ready-mix plant or jobsite, and thus aggregate transportation costs are low compared with those for heavy materials, which are usually available only quite far from the jobsite. Also, because commercially available crushing and sizing equipment is geared to aggregates of normal weight, its rate of wear is increased by heavy aggregate, and the volume handled may be inversely proportional to the specific gravities of the aggregates. Also, whereas a concrete mixer can handle 5 cu yd of concrete totaling about 20,000 lb, it is likely that 2 to 3 cu yd of heavy concrete would constitute an overload. Thus normal-weight concrete is often used in applications where mass or weight is of prime consideration even though the volume taken up by it is much greater than it would be with heavy concrete. Nevertheless, a heavier weight of concrete may be necessary or desirable in the design of many structures or facilities for radiation shielding, in counterweights, or where an increased dead weight is required and especially where space is at a premium. The thickness of a wall or floor can be cut in half merely by doubling the density of the concrete used in its fabrication. The increase in use of heavy concrete over the past 10 years is certainly indicative of its need in the field of nuclear energy and in the shielding problems that have accompanied advances in nuclear technology. The technician now finds it necessary to manipulate the various components within a nuclear furnace or chamber, and thus bulk is rapidly becoming a most objectionable attribute of concrete.

1 Nuclear or Radiation Shielding

Radiation shielding is provided and is necessary primarily for the protection of personnel operating in and about facilities which emit nuclear particles (neutron, proton, alpha, and beta) and x-rays or gamma rays. These particles or rays in general are stopped, deflected, transformed, or attenuated merely by mass, i.e., by the weight of concrete (pounds per square foot) which lies between the source of energy and the persons concerned. Neutrons, however, require in addition to mass particular substances such as appreciable amounts of iron, hydrogen, and boron or cadmium. In some instances, the hydrogen contained in moisture present in the concrete is sufficient for neutron attenuation. It must be cautioned, however, that this moisture may be dissipated if and when there is a subsequent rise in the temperature of the concrete in the shield. This moisture may be in the

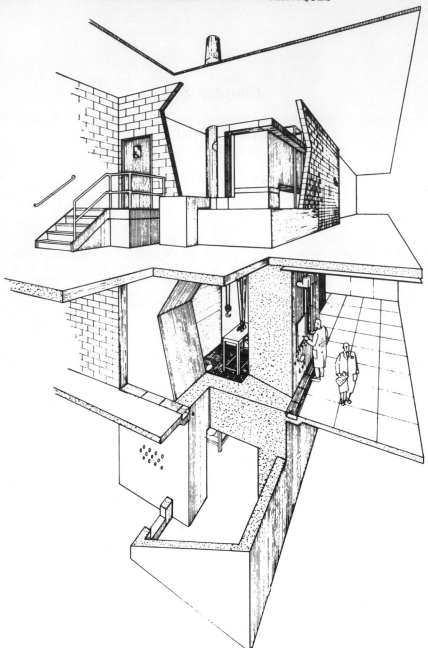

FIG. 1.   Cutaway drawing of hot cell shows all three levels that were surrounded by 66-in.-thick concrete walls. Isotope-handling room is located on second-floor cell, has 4-ft-thick floor and ceiling. The concrete was placed in three separate lifts of one story each, the separate sections being joined by 3- by 10-in. keys at each floor line. (*Courtesy of Master Builders Company*.)

form of absorbed water, water of hydration chemically bound by the portland cement, or water of combination available in certain aggregates (hydrous ores, limonite, and serpentine). Boron or cadmium, on the other hand, must be intentionally introduced as an aggregate or as an admixture.

A typical hot cell is shown in Fig. 1.

Strength of shielding concrete is much dependent upon the aggregate quality and gradation as well as on the water-cement ratio. Strengths as high as 10,000 psi have been obtained, although many shielding concretes have been made that are just about strong enough to hold their shape. Strengths of 1,000 to 2,000 psi are readily attained for use in massive slabs on grade, walls, or deep wells where strength is secondary to shielding capacity. Some choice of aggregate may be necessary where strengths of 4,000 psi or higher are required. Furthermore, volume change and freedom from cracking are also items of prime concern, especially where structural members form part or all of the shield. Thus it is concluded that the manufacture of shielding concrete is nowhere so simple as for normal-weight aggregate concrete. First, one must know something about the source, nature, and intensity of the nuclear particles and rays which are to be stopped or attenuated at least to some acceptable limit. Secondly, the aforementioned choice must be made as to the aggregate and possibly to one or more admixtures which will result in the attainment of the concrete necessary to do the job. Normal Type I cement is regularly used as the binder, although in some instances Types II, IV, or V are chosen for their particular attributes. Likewise, low-alkali cement (less than 0.60 total alkalies as Na_2O) is sometimes required because the chosen aggregate is susceptible to alkali reactivity. In this same regard, an alkali-reactive aggregate should not be employed where the resultant concrete will be exposed to liquors or other substances high in alkali regardless of the use of low-alkali cement in such concrete.

2 Counterweight Concrete

Heavy concrete is often employed in the fabrication of counterweights or simply as a means to increase the dead weight of some facility economically and yet without the bulk volume which normal-weight-aggregate concrete would occupy. The aggregates used for these purposes can be the same as are employed in shielding concrete except that the exposure of counterweight concrete may be even more critical from a different viewpoint. Thus additional stipulations may be required as to the quality of the concrete and the aggregate. If freezing and thawing conditions are to be encountered, the concrete should be made of a durable aggregate and should contain the usual recommended amount of entrained air commensurate with the maximum size of aggregate. Even though air entrainment causes a reduction in dead weight, it usually has less influence on weight than would a change in the aggregates, and it is necessary in concrete which is to be subject to freezing and thawing.

If the concrete is to be exposed to salt water, deicing salts, or other chloride-ion exposures, care must be taken to avoid rusting and subsequent detrimental expansion of the concrete made with steel punchings and the like as aggregate. In general, the use of high cement factors, low water-cement ratios, and 3 to 4% entrained air is conducive to the making of an impermeable cement paste or mortar which will satisfactorily encase the iron aggregate in concrete subjected to almost any exposure. No chlorides should be used. Submerged pipelines for gas, air, and even certain liquids are often weighted by attaching to them concrete saddles or by encasing them in concrete, heavy or normal weight. Some heavy-industrial wear-resistant floor-surfacing mortars or concretes have densities of up to 240 pcf inasmuch as they employ a specially graded and prepared iron aggregate as one of their components; these aggregates are not used because of their influence on weight but rather because of the accompanying high resistance to abrasion of the iron-aggregate floor topping.

3 Heavy Aggregates

In this category, and primarily because shielding is given prime consideration here, one must include in the range of aggregates the heaviest available (iron or steel punchings, etc.) and some that are even lighter in weight than is normal-weight aggregate, namely, the hydrous ores (bauxite) and the borates, which are sometimes employed as additives because of their special attributes. A list of those commonly used is given in Table 1. The ranges in specific gravity indicated

Table 1. Aggregates in Common Use in Nuclear-shielding Concrete

Concrete aggregates	Specific gravity	Concrete density, pcf
Steel punchings, shot, and sand........	7.5–7.8	340–380
Ferrophosphorus.....................	5.8–6.5	280–300
Magnetite, ilmenite..................	4.2–4.8	210–250
Barite.............................	4.0–4.5	210–230
Limonite...........................	3.4–3.8	180–200
Crushed stone......................	2.6–2.9	150–165
Sand and gravel.....................	2.5–2.7	140–150
Serpentine.........................	2.4–2.6	*
Boron frit..........................	2.4–2.6	*
Borates, colemanite.................	2.0–2.4	*
Bauxite............................	1.8–2.2	*

* Additive materials.

for each are much dependent upon the purity of the ore or that of the processed or manufactured material. Ferrophosphorus is a slag and thus is subject to variations in density as are the iron or ferrous aggregates. Serpentine and limonite are both hydrous ores and contain appreciable amounts of combined water over and above any absorbed moisture; this characteristic is of special merit in that this combined water will not be driven off until the temperature of the concretes in which they are used attains a level somewhat higher than normal.

Magnetite and ilmenite are the most commonly used aggregates in the making of shielding concrete, with barite, ferrophosphorus, and steel aggregate being used in much of the balance. Combinations of these aggregates are employed such as magnetite coarse aggregate and steel-shot fine aggregate. The borates, bauxite, and boron frits are usually introduced into the mix so as to better provide for neutron attenuation, especially when the concrete temperature in service is expected to be high and when most of the water will be driven out. The borates (colemanite included) are quite fragile and are also somewhat soluble in concrete; the primary objection to this is the marked and uncontrollable retardation of set of the concrete which often accompanies their use in portland-cement concretes. The tendency is therefore to use the boron frits in lieu of the commoner borates.

4 Proportioning and Mix Designs

Essentially the same principles hold for proportioning of heavy concrete as are recommended for normal-weight concrete as described in Chap. 11. The recommendations of ACI 211.1 can be followed with few modifications. It may be that more trial mixtures will be necessary to arrive at optimum amounts of coarse and fine aggregate inasmuch as the heavier and harsher aggregates behave somewhat differently from the normal-weight aggregates. Also, with regard to workability, the slump should never be greater than 2 to 3 in., the mixture should be somewhat on the harsh side, and the concrete is best consolidated by vibration. The use of a minimum water content will aid measurably in reducing bleeding and segregation; water pockets adjacent to flat aggregate surfaces subsequently will serve as undesirable windows or openings through which unwanted nuclear rays or particles pass unhindered. Since the aggregates and the mortar have specific gravities that differ considerably from each other, the denser aggregate tends to settle to the bottom of the concrete. The use of a water reducer or water-reducer retarder is recommended primarily because the concrete is made more workable and homogeneous through its use, and with less water. Air entrainment up to a maximum of 2 to 3% aids in reducing bleeding and in the attainment of a more homogeneous concrete; for these reasons, it is used even in spite of its adverse effect on density. Essentially all the test methods stipulated for the control and evaluation of normal-weight concrete are likewise applicable to heavy concrete. Field inspection should include slump, air content, yield, and the making and curing of cylinders for strength tests. Tests made prior to concreting should include the usual ones on cement and particularly those pertaining to the density, grading, abrasion resistance, and potential alkali reactivity of the aggregates. Admixtures as well as aggregate-like additives should also conform to established ASTM standards and to any particular project-specification requirements.

5 Mixing and Placing

As with proportioning, heavy concrete may be mixed and placed by conventional means as recommended in ACI 211.1 for normal-weight-aggregate concrete. One must remember, however, that this concrete is denser and that the number of cubic yards that a mixer will handle is cut in half or some other amount inversely proportional to the density. This caution in handling applies also to the concrete chute supports, the capacity of cranes, the size of concrete buckets, the strength of the forms, as well as other points of similar concern.

Heavy concreting is also accomplished by the grout-intrusion method or by puddling. In the grout-intrusion method, coarse aggregate is placed and consolidated so as to fill the forms completely; the grout is then pumped in beginning at the bottom, and in amount, manner, and sequence sufficient to fill all the voids in the mass of aggregate (see Chap. 38). To make this possible, the coarse aggregate ($1\frac{1}{2}$- or 2-in. maximum size) is devoid of the finer-size fractions; for instance, all aggregate smaller than $\frac{1}{2}$ or $\frac{3}{4}$ in. is eliminated to make it ready for use. Similarly, the sand for the grout is screened over a No. 16 sieve, all the oversize material being discarded. Thus the No. 16 sieve-size sand grout is capable of being readily pumped into and through the voids of the processed coarse aggregate already in place. To make this intrusion of grout as complete and as thorough as possible, admixtures are invariably used, and these may include plasticizers, pozzolans, air-entraining agents, water-reducer retarders, and possibly aluminum powder. The aluminum powder generates bubbles of hydrogen gas in the alkaline cement-paste environment, the resultant gas pressure being enough to assure greater contact and bond of the grout and the aggregate. This type of concrete has much merit in its application to the field of nuclear shielding, primarily because the aggregate concentration is uniform throughout the mass; settling of coarse aggregate is hardly possible unless the coarse aggregate is lifted out of place by the pressure of the intruding grout.

Puddling as contrasted to the other methods is accomplished mainly by placing a fluid heavy grout in the forms into which coarse heavy aggregate is puddled, pushed, or embedded. This is generally done in stages of 1 ft in height. Puddling is the less favored process because there is little assurance that the amount of coarse heavyweight-aggregate concrete is as great as it should or could be. The key to obtaining good shielding concrete by puddling lies in the insistence of the operator in adding and puddling the coarse aggregate until positive resistance is met and the grout will not take more aggregate.

Section 9

ADVANCED BUILDING-CONSTRUCTION SYSTEMS

Chapter 33

ON-SITE PRECASTING AND TILT-UP

CHARLES J. PANKOW

Individuality of a building results from a design developed by the architect, expressing the concepts of both the architect and the owner and incorporating necessary structural provisions evolved by the structural engineer. Careful preplanning and cooperation of the architect and builder can permit the contractor to explore the most economical and efficient methods of achieving the structural and aesthetic values sought by the architect.

BASIC PRECEPTS

For example, some types of building units lend themselves well to precasting and prestressing on the site. Where there are many identical beams, panels, or similar units, the builder should investigate the possible economies resulting from on-site precasting. Beams and girders can be precast and tensioned, and wall panels can be cast in a flat position, using floor-construction techniques, and then tilted into position. Many other units can be cast on the site and lifted into position by crane or derrick.

In many cases, slight changes in structural design of cast-in-place members will permit precasting operations for the contractor. The economy of such a change, which is usually initiated by the contractor, is generally beneficial. However, each case must be considered individually in making a comparative cost analysis. Items for consideration on such a change, other than the costs of the members themselves, may include either more or less reinforcing steel, shoring and reshoring changes, cast-in-place joinery connections, sandblasting of beam tops, and tensioning procedures, as well as design features, erection operations, and savings or added costs in deck-forming methods that may be adaptable to change. With respect to this last item, should deck-framing beams be changed from cast-in-place to precast, the cost of deck forming may be significantly reduced by using members to span across the precast beams. Such a forming system can completely eliminate shoring requirements other than those for the beams themselves.

A typical situation might be a multistory office building, preferably at a site with plenty of casting area adjacent. Ideal circumstances would permit the construction plant designer to incorporate numbers, weight, reach, distance, and other factors into the automating considerations. For example, a traveling tower crane of a certain capacity may be committed to a design that will employ the ultimate capability of that equipment. Such preplanning is becoming more necessary as costs of field production continue to go up.

A 9- to 13-story building will require as many as four, or even more, form elements in order to accomplish the proper cycle of production. The number of form elements, however, is directly related to what other items are being precast. For example, the decision to precast wall units might demand similar consideration for structural units, or vice versa. Therefore, either selecting equipment is a function of what the designed units are, or design of the units depends to some extent upon the equipment selected or available. It can thus be seen that design and methods coordination are a large part of preplanning and can be an important factor in the final economic results.

Because of the many variations in shape, size, finish, and other properties of building units that can be precast, this discussion will dwell on generalities as they relate to a typical set of circumstances, rather than attempt to confine itself to any one form or unit.

In high-rise building construction, concrete curtain walls are frequently nonstructural. However, even though they may be structural, there should be no special problems in construction or erection because of this. The physical problems of making and handling precast units, on the other hand, are a function of the configuration. Long, narrow, and deep sections are literally impossible to cast in place; but such members can be built, utilizing precast concrete forming elements, thereby giving the architect or designer the opportunity to incorporate very delicate lines on exterior architectural wall surfaces.

It has been the author's experience that architectural panels generally require removable form elements or parts, and the positive anchorage and seating of these elements are best obtained with either steel or fiber-glass forms. Being heavily customized, these forms can be expensive as compared with contractor-built wood or concrete forms. However, contractor-built fiber-glass forms, though more costly than wood or concrete forms, should be considered because of better use factors, ease of duplicating additional form units, and configuration advantages, which are more limited in wood and concrete. A concrete or fiber-glass form built on the site by the contractor will probably be ready for use before a steel form can be ordered and delivered—a factor that must be considered if speed of construction is important (as it usually is).

Cost, timing, and the technicalities of efficient use and reuse must all be considered. Whatever the form material, it is essential that workmanship be of the highest quality. Watertightness is essential, as are accuracy of lines affecting draft and alignment. One must remember that the finished units will be adjacent to each other on the building, both horizontally and vertically, thus requiring tolerances of a precise nature in dimensions and, for exposed architectural units, in finish as well.

It is basically important that equipment and duplication of operations should be employed to the fullest extent. Therefore, maximum capacity, reach, line speed, etc., will serve to determine the type of equipment. Repetitive production-line methods will simplify educational and supervisory activities. Whether the forms are of steel, wood, concrete, or some other material is not relevant at this time.

Most builders are, of course, constantly reviewing new work opportunities, some of which represent "fixed-plan" competitive-bid situations that generally do not lend themselves to the type of work discussed here. Other occasions will convey opportunities for the builder to express himself with his experience and knowledge. The willingness on the part of the owner and architect, and the contractor's ability to contribute to owner and designer matters of cost control, will be truly represented when the factors discussed in this chapter can be realistically incorporated into the work during the preparation of final design and drawings.

Of fundamental and everlasting importance is the experience of the contractor. The architect will probably not permit his job to be a laboratory for experimenting with new and untried methods, nor is he likely to make it a training ground for an inexperienced builder. Conferences between the owner, architect, and contractor during the preliminary design stages will enable each to contribute his share to the successful and economical completion of the project.

To summarize then, this discussion is based on three fundamental concepts: first, that the daring and imagination of both the designer and builder constitute a dynamic mutual interest that can be used effectively in developing the final design and execution of the plans for the building; second, that the builder has the necessary experience to enable him to engage in the fabrication and erection of precast and sculptured concrete; and third, that the designer feels that a limited budget makes it desirable to utilize the unique experience of the builder to take advantage of the economies of the on-site precasting.

1   Fiber-Glass Molds

Where the job schedule permits time to make fiber-glass molds for production of units, it is generally economical to do so, especially when many forms are needed. Reinforced fiber-glass is very durable, and forms made from it have an almost unlimited life. Fiber-glass molds resist abrasion well and are not affected by weather. When forms of wood are not feasible because of unusual architectural configurations, curved surfaces, or other special problems, steel forms and fiber-glass forms should be carefully considered. The use of steel forms is limited, however, because of the physical properties of steel itself. Architectural precast units that involve small-radius curves and very few flat surfaces may be extremely expensive to form in steel, if not virtually impossible. In such cases, the use of reinforced fiber-glass molds, as described in the following paragraphs and in Fig. 1, is recommended.

The master pattern must be dimensionally correct in every respect, as the production units will be exact copies. In very unusual architectural units where the design contains curvilinear contours, the master pattern will have to be made in sculptured plaster or a similar medium. Otherwise, straight-line contour patterns can easily be fabricated from prefinished plywoods. Extreme care must be exercised to ensure that the proper amounts of draft or taper are present on this master pattern, because in some cases the pattern will be used only once and will be destroyed during the process of stripping the first negative mold. If there is no draft, or very little draft, present for deep-sectioned units, the production units will not strip from the molds. Draft may be hardly measurable; but in deep sections it must be proportionately increased in order to overcome the increased friction and suction of the deeper form. In sections 24 in. deep, a draft of ½ in. has been satisfactory in the use of both hard-rock and lightweight concrete. The finished surfaces of the master pattern should have a smooth and high-gloss finish, for the fiber-glass overlay will conform to these surfaces.

For illustration only, we have chosen a 12 $\times$ 12-in. column showing a mold in simplest form. The master pattern as shown in step 1, Fig. 1, contains straight-line contours; so ½-in. preglazed plywood has been selected as the material. The succeeding procedure will be to apply the releasing agents. To assure a complete separation, a minimum of two coats of mold wax should be applied with one sprayed-on coat of polyvinyl alcohol. At this time it is important to mention that the correct resin materials be selected for this type of open-mold fabrication. Some resin manufacturers are designing their products to satisfy jobsite needs. Depending on climatic conditions, a choice should be made between winter or summer resins and for the proper degree of hardness. The hard resins are normally employed as outer skins for concrete resistance; the softer materials are more economical and can be used as back-up laminations. Figure 1 illustrates the procedure of one lamination. This procedure can be repeated until the desired thickness is reached. Assuming that for this particular mold we have elected to apply three laminations of glass or ⅜-in. thickness, then the wood reinforcing material, as shown in step 2, will be installed after the second lamination. The last lamination, producing the required thickness of glass, will adhere to the wood, completely encasing it with glass. The pre-assembled wood collar, as shown in step 2, should be installed while the second lamination is still tacky, causing an adherence there also. It is recommended that the wood joints be glassed for rigidity, as shown in step 3.

Air orifices are of great importance for stripping purposes. In some cases, air will be sufficient to create the initial separation between concrete and glass. The crane will then make the final liftoff to the stacking area. In other cases, where unusually large and deep sections are being produced, a combination of air and hydraulic jacks is used for the initial liftoff. For large units, strongback systems are used. Steel I-beams as strongback members can be bolted to inserts cast in the units so that hydraulic hand jacks can be placed between the beam and

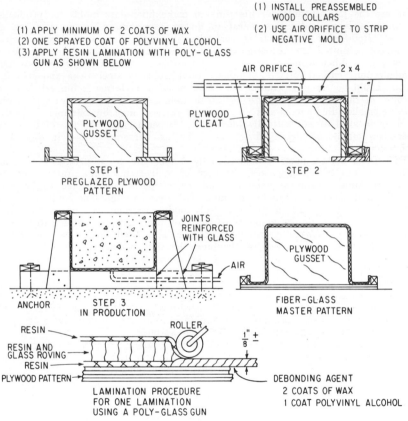

Fig. 1. Sequence for making fiber-glass mold for repeated production of a simple 12×12-in. column.

the mold itself. For a positive stripping system, jacks must be placed to exert force directly between the unit and the mold. If the jacks bear against the ground or any place other than the mold, it is likely that they will pick up the unit and the mold together.

Depending on the complexity of the pattern and the number of negative molds required, it is sometimes more economical to produce a fiber-glass master pattern, as shown in Fig. 1, and then use this pattern to make negative molds. This is especially desirable when repeated stripping of negative molds causes extensive damage to the wood pattern. The procedure is to follow through with steps 1 and 2. Using the first negative mold, as shown in step 3, two laminations should then be applied. Plywood cross ribbing is added for rigidity. This type of pattern will give an infinite number of negative molds.

For equipment, it is necessary to have an air compressor capable of delivering 100 cfm. The poly-glass gun is a simple and reliable piece of equipment that is easy to maintain and to use. However, its workings should be understood

FIG. 2. One type of fiber-glass mold. FIG. 3. Precast panels cast from fiber-glass mold.

FIG. 4. Fiber-glass mold for precast wall units.

by all operators who are using it. It is also extremely important that the manufacturer's preventive maintenance program be carefully followed.

The fiber-glass mold shown in Fig. 2 was used for forming the two-story-high panels shown in Fig. 3. Ten of these molds were moved to each building site for casting the panels. Figure 4 shows a fiber-glass mold used in an off-site casting yard in the Midwest. The panels cast from molds of this type are illustrated in Fig. 5. Another use of fiber-glass molds is shown in Figs. 3 and 4 of Chap. 14.

FIG. 5. Exterior of office building enclosed with wall units cast in the molds shown in Fig. 4.

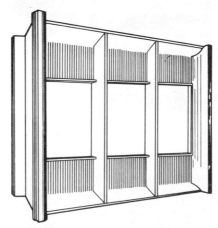

FIG. 6. Type of precast wall panel made with steel form. Approximate weight 8,200 lb. (See also Figs. 7 through 9.)

2   Steel Forms

As mentioned in Art. 1, steel forms are most economical when many precast units can be cast from very few forms. Also economically, their best use is for architectural units involving curved surfaces and unusual configurations. Another advantage is in having a rigid steel member for screeding the concrete, which results in exact alignment of interior edges of adjacent panels, eliminating the need for possible grinding, etc., to present a smooth interior finish at the joints.

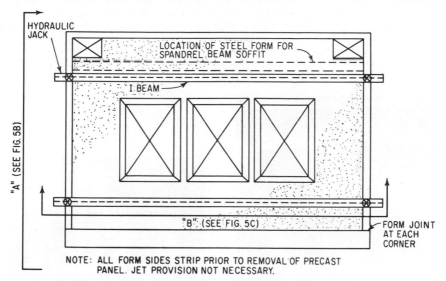

NOTE: ALL FORM SIDES STRIP PRIOR TO REMOVAL OF PRECAST PANEL. JET PROVISION NOT NECESSARY.

Fɪɢ. 7A.   Precast wall panel in form ready for stripping.

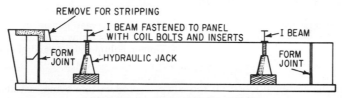

NOTE: DROP ALL SIDES PRIOR TO JACKING.

VIEW "A"

Fɪɢ. 7B.   Precast-wall-panel form.

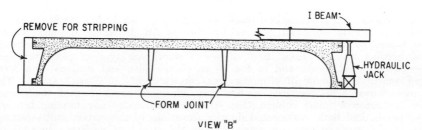

VIEW "B"

Fɪɢ. 7C.   Wall-panel form containing precast panel.

In the construction of the forms themselves, it may be necessary to provide holes and fittings for air jets to aid in stripping the forms from the concrete castings, especially if the precast unit sections are very deep and thin or if parts of the form are surrounded by concrete on four sides as an inverted tub to form a recess in the unit. The placing of such jets, of course, depends on the shape of the form itself, but should any difficulty be anticipated in stripping the precast units, it is wise to provide these means beforehand. Difficulty in stripping will also be increased should the forms be fabricated without the proper amounts of draft on all vertical surfaces, or with the steel surfaces or joints left rough. The joints between the steel plates should be welded and then ground off flush and checked with a straightedge to be sure there are no small humps present. This will also eliminate the possibility of mechanically locking in the concrete unit and will also eliminate the joint lines showing on the finished concrete product. Caution is emphasized to make sure the forms have the proper draft

FIG. 8. Strongbacks and jack used in stripping the panel from the form.

FIG. 9. Precast panels in place in the building, showing method of forming joiner column and spandrel beam.

and no mechanical-bind spots. If the concrete products are being cast and stripped each day, the product will normally be warm because of retained heat of hydration when stripping commences, whereas the steel form may be cool because of the air temperature at the same time. This will cause an additional tightness. The combination of this temperature differential with any lock of forms due to draft or bind spots mentioned above can prohibit stripping altogether without damage to the concrete product.

In stripping the products, best results are usually obtained using the same strong-back-type system and hand jacks as described in Art. 1. Again, the jacks should be located so that the force to lift the product from the form is exerted between the product and the form itself. Figure 6 shows one type of panel that can be cast in steel forms, and in Fig. 7 are shown plan and sections of a form and panel. Positioning of stripping jacks shows clearly in Figs. 7 and 8. The panels shown here were cast so that two adjoining panels formed the closure for the concrete joiner column (Fig. 9). This eliminated any forming between the panels, and both exterior and interior elevations of the curtain wall presented the smooth surfaces that are obtained from the steel-form surfaces. In addition to the panels forming the column closure, the panels were cast so that the spandrel-beam soffit and sides were also formed by the panel, thereby eliminating any need for conventional beam forming and shoring.

Another wall panel and forms are shown in Figs. 10, 11, and 12. Procedure in manufacturing this panel is almost identical to that for the panel in Figs. 4 and 5. After panels have cured, they can be stacked and then lifted and set in place (Figs. 13, 14, and 15). On this particular job, erection of the panels was done with a tower crane, which proved to be the most economical piece of equipment for the purpose.

3 Precast Beams

Precast-beam forms for repeated usage are usually fabricated with the soffit and one side fixed and the other side either hinged or removable (see Fig. 16). With the fixed side being aligned and plumb before any units are cast the removable

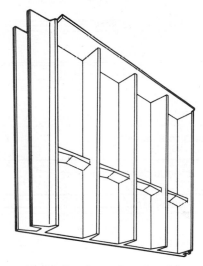

Fig. 10. Another type of precast wall panel made with steel forms. Approximate weight 8,400 lb. (See also Fig. 11.)

side can easily be lined to it and spaced for each casting operation. In preparing for the stripping operation then, the removable side is stripped first and the precast beams can then easily be removed with a crane. This forming method, however, applies only to rectangular-sectioned beams. Should the beam section be an I configuration or other, it will be necessary to have both form sides hinged or removable.

Types of precast beams fall into three categories—steel-reinforced beams, posttensioned beams, and pretensioned beams. For each type, the precasting sequence of form setting, placing concrete, and stripping varies only slightly. If reinforcing steel requirements for the beams are light, the steel cages are fabricated and stockpiled ahead and then placed in the form as one unit. The sequence starts with cleaning and reoiling of the form in preparation for the next placement. Next the reinforcing is placed and then posttensioning cables or prestressing strands are inserted if the beams are one of these types. In the case of prestressed beams, the following operation is stressing of the strands. Next the removable or hinged form side is placed, the beam-end bulkheads set, and the concrete placed.

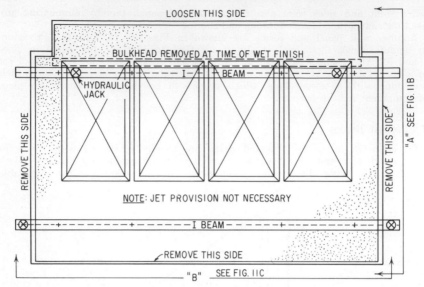

FIG. 11A

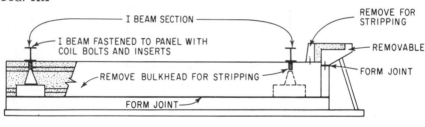

FIG. 11B   "A" VIEW

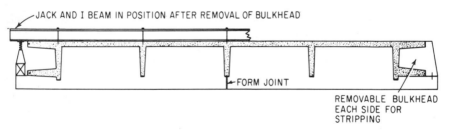

FIG. 11C   "B" VIEW

Fig. 11. Forming for the panel shown in Fig. 10.

After the concrete has attained sufficient strength, the hinged or removable side form is removed and the beam is lifted out of the form. The beams are usually stockpiled then prior to erection, where their specified strength is attained. If they are of the posttensioned type, they are easily posttensioned in the stockpile unless structural requirements dictate that they be stressed in place. Usually the reinforcing steel and the pickup points can be arranged so that the beams can be removed and stockpiled prior to posttensioning.

Details of prestressed-concrete construction are discussed in Chap. **41.** Methods of pretensioning and posttensioning that are used in central casting yards are, in general, applicable to on-site operations and need not be repeated here.

Placing and consolidating of the concrete, finishing, and curing all follow standard construction practices. Beams are frequently steam-cured in order to expedite strength development as they can be moved sooner, thus making the casting beds available for the next round of girders.

FIG. 12.   Photograph of the panel form shown in Fig. 11.

FIG. 13.   After the panel has been jacked loose from the form, it is lifted out to be stacked prior to erection.

FIG. 14.   Panel being lifted to storage area.

4   Precast Prestressed Floor Slab Elements (Planks)

In addition to precast beams (pretensioned or posttensioned), precast, prestressed floor planks can be job-cast. These, in conjunction with the beams, form the structural components of the floor slab over which a topping of 2 to 3 in. of concrete is placed. The availability of pretensioned planks and beams provides a great deal of flexibility to a building project, as this material can be easily installed, and, in conjunction with high-strength, light-weight concrete, it permits substantial reduction in the total weight of materials. Figure 17 is a plan of

FIG. 15. After the panels are situated in the building, concrete is placed in the columns.

an on-site casting bed for beams, planks, and wall panels. One traveling crane was used to service the casting operation, and a climbing crane was used in the building to erect the units. Figure 18 is a view of the precast beam and plank application in a 30-story high-rise office building.

Figures 19 and 20 are general overall views of an off-site yard used for casting elements for an office building and adjacent parking garage. Elements consisted of precast, prestressed beams; precast columns; and precast architectural wall panels for both jobs.

5 Precast Tilt-up Panels

This is a type of precast construction in which wall panels are cast in a horizontal position at the site, tilted to a vertical position, and moved into final location to become the building walls. Panels may be either solid concrete or of sandwich construction in which relatively thin, high-strength conventional concrete surface layers are separated by a core of low-density insulating material. Generally, tilt-up wall panels are cast directly on the floor slab of the building. Where there is not enough room available, they are usually stacked—that is, cast one on top of the other.

In considering a building for tilt-up construction, the floor must be designed to support the loads of cranes, ready-mix trucks, and similar equipment that may impose heavier loads than the proposed occupancy of the building. A well-compacted base to support the slab is essential to prevent movement and cracking of the slab. Sometimes it may even be desirable to increase the slab thickness, at least in areas where heavy equipment will be operating.

The casting floor should be perfectly smooth and uniform, as any imperfection in the floor will show on the panel cast against it. Imperfections in the floor can be smoothed off by filling them with stiff mortar, after thorough removal of all curing compound or dirt in the area being patched. Temporary fillings such as would be required in a joint in the floor, and which are to be removed from the floor after the panels have been removed, can be made with plaster of paris.

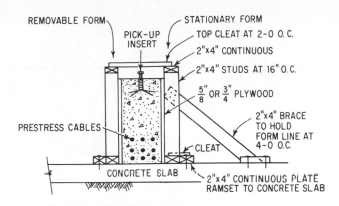

REMOVABLE FORM
PICK-UP INSERT
STATIONARY FORM
TOP CLEAT AT 2-0 O.C.
2"x4" CONTINUOUS
2"x4" STUDS AT 16" O.C.
$\frac{5}{8}$" OR $\frac{3}{4}$" PLYWOOD
2"x4" BRACE TO HOLD FORM LINE AT 4-0 O.C.
PRESTRESS CABLES
CLEAT
CONCRETE SLAB
2"x4" CONTINUOUS PLATE RAMSET TO CONCRETE SLAB

(A) STANDARD PRECAST BEAM FORM ON CONCRETE SLAB

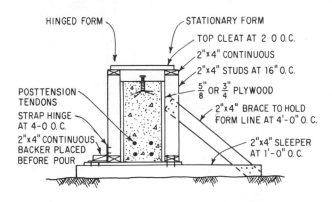

HINGED FORM
STATIONARY FORM
TOP CLEAT AT 2-0 O.C.
2"x4" CONTINUOUS
2"x4" STUDS AT 16" O.C.
$\frac{5}{8}$" OR $\frac{3}{4}$" PLYWOOD
2"x4" BRACE TO HOLD FORM LINE AT 4'-0" O.C.
POSTTENSION TENDONS
STRAP HINGE AT 4-0 O.C.
2"x4" CONTINUOUS BACKER PLACED BEFORE POUR
2"x4" SLEEPER AT 1'-0" O.C.

(B) STANDARD PRECAST BEAM FORM ON SOIL

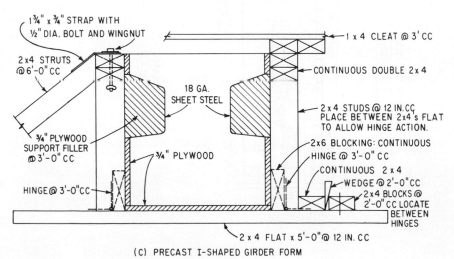

$1\frac{3}{4}$" x $\frac{3}{4}$" STRAP WITH $\frac{1}{2}$" DIA. BOLT AND WINGNUT
1 x 4 CLEAT @ 3' CC
2 x 4 STRUTS @ 6'-0" CC
CONTINUOUS DOUBLE 2 x 4
18 GA. SHEET STEEL
2 x 4 STUDS @ 12 IN. CC PLACE BETWEEN 2x4's FLAT TO ALLOW HINGE ACTION.
$\frac{3}{4}$" PLYWOOD SUPPORT FILLER @ 3'-0" CC
2x6 BLOCKING: CONTINUOUS
$\frac{3}{4}$" PLYWOOD
HINGE @ 3'-0" CC
CONTINUOUS 2 x 4
HINGE @ 3'-0" CC
WEDGE @ 2'-0" CC
2x4 BLOCKS @ 2'-0" CC LOCATE BETWEEN HINGES
2 x 4 FLAT x 5'-0" @ 12 IN. CC

(C) PRECAST I-SHAPED GIRDER FORM

FIG. 16. Beam forms for repeated use showing: (*A*) Removable side; (*B*) hinged side; and (*C*) precast I-shaped girder form.

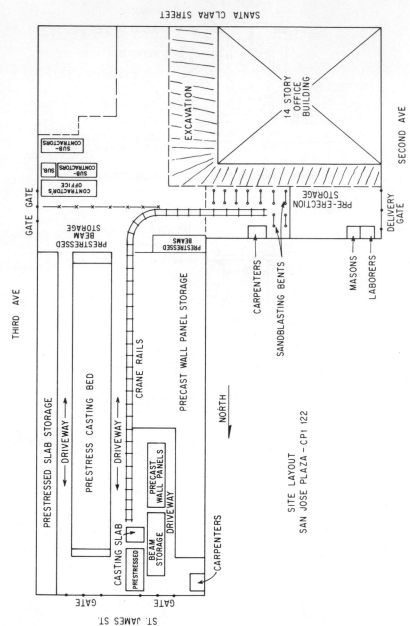

Fig. 17. Typical plan of an on-site casting yard.

FIG. 18. Installing precast planks on precast beams in a 30-story office building.

FIG. 19. A bird's-eye view of an on-site casting yard. A portion of the 400-ft stressing bed is shown to the left of the crane rails. Prestressed beams are stored adjacent to the bed, with precast columns on the extreme left.

Fig. 20. Another view of the yard shown in Fig. 19. Crane rails and traveling tower crane are on extreme left. Two fiber-glass garage-wall forms are in the center of the figure; completed garage-wall panels are stored at the right. Forms and completed wall panels for the main building are in center background.

Fabrication is accomplished by first placing a bond breaker on the casting floor. Liquids of various types are generally used, although sheets of plywood, metal, or paper can be used. Liquids consist of special formulations for this purpose, curing compound, and waxes, which are applied in two coats, the second coat being applied shortly before the panel concrete is placed. Uniformity of application is important. Paper and felt bond breakers nearly always wrinkle, and asphalt materials will stain the concrete and hence should be avoided.

A good type of bond breaker is one that combines curing with bond breaking. This material is applied to the floor as soon as final troweling is completed. A second coat is applied to the cleaned surface after forms have been placed but before the reinforcement and inserts have been set in place. Silicone-base materials have proved to be very satisfactory.

Side forms are usually of lumber. Forms for window and other openings may be metal or wood, with metal preferred because the swelling of wood frames makes them difficult to remove and might crack the concrete. Figure 21 shows a section of typical edge form for a flat wall panel. Dowels extending through the edges of panels fit into cast-in-place columns that are constructed after the panels have been braced in their final position. The presence of these dowels makes necessary the use of split forms on the vertical edges of the panels, as shown in Fig. 21.

Reinforcing steel may be fabricated in place, steel mats may be prefabricated on the site outside the forms, or heavy welded steel mesh can be used, depending on the relative cost at any one building site.

The panels are cast inside face up if they are to be erected from the inside, and outside face up if they are to be erected from the outside. However, should there be beam protrusions monolithic with the panels, or should a type of concrete finish on one face or the other dictate that that face be cast up, it may not be possible to erect in this manner. For this case, the erection of each panel is called a "suicide pick" as the panel would then fall into the crane. This situation is therefore to be avoided if at all possible.

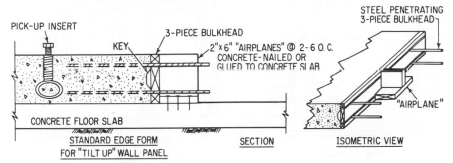

FIG. 21. Details of edge form for tilt-up panel.

FIG. 22. Tilting panel into place with a motor crane.

Placing and consolidating the concrete should be done in accordance with good construction practices. Attachment fittings for lifting hardware are inserted in the form prior to concreting. Edge forms serve as screeds. Because of the presence of inserts and other hardware, it is frequently not possible to use a vibrating strike-off, and immersion vibrators must be used to consolidate the concrete thoroughly. Many special finishes can be applied while the concrete is still plastic, including the embedment of architectural details and ornamentation, such as exposed aggregate.

In finishing a tilt-up panel, it should be remembered that there will be more bleed water rising to the surface than there would be for a slab on grade, as there is no absorption of water by the subgrade. This feature will probably make necessary somewhat more wood floating than is normally required.

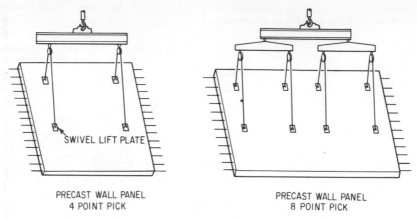

SWIVEL LIFT PLATE

PRECAST WALL PANEL
4 POINT PICK

PRECAST WALL PANEL
8 POINT PICK

Fig. 23. Two types of rigging for tilting panels.

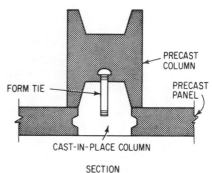

PRECAST COLUMN

FORM TIE

PRECAST PANEL

CAST-IN-PLACE COLUMN

SECTION

Fig. 24. Precast panels in erected position, held in position by temporary adjustable braces. Note also precast beams ready for posttensioning.

Fig. 25. Precast panel and column joinery.

Various types of cranes and gin poles have been used to lift the panels into place but motor cranes are normally used (Fig. 22). The panels are tilted onto a layer of mortar on the foundation and braced temporarily until the columns have been constructed.

Tilting places unusual stresses on the panels which should be considered in the design. The panel should reach the required design strength with a factor of safety, before the panel is tilted. Field-cured cylinders will provide this information. Pickup points must be carefully located and lifting equipment designed so as to avoid high localized stresses in the panel that might cause cracking, splitting, or spalling of the concrete. Vacuum lifting attachments are sometimes used. Figure 23 shows two rigging arrangements for erecting panels. The location of the pickup points to erect the panels is very important. If they are not located correctly in relation to the center of gravity of the panel, the panel will not hang in the proper position for erection. The normal position for easy erection is slightly out of plumb in a direction away from the crane. Computations for the pickup-point locations are usually done by the structural engineer or the engineering staffs of the concrete-accessory companies.

Considerably more lifting force is required to break the panel loose from the casting floor than is necessary to lift the panel after movement has started. For this reason, it is a good idea to move the panel slightly, if possible, before lifting. This can be accomplished with jacks operating in a horizontal direction, sliding the panel a fraction of an inch, to break the bond.

Lifting hardware and braces are attached to fittings that are embedded in the concrete when the panel is cast. Temporary bracing is best provided by tubular braces, including turnbuckles, for adjustment, that can be attached to the panel before it is erected (Fig. 24). After the panel is in place, the braces are attached to floor anchors and final adjustment is made to the panel location. Proper coordination of lifting and bracing will help to free the crane for other work.

After the panels have been erected and plumbed in their final positions, the joinery columns are formed and concrete is placed in them. In some cases the joinery columns can also be precast, which gives an exterior elevation free from form markings of any kind. Figure 25 is an example of a job where this method was followed and the precast column formed the exterior face of the poured column joinery. Both the column and panel can be given different treatments of architectural finish including colored concrete and/or contrasting texture by sandblasting either lightly or heavily.

Chapter 34

SLIPFORM CONSTRUCTION OF BUILDINGS

CHARLES J. PANKOW

The purpose of this chapter is to further the reader's understanding of slipform construction and bring to his attention methods by which the full efficiency of the slipform can be utilized. In the area of office building construction slipforming has been limited almost exclusively to the "core" portion of the building, or that portion which generally contains the elevators, stairs, toilet rooms, etc. This type of form is usually small in area and can be handled as a secondary operation in relation to construction of the balance of the building. In apartment, hotel, and similar residential-type multistory buildings, the slipform method of construction is highly effective, if architectural and structural design features are properly followed.

The ultimate approach in incorporating a slipform into building construction is to slipform all vertical concrete. This makes the slipform the primary operation controlling the other phases of construction. If the design of the building is for bearing-wall construction, then all the vertical elements can be slipformed.

First of all, before going into details of a slipform operation, we should evaluate the advantages and disadvantages of the basic system in order to determine if a job should be slipformed. It is generally recognized that slipforming is a faster method of construction than conventional forming; so it follows that if the construction cycle is geared to the slipform as the controlling item, an earlier completion date can be obtained. From the owner's standpoint, an earlier completion date means an earlier return on his invested capital. From the contractor's standpoint, an earlier completion date means lower overhead costs and more jobs that can be completed with his organization in a given period of time.

Under ideal conditions a slipform will provide substantially lower construction costs which directly result in lower completed-building costs. This may make possible the building of projects that otherwise could never have been built because of limited financing.

Slipforming is one of the systems of automation which can be more broadly included in building-design systems. The efficiency of its use depends on the building design and its adaptability to construction details that are necessary to slipforming and which are discussed later in this chapter.

In order to achieve this automation, work of all the various trades must be organized and coordinated so that they mesh with each other with a minimum of conflict. The materials deliveries, such as concrete and reinforcing steel, should be made on schedule in order to develop a maximum rate of slip. Specialized help is required for the duration of the slipform work, much more so than in conventional construction. It is imperative that some person thoroughly familiar with the overall operation be on the form at all times. If care is taken in the planning stage and carried through the complete operation, many of the normal pitfalls can be avoided and an efficient operation can be achieved.

Not all jobs are suitable for slipforming, and each should be evaluated to see if a slipform is feasible. This should be done in consultation with the architect and structural engineer, for sometimes limitations of structural or architectural design preclude the possibility of making the changes necessary to set up a slipform operation. If the design is in an early stage, it is sometimes possible to design around a slipform or, in other words, make the basic design one that can be slipformed. Many owners have gone to this method of architect-engineer-contractor coordination and cooperation in order to obtain the most for their money.

1 Design Requirements

Following is a discussion of the general design requirements that are necessary in order to develop an economical and efficient slipform. Details vary but these requirements remain constant.

Number one on the requirement list is a layout that remains typical from floor to floor. This will enable a form to be built that will not have to be modified to any great extent as the slips progress. If possible, the wall thicknesses should remain constant throughout their full height, although walls can be narrowed down by the use of filler panels inserted in the form, if necessary. Sometimes the saving in concrete is more than offset by the additional labor spent in modifying the form, not to mention the time lost. Minimum wall thicknesses may vary depending on whether lightweight or hard-rock concrete is used, but generally they should be at least 7 in.

The reinforcing-steel design is the number two item that will govern the efficiency of a slipform operation. Large concentrations of steel, such as found in spandrel beams and interior beams, should be kept to an absolute minimum, for these concentrations make it difficult if not impossible to place the steel while the slip is in progress. In some cases horizontal construction joints have to be located with this in mind. This also causes the work load of the ironworkers to vary greatly, thus making it difficult to maintain an efficient crew size. If the slip is not to be a continuous one, then the vertical steel should be detailed so that it can be placed while the form is at rest. If continuous, a pattern of splice points satisfactory to the engineer should be worked out to enable the vertical steel to be placed while the form is moving. Again, consideration should be given to providing as even a work load as possible for the ironworkers by establishing the pattern of splices so that only part of the steel has to be spliced at any one elevation. Horizontal steel for slab doweling, No. 5 bars or smaller, may be bent to lay along the face of the form and straightened out after the form has slipped by, or if large in diameter, the bars can be welded. The ends of the bars to be welded are normally encased in a blockout that is set in the form and therefore exposed after the slip. If bars are to be welded, only weldable steel should be used. This should be settled before construction starts (see Chap. 3).

In order to achieve a satisfactory finished product, in this case a completed structure incorporating all required architectural details, care must be taken to see that all of the items to be installed within or connecting to the slipformed walls have tolerances in their design to ensure proper fit. This is no different from conventional construction, but the difficulty of setting work to an exact position from a moving deck accents this importance.

ACI Committee 347[1] recommends that the total deviation of any point on the slipform, measured in a horizontal plane with respect to the projection of a corresponding reference point at the base of the structure, should not exceed 1 in. per 50 ft of height. From this it can be seen that on tall buildings where the elevator shafts are of slipformed-concrete construction, the elevator door frame and sill are constant, and so the profile of the shafts in relation to plumb should be plotted to enable the rail line to be set and fit properly within the shaft. In this manner, the proper architectural detail is maintained at each floor. Using a plaster skim coat on the lobby side of the elevator walls allows more flexibility in setting the elevator frames to the exactness required. Differences in projection of the frames from the concrete wall can be adjusted for appearance by varying the thickness of the skim coat.

When framed openings are required in slipformed walls, they can be handled in several ways: The first method, which can be used in the case of pressed steel or channel frames, is to set the frame in its final position similar to the way it would be in conventional construction, then place the concrete around it as the slip progresses. If this method is used, it is necessary to position the frame firmly by welding or some other means to make sure that it does not shift during the slide. The second method is to provide a blockout in the wall slightly greater than the out-to-out frame dimension and install the frame after the slip has moved out of the way. The third method is to provide oversized blockouts in the walls and then set the frames after the slip has passed, and

FIG. 1.   In the background, the slipform has started up the core of the building. Precast panels and columns are visible in the foreground.

either grout them in or plaster around them.   Any of the above-mentioned methods is satisfactory, but the method selected should be dependent upon the architectural requirements of the job.   The cast-in-place frame method is least desirable.

2   Finishing

Another basic consideration in setting up a slipform is the type of concrete finish that is going to be required. The normal "slipform finish" is a rubber-float finish that is applied to the wet surface of the concrete as the slip progresses or applied dry after the slip is complete and the form stripped off.

In the first method, a finisher's platform is suspended below the form and the finishing crew works with the freshly exposed concrete to give it the texture required. A membrane cure is sprayed on the surface after the finishing operation is complete, and it is usually unnecessary to go over the surface again once the slip is complete. Figure 1 shows a slipform in the background, and the finisher's platform can be seen below the form.   Another slipform is shown in Fig. 2.

The second method, that of applying a "dry" finish, is not to touch the surface to be finished until after the slipping operation is completed full height. When complete and the form has been removed, the walls are float-finished from a swinging stage by applying either a sand finish or a cement-plaster finish. If this method of finish is to be used, the finisher's scaffold is usually omitted except at critical areas where it is, or may be, necessary to have access under the form. Thus, the cost of fabricating and installing a finisher's platform is saved and can be applied to the higher unit cost of applying the finish in this manner. If a smooth finish is desired on the exterior of the building, a plaster skim coat

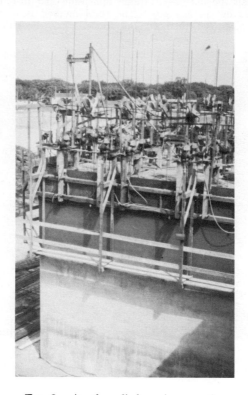

Fig. 2. Another slipform in operation.

can be applied over the rubber-float finish. Concrete finishing is discussed in Chap. 27, and plastering is covered in Chap. 39. Sandblasting of slipformed concrete is not recommended. The possibilities of uniform texture are practically nil. Furthermore, if sandblasting is attempted and found unacceptable, corrective procedures are costly in both time and money.

3 Elevators

Since we have been discussing basic considerations in order to determine whether a slipform is feasible or not, we should include the elevator requirement in arriving at a conclusion. One advantage of slipforming, as mentioned before, is that the elevators can be started before the building is topped out. This is especially true where the portion of the building containing the elevator shafts and equipment rooms is a stable structure within itself and can be slipped to the top and completed very early in the job. On tall buildings where the shafts cannot be slipped to the top ahead of the balance of the building, it is possible to get a head start with the elevator installation by establishing a rail line based upon the first seven or eight floors and projecting it upward in similar increments as the shafts are completed. It is also necessary to block off the shaft at some point below the slipform to protect the elevator constructors. The most important thing to take into consideration is the alignment of the shafts for, once the rail line for the first segment is established, it is necessary to continue this line for segments above. This makes the elevator-front design more important than ever.

4   Jacking Systems

After it has been decided to slipform a portion or all of the work, the details necessary to put the system into operation must be worked out. Of course, the initial decision must be that of selecting which jacking system to use. At present there are four primary systems: hydraulic, pneumatic, electric, and manual. Of these, the hydraulic and pneumatic systems seem to be the commonest, and we shall discuss their details.

The hydraulic jacking system is a patented system employing a network of hydraulic jacks connected by oil lines to a central reservoir and powered by an electric pump. The jacks usually climb on pipe that has a diameter of 1 in. and a wall thickness of ⅛ in. Each jack is calibrated to climb approximately 1 in. each time the pump is activated. It is necessary to set the pump pressure high enough to make sure all jacks have raised before the pump is turned off. If this is not done, that portion of the form which is lightly loaded will progressively gain on the balance of the form and throw the deck out of level. The yokes that are used with this system are made of steel and are furnished by the slip-form-equipment supplier as part of the equipment leased to the job. The clearance from the working deck to the underside of the yoke is approximately 12 to 14 in., depending on the location of the bottom waler. Normal spacing between the yokes is 6 to 8 ft, with 7 ft considered to be optimum. If this spacing has to be increased because of architectural or structural requirements, special steps should be taken to ensure that the deflection of the walers between the yokes is kept within acceptable limits.

The pneumatic system of slipforming is also a patented system employing a network of jacks connected by air lines to an air compressor generally located near the base of the slipform. The control is simply a pressure exhaust valve located on the form and operated manually to raise the form in ½-in. increments. The rods that these jacks climb on are solid 1 in. in diameter, drilled and tapped at each end for a coupling stud. The yokes are made of wood and fabricated on the job, which gives an added degree of flexibility in the form designing.

As mentioned before, there are systems employing manual and electric jacks, but these are not employed to any great extent in building-construction work. Figure 3 illustrates the wood yokes and shows the pneumatic jacks in place.

5   Form Design

After the jacking system has been selected, the next step is to design a form around this system making proper provisions for the jacks, yokes, etc. Slipforms for building construction are usually built from 4 to 5 ft high with the outside form being about 6 in. higher than the inside in order to act as a splash board to prevent concrete and other material from falling from the form. Yoke spacing should be held to a maximum of 7 ft with "dummy" yokes (nonjacking) added as necessary to provide additional support. Where the load of the form is high, jacks will have to be spaced closer together or doubled up to provide the required lifting capacity. The weight of concentrated loads, such as concrete or reinforcing, is an example.

Great care should be taken during the initial fabrication and setup period to see that design dimensions are maintained and that the form is strong enough to stand the stresses of the slipping operation. The forms should be battered about ⅛ in. in a 4-ft height so that they tend to clear themselves as the slide

Fig. 3. Operating deck of a slipform, showing yokes, steel template, and jacks.

Fig. 4. Slipform being set up, showing inside form and reinforcing steel in place before placing outside form.

progresses. In some cases, the batter is built into one side only with the other side remaining vertical.

The form panels can be built from ¾-in. plywood or some other material such as sheet steel or 1-in. boards, and may utilize a two- or three-waler system. The size of the walers will depend upon the spacing between yokes and may be 2 × 6 or 2 × 8 timber or, in some cases, structural steel. Blocking should be placed between the walers at each jacking point to transmit the load from the yoke into the form. The two-waler form usually will incorporate a herringbone bracing system between the walers continuous along the form. Recommended form surfacing is 1×4 tongue and groove placed vertically. The vertical assembly permits slippage along the way between yoke locations where vertical adjustments are occasionally made; secondly, 1×4 will not "cup" as much as wider boards. If

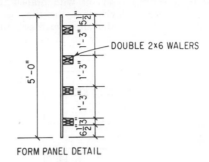

FORM PANEL DETAIL

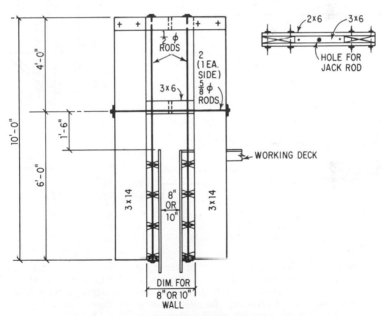

FIG. 5. Timber yoke for a 5-ft slipform.

plywood is used, a "plate girder" effect tends to occur, thereby inducing "corrective" action into adjacent jacking location where correction in fact is not required.

The yoke design is usually standardized to allow for walls of various thicknesses to be obtained from one size of yoke, and as much clearance as practical is left between the bottom of the yoke and the working deck. Large concentrations of horizontal steel (especially in earthquake zones) make it necessary to delay the slip or, in some cases, stop the slip in order to place the steel properly. There has been at least one attempt to increase the time available for placing this steel by increasing the clear distance between the bottom of the yoke to the working platform from 12 to 36 in. The results of this high-yoke trial were quite satisfactory and can be helpful in critical areas of heavy reinforcing. Figure 6 illustrates a simple slipform layout and shows how a template is used to aid in the reinforcing-steel placement.

The deck-leveling method used for each of these jacking systems is a water level connected to a central reservoir with branch lines running to each jacking point. The working deck is leveled at the time of initial setup and the water level is referenced so that as the slip progresses the level of the deck can be adjusted as desired. It is necessary to know relative elevations of the deck in

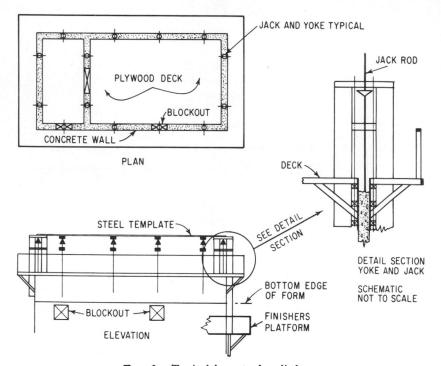

Fig. 6. Typical layout of a slipform.

order to control the lateral movement of the form. Plumb bobs and/or optical plummets are used and referenced to a line at the bottom of the slip. These should be checked continuously so that the slip can be adjusted quickly to compensate for deviations. Correction for deviation from line is done by changing the relative elevation of portions of the slipform deck. This adjustment is done by means of shutting down certain jacks during several lifting operations until the error has been compensated. This is a difficult operation and should be under the direction of a supervisor experienced in slipform work.

The elevation of the form is registered by a steel tape, or similar device, attached to the base of the slipformed walls at a known elevation. The point where the tape is read at the form level is usually adjacent to the water level, therefore providing a direct method of establishing elevation anywhere on the form.

A good reference book giving design loads, methods of calculating pressures, and additional slipforming information is published by the American Concrete Institute.[2]

In order to slipform a portion or all of a building a method must be selected for hoisting concrete, reinforcing steel, and the rest of the construction material required. This can be a material hoist, a mobile crane, a tower crane, a gin pole, or any other method that will satisfy the hoisting requirements of the slipform. In Europe, tower cranes have been used for many years to speed up the construction of high-rise buildings and are now very popular in the United States. In a slipform application the tower crane may be carried to the top with the form and used to satisfy the hoisting requirements, or a portion of the hoisting requirements, for the whole building after the slipform operation is complete. Whatever method is selected it must be based upon an objective view of the whole job with the ultimate goal of satisfying the needs of all phases of the construction. Figure 7 shows a typical slipform which was poured using a tower crane. Tight scheduling

Fig. 7. A later stage of the building shown in Fig. 1, showing typical slipform, blockouts, and finisher's platform.

enabled the tower crane to service the slipform in addition to erecting the precast members.

A portion of another central core slipform, which was slipped over 400 ft, is shown in Fig. 8. All the precast structural elements and architectural wall panels on this job were hoisted by use of a climbing tower crane. Figure 9 shows the slipform on a 23-story apartment building in which all the vertical concrete was slipformed.

6 Slipping

The last major division of slipforming to be considered is the actual operation of the form during and between the slip. The quality of this portion of the work is dependent upon the proper preparation of the details prior to starting to slip, scheduling of material to be delivered to the job, primarily concrete, coordination of subcontractors, and a thorough knowledge of the slipform work including all details of all work to be incorporated into the slip. Figure 10 shows a typical detail sheet which is made for the slipform operation. This type of

FIG. 8. A portion of a central-core slipform. Note the base of the tower crane in the center of the form.

FIG. 9. All the vertical concrete on this building was slipformed.

drawing is made for each wall at each floor and shows all items to be incorporated into the slip. The responsibility for this should be given to one person who stays on the form at all times and has the authority to see that the necessary work is completed as the slip progresses. A special eye should be kept for any indication of form trouble, which often shows up as a change in wall thickness, errant plumb-bob readings, form noise, etc. Concrete should be placed in a definite pattern in lifts of 6 to 8 in. and about 2- and 3-in. slump with the vibrator penetrating just into the lift below. By keeping the form full, the maximum rate of slip can be obtained consistent with installation of other work. Rate of slipping will vary according to concrete slump, weather, and installation of other work, but the average is about 12 in. per hr with some slips approaching 24 in. per hr at certain times.

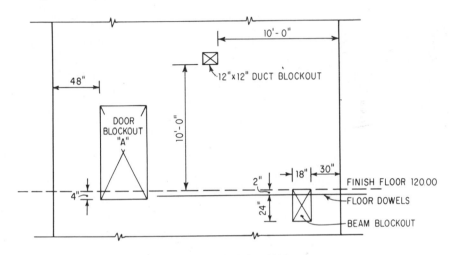

SOUTH WALL AT ELEV. 120.00

FIG. 10. Detail sheet for a slipform operation.

To summarize, the single most important thing necessary in order to obtain a satisfactory slipforming operation is thorough planning. From conception, through the structural design or modification thereof, through the form design and construction, to the actual operation of slipping, planning will prove the difference between a good low-cost job and one where troubles are prevalent and costs are high.

REFERENCES

1. ACI Committee 347 Report, Recommended Practice for Concrete Formwork, ACI Book of Standards, 1972.
2. Hurd, M. K.: Formwork for Concrete, *ACI Spec. Publ.* 4, American Concrete Institute, 1963.

Chapter 35

LIFT-SLAB CONSTRUCTION

JOSEPH J. WADDELL

In the lift-slab technique, the building floor usually serves as the casting floor on which the roof and upper floor slabs are cast at ground level. Several slabs may be involved, stacked one on top of the other. After the slabs have cured and have been posttensioned, they are raised, each to its required elevation, by means of jacks mounted on top of the preerected building columns, opening having been left in the slabs to enable them to rise along the columns.

Costly equipment, special experience, and the critical nature of many of the steps make lift-slab construction a job for the specialist. Preconstruction planning and consultation between the architect, engineer, builder, and lift-slab specialist will enable full advantage to be taken of this unique type of construction to effect the greatest time and money savings of which the system is potentially capable.

Many contractors have selected the lift-slab method to enable them to construct buildings in the most efficient manner to meet the requirements of safety and structural adequacy. Almost all types of buildings of medium to large size with an area of about 25,000 sq ft or more are appropriate to this construction system, as shown in Fig. 1. Most structures have been three to five stories in height, but the method has been used in constructing a number of high-rise apartment and office buildings (Fig. 2). Advantages claimed are the elimination of costly formwork and shoring, as only edge forms are necessary; slabs can be cast, finished, and tensioned faster and more easily because the work can be done at ground level; electrical and plumbing services are easily installed before the concrete is placed; curing and weather protection are necessary only at ground-level area where the slabs are being cast; and tile, flooring, or other finishing materials can be applied directly to the slab concrete. Posttensioning the units minimizes cracking because shrinkage is restrained, and it is not necessary to compensate for deflection of the slab when installing partitions.

The foundation for a lift-slab structure is usually the ground-floor slab of the building. Special care is necessary to make sure that the foundation slab is smooth and free of blemishes, as the underside of the slab cast on it will reflect any such blemishes.

Before casting the lift slabs, it is necessary to erect the building columns, which may be of precast concrete, cast-in-place concrete, or steel. Columns are frequently spliced, the upper portions being installed after lifting of the slabs is commenced. For example, slabs may be parked at the first three floor levels while the columns are being spliced; then the slabs are raised to their final elevations for attachment to the columns. Figure 3 shows slabs temporarily parked while columns are being extended. Column design must take into account the stresses imposed on them during slab lifting, considering among other things wind loads, stability, and stiffness during construction. Because of the design of the lifting equipment, there is small likelihood of eccentric loading on the columns. Steel lifting collars, cast into the slab later, are stacked on each column at the time the columns are erected. There should be clearance between the collar and column to ease lifting and permit relative movement caused by shrinkage and other factors.

FIG. 1.   Lifting is not confined to flat slabs, as this domed auditorium roof was lifted in one section.   (*Vagtborg Lift Slab Corporation.*)

FIG. 2.   Architect's rendering of an 11-story hotel that was lifted.   (*Vagtborg Lift Slab Corporation.*)

Casting, finishing, and curing follow closely the procedures for any first-class slab production.   Slabs may be up to 10 in. thick, frequently of lightweight concrete.   Conventional reinforcement, stressing cables, conduit piping, and inserts are all installed in the usual manner.   Concrete with a slump not exceeding 4 in. is placed and consolidated by vibration to eliminate all rock pockets, then is finished to a smooth finish with power tools.   Curing is best done with a resin-base liquid material that also serves as a parting compound on which concrete for the succeeding slab can be placed, after the first slab has developed sufficient strength.

After a suitable curing period the slabs are posttensioned (see Chap. 41) usually in both directions; that is, tendons are positioned and cast in the concrete at right angles to each other. Tendons should not be deflected or diverted in plan around openings, as erratic friction losses are apt to be introduced. Stressing heads should be placed in wooden box forms recessed several inches along the edge of the form. Short lengths of heavy wire for pulling can be stapled or nailed in the boxes and, after removal of the boxes and completion of stressing, the cavities can be dry-packed, thus protecting the stressing heads from corrosion.

Fig. 3. Slabs temporarily parked while another section is attached to the columns. Note the recesses and prestressing connections in the slab edges. (*Vagtborg Lift Slab Corporation.*)

Tendons less than about 70 ft long can be stressed from one end; those longer should be stressed from both ends. Usually adjacent short tendons are stressed alternatively from opposite ends, thus tending to equalize stresses in the slab.

After the columns have been erected, special hydraulic jacks are placed on top of the columns, one to each column (see Fig. 4). The jack illustrated has a large-diameter ram in the center that has a stroke of about ½ in. Follower nuts, following closely behind the jack movement, automatically hold the slab when the jack is retracted. A jack of this type may have a capacity as high as 100 tons under full hydraulic pressure of 2,000 psi. When the slabs are ready for raising, suspension rods, reaching down from the jacks on top of the columns, are attached to the collars that were previously cast in the top slab (usually the roof slab), one collar encircling each column. All jacks are synchronized, operating from a central control console (Fig. 5). Should one of the jacks move ahead or lag behind by ½ in. in elevation, a signal light on the console gives a warning. The slab is maintained level within a tolerance of about ¼ in. Usual jacking rate is between 5 and 15 ft per hr, with 7 to 10 about normal. If no

Fig. 4. A typical hydraulic lifting jack atop a steel H column. Follower nuts are rotated through the chain drives that are actuated by small hydraulic motors. (*Vagtborg Lift Slab Corporation.*)

Fig. 5. Regulation of all jacks on the slab is effected at the operating console. (*Vagtborg Lift Slab Corporation.*)

splicing of the columns is required, the roof slab is first raised to its final position, followed by the floor slabs, normally one at a time. On a small building more than one slab might be lifted simultaneously.

Height of the building and the number of slabs to be lifted determine the exact lifting sequence to be followed in each case. For a building requiring more

FIG. 6. Two slabs have been raised to their final elevations, and the two top slabs have been parked temporarily while columns are extended. (*Vagtborg Lift Slab Corporation.*)

FIG. 7. The cast-steel collar which is embedded in the slab concrete is lifted by two jack rods, one of which is shown just above the collar. (*Vagtborg Lift Slab Corporation.*)

than one column section (because of height) in the interest of providing column stability during construction, it is sometimes prudent to park several slabs at the top of the first section of columns as shown in Fig. 6. Part of the slabs can be lifted and braced, then columns extended, and finally the upper slabs lifted to their final elevations.

Slabs are connected to the columns by means of metal collars previously cast into the concrete of the slabs. Collars are sometimes steel weldments but more often are stress-relieved steel castings, as shown in Fig. 7. Provision is made for insertion of a steel shear key or bar that passes through the column and provides permanent attachment of the slab to the column. As soon as the slab has been lifted, the shear key is inserted, coming to rest on a steel bearing plate in the column, with the slab collar resting in turn on the key. Accuracy in fabricating is necessary so that there is an even bearing between the collar and the key, and between the key and the bearing plate. Steel shims can be used to equalize bearing, if necessary.

If steel columns are being used, satisfactory attachment can be achieved by welding the collar to the column.

After the slabs have been leveled and permanently attached to the columns, the space between the collar and column can be dry-packed or filled with grout.

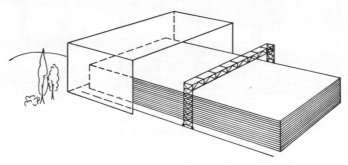

Fig. 8. Floor and roof slabs for the MLC System are cast, one on top of the other, on the ground-floor concrete slab. (*Kolbjorn Saether & Associates, Inc.*)

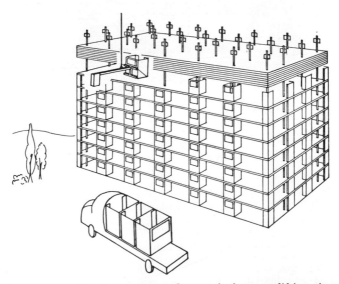

Fig. 9. The stack of slabs is raised one floor each day, overlifting about 5 in. so angular wall columns and prefabricated units can be installed; then the lowermost slab is lowered to rest on the columns. (*Kolbjorn Saether & Associates, Inc.*)

FIG. 10. Second- and third-floor slabs have been landed on the columns, and the precast columns will now be placed in the third floor to support the next higher slab. (*Kolbjorn Saether & Associates, Inc.*)

Connections between slabs and walls should not be made until the slab is at least 30 days old to reduce the possibility of horizontal movement of the slab relative to the shear walls resulting from creep of the slabs.

The Multileveling Component System involves casting a stack of slabs in the usual manner (Fig. 8). After the concrete has been placed in all the slabs, temporary lifting columns are erected, each with two hydraulic lifting jacks riding on the sides of the columns with steel rods connecting the slabs to the jacks. All slabs are now raised simultaneously until the bottom slab is 5 or 6 in. above its final elevation, where it is separated from the others and brought to rest on temporary holding brackets. Precast permanent columns, called "wall columns," previously cast to close tolerances, are now set in place, and the slab is lowered onto neoprene gaskets on the wall columns. With the slab attached to the wall columns, the resulting frame provides stability to the structure. Figures 9 and 10 illustrate these operations.

Overlifting the slab permits full floor-to-ceiling-high prefabricated components to be placed in the building before the slabs are lowered to rest on the wall columns. Lifting and placing the precast wall columns and prefabricated units are accomplished with a special crane-lifted balanced beam that places the unit within the building frame.

Section 10

SPECIALIZED PRACTICES

Section 16

SPECIALIZED PRACTICE

Chapter 36

PUMPED AND SPRAYED CONCRETE
AND MORTAR

JOSEPH J. WADDELL

Conveying of mortar and concrete through a pipeline has been a highly satisfactory method of transportation for many years. Two common methods are employed, pumping by mechanical pumps and conveying the material in a stream of air.

Pumped concrete or mortar is premixed, including the water, and conveyed through a pipeline or hose by the mechanical force exerted by a piston, pneumatic pressure, or a roller acting on the material. The material moves slowly through the conduit and is discharged by dropping into the form or a receiving hopper (see Figs. 1 and 2).

Shotcrete, or gunite, on the other hand, consists of either a dry mix, to which water is added at the nozzle, or a wet mix containing the necessary mixing water. Both are conveyed at relatively high velocity by pneumatic pressure through the conduit and are discharged at high velocity and forcibly deposited on the receiving surface.

There is some overlapping of methods and equipment, and some of the pneumatic machines perform as pumps as well as shotcrete guns; so no hard and fast rules govern the classifying of some of the equipment in one category or the other. It may fit in both.

PUMPED CONCRETE AND MORTAR

Conditions that lend themselves especially well to pumping exist on those sites and areas where access is limited and the site is crowded with materials and equipment, such as many city building sites.[1] Nearly all pumps, large and small, can pump 100 to 150 ft vertically, thus lending themselves very well to high-rise building construction. A pump takes small space and can be located any place that the ready-mix truck can reach. Conveying hose and pipe are easily placed out of the way and take little room. In locations difficult to reach with the ready-mix truck, such as on a steep hillside, a pump can easily move the concrete over obstructions that would be exceedingly difficult for the trucks to overcome. In many cases, the cost of pumping concrete is less than that of other methods of transporting. Each job, of course, must be analyzed on its own requirements and conditions to determine the most economical method of moving concrete.

1 Equipment

Heavy Equipment. For use when relatively large quantities of concrete are to be placed, heavy mechanical pumps with a rated capacity of up to 65 cu

FIG. 1. Discharge of concrete from hose of small-line pump. (*Ridley and Company, Inc.*)

FIG. 2. A boom arrangement such as this enables the pipeline to clear obstacles and facilitates concrete placing. (*Transit Mixed Concrete Company.*)

yd per hr are used. As shown in Table 1, these machines can pump 3-in.-slump concrete through a 6- or 8-in. pipeline up to 1,000 ft long, raising it as much as 120 ft vertically. Concrete frequently contains aggregate as large as 2 or 3 in.

One of the largest of these machines, shown in Fig. 3, features two separate outlets that can be operated either simultaneously or individually, permitting the use of the machine with two separate discharge lines to different forms if desired. Pistons and valves are mechanically operated as shown in Fig. 4. The receiving

Table 1. Specifications for Heavy-duty Concrete Pumps*

Model	160 single	200 single	200 double
Capacity, cu yd per hr...........................	15–20	25–33	50–65
Bore and stroke, in..............................	6.3 × 12	7.9 × 12	7.9 × 12
Approx rpm....................................	50	50	50
Max size aggregate, in..........................	2	3	3
Pipeline, in....................................	6	8	8
Remixer capacity, cu yd concrete.................	¾	2	3
Avg pumping distance, straight pipe, ft:			
Horizontal...................................	800	1,000	1,000
Vertical.....................................	100	120	120
Weight on skids, complete machine, lb.............	6,850	13,950	25,050

* Rex Chainbelt, Inc.

Fig. 3. Heavy-duty duplex concrete pump. Note the two discharge lines. (*Rex Chainbelt, Inc.*)

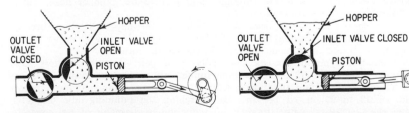

INLET VALVE BETWEEN HOPPER AND CHAMBER IS OPENED TO ALLOW CHARGE OF CONCRETE TO BE DRAWN INTO THE CYLINDER BY GRAVITY ASSISTED BY THE SUCTION OF THE PISTON. AT THIS TIME, THE OUTLET VALVE IS CLOSED.

OUTLET VALVE IS OPENED, AND WITH INLET VALVE CLOSED, PISTON FORCES CHARGE OF CONCRETE THROUGH OUTLET PASSAGE. VALVES, WHICH ARE MODIFIED "PLUG COCKS," ARE OPENED AND CLOSED IN TIMED RELATIONSHIP TO THE MOVEMENTS OF THE PISTON.

Fig. 4. Operation of heavy-duty concrete pump. (*Rex Chainbelt, Inc.*)

hopper is a pugmill-type mixer that helps to maintain uniformity of the concrete and minimize segregation.

Most of the concrete for the Bartlett Dam in Arizona, built in the late 1930s, was pumped into place. The pump was set up adjacent to the concrete-mixing plant and handled a mix containing 3-in. maximum-size aggregate containing about 40% crushed material. Slump of the concrete, which contained 34% sand, averaged 3 in., and the cement content was about 470 lb per cu yd. Pumping rate was about 40 cu yd per hr.

Small-line Pumps. A large variety of pumping equipment is available suitable for almost every concreting job. These pumps owe their name to the fact that the concrete is pumped through conduit 5 in. or less in diameter, this being small when compared with the 6- and 8-in. lines of the heavy pumps. Many of these rigs are small and portable enough to be mounted on a trailer that can be hauled behind a pickup truck (Fig. 5). Others are truck-mounted. Capacities up to

Fig. 5. Small-line concrete pump with a rated capacity of 55 cu yd per hr. This pump can handle concrete containing 1½-in. aggregate. (*Ridley and Company, Inc.*)

50 cu yd per hr are claimed, the actual rate depending on length of hose, vertical lift, maximum size of aggregate, mix proportions, and slump. In general, these small-line pumps evolved from grout and plaster pumps. There are several makes of piston pumps, either hydraulically or mechanically driven, most of them with two pistons, alternating on the power stroke. The large (6- to 8-in.-diameter) low-velocity pistons force concrete through reductions to the pipe or hose, which may be from 2 to 4 in. in diameter.

Concrete from the ready-mix truck is deposited in the hopper leading directly to the loading chamber, passing through valves into the unloading chamber where the piston forces it into the pipe or hose for delivery to the forms.

Another small-line pump consists essentially of a pair of rubber rollers that squeeze the concrete out of a rubber tube lining half of a drum-shaped vacuum chamber (Fig. 6). Pumping tube sizes, from 2 to 5 in., provide maximum pump capacities of 12 to 90 cu yd per hr. By matching the pumping-tube diameter to that of the placing pipe line, there is no reduction in diameter of the concrete stream being pumped, and therefore no increase in pressure as the concrete enters the pipe. There are no valves, pistons, or other mechanisms in direct contact with the concrete.

A third type of small-line pump is the pneumatic type in which the concrete is carried through the pipe by air pressure in a manner similar to shotcrete, but discharge is at low velocity.

Remote control is sometimes provided, enabling the man at the hose discharge to start or stop the pump. This feature is especially desirable when visual signals cannot be used because the hose discharge is not visible from the pump. Telephone or radio communication is sometimes used.

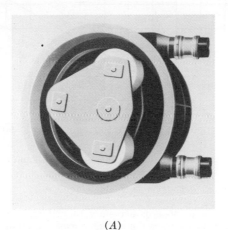

(A)

ROTATING ROLLERS   MATERIAL HOSE

PRESSURE

RE-MIXER HOPPER

PUMPING CHAMBER

SUCTION

PUMPING TUBE

(B)

Fig. 6. *A,* interior of the vacuum-squeeze pump; *B,* a schematic cross section of the pump. A vacuum in the pumping chamber causes the pumping tube to open, drawing the concrete into the tube immediately behind the roller at the bottom, while the roller at the top is forcing the concrete out through the pressure line. (*Challenge-Cook Bros., Inc.*)

Pipes. Large-diameter pipe for heavy-duty machines may be about 8 in. in diameter. It is quite apparent that a fairly large amount of concrete may be contained in a long line. Figure 7 will aid in determining the amount of concrete thus contained, and is of value in holding concrete loss to a minimum when nearing the end of a run. This chart can also be used to determine the amount of water required to clean out the line with the go-devil (a plug that can be introduced into the line at the pump).

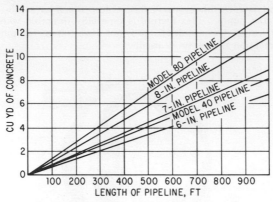

Fig. 7. Chart for computing amount of concrete in pipelines of various diameters and lengths. (*Rex Chainbelt, Inc.*)

In figuring pipeline for a job it is necessary to make an adjustment for vertical lifts and bends, converting them to equivalent horizontal pumping. Rex Chainbelt, Inc., recommends the following equivalents for their Pumpcrete machines:

    1 ft of vertical pipe = 8 ft horizontal
    One 90° bend = 40 ft horizontal
    One 45° bend = 20 ft horizontal
    One 30° bend = 13 ft horizontal

For example, assume that a line has an actual length of 360 ft which is made up of the following sections:

    320 ft straight pipe
    Two 90° bends
    Four 45° bends

There is a 40-ft vertical lift at the end of the pipeline.
The equivalent length of straight horizontal pipe is determined as follows:

320 ft straight pipe equals....................................	320 ft
Two 90° bends (each equivalent to 40 ft straight pipe)............	80 ft
Four 45° bends (each equivalent to 20 ft straight pipe)...........	80 ft
40 ft vertical lift (1 ft vertical is equivalent to 8 ft horizontal).....	320 ft
Total	800 ft

Layout of the pipeline (called the "slickline") for any size of pump is important, keeping in mind the fact that bends introduce additional frictional resistance. Alignment from the pump to discharge should be as direct as possible. Large areas of placement are most efficiently covered by adding sections of pipe as the work progresses. In some cases, however, such as on a large slab, it may be more desirable to work in the opposite direction, progressively removing sections of pipe. The sections removed should be cleaned immediately. Normally, a length of hose is used at the discharge end to facilitate placing the concrete exactly where required.

When it is necessary to run the pipe over a large area where a considerable amount of concrete is to be placed, it is sometimes feasible to run the pipe on steel supports that can be left in the concrete upon completion of the placement. In this situation the steel supports are cut off just below finish grade before the concrete is finished.

In general, the larger the inside diameter of the pipe or hose, the less pressure is required to move a given quantity of concrete in a certain time interval. (Hose, however, requires more pressure than pipe because of higher frictional resistance.) Larger pipe requires more labor to handle the pipe sections and may require heavier bracing and supports. The total weight of a 3-in. ID hose full of normal-weight concrete is 250 lb for a 25-ft length; a similar length of 5-in. hose full of concrete weighs 645 lb.

Aluminum pipe must not be used because of a reaction of the aluminum with the cement in the concrete that produces an expansion of the concrete, causing the concrete to suffer a significant loss of strength. A number of cases have been reported in which this happened on the job.

2 Materials

Aggregates. The size, shape, grading, and proportions of aggregates are all important in obtaining a pumpable concrete. Some operators suggest that the maximum size of coarse aggregate should be no larger than about 40% of the conduit diameter, as shown in Table 2.

Table 2. Maximum Aggregate Sizes for Different Conduit Sizes

Pipe or hose ID, in.	Max aggregate size, in.
1½	⅝
2	¾
2½	1
3	1¼
4	1½

Rounded or subrounded aggregates make better mixes for pumping than aggregates containing a large proportion of crushed material, although the latter can be used satisfactorily. If it becomes necessary to use crushed-rock coarse aggregate, every measure should be taken to assure maximum plasticity of the concrete as described in the chapter on mixes.

Grading of the aggregates should conform to the requirements of the code or specifications under which the work is being performed. Sand must contain adequate fines, with 15 to 20% passing the No. 50 screen and at least 3% passing the 100-mesh. In 1- or 1½-in. mixes, the total aggregate should contain about 10 to 15% pea gravel.

Poorly graded aggregates are apt to result in concrete that is difficult to pump, the friction of such concrete in the conduit offering so much resistance that maximum pumping efficiency is not achieved. Skip or gap gradings are not suitable.

Cement. Any of the usual types of portland cement can be used in concrete to be pumped.

Admixtures. Good construction practices govern the use of admixtures, and neither special limitations nor tolerances need to be applied so far as pumping is concerned. During hot weather a set retarder may be desirable if concrete is being pumped a long distance. Water reducers can be used to advantage. Pumping-aid admixtures are available and can be helpful in difficult pumping, such as a long pipe line, high lift, harsh mix, or poor aggregate grading. Among the latter are finely ground minerals, such as hydrated lime, pulverized quartz or limestone, and pozzolans. Some users report favorable results from certain water-soluble cellulose polymers.

3 Mixes

Concrete for pumping must be plastic and workable. Because of this many persons have felt that a very high percentage of sand, as much as 65% of total aggregate for a 1-in. maximum-aggregate concrete, is necessary. However, the heavy pumps can handle concrete with far less sand, and the small-line pumps have now been improved to the point where they can handle mixes with a higher proportion of coarse aggregate. Heavy pumps handle concrete with 1½-in. maximum aggregate containing 33% sand, 3 to 5% entrained air, and cement contents of about 580 lb per cu yd. While a slightly oversanded mix is more pumpable, other things remaining the same, excessive oversanding is not considered necessary. The percentage of sand in the mix should be based on the void content of the coarse aggregate, a properly proportioned mixture containing sufficient sand to fill the voids in the coarse aggregate and enough paste to coat the aggregate particles. If this condition is obtained, a pumpable concrete can be realized even with relatively poorly graded coarse aggregate. However, the better the overall grading, the better the pumping results will be.

If harsh, angular coarse aggregate must be used, it is especially important to make sure that both the fine and coarse are well graded, as previously described. A somewhat higher cement content and percentage of sand than normal will be helpful, and the maximum size of coarse aggregate should be about ¾ in. Entrainment of about 4% air will be beneficial.

Mixes handled by the small-line pumps may contain coarse aggregate as large as 1½ in. One manufacturer reports typical mixes pumped by their equipment as shown in Table 3.

Table 3. Mixes for Small-line Concrete Pump*

1. Cement, 440 lb/cu yd
   35% 1½-in. aggregate
   27% 1-in. aggregate
   38% sand
   Water-reducing retarder
   Slump: 3–5 in.
2. Cement, 500 lb/cu yd
   57% 1-in. aggregate
   43% concrete sand
   Slump: 3–5 in.

* Ridley and Company, Inc.

A plastic, workable mix of about 2½- to 5-in. slump is best. Visual inspection of the concrete as it emerges from the line assists in evaluating plasticity. The above slump limits are broad enough to cover any normal material and job conditions but should not be used as limits on any one job. Once a plastic, workable mix has been developed every effort must be made to maintain all conditions within reasonably close limits so as to keep a supply of suitable concrete coming to the pump. A water-reducing admixture, if used under careful technical control, is desirable in order to keep the amount of water as low as possible. If the slump is too low, the friction of pumping is higher, resulting in less volume of concrete per unit of pump horsepower. However, segregation tendencies are lessened. Mixes that are too wet, on the other hand, are more apt to segregate, a condition that is liable to result in line blocks or plugs.

Cement content is usually around 470 or 520 lb per cu yd, with a lower limit in the neighborhood of 420 lb per cu yd. Richer mixes can be pumped successfully. Pumps can handle 1½-in.-aggregate mixes, but ¾- or 1-in. mixes are commonly used.

In general, pumping difficulties that are caused by the concrete mix result from attempting to pump concrete with too much or too little slump; harsh, unworkable mixes caused by poorly graded or angular aggregates; concrete containing porous, highly absorptive aggregate, or aggregate too large for the machine or conduit.

A loss of slump during pumping is normal and should be taken into consideration when proportioning the concrete mixture. A loss of 1 in. per 1,000 ft of conduit is not unusual, the amount depending upon ambient temperature, length of line, pressure used to move the concrete, and moisture content of aggregate at time of mixing. The loss is greater for hose than for pipe, and is sometimes as high as ¾ in. per 100 ft.

Air-entrained mixes can be successfully pumped, although some operators recommend slightly less air than is normally desired for durability of the concrete. There is probably a slight loss of air during pumping, and evidence indicates that the concrete, while under pressure in the pipeline, suffers a temporary loss of workability because of compression of the entrained air.

4 Pumping

Before pumping of concrete is started, the conduit should be primed by pumping a batch of mortar through the line to lubricate it. A rule of thumb is to pump 5 gal of mortar for each 50 ft of 4-in. hose, using smaller amounts for smaller sizes of hose or pipe. Dump concrete into the pump hopper before the last of the mortar disappears into the pump loading chamber, pump at slow speed until concrete comes out the end of the discharge hose, and then speed up to normal pumping speed. Once pumping has started, it should not be interrupted if at all possible as concrete standing idle in the line is liable to cause a plug. Of great importance is to keep concrete in the pump receiving hopper at all times, which makes necessary the careful dispatching and spacing of ready-mix trucks. From the standpoint of the pumper, it is better to have a ready-mix truck stand by for a few minutes rather than to shut down the pump to wait for concrete. However, having ready-mix trucks standing by waiting to discharge their concrete is neither efficient nor desirable.

Other causes of line blocks are slump too high; harsh, unworkable mix resulting from poor aggregate grading; mix too dry or undersanded; bleeding of the concrete; a long line exposed to the hot sun; improper adjustment of the pump valves; dirty and dented pipe sections; or a kinked hose.

With heavy equipment, a line plug can cause a delay of an hour or more while the pipe sections are separated and the plug is located. With a 10-ft section of 8-in. pipe full of concrete weighing over 200 lb, the amount of labor involved can be appreciable. During an extended delay at the placing end, it is good practice to run the pump for a few strokes every few minutes, even though the concrete has to be wasted, in order to avoid a plug in the line. This is especially necessary in hot weather. If the pump can be reversed, thus reversing the flow of concrete, an attempt can be made to free the blockage in this manner. If the blockage cannot be freed after one or two attempts, the effort should be abandoned and the blockage cleaned from the system by other means. The plug in the pipe can be located by tapping on the pipe, starting at the pump. The dull thud of the tapping changes to a more hollow sound after the plug has been passed. Another method is to loosen but not release the couplings between pipe sections. Grout or mortar will spurt out if the coupling is between the pump and the blockage; after the blockage has been passed, the grout will not escape.

Concrete can be pumped upward, but downhill pumping has a few special problems, because the concrete is apt to separate or segregate in the pipe unless there is resistance to pump against. Resistance can be provided by a valve at the discharge end that can be adjusted to restrict the flow of concrete, or by inclining the final lengths of pipe upward.

If it is necessary to run the line up and over an obstruction, it is advisable to install an air-release valve at the highest point of the line to prevent an accumulation of air that could result in a line plug.

When nearing the end of a placement, with a heavy pump, the amount of concrete in the line should be computed, using Fig. 7 for this purpose, and the supply of concrete cut off in sufficient time to avoid excessive waste of a line full of concrete. Cleaning of the last of the concrete out of the line is accomplished with a "go-devil," a plug that is inserted in the line at the pump and is forced through the pipe by air or water pressure. When using air, there is no problem of disposal of the water following the go-devil, but care is necessary to prevent the go-devil from coming out of the end of the pipe like a bullet. As the go-devil nears the end of the line (determined by tapping on the pipe) the air pressure should be progressively lowered. Near the end, the resistance to the go-devil decreases to the point where the air can be shut off entirely, and the expanding air will force it out.

The use of water under pressure does not present the safety problem that compressed air does. However, arrangements must be made at the forms to dispose of the water outside the forms. This can sometimes be done by breaking the line before it comes over the forms.

Small-line pumps are also cleaned by pumping clean water. This is accomplished by placing a large sponge or rubber flush plug in the feed end of the conduit. (The sponge is inserted in the pump section line of the vacuum roller machine.) After being washed out, the hopper is filled with water and the pump is started, forcing the remaining concrete out of the conduit followed by the sponge or plug. One type of pump has a separate washout pump that supplies water to force the plug through the conduit. All parts of the pump that come in contact with the concrete, including the hopper, heads, and cylinders, must be cleaned of concrete and mortar.

Users of the vacuum squeeze pump have developed a booster system for pumping concrete for high-rise construction in which the suction line of one pump is connected directly to the discharge line of the previous pump, all pumps in the system being controlled from one central panel. In this way, it is possible to maintain a steady maximum output of concrete regardless of height or distance. There are no hoppers on the booster pumps. The entire system can be reversed to simplify cleanup and discharge of wash water at ground level.

Lightweight structural concrete can be successfully pumped. The main problem in pumping lightweight concrete has been the loss of plasticity of the concrete resulting from absorption of water by the dry, porous aggregate particles when the concrete undergoes pressure in the line. Some operators report a slump loss of about 3 in. between the time the concrete is dumped into the pump hopper and when it comes out of the hose. This has made necessary a slump of 7 in. or more at the pump. Loss of slump can be minimized by prewetting the coarse aggregate, accomplished by some aggregate producers by vacuum and other treatments at the manufacturing plant. Success is more likely if natural sand can be used for the fine aggregate.

PNEUMATICALLY APPLIED CONCRETE AND MORTAR

Sprayed dry-mix mortar has been used for many years for both new construction and repair of old structures. Equipment is also available for the pneumatic delivery and application of wet concrete mixes containing coarse aggregate as large as 1 in. No formwork is necessary in many instances; yet intricate shapes and thin overlays can be successfully constructed, provided good materials are used and proper procedures are followed.

5 Shotcrete

Description. When cement mortar is sprayed on a surface, the product is variously known as pneumatically applied mortar, shotcrete, or gunite. The process consists essentially of mixing dry sand and cement in a mixer, then placing this mixture in the delivering equipment, the first part of which is a vertical double-chambered vessel (known as the gun) wherein the mixture is placed under pneumatic pressure (Fig. 8). Under pressure, the mixture flows through a rubber hose to the nozzle, where water joins the material, wetting the mixture as it leaves the nozzle under high velocity. Compaction is achieved by the force exerted by the impact of the mortar on the receiving surface.

Fig. 8.   A small mortar-gunning machine with a rated capacity of 2½ cu yd per hr. (*Air Placement Equipment Company.*)

Shotcrete produces a high-quality material with the following desirable properties:
1. Low water-cement ratio, resulting in high strength and low permeability.
2. Dense concrete, as a consequence of low water content and high impact velocity.
3. Superior bonding ability, making it especially suitable for repair work on many types of structures.
4. Relatively simple work-area requirements. An air compressor and source of water are necessary, but buckets, cranes, elevators, trucks, and similar equipment are not necessary. Concrete (or mortar) is easily transported through a hose from the gun to the site of application.
5. Shotcrete lends itself to the production of many shapes and thin sections with a minimum of formwork, or no formwork at all.
6. Resistant to weathering and many types of chemical attack. Good abrasion resistance.
7. With proper aggregates, has good refractory properties.

Uses. Application of mortar by the pneumatic method is adaptable to either new construction, the application of a coating or covering, or repair of existing structures.

New construction includes linings for canals, reservoirs, tunnels, and pipe; thin slabs, walls, and domes; erosion control on earth slopes; swimming pools.

Coatings may be applied to deteriorated concrete or masonry, after removal of unsound, deteriorated material; to rock surfaces to prevent scaling or disintegration of newly exposed surfaces; to steel and timber for fireproofing; to pilings for encasement. Coatings are used for refacing dams and for roof protection in mines and tunnels.

Materials. Any standard cement conforming to ASTM Designation: C 150 may be used. Normally Type I or Type II is used, unless another type is specified. Aluminous cement is used for refractory applications.

Aggregate should conform to the requirements of ASTM Designation: C 33, except for grading. The grading suggested by the Gunite Contractors Association is shown in Table 4. (Some operators use plaster sand for the finish cover coat.)

Table 4. Sand for Gunite*

Sieve size	% passing	
	Min	Max
⅜-in.	100	
No. 4	95	100
8	65	90
16	45	75
30	30	50
50	10	22
100	2	8

Moisture content: 3–6%.
Fineness modulus: 2.70–3.30.
* Gunite Contractors Association Brochure G55–66.

To prevent segregation and to assure efficient mixing and application, the sand should have between 3 and 6% moisture.

Any potable water suitable for use in concrete can be used in shotcrete. A uniform supply at steady pressure is necessary.

Reinforcement consists of bars, welded mesh, or expanded metal lath, depending on the nature of the structure, thickness of the coating, and extent of the work. Mesh and lath may be self-furring, or used with chairs. Steel must be free of loose scale, heavy rust, oil, paint, or any material that will interfere with bond.

Mixes. Usual mixes consist of 1 part cement to 4 or 4½ parts of sand by dry loose volume. Strength of these mixes will normally be between 3,000 and 4,000 psi at 28 days, or perhaps slightly higher for the 1:4 mix. Leaner mixes are sometimes used, but the amount of rebound is considerably higher with the lean mixes. Water-cement ratio is from 0.35 to 0.50 by weight.

Preparation of Surface. Forms, when required, must be substantial, true to line and grade, and so constructed as to permit the escape of air and rebound. When constructing a wall, only one side needs to be formed. Columns are formed on two adjacent sides, and beam forms usually consist of one side and the soffit. Sufficient grounds must be provided to establish the surface and finish lines of the work. Ground wires should be taut and true to line and surface, and must be maintained in a taut condition until the finish coat has been applied.

In repairing old concrete or masonry, all the old unsound material must be removed so the shotcrete can be applied to a sound surface. All weathered and unsound concrete or masonry should be chipped off with pneumatic tools, using sawtooth bits and gads instead of chisels or sharp points. The entire surface should be sandblasted in areas where chipping is not required. Finally, the surface is cleaned with compressed air and water. Heavily corroded steel should be sand-blasted. About an hour before application of mortar, concrete and masonry should be wetted, but not saturated; otherwise suction will be reduced. Surface should be moist, but not wet, when the shotcrete is applied.

Earth surfaces to receive shotcrete should be well compacted, trimmed to line and grade, moist, and free of frost.

Water seepage or leakage must be removed by installation of drainpipes to take the water away from the surface being shot. Seeps in rock or old concrete can be sealed with a quick-setting mortar, such as a mixture of portland and aluminous cement with 2 parts sand, portland-cement mortar with 5 to 10% (by weight of cement) of sodium carbonate, or certain proprietary compounds.

Reinforcement must be rigidly anchored in place so it will not move. In placing reinforcement for a repair, the steel can be wired or welded to existing reinforcement, or anchored to dowels secured in holes drilled into the concrete. Cement mortar or epoxy mortar can be used to grout the dowels in the concrete. Joints in reinforcing bars should be lapped 40 diameters, and mesh should be lapped one full mesh. Reinforcement should be furred at least ½ in. from the surface and covered with a minimum of ¾ in. of mortar on surfaces not subject to fire hazard. Fire-hazard exposure should have the same cover as cast-in-place concrete.

Equipment. Mixers are of either the rotating-drum or paddle type. Some rigs include a mixer as part of a self-contained unit. More elaborate rigs include mixer, elevator, air compressor, gun, and necessary auxiliary parts, such as a hose reel, all mounted on a truck or trailer. Placement capacities of 20 cu yd per hr are common.

The commonest gun is the vertical double-chambered type (Fig. 8). The dry-mixed mortar or concrete is admitted to the upper pot while the gate between the two pots is closed. The charging port is then closed, the upper pot is pressurized, and the transfer gate is opened, allowing the material to flow by gravity into the lower chamber. By means of a feed wheel or similar device the mix is fed into the delivery hose where a stream of air carries the material to the nozzle. As soon as the upper pot is empty the transfer gate is closed, pressure is released on the upper chamber, a new batch of material is admitted through the charging port, and the cycle is repeated. Meanwhile, pressure is maintained on the lower pot, which continues to deliver an uninterrupted flow of material to the hose.

Another type of gun makes use of a screw to feed the mixture to the delivery hose, where the air stream picks up the mixture and delivers it to the nozzle.

Special hose in 1¼ to 2 in. inside diameter is available. Most manufacturers rate their machines with 150 or 200 ft hose. However, hose lengths of 500 ft are frequent, and lengths of 1,000 ft have been used on occasion. The Gunite Contractors Association recommends not less than 365 cu ft of free air per minute at a minimum pressure of 45 psi in the gun chamber for proper placement and adequate blow-out jet requirements. If more than 100 ft of hose is used, air pressure should be increased by 5 psi for each additional 50 ft of hose. However, a maximum of 75 psi is usually adequate.

Water is conveyed through a separate hose to the nozzle under a pressure of about 15 psi higher than the air pressure in the hose at the gun. Radial perforations in the water ring at the nozzle spray water into the cement-sand mixture as it passes through the nozzle.

Application. Before starting to shoot shotcrete, precautions should be taken to protect property in the area. Adjacent construction, openings, shrubbery, and all areas that might be discolored or damaged by rebound, cement, water, or dust must be covered with tarpaulins or plastic sheets to protect them from damage.

The amount of water added at the nozzle should be a minimum but should

be sufficient to prevent excessive rebound and assure hydration of the cement.
The nozzle should be held as nearly perpendicular as is possible to the surface
being treated and should be held uniformly the same distance away from the
surface at all times. For a large nozzle, the distance is about 3 ft, ranging down
to as close as 12 in. for a small nozzle in close quarters.

When enclosing reinforcing steel, the stream from the nozzle should be directed
at an angle so as to fill the space behind the bars, shooting from each side
of each bar individually. The nozzleman's helper should use an air jet to blow
out rebound ahead of the application of shotcrete.

The flow of shotcrete out of the nozzle should be uniform, without slugs of
wet or dry material. If the nozzle starts slugging, it must be turned away from
the work until the flow becomes steady again. Slugs are usually caused by insufficient air for the amount of material being handled or the length of hose in
use.

Guniting or shotcreting should not be done during a rainstorm, as the rain
will wash cement out of the material, nor should it be done during a high wind
that blows the nozzle spray and makes control difficult. Work should be suspended
during freezing weather, unless all operations can be protected from freezing.

Shotcrete requires expert and conscientious workmanship and should be done
under careful supervision and inspection, employing only experienced workmen.
The ACI Recommended Practice[3] points out the need for skill and experience,
as follows:

The foreman should have good personal experience, preferably including not less than
two years as a shotcrete nozzleman.

The nozzleman should have served at least six months apprenticeship on similar applications and should be able to demonstrate by tests his ability to perform satisfactorily his
duties and to gun shotcrete of the required quality.

The nozzleman's duties are to:

1. Insure that all surfaces to be shot are clean and free of laitance or loose material,
using air and air-and-water blast from the nozzle as required.

2. Insure that the operating air pressure is uniform and provides proper nozzle velocity
for good compaction.

3. Regulate the water content so that the mix will be plastic enough to give good compaction and a low percentage of rebound, but stiff enough not to sag. (In the dry-mix
process the nozzleman actually controls the mixing water, while in the wet-mix process he
directs changes in consistency as required.)

4. Hold the nozzle at the proper distance and as nearly normal to the surface as the type
of work will permit, in order to secure maximum compaction with minimum rebound.

5. Follow a sequence routine that will fill corners with sound shotcrete and encase reinforcement without porous material behind the steel, using the maximum practicable layer
thickness.

6. Determine necessary operating procedures for placement in close quarters, extended
distances or around unusual obstructions where placement velocities and mix consistency
must be adjusted.

7. Direct the crew when to start and stop the flow of material, and stop the work when
material is not arriving uniformly at the nozzle.

8. Insure that sand or slough pockets are cut out for replacement.

9. Bring the shotcrete to finished lines in a neat and workmanlike manner.

The delivery equipment operator operates the mechanical feeder (gun) and directs the
work of the mixing crew. He receives instructions or signals from the nozzleman to deliver
material as required. His most important duty is to see that the material flow to the nozzle
is at a uniform rate and at the pressure requested by the nozzleman. It is also his responsibility to see that no premixed material that stands more than 45 minutes is permitted to
get to the nozzleman.

The nozzleman's apprentice or helper operates an air blow pipe at least ¾ inches in
diameter, to assist the nozzleman in keeping all rebound and other loose or porous material
out of the new construction (except in classes of work where the trapped rebound can
readily be removed by the nozzleman). He also assists the nozzleman in other assignments
as required.

The nozzleman's other assistant is the hoseman. He helps with the material hose, keeping it advanced for efficient progress. On some jobs the nozzleman's helper also acts as
hoseman.

On vertical or overhanging surfaces, the mortar should be applied in layers not exceeding ¾ in. in thickness. On horizontal or nearly horizontal surfaces (when shooting down), the thickness may be as much as 3 in. Excessively thick layers will cause the mortar to slough, or sag. Sagging may also occur if insufficient time elapses between layers. Time between successive layers should be at least 30 min but not so long as to permit the previous layer to set completely, as the surface is apt to glaze, with a resulting loss of bond between the layers. Light brooming of the surface will remove any rebound and assist in bonding the next layer of mortar.

Rebound. A portion of the mortar bounces from the surface where it is being applied, the amount varying with air pressure, quality of sand, placement conditions, and the cement and water contents. The amount of rebound varies from 25 to 50% when shooting overhead, 15% of less when shooting down, and 15 to 30% on vertical or sloping surfaces. It is higher with high nozzle velocity and decreases with higher water content. This rebound material, consisting mainly of coarse sand particles, should not be reused because of its variable quality and low cement content, although a part of it can be used as sand provided it does not exceed 25% of the total sand in the batch.

Rebound is higher when a coarse sand is used, when a lean mix is used, or when the water content of the mix is too low. Frequent moving or "waving" of the nozzle, or shooting at an angle, increases rebound. Sometimes holding the nozzle close to the surface being shot decreases rebound, but this is apt to result in a rough or irregular surface. Use of a wetting agent or water-reducing admixture, introduced with the sand, sometimes reduces rebound.

Finishes. In many applications, the surface produced by the shooting is satisfactory. Tunnels, channel linings, ditches, and dam facings are a few of the structures frequently treated in this manner. Usual practice is to apply the finish coat by starting at the top and working down to avoid shooting over the finished work.

If finishing is required, the first operation after shooting is to rod (screed) off to the ground wires or other guides, then remove the grounds, after filling in low spots revealed by the rodding. The thin-edged rod is worked back and forth with a slicing motion, starting at the bottom of the area. Sometimes this is all the finishing that is required. Additional finishing might consist of brooming, sacking, floating with a rubber or wood float, or troweling. The amount of pressure required is normally less than that required for regular concrete finishing.

Curing. A fine water spray, or fog, should be applied to the surface of the shotcrete as soon as possible, to prevent drying out. The initial application of fog should be just enough to maintain the moisture in the mortar without wetting the surface. As soon as the surface has hardened, a heavier application of water is permissible, keeping the shotcrete wet for a period of from 5 to 7 days, depending on weather conditions, or as required by the specifications.

Hot, dry weather, especially when accompanied by wind, causes the surface of shotcrete to dry very rapidly, leading to cracking, spalling, scaling, and low strength. Under these conditions the fresh shotcrete must be covered with burlap or similar material kept damp for the required curing period.

Liquid membrane-forming compounds are satisfactory for curing, and should be applied after first dampening the surface with water. The surface must be dark in color but not shiny-wet when the compound is applied.

Testing. Cylinders for compressive-strength tests are made by shooting the mortar vertically into molds made of ½-in.-mesh hardware cloth. The excess material is trimmed off the outside of the mold immediately after filling. Cylinders should be handled and stored under standard conditions. After several days, when the shotcrete has developed sufficient strength, the hardware cloth is removed. Cylinders can be standard 6 × 12-in. size, although some agencies prefer a smaller size. Because strengths indicated by cylinders are liable to be somewhat erratic, strength of shotcrete is sometimes determined by coring the finished work. In the event that the structure cannot be cored, cores may be cut from test panels made especially for that purpose.

The presence of hollow spots or areas containing an excess of rebound included in the shotcrete can be detected by sounding the surface of the finished work with a hammer. Imperfections should be removed and replaced with new shotcrete.

6   Refractory Applications

Shotcrete is sometimes used for lining smokestacks, breechings, and process equipment in furnaces, steel mills, oil refineries, and similar installations, making use of aluminous cement and certain selected aggregates (Table 5), where resistance

Table 5. Aggregates for High-temperature Exposures

Material	Max temp, °F	Strength	Conductivity	Abrasion resistance	Insulating value	Corrosion resistance	Volume stability
Siliceous sand.	500	High	High	Good	Poor	Good	Poor
Calcined clay or shale	2,000	High	Medium to low	Good	Good	Good	Fair to good
Vermiculite and perlite..	2,000	Low	Low	Poor	Excellent	Good	Good
Traprock (igneous)...	1,800	High	High	Good	Poor	Good	Fair
Crushed firebrick.......	2,500	Medium	Medium	Fair	Fair	Fair	Good
Granulated blast-furnace slag........	1,000	Low	Low	Good	Good	Good	Fair

Note.   Ratings are comparative only and may vary widely for different materials.

to high temperature is an important factor.   Mixes are usually about 1:3 or 1:4 by dry loose volume. A small amount of plastic fireclay (not over 12 lb per cwt of cement) may be added to improve workability. There are available on the market premixed dry-packaged castable refractories, consisting of aluminous cement and aggregate, suitable for gun application.

Steel must be clean, and a reinforcing mesh should be placed about 1 in. away from the steel shell, covered with at least 1 in. of shotcrete. Mesh is usually 3 × 3, 10–10, or 2 × 2, 12–12, rigidly supported.

After application, the shotcrete should be permitted to set for a maximum of 8 or 10 hr; then water curing is applied for 24 hr. In lieu of water curing, a liquid membrane compound can be applied.

7   Wet-mix Shotcrete

The previous discussion in this chapter has dealt with dry-mix shotcrete, or gunite, a product that has been used for many years. Equipment for pneumatic application of premixed mortar and concrete is also commonly available. The principal advantage of wet-mix shotcrete over dry-mix is that the amount of water can be established beforehand and maintained at this value during the gunning operation. Gunning capacity is about the same as for dry-mix, a maximum of about 20 cu yd per hr (Fig. 9). Pumps are of the auger type, in which the mixture is fed into the hose by means of a screw. A high-velocity flow of air conveys the mixture to the nozzle where it is shot onto the receiving surface. Another type consists of a pressurized tank in which rotating mixing paddles inter-

Fig. 9. Application of wet-mix shotcrete to a ditch lining. (*Transit Mixed Concrete Company.*)

mittently introduce air with the material into the hose or pipe. These machines handle concrete containing 470 lb of cement per cu yd and ¾-in. aggregate, with a zero slump.

Much of the previous discussion covering surface preparation, application, finishing, and curing of dry-mix shotcrete applies to wet-mix also.

REFERENCES

1. ACI Committee 304 Report, Placing Concrete by Pumping Methods, *J. ACI*, May, 1971.
2. A series of five booklets covering equipment, materials, mixes, and application of concrete pumps. Challenge-Cook Bros., Inc. Industry, Calif., 1971.
3. Recommended Practice for Shortcreting, ACI 506-66, American Concrete Institute, Detroit, 1966.
4. "Gunite Specifications and Recommended Practice," Brochure G55-66, Gunite Contractors Association, Burbank, Calif., 1966.

Chapter 37

VACUUM PROCESSING OF CONCRETE

JOSEPH J. WADDELL

Vacuum processing of concrete is accomplished by applying a vacuum to surfaces of fresh concrete, either formed or unformed. This patented process removes as much as 40% of the water from a few inches of the concrete surface, producing in effect a "casehardened" concrete. At a depth of 6 in. it is common to remove 20% of the water. In addition, air bubbles or "bug holes" are removed from the surface, as shown in Fig. 1.

FIG. 1. Surface finish of vacuum concrete cast against forms. The lower portion was cast against ordinary lumber forms, and the upper portion was vacuum-processed. Note absence of pits in the vacuumed surface. (*Courtesy of Aerovac Corp.*)

The effect of vacuum processing is one of compaction; the cement paste is consolidated or densified upon the removal of water from the mass. Removal of air is a surface phenomenon only, but the effect of vacuum in removing water is effective for a depth of as much as 12 in. This compaction and lowering of the water-cement ratio results in markedly greater compressive strength; up to an age of 7 days the strength of vacuumed concrete is as much as 100% higher than unprocessed concrete, and normal 28-day strength will be reached in 8 or 10 days. Final strength will also be higher. The curves in Fig. 2 compare 28-day strengths for vacuum and plain concrete at different water-cement ratios. Durability is improved, the resistance to erosion, abrasion (see Fig. 3), and cycles of freezing and thawing being greater than for plain concrete. Shrinkage and cracking tendencies are reduced.

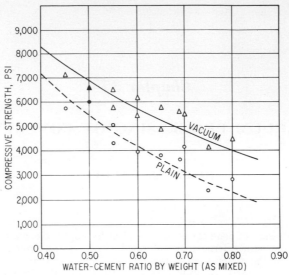

Fig. 2. Vacuum treatment improves compressive strength of concrete at all water-cement ratios. (*Courtesy of Aerovac Corp.*)

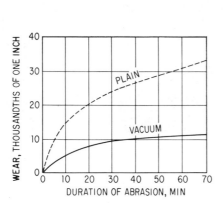

Fig. 3. Abrasion resistance of concrete is improved by vacuum processing. (*Courtesy of Aerovac Corp.*)

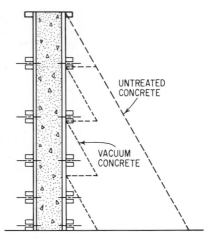

Fig. 4. Ten minutes after placing each lift of concrete, vacuum processing reduces pressure against the form to zero by reducing slump to zero. (*Courtesy of Aerovac Corp.*)

Another advantage of vacuum concrete is an early solidification or apparent hardening of the concrete that in some instances permits removal of forms in a matter of minutes instead of hours. Figure 4 shows that vacuum processing reduces pressure on the form practically to zero for each lift processed, but the untreated concrete is still exerting pressure on the entire height of form after completion of placing. Similarly, finishing of slabs, as illustrated in Fig. 5, can start within

FIG. 5. Vacuum pads, shown in the center, have been applied to the fresh concrete. After processing, the concrete is immediately finished. (*Courtesy of Aerovac Corp.*)

a few minutes after placing the concrete and application of vacuum. Early strength is especially desirable in precast work, as it permits removal of forms as soon as processing has been completed, thus releasing the equipment for the next round of casting.

The value and extent of early form removal and early finishing must be considered in connection with any other advantages when determining whether to use the vacuum process, as there are some limitations. If careful formwork and proper vibration will do the job, there is no advantage in applying a vacuum treatment. It is difficult to justify vacuuming merely for improving the appearance of concrete, and usually proper air entrainment will provide all the durability that is required, although it has been reported that entrained air and vacuum processing together give better durability than either treatment alone. Because of the time involved, there is little benefit in attempting to process to a depth greater than 12 in.

Normal concrete mixes can be successfully vacuumed, although best results are obtained when the fines are at a minimum, that is, for relatively lean mixes containing the minimum practicable amount of sand, and with the coarser sands.

FIG. 6. Vacuum mat for horizontal surface. The surface of the pad that contacts the concrete is open at corner to show lath, screen, and filter cloth. (*Courtesy of Aerovac Corp.*)

FIG. 7. Vacuum mat for processing inclined or horizontal formed surface. Lining of lath, screen, and filter cloth is applied to the rectangular areas. Vertical and horizontal seal strips limit height and width of lifts subjected to vacuum processing. (*Courtesy of Aerovac Corp.*)

Uniformity of slump from batch to batch is important, and variations should not exceed 1 in. Vacuum should be applied as soon as possible after the concrete has been placed, while it is still plastic. Vibration of the concrete during the first few minutes of processing is desirable, as this tends to close the small channels in the concrete that are formed as the vacuum draws the water out, thus providing improved watertightness. Efficiency of the vacuum system is somewhat higher at lower temperatures.

On unformed surfaces the fresh concrete is struck off to grade; then the mats are laid on the surface and vacuum is applied. Processing usually requires between 10 and 20 min per panel. Immediately after processing workmen can walk on the concrete without leaving footprints in it, to remove the mats. Floating of the concrete is then commenced, and finishing proceeds in the normal manner.

Vacuum is applied to the concrete surface by means of special form liners, or by mats. These liners and mats are shown in Figs. 6, 7, and 8. On vertical

Fig. 8. Attaching filter cloth to form which has been covered with metal lath and screen. Form is being prepared for the next lift on the interior of the training wall.

surfaces the form is lined with two layers of screen cloth covered with muslin or plastic, and vacuum is applied to fittings on the exterior of the form. Usual practice is to make the panels long and narrow, about 12 to 18 in. high, so that each area will be covered with concrete in as short a time as possible and the vacuum can be applied.

On horizontal surfaces mats, covering about 12 sq ft, are made of reinforced plywood lined with screen and muslin or plastic. For curved, unformed surfaces, the mats are made of light-gage flexible steel instead of plywood.

A vacuum of between 15 and 20 in. of mercury is drawn by a special vacuum pump. On a small job, it is sometimes possible to obtain sufficient vacuum by connecting the intake of an air compressor to the vacuum system, exhausting the compressor discharge.

Chapter 38

PREPLACED-AGGREGATE CONCRETE

JOSEPH J. WADDELL

Originally developed for structural concrete repairs, preplaced- or prepacked-aggregate concrete, or prepacked concrete, which is composed of graded coarse aggregate solidified with a special portland-cement mortar, has been found to be suitable for use in certain types of new construction as well. Preplaced-aggregate concrete is a system in which coarse aggregate is placed in the form and a special mortar is then pumped into the form, starting at the bottom, to fill the voids and create a solid concrete unit. Some of the equipment, procedures, and materials are patented, and the user should check with authorized installers before using this method.

One of the main advantages claimed for preplaced-aggregate concrete is the ease with which the concrete can be placed in locations where conventional concrete could be placed only with extreme difficulty. In some cases on jobs where ready-mixed concrete is not readily available, preplaced-aggregate concrete can effect a saving in time and money. The point-to-point contact of the preplaced-aggregate particles and the special properties of the mortar reduce shrinkage to a negligible amount. Drying shrinkage, following proper curing, is normally in the range of 200 to 400 millionths as compared with ordinary concrete, which ranges between 400 and 600 millionths.

Preplaced-aggregate concrete has been used for repairing damaged or deteriorated bridge piers, for refacing dams and spillways, for repairing structural concrete, and for pile encasements. It is well suited to underwater construction and cofferdam seals. In some usages, such as in mines and tunnels, waste rock can serve as coarse aggregate. Special procedures for repair methods are described in Chap. 48.

A common usage of preplaced-aggregate concrete is in constructing high-density biological shielding of nuclear reactors and other equipment making use of radioactive substances. A portion of one of these structures, with preplaced aggregate in place, is shown in Fig. 1. Heavy aggregates such as barite or iron ore are required which, because of their heavy weight, angular shape, and friable nature, are extremely hard to handle. Hand placing of the aggregate is sometimes necessary because of the congestion of pipes, instruments, reinforcement, and similar gear within the form.

About 450,000 cu yd of prepacked concrete was used in the substructure for the Mackinac Bridge. All 32 piers below water, and much of the above-water concrete in the principal piers, employed prepacked concrete. Figures 2, 3, and 4 show some of the details of this construction, which is typical of the large-scale work sometimes performed.

Figure 5 shows a cofferdam cell for the Columbia River Bridge at Astoria, Ore. This was typical cofferdam construction, in which washed coarse aggregate was placed below water inside the cofferdam, and mortar was then injected from

Fig. 1. Preplaced aggregate in an atomic-reactor shield, ready to receive mortar. (*Courtesy of Lee Turzillo Contracting Co.*)

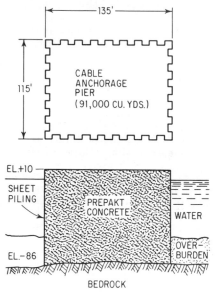

Fig. 2. One of the large piers for the Mackinac Bridge which was placed in one continuous operation by the prepacked method. (*Courtesy of Intrusion-Prepakt, Inc.*)

the bottom up through the grout pipes shown, displacing the water upward and out.

A typical repair job was the refacing of a railroad grade separation in which scaled and disintegrated concrete was first removed from the wing wall, forms were erected (Fig. 6), aggregate was placed, and finally mortar was pumped into the bottom of the form. As the mortar rose inside the form, it filled the voids in the aggregate, resulting in a solid concrete facing for the wall. Bond with

FIG. 3. Self-unloading boat depositing aggregate for the large pier. Note that grout pipes are in position. (*Courtesy of Intrusion-Prepakt, Inc.*)

FIG. 4. Aggregate in place ready for injection of mortar through the grout pipes extending above the aggregates. (*Courtesy of Intrusion-Prepakt, Inc.*)

FIG. 5. Cofferdam for Pier 169 for the Columbia River Bridge. Mortar will be injected through the long pipes that protrude above the river. The shorter pipes serve as gages to measure the elevation of mortar as it is pumped upward in the aggregate. A floating batch plant is in the background to the right. (*Courtesy of Lee Turzillo Contracting Co.*)

FIG. 6. Forms erected for refacing of a deteriorated railroad wing wall. (*Courtesy of Lee Turzillo Contracting Co.*)

the old concrete was assured by removing all old disintegrated concrete down to solid, sound concrete, and by virtue of the small amount of shrinkage of the preplaced aggregate facing.

Any of the standard types of portland cement can be used. Fine aggregate for the mortar should conform to the requirements of ASTM Designation: C 33, except that the grading is somewhat finer, as shown in Table 1, which is an average of commonly recommended gradings. Fineness modulus should not be below 1.30 and the maximum should be about 2.00 or 2.10. A well-graded sand near the middle of the limits is best. Uniformity of grading is very important, and specifications usually place a limit on the amount that the fineness modulus of a single sample taken at the mixer can vary from the average fineness modulus.

Coarse aggregate, which can be either gravel or crushed stone, must meet ASTM C 33, and should be so graded as to have a minimum void content, usually specified to be between 35 and 40% after compaction, being least for a large-maximum-size smooth, well-graded, well-compacted gravel and highest for small-size, poorly graded crushed stone. The maximum size can be as large as convenient to handle, provided that the maximum size does not exceed one-fourth of the minimum dimension of the portion of the structure in which the aggregate is to be used, or two-thirds of the minimum distance between reinforcing bars. The minimum size should be no smaller than ½ in. when the concrete section is not more than 12 in. thick, or ¾ in. for thicker sections, in order to provide ample space for the mortar to flow into interstices and become distributed throughout the form. The specifications should allow a tolerance of 5 to 10% undersize passing the nominal minimum-size screen.

Table 1. Suggested Sand Grading for Preplaced-aggregate Concrete

Sieve size No.	Cumulative % retained	
	Min	Max
8	0	
16	0	5
30	17	43
50	52	75
100	70	90
200	90	100

Admixtures consist of a pozzolanic filler which imparts a degree of workability and contributes to ultimate strength of the mortar. Fly ash and similar fine materials have been used for this purpose. Corps of Engineers Specification CRD-C-262, Pozzolan for Use in Portland Cement Concrete, is sometimes referred to with respect to this material. A proprietary intrusion aid or fluidifier, used in the mortar to inhibit settling, lower water requirement, delay early stiffening, and entrain air, usually contains a small amount of aluminum powder or similar material that reduces settlement shrinkage by causing a slight expansion of the mortar just before initial set. It is sometimes required to meet the requirements of Corps of Engineers Specification CRD-C-566, Grout Fluidifier.

Forms are constructed in much the same manner as good formwork for conventional concrete, but absorptive form lining cannot be used. Because of the fluid nature of the mortar, joints between panels must be tight to prevent leakage that would not be a problem with conventional concrete. In some cases the joints are sealed with putty. Forms should be adequately vented for the escape of air and water during pumping of mortar.

Proportions of the mortar vary, depending upon the type of construction or repair being performed and the materials available. Homogeneity and uniformity from batch to batch are important. One typical mix consisted of portland cement, 190 lb; sand, 3 cu ft; pozzolanic filler, 75 lb. Preplaced aggregate was ¾- to 6-in. gravel and ¾- to 2½-in. crushed granite. Other projects have used mixes ranging from about 165 to 750 lb of cement per cu yd.

The washed and drained coarse aggregate should be handled and placed in the forms in such a manner that segregation and breakage are minimized. Sometimes an 8- or 10-in. pipe chute can be used, with rubber-belting baffles. Reinforcement and items to be embedded in the concrete are fixed in place, and vertical grout pipes, usually ¾ in. diameter, spaced about 5 ft apart, are installed as placement of the aggregate starts and are extended above the top of the form. Actual spacing of the mortar-injection pipes is controlled by the characteristics of the section being constructed, such as size and depth of the section, size of aggregate, and proximity of forms or embedded items. In an extremely tight and congested location aggregate may be placed by hand to ensure closure around pipes and other items. The aggregate, when placed above water, can be lightly tamped or vibrated during placing to minimize voids. Vent pipes must be provided in certain types of construction where the aggregate is completely enclosed.

Mortar is pumped through the injection pipes starting at the lowest point in the form. Pumping should be done slowly and at a uniform rate without interruption so the mortar, as it rises in the aggregate-filled form, will completely fill all voids in the coarse aggregate. Vibration of the forms in the area of the mortar surface during pumping helps to remove air bubbles and improve appearance of the completed concrete.

If the top surface is to be finished, mortar is pumped into the form until the prepacked aggregate is flooded, any diluted mortar is removed by brooming, and a thin layer of pea gravel is spread over the surface. After the pea gravel has been tamped down into the mortar, the surface can be struck off and floated or troweled, as required.

Stripping of forms and curing are accomplished in accordance with good construction practices described elsewhere in this handbook.

Chapter 39

PORTLAND-CEMENT PLASTER

JOSEPH J. WADDELL

Portland-cement plaster is a facing material for either interior or exterior application to almost any building surface. When properly prepared and applied, it has properties akin to those of good concrete. Stucco is sometimes meant to designate exterior plaster. Strictly speaking, the term stucco is usually accepted as meaning the final coat of exterior plaster in which the color and texture of the plaster are obtained.

Plaster is mixed on the job by combining plastering sand, portland cement, and water to produce a mixture with a consistency suitable for either hand or machine application.

The chief advantage of portland-cement plaster, in addition to the strength and durability that make it especially suitable for exterior exposure or for interiors to be exposed to wetting or severe dampness, is its versatility. It can be used to construct free-form curved and irregular surfaces without formwork by using the reinforcement itself as a form. In addition to form and texture, plaster lends itself to such artistic treatments as fresco and graffito.

1 Materials

Cement is usually specified as Type I or II conforming to the requirements of ASTM Designation: C 150 except in respect to the limitations on insoluble residue, air entrainment, and additions subsequent to calcination. Plasticizing agents in an amount not exceeding 12% by volume may be added to Type 1 or Type II cement in the manufacturing process. This cement is known as plastic cement and is commonly used for hand-applied plaster. For gun or machine application, plastic cement is modifiied by the addition of 2 or 3% of asbestos fibers, added at the mill. In some parts of the country masonry cement, conforming to ASTM Designation: C 91, Type II, is used. In other areas regular Type I or II cement is employed, requiring the jobsite addition of a plasticizer such as hydrated lime or short-fiber asbestos. When lime is used with regular cement, its amount should not exceed 10% by weight of the cement.

Aggregate may be normal-weight sand (ASTM: C 33) or, where weight or fireproofing is of prime importance, vermiculite or perlite (ASTM: C 332). Sand should be washed clean, free of organic matter, clay, and loam, and well graded, conforming to the specified grading, as shown in Table 1. The portion retained on (or passing) each sieve should be near the midpoint of the specifications. Especially to be avoided is an excess of fines. A well-graded sand has a minimum of voids, making a plaster with a minimum water requirement, other things being the same.

Quality and type of aggregate for finish-coat stucco are usually determined by the stucco manufacturer. In most cases color of aggregate is important. Sometimes special materials, such as perlite fines or mica flakes, are added for special effects.

Table 1. Sand Grading
Cumulative percents retained

Sieve size	USAS[a] A42.2 and A42.3		UBC[b] and ASTM: C 35		ACI[c] Committee 524		Stucco Manufacturers Association[d]		PCA[e]	
	Min	Max	Min	Max	Min	Max	Min	Max	Min	Max
4	0		0		0		0		0	
8	0	10	0	5	2	8	0[f]	5[f]	0	
16	10	40	5	30	22	38	10	35	10	40
30	30	65	30	65	52	78	30	65	30	65
50	70	90	65	95	82	100	70	90	70	90
100	95	100	90	100	97	100	90	95	95	100

[a] United States of America Standards Institute Specification for Portland Cement Stucco and Portland Cement Plastering.
[b] Uniform Building Code.
[c] Report of ACI Committee 524, *J. ACI*, July, 1963.
[d] Proposed Specifications for Finish Coat Stucco.
[e] "Plasterers Manual," Portland Cement Association.
[f] No. 12 sieve.

Perlite aggregate is a volcanic glass that has been expanded by heating to at least 1400°F. Perlite aggregate weighs less than 15 pcf and produces plaster weighing approximately 60 pcf. It is used for interior plaster where insulating value and minimum weight are of importance. It is not recommended for exterior exposure.

Vermiculite is an expanded mica weighing 10 pcf or less and is produced by heating the mineral in a manner similar to perlite. It is used in the same manner as perlite.

In some areas of the country pumice, a natural lightweight aggregate, is available. It weighs about 30 pcf.

Water should be of a quality suitable for concrete. Hydrant or tap water is ordinarily suitable. USA Standards[1] A42.2 and A42.3 state that "water shall be clean, fresh, suitable for domestic consumption, and free from such amounts of mineral and organic substances as would affect hardening of the mortar." Avoid the use of stagnant water, as organic impurities are often present.

Stucco finish coat is a factory-produced finish-coat material consisting of standard portland cement (usually white), fine aggregate, hydrated lime, pigment, and plasticizing agent, all mixed and sacked ready for use, requiring only the addition of water on the jobsite. It conforms to the building codes and is superior to job-mixed finish coat because of its greater uniformity in properties and color.

Color in finish coats is most frequently obtained by the use of premixed stucco. Usually white cement is used in formulating stucco, instead of common gray cement, because of the clean and pure colors obtained, especially in the light pastel shades. Pigments should consist of metallic oxides because of their stable and nonfading properties. Iron oxides are used for black, gray, brown, red, and yellow; chromium oxide for greens; and cobalt oxide for blues.

Admixtures are not ordinarily added at the mixer. Plastic cement, gun plastic cement, masonry cement, and stucco all contain the necessary plasticizing agents for their respective uses. The use of waterproofers or dampproofers is not recommended. Air-entraining agents are usually not necessary as these cements entrain sufficient air for workability and durability.

Calcium chloride should conform to ASTM Designation: D 98. It is highly hygroscopic and must be kept in sealed containers at all times. It should not be used if it becomes sticky or caked. Furthermore, calcium chloride can be used only during very cold weather to hasten the hydration and strength development of exterior plaster. It ought not to be used for interior work, because the work space can be heated.

Lime is necessary only if regular cement is used. For this reason, lime is rarely added at the mixer. If used, it should conform to the specifications for special finishing hydrated lime, ASTM: C 206, Type S.

Asbestos must be free of unhydrated oxides or other substances that might be harmful to the plaster.

Lath consists of several different forms of metal reinforcement. (Wood lath has been almost completely supplanted by metal lath.) One common form of reinforcement is expanded metal lath, which is made by cutting parallel, staggered slits in sheet metal, then expanding or stretching it to form diamond-shaped openings. A modification is a self-furring type in which dimples hold it about ¼ in. away from the backing, when it is applied to a smooth surface, such as concrete. V-rib lath is expanded metal lath with longitudinal ribs about 6 in. apart. The V ribs, which are about ⅜ in. high, impart stiffness to the lath and permit wider spacing of lath supports. It is especially desirable for use on ceilings and soffits. However, it must not be used in areas exposed to the weather or to moisture as it may rust because it is nearly impossible to cover the ribs completely with plaster. Standard sheet sizes are 24 or 27 in. wide by 96 in. long. Expanded metal lath is made from copper-bearing steel sheets, painted with rust-inhibitive paint, or from zinc-coated steel sheets. It should weigh at least 2.5 lb per sq yd for use on vertical surfaces and 3.4 lb for horizontal surfaces.

Stucco mesh is similar to diamond mesh, except that it has openings about $3 \times 1\frac{1}{2}$ in. It is used for exterior plastering. It should weigh at least 1.8 lb per sq yd. Stucco mesh and expanded metal lath should conform to the requirements of Federal Specifications QQ-B-101.

Paper-backed wire-fabric lath consists of square or rectangular mesh of No. 16 gage (minimum) galvanized wire with a maximum spacing of 2 in. welded at the intersections. The paper backing is integrally woven with the wire mesh, thus providing a backing for machine-applied plaster. Furring crimps are usually woven into the mesh.

Stucco netting is a hexagonal woven-wire mesh (chicken wire) of 17- or 18-gage galvanized wire. It is fastened with furring nails over waterproof building paper. Mesh is either 1 or 1½ in. depending on the gage. Paper-backed mesh is also available.

Miscellaneous materials include waterproof building paper or felt meeting the requirements of Federal Specifications UU-P-147, Type II, Class D. It should be free of holes and breaks, and should weigh at least 14 lb per 108 sq ft roll. Flameproof paper for certain specified installations should comply with ASTM: D 777.

There is a large assortment of base screeds, casing beads, corner reinforcements, clips, furring nails, tie wires, and other accessories for various special uses.

Grounds and screeds are devices for establishing the proper thickness of the plaster membrane. Grounds are narrow strips of metal or wood attached to the base adjacent to edges and openings in the plastered area. Door and window frames are frequently used for this purpose. Screeds are narrow strips of plaster within large areas to be plastered, built up to the thickness of the base coat.

2   Bases for Plaster

A fundamental requirement is that the base for plaster must be rigid and securely braced to prevent movement either during plastering or later. It must be kept in mind that plaster is a thin membrane and it cannot be expected to assume any significant part of the structural load. Faulty design or construction may transfer part of the load to the plaster, resulting in cracks. Table 2 shows typical approved methods of attaching lath and appurtenant fixtures.

Wood-frame structures are either sheathed or open-frame. In open-frame construction, soft annealed wires of No. 18 gage or heavier, called "line wire" or "string wire," are stretched horizontally across the studs at a vertical spacing of about 6 in., being attached at every fifth stud. The wire is drawn taut by raising or lowering attachments to intermediate studs. Waterproof felt or paper is next nailed in place, with edges lapped at least 2 in. In sheathed construction, the felt is applied directly to the sheathing, without the string wires.

Table 2. Attachments for Exterior Plaster Reinforcement, String Wires, and Paper Backing

Type of construction	Attachments for[a]		
	String wires 6 in. apart vertically	Paper backing[b]	Reinforcement (furred out ¼ in. min)[c]
Wood frame sheathed..	. . . .	Nails or approved staples[d]	Furring nails, approved staples, or other furring device, 6 in. apart vertically on supports
Wood frame open......	32 in. on center horizontally[e]	Nails or approved staples[d]	Furring nails, approved staples, or other furring device, 6 in. apart vertically on supports
Steel frame open.......	. . . .	Clip or other attachment	Clip or other attachment 6 in. apart vertically on supports
Masonry or concrete (when reinforcement is used)	. . . .	. . . .	Furring device 6 in. apart vertically and 16 in. apart horizontally

[a] All nails, staples, or other metal attachments of lath and reinforcements shall be corrosion-resistant.

[b] Paper backing may be omitted in the following cases:
1. When exterior covering is of approved weatherproof panels
2. In back-plastered construction
3. When there is no human occupancy
4. Over water-repellent panel sheathing
5. Under approved paper-backed metal or wire-fabric lath
6. Under metal lath, wire lath, or wire-fabric lath on incombustible construction

[c] Self-furring metal lath or self-furring wire-fabric lath meets requirements. In addition furring shall not be required on steel members having a flange width of 1 in. or less.

[d] Attachment for paper backing shall be of such nature that it will not tear paper.

[e] String wires must be stretched in taut position either by staggering the attachments up and down (herringbone fashion) or by wrapping the wire around the attachments; wires must be securely fastened in place.

Metal lath or reinforcement is applied over the felt. Usually for exterior plaster this consists of hexagonal stucco netting, applied with furring nails so as to space the netting about ¼ in. clear of the paper. Other types of metal lath also can be used with this system. Paper-backed wire-fabric lath can be used on wood-frame structures, applied directly to the studs or sheathing without first attaching string wire and felt.

Metal reinforcement must not be jointed at corners. Instead, it should be returned 6 in. on wood sheathing, and to the first stud on open construction. Metal reinforcement should be installed with the long dimension across the studs, using galvanized or rust-resistant nails and staples. Aluminum nails should not be used where they will be in contact with fresh plaster, because of a chemical reaction. Spacing of fasteners should not exceed 6 in. using 1½-in. barbed roofing nails on horizontal supports. One-inch roofing nails or 1-in. 4-gage staples are used on vertical supports.

Steel-frame construction is similar to wood-frame in that the lath is attached to studs, in this case steel studs. Attachment is made by means of wire ties or "hog rings," the latter being open wire rings that are attached with a special tool. Wire should be of 18 gage and spacing should be 6 in.

Masonry surfaces must be clean for the application of plaster. Concrete block, clay brick, or clay tile should be free of paint, oil, dirt, or other material that might interfere with bond. Bond depends on two things: mechanical key and suction. Mechanical key is obtained when the surface is rough enough for plaster to adhere, of the kind found on rough porous concrete blocks. Suction is the ability to absorb moisture. Lightweight concrete block, common brick, and porous tile have high suction; dense stones such as granite, hard-burned brick, and glazed tile have low suction rates.

Old masonry surfaces can be cleaned by sandblasting, brushing, or washing, depending on the amount and kind of dirt. Efflorescence is removed by washing the surface with a 10% muriatic acid solution, then rinsing with clean water. Grease can be removed by scraping and washing with a strong detergent, or by burning with a blowtorch. Disintegrated or weathered masonry surfaces must be prepared by removing all unsound material.

Uniform suction is necessary for a satisfactory plaster job. It may be necessary to moisten the surface several times starting about 48 hr ahead of plastering, to control areas of high suction. It should not be saturated.

On dense surfaces, if there is any doubt about obtaining a good bond, the clean surface can be improved by the application of a dash coat of plaster, or by a bonding agent. The dash bond coat, consisting of 1 part cement to 1 or 2 parts sand and water, is dashed or thrown on the surface with a whitewash brush or small broom, or by machine. It is not floated or troweled and is permitted to set before the next coat is applied. It should be kept damp for a day or two, then permitted to dry before the next coat is applied. Bonding agents are proprietary compounds that are applied to the masonry just ahead of application of the first plaster coat. Manufacturer's instructions should be followed in using these materials.

Surfaces on which it is not possible to obtain bond should be covered with felt, over which metal reinforcement is attached, furred out ½ in. Paper-backed wire-fabric lath can be used instead.

Chimneys are covered with metal lath, properly furred, without waterproof paper or felt.

Concrete surfaces are treated in much the same manner as masonry. Smooth surfaces lacking mechanical bond should be roughened by sandblasting, bushhammering, or chipping. In some cases wire brushing or acid etching may be satisfactory. Disintegrated concrete must be removed down to a sound surface. On new work, when it is known that plaster is to be applied, oil should not be used on the forms. If form oil is present on the surface, it can be removed by scrubbing with soap and water, or by sandblasting.

Old plaster or stucco is treated in the same manner as masonry. Unsound plaster must be removed, or covered with felt and lath.

Gypsum products such as gypsum lath, plaster, or block, and magnesite plaster, should not be used as a base for portland-cement plaster because of the possibility of bond failure due to chemical reaction or other causes.

3 Proportioning and Mixing

Mixes. Codes and specifications permit volumetric proportioning (Table 3). Measurement of sand is usually accomplished by counting the number of shovelfuls of sand per sack of cement, 7 No. 2 shovelfuls being considered equivalent to 1 cu ft. The amount of water is controlled by the appearance of the plaster in the mixer, although some agencies are now requiring that the slump of the plaster be measured, using a small cone 6 in. high, 4 in. in diameter at the bottom, and 2 in. in diameter at the top. Slump should be 1½ to 3 in. for either hand- or gun-applied plaster, at the point of application.

The use of plastic cement or masonry cement greatly simplifies jobsite mixing because no plasticizing agent is necessary with these cements. Hydrated lime may be added to the mix as a plasticizer when standard portland cement is used, in an amount not exceeding 10% by weight or 25% by volume of the cement. Other plasticizers may be used if approved by the engineer or architect, the amount depending on the agent used and the degree of plasticity required, using the smallest amount possible. The plasticizer should not reduce the compressive strength of the plaster more than 15% below the strength of the plaster without the plasticizer.

Table 3. Exterior Portland-cement Plaster

Coat	Max volume of sand per volume of cement	Min thickness, in.	Min period moist curing, hr	Min interval before applying succeeding coat
First, or scratch...	4	½*	48	48 hr†‡
Second, or brown..	5	First and second coats, ¾	48	7 days‡
Third, or finish....	3§	⅛		

* Measured from backing to crest of scored plaster.

† To ensure suction there should be a drying period after the moist-curing period.

‡ When applied over gypsum lath backing, the brown coat of reinforced exterior plaster may be applied as soon as the first coat has become sufficiently hard. In some areas when approved by local building officials, brown coat can be applied the day after scratch coat is applied. In such cases, the interval between brown and finish coats is extended to a minimum of 14 days. Shrinkage cracks, if any, in the brown coat should be filled in approximately 7 days after application of brown coat.

§ If proprietary color-coat stucco is not used.

Dry materials should be mixed to a uniform color before adding water, using a power mixer, then mixed for a total of 5 min after all materials including water are in the mixer. If the mixer is progressively charged by one man shoveling while mixing, allow at least 3 min after all materials are in the mixer.

Avoid sloppy, overwatered mixes, as they can result in segregation problems during temporary shutdowns, as well as create a number of "on-the-wall" problems—glazing, shrinkage cracking, slow take-up, drop-outs, low strength, efflorescence, and crazing, to name a few. Do not retemper batches that are delayed and become too stiff to apply, as this results in low strength and color variations. Incomplete mixing results in variations in workability of the plaster.

The finish coat is best made with prepared stucco, following the recommendations of the stucco manufacturer. If finish coat is mixed on the job, better color results will be obtained with white cement and light-colored sand. Use a mix consisting of 1 part white cement, ½ part (maximum) by volume of hydrated lime, 2 to 3 parts sand, and the necessary mineral-oxide pigment to give the correct color. Pigment must be weighed for each batch in order to obtain uniform color for all batches. Sand should be clean, well graded, and all passing the No. 16 sieve.

4 Application of Plaster

Hand Application. On many jobs, plaster is applied by hand methods, using the familiar hawk and trowel. Hand methods are especially suitable for the small, isolated job, such as a single dwelling, where it is not economical to bring in the plastering machine.

Plaster should be machine-mixed. Entirely satisfactory results are obtainable in hand-applied plaster, and the choice of hand or machine application is purely one of economics.

Machine Application. Plastering machines are of either the piston or worm-drive type (Fig. 1). In the piston-type machine, plaster flows by gravity or suction from the hopper to the pump cylinder. The pump piston is usually mechanically driven, although some machines have pneumatic or hydraulic power. Large-piston

PUMP DISCHARGE
TO HOSE

Fig. 1. A piston-type plastering machine or "gun." (*Courtesy of Thomsen Equipment Div., Royal Industries, Inc.*)

Fig. 2. Brown-coat plaster is machine-applied to the hardened scratch coat.

Fig. 3. As soon as the first coat of plaster has been applied it is scratched with special tools, hence the name "scratch coat."

machines are frequently used to lift or hoist plaster to the upper floors of a building where the plaster either is applied directly to the wall or is fed to another pump that delivers the plaster to the nozzle. The worm-drive machine consists of a rotor and stator similar in action to a meat grinder. Action of the screw in the rubber-lined stator forces plaster into the hose. Machines can be used for application of all plaster coats. Air joins the material at the nozzle and ejects the plaster at high velocity.

Inadequate maintenance of equipment, improper operation, or unsuitable materials cause operating and plastering difficulties. A sand block, which is a stoppage of the hose between the plastering machine or "gun" and the nozzle, sometimes results, as shown in Table 4.

General. Plaster is normally applied in three coats—scratch, brown, and finish (Fig. 2). Some agencies permit two-coat application on solid backing such as concrete or masonry if no reinforcement is used, in which case the first coat is a cement-grout dash coat, applied as previously described under Masonry Bases.

Concrete or masonry to which plaster is to be applied should be completely and evenly dampened, but not saturated, just prior to application of plaster. There should be no free water on the damp surface when the plaster is applied.

When applying scratch coat to metal reinforcement, sufficient material and pressure should be used to force the plaster through and completely embed the reinforcement. Not more than 10% of the area of the reinforcement should be exposed. After application, the plaster is scratched or scored horizontally, using a special scarifier, to provide good mechanical bond with the second coat (Fig. 3).

The second, or brown, coat is applied as soon as permitted by local building regulations (Table 3). The scratch coat should be evenly dampened but not saturated, to provide uniform suction, immediately prior to applying the brown coat.

Table 4. Sand Blocks

Cause	Cure
Dirty pump, hose, or nozzle. Gravel or sand left in hose after inadequate cleaning. Cement buildup in pump, hose, or nozzle	Pump, hose, and nozzle should be thoroughly cleaned after every day's use, or after a long delay. Cleaning is done by filling the hopper with clean water and running it through the hose until it comes out clear. Disconnect the hose at the feed end, insert a rolled-up sponge, reconnect the hose, and force the sponge through the hose under water pressure
Lumps clogging hose or nozzle resulting from lumps in aggregate or hardened plaster breaking off from pump hopper, hose, or mixer. Soft lumps may be due to insufficient mixing	A screen on the pump hopper will catch most of these. Keep equipment clean
Loose or leaky hose connections. This permits the water or cement paste to separate out of the plaster, leaving a sand block	Provide regular preventive maintenance. Repair or discard worn hose or fittings. Keep all joints tight
Hot, dry hose at start of pumping	Run water through hose before starting to pump plaster
Mortar variability, caused by insufficient mixing	Mix dry materials to a uniform color before adding water; then mix for at least 5 min after all materials are in the mixer
Mortar variability, caused by irregularities in batching	The presently accepted method of batching by the shovelful is variable and inaccurate and is the source of variations in the plaster mixture. These variations can be minimized by careful attention to the batching or measuring operation. Keep sand piles close to mixer; use standard shovels
Oversanded mix	Use a richer mix
Mix too stiff	Slump, as measured by the 6-in. cone, should be about $2\frac{1}{2}$ in. at the nozzle
Hose too long, or passing through and over buildings and corners, or kinked hose. For example, a hose passing up and over a railing or window opening tends to flatten out and restrict the flow of plaster	Keep hose as short and straight as conditions permit. Use $1\frac{1}{2}$-in. pipe on long runs, substituting for part of the hose
Exceeding capacity of machine. Overworking of machine. May cause loss of efficiency and pressure, possibility of stalling	Avoid such practices

Thickness of the brown coat should be about ⅜ in. The brown coat is brought to an even flat surface by smoothing or straightening with the rod and darby, and is left in a rough floated condition for application of the finish coat. Defects such as rough spots or deep scratches must be removed, as they will show up in the finish coat. Uniformity of surface is essential. Apply brown coat to an entire panel without stopping, if possible, as stopping and starting joints in the brown coat will show through the finish as irregularities or discoloration in the surface. Make stops at corners, columns, pilasters, doorway or other openings, downspouts, belt courses, control joints, or other locations where a slight change in appearance of the finish coat will not be obvious. The joinings resulting from scaffold levels and locations should be planned accordingly, and should be kept to a minimum. Horizontal joinings in the brown coat are made at least 6 in. below joinings in the scratch coat, except as described above at corners and similar locations.

After curing the brown coat, it is good practice to permit the plaster to dry out and take its shrinkage before applying the finish coat. Most codes require a period of at least 7 days. More uniform suction results from the delay also. Just prior to application of the finish coat, the brown coat is dampened, but not saturated. Finish coat is applied by either hand or machine to a thickness of ⅛ in. or more. Uniformity of materials and methods is essential for obtaining a finish that is free of color and texture variations.

5 Special Finishes

A great variety of finishes are obtainable by using different tools during finishing. Highly textured finishes are common, especially on exterior plaster. Special effects

Fig. 4. A small hand gun is used to lay the marblecrete aggregate on the still-plastic bedding coat.

are obtained by the use of brushes, brooms, or sponges, applied to the surface, sometimes brushed or troweled afterward. An even, sanded texture is obtained by floating with a carpet or sponge-rubber float. Other finishes are obtained by spraying the color coat with bits of mica, colored metallic tinsel, or ground glass. Patterns may be applied by various tooling methods, such as combing or scoring.

Marblecrete (Fig. 4) consists of exposed natural or integrally colored aggregate, partially embedded in a natural or colored bedding coat of plaster. It is frequently used for making colored and textured murals, either interior or exterior. Aggregates consist of chips or pebbles of marble, quartz, crushed glass, crushed china, seashells, cinders, or other material having a hardness of at least 3 Mohs' scale. They must be durable, alkali-resistant, and inert, and must be free of all dust and dirt at time of application.

Marblecrete can be applied to:

1. Unset plaster base that has been allowed to take up until sufficiently firm to carry the marblecrete

2. Hardened plaster that has been moistened to create even suction, or has been treated with a liquid bonding agent

3. Concrete that has been treated with a bonding agent (or a leveling coat may first be applied to the treated concrete before bedding coat is applied)

4. Concrete masonry that has been coated with a plaster dash coat and allowed to cure for 48 hr

Thickness of the bedding coat depends on the size of aggregate used, and should be at least 80% of the nominal aggregate dimension (see Table 5). After the bedding coat has taken up to the proper consistency, the aggregate is laid on by throwing it onto the bedding coat by hand or by mechanical hand-hopper machine that embeds the aggregate slightly; then the aggregate is tamped gently and evenly so as to embed the aggregate from one-half to three-fourths of the aggregate dimension. Aggregate smaller than No. 4 can be applied by machine, but larger particles must be hand-placed. After 24 hr, if weather conditions make it necessary, the marblecrete can be cured by keeping it damp with a light fog of water. After the surface dries out, a clear nonstaining, nonpenetrating liquid glaze coating may be applied, or a clear nonstaining waterproofing sealer can be used. Either material should resist deterioration and discoloration from weathering.

6 Curing

Scratch and brown coats must be cured, as shown in Table 3, by careful spraying with water. Most codes are silent on curing of the finish coat, there being some objection because of the danger of causing streaks or color and texture variations if the surface becomes too wet. However, curing improves the durability and strength of plaster, especially during hot, dry weather, and if properly done will not cause discoloration. After the finish plaster has hardened for 24 hr a light fog is applied, taking care to apply just sufficient fog so the plaster does not

Table 5. Bedding-coat and Aggregate Sizes

Chip No.	Sieve size, in.		Min bedding-coat thickness, in.
	Passing	Retained on	
0	1/8	1/16	3/8
1	1/4	1/8	3/8
2	3/8	1/4	3/8
3	1/2	3/8	3/8
4	5/8	1/2	1/2
5	3/4	5/8	5/8
6	7/8	3/4	3/4
7	1	7/8	7/8
8	1 1/8	1	1

Chip sizes conform to National Terrazzo and Mosaic Association Standards.

dry out, keeping the surface damp for about a day, but avoiding a heavy application of water that will wash the surface and cause streaks.

Do not spray water on any coat of dry, hot plaster, and do not permit the plaster to dry out between applications of water; otherwise cracking and discoloration will develop. A fogging nozzle of the type used in hothouses or nurseries is especially well adapted to curing plaster.

7 Plaster Failures and Distress

The proper choice and use of materials and methods will result in a good job of plastering. Problems with plaster can invariably be traced to the failure to follow good plastering practices.

Discoloration. Causes of nonuniform appearance are variations in mix proportions, especially the water content; too much water in the mix; color coat spread too thin; incomplete mixing of job-mixed finish coat; inferior pigments; dirt, clay, or organic material in the sand; dirty tools, including pump and accessories, mortar boards, and hand tools; retempering batches that have dried out; variations in surface, suction, or moisture content of base coat; careless workmanship, including variations in trowel or float angle and pressure; overtroweling; dry floating; starting and stopping joints or joinings; rain on fresh color coat; and variations in curing.

Discoloration can occur at any time after application of the stucco because of a number of causes. Rusting of metal brackets, trim, etc., will cause staining. Improper design or installation of flashings, drips, sills, etc., permits water to enter the hardened stucco. Staining material originating in the backing material is brought to the surface by migrating water. The obvious remedy is proper and adequate design and installation of all appurtenances to stucco as well as proper application of the plaster and stucco.

Efflorescence. A whitish bloom, film, or crystal formation resulting from the deposition of mineral salts on the surface by water migrating from the interior of the wall and evaporating from the surface is called efflorescence, for example, mineral salts present in the sand or mixing water; organic matter in aggregate or mixing water; soluble calcium hydroxide leached out of the plaster; efflorescence already present on concrete, brick, or masonry base.

Efflorescence can be prevented by attention to certain details. Use only clean materials conforming to applicable specifications. Do not use unwashed sand or brackish, stagnant water. Do not apply plaster directly to masonry or concrete surfaces showing evidence of efflorescence. Prevent water from entering the wall by proper design, such as the use of an impervious membrane. Stop plaster above the soil line to prevent movement of salt-carrying water from the soil into the plaster.

Cracking. In general, cracking is caused by structural movement of the building resulting from settlement or swelling of foundation soil; failure to provide control joints about every 25 ft on large areas; inadequate joining of lath to studs or backing; failure to lap and fasten lath to each other; shrinkage or warping of wood framing; excessive loading of structure; weak sections such as at changes in cross section or at openings; continuous plaster membrane over two different backing materials; too much suction resulting from application of plaster to a dry base (including previous coat of plaster); shrinkage of plaster caused by wet mixes, inferior, poorly graded, or dirty aggregate; mix too rich; rapid drying of the fresh plaster; plaster membrane too thin, especially over studs; inadequate curing.

Other conditions that contribute to the formation of cracks are hot, dry, windy weather; overfinishing, or finishing too soon; loose line wires in open-frame construction; bond failure between the plaster coats or between plaster and substrate; and thermal shock, or sudden and large differences in temperature.

Low Strength. Properly applied plaster possesses sufficient strength for any exposure condition normally encountered. Conditions that lead to poor strength performance are much the same as those affecting concrete. Low strength may be caused by inferior aggregates, poor aggregate grading (especially excessive fines), insufficient cement in the mix, too much water (sloppy mix), incomplete mixing, retempering batches retained too long after mixing and before using, lack of curing, or hot, dry weather.

Softness of the finished plaster is evidence of low strength. Improper use of such admixtures as detergents and soap, low temperatures that retard hydration, freezing, and impurities in the water or aggregate all lead to low strength and soft plaster.

8   Plastering in Cold Weather

Because exterior plaster is a thin membrane on an exposed structure that is difficult or impossible to protect from low temperature, especially if accompanied by wind, plastering during cold weather should be avoided. During cold weather setting and strength gain of mortar are slower than at more moderate temperatures, causing long delays between application of the plaster and floating or finishing. Particularly risky is plastering when nighttime temperatures are below freezing as there is great danger that the fresh plaster will suffer irreparable damage if it freezes before it has hardened, resulting in low strength, softness, and unsoundness.

If it is necessary to plaster during cold weather when the minimum temperature is above freezing, setting of the cement should be accelerated so the plaster will take up faster and floating can be accomplished within a reasonable time. Sometimes all that is necessary is to heat the mixing water to about body temperature, easily accomplished by directing the heat from a salamander or other heater against the water barrel. Water should be warm but not hot, as hot water might cause a flash set.

Acceleration of setting and strength development can be accomplished by using a small amount of calcium chloride in the plaster. Calcium chloride, which is available from most building-material dealers, should be added to the batch in solution, using about 1 lb of the salt for each sack of cement in the mix. (Because specialty cements are frequently used in plaster, the user should check with the cement manufacturer regarding the use of any admixture.) Calcium chloride is not an antifreeze and does not protect the plaster from freezing, but it does accelerate hardening of the plaster, thus enabling it to withstand freezing temperatures better.

The reader is referred to Chap. 30 for a detailed discussion of cold-weather concreting and the use of calcium chloride.

REFERENCES

1. Standard Specifications for Portland Cement Stucco and Portland Cement Plastering, USAS A42.2 and A42.3, 1946, rev. 1960, American Society for Testing and Materials, Philadelphia, Pa.
2. Uniform Building Code, International Conference of Building Officials, Whittier, Calif., 1973.
3. Diehl, John R.: "Manual of Lathing and Plastering," National Bureau for Lathing and Plastering, Washington, D.C., 1960.
4. Lathing and Plastering Reference Specifications, California Lathing and Plastering Contractors Assn., Los Angeles, Calif., 1965.

Chapter 40

CONCRETE MASONRY

JOSEPH J. WADDELL

Masonry is used in all types of buildings: dwellings, schools, churches, industrial, commercial, and farm buildings; from small single garages to high-rise apartments. Retaining walls are frequently built of concrete masonry.

Concrete masonry is used extensively as a backup for other material (stone, brick, tile, plaster), for strictly utilitarian walls without adornment, as a finish wall material for both exterior and interior applications, and for ornamental grilles, solar screens, and garden walls. Masonry building walls may support part of the structural load of the building, in which case they are called load-bearing, or they may be non-load-bearing space dividers.

1 Masonry Building Construction

The six types of masonry building construction are hollow-unit masonry, solid masonry, grouted masonry, reinforced hollow-unit masonry, reinforced grouted masonry, and cavity wall masonry. A seventh type, used for architectural and decorative purposes, is the screen wall used for grilles, garden walls, and solar screens.

Hollow-unit masonry consists of hollow units all laid and set in mortar. When two or more units are used to make up the thickness of the wall, header courses or metal ties are required.

Solid masonry is made of concrete brick or solid load-bearing concrete-masonry units laid contiguously in mortar. All head, bed, and wall joints are required to be solidly filled with mortar. Adjacent withes must be bonded by means of header courses or metal ties.

Grouted masonry is made with solid brick units or hollow units in which interior masonry joints and all cells or cavities are filled by pouring grout therein as the work progresses. Type S mortar is specified. Wall may be of either low-lift or high-lift construction. High-lift construction requires special wire wall ties and permits grouting in 4-ft lifts.

Reinforced hollow-unit masonry is made with hollow masonry units in which some of the cells, containing vertical reinforcement, are continuously filled with concrete or grout. Type S mortar is specified. A cleanout opening is required at the bottom of each cell to be filled where each lift exceeds 4 ft in height.

Reinforced grouted masonry is the same as grouted masonry except that it is reinforced and the thickness of grout or mortar between masonry units and reinforcement must not be less than $1/4$ in. However, $1/4$-in. bars can be laid in horizontal joints at least $1/2$ in. thick and steel-wire reinforcement can be laid in horizontal joints at least twice the thickness of the wire diameter.

Cavity wall is made with units in which facing and backing are completely separated except for metal ties that serve to bond the two tiers together. The

Uniform Building Code[1] requires a minimum net thickness of 3½ in. for both the facing and the backing, and a net width of the cavity between 1 and 3 in.

Screen walls are made of special patterned, pierced, or open block. Besides their ornamental value, they are used to separate space, both interior and exterior, as solar screens to shade windows and other areas, and to provide privacy yet permit passage of air and light.

2 Masonry Units

Dimensions. Concrete-masonry units are usually referred to by their nominal dimensions. Actual dimensions are smaller to allow for the thickness of mortar

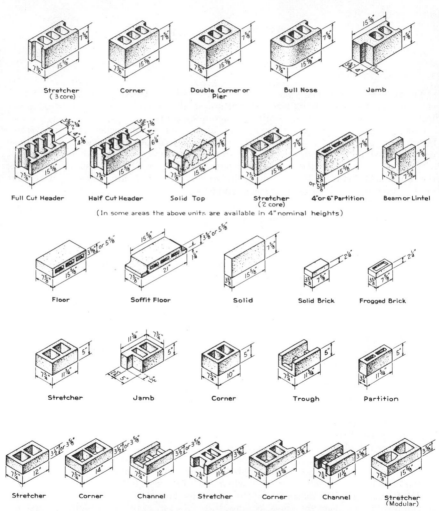

Fig. 1. Typical sizes and shapes of masonry units. Dimensions shown are actual unit sizes. A 7⅝ × 7⅝ × 15⅝-in. unit is commonly known as an 8 × 8 × 16-in. concrete block. Half-length units are usually available for most of the units shown below. See concrete-products manufacturer for shapes and sizes of units locally available. (*From "PCA Concrete Masonry Handbook," Portland Cement Association, Chicago.*)

joints. For example, an 8 × 8 × 16-in. unit actually measures $7\frac{5}{8} \times 7\frac{5}{8} \times 15\frac{5}{8}$ in., allowing for $\frac{3}{8}$-in. mortar joint. Common sizes and types are shown in Fig. 1.

Pierced or screen blocks come in a variety of sizes that should be ascertained locally before use.

Modular Construction. If advantage is taken of the nominal sizes of units, construction can be greatly simplified and cutting or breaking of units is minimized. Half-length blocks are available for further simplification. Tables 1 and 2 show the nominal lengths and heights of walls built with the two commonest sizes of blocks. All vertical dimensions should be in multiples of nominal full-height units, and horizontal dimensions in multiples of half-length units. Sizes and location of doorways and window openings should be planned to accommodate these dimensions, making proper allowance for sills, jambs, lintels, and frames.

Table 1. Nominal Length of Concrete-masonry Walls by Stretchers[6]

No. of stretchers	Nominal length of concrete-masonry walls	
	Units $15\frac{5}{8}$ in. long and half units $7\frac{5}{8}$ in. long with $\frac{3}{8}$-in.-thick head joints	Units $11\frac{5}{8}$ in. long and half units $5\frac{5}{8}$ in. long with $\frac{3}{8}$-in.-thick head joints
1	1 ft 4 in.	1 ft 0 in.
$1\frac{1}{2}$	2 ft 0 in.	1 ft 6 in.
2	2 ft 8 in.	2 ft 0 in.
$2\frac{1}{2}$	3 ft 4 in.	2 ft 6 in.
3	4 ft 0 in.	3 ft 0 in.
$3\frac{1}{2}$	4 ft 8 in.	3 ft 6 in.
4	5 ft 4 in.	4 ft 0 in.
$4\frac{1}{2}$	6 ft 0 in.	4 ft 6 in.
5	6 ft 8 in.	5 ft 0 in.
$5\frac{1}{2}$	7 ft 4 in.	5 ft 6 in.
6	8 ft 0 in.	6 ft 0 in.
$6\frac{1}{2}$	8 ft 8 in.	6 ft 6 in.
7	9 ft 4 in.	7 ft 0 in.
$7\frac{1}{2}$	10 ft 0 in.	7 ft 6 in.
8	10 ft 8 in.	8 ft 0 in.
$8\frac{1}{2}$	11 ft 4 in.	8 ft 6 in.
9	12 ft 0 in.	9 ft 0 in.
$9\frac{1}{2}$	12 ft 8 in.	9 ft 6 in.
10	13 ft 4 in.	10 ft 0 in.
$10\frac{1}{2}$	14 ft 0 in.	10 ft 6 in.
11	14 ft 8 in.	11 ft 0 in.
$11\frac{1}{2}$	15 ft 4 in.	11 ft 6 in.
12	16 ft 0 in.	12 ft 0 in.
$12\frac{1}{2}$	16 ft 8 in.	12 ft 6 in.
13	17 ft 4 in.	13 ft 0 in.
$13\frac{1}{2}$	18 ft 0 in.	13 ft 6 in.
14	18 ft 8 in.	14 ft 0 in.
$14\frac{1}{2}$	19 ft 4 in.	14 ft 6 in.
15	20 ft 0 in.	15 ft 0 in.
20	26 ft 8 in.	20 ft 0 in.

Actual length of wall is measured from outside edge to outside edge of units and is equal to the nominal length minus $\frac{3}{8}$ in. (one mortar joint).

* From PCA "Concrete Masonry Handbook."[6]

Table 2. Nominal Height of Concrete-masonry Walls by Courses°

No. of courses	Nominal height of concrete-masonry walls	
	Units 7⅝ in. high and ⅜-in.-thick bed joint	Units 3⅝ in. high and ⅜-in.-thick bed joint
1	8 in.	4 in.
2	1 ft 4 in.	8 in.
3	2 ft 0 in.	1 ft 0 in.
4	2 ft 8 in.	1 ft 4 in.
5	3 ft 4 in.	1 ft 8 in.
6	4 ft 0 in.	2 ft 0 in.
7	4 ft 8 in.	2 ft 4 in.
8	5 ft 4 in.	2 ft 8 in.
9	6 ft 0 in.	3 ft 0 in.
10	6 ft 8 in.	3 ft 4 in.
15	10 ft 0 in.	5 ft 0 in.
20	13 ft 4 in.	6 ft 8 in.
25	16 ft 8 in.	8 ft 4 in.
30	20 ft 0 in.	10 ft 0 in.
35	23 ft 4 in.	11 ft 8 in.
40	26 ft 8 in.	13 ft 4 in.
45	30 ft 0 in.	15 ft 0 in.
50	33 ft 4 in.	16 ft 8 in.

For concrete-masonry units 7⅝ in. and 3⅝ in. in height laid with ⅜-in. mortar joints. Height is measured from center to center of mortar joints.

* From PCA "Concrete Masonry Handbook."[3]

Kinds of Block. Blocks are designated as hollow load-bearing, solid load-bearing, hollow non-load-bearing, and brick. These may be either normal-weight concrete or lightweight concrete (about 105 pcf), with the latter usually preferred because they weigh about one-third less than the heavy ones. Lightweight blocks weigh 24 to 30 lb each.

Many types of building units are ornamental by virtue of special textures, patterns, or colors incorporated in them at the time of manufacture. Other units are specifically designed for aesthetic value such as *grille* and *screen* block cast in special molds that give a pierced or open effect, especially desirable for screen walls, garden enclosures, and fences; *slump* block, a rustic-finish block resembling adobe brick in appearance, made with a mix of such consistency that it slumps or sags when the form is removed, resulting in blocks of variable appearance, texture, and height; *split* block that looks like rough stone, made by splitting a hardened block; *shadow* block, with embossed or recessed face patterns that give a three-dimensional effect; and *faced* or *glazed* blocks that are coated with a ceramic or plastic material giving a smooth, sanitary surface for use in bathrooms, kitchens, and similar locations.

Construction Requirements. Blocks must be dry when they are laid in the wall; otherwise the units will shrink and the wall will tend to crack. For this reason blocks must be permitted to age for about 4 weeks before they are used. Aging periods may be shorter in a hot, dry climate, or when autoclaved blocks are used. Blocks are classified by ASTM as follows:

ASTM C-55: Concrete building brick
ASTM C-90: Hollow load-bearing concrete masonry units
ASTM C-129: Hollow nonload-bearing concrete masonry units

ASTM C-139: Concrete masonry units for construction of catch basins and manholes

ASTM C-145: Solid load-bearing concrete masonry units

Blocks in each of the classifications (except C-139) are further designated as Type I, moisture-controlled units, and Type II, nonmoisture-controlled units. Blocks conforming to Type I, at the time of delivery to the site, shall have a maximum moisture content ranging in value from 25 to 45% of the total absorption, depending on the average annual relative humidity of the geographical area in which the blocks are to be used and the allowable shrinkage of the blocks. Blocks are further classified as to grades as shown in Table 3.

Table 3. ASTM Requirements for Concrete-masonry Units

ASTM designation	Grades
C-55	Grade U: For use as architectural veneer and facing units in exterior walls and for use where high strength and resistance to moisture penetration and frost action are desired
	Grade P: For general use where moderate strength and resistance to frost action and moisture penetration are desired
	Grade G: For use in back-up or interior masonry, or where effectively protected against moisture penetration
C-90	Grade N: For general use, as in exterior walls below and above grade that may or may not be exposed to moisture penetration or the weather, and for interior walls and back-up
	Grade S: Limited to use above grade in exterior walls with weather-protective coatings and in walls not exposed to the weather
C-129	No requirements
C-145	Grade U: For use in unprotected exterior walls below grade and also for unprotected exterior walls above grade which may be exposed to frost action
	Grade P: For use in protected exterior walls below grade and also for protected exterior walls above grade that may be exposed to frost action
	Grade G: For general use above grade in walls not subjected to frost action, or in walls protected from the weather

Strength, absorption, density, and allowable moisture contents vary considerably for different classifications of blocks. For these reasons, the producer and user should check local code provisions concerning requirements for blocks and should also be sure that the latest applicable ASTM specifications are being used, since these are subject to revision. Materials, dimensions, and appearance are all covered in the standard specifications.

Block, upon delivery to the site, must be stored on planks or other supports to keep them off the ground, and should be covered with canvas, plastic, or waterproof paper to protect them from moisture, if atmospheric conditions require.

Estimating Quantities. The number of block and the amount of mortar required for a plain masonry wall can be estimated from Table 4.

3 Foundations

Concrete-block walls and structures must be constructed on concrete footings of adequate width and thickness to carry the expected loads, situated on undisturbed or well-compacted earth, in accordance with the design and local building requirements. The foundation should be constructed in accordance with good construction practices and screeded level. At the time blocks are laid on it the concrete should be at least 3 days old, clean and free of all laitance and dirt.

Table 4. Weights and Quantities of Materials for Concrete-masonry Walls°

Actual unit sizes (width × height × length), in.	Nominal wall thickness, in.	For 100 sq ft of wall					For 100 concrete units, § mortar, § cu ft
		No. of units	Avg weight of finished wall			Mortar, § cu ft	
			Heavy-weight aggregate, lb†	Light-weight aggregate, lb‡			
3⅝ × 3⅝ × 15⅝	4	225	3,050	2,150		13.5	6.0
5⅝ × 3⅝ × 15⅝	6	225	4,550	3,050		13.5	6.0
7⅝ × 3⅝ × 15⅝	8	225	5,700	3,700		13.5	6.0
3⅝ × 7⅝ × 15⅝	4	112.5	2,850	2,050		8.5	7.5
5⅝ × 7⅝ × 15⅝	6	112.5	4,350	2,950		8.5	7.5
7⅝ × 7⅝ × 15⅝	8	112.5	5,500	3,600		8.5	7.5
11⅝ × 7⅝ × 15⅝	12	112.5	7,950	4,900		8.5	7.5

Table based on ⅜-in. mortar joints.
* From PCA "Concrete Masonry Handbook."[3]
† Actual weight within ±7% of average weight.
‡ Actual weight within ±17% of average weight.
§ With face-shell mortar bedding.   Mortar quantities include 10% allowance for waste. Actual weight of 100 sq ft of wall can be computed by formula $WN + 150M$ where W = actual weight of a single unit, N = number of units for 100 sq ft of wall, M = cu ft of mortar for 100 sq ft of wall.

Size and depth of footings is a function of design, and the builder is concerned with making the footing in accordance with the plans. A footing width twice the wall thickness is usually adequate for light buildings. Base of the footing should be at least 12 in. below ground line or below the frost line, whichever is deeper. In areas of expansive or unstable soil, the designer may impose additional requirements. Exterior doorsills can frequently be cast as part of the footing. Footings for garden walls are made at least 12 in. wide and 8 in. deep.

4   Reinforcement

The amount and location of reinforcement is a design consideration, and the builder must be governed by the plans and specifications in this respect. The following comments are offered to assist in placing reinforcement in accordance with the design requirements.

Footings. Usually footings for light buildings and walls are unreinforced, but the plans and local code should be consulted in this respect. Heavy loads, poor soil conditions, and areas subject to high winds or earthquakes may make reinforcement necessary. For example, in earthquake zones, it is customary to place two horizontal ½-in. bars in the footing, one 3 in. from the bottom and one a like distance down from the top. Minimum steel to tie walls to footings may consist of ½-in. vertical dowels adjacent to door openings. In earthquake areas two vertical dowels may be required adjacent to door openings in exterior and bearing walls and at corners and intersections, with additional dowels in the footing.

Wall and Joint Reinforcement. The amount and location of vertical steel is usually determined by design considerations for vertical structural loads, and seismic or wind conditions. Vertical bars, when required, are placed in the cores of the blocks at each foundation dowel, lapping the dowel 30 bar diameters. Bars should be wired to the dowels or spaced one bar diameter, at splices, and placed at the centerline of the wall, at least ¼ in. clear of the masonry. Bars must be held at sufficient points to prohibit movement of the bars while placing masonry and grouting, usually at intervals not exceeding 192 bar diameters. Where two

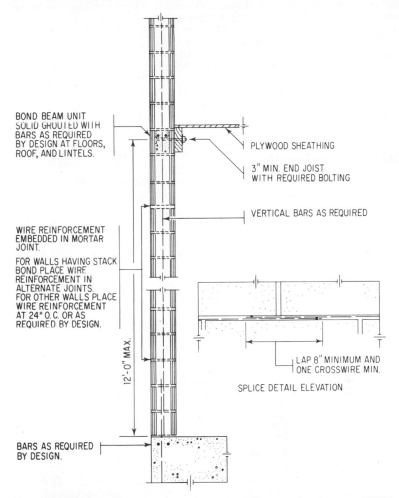

BOND BEAM UNIT SOLID GROUTED WITH BARS AS REQUIRED BY DESIGN AT FLOORS, ROOF, AND LINTELS.

PLYWOOD SHEATHING

3" MIN. END JOIST WITH REQUIRED BOLTING

VERTICAL BARS AS REQUIRED

WIRE REINFORCEMENT EMBEDDED IN MORTAR JOINT.

FOR WALLS HAVING STACK BOND PLACE WIRE REINFORCEMENT IN ALTERNATE JOINTS. FOR OTHER WALLS PLACE WIRE REINFORCEMENT AT 24" O.C. OR AS REQUIRED BY DESIGN.

12'-0" MAX.

LAP 8" MINIMUM AND ONE CROSSWIRE MIN.

SPLICE DETAIL ELEVATION

BARS AS REQUIRED BY DESIGN.

FIG. 2. Wire reinforcement in mortar joints. Wire conforming to ASTM A82 may be used as temperature steel or to replace running bond and may be added to bond beam and foundation bars for required minimum reinforcement. Spacer wires welded to longitudinal wires shall be spaced at 16 in. on centers maximum. Wire reinforcement is usually specified as No. 9 wire or ³⁄₁₆-in.-diameter wire. *(From "Concrete Masonry Design Manual," Concrete Masonry Association of California, Los Angeles, 1964.)*

bars are called for at a jamb, the individual bars can be placed in adjacent cores. If the cores are to be grouted in lifts exceeding 4 ft, a cleanout hole should be provided at the bottom of the lift in the cell or core containing the steel.

Where horizontal reinforcement is required, it usually consists of ¼-in. rods or commercially available fabricated assemblies of longitudinal wires joined by means of diagonal cross wires, laid on the webs of the masonry, and placed in the face-shell mortar. Splices in wire-reinforcement assemblies should be lapped at least 8 in., with one cross wire in each assembly included in the lap. If crossties are not in the same plane as the longitudinal reinforcing wires, the assembly should be laid with the crossties down, thus providing better embedment by holding the wires up from the surface of the lower block.

Many walls are built without horizontal joint steel. However, the steel strengthens the wall and distributes stresses that might occur in the wall. If the wall shrinks and cracks, the cracks will be much finer and closer together than in a similar wall without the steel. Horizontal steel is necessary in bond beams at roof and floor levels, at the top of wall openings, in parapets, and in walls of stacked bond. See Fig. 2 for typical details.

Special fabricated wire assemblies are available for use as horizontal steel to tie together the two withes or tiers of cavity walls. Cross rods are provided with a drip to prevent migration of moisture.

5   Mortars for Concrete Masonry

Materials.   Cementitious material for masonry mortar consists of masonry cement, plastic cement, or common portland cement with hydrated lime or lime putty. Cements are described in Chap. 1. Hydrated lime should conform to ASTM Designation: C 207, Type S, and quicklime to ASTM: C 5.

Table 5. Sand Grading for Masonry Mortar, ASTM C 144

Sieve size No.	% passing	
	Natural sand	Manufactured sand
4	100	100
8	95–100	95–100
16	70–100	70–100
30	40– 75	40– 75
50	10– 35	20– 40
100	2– 15	10– 25
200		0– 10

Aggregate consists of sand conforming to the grading shown in Table 5, this being the grading shown in ASTM: C 144. Neither ASTM nor the Uniform Building Code has a limitation on the amount of natural sand passing the 200-mesh sieve. However, the fines should be kept below 5%, as a sand that is too fine requires more water for workability and is apt to produce weak mortar. On the other hand, a sand deficient in fines passing the No. 50 and 100 sieves will probably be harsh and unworkable, making it difficult to produce weathertight joints. Sand should be washed and free of clay, silt, and organic matter.

Admixtures need not be added at the mixer, except that 1 lb of calcium chloride per cwt of cement can be used as an accelerator during cold weather.

Cement and aggregates should be stored where they will not become contaminated but should be close to the mixer.

Properties of Mortar.   Mortar should be workable, or "buttery," have good water retentivity, and be capable of developing the specified strength and bond. These properties are achieved by the use of well-graded sand, proper mix proportions, and the minimum amount of water to give the required plasticity.

Because masonry units are dry when laid in the wall, their suction draws water out of the mortar, resulting in a loss of plasticity of the mortar and sometimes affecting the bond. Water retentivity is the property of the mortar that resists rapid loss of water to the masonry as determined by the method of ASTM: C 91.[2] Mortar with a low water-retention value (less than 70%) undergoes a rapid loss of water when placed in the wall, resulting in loss of plasticity and the danger of serious loss of bond.

The property of mortar that unites it to the masonry is the bond. Bond is affected by the amount and kind of cementitious material, workability and plasticity of the mortar, texture and suction of the masonry units, and the workmanship used in laying the masonry.

The specifications provide for several grades of mortar as shown in Table 6, and the class of mortar used for any structure depends upon design and exposure conditions. The Uniform Building Code designates the types of construction and

Table 6. Mortar Proportions*
Parts by volume

Mortar type	Minimum compressive strength at 28 days, psi	Port- land cement	Hydrated limes or lime putty†		Masonry cements	Damp loose aggregate
			Min.	Max.		
M	2,500	1	—	¼	—	Not less than 2¼ and not more than 3 times the sum of the volumes of the cement and lime used
		1	—	—	1	
S	1,800	1	¼	½	—	
		½	—	—	1	
N	750	1	½	1¼	—	
		—	—	—	1	
O	350	1	1¼	2½	—	

* From Table 24-A, Uniform Building Code.

† When plastic or waterproof cement is used as specified in Section 2403 (p), hydrated lime or putty may be added, but not in excess of one-tenth the volume of cement.

allowable unit stresses for the different classes of mortar. Code requirements should always be ascertained before starting construction.

Mixes. The strength of masonry mortar depends on the quantity and kind of cementitious material, amount of water in the mix, quality of aggregate, water retentivity, plasticity, and workmanship. Mortar should have good plasticity but overly rich mixes should be avoided as they are subject to volume change. Fireclay and similar materials should not be used.

Mortar should be mixed in a power mixer. Half of the water and sand are put in the mixer, then the cement (and lime, if any), followed by the balance of the water and sand. Mixing is continued for 3 min. Mortar should be used within 2½ hr after mixing when the air temperature is 80°F or higher, or 3½ hr when the air temperature is below 80°F. Water can be added to the mortar within these time limits to maintain plasticity if the mortar has stiffened because of evaporation of water.

Color. In joint mortar, color can be obtained by the use of pigments as described in Chap. 4. Experiment with the mix and materials to be used on the job by making small test panels and storing them on the job for about a week. Brilliant and dark colors are difficult to obtain. What is sometimes thought to be fading is actually a deposit of efflorescence on the surface that hides the color. A faded appearance also results from weathering in which the aggregates are exposed by erosion of the surface.

6 Laying Masonry

The First Course. Alignment of the first course is extremely important in order that the finished wall will be true and straight. The concrete footing should be swept clean of all sand, dirt, and other debris. Block laying starts from the corners, using a chalked snap line to mark the footing. Lay blocks with the thick end of face shells up, using full bedding; that is, the webs of the units as well as the face shells are embedded in mortar. Mortar should not extend into the cores more than $\frac{1}{2}$ in. If the wall is more than 4 ft high, provide cleanouts of about 12 sq in. area at the bottom cell containing vertical reinforcement. Frequent use of the straightedge and level assures a straight course of masonry.

Laying the Wall. Corners are kept higher than the rest of the wall. Ends must not be toothed, but racking a maximum of five tiers is permitted. Unless specified otherwise, the blocks are laid in common or running bond.

The method of bedding depends on the type of wall under construction. Load-bearing building walls, columns, piers, and pilasters, especially those for heavy loads, and the first or starting course on the footing or foundation wall should have full bedding; that is, mortar should be applied to the webs as well as the face shells of the units. All other hollow concrete masonry can be laid in face-shell bedding.

Use a mason's line stretched between the corner blocks to keep the units on the proper line and grade. Place the bed-joint mortar on two or three blocks in the previous course of blocks just ahead of the unit being placed, but not so far ahead that it loses moisture and plasticity before the block is placed on it. For vertical joints, mortar is usually applied only to the face shells. Some masons butter the end of the block before setting it in the wall, while others apply mortar to the end of the block already in place. Either method is satisfactory. When the block is placed, shove it against the previously laid block so as to compact the vertical mortar joint. Once a block has been set into the wall it should not be moved after the mortar has started to stiffen, as bond will be broken and a leaky wall will result. Instead remove the block and mortar and apply fresh mortar; then reset the block.

The mason should use the level, straightedge, and story pole to maintain line and grade and to assist him in keeping alternate head joints (in common bond) in vertical alignment. Cores containing vertical reinforcement must be carefully kept in alignment. It is usually best to lay up pilasters at the same time as the wall.

When building up the corners first and working from both ends toward the middle of the wall, the last block laid in each course is the closure block. The closure block is laid by buttering all edges of the opening and all four vertical edges of the block. If any of the mortar falls out before the block is tapped into position, remove the block, apply new mortar, and reset it. When work is interrupted for any reason, cover the top of the wall with canvas, building paper, or polyethylene to keep out rain and snow. Be sure the covering is anchored so it will not blow away.

Solid grouting for bond beams, as shown in Fig. 2, is accomplished by using special bond-beam blocks that are filled with concrete or mortar after installation of specified reinforcing steel.

Openings and Intersections. Depending upon personal preference and site conditions, either of two methods can be used for door openings. One method is to square and brace the door frame in its final position before laying blocks. Anchor bolts or ties are then attached and extended into the joint mortar as the wall is laid up. Another method is to set in a temporary wood frame of the exact size and shape and lay the blocks around it, removing the frame later. The latter method is normally used when special jamb blocks are laid, and is the preferred method. Window openings are constructed in the same manner. Metal sash requires special attention, as metal-sash blocks, grooved to accept the metal frame, are used, and it is sometimes necessary to set the metal frame before the lintel is laid.

Doorsills are sometimes cast in the concrete of the foundation. Windowsills are of two types, both of which have sloping tops and project 1½ in. outside the wall, with a drip groove on the bottom to prevent rainwater from running back into the wall. The lug sill is set in place as the wall is built up, and the slip sill is mortared in place after the wall has been completed.

Lintels may be precast or cast-in-place concrete, steel angles supporting concrete block, or lintel block. Precast-concrete lintels may be either one-piece or split, i.e., consisting of two separate beams side by side. One-piece lintels are preferred. Lintel blocks are U-shaped blocks, placed with the open end up. They must be well supported during construction. After the blocks are placed, two reinforcing bars are placed in the bottom of the channel formed by the blocks and the channel is filled with mortar or concrete. The use of lintel blocks avoids the difference in texture that results when precast is used.

Usual practice is to tie corners of bearing walls with a masonry bond, but other intersections are not bonded; instead the intersecting wall terminates and is tied to the other by means of metal ties. Nonbearing wall intersections are tied by means of strips of metal lath or hardware cloth in the mortar joint in alternate courses. The vertical joint at the intersection should be raked and caulked. Metal ties of this type, used for tying interior partitions to exterior walls, are slightly flexible and permit slight differential movement.

Anchors and Attachments. Anchors and inserts must be set in mortar or grout. Plate anchor bolts can be set in the mortar or concrete in the bond beam. Where an anchor or insert extends into the core of a block, the core is filled with mortar after first placing a piece of metal lath in the horizontal joint just below. If the anchor extends into a core containing vertical steel it should be solidly wired to the steel to prevent displacement when the cores are grouted.

When interior finish requires the use of furring strips, these can be attached to masonry by the use of special nails designed for this purpose.

Flashings over door and window openings, on parapets, and on roofs are shown on the drawings. Where the flashing extends into the mortar joint, it should not penetrate into the joint more than 1 in.

Screen Walls. Laying block in screen walls is done in virtually the same manner as in other types of masonry. Many of these walls are laid in stacked bond, requiring horizontal steel in every joint.

Cold Weather. During cold weather the work area should be enclosed, and the newly laid wall protected from low temperatures for at least 48 hr. When the air temperature drops to 40°F the water should be heated to not over 160°F in order that the mortar temperature will be between 75 and 100°F. When the air temperature drops below freezing the sand should be heated also, taking special care that frozen lumps are thawed out. Masonry units should be stored indoors where they can be protected from frost and snow. If it is necessary to lay masonry when the air temperature is below 20° the units should be warmed to about 40 to 50°F.

Block should not be laid on a frozen foundation or on frozen block. When cold weather makes protection necessary, both sides of the wall must be protected.

7 Mortar Joints

The success of a masonry wall depends upon the quality of the joints, hence the importance of good workmanship and materials. The properties of mortar that lead to good joints are clean sand of the proper grading, adequate cement of the proper type, minimum water to give the necessary plasticity and workability to the mortar, and adequate water retentivity of the mortar.

Making the Joint. Mortar joints should be ⅜ in. thick, plus or minus ⅛ in. The starting bed joint on the foundation should be laid with full mortar coverage on webs as well as face shells. All other joints are laid in face-shell bedding except in pilasters, columns, cores at vertical steel, and as shown on the plans. Bedding mortar is spread on the previously laid blocks ahead of the one being laid but should not be permitted to dry out or stiffen before the block is laid. Joints must be full and level. Head joints must be carefully buttered, laid up to the full width of the face shell. Excess mortar is struck off the face of the blocks as soon as it is squeezed out of the joint by the block being shoved into place.

When laying a cavity wall, droppings can be kept out of the cavity by laying a board across the metal ties. When the masonry reaches the level of the next row of ties the board is removed by wires attached to it, cleaned, and reset on the new level of ties.

Tooling Joints. Customary practice is to tool the joints to dress them up and make them watertight. This is accomplished with a jointing tool, going over the joint to compact the mortar and seal it tightly against the masonry units. Excess mortar is then trimmed off. Tooling should be done when the mortar has stiffened somewhat but before it sets. All exterior joints must be tooled if a watertight wall is to be obtained. Joint finishes of the type shown in Nos. 1 through 5 in Fig. 3 are necessary for exposed walls, as they are watertight when properly made. Unexposed and interior walls, or exterior walls in a mild dry climate, can be of the types shown in Nos. 6, 7, and 8 of Fig. 3.

Tooling makes a dense surface texture that sheds water, gives better bond by forcing the mortar against the masonry units, and reveals places where mortar is thin or lacking. An untooled joint, which is made by striking off the fresh mortar with a trowel, is apt to be porous and permeable, with the mortar torn and drawn away from the units.

When a smooth, uniform texture is desired, as on a wall to be painted, the joints can be struck flush and then sack-rubbed (No. 5 in Fig. 3).

8   Grouting

Mixes. Grout for grouted masonry, unless otherwise specified, consists of 1 part cement, 3 sand, and 2 pea gravel mixed to a consistency as fluid as possible for pouring without segregation. Where the grout space is less than 3 in. in its least dimension the grout consists of 1 part cement with 2¼ to 3 parts sand. All proportions are by damp loose volume. Mixing should be continued for at least 3 min after all ingredients including water are in the mixer.

Grout to be pumped is usually delivered to the job in ready-mix trucks and should contain at least 660 lb of cement per cu yd and sufficient water to be of fluid consistency, as described above.

Grouting. Vertical cells to be filled must be aligned so as to have a continuous vertical cell measuring at least 2×3 in. in area. Cells or cores not to be grouted should be covered with wire lath, a broken piece of block, or similar material to keep grout from entering. All mortar droppings and other foreign material should be cleaned out of the cell and off the reinforcement; then the cleanout hole is closed.

All bolts and anchors should be tightly grouted, as well as spaces around metal door frames and similar items. Cells containing reinforcement should be carefully filled with grout. When it is necessary to stop grouting, a key is formed by stopping 1½ in. below the top of the course of masonry. Beams over doors and other openings should be grouted in one continuous operation. All grout must be consolidated by vibration or puddling.

High-lift Grouting. In high-lift grouted masonry vertical grout barriers of solid masonry are built across the grout space between the withes at a horizontal spacing of not more than 25 ft. Walls of the two withes of masonry should be permitted

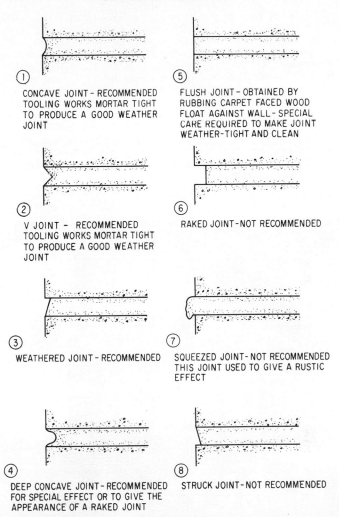

① CONCAVE JOINT - RECOMMENDED TOOLING WORKS MORTAR TIGHT TO PRODUCE A GOOD WEATHER JOINT

⑤ FLUSH JOINT - OBTAINED BY RUBBING CARPET FACED WOOD FLOAT AGAINST WALL - SPECIAL CARE REQUIRED TO MAKE JOINT WEATHER-TIGHT AND CLEAN

② V JOINT - RECOMMENDED TOOLING WORKS MORTAR TIGHT TO PRODUCE A GOOD WEATHER JOINT

⑥ RAKED JOINT - NOT RECOMMENDED

③ WEATHERED JOINT - RECOMMENDED

⑦ SQUEEZED JOINT - NOT RECOMMENDED THIS JOINT USED TO GIVE A RUSTIC EFFECT

④ DEEP CONCAVE JOINT - RECOMMENDED FOR SPECIAL EFFECT OR TO GIVE THE APPEARANCE OF A RAKED JOINT

⑧ STRUCK JOINT - NOT RECOMMENDED

FIG. 3. Masonry joints. (*From "Concrete Masonry Design Manual," Concrete Masonry Association of California, Los Angeles, 1964.*)

to cure for a minimum of 3 days. Hollow-unit masonry should cure for a minimum of 24 hr. Grout barriers or dams are, of course, not necessary when the grouting consists only of filling the cells in hollow units.

Grout should be puddled or vibrated at the time it is poured, and reconsolidated a second time just before it loses its plasticity.

9   Cracking of Masonry

Causes of Cracking. In most instances, cracking of the wall can be traced to shrinkage of the blocks. Other causes of cracking are improper design or workmanship in the footing that leads to settlement, or movement caused by frost, or soils that expand when wet. Temperature changes may cause movements. Expansion or contraction of floor or roof elements can cause abnormal distortion. Overloads on the wall might result from improper design, vibration, earthquakes, or extremely high winds. Substandard units, such as thin webs and shells, or weak concrete, or poor workmanship in laying the units can cause cracking.

Crack Prevention. Of greatest importance is to use blocks that have dried out adequately. Specifications for Type I blocks limit the total moisture content of blocks at the time of delivery. These are maximum values, and lower ones are preferred. Blocks that have been stockpiled and permitted to dry out for 28 days will satisfy these requirements. In very dry climates, more rapid drying permits using the block sooner. Autoclaved blocks dry out more rapidly than water- or steam-cured ones. Moisture content and absorption tests should be made to determine when the blocks have dried out sufficiently. The drier the block, the less shrinkage it will undergo after construction of the wall, and the less cracking the wall will suffer.

Fig. 4. Examples of concrete-masonry construction, showing structural walls of split block, typical one-story construction, screen walls, and fences.

FIG. 4. (*Continued*)

Mortar must be made of sound materials, of the correct proportions, properly mixed and applied. The use of reinforcement in the joints in accordance with good design practices is desirable.

Contraction joints should be installed at intervals of about 25 ft in long walls. The joint should be entirely through the wall, plumb, and of the same thickness as the other mortar joints so it will not be conspicuous. Sometimes it can be hidden behind a downspout or other feature of the building.

The designer has a wide range of joint designs available, including special rubber strips that are inserted in grooves in the block, the use of building paper to break bond, installing Z-bar reinforcing on weakened joints, and others. All joints must be caulked with a flexible or plastic caulking compound.

10   Testing

Methods of sampling, inspecting, and testing masonry units are described in the ASTM specifications referred to in Table 3.

Grout and mortar should be sampled twice a day, or whenever there is a change in materials or mix. Specimens are handled in the standard manner and are tested in compression.

Grout. The method prescribed in the Uniform Building Code[1] is as follows: On a flat nonabsorbent base, form a space approximately $3 \times 3 \times 6$ in. high, i.e., twice as high as it is wide, using masonry units having the same moisture condition as those being laid. Line the space with a permeable paper or porous separator, such as paper toweling, so that water may pass through the liner into the masonry units. Thoroughly mix or agitate grout to obtain a fully representative sample and place into mold in two layers, and puddle each layer with a 1×2-in. puddling stick to eliminate air bubbles. Level off, immediately cover mold, and keep it damp until taken to the laboratory. After 48 hr, carefully remove the masonry units and place the sample in the fog room until tested in a damp condition.

Mortar. Again quoting the Uniform Building Code, the recommended method is to spread mortar on the masonry units $\frac{1}{2}$ to $\frac{5}{8}$ in. thick, and allow to stand for 1 min, then remove mortar and place in a 2×4-in. cylinder in two layers, compressing the mortar into the cylinder using a flat-end stick or fingers. Lightly tap mold on opposite sides, level off, and immediately cover mold and keep it damp until taken to the laboratory. After 48 hr, remove the mold and place the sample in the fog room until tested in a damp condition.

REFERENCES

1. Uniform Building Code, International Conference of Building Officials, Whittier, Calif., 1973.
2. ASTM Specifications C 91, Specifications for Masonry Cement, American Society for Testing and Materials, Philadelphia, 1973.
3. "Concrete Masonry Handbook," Portland Cement Association, Chicago, Ill., 1951.
4. "Concrete Masonry Design Manual," Concrete Masonry Association of California, Los Angeles, 1964.

Section 11

PRECAST AND PRESTRESSED CONCRETE

Chapter 41

PRESTRESSED CONCRETE

JOSEPH J. WADDELL

Prestressing is accomplished when two stressed materials are joined in such a manner that a force acting in one is balanced by an opposite force in the other. Thus in prestressed concrete a tensile force applied to steel tendons embedded in the concrete generates a compressive force in the concrete. The reason for prestressing concrete is to enable it to withstand tensile stresses, as concrete is weak in tension, its tensile strength being roughly one-tenth of its compressive strength. This permits the use of smaller, lighter concrete members. Prestressing imparts greater stiffness to the member and enables the designer to take advantage of high-strength steel. Because the concrete is in compression, applied loads can be greater than for reinforced-concrete members of the same size before actual tensile stresses are set up in the concrete.

In any prestressing operation, the procedure follows good concreting practices in regard to materials, mixing, placing, curing, and other phases of concrete making. Additional work involves placing and tensioning strands or wires and, in the case of pretensioning, release of tension at the proper time.

Control of prestressed concrete includes the usual inspection and supervisory procedures for any good concrete. Uniformity of the concrete is especially important, as nonuniform concrete results in variations in camber and other features of the finished units. Concrete strength as high as 8,000 psi is frequently specified.

Steel forms are almost universally used, although wood is acceptable for a limited number of reuses. They should be stoutly built and should close tightly to prevent leakage of mortar when the concrete is placed in them. Because forms are used repeatedly, they are apt to warp and bend, resulting in mortar leaks and uneven surfaces.

Prestressed units require conventional reinforcing steel, the same as nonprestressed concrete. Stirrups in girders and beams should be accurately placed as shown on the drawings. Frequently special steel is installed to control certain types of cracking. A relatively small amount of extra stirrup reinforcement near the end of a pretensioned girder is effective in minimizing the occurrence of cracks.

GENERAL PROCEDURES

1 Basic Methods

There are two basic methods of prestressing concrete, pretensioning and posttensioning. Pretensioning lends itself particularly well to mass production of members in a central casting yard. In pretensioning, the stressing strands are tensioned in long beds before the concrete is placed in the forms. After the concrete hardens and has cured, the tensile load on the strands is relieved, transferring the stress to the concrete by means of bond between the concrete and steel. Forms are usually continuous, separated by bulkheads to give the requisite length of members. In some yards, concrete is placed continuously in a long form and the hardened concrete section is sawed into appropriate lengths.

In posttensioning, which is especially suitable for on-site construction, ducts are formed by embedding hollow conduits in the concrete when the member is cast. After the concrete has reached the required compressive strength, the strands or wires in the ducts are tensioned, thus putting the concrete in compression, and the ducts are (usually) grouted.

Whether to use one method or the other is largely a matter of economics. Pretensioning usually presents greater economy when there are many identical or near identical units. However, if units exceed about 80 ft in length, there are apt to be problems in transporting them to the jobsite, unless rail or barge transportation can be used. Pretensioning beds require heavy abutments to withstand the loads imposed by the stressed strands, but posttensioning loads are carried by the concrete of the member being stressed. Posttensioning can sometimes increase the effective spans of cast-in-place slabs or beam and slab systems, at the same time offering good deflection control, and aids in developing continuity in precast construction.

2 Materials

Prestressing Steel. Steel for prestressed concrete consists of stress-relieved high-tensile-strength wire or strand or, less frequently, high-strength rods, as described in Chap. 3. The following types are regularly used:

1. Small-diameter strand ($\frac{1}{2}$ in. or less in diameter) made up of six uncoated wires wrapped helically about a center wire. Used mostly for pretensioned concrete. ASTM Designation: A 416.

2. Cold-drawn single wire, uncoated and stress-relieved, or hot-dip-galvanized, used in groups of two or more essentially parallel wires for posttensioning. ASTM Designation: A 421.

3. Cold-stretched alloy steel bars. Used mostly for posttensioning.

4. Large-diameter strand consisting of 37 or more wires. For posttensioning.

Detail requirements differ for different users, but the reference ASTM specifications provide sufficient detail for the applicable types. Strength of the wire and strand is usually specified as 250,000 psi or more. The elastic modulus is about 28 million psi.

Many of the comments relative to reinforcing steel apply to stressing steel also. For example, a small amount of rust has been found to be beneficial to bond, but severe corrosion should not be permitted. Avoidance of corrosion requires somewhat more care for stressing steel than for reinforcing steel, as high-strength steel is more susceptible to corrosion. Strand should be protected from the weather during storage, preferably by keeping it inside a building. Severe corrosion may occur if the steel is exposed to galvanic action while in storage.

In pretensioned concrete, and most types of posttensioned concrete, the prestress forces are maintained exclusively by bond between the steel and the hardened concrete, hence the importance of maintaining the steel free of deleterious coatings and contamination.

Concrete. Materials and mixes are covered elsewhere in this handbook. Only clean, well-graded aggregate of high quality should be used. Natural or artificial lightweight aggregates can be used to reduce the weight of members. Any of the standard types of cement, including Type III, are satisfactory. Admixtures that can be employed are air-entraining agents and water-reducing retarders. Admixtures containing calcium chloride must be avoided, because of the subsequent danger of serious corrosion of the stressing strands. Air entrainment has a tendency to lower strength values for rich mixes but is desirable from the standpoint of improved workability and durability. A water reducer permits improved strength development by reducing the amount of water for the same slump.

Compressive strength for detensioning (transfer of stress from the steel to the concrete) and ultimate strength are specified by the designer. Detensioning strength must be adequate for the requirements of the anchorages or transfer of stress through bond and must meet camber or deflection requirements. The ACI Building Code specifies a minimum transfer strength for seven-wire strands of 3,000 psi for ⅜-in. or smaller strand, and 3,500 psi for 7⁄16- and ½-in. strands. Most engineers specify considerably higher strengths.

Mixes are usually quite rich, containing as many as 8 sacks (752 lb) of cement per cu yd. Type III high-early-strength cement is frequently specified. Slump is low, in some plants less than 1 in. Consolidation is always by vibration, either internal, external, or both.

3 Camber and Cracks

Camber. Prestressed-concrete members have a natural tendency to bend in a vertical plane so the midpoint is higher than the ends. This curvature is called *camber* and results from the eccentric application of the prestressed load. Both pretensioned and posttensioned members will camber.

It is difficult to forecast exactly how much a member will camber, since camber depends largely on the modulus of elasticity of the concrete, and increases with time because of creep. The main quality-control consideration is the difference in camber between adjacent members in the structure. For example, box beams in a bridge deck, to be covered with a cast-in-place composite slab, could tolerate a difference of 1 in., while adjacent tees in a building could accommodate perhaps half as much.

Camber can be measured while the member is still on the stressing bed, just after detensioning of pretensioned members, or stressing of posttensioned members, using the soffit plate or bed as a line of reference.

Cracking. One of the principal reasons for prestressing is the lessened occurrence of diagonal tension and flexure cracks under load. However, some cracking does occur during the manufacturing process which, in moderation, has no effect on the structural integrity of the member.

Some minor cracking results from shrinkage of the concrete. Because of the very low-slump concrete normally used, shrinkage cracking is minor and of little or no consequence. Short discontinuous cracks, commonly not over 3 in. long, have been observed occurring horizontally at the angle of the web and bottom flange of I-beam girders. These are shallow shrinkage cracks and are not serious. Cracking may occur as a result of failure to follow proper curing and detensioning procedures. Vertical cracks can result from shrinkage and cooling before the member is detensioned. Once formal steam curing is discontinued, holddowns and strand anchorages should be released immediately.

Some cracks occur at the ends of pretensioned members that can be controlled by proper conventional reinforcement. This will not prevent the cracking, but it will minimize the size and frequency of cracks. Stress concentrations occur at the ends of posttensioned members that can lead to cracking. Transverse reinforcement will reduce the size of such cracks.

PRETENSIONING

In pretensioned concrete, the steel tendons are placed in the forms and a tensile load is applied to them by means of jacks. The concrete is then placed, and after it reaches a certain specified strength, the tendons are cut loose at the ends. This transfers the load to the concrete as a compressive force, and it is held by means of the bond between the concrete and the steel.

Steel for pretensioning consists of twisted strands that are anchored in grillages at the ends of the casting bed and are elongated by means of hydraulic jacks. All the strands in a member may be stressed simultaneously, or the strands may be stressed one at a time. The trend is now toward single-strand tensioning.

4 Casting Beds

Casting, or stressing, beds may be less than 100 ft long, ranging up to 650 ft. The reason for the long beds is to permit several forms to be set up in a line when cross section and tendon pattern in the units permit. One typical plant had four beds ranging from 20 to 450 ft in length. In this plant, the bed consisted of a raised soffit plate supported on two pilot liners, or I-beam rails set in the

Fig. 1. The soffit plate rests on pilot liners and steel beam sections.

Fig. 2. End anchorage for a small duplex bed designed for a maximum of twelve ⅜-in. strands on each side.

concrete base. The raised soffit plate provided access to the underside of the form for installation and release of holddowns for draped, or deflected, strands. Additional support for the soffit plate was provided by short lengths of I beam spaced at 30-in. intervals and set at right angles to the long axis of the bed. This was a specialized bed, designed for large bridge beams of either I section or hollow-box section (see Fig. 1).

The bed itself should be of concrete, finished to a smooth surface to support the steel forms, with end anchorages built into each end. Horizontal loads imposed on the end anchorages by the fully stressed strands in a bridge girder may approach 2 million lb, requiring extremely heavy construction to resist the moments induced by such loads. End anchorage may be constructed of reinforced concrete or structural steel, extending down into the ground so as to transmit the force into the ground or into the concrete of the casting bed. The strands pass through openings

FIG. 3. The end anchorage for a large bed for heavy bridge girders. Twenty ½-in. strands are shown. Some girders have 60 or more strands.

FIG. 4. Another view of the anchorage shown in Fig. 3. The single-strand jack is shown in a movable framework.

in the end anchorage to the stressing equipment beyond, as shown in Figs. 2, 3, and 4. Instead of the heavy end anchorage, the bed can sometimes be designed so the forms take the reaction to the stressing load.

Many prestressed units are designed with deflected or draped strands, sometimes called "harped" strands, in which part of the strands are raised at each end of the girder. This construction makes necessary the introduction of holddowns in the forms to guide the strands into the proper pattern.

Members in a long bed are separated by means of bulkheads installed in the forms prior to placing the concrete (Fig. 5). A number of bulkheads are available on the market, and some precasters make their own. For a limited number of uses plywood can be used, but most bulkheads are of steel. One type is segmented so it can be fitted around the strands at the proper location in the form. Another type can be threaded onto the strands in a bundle, then separated and each bulkhead moved to its proper location. Usual practice is to anchor the bulkhead to the strands to prevent movement during concrete placing. When draped strands are used the bulkheads serve also as a "holdup" for the strands.

Curing facilities must also be provided and may consist of steam, hot water, or hot-oil piping (see Curing in Art. 7).

5 Forming

Usually only steel side forms and steel or concrete bottom forms are recommended. Forms should be of sufficiently thick material, rigidly braced, and anchored, to withstand vibratory placing of fairly stiff concrete without bulging or deflecting.

Forms must be easily cleaned and should be cleaned after each usage.

FIG. 5. Tendons have been pulled through the bed and bulkheads set, indicating the ends of the girders.

Joints in forms must be smooth and tight to prevent unsightly offsets, sand streaks, or other blemishes on the concrete. Joints in the surface of a form must be ground smooth. Joints between form sections, soffits, and bulkheads must be tight, and gasketed if necessary. Certain types of chamfer strips of rubber or plastic can double as gaskets. All edges should be chamfered.

Most forms are externally supported, and no form ties are necessary. If form ties are necessary, they must be snap ties or she-bolts that break away beneath the surface of the concrete, leaving no pieces of metal in the surface of the concrete.

A form oil or parting compound must be applied to the forms to prevent adhesion of the concrete. There are many good materials on the market especially compounded for high-temperature curing in steel forms. Special care must be exercised, when applying the form oil, to avoid getting any of it on the prestressing strands. If the bond breaker is applied to the form before the strand is placed, the form must be covered with paper, plastic, or other protective materials so the strands will not be contaminated when they are installed. If the form oil is applied after the strands are in place, special care is likewise needed to keep the strands clean. Any form oil that gets on the stressing wires should be removed immediately with a solvent.

Most forms include provisions for the attachment of heating pipes for curing.

Some designs, especially heavy bridge girders, require bearing plates at the girder ends. These bearing plates must be placed accurately on the casting bed before the forms are set. Other inserts may be required, and the plans should be checked to determine type and location.

Fig. 6. The tendons have been stressed and conventional reinforcement, consisting of longitudinal bars and stirrups, has been placed. Note that some of the tendons are draped or deflected.

Fig. 7. The bed is now ready for the forms to be set in place. Note the end anchorages, stressing strands, and reinforcement.

Prestressed units contain conventional reinforcement also (Figs. 6 and 7). Stirrups are especially important in controlling some types of cracking. Reinforcement is usually prefabricated into cages that are inserted in the forms either before or after stressing, depending on the type of unit. Some forms, especially those for I beams, are so arranged that the entire assembly of stressed tendons and conventional reinforcement is in place before the side forms are set up. Bulkheads and deflecting hardware can be placed also before setting forms.

Void formers, discussed in Chap. 14, are placed in the forms to displace concrete near the neutral axis of a member to reduce weight with no loss in load-carrying capacity. Small, cylindrical voids are formed in roofing and deck units and large voids of square or rectangular section are used in making hollow box girders for bridges. Installation of formers is shown in Figs. 8 and 9.

FIG. 8. The same bed shown in Fig. 7. Forms, spreaders, and bulkheads have been set, part of the concrete placed, and void formers installed. Balance of the concrete will be placed immediately, before a cold joint develops.

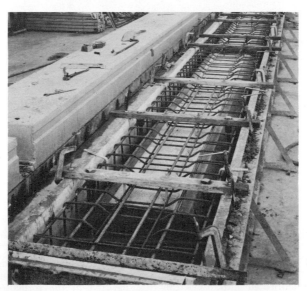

FIG. 9. A form for a hollow box girder. All steel is in place and a large void former tied in place.

6 Tensioning

When a tensile load is applied to a piece of steel, the steel starts to stretch or elongate. This elongation (called strain) is proportional to the applied load (called stress) up to a point called the proportional elastic limit. If the load is now released, the wire will resume its former length. If the load is carried beyond the elastic limit, the wire is permanently stretched to a degree depending on the load. This is called permanent set. In prestressed concrete, it is desirable to elongate the strands to a stress just below the elastic limit, generally about 175,000 psi (see Fig. 10).

The modulus of elasticity E is a measure of the elasticity of a material. It is equal to the unit stress divided by the unit deformation or strain. The modulus of elasticity of steel for prestressing is in the neighborhood of 28 million psi. This is apt to vary as much as 7 or 8% from lot to lot, thus introducing a possible error in stress values computed from strain measurements. By using jack pressure for computing stress, a check is obtained. As long as the two are within reasonably close agreement, there is no question of the accuracy of the stress measurement. Each job will have to set up its own standards and tolerances, but a difference of 5% between stress computed from jack pressure and that computed from measured elongation is acceptable. The ACI Building Code states that the prestressing force shall be determined by measuring tendon elongation and also either by checking jack pressure on a recently calibrated gage or by the use of a recently calibrated dynamometer.

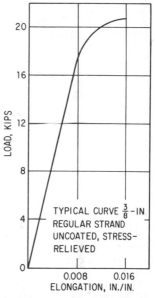

Knowing the properties of the steel, tensioning requirements, and dimensions of the stressing bed, it is not difficult to compute the required pressure to apply to the stressing jacks, or the elongation to be obtained. Differences in the modulus of elasticity of different lots of steel, or inaccuracies in pressure gages on the stressing jacks, are sources of error in measuring stressing loads. These measurements are complicated when deflected or draped strands are used. All equipment should be calibrated at regular intervals, preferably under operating conditions. Any measurements should be checked occasionally with a dial gage extensometer applied to the wire during the stressing operation.

Fig. 10. A typical stress-strain curve for prestressing strand.

As a check on stressing, for either straight or deflected strands, the occasional use of load cells is recommended at the beginning of a job. A load cell makes use of electric strain gages (SR-4) for measurement of loads.

Strain gages applied directly to the strand are useful to determine the uniformity of tension along a deflected strand. This procedure is used when checking a new tensioning system but is impractical as a routine check.

The sudden release of energy that occurs when a strand breaks or an anchor fails causes rapid recoil of the strand and other hardware in the stressing bed. For this reason, inspectors and workmen must be alert for any unusual occurrences during stressing.

Tensioning Methods. In pretensioning, strands are anchored at one end of the bed and are attached to the tensioning device at the other end. In both multiple-strand and single-strand stressing, special hydraulic jacks with long ram travel are used.

It is common practice to apply a small preload to the tendons (strands) by tensioning them to a low value which is the minimum force that will take the slack out of the tendons and equalize stresses. The load required depends on bed length, size of strands, and whether draped strands are used. Values commonly used are between 500 and 2,000 lb. Preload can be applied by using the stressing jack or other means to apply a measured load. After application of the preload, reference points must be established from which total elongation can be measured. Elongation must be measured within a tolerance of ⅛ in. or 1% of theoretical elongation, whichever is smaller.

In multiple-strand tensioning, the jacks push a crosshead that pulls the template to which the strands are attached. When the proper tension is reached, as indicated by jack pressure and elongation, the strands are anchored in the grillage and the jacks are removed to the next bed. When a large number of strands are being stressed, more than one jack might be used.

Single-strand tensioning permits the use of smaller, more portable jacks. A center-hole jack may be used, in which the strand to be tensioned is fed through the center hole of the jack and anchored on the outboard end. After stressing, another anchor on the strand holds the strand against the grillage, and the jack is released and moved to the next strand.

Temperature Effect. Normally, the difference in temperature between the steel at time of tensioning and that of the concrete is not significant in causing stress change in the tendons. If steel is stressed at a low temperature and warm concrete is placed around it, the steel will expand and reduce the tension. Of course, the opposite effect results if temperatures are reversed.

Total length change in the steel resulting from temperature change is expressed by the equation (all length measurements in inches)

$$\Delta_t = etl \qquad (1)$$

in which Δ_t = length change of steel due to temperature change
$\quad e$ = linear coefficient of expansion of steel = 0.0000065 in. per in. per °F
$\quad t$ = temperature difference
$\quad l$ = length of strand subject to temperature change

Unless there is a large temperature difference, the effect can usually be neglected.

Discrepancies and Losses. Small discrepancies can be expected between calculated and measured elongations. As long as these are recognized and evaluated, there is no cause for concern. One source of discrepancy is slippage in the gripping devices or anchors on the ends of the strand. This can easily be determined and included in the computations, as shown in the paragraph on computing stresses. At the start of any prestressing job, the movement of anchoring abutments and elongation of anchor bolts should be evaluated. Usually this can be ignored, but it should be checked if there is an otherwise significant discrepancy.

In single-strand stressing, the strand will have a tendency to untwist, causing the jacking ram to rotate. A rotation of one turn is permissible.

Occasionally tendons will become twisted or crossed. An occasional crossing is not serious and need not require removal and restringing of the tendons. Sometimes a wire will break during stressing. If the number of broken wires is less than 2% of the total number of wires, the stressing is still acceptable.

Computing Prestress. The size, number, and distribution of tendons, and the stress to be applied, will be computed by the designer, but this information has to be translated into pressure and elongation values for field use.

The following computation shows how to compute the tensioning load in a group of strands being stressed simultaneously. If single-strand stressing is being used, the equations are still valid, in which case $N = 1$ and $J = 1$.

Let A = area of one strand, sq in.
$\quad a$ = net area of jack ram, sq in.
$\quad d$ = unit deformation or strain, in. per in.
$\quad E_s$ = modulus of elasticity of strand, psi

Δ = total change in length l, in. (elongation)
J = number of jacks
L = net length of stressing bed, ft
l = length, in.
M = measured movement of jack-end anchors, in.
N = number of strands in one group
P = jack pressure, psi
S = unit stress applied to strand or required, psi
T = total slippage, in.
W = total load on N strands, lb

The number of strands N is the group of strands being stressed at one time by a group J of jacks. This will vary with different plants but will probably be the number of strands in one casting, assuming that the bed consists of several castings in tandem. The unit stress S is determined on the basis of the physical properties of the steel being used, and this will be the basis for the field computations. The modulus of elasticity E_s is determined from laboratory test and is the basis for determining the change in length of the strands in the bed.

To obtain the jack pressure reading P required to produce the unit stress S in the strand, we have

$$W = ANS$$

then
$$P = \frac{W}{Ja} = \frac{ANS}{Ja} \qquad (2)$$

In using E_s to determine the total elongation of the strands, it is necessary to consider the amount of slippage of the strands in the anchors, both at the dead end of the bed and at the jack end. As the load is applied, there will be a slight movement in the anchors at each end as they grip the strand. This is normally of the magnitude of $\frac{1}{4}$ in. To find what it is, mark each strand, under no load, just where it emerges from the anchor, using crayon, soapstone, or similar material. After the bed has been stressed, measure the distance the marks have moved away from the ends of the anchors. The sum of the slippages at each end is the total slippage T.

From the relationships,

$$d = \frac{S}{E} \qquad \text{and} \qquad d = \frac{\Delta}{l}$$

we get
$$\Delta = \frac{Sl}{E_s} = \frac{12SL}{E_s} \qquad (3)$$

Δ being the total required elongation in a bed of length L. Movement of the ram

$$M = \frac{12SL}{E_s} + T \qquad (4)$$

The slippage T should be checked for each operation. Once the pressure and elongation have been set up, there will be no change unless a change is made in the other constants. In some plants it will be possible to measure the elongation Δ directly, without measuring slippage. In this case, use Eq. (3).

The following example illustrates the computations.

A certain plant has a stressing bed 400 ft long using one jack with a net ram area of 32 sq in. It is planned to stress a string of beams using 16 strands of 0.035 sq in. area. The steel has an elastic modulus of 28 million psi, and the strands will be stressed to 175,000 psi. Slippage in the anchors is $\frac{1}{4}$ in. at the jack end and $\frac{1}{4}$ in. at the dead end. Compute required jack pressure and movement of jack-end anchors.

From the above,

$A = 0.035$ $N = 16$
$a = 32$ $J = 1$
$E_s = 28{,}000{,}000$ $S = 175{,}000$
$L = 400$ $T = \frac{1}{2}$

$$P = \frac{ANS}{Ja} = \frac{0.035 \times 16 \times 175{,}000}{1 \times 32} = 3{,}062 \text{ psi}$$

$$M = \frac{12SL}{E_s} + T = \frac{12 \times 175{,}000 \times 400}{28{,}000{,}000} + \frac{1}{2} = 30.5 \text{ in.}$$

In practice it is necessary to apply a slight load to the strands in order to take up all the slack and give a reliable starting joint for measuring elongation. This is done by applying a small load, say 100 psi, to the jack which will take the catenary or droop out of the strands through the forms. The stress in the strands can be computed from Eq. (2) and the initial elongation by proportion.

Using values from the above example, for a 100-psi initial jack load, from Eq. (2),

$$S = \frac{PJa}{AN} = \frac{100 \times 1 \times 32}{0.035 \times 16} = 5{,}714 \text{ psi}$$

$$\frac{5{,}714}{175{,}000} = 0.0325$$

Elongation at 100 psi $= (M - T) \times 0.0325 = 30 \times 0.0325 = 0.98 \text{ in.}$

Deflected Strands. The foregoing computations apply to a bed with straight strands. Many prestressed units, on the other hand, contain deflected, or draped, strands. A draped strand is one that extends a predetermined distance near the bottom of the beam or unit being made, equal distances each side of the center of the span, then rises and emerges from the ends of the beam near the top. The purpose of deflecting strands is to provide a better stress distribution in the pre-stressed member and to reduce eccentricity of the prestress force at the ends of the member.

Elongation and load computations are made as for a straight strand. However, it is sometimes not possible to stress a deflected strand from one end only, because of friction in the hardware holding the strand down and up at several points through the length of the bed. In this case, the full load is applied to one end of the strand and the elongation noted. The jack is then moved to the other end of the bed and the full load applied at that end, the elongation there being noted. The sum of the two elongations should equal the computed total elongation within 5%.

There are two common methods of stressing deflected strands, both of which appear to give satisfactory results. In one system, the strands are tensioned the same as straight tendons, as described above. If the bed contains more than one or two members in tandem, it is usually necessary to stress from both ends. In the second system, all the tendons, running straight through the forms, are partially stressed a predetermined amount with the jack. The balance of the stress is applied by pulling the strands up or pushing them down to final deflected position.

A great variety of hardware is available for holddown and holdup units. Hold-downs in some cases extend through the bottom form and are held in the anchor plates by means of pins that can be easily and quickly removed. Holdups consist of various assemblies, some of them with small grooved wheels to carry the strands to minimize friction.

An effect similar to that obtained with deflected strands can be achieved by breaking the bond between the concrete and the end of the tendons. The distance to be unbonded is determined in the design office. Bond is prevented by enclosing the tendon in a paper or plastic sleeve, or by coating the strand with a bond-inhibit-ing compound.

7 Concreting

After the tendons have been tensioned, the reinforcement placed, and the forms buttoned up (Fig. 11), concrete is placed in the conventional manner, following good construction and control practices.

Some plants use ready-mixed concrete, but most have their own batching and mixing plants, frequently with a turbine-type mixer because of the low-slump concrete used. Distribution of concrete in the yard is by practically any kind of handling equipment as shown in Fig. 12. Concrete should be distributed in the forms in lifts not over about 16 in. in depth, thoroughly consolidated by vibration, both internal and external. Cold joints must be avoided.

Concrete test cylinders should be made during the time that concrete is being placed. The exact number will depend on the manufacturing schedule, but a minimum of six cylinders should be field-cured with the concrete members, and the time of releasing stress on the tendons determined by the strength developed

Fig. 11. A small bed for thin concrete planks, ready to receive concrete.

Fig. 12. Various methods of transporting concrete are used. In this plant, they hauled the concrete in a front-end loader.

FIG. 13. Heavy canvas is spread over the finished girder, and then steam is admitted through perforated pipes. The canvas is on frames so there is circulation of steam around the beam.

in these specimens. Cylinders should be made for standard curing also. Specimens for standard curing may be cast in either cardboard or metal molds, but the field-cured ones should be cast in metal molds.

Curing. Some sort of accelerated curing is practiced in nearly every casting yard, because of the requirement for 4,000 psi compressive strength, or more, in about 16 hr. Many of the forms have incorporated in them pipes for steam, hot water, or hot oil. Means must be provided to keep the units moist during the curing period. Beds can be covered with heavy tarpaulins or boxes to prevent loss of moisture and heat. When heat is supplied by wet steam, sufficient moisture is usually present. If heat is supplied by hot water or hot oil in pipes, moisture must be supplied to the top exposed areas of concrete by soaker hoses, fog spray, wet burlap, or other means. Water must be hot. In general, the best curing is obtained with wet steam under tarpaulins, except in exposed outdoor beds in cold weather.

FIG. 14. This beam, exposed to normal atmospheric temperatures, was covered with burlap kept wet with soaker hoses. Such a procedure could not be used during cold weather.

Start of high-temperature curing should be delayed 3 to 4 hr after the last concrete is placed, and the temperature should not exceed 160°F. After the expiration of the curing period (determined by field cured cylinders) the concrete is permitted to cool to ambient temperature. When the air temperature is below 50°F, the concrete should be cooled at a rate not exceeding 5° per hr.

Curing methods are depicted in Figs. 13 and 14.

8 Detensioning

As soon as the concrete reaches the specified strength for transfer of stress to the concrete, the strands are released, or detensioned. This may be accomplished by releasing the jacks when multiple-strand tensioning has been applied to straight strands. Single-strand tensioning, and the presence of draped strands, make other methods necessary. One method is to cut the strands individually with an acetylene torch, using a sequence or pattern that has to be developed for each type of unit and stressing bed.

The relationship between curing, release of tension in the strands, stripping of forms, and release of strand holddowns is important in the control of cracking in the units. Forms should be loosened as early as possible in the curing cycle, but in any event while the concrete is warm and before the detensioning is done.

This should be done as rapidly as possible to minimize cooling of the concrete before the bed has been completely released; otherwise undesirable cracking may result.

Multiple-strand Release. The entire stressing load is picked up by the jack, or jacks, and is then gradually released. There will be some sliding of the stressed members on the bed as stress is released, the amount depending on the length of the tendon exposed between members and between the last member and the dead end. For this reason, there should be no restraint of movement of the members, except friction on the bed.

Single-strand Release. Using a sequence that keeps the stresses nearly symmetrical about the axis of the members, the strands are heated with a low-oxygen flame until the metal softens and loses its strength, causing the strands to part gradually, in a matter of several seconds for each one. Best results are obtained if heating can be done at both ends of the bed simultaneously.

Deflected Strands. For heavy members, in which the weight of each member is at least twice as much as the holddown force within the member, the holddown devices are released; then the strands, both deflected and straight, can be detensioned by either multiple- or single-strand release method.

In the case of light members weighing less than twice the holddown force, the draped tendons are first released by heating the tendons at each uplift point (between the bulkheads), and then the holddowns are released. Finally the straight strands are released by either single- or multiple-strand method. Light members can also be detensioned in the same manner as heavy ones if sufficient weight or restraint is applied to the members directly over the holddown points.

9 Handling and Transporting

Types of handling equipment include cranes, lift trucks, and straddle carriers of various types. Rubber-tired or rail-mounted gantry cranes are sometimes used. In enclosed plants, overhead-cab-operated bridge cranes are suitable.

Transportation of girders to the job is usually accomplished with motor trucks. Long girders are hauled on a tractor and bogie combination. Frequently the bogie is equipped with a steering device similar to those used on long fire trucks.

Prestressed members must be maintained in an upright position at all times, and supported or picked up only at designated pickup points. Beams and girders must be picked up at the ends, over the bearings. Disregard of these precautions will result in a damaged member.

Figures 15, 16, and 17 illustrate handling of members.

Fig. 15. Straddle carriers are frequently used for moving girders short distances.

Fig. 16. On the Lake Pontchartrain Bridge job, a special 200-ton gantry picked the precast deck unit out of the stressing bed and carried it to a barge for transportation to the construction site.

Fig. 17. Cranes are used to move large girders about the yard and to erect them in the structure. Pickup must be made at the bearings, with the member in an upright position at all times. (*Illinois Tollway Photograph.*)

POSTTENSIONING

Prestessing by the posttensioning process is accomplished by forming ducts through the concrete at the time it is cast. When the concrete has reached the required strength, as indicated by field-cured cylinders, the stressing wires are inserted in ducts, tensioned, anchored, and grouted. Whereas pretensioned members are manufactured at a central casting yard or plant, posttensioned members are frequently made at the site. One reason for posttensioning is that there is no need to transport the units over highways and streets; hence the logistics of getting the units in place does not impose a limitation on size. Pretensioned girders are rarely over 80 ft long for this reason.

10   General

The same care in production as for pretensioning is necessary. Frequently the ducts are curved or draped, a condition that is apt to lead to friction on the wires while they are being stressed. For this reason, it is especially important to check jack pressure and elongation at both ends. The ACI Building Code states that friction losses in posttensioned steel should be based on experimentally determined wobble and curvature coefficient, verified during stressing operations. The values of coefficients assumed for design and the acceptable ranges of jacking forces and steel elongations should be shown on the plans. These friction losses can be calculated by the equation

$$T_0 = T_x \epsilon^{KL + \mu\alpha} \tag{5}$$

When $KL + \mu\alpha$ is not greater than 0.3, Eq. (6) may be used.

$$T_0 = T_x(1 + KL + \mu\alpha) \tag{6}$$

in which K = wobble friction coefficient per ft of prestressing steel
L = length of prestressing steel element from jacking end to any point x
T_0 = steel force at jacking end
T_x = steel force at any point x
ϵ = base of Naperian logarithms
α = total angular change of prestressing steel profile, radians from jacking end to any point x (Fig. 18)
μ = curvature friction coefficient

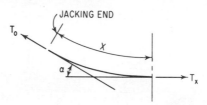

Fig. 18.   Angle α, expressed in radians and measured as shown here, is a maximum at the midpoint of the draped strand.

Values of K and μ can be taken from Table 1, which shows values for metal sheathing that can be used as a guide. Values of K (per lineal foot) and μ vary appreciably with duct material and type of construction.

Ducts or core holes for wire can be formed in the concrete by casting in flexible metallic tubing which is left in the concrete, or by flexible rubber tubing, hose, or fiber conduit that can be removed. Conduit or tubing should be securely tied to the reinforcing cage at intervals not exceeding one-tenth of the length of the member, with maximum intervals of 10 ft. In making straight ducts in lengths of hollow piles that are subsequently joined together, the ducts are formed by inserting steel rods, covered with rubber hose, in the form. Removal of the rods loosens the rubber, which can then be pulled out.

Individual wires are usually used in posttensioning, instead of the strand commonly used in pretensioning. By means of special equipment illustrated in Fig. 19, a group of wires is forced through the duct and anchored by means of special wedges. In another type of posttensioning, the wires are accurately measured and assembled into groups in a length of flexible metallic tubing. Each wire is then upset on each end, forming small buttons that bear against a special perforated plate. The entire tube and wire assembly is located in the forms and the concrete is placed. After the concrete has developed the required strength, a special jack grips the perforated plate and elongates the wire, which is then anchored, and the tube is grouted.

Table 1. Values of K (per Lineal Foot) and for μ Metal Sheathing

Type of steel	Suggested design values	
	K	μ
Wire cable............	0.0015	0.25
High-strength bars.....	0.0003	0.20
Galvanized strand......	0.0015	0.25

FIG. 19. Around the edge of the 54-in. concrete cylinder (right) are 12 holes extending the length of the section. One man operating the machine collects 12 wires from turning spools, right background, and shoves the wires through each hole to join six 16-ft sections into one 96-ft pile.

11 Stressing and Grouting

A small preload is usually applied to take the slack out of the tendons, as described under pretensioning, and then the design stress is applied, with proper allowances for friction. The tendons are next fixed in the anchoring devices, and the elongation is accurately measured, as shown in Figs. 20 and 21. It is important that both elongation and jack pressure be measured and the discrepancy noted (agreement must be within 5%).

Grouting. The purpose of grouting is to protect the tendon from corrosion and to provide bond between the tendon and the concrete. Grouting should be done as soon after stressing as possible, but in no event more than 48 hr later. Ducts to be grouted must be provided with ports for entrance and discharge of the grout and may require air bleeders at the high points of the profile.

Ducts must be clean when the grout is admitted, which may require flushing with water, after which the duct is blown out with air. Grout may be admitted

FIG. 20. Two posttensioning jacks on a 36-in. pile. The wires in part of the ducts have been stressed, anchored, and burned off. As soon as wires in all ducts have been tensioned, grout will be pumped in. After a suitable curing time, the wires will be burned off, releasing the tensioning ring.

Fɪɢ. 21.   Marking reference points for measuring elongation.

in one or more points, depending on the length and configuration of the duct. Grout is pumped continuously until it flows steadily out the discharge port. The discharge port is then closed and the grout pressure gradually increased to a pressure of 75 to 125 psi and held for 10 or 15 sec. The entrance vent is then closed.

Mixing, agitating, and pumping units are available for handling grout. The grout used for this purpose is quite fluid, sometimes containing a small amount of fine sand. Materials for grout must be accurately measured. After mixing, the grout is passed through a screen before it is admitted to the pump. Some specifications will allow a small amount of unpolished aluminum powder (about 1 teaspoonful per sack of cement) to alleviate shrinkage of the grout.

CIRCUMFERENTIAL PRESTRESSING

Another type of posttensioning is applied to cylindrical objects, such as large concrete pipe and concrete tanks. After the pipe or tank has been cast in a conventional manner and has attained the necessary strength, special machines wrap high-tensile wire around the object, stressing the wire to the required tension. The entire outer surface is then coated with shotcrete.

Manufacture of concrete pipe by this process is briefly described in Chap. 42.

Tanks are made by first constructing a conventional cylindrical tank on a concrete slab. The joint between the slab and the tank walls is usually flexible; that is, some movement is possible. Leakage is prevented by installing a rubber or plastic water stop in the joint. A special machine, suspended from the top of the tank and traveling in a circular course, wraps high-tensile-strength wire around the tank under the proper degree of prestress. The machine is operated so as to space the wire wrapping properly from bottom to top of the tank. After the prestressed wire has been installed, the entire tank is coated with shotcrete to protect the wire.

CHEMICAL PRESTRESSING

Portland cement can be formulated in such a way that an expansive force is generated during the hydration process. Called expansive, shrinkage-compensating, or self-stressing, this is a calcium sulfoaluminate cement.

Experimental use has been made of expansive cement to produce a self-stressing concrete. It has been found that, under favorable circumstances, when concrete is properly restrained with adequate reinforcement, a prestressing load can be induced in the concrete and steel similar to the stresses induced by mechanical prestressing. Although still in an experimental stage, the process shows promise. Dependent to a large extent on the amounts of normal cement and expansive cement in the mix, results depend also on the particular type of cement being used, the amount and distribution of reinforcing steel, and curing conditions.

Chapter 42

PRECAST CONCRETE

JOSEPH J. WADDELL

Precast concrete, as discussed in this chapter, includes precast building units, concrete masonry, pipe, and various specialty items such as railroad ties and poles.

PRECAST BUILDING UNITS

Precasting of building units can be divided into two general categories: (1) precasting on the site, using tilt-up, lift-slab, or other on-site construction methods (see Chap. 33), and (2) plant precasting, using manufacturing assembly-line techniques (Fig. 1). Units made by either method may or may not be prestressed. Pre-

FIG. 1. Modern precasting plants require adequate space for storage of finished members. This is the Schokbeton yard of Rockwin Prestressed Concrete Corp., Los Angeles.

stressed units may be either pretensioned or posttensioned (Chap. 41). This chapter is concerned with off-site plant precasting.

Concrete lends itself well to prefabrication of building units, not only because of its structural qualities but also because it can be cast into intricate shapes and details (Figs. 2 and 3). Off-site production-line manufacture of building components is an important part of the "system building" concept, which is merely a term to describe a system of construction in which design, production, erection,

FIG. 2. The contrast between the precast white concrete and gray concrete lends interest to this office building. White-concrete units were manufactured by the Schokbeton process. (*Rockwin Prestressed Concrete Corp., Los Angeles.*)

and overall administration of the building and its construction are coordinated, or preengineered (see also Chap. 33).

The same basic design and construction requirements that apply to any on-site concrete work apply also to precast work. In addition, there are several special requirements peculiar to precast units. Concrete wall panels, for example, may be of regular weight or lightweight concrete. Some are solid; some consist of a "sandwich" of concrete on the outside with a center layer of foamed, lightweight insulating material.

1 Shop Drawings

The manufacturer of precast concrete is usually required to furnish shop drawings showing complete information for making and installing units. Drawings should include an erection plan with all pieces marked, showing the necessary bracing that will be required during erection. Shop drawings should also give all dimensions, including tolerances. If units are to be prestressed, the allowance for elastic shortening should be designated. Drawings should show size, amount, and location of reinforcement; details and location of connections and method of adjusting; location and details of erection devices and special reinforcement for same; details on all inserts, reglets, attachments, etc., and method of anchoring; joint details and materials; concrete finishes; and methods of protecting exposed surfaces during storage and erection.

2 Forms

Steel, wood, fiber-glass-reinforced polyurethane, concrete, and plaster can be used for forming material, depending on the configuration of the units, number to be made, and preference of the manufacturer (see also Chap. 14).

Steel forms are desirable when many reuses of the form are required and where complicated irregular and reentrant surfaces are not necessary. Rivet heads and joints between sheets must be carefully ground smooth. Steel forms must be opened up or disassembled to permit removal of casting. Sheet-steel contact areas must be checked occasionally to detect buckling or dimpling, and must be coated with a light form oil before each usage. Steel forms are especially desirable when the product is to be steamed.

Fiber glass can be molded to complicated shapes on a master pattern and can be used many times (Chap. 33). The material is strong but requires support to give it rigidity. Concrete surfaces cast against new fiber glass are apt to be very smooth and glossy.

Wood forms are easily fabricated into intricate details but are not so durable as steel or fiber glass for many reuses. Wood forms should be treated with a sealant to protect the wood and should be coated with a parting compound when used.

A concrete form is made by casting against pattern of concrete, wood, or plaster, making an allowance for shrinkage. Waxes and sealants should be applied to the concrete.

FIG. 3. Precast panels may be two or more stories in height. Large units require fewer joints than small ones but usually entail little or no additional work in handling. Instead, there being fewer units to handle, erection costs are less.

3 Fabrication

Many times a casting yard is combined with a prestressing yard, thus permitting maximum utilization of such facilities as batching plant, mixers, and cranes. Forming, casting, curing, and handling are similar in the two plants. Mass-production assembly-line techniques are common. Individual reinforcing bars are cut, bent, bundled, and tagged, and bar mats and mesh can be cut to size. By setting aside an area of the yard for steel fabrication, jigs can be set up on benches at a convenient working height, and the individual steel components assembled into complete reinforcing units, or cages, for the precast units. The completed cages are stockpiled for later installation in the forms. Frequently inserts, such as lifting eyes, connecting hardware, and anchor bolts, can be attached to the steel cage, and electrical conduit can be tied in. Where similar but different units are being produced, a system of numbering must be established before manufacturing commences, and the reinforcing cages identified with these numbers.

While the steel crew is making reinforcing cages, another crew prepares the forms for concreting. The previous castings are stripped out and moved to storage for additional curing, using a motor crane, gantry, or straddle buggy. Forms are then cleaned of all loose or adhering concrete and other material, oiled, buttoned up, and the steel cage set in place. Small and light cages can be handled manually; heavier ones require the use of a crane. Cages must be accurately positioned in the forms and tied in such a manner as to prevent any movement during concrete placing. The type of concrete surface specified determines the type of chairs to be used, if any, or other means of fixing the steel in place.

Before placing concrete, the foreman and inspector should inspect the forms carefully to make sure that reinforcement and inserts are accurately located, well braced, and of the correct size, type, and quantity; that the forms are clean, well braced, and properly "buttoned up"; that the parting compound or form oil has been applied to the form; that void formers, if required, are well tied down to prevent movement or floating. Void formers will float in the plastic concrete during vibration, carrying reinforcement with them, unless they are well tied down to the form. Many of these units are prestressed, and the reader is referred to Chap. 41 covering this phase.

Concreting follows conventional methods. Slump of concrete is 2 in. or less. Immersion vibrators consolidate the concrete, frequently assisted by form vibrators. The desire for high-early-strength concrete to permit rapid turnover of the forms and casting beds makes necessary the use of rich mixes, steam curing, and sometimes Type III cement. Compressive strength of 5,000 psi is commonly specified for facing mixes, and 4,000 psi for other mixes. Air entrainment is recommended for attainment of maximum durability and to minimize permeability.

Curing, discussed in detail in Chap. 29, can be by moist curing at a temperature above 50°F for 5 days if regular cement is used, or 3 days for Type III cement. Many plants steam-cure at temperatures up to 160°F. Curing compounds are rarely used.

After curing, units are sometimes stored for a period of time. Storage should be in such a manner as to minimize warping or distortion. Panels can be stored flat if they are fully supported on a flat surface; however, there is less danger of warping if they can be stored on edge, provided the panel edge is fully supported and the panels rest against a rigid nonstaining frame. Beams should be supported at their normal support points. If storage is for more than a few days, units must be covered with polyethylene to protect them from atmospheric staining.

All repair to minor spalls and similar defects must be made as early as possible.

In handling and transporting units, care is necessary to avoid staining or marring. Edges and corners are especially vulnerable and should be protected with timber lagging if it is necessary to lift the unit with a sling or tie it down to a truck or car bed. Ropes are less apt to damage surfaces than cables or chains.

4   Sandwich Panels

In this type of construction the wall panel, cast in a horizontal or flat position, consists of a layer of structural concrete, a layer of lightweight insulating material, then another layer of structural concrete. By this method, the panel has the advantage of relatively high structural strength, at the same time possessing fairly high insulating value.

Besides possessing good thermal-insulating properties, the insulating material must be resistant to the absorption of moisture, as the insulating ability diminishes when moisture is absorbed. A desirable material is one having low absorption, no capillarity, and capable of little or no vapor transmission. In addition, the material must be capable of developing good bond with the enveloping concrete, and must develop a reasonably high modulus of elasticity and compressive strength. These strength properties are necessary to provide a satisfactory degree of interaction between the concrete and the insulation, and to give the unit the required load-carrying capacity.

Materials that meet these requirements are portland-cement concretes using expanded mineral aggregates (perlite, vermiculite); foamed concrete, foamed glass, certain foamed or expanded plastics such as polystyrene and polyurethane, and paper honeycomb. Preformed insulation must be protected from damage during handling and storage, and protected from absorption of moisture. Densities of these materials range from less than 15 pcf to a little over 50 pcf.

Fig. 4. Well-constructed molds are necessary for the Schokbeton process. Note the extremely dry white concrete in the background, under the conveyor, that is being placed in the vibrating mold. (*Rockwin Prestressed Concrete Corp., Los Angeles.*)

The outside concrete layers of sandwich panels are interconnected by means of mechanical shear ties, or by concrete spacers or ties. In general, properly located concrete ribs are more efficient than metal ties.

5 The Schokbeton Process

In this plant precasting process, which is patented, zero-slump concrete is placed in the mold and the entire assembly is subjected to a low-frequency vertical shock or impact that consolidates the concrete without segregating it. Speed of vibration is of the magnitude of 250 high-amplitude blows per minute. Since the entire element is subjected to the same processing simultaneously, there are no conflicting impulses (Fig. 4).

Molds must be accurately made, rigid, and rugged to withstand the casting process. Details must be accurate, with precise corners and arises, so as to assure excellence in the finished product. Mold dimensions are usually required to maintain a tolerance of plus 0, minus $\frac{3}{32}$ in. for 10 ft or over. Resulting concrete possesses high strength, high density, and low water absorption, making members with exact tolerances and sharp arrises.

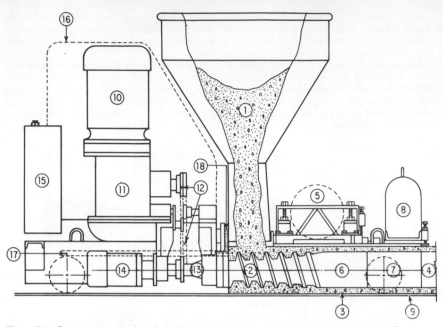

Fig. 5. Operation of the Spiroll extruder starts with introduction of the concrete mix into the machine hopper (1). The process continues as follows: (2) A set of rotating augers collects the concrete mix from the discharge end of the hopper and forces it toward the rear of the extruder. (3) The concrete, which has been forced to the rear by the auger flights, enters a molding chamber where the bottom, top, sides, and cores of the slab are finally finished before it is extruded from the machine. (4) The extruded slab, as soon as it leaves the extruder, is strong enough to support a man's weight if he stands upon it. (5) In order to assist the pressure generated by the augers to consolidate the concrete mix, vibrators are affixed to the top forming plate adjacent to the hopper. (6) For the same reason, a vibrator is mounted inside the hollow core of each auger. (7) The final finish is applied by the separate follower tubes which are rubber-mounted to the ends of the augers so that extreme vibration is not transmitted this far. Vibration at this point of the extrusion tends to deteriorate rather than improve slab compaction. (8) A counterweight is mounted on the rear of the machine just over the final troweling plate to prevent it and the machine from rising. Naturally, there will be a tendency for this to occur as the resultant force of extrusion is downward as well as rearward. (9) The extruder travels on, and the slab is cast into polished steel pans. These, of course, are laid between the stressing abutments, and the extruder starts at one abutment and continues until it reaches the other. (10) The electric drive motors impart their torque through (11) a gearbox and (12) chain-and-sprocket drive to (13) the auger drive shaft, which is mounted in the main bearing housing. This bearing housing is welded solidly to the extruder's frame and forms the very core of the machine. (14) The vibrator, which is mounted inside each auger, is driven through the hollow core of the auger and its drive shaft by a high-speed electric motor mounted on the front side of the bearing housing. (15) The extruder controls are electrical and are operated from a box at the front of the machine by using push buttons on the top of the box. (16) The drive motors and other working parts are covered by a lightweight fiber-glass housing. (17) Hydraulically powered rods may be used to insert transverse steel in the top of the slab when required. (18) The transverse steel can be stacked at the front of the hopper where it is picked up by the plunger rods.

FIG. 6. An extruder for a 6 × 48-in. slab. The eight tubes at the bottom of the machine form voids in the slab. (*Spiroll Corp., Ltd., Winnipeg.*)

Reinforcing steel must be accurately placed and rigidly tied. Manufacturer's specifications require a minimum coverage of ¾ in. of concrete for smooth surfaces and 1 in. for exposed-aggregate surfaces, unless the local code specifies other requirements.

The concrete is specified to have a compressive strength of not less than 6,000 psi at 28 days.

6 Extrusion Process

Hollow-core prestressed precast slabs are made by a machine that extrudes concrete as it moves along the casting bed. One of these machines, shown in Figs. 5, 6, and 7, generates its own motive force as a reaction to the force that extrudes the concrete. These extruders are particularly well adapted to long-line continuous manufacture of prestressed concrete.

Casting surfaces are metal pallets, permanently attached to the bed. The first operation is to clean the pallets and apply a parting compound. Stressing strands are then inserted in the bed and tensioned. Tracks attached to the pallets provide a smooth path for the extrusion machine in which zero-slump concrete feeds by gravity from the receiving hopper to the rotating augers in the molding chamber. As the slab emerges from the machine, it is covered with wet burlap or curing compound to keep it from drying out during the preset time. After steaming, the slabs are cut to the required lengths, usually on the following day, using a special saw, and moved to the storage area.

FIG. 7. An 8 × 48-in. hollow-core slab extruder operating in an outdoor plant. (*Spiroll Corp., Ltd., Winnipeg.*)

Speed of the machine along the bed varies in accordance with the mix proportions, type of aggregate, moisture content of the mix, size of slabs being extruded, and the setting of the plates on the molding chamber. Speeds as high as 5 lin ft per min have been reported, with normal production about 3½ fpm.

Fig. 8. Rigging for a wall panel with a horizontal axis much longer than vertical.

Conventional methods of mixing and handling zero-slump concrete are used. Some plants use a gantry spanning several beds to feed the extruder hopper. Cranes and front-end loaders have been used.

7 Finishes

Materials for precast units are covered in detail in Sec. 1 of this handbook. Because precast units, especially wall panels, are frequently exposed, special architectural finishes may be specified requiring special aggregates such as selected gravel or crushed rock, marble, white limestone, quartzite, granite, or artificial aggregates, such as expanded shales and clays, ceramics, and crushed vitreous aggregate. If a proposed aggregate does not have a satisfactory service history in a similar usage and exposure a careful investigation of that aggregate must be made before it is used. White cement may be specified, sometimes with color added. Finishes may be smooth formed surfaces, or aggregates exposed by retarded surface, acid etching, sandblasting, or bushhammering (see Chap. 27). Sometimes individual aggregate particles are hand-placed on the surface of the fresh concrete.

Uniformity of materials and finish is very important. Besides color of materials, the aggregate must be unvarying in size and grading if it is to be exposed.

Liners of many varieties are available to be used in combination with the forms to provide special finishes. Rubber and plastic sheets are commonly used. One technique that is used on panels for a deep reveal is to prepare an absorbent bed of sand within the form, hand-place the surface aggregate, and then place the matrix mortar, taking care that the mortar fills all interstices between the aggregate particles. Finally the reinforcing steel, inserts, and the backup concrete are placed. See Chap. 27 for finishing techniques.

8 Erection

Handling of precast members should be kept to a minimum. Working loads on lifting insets should be based on a factor of safety of at least 3, and the design engineer of record should specify or approve location and type of inserts. Many precast units are subjected to greater loads, and sometimes different loadings, during erection, from those to be experienced after the units are in place in the building, and this fact should be taken into consideration when designing the structure. Lift points should be located so the crane hook will be directly above the center of gravity of the unit while it is being lifted, thus keeping the bottom edge level. Figures 3 and 8 illustrate good rigging for wall panels.

Large wall members should not be laid flat after they have been removed from the molds. If they must be stored, it should be in a vertical position, with adequate support under the bottom edge and corners. Units should not be stored so the weight of one bears against one previously stored.

As soon as a unit has been erected in place in the structure, shores and braces must be installed while temporary connections are being made. Whether the temporary connections are adequate without bracing must be given very careful engineering analysis. Serious accidents have been caused by failures of temporary connections and bracing, and a little extra bracing is good insurance against failure. In addition, good rigging practices must be followed when handling units.

9 Joints

As a rule, joints involving precast concrete units are not intended to be weathertight until they have been caulked. Usual construction is to insert a backup filler in the joint to restrict the depth of sealant and prevent the sealant from adhering to the back of the joint. Foamed polymers that change volume with changes in joint width are commonly specified. These include certain urethanes, ethylenes, vinyl chloride, and others.

After insertion of the backup filler, the sealant is applied, preferably with a power-actuated gun, in accordance with the manufacturers' instructions. Thermosetting plastics, such as a polysulfide, are preferred as joint sealants, although some thermoplastics and mastics have been found suitable. Oil-base compounds eventually dry out and are not suitable. Cement mortars do not possess the extensibility necessary for a good joint.

The minimum amount of sealant to make a satisfactory joint should be used. Usually most of the joint is filled with the premolded elastic filler, with the sealant used on the exterior ½ in. of the joint. Sealant should be applied to a clean and dry joint. Do not apply during freezing weather.

10 Connections

Connections are defined as the hardware used for joining the precast elements to the building frame and to each other. Design of connections, a function of the structural engineer, should be taken care of when the members are designed, as both members and connections share the responsibility for structural adequacy with the building frame. Analyses of building failures resulting from earthquakes have shown that structural weaknesses leading to the failure were the result of improperly designed or installed connections, rather than any weakness of the building units.

Reference 1 is a good source of information on design of connections. This article points out that structural adequacy is of greatest importance, followed by architectural function, which requires that connections be neat and clean, essentially invisible, and compact. Finally, economy requires both ease of erection and ease of fabrication. For example, it is well to design a connection so it can be temporarily fastened with erection bolts, thus freeing the crane for other work while the permanent connection is made. Permanent fastening is accomplished by welding, bolting, or riveting.

CONCRETE BLOCK

Concrete blocks, or concrete-masonry units, are made in a variety of machines that vibrate, jolt, press, and otherwise consolidate a concrete of barely moist consistency. According to the Concrete Industries Yarbook:*

The manufacture of concrete masonry units involves six basic steps: (1) Selection of raw materials of a type and quality, and of proper gradation in the case of aggregates, that

* Much of the information in this chapter is from that publication.[2]

will permit production of finished units meeting all pertinent specifications; (2) Accurate batching of the raw materials to meet the requirements of selected mix designs; (3) Thorough mixing of the batched materials to produce concrete of suitable workability; (4) Molding in a machine properly adjusted to produce units of consistent quality from an adequate range of mix designs; (5) Proper curing to develop product qualities that will minimize problems in initial handling and ultimately meet pertinent specifications with respect to strength and other characteristics; (6) Storage of finished units under methods that will promote the development of ultimate strengths and moisture content requirements.

Each of these six steps, except step 4, has been discussed in detail in other parts of this handbook, and will now be considered as especially applied to concrete block.

11 Materials

All masonry units contain portland cement, aggregate, and water. In addition certain admixtures are frequently used, such as coloring pigments, air entrainers, pozzolans, accelerators, and retarders (see Sec. 1).

Cement can be standard Type I or II, depending mainly on location of usage. Where early strength is desirable, or to reduce curing time, Type III high-early-strength cement may be used. Most cement companies now produce a block cement which is of a lighter color than the standard or normal types, and approaches Type III in strength development. White cement is commonly used in such specialty block as split block, where the mortar color is especially important, and in decorative grille and screen units.

Aggregate should (1) be clean, and free of particles or substances that are unsound, dirty, or would interfere with strength and durability, or would cause surface imperfections; (2) have sufficient strength, toughness, and hardness to resist the expected loading; (3) be durable so as to resist freezing and thawing, temperature fluctuations, and changes in moisture content. and (4) have a uniform grading from finest to coarsest size to produce a workable, economical, and uniform mix.

Normal or heavy aggregates include sand, gravel, crushed stone, and blast-furnace slag. Sand should all pass the No. 4 mesh screen, have 15 to 25% pass the 50-mesh and 5% minus 100-mesh, with a fineness modulus between 2.60 and 3.00. A maximum of 5% silt is usually not objectionable. Coarse aggregate passes the $\frac{3}{8}$-in. screen and is retained on the No. 8. with a tolerance of not more than 5% passing the No. 8. About 25 to 40% should pass the No. 4 screen.

Many block manufacturers use lightweight aggregates. Natural lightweight aggregates are pumice, diatomite, and scoria. Manufactured lightweight aggregates are expanded shale, expanded clay, water-quenched blast-furnace slag, and cinders. Grading for lightweight aggregate should be about the same as for regular aggregate. Other important properties are unit weight and absorption, which should be maintained at uniform values. Many lightweight aggregates are quite angular and harsh, and some, especially the very lightweight ones, are apt to be somewhat fragile. The desirable properties listed above for normal-weight aggregates apply also to lightweight. Aggregate should be free of sulfur compounds, excessive lime, iron particles, pyrites, and any material that might cause staining and popping.

Control of segregation is important for all types of aggregates, as segregation results in variations in grading. Grading variations cause variations in workability of the mix, water demand, texture and strength of the finished block, and yield (number of blocks) per yard of aggregate.

Water should conform to the requirements for mixing water for plastic concrete (Chap. 4).

Admixtures play an important part in block production. Whenever a block producer is considering the use of any admixture, he should carefully evaluate it in his own plant, using his normal materials and methods, observing the effect of the admixture on all manufacturing operations and all properties of the block. The amount to use and the conditions under which it can be used depend on the mix, materials, equipment, weather, and many other factors.

An air-entraining agent causes the formation of many microscopic air bubbles in the mix, greatly improving durability of the block. It improves workability of the mix, especially with aggregates lacking fines, and improves surface texture and density of the block. Absorption of the block is usually decreased, and breakage or damage to green block is lessened.

Calcium chloride, used as an accelerator, hastens cement hydration, thus developing early strength during cold weather, in much the way of Type III cement. Not more than 2% by weight of cement should be used, and usually less suffices. By making a solution of 25 gal containing 100 lb of the salt, each quart contains 1 lb of chloride. The solution should be introduced into the mixer along with the mix water, keeping the total amount of water the same.

Some of the metallic stearates perform as waterproofing or capillarity-reducing agents, but their effectiveness is limited.

Pozzolans, such as fly ash, burnt and pulverized shale, or pumicite, can be successfully used with some aggregates to improve the quality of the block. Pozzolans should be carefully evaluated by laboratory and plant tests before they are used in production.

12 Materials Handling

Most plants use bulk cement, delivered in special tank rail cars or trucks equipped with pneumatic pumps that convey the cement through a pipe to the top of the silo or bin. Some plants receive the cement in a ground-level hopper from which screw conveyors or bucket elevators carry it to the silo. Sacked cement is rarely used except in very small plants.

Cars or trucks discharge aggregates into ground-level hoppers from which inclined-belt conveyors or vertical bucket elevators lift the aggregate to overhead storage bins. Occasionally a clamshell unloads cars into stockpiles or moves aggregate from cars and stockpiles into bins, but this system is seldom used any more because it is costly and inefficient.

Movement of materials, especially cement and other fine bulk material, is aided considerably by the proper instal-

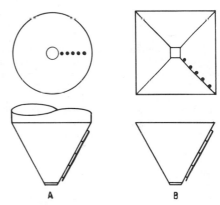

FIG. 9. Installation of low-pressure air jets in the bottom of a cement silo fluffs the cement and keeps it in a free-flowing condition. Air pressure should not exceed 5 psi. The Remco aerator fits on the outside of the silo and admits a large volume of air through a fabric held between a circular housing and a grill. (*A*) In a conical-bottom silo, one row of jets is usually adequate. (*B*) In a pyramidal-bottom square bin, a row of jets should be installed adjacent to each of the valleys. One row of jets is shown.

lation of external vibrators on hoppers and bins. Low-pressure air jets, installed in a conical silo bottom just above the gate opening, aid in preventing arching and sticking of cement (Fig. 9).

13 Mixes

Combined aggregate, that is, minus ⅜-in. gravel or other coarse aggregate plus the sand, must be so proportioned that the fineness modulus is between 3.5 and 4.0. Usually a mix will consist of 55 to 70% sand with 45 to 30% gravel. These proportions might be changed to 40 to 75% sand and 60 to 25% of crushed rock. Strength of the block for any given cement content is higher when the aggregate grading is coarser. However, sufficient fines are always necessary to produce a workable mix that will give regular and uniform surfaces, and minimize breakage.

Proportions of aggregate and the cement content have to be determined by trial mixes in the plant. Workability in the machine, cracking, surface texture, green strength, density, absorption, strength of the cured block, economy, and general appearance of the block all have to be considered. The maximum proportion of coarse aggregate that the mix will tolerate should be used. Sometimes this may be as little as 25% of the total aggregate but usually it is around 40 to 50%. Trial mixes should consist of varying proportions of coarse to fine aggregate, combined with different amounts of cement. Richest mix will probably be about 1 part cement to 6 parts of damp, loose mixed aggregate, ranging down to 1:10. Because of the very dry nature of the concrete, the water-cement ratio is not critical from a strength and durability standpoint.

Behavior of the mixture in the machine, appearance of the green block, and the ease of handling the green block all give clues as to the final quality of the block. Once a mix has been decided upon, every effort should be made to keep all operations and materials the same and equipment in good operating condition.

When using lightweight aggregate it may be necessary to use less fine material in the mix than for normal-weight aggregate in order to obtain the required light weight, texture, insulating value, and acoustical properties. This is accomplished at a sacrifice of some of the strength.

Appearance of the blocks as they come off the machine is important in judging whether the amount of water in the mix is correct. Differences in machines and materials affect the amount of water. It is best to use the maximum amount permissible without causing the block to slump. If insufficient water is used, the

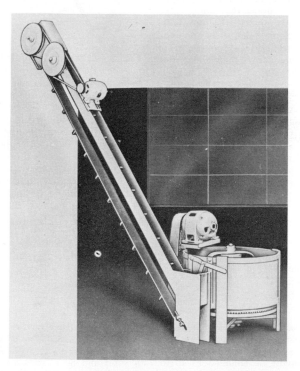

Fig. 10. A small turbine mixer can be installed flush with the floor to facilitate charging. The inclined conveyor elevates mixed concrete to the block machine. (*W. E. Dunn Mfg. Co.*)

block will have a dull gray color, with a "dry" appearing texture. Too much water will cause the block to slump. The correct amount produces a water web or sheen on the molded faces of the block when it is demolded.

14   Batching and Mixing

Section 5 of this handbook covers descriptions of most types of batching and mixing equipment. In most modern plants, batching is done in automatic or semiautomatic batchers. The weigh batcher is positioned under the storage bins and a measured amount of each ingredient except water is discharged into the batcher. The batcher then moves to a position over the mixer and discharges the batch into the mixer while water is measured into the mixer. The admixture, if any, is introduced with the water. One mixer installation is shown in Fig. 10. Other mixers, both large and small, are frequently horizontal-shaft pugmill units.

Control of mix water is best accomplished by means of some type of moisture meter. One type has probes in the mixer that sense the moisture content of the batch, usually by electrical-resistance methods. A dial gage shows the moisture content and enables the operator to adjust the amount of water. The meter can be made to actuate automatic controls that regulate the amounts of water and sand in the batch. Probes should be cleaned regularly and the equipment kept in adjustment to assure proper operation.

Adequate mixing is essential and results in better block than short mixing. Adequate mixing results when the mixer is not overloaded, the time period is sufficiently long, and the proper consistency of the batch is obtained through control of the water. The mixer is overloaded if the batch in the mixer covers the mixer shaft; length of time period depends on the type of machine, batch proportions, and materials. It must be determined by trial. With lightweight aggregates, it may run about $7\frac{1}{2}$ min, including about 1 min prewetting of the aggregate before cement is added. Water control is best achieved by use of automatic equipment.

Fig. 11. This semiautomatic block-making machine makes two 8-in.-high blocks at a time and a speed of up to 8 cycles per minute. (*The Besser Company.*)

FIG. 12. A large automatic block-making machine of this type can make **1,400** $8 \times 8 \times 16$ blocks per hour. (*The Besser Company.*)

15   Making the Block

After mixing, the concrete batch is discharged into the block-machine hopper, from which controlled quantities of concrete are fed into the molds. Vibration and pressure consolidate the concrete in the molds, which rest on a steel or wood pallet during molding of the block. Output of the machine is based on the number of $8 \times 8 \times 16$-in. blocks, or equivalent amount of concrete, produced per machine cycle, the machine speed usually being about **5** cycles per minute. Figures 11, 12, and 13 are views of block machines and plant layouts.

The pallet, usually of steel, but occasionally of wood, measures $18\frac{1}{2} \times 26$ in., the 26-in. length permitting working space to mold three 8-in. blocks, two 12-in., four 6-in., six 4-in., or 12 bricks. Thus the machine, by changing the mold box, can produce any of the common modular units.

As soon as the concrete is compacted and struck off, the mold box lifts and the pallet holding the green blocks emerges from the machine, while a new pallet is pushed into the other side of the machine. The pallet containing the green blocks then travels to the curing area (Fig. 13).

Many of the steps are automated, even on some of the small machines. Pushbutton controls permit one-man operation of machines with a capacity exceeding a thousand 8-in. blocks per hr. Block machines are ruggedly built, as they are subject to moisture, vibration, and exposure to abrasive materials, and proper maintenance is essential to keep them operating efficiently. Thorough cleaning and lubrication every shift are very important. As an aid in this respect, compounds are available that can be sprayed or brushed on surfaces to prevent adhesion of concrete. Pallets must be kept clean and mold boxes checked for wear; those showing excessive wear should be discarded. Blocks should be checked for density and dimensions occasionally to be sure they are maintaining the proper quality.

Offbearing, the process of moving the block off the machine, in many plants is either semiautomatic or automatic. In some plants this operation is performed

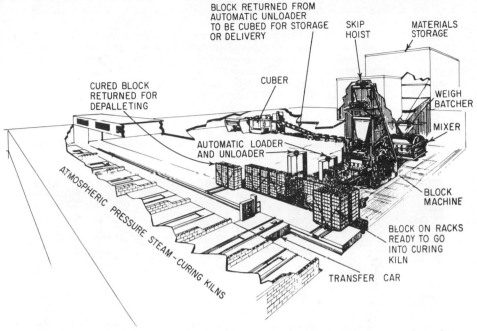

BLOCK RETURNED FROM
AUTOMATIC UNLOADER
TO BE CUBED FOR STORAGE
OR DELIVERY

SKIP
HOIST

MATERIALS
STORAGE

CURED BLOCK
RETURNED FOR
DEPALLETING

CUBER

WEIGH
BATCHER

MIXER

AUTOMATIC LOADER
AND UNLOADER

BLOCK
MACHINE

BLOCK ON RACKS
READY TO GO
INTO CURING
KILN

ATMOSPHERIC PRESSURE STEAM-CURING KILNS

TRANSFER CAR

FIG. 13. Mechanical equipment in the several stages of block manufacturing serves to automate this one-car automatic plant with atmospheric curing. Many combinations of equipment are available for automation of almost any type of plant. (*The Besser Company.*)

manually. The loaded pallets are placed in steel curing racks which are now taken to the curing kiln or autoclave. Transporting of racks is accomplished with lift trucks, casters, small rail cars, or monorail.

16 Curing

High-temperature steam curing, almost universally practiced, can be accomplished with saturated steam at atmospheric pressure, or can be accomplished by means of high-temperature high-pressure steam in a pressure vessel called an autoclave. Curing is discussed in Chap. 29.

Atmospheric steam curing rooms, called kilns, should be tight so as to minimize heat loss. Either hinged or vertical sliding doors can be used, or sometimes heavy drop curtains. Walls and roofs should be made of insulating material, such as lightweight-aggregate concrete or block, and the floor should be concrete. Adequate drains must be provided. The ideal situation is one in which every block in the room is exposed to exactly the same curing cycle. This ideal is seldom achieved, but adequate space around all racks of block, to permit free movement of the saturated warm air, will help to approach the ideal. There should be a clearance of 18 in. above the highest block in the racks, and 6 in. on all sides of each rack.

According to "Modern Steam for the Concrete Industries,"[3] most kilns and steam-curing equipment are improvisations developed by the block maker himself. There apparently is no general standard for the construction of kilns. They are made from all types of materials, from a canvas-covered frame and canvas flap door to a concrete or concrete-block building with seal-tight quick-opening doors. However, a good kiln has walls, top, and door correctly insulated against heat losses and has a door that will give a tight seal, preventing cold air from entering the kiln and steam from escaping.

The best method in block curing is to have a set of kilns in which the various operations of loading, steaming, soaking, and unloading can be carried on separately. In this way, production is in chainlike manner; that is, loading one kiln, unloading another, steaming another, and soaking still another, all at the same time.

Some block makers load an entire 8-hr output of blocks in one large curing room. Then they apply steam to the entire lot at one time. This apparently is not the ideal method, as it requires the continuous-chain production method without any increase in output.

The most important part of steam curing is to know the temperature of the block throughout the entire curing cycle. If heat is controlled during the entire curing cycle, then the block maker knows that each batch of blocks is precisely

Fig. 14. Autoclaves are used for high-pressure high-temperature steam curing of block, pipe, and similar precast materials. (*Clayton Mfg. Company.*)

the same quality and color as the preceding one. Since a temperature rise of 60° per hr is permissible in a block, it is possible with the proper steam equipment to carry out the steaming operation in a matter of 3 or 4 hr.

Steam is evenly distributed in the kiln by means of a perforated steam header. Condensate can be allowed to drain out the door of the kiln, or through special drains. Thermometers are usually installed in the kiln so they can be read from outside the kiln.

High-pressure steam curing in an autoclave produces blocks which are lighter in color, which exhibit greater dimensional stability, that is, less shrinkage, and which are considerably drier when removed from the kiln, compared with blocks cured under atmospheric conditions. One-day strength of autoclaved block is about equal to standard 28-day strength.

The autoclave, shown in Fig. 14, is a long, tanklike steel cylinder of diameter large enough to accept racks of blocks. Blocks from the machine are held from 2 to 5 hr before placing in the autoclave to permit initial hydration of the cement and hardening of the blocks. Next the blocks are heated gradually in the autoclave so that full pressure is reached in not less than 3 hr. Units are then steamed for 5 or 6 hr at a temperature of 350°F. Thick units require slightly more time. After the steaming, or soaking, period, pressure is rapidly released over a period of ½ hr or less so the blocks will dry as much as possible. This rapid pressure release removes sufficient moisture from normal-weight blocks to bring them to a relatively stable air-dry condition. Additional drying is usually necessary for lightweight block.

Heat Computations. Augmenting the discussion of steam curing in Chap. 29, the following computations, adapted from "Modern Steam for the Concrete Industries,"[3] illustrate the steps to determine the necessary boiler capacity for a high-temperature steam-curing system in a block plant.

In any high-temperature curing system, the following information is required:

1. Mean average low outside air temperature (if autoclave or kiln is set outside in the weather).

2. Mean average low air temperature of the room (if autoclave or kiln is set inside building).

3. Desired curing temperature of the blocks.

4. Time required to raise temperature to curing temperature.

5. Total weight of blocks to fill autoclave or kiln.

6. Total weight of racks for holding blocks.

7. Total weight of pallets for holding blocks.

8. If more than one autoclave or kiln is used, determine cycle for peak steam load.

9. Operating pressure and temperature of autoclave. Operating temperature of kiln.

10. If autoclave or kiln is insulated, determine K factor for radiation loss.

11. If autoclave or kiln is not insulated, determine K factor for radiation loss.

12. Complete dimensions and weight of autoclave (diameter, length, and thickness). Dimensions of kiln.

13. Other steam requirements—such as water heating, heating aggregate, and space heating.

Kiln Curing. By far the largest steam requirement occurs during the time the loaded kiln is brought up to the curing temperature. This includes bringing up to temperature the kiln, the concrete blocks, the concrete-block racks, the pallets, and any other handling equipment that is contained within the kiln during curing.

The general formulas used are as follows:

$$\text{hp} = \frac{\text{weight of all items} \times \text{specific heat} \times \text{temp rise (per hr)}}{33,500}$$

(temperature rise should be at the rate of 60°/hr)

$$+ \text{ steam load to fill kiln} = \frac{\text{cu ft of kiln} \times \text{Btu/lb of steam}}{33,500 \times \text{volume/lb of steam}} + \text{radiation loss of kiln}$$

$$= \frac{\text{heat-transfer coefficient} \times \text{surface area} \times \text{temp difference}}{33,500}$$

Because there is a difference of only 2 or 3 hp between heat-up and holding rates, the full radiation rate has been used, to simplify computations

General formula:

$$\text{hp} = \frac{W(60°) \text{ specific heat}}{33,500} + \frac{\text{volume (Btu/lb)}}{\text{(cu ft/lb)} \times 33,500}$$
$$+ \frac{(K \text{ factor})(\text{surface area})(T_1 - T_2)}{33,500}$$

Example—Kiln Stem-curing Concrete Blocks. Assume the following:

1. Kiln located outside building

2. Minimum low outside temperature +20°F

3. Initial temperature of all materials including concrete blocks, racks, etc., +40°F

4. Final temperature of above material +165°F

5. Size of kiln, 10 ft wide × 7 ft 6 in. high × 30 ft long. 2 doors 10 × 7 = 140 sq ft. Exposed walls and roof = 780 sq ft. Cubical contents, 2,250 cu ft

6. Racks (14) for holding concrete blocks, 700 lb each $\times$ 14 = 9,800 lb
7. 350 pallets at 50 lb each = 17,500 lb
8. Weight of steam and condensate distributing pipe, say, 700 lb
9. Weight and number of concrete blocks, 1,000 at 44 lb each = 44,000 lb
10. Specific heat of concrete blocks, 0.25
11. K factor for exposed walls for kiln 0.25. K factor for doors for kiln 1.00. K factor for cubical contents of kiln 0.02
12. Temperature rise $(T_1 - T_2) = 165 - 40 = 125°$

13. Heat-up time at 60° per hr = $\dfrac{125}{60}$ = 2.08 hr (some authorities recommend a maximum rate of temperature rise not to exceed 40°F per hr)

Estimate:

$$
\begin{array}{lr}
\text{Weight of racks}\ldots\ldots\ldots & \text{9,800 lb} \\
\text{Weight of pallets}\ldots\ldots\ldots & \text{17,500 lb} \\
\text{Weight of pipe}\ldots\ldots\ldots & \text{700 lb} \\
\hline
\text{Total}\ldots\ldots\ldots\ldots & \text{28,000 lb}
\end{array}
$$

$$\frac{28,000 \times 0.12 \times 60° \text{ temp rise/hr}}{33,500} = \text{say 6 bhp}$$

Concrete blocks:

$$\frac{44,000 \times 0.25 \times 60° \text{ temp rise/hr}}{33,500} = \text{say 19.7 bhp}$$

Steam to fill kiln:

Operating pressure = 75 psi
Latent heat, Btu/lb of steam at 75 psi = 894.7 Btu
Volume of steam at 75 psi = 4.896 cu ft/lb of steam
cu ft of kiln = 2,250 less load, say, 50% = 1,125 cu ft

$$\frac{1,125 \times 894.7}{33,500 \times 4.896} = \text{say 6.2 hp}$$

Radiation loss kiln:

$$T_1 - T_2 = 165°F - 20°F = 145° = \text{temp rise}$$

Doors

$$\frac{140 \text{ sq ft} \times 145° \times 1.0}{33,500} = \text{say 0.61 hp}$$

Exposed walls

$$\frac{780 \text{ sq ft} \times 145° \times 0.25}{33,500} = \text{say 0.85 hp}$$

Cubic contents

$$\frac{1,125 \text{ cu ft} \times 145° \times 0.02}{33,500} = \text{say} \quad \begin{array}{l} 0.10 \text{ hp} \\ \hline 1.56 \text{ hp} \end{array}$$

$$\text{Plus 10\% safety factor} \quad \begin{array}{l} 0.16 \text{ hp} \\ \hline 1.72 \text{ hp} \end{array}$$

$$
\begin{array}{lr}
\text{Weight racks}\ldots\ldots\ldots\ldots & \text{6 hp} \\
\text{Concrete blocks}\ldots\ldots\ldots\ldots & \text{19.7 hp} \\
\text{Steam to fill kiln}\ldots\ldots\ldots\ldots & \text{6.2 hp} \\
\text{Radiation loss}\ldots\ldots\ldots\ldots & \text{1.7 hp} \\
\hline
& \text{33.6 bhp/hr required during heat-up} \\
\text{Say 10\% for transmission-line losses}\ldots & \text{3.4 hp} \\
\hline
\text{Total}\ldots\ldots\ldots\ldots & \text{37.0 bhp/1,000 block/hr during heat-up}
\end{array}
$$

After 3 hr heating time, soaking time requires

$$
\begin{array}{lr}
\text{Radiation loss} & 1.7 \text{ hp} \\
\text{Transmission-line loss} & \underline{2} \text{ hp} \\
& 3.7 \text{ hp}
\end{array}
$$

To the above estimates, any additional steam load such as space heating, heating stockpiles, heating water, steam jetting hoppers or railroad cars, or steam cleaning must be added to the total horsepower load. Also, future plant expansion must be considered; the steam generator should be large enough, or provisions should be made for the installation of additional units.

Autoclave Curing. As in the kiln method, the largest steam requirement occurs during the time the loaded autoclave is brought up to curing temperature. This includes bringing up to temperature the autoclave, the concrete blocks, concrete-block racks, pallets, and any other handling equipment that is contained within the autoclave during curing. The general formulas used are as follows:

$$
\text{hp} = \frac{\text{weight of all items including autoclave} \times \text{specific heat} \times \text{temp rise}}{33,500 \times \text{min temp raising time, hr}}
$$

$$
+ \text{ steam load to fill autoclave} = \frac{\text{cu ft of autoclave} \times \text{Btu/lb of steam}}{33,500 \times \text{volume/lb of steam}}
$$

$$
+ \text{ radiation loss of autoclave}
$$

$$
= \frac{\text{heat-transfer coefficient} \times \text{surface area} \times \text{temp difference}}{33,500}
$$

General formula:

$$
\text{hp} = \underset{\text{Items}}{\frac{W(T_1 - T_2)\ \text{specific heat}}{33,500 \times \text{hr}}} + \underset{\text{Autoclave}}{\frac{\text{volume (Btu/lb)}}{(\text{cu ft/lb}) \times 33,500}}
$$

$$
+ \underset{\text{Radiation loss}}{\frac{(K\ \text{factor})(\text{surface area})(T_1 - T_2)}{33,500}}
$$

Note. To simplify estimate, 33,500 Btu per hp was used in place of 33,475 Btu per hp.

Example—Autoclave Steam-curing Concrete Blocks.   Assuming the following:

1. Autoclave located outside building
2. Minimum low temperature $+40°F$
3. Initial temperature all materials including blocks, racks, etc., $+40°F$
4. Final temperature of above material $+350°F$
5. Size and weight of autoclave, 8 ft 0 in. diameter $\times$ 90 ft 0 in. long, $\frac{7}{8}$ in. thick, specific heat 0.12, weight of steel per sq ft = 35 lb.   Two heads 8 ft 0 in. diameter $(\pi R^2) \times 2 = 100.5$ sq ft.   Shell, 90 ft 0 in. long $(\pi D \times 90) - 25.12 \times 90 = 2,261$ sq ft $+ 101 = 2,362$ sq ft. $2,362 \times 35 = 82,670$ lb
6. Racks (28) for holding concrete blocks 700 lb each $(28 \times 700 = 19,600$ lb)
7. 700 pallets at 50 lb each = 35,000 lb
8. Weight steam and condensate distributing pipe (assume 1,000 lb)
9. Weight and number of blocks, 2,000 at 44 lb each
10. K factor for insulated autoclave = 0.29.   0.29 Btu/hr/sq ft/temp difference = $0.29 \times 310 = 90$ Btu/loss/sq ft
11. Specific heat of concrete block -0.25
12. Heat-up time 3 hr
13. $T_1 - T_2 = 350 - 40 = 310°F$

Estimate:

Weight autoclave.........	82,670 lb
Weight racks.............	19,600 lb
Weight pallets...........	35,000 lb
Weight piping...........	1,000 lb
Total.................	138,270 lb

$$hp = \frac{138,270 \text{ lb} \times 0.12 \times 310}{33,500 \times 3} = \text{say } 51 \text{ hp}$$

Concrete blocks:

$$hp = \frac{2,000 \times 44 \times 0.25 \times 310}{33,500 \times 3} = \text{say } 68 \text{ hp}$$

Radiation loss:

$$\frac{90 \times 2,362}{33,500} = \text{say } 6 \text{ hp}$$

Steam to load autoclaves:

Latent heat of steam at 150 psi = 857.0 Btu
Volume of steam at 150 psi = 2.752 cu ft/lb

Cu ft of autoclave = area × length = 50 × 90 = 4,500 cu ft (deduct for block load)
= say approx 50% or 2,250 cu ft

$$hp = \frac{\text{volume (Btu/lb)}}{\text{(cu ft/lb)(hr) } 33,500} = \frac{2,250 \times 857.0}{2.752 \times 3 \times 33,500} = \text{say } 7 \text{ hp}$$

Total hp required to heat autoclave in 3 hr:

Heat handling material, say.................	51 hp
Heat product, say..........................	68 hp
Radiation loss, say........................	6 hp
Load autoclave...........................	7 hp
Total.................................	132 hp

Say plus 10% for transmission-line losses......	13 hp
Total hp required per hr for first 3 hr...........	145 hp

After 3 hr, steam required for heating time, soaking time requires per hour

Radiation loss................	6 hp
Transmission-line loss..........	13 hp
Total.....................	19 hp

145 hp − 19 hp = 126 hp available for the second autoclave or other steam requirements.

17   Carbonation

It has been demonstrated that exposure of concrete to carbon dioxide in a normal atmosphere will result in shrinkage of the concrete. Tests by the Portland Cement Association and the National Concrete Masonry Association have shown that precarbonation of concrete-masonry units reduces shrinkage by as much as 50% under subsequent exposures to carbon dioxide or cycles of wetting and drying. Conclusions based on the PCA-NCMA tests[4] were:

1. The relative humidity in the kiln should be between 15 and 35%.
2. The CO_2 content should be as high as possible but not less than 1.5%.
3. The temperature should be between 150 and 212°F.
4. Generally a 24-hr treatment will be required, but under optimum conditions a shorter period may provide considerable reduction in shrinkage.

5. Potential shrinkage reduction of more than 30% may be obtained under favorable conditions.

Practice in the industry has been to introduce waste flue gas into the kiln after the blocks have been cured and are in the drying period. During this period, the blocks must be exposed to conditions that dry the blocks to a favorable moisture level, maintaining that level during treatment, and that provide the correct CO_2 concentration to effect rapid carbonation.

Equipment is available that provides an automatic cycle to cure the block in this manner. The system includes equipment that steams, dries, and carbonates the concrete in accordance with the following program:

a. Preset: Required of all concrete to be cured at any temperature above normal atmospheric temperature. During this period, air is circulated through the kiln to maintain a uniform atmosphere.

b. Steam: A gas-burning heater is activated to raise the kiln temperature, evaporate water for steaming, and introduce the products of combustion into the kiln to provide carbon dioxide.

c. Drying: Steam generation is stopped, but dry heat and carbon dioxide continue to enter the kiln, drying and carbonizing the block.

d. Exhaust: The kiln is now vented to remove moisture and carbon dioxide.

An installation of equipment of this type is shown in Fig. 29-5.

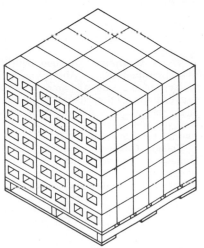

Fig. 15. Finished blocks are cubed and palletized for ease in handling. This pallet contains 108 8 × 8 × 16-in. blocks. The pallet shown here is not the type of pallet that is used in the block machine, but constructed of 2 × 4's and 1-in. boards; it is commonly used for palletizing many kinds of merchandise.

18 Handling and Storage

Cubing is the process of assembling blocks into convenient groupings for further handling. Upon completion of curing, racks of block are conveyed to the cubing station, where the cubes are assembled either by hand or by automatic machinery. A cube consists of five or six layers of 12 to 18 blocks, 8 × 8 × 16 in., or an equivalent volume of other size of block (Fig. 15). Other sizes of cubes may be used, depending on local custom and facilities. It usually requires four laborers to cube the production of one three-at-a-time block machine, but a semiautomatic cuber can do the job with two men. After cubing, the blocks are moved to storage.

Lift trucks, or fork lifts, are extensively used for moving racks and cubes of block in the yard. These machines, normally operating on pneumatic tires, are very versatile and can operate in unpaved yards. By cubing the blocks on pallets, or cubing them so the tines of the fork lift can slide through the cores of the bottom course, the entire cube can be lifted into the stockpile or onto a truck.

19 Properties of Masonry Units

Types of block are briefly discussed in Chap. 40, and common types are shown in Fig. 1, Chap. 40. Screen and grille blocks offer an almost infinite number of interesting and artistic designs, and the introduction of textures and color (including pure white) has made available a wide array of specialty precast products including bas-relief surfaces, garden accessories, and other special-use units. Different textures can be obtained by use of different aggregate gradations, by changing the mix consistency, or by mechanical means.

While concrete blocks are similar to structural concrete in many respects, there are important differences. In general, quality of the product depends upon the qualities of materials, mix, and workmanship, the same as structural concrete. However, block mixes contain less cement (260 to 375 lb per cu yd) and less water. Aggregate rarely is larger than ⅜ in. and is frequently lightweight. Curing is nearly always at elevated temperatures.

Strength. Compressive strength is determined on the entire block and is computed on the gross area of the block, that is, overall dimensions, including core space. Compressive strength based on net area, excluding core space, is about 1.8 times the gross-area value. Compressive strength is influenced principally by the type and amount of cement per unit, type of aggregate, grading of aggregate, degree of compaction attained in the molding machine, age of the specimen, curing, and moisture content at time of test. Lightweight blocks are usually not as strong as heavy ones.

Water Absorption. A good indication of the density of block is obtained from the absorption value. Absorption varies widely, from about 4 to 20 pcf depending on type of aggregate and amount of compaction of the block. Permeability, thermal conductivity, and acoustical properties are also influenced by absorption. Although high absorption is not desirable, it of necessity accompanies lightness in weight and desirable thermal and acoustical properties. Block for exterior exposed walls should have low absorption.

Volume Changes. Masonry, the same as any concrete, undergoes changes in dimensions due to changes in temperatures and moisture content. One of these changes is the original drying shrinkage of green concrete. If blocks are laid up in a wall before they have taken their drying shrinkage, tensile stresses are developed that are apt to lead to cracks in the wall. By proper curing and drying of block in the yard, the moisture content of the block can be reduced to an equilibrium condition with the surrounding exposure and cracking tendencies are minimized.

Texture and Color. Texture is the disposition of particles and voids in the surface of the units. The desired texture can be obtained by adjusting (1) the aggregate gradation, (2) the amount of mixing water, and (3) the degree of compaction in the mold. In addition, special face molds and rubber or plastic mold liners are available for special finishes, and the units can be ground with an abrasive, or the aggregate can be exposed by chemical means. The desire for color and texture has led to the use of bonded facings, such as those obtained with thermosetting resin binders with silica sand, and vitreous glazed surfaces.

Color depends mainly on the materials, although autoclaving results in a lighter color. Different cements are of different shades of gray. Block cement is lighter in color than regular cement, and white cement results in white units. Pigments, available from reputable producers, make possible a veritable rainbow of colors. The use of white cement with pigments results in clean, pastel colors, and gray cement gives the dark colors. The choice of cement, aggregate, and colors is especially critical for split block. The many types of standard and specialty blocks are described and illustrated in Chap. 40.

20   Specifications and Quality

The most commonly accepted specifications are those of the American Society for Testing and Materials (Table 3, Chap. 40). The Federal government and other agencies also have their own specifications. Most building codes, including the Uniform Building Code, refer to ASTM specifications. Testing of block is covered by ASTM Designation: C 140. The number of samples required for testing is shown in Table 1.

An industry program is the "Q block quality-control program,"[5] sponsored by the National Concrete Masonry Association. Manufacturers participating in this program are required to have their products tested by an independent testing laboratory. The NCMA reviews the laboratory reports and issues a certificate

Table 1. Number of Concrete-masonry Samples

No. of units in lot	No. of specimens	
	For strength test only	For strength, absorption, and moisture content
10,000 or less	5	10
10,000–100,000	10	20
Over 100,000	5 per 50,000	10 per 50,000

of quality to the qualifying producer, authorizing him to market his product as "Q block."

CONCRETE PIPE

Methods for making precast-concrete pipe include tamping, compression by means of a revolving packer head, centrifugation, vibration, and combinations of certain of these processes. The pipe can be broadly classified into pressure pipe and nonpressure pipe, for which there are a number of specifications, as shown in Table 2.

Nonreinforced pipe up to 24 in. diameter, classed as agricultural pipe, is used for irrigation and drainage of farmland. Irrigation pipe may have tongue-and-

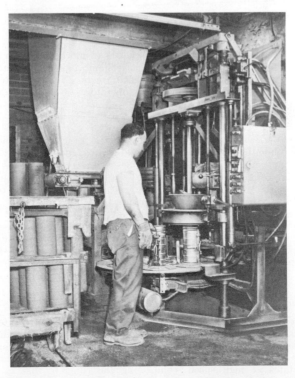

Fig. 16. Short lengths of small drain tile can be made on this machine, shown making 6-in. drain tile with butt joints. (*W. E. Dunn Mfg. Company.*)

FIG. 17.   A medium-sized high-capacity packer-head machine with a size range of 4 to 18 in. in 4-ft lengths, for producing gasket-joint pipe using machined socket curing pallets or mortar-joint pipe using socket formers.   The complete pipe-making cycle can be automatic or manual with an operator.   (*Concrete Pipe Machinery Company*.)

groove joints sealed with cement mortar or some type of rubber coupling.   Drain tile may have plain butt ends or tongue-and-groove ends that are left open to admit drainage water.

Culverts and storm sewers, up to 24 in. diameter, are usually nonreinforced; large pipe is reinforced.   Culvert joints may be cement mortar or rubber gasket. Storm sewers usually have rubber-gasketed joints.

Nearly all pressure pipe is laid with a gasket joint of some type.   All pressure pipe is reinforced, some is prestressed circumferentially, and some consists of a steel cylinder with mortar on both the inside and outside.   Pressure pipe is used for municipal water supply, pressure lines for irrigation, pressure sewer mains and outfalls, siphons, and in any installation where there is internal hydraulic pressure on the pipe.

21   Packer-head Process

In the packer-head process a no-slump, moist concrete mix is fed into a stationary split cylindrical form, or jacket, and is packed or compacted radially outward against the inside of the form by the revolving and rising shoe, called a packer head or roller head.   Starting at the bottom of the form, the packer head slowly rises as it rotates at high speed, compacting the concrete that is fed into the open top of the mold as the packer head rises.   Speed of revolution is about

Fig. 18. One of the largest packer-head machines available, this one makes 72-in. pipe in 8-ft lengths. (*Concrete Pipe Machinery Company.*)

250 rpm. This method is restricted to cylindrical pipe, and is adaptable to making both reinforced and nonreinforced pipe. Range of sizes produced is from 4 to 48 in. in diameter with lengths up to 8 ft. Pipe can be made in either the bell-up or bell-down position. Figures 16, 17, and 18 are views of three sizes of packer-head machines.

As soon as the packer head travels its course, the jacket containing the pipe is taken to the curing area and the jacket is immediately removed. Because of the dryness of the mix and degree of compaction, the freshly made pipe has sufficient "green strength" to stand without support.

22   Tamped Process

Both reinforced and nonreinforced pipe can be made by the tamped process, in which a split jack and base plate, resting on a turntable, rotate about a stationary core. Dry-mix concrete is fed into the top of the annular space and is tamped by metal- or steel-shod wooden tamping bars striking at a rate of 430 to 600 blows per minute. As the pipe wall is built up, the tampers move upward also, at the same time striking the surface of the concrete with a constant amount of energy to compact the concrete uniformly from bottom to top of the pipe being made. Rotation of the jacket rotates the concrete about the stationary core, providing a troweling action that smooths the interior of the pipe. Upon completion of tamping, the core is withdrawn and the new piece of pipe is taken to the curing area, where the jacket is immediately removed, as shown in Fig. 19.

Table 2. Listing of Standard Specifications and Sizes of Concrete Pipe*

Specification designation	Type of pipe	Diameter, range, in.

ASTM C412
  Standard quality
  Extra quality, special quality
	Concrete drain tile	4–12
		4–24

ASTM C-14
  Standard strength
  Extra strength
	Concrete sewer, storm drain, and culvert pipe	4–24
		4–24

ASTM C76
  Class I
  Class II, III, IV, V
	Reinforced concrete culvert, storm, drain, and sewer pipe	60–108
		12–108

ASTM C118	Concrete irrigation and drainage pipe	4–24
ASTM C361	Reinforced concrete, low-head pressure pipe	12–108

ASTM C444
  Standard strength
  Extra strength
	Perforated concrete pipe	4–24
		4–24

ASTM C443	Joints for circular concrete sewer and culvert pipe using rubber type gaskets	
ASTM C478	Precast concrete manhole sections	Up to 72
ASTM C505	Nonreinforced irrigation pipe with rubber type gasket joints	6–24
ASTM C506	Reinforced arch culvert storm drain and sewer pipe	Equivalent round 15–108
ASTM C507	Reinforced elliptical culvert storm drain and sewer pipe	Equivalent round 18–108 36–108
ASTM C655	Reinforced concrete D-load culvert, storm, drain and sewer pipe	
ASTM C497	Method of tests	

ASTM C654-70T
  Standard strength
  Extra strength
	Porous concrete pipe	

AASHO M178
  Standard quality
  Extra quality, special quality
	Concrete drain tile	4–12
		4–24

AASHO M86
  Standard strength
  Extra strength
	Concrete sewer, storm drain pipe	4–24
		4–24

AASHO M170
  Class I
  Class II, III, IV, V
	Reinforced concrete culvert, storm drain and sewer pipe	60–108
		12–108, 12–84, 12–72

Table 2. Listing of Standard Specifications and Sizes of Concrete Pipe* (*Continued*)

Specification designation	Type of pipe	Diameter, range, in.

AASHO M175
  Standard strength
  Extra strength
	Perforated concrete pipe for underdrainage	
		4–24
		4–24

AASHO M176-63 1
  Class II
  Standard strength
  Extra strength
	Porous concrete pipe for underdrainage	
		4–24
		4–24
		4–24

AASHO M 198	Joints for circular concrete sewer and culvert pipe using rubber type gasket	
AASHO M 199	Precast reinforced concrete manhole sections	
AASHO M 206	Reinforced concrete arch culvert, storm drain, and sewer pipe	Equivalent round 15–108
AASHO M 207	Reinforced concrete elliptical culvert, storm drain, and sewer pipe	Equivalent round 18–108

Federal
  SS-P-371
  Standard strength
  Extra strength
	Concrete, nonreinforced, sewer storm drain, and culvert pipe	
		
		4–24
		4–24

Federal SS-P-375	Reinforced concrete sewer storm drain and culvert pipe	12–108
SCS 123	Nonreinforced concrete irrigation pipe with rubber-type gasket joints	6–24
AWWA C300	Reinforced concrete water pipe—steel cylinder type, not prestressed	20–96
AWWA C301	Reinforced concrete water pipe—steel cylinder type, prestressed	16–96
AWWA C302	Reinforced concrete water pipe—noncylinder type, not prestressed	12–96
Federal SS-P-381	Pressure reinforced concrete pipe—pretensioned reinforcement (steel cylinder type)	10–42

* "Concrete Industries Yearbook."[2]

Fig. 19. While the man with the dolly picks up the empty mold and returns it to the pipe machine, the other workman carefully removes the split jacket, setting it aside to be "buttoned up," after he removes any ragged pieces of concrete and smooths the groove end of the pipe.

23   Centrifugal Process

In the centrifugal process, low-slump concrete is placed in a horizontal mold rotating at slow speed, just fast enough so centrifugal force holds the concrete against the mold. The rotational speed is slowly increased, causing centrifugal force to compact the concrete and drive out the excess water, which, being lighter than the aggregate and cement, migrates to the interior of the pipe where it can be removed by light troweling or brushing.

High-strength dense concrete results from the centrifugal process. Slump of the concrete does not exceed 2 in. when it is placed in the form. During spinning a large portion of the water is removed, leaving a concrete with a very low

Fig. 20. A Cenviro machine for making pipe by a combination of centrifugal force, vibration, and pressure.

water-cement ratio. Pipes from 12 to 78 in. in diameter and from 8 to 16 ft long are made by centrifugation. Pipe may be either reinforced or nonreinforced.

The so-called cylinder pipe consists of a steel cylinder in which the pipe concrete is spun, the cylinder thus becoming a part of the pipe. After the interior concrete or mortar lining has been made, a cement-mortar covering is shot onto the exterior of the steel shell. For high-pressure service (up to 350 psi) the exterior of the shell is wrapped with high-tensile-strength steel wire which is then stressed to about 135,000 psi, in effect applying a circumferential prestressing to the pipe. After the wire is stressed, the mortar coating is applied.

Steel rings for the bell-and-spigot joint are welded to the steel cylinder prior to spinning. To assure accuracy of fit of the steel joint rings, they are stretched on a mandrel just beyond their elastic limit. Rubber gaskets are used in the joints.

A variation of the centrifugal process is one in which dry-mix concrete is placed in the rotating mold, compaction being achieved by means of a combination of centrifugation, vibration, and compaction with a roller as shown in Fig. 20. After the requisite amount of concrete has been deposited in the form, vibration is applied to the form, and a heavy roller is brought to bear on the concrete in the mold.

Defects in spun pipe are rare when proper manufacturing techniques are followed. When they do occur, the manufacturer can eliminate them, as in any other process, by eliminating the cause. Occasionally a seam crack may occur, caused by leakage of mortar through the jacket gate during spinning. Other defects might include exposed steel, usually due to faulty alignment of the cage, rather than to faulty cage dimensions; inside diameters not within allowable tolerances; rough interior, frequently caused by attempting to fill an underfilled pipe after it has spun several minutes, or lightweight-aggregate particles; crooked gasket ring in spigot, misalignment of the ring ends, or ring not perpendicular to axis of pipe; and rough, sandy spots resulting from excessive form oil.

At times blisters, or drummy areas, in which the interior concrete separates from the main body of the pipe wall, occur in spun pipe. These blisters are attributed to a number of conditions, including concrete too soft or too wet (high slump), insufficient spinning, starting steam curing too soon after spinning, and steaming at too high a temperature (above 155°F).

Blisters are usually worse during wet, cold weather than at other times. Probably all the above factors contribute to their formation at times. It has been found that, by giving the pipe its initial steam curing in a vertical position, or at least steeply inclined, the blisters are largely eliminated. The effect of lower-slump concrete is definite in reducing the incidence of blisters.

24 Cast Pipe

Large pipe, usually larger than 48 in. diameter, is sometimes cast in vertical forms, using plastic concrete mixes and following normal good practices for structural concrete.

Forms must be tight. Joints and gates in the form jackets and inside liners, and the joint with the base ring must be sealed with gaskets or tape to prevent the loss of mortar that will result in sand streaks. To facilitate placing concrete in the form, a cone or plate should be placed on top of the form so concrete will feed into the annular space between the form jacket and inner liner. Usually several forms are set up at one time to give maximum production.

Concrete should be placed in shallow lifts and vibrated with high-speed form vibrators operating at speeds in excess of 8,000 rpm. Speeds of this magnitude dictate the use of pneumatic or high-cycle electric vibrators. Vibrators should not be operated on empty areas of forms but should run while concrete is being placed in their vicinity and until the level of the concrete is well above them; then they should be stopped. Usually pipe is cast spigot end up. As the level of concrete rises in the form and approaches the top, particular care is necessary to avoid overvibration that produces weak concrete in the spigot. Slump should be kept low and the amount of vibration reduced.

Revibration is sometimes helpful in reducing the incidence of horizontal settlement cracks occasionally found near the top of the pipe. Revibration should be delayed as long as possible. The best time is just before the concrete loses its plasticity, and depends on temperature, slump, presence of admixtures, and other factors.

Soft spots in the end of the pipe cast against the base ring will result if an excess of form oil is allowed to accumulate and is not removed before placing concrete.

25 Mixes

Most manufacturers are unwilling to divulge information relative to the mixes they use, and it is necessary to proportion mixes based on commonly known principles within the limitations of specifications and pipe-making processes.

Some of the ASTM specifications set forth certain limits for some mix properties, and these limits can be used as a starting point. Trial mixes are necessary in order to achieve the most economical mix that will perform satisfactorily. Maximum size of aggregate depends on the size of pipe and method of manufacture. Packer-head and tamped pipes rarely contain aggregate larger than ⅜ or ½ in. Although ASTM C 76 and similar specifications require a minimum cement content of 6 sacks (564 lb) per cu yd, the need for plasticity, especially of packer-head and tamped pipe, makes more cement necessary. Some producers claim that the use of an air-entraining agent improves plasticity. The amount of air-entraining agent required, in the case of tamped pipe, is about half the amount normally used in concrete, but for packer-head pipe the amount may be twice as much as normal. Normal air entrainment would be of benefit for cast pipe, but there is no value in using air-entraining agents in centrifugal pipe.

As previously pointed out, pipe concrete is dense concrete. Density is achieved mechanically by the several methods of compacting and forming the pipe. In all methods, the mix is very dry. Cast and centrifuged pipe mixes have a slump of 2 in., ranging down to almost zero. Packer-head and tampered mixes are barely damp, sometimes called less than zero slump because considerably more water can be used in the mix before any indication of slump is obtained in a slump test. Water content may be as low as 3½ gal (30 lb) per cwt of cement.

Materials should conform to applicable ASTM or local code requirements, should be measured accurately, preferably by weight, and should be well mixed. Plasticity of the fresh concrete is essential in all processes, but too much plasticity can lead to trouble. It might create a mushy mix which leads to broken tamping sticks, or it might cause the freshly stripped pipes to slough. The maximum proportion of coarse aggregate gives the best strength, but this has to be balanced against density and permeability.

26 Reinforcement

All pipe of 24 in. diameter and over is reinforced with longitudinal bars and heavy wire wound into a cage. Pipe as small as 12 in., when subjected to internal hydrostatic pressure or heavy fill loading, is reinforced.

Reinforcing steel consists of longitudinal bars around which the circumferential steel is wrapped in a continuous spiral as shown in Fig. 21. Resistance welds are made at each crossing of the longitudinal bar and circumferential steel. In pipe designed to bear especially heavy fill loads, the circumferential steel may consist of two concentric cages, or a single cage may be formed into an elliptical shape. When elliptical cages are used, it is essential that some positive means of identifying the axes of the ellipse on the finished pipe be provided, as the long axis of the ellipse must be horizontal whenever the pipe is laid. Cases have been reported in which large pipe cracked in storage because it was laid down with the long axis of the cage in a vertical orientation.

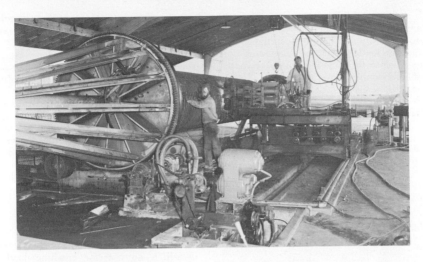

Fig. 21. Reinforcing cages are made on some sort of mandrel. Shown is a large automatic cage machine that automatically wraps circumferential steel over the previously placed longitudinal bars, spacing the wraps correctly and welding each intersection. As the mandrel rotates on its longitudinal axis, the carriage moves along its track to space the circumferential spiral.

Reinforced-concrete pipe may crack if the reinforcing steel is too close to the surface. The amount of cover must be determined by the manufacturing process, type and size of pipe, type of reinforcing cage, and experience. As little as ½ in. of cover is adequate in some cases. (Note that this cover is considerably less than the 2 to 3 in. recommended for structural concrete.) However, careless placement of the cage may result in exposure of the steel on either the outside or the inside of the pipe.

27 Curing and Handling

Curing. Machine-made pipe is handled and cured in a manner similar to concrete block, practically all pipe being low-pressure steam-cured in kilns as shown in Fig. 22 (see also Fig. 19). Large cast pipe is frequently steam-cured by covering individual pipes, or groups of pipes, with an enclosure of wood, metal, or canvas and admitting steam into the enclosure.

Fig. 22. Pipe 36 in. in diameter by 6 ft long in curing kiln awaiting the normal delay before the kiln is closed and steam admitted. (*Concrete Pipe Machinery Co.*)

Some producers provide fog nozzles in the curing kilns. If the fog nozzles are turned on progressively across the room as the kiln is filled with pipe, the pipe does not dry out before steaming.

Pipe sections, when steam-cured, should have both the exterior and interior exposed to the elevated temperature. Admitting steam to the interior only, especially if the ambient-air temperature is relatively low, might cause fine longitudinal cracks on the wall of the pipe.

Handling Pipe. Nearly as varied as the plants making pipe are the methods and equipment for handling. Figure 23 shows a fork lift with a special hydraulically powered fitting for gripping large pipe. Dollies of numerous types, one of which appears in Fig. 19, are frequently used, especially for small sizes, with crawler and motor cranes to handle the large ones.

FIG. 23. Special grips or fittings on fork lifts can be adapted to handle heavy sections of pipe.

28   Properties and Tests

Most specifications require a load test, in which an exterior crushing load is applied to a pipe and the strength is determined in pounds per lineal foot of pipe when a 0.01-in. crack appears along a length of 1 ft or more, or at ultimate failure. Figure 24 depicts a "D-load," as it is called, being applied to a piece of pipe, and Fig. 25 indicates the method of supporting the pipe and applying the load. In the case of large pipe, it is usually specified to withstand a certain load, and the 0.01-in. crack load is not specified.

Absorption tests are made on pieces of pipe broken during a load test.

Some pipe, especially for pressure lines, is required to pass a hydrostatic test, in which ends of the pipe are bulkheaded off (frequently two pieces of pipe are joined together so as to test the joint) and an internal hydrostatic load is applied.

Inasmuch as pipe is covered by standard ASTM specifications, structural analysis is not normally necessary. The D-load, or crushing, test is relied upon for design purposes based on determination of pipe bedding and trench load factors in the installation.

In making tamped and packer-head reinforced pipe, reasonable care must be exercised to avoid displacement of the reinforcement, especially in the bells and spigots where it might be exposed. Displacement of the steel weakens the pipe and interferes with adequate compaction of the concrete.

Pipe may be rejected because of circumferential cracks, longitudinal cracks, ragged tongues, or rough interiors.

Besides cracks and spalls due to rough handling, other defects are circumferential cracking of reinforced tamped pipe. Sometimes these cracks are in the form of a helix, caused by reinforcing cage hoops poorly welded to longitudinals, inexperienced or careless strippers, poorly graded mix containing insufficient gravel, or twisting of cage during tamping released when the pipe is stripped. Cracks in general may be caused by drying of pipe before initial steaming, careless stripping, mix too wet, or length-diameter ratio too great (as 4-ft lengths of 15-in. pipe).

Rough interiors result from poor aggregate grading or using a mix that is too

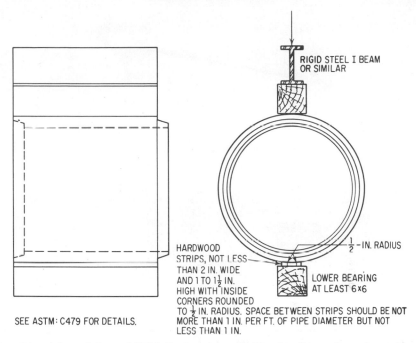

RIGID STEEL I BEAM
OR SIMILAR

$\frac{1}{2}$ – IN. RADIUS

HARDWOOD
STRIPS, NOT LESS
THAN 2 IN. WIDE
AND 1 TO 1$\frac{1}{2}$ IN.
HIGH WITH INSIDE
CORNERS ROUNDED
TO $\frac{1}{2}$ IN. RADIUS. SPACE BETWEEN STRIPS SHOULD BE NOT
MORE THAN 1 IN. PER FT. OF PIPE DIAMETER BUT NOT
LESS THAN 1 IN.

LOWER BEARING
AT LEAST 6×6

SEE ASTM: C479 FOR DETAILS.

Fig. 24. Adapted from ASTM Designation: C 497, this figure shows how pipe is supported and the load applied. Properly called the crushing test, this procedure is frequently called the D-load, or three-edge bearing test. Figure shows a length of tongue-and-groove pipe under test. The same arrangement is used for bell-and-spigot pipe except that load is not applied to the bell.

wet or too dry. Pinholes, causing hydrostatic failures, sometimes result from using sand lacking in fines.

29 Joints

Joints in pipe are either tongue-and-groove or bell-and-spigot (Fig. 26). Sealing can be either with cement mortar or by means of a rubber gasket as shown in Fig. 27. There are many types of gasket joints, some of which are patented. The advantage of gasketed joints is a slight degree of flexibility.

Concrete drain-tile lines have open joints to admit drainage water. In order to prevent infiltration of soil into the pipe, the pipe is laid in a specially proportioned bed of filter gravel.

CONCRETE PILES

30 Precast Piles

Piles can be manufactured in a casting yard adjacent to the construction site for a large project, or they may be made in a central plant and transported to the job by rail, barge, or truck. An example of a jobsite plant is the casting yard for the Chesapeake Bay Bridge and Tunnel. A completed pile structure is shown in Fig. 28.

Square and hexagonal (or octagonal) piles are the commonest cross sections, although cylindrical spun piles, usually hollow, are frequently used also. .

Square and Octagonal Piles. Except for a taper at the point, piles are normally of uniform section throughout their length. Whether to make square or octagonal piles depends mainly on personal preference. (In this discussion, "octagonal" is meant to include hexagonal also. Octagonal section is more prevalent than hex-

Fig. 25. In the three-edge bearing test, a crushing load is applied to the pipe by means of hydraulic jacks and the load per lineal foot of pipe is computed. Note the rejected reinforcing cages serving as wells for the counterweights, and the two hydraulic jacks for applying the load.

agonal.) One advantage of an octagonal pile is that its flexural strength is the same in all directions. Octagonal piles in a group present a pleasing appearance because rotation of any of them during driving is not noticeable. Other advantages are that the lateral ties can easily be made in the form of a continuous spiral. They can be made in wood or metal forms and edge chamfering is not necessary.

On the other hand, longitudinal steel in a square pile is better located to resist flexure. Square piles are easy to form, especially in banks or tiers, and concrete placing is easier than in octagonal ones. Also, there is more surface area per volume of concrete.

Size of piles can vary to suit almost any conditions. Piles 24 in. square over 100 ft long are fairly common and piles as small as 6 in. square are also made. Reinforcement, consisting of longitudinal bars in combination with spiral winding or hoops as detailed on the plans, is designed to resist handling and driving loads as well as service loads. Coverage over steel should be at least 2 in. except for small piles of special dense concrete when slightly less coverage is permissible. For piles exposed to seawater or freezing and thawing cycles while wet, coverage should be 3 in. Splices in longitudinal steel, if required, should be staggered.

The ends of the pile are called the head and the point (Fig. 29). The head should be carefully made, smooth, and at right angles to the axis of the pile. It is deeply chamfered on all sides, and extra lateral reinforcement is provided for a distance at least equal to the diameter of the pile. Shape of the point depends on expected soil and driving conditions. For hard strata and cohesionless sand and gravel, the tip may be about one-fourth the pile diameter with a length three times the pile diameter. In plastic soil the diameter may be $\frac{1}{4}D$ and the length $1\frac{1}{2}D$, or there may be no taper at all. In particularly difficult driving a metal shoe is sometimes provided. Longitudinal bars should follow the taper by drawing together in the center of the pile. Avoid bunching them together at one side of the point. Extra lateral reinforcement should be inserted in the tip.

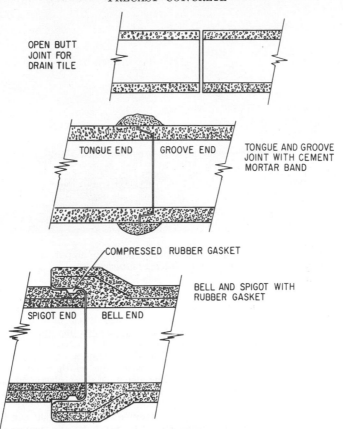

OPEN BUTT
JOINT FOR
DRAIN TILE

TONGUE END   GROOVE END

TONGUE AND GROOVE
JOINT WITH CEMENT
MORTAR BAND

COMPRESSED RUBBER GASKET

BELL AND SPIGOT WITH
RUBBER GASKET

SPIGOT END   BELL END

Fig. 26. Concrete pipe can be joined by a number of joint systems, as shown. There are a number of modifications available, especially of the gasketed joints.

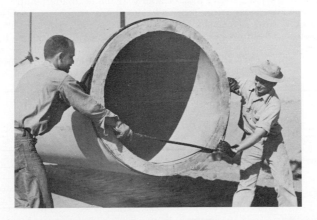

Fig. 27. The workmen are attaching a rubber joint gasket to the spigot end of a pipe. Besides being made of a ring of circular cross section as shown here, gaskets may be of different cross sections.

FIG. 28. Vertical and batter piles have been driven to support a waterfront warehouse and berth. (*Los Angeles Harbor Dept. photograph.*)

Manufacture of piles should follow the fundamentals of good concrete construction explained elsewhere in this handbook. Casting-yard facilities can range from very simple hand operations using ready-mixed concrete, when only a few piles are required, to elaborate casting yards of the type shown in Fig. 30. Regardless of the size of the casting yard, it should be designed to permit logical movement of materials and finished piles with a minimum of overlapping.

Square forms can be arranged on centers twice the width of the piles. Sides of the forms are covered with heavy building paper or sheet plastic which becomes the form for the second group of piles after the first ones have been cast and stripped.

Common practice is to assemble the reinforcement into cages that can be handled with a crane and set into the forms so as to preclude any movement during placing and vibration of the concrete. Small concrete or mortar blocks can be wired to the steel to space the steel on the sides as well as on the bottom of the form, or metal or plastic chairs can be used. The finished cage should be 6 in. shorter than the pile, to give 3 in. clearance at the head and point.

Mixes are usually fairly rich, especially for piles to be exposed to seawater, containing 6 or 7 sacks of cement (564 to 658 lb) per cu yd. Low-alkali low-C_3A cement is desirable. Slump need not exceed 2 in. for vibrated concrete containing entrained air. After the water sheen disappears from the top surface of the fresh concrete in the form a wood float finish is applied. Finally, good curing practices

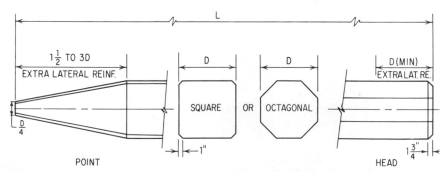

FIG. 29. Solid concrete piles can be square, octagonal, or hexagonal.

Fig. 30. Precasting yard of Bayshore Concrete Products Corp. Besides precasting 54-in. prestressed cylinder and other kinds of piles, this yard is used for precasting pile caps and bridge-deck units.

must be followed. Concrete should normally have a compressive strength of not less than 4,000 psi when the piles are driven, but the plans should be checked to determine what is actually specified for any certain job.

Handling imposes high stress in a pile and should be done carefully. The pile should be picked up at the proper designated points to avoid bending stresses and shocks and jolts should be avoided. Field-cured cylinders, exposed to the

Fig. 31. This pile hammer, used for driving 54-in. prestressed cylinder piles, is capable of delivering 50,000 ft-lb of energy with each blow. Most pile hammers are steam-operated, but many are also air-operated.

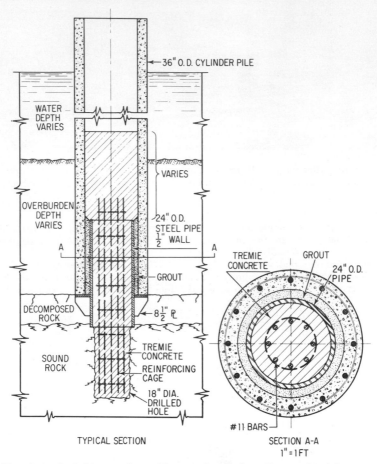

FIG. 32. A method for securing a Raymond cylinder pile to a hard bottom where there may be insufficient overburden or where the pile may be subject to tensile or pull-out forces. By concreting a reinforcing cage into a socket, or by driving a steel pin into soft rock and concreting the annular space between pin and pile, an effective method for anchoring the pile can be achieved.

same conditions as the piles, give a reliable indication of strength of the concrete in the piles to determine when the piles can be moved. Straightness of the piles is important, and specifications require that the maximum allowable deviation of the longitudinal axis from a straight line drawn from the center of the tip to the center of the head shall not exceed a certain amount, usually ¼ in. per 35 ft of length of pile.

Hollow Cylindrical Piles. These piles are made under a licensing arrangement with Raymond International, Inc. They were used on the Lake Pontchartrain causeway, numerous bridges on the Northern Illinois Tollway, the Chesapeake Bay Bridge, and other structures. This cylinder pile, as it is called, is a hollow, cylindrical, precast, posttensioned concrete pile made up of a series of sections placed end to end and held together by posttensioned cables. The pile sections are manufactured by centrifugal spinning, similar to the manner in which concrete pipe is made. Each section is reinforced with a small amount of longitudinal and spiral steel to facilitate handling. Longitudinal holes for the prestressing wires are formed in the wall of the section by means of a mandrel enclosed in rubber hose.

After curing, the sections are assembled end to end on the stressing racks with the ducts lined up, and a high-strength polyester resin is applied to the joint surfaces. Posttensioning wires are inserted and tensioned, ducts grouted, and the sections allowed to rest while the grout cures. Next the stressing wires are burned off and the pile is ready for driving.

Cylinder piles are generally made in standard diameters of 36 and 54 in. with wall thickness and number of stressing cables varied, depending on job requirements. The 36-in. piles have wall thickness from 4 to 5 in. containing 8, 12, or 16 prestressing cables; the 54-in. piles are 4½, 5, or 6 in. thick with 12, 16, or 24 cables.

Fabrication is basically the same at all plants licensed to manufacture these

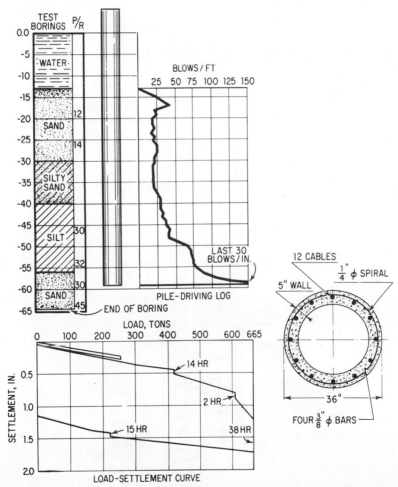

Fig. 33. Pile-driving log and load-settlement curve of a 36-in.-OD Raymond cylinder pile driven in Lake Maracaibo in 1959. The 80-ft pile was fitted with a flat point by filling the lower 4 ft with concrete. Driving was done with a Raymond 5/0 hammer delivering 57,000 ft-lb of energy.

piles. There are variations in methods of spinning, form stripping, and curing, but the variations are minor and the piles made are essentially identical. These large piles require a hammer capable of delivering a large amount of energy, of the type shown in Fig. 31. Figures 32, 33, and 34 by Raymond International show typical construction details.

Sheet Piles. Made in the shape of interlocking planks, sheet piles are a special type of precast piles. Planks vary in thickness from 6 to 12 in. and in width from 15 to 24 in. Interlocking is provided by a tongue-and-groove design, as shown in Fig. 35. The point of the pile is beveled so that it will be forced against the previously driven adjacent pile, thus assuring continued alignment of the wall under construction.

Handling Piles. Piles must be rolled over, picked up, loaded, transported, unloaded, picked up, turned to a vertical position, and finally driven. All this handling imposes loads on the pile which must be considered when the pile is designed. Pickup points should be indicated on the finished pile, if they are critical, and

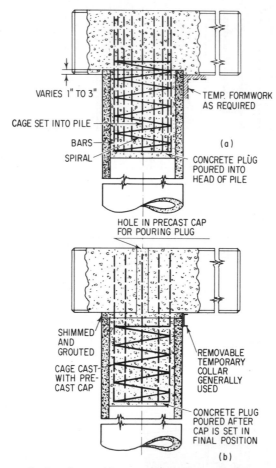

FIG. 34. Two methods of attaching a pile cap are illustrated. A cast-in-place pile cap is applied to a cylinder pile following method *a*. For a structure utilizing a precast cap, method *b* is followed. In each case an expendable platform is inserted just below the cage to support the fresh concrete inside the pile. (*Raymond International, Inc.*)

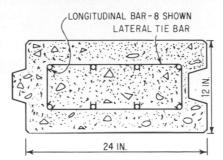

LONGITUDINAL BAR – 8 SHOWN
LATERAL TIE BAR
12 IN.
24 IN.

FIG. 35. Sheetpiling comes in a number of sizes and shapes. This one is typical. The amount of reinforcing depends on the length of the pile and anticipated lateral loading.

the pile should be picked up at these points. Short piles less than 25 ft long can usually be handled with one pickup, but longer piles may require a two-, three-, or even four-point pickup, except that most prestressed piles can be handled with one pickup, as depicted in Fig. 36. Each suspension system must be designed for each situation.

31 Cast-in-place Piles

These members are constructed either by drilling a shaft in the earth and filling the shaft with concrete, or by driving a hollow sheet-metal shell by means of a mandrel. After driving, the mandrel is withdrawn and the shell is filled with concrete. Another type consists of a heavy shell (7 gage or heavier) driven without

FIG. 36. Special heavy equipment, including a self-elevating driver platform with elaborate leads, jets, and whirler crane, is used for handling of 54-in. prestressed cylinder piles. Note single-point pickup. (*Raymond International, Inc.*)

FIG. 37. Vertical leads, to hold the pile in a vertical position while it is being driven, can be mounted on skids, a barge, or railroad car or, as shown here, can be attached to the boom of a crane. Hollow steel shells for cast-in-place piles are being driven here.

a mandrel as shown in Fig. 37. A special type of cast-in-place pile is the caisson footing in which a shaft is drilled and the bottom is belled to a larger diameter to provide an extended bearing area for the concrete.

While these piles may come in the class of "rough" concrete, they are vitally important to the safety and success of the structure they support, and there is no reason to abuse usual good construction practices. They are not subject to weathering exposure such as may be the case with precast piles, but they may be subject to attack by aggressive waters in the soil. It is usually permissible to use a slightly leaner mix in cast-in-place piles than in precast piles because they receive vertical loading only. However, there are uncertainties in soils and foundation work, and the extra cost for good concrete is negligible. Water-cement ratio should not exceed 6½ gal per sack of cement (0.58 by weight). The hole should not contain free water when the concrete is placed in it. Tops of piles, if exposed, should be given proper curing and protection.

Shell piles may be either reinforced or unreinforced. If reinforcement is used, it is usually assembled into a cage that is lowered into the shell after the shell has been inspected. Inspection is accomplished by lowering a light into the shell. A shell that is not watertight, or that shows kinks, bends, or other deformation resulting from driving that would impair the quality of the pile, should be repaired or removed. The reinforcing cage should be provided with chairs or other devices to assure clearance from the sides of the shell. It is usual to place a foot or two of mortar in the bottom of the shell just before filling with concrete. The

Fig. 38. Precast-concrete roof joist units in storage in the casting yard. On the trailer are four dome units of precast concrete.

concrete should be vibrated. If practicable, all the piles in one construction unit, such as a pier, bent, or abutment, should be driven before concrete is placed in any of them. This provides better uniformity in construction, avoids damaging completed piles, and permits removal of a shell if it becomes displaced or damaged because of subsequent driving operations.

MISCELLANEOUS PRECAST ITEMS

Concrete roofing tiles are made on a variety of machines that extrude, press, or vibrate the concrete mixture. In one machine, mixed material in a hopper feeds by gravity onto a series of pallets moving forward on an endless chain, passing under vibrating tampers and trowels that densify and smooth the tiles. Color is achieved either through the use of integral coloring material that colors all the concrete and extends entirely through the tile, or by a cementitious glaze that is applied to the surface of the tile with a pneumatic gun. Each tile is $10\frac{1}{2} \times 15$ in. with an exposed area 8×12 in.

Railroad crossties have been installed on a number of test installations in the United States and Canada, after favorable reception in Europe. The ties that have been used in North America have been prestressed, using high-quality low-slump concrete; most are steam-cured.

Other precast items include manhole rings for constructing manholes for sewers and utility installations underground; septic tanks; parking bumpers; meter boxes and covers; burial vaults; garden furniture; laundry trays; fireplaces; silo staves; cribbing for retaining walls; fence posts and rails; steps; multiple-duct electrical conduit; and many others, apparently limited only by the ingenuity of the precaster.

Many of these items use rich mixes, low slump, or no slump at all, and compaction methods to give a very dense concrete. Precast roof joists are shown in Fig. 38.

REFERENCES

1. Birkeland, Philip W., and Halvord W. Birkeland: Connections in Precast Concrete Construction, *J. ACI*, March, 1960, pp. 345–368.
2. "Concrete Industries Yearbook," Pit and Quarry Publications, Chicago, published annually.
3. "Modern Steam for the Concrete Industries," Clayton Manufacturing Co., El Monte, Calif.
4. Toennies, H. T., and J. J. Shideler: Plant Drying and Carbonation of Concrete Block—NCMA-PCA Cooperative Program, *J. ACI*, May, 1963, pp. 617–633.
5. Q Block Specifications, National Concrete Masonry Association, Washington, D.C.

Chapter 43

ESTHETICS IN CONCRETE

ROBERT L. JONES

INTEGRATING RELIEF PATTERNS IN CONCRETE CONSTRUCTION

Incorporation of artistic values in concrete construction at low cost has been made possible by space-age technology. One established method uses patented disposable polystyrene form liners marketed commercially as Labrado forms. This method integrates relief patterns in the concrete structure.

The liners have proved themselves in Southern California in the construction of public, commercial, and industrial buildings where esthetic benefits are considered desirable or essential. Their success is due to low costs and control of results. Use of the forms involves little extra effort in the construction of attractive exterior and interior walls. Designs for walls are limited only by the imagination and can either conform to the environment or create the environment. Standard patterns, however, are more economical. Some examples are shown in Fig. 1.

The general characteristics of the liners make them easy to use. They are easy to cut or saw, lightweight, rigid, noncompressible, easily transported, and readily stripped from the finished wall. Moreover, they withstand the considerable abuse peculiar to a rough construction environment.

Manufacture of Disposable Liner. A hot-wire stylus is used to carve predetermined designs in a slab of foam Dylite. From that a plaster-of-paris cast is formed to make a sand casting that will shape an aluminum mold. Expandable polystyrene beads are processed by a preheating method prior to being forced into the aluminum mold, which is also preheated. The platens are clamped together; the aluminum mold cavity is filled uniformly, and steam is injected into the cavity; this causes the beads to expand further and fuse into a homogeneous form. The mold is then cooled and opened, and the finished artistic liner is ejected.

Although other plastic forms are used, polystyrene foam possesses maximum desirable properties, such as low cost, low density, high water repellency, high compressive and bending strengths, and a surface of variable texture to permit control of the texture of the concrete wall or ceiling. The surface of polystyrene foam can be varied from smooth to coarse through the control of temperature, pressure, and size of the polystyrene expandable beads.

One mold can produce as many liners as will be needed at the rate of one every 3 min. Patterns and effects desired, as well as construction ease, determine the size and shape of the form liner.

Labrado form liners have a depth slightly greater than the depth of desired relief and bear an intaglio impression of the artistic relief on one face. The opposite face of the liner is smooth for resting upon a floor or other support. Liners have a compressive strength sufficient to withstand up to about 150 psf to accommodate tilt-up walls with a thickness of 12 in. of concrete. For cast-in-place walls, a compressive strength of from 10 to 20 psi is used.

How Liners Are Used. In the construction of tilt-up walls, form liners are fitted within the wall form, smooth side down, intaglio side up to imprint the concrete. In some instances, fasteners such as concrete nails, or preferably an adhesive or mastic, on the smooth side may be indicated to prevent movement. Usually liners are cut to fit snugly inside the wall form, as shown in Fig. 2, so that accidental movement is negated.

Fig. 1. Typical designs that have been integrated into the construction of concrete walls by use of Labrado form liners.

Appropriate reinforcing members (rods) are then positioned within the wall form in their customary location; the concrete is placed on top of the liner (Fig. 3), filling the form to its upper edge in the usual manner. To assure filling out the intaglio, the concrete should be compacted by vibration during placing. The upper surface of the concrete is finished, and the wall panel is allowed to cure.

An extra step is added in laying veneer brick walls horizontally for tilt-up. Specifically, the liners are recessed to hold individual bricks firmly in place during concrete placing. Acting as templates, the liners are placed on the floor, smooth side down, within the wall form. Bricks are inserted into the liner recesses by hand (Fig. 4). Reinforcing rods are fixed in position; concrete is placed in the form, finished, and left to cure.

After curing, the form is removed, and the concrete panel, with liner intact, is tilted up into position as part of the wall. In the case of ceilings, the panel is elevated into position and secured to structural beams or rafters.

For all patterns, once the wall panels are secured in their position, the polystyrene form liners are stripped from the concrete (or brick and concrete) by hand or by high-pressure water hose, leaving the design in relief on the concrete.

For concrete walls to be cast in place, Labrado forms are positioned inside

FIG. 2. Placing 2 × 7-ft liners on a concrete floor within wall-panel form. Liner lengths are alternated to break joint seam.

FIG. 3. Concrete being placed on liners that were fitted into the wall form shown in Fig. 2.

FIG. 4. Veneer brick wall in background was laid horizontally for tilt-up. White polystyrene form liner, leaning against the brick wall, shows how it is recessed to hold bricks uniformly within wall form while concrete is being placed and cured.

the outer gang form. They are then secured with mechanical fasteners, such as staples or adhesives, as mentioned earlier. An alternative method may be used to avoid reduction of wall thickness. In this method, a panel is cut out of the gang form, over which an offset panel is attached to house the form liner, so that the design will be added to the thickness of the wall. When the concrete has been cured, the forms and liners are stripped simultaneously, leaving the design as part of the wall as shown in Fig. 5.

WALL TILES

Thin panels, or tiles, for ornamental veneer are sometimes made of portland-cement mortar or concrete. These should not be confused with precast panels, as the items presently under discussion come under the general classification of tiles normally not more than 1 ft square, used strictly for ornamentation. One type, called the "Labrado," is especially attractive when used in conjunction with

Fig. 5. Cast-in-place wall that incorporated polystyrene liner pattern in the outer gang form to produce the design in the wall.

the Spanish style of architecture. By using white cement and light-colored aggregate, delicate pastels as well as bold, strong colors are possible.

There are several specialized methods of making these units, one of which consists of pressing a very dry (earth-moist) mixture into a metal mold under several tons of pressure. The unit is immediately removed from the mold for curing. Color and design can be obtained by the use of a thin layer of colored slurry placed in the mold before the backing mix is introduced and pressure applied. A very smooth and dense face is obtained by this method. Heavy machined-steel molds are required to withstand molding pressure and to produce a smooth surface on the tile.

Labrado tiles are cast in rubber molds which allow an almost limitless range of textures and surface relief to provide interesting artistic effects. These sculptured concrete units are $11\frac{5}{8} \times 11\frac{5}{8}$ in. in size and 2 to 4 in. thick, depending on design and amount of relief. The $11\frac{5}{8}$-in. dimension permits modular construction in 1-ft increments, allowing $\frac{3}{8}$ in. for the mortar joint.

Molds. For each design there must be a model from which to construct a mold. This is done by casting a block of plaster 11⅝ × 11⅝ × 3 in. in size, in which to sculpt or carve the desired pattern. Carving should be done very accurately, when the casting plaster is damp or cold to touch, as it is easier to carve in this condition, and the danger of chipping the edges is lessened. Draft, or taper, should be provided so the completed units will release from the mold. Surface of the plaster model is coated with a mixture of half kerosene and half petroleum jelly, applied with a brush and allowed to penetrate into the plaster, after which the excess is wiped off.

A mold frame or casing made of ½-in. steel plate having been prepared as shown in Fig. 6, the model is now covered with a minimum of ½ in. of modeling clay, or plasticene, after which the steel mold frame is set in place as shown in Fig. 7a. Tape the edges of the plasticene and cast the mold backing by pouring casting plaster through the sprue until it reaches the top of the mold frame at the sprue and vent holes. Allow the plaster to harden; then remove the model

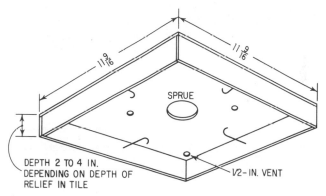

FIG. 6. The mold frame, or casing. A 1½-in. sprue is shown in the center of the cover plate. Vent holes ½ in. in diameter should be bored through the cover plate opposite the high points in the plaster model. Heavy wire hooks are welded to the ½-in. steel sides to hold the casting-plaster backing. Cover plate of ½-in. steel is welded to the side plates.

and plasticene layer. Drill out the plaster through the holes in the top of the mold frame; then let the mold backing set for 48 hr.

Reposition the model and plaster mold backing in the shell, leaving a void where the plasticene had formerly covered the model (Fig. 7b). The unit is now ready for pouring the rubber mold liner.

The rubber should be a cold-setting two-component liquid system consisting of a combination of liquid polymers and fillers which are combined with a curing agent at the time of use. After curing, the material should have a Shore A durometer of about 30 and a tensile strength of 125 psi. Follow the manufacturer's instructions on mixing, pouring, and curing.

Pour the rubber into the mold in the same manner as the casting plaster had been poured, and keep the assembly at a temperature of at least 70°F until the rubber has cured. The assembly now appears as in Fig. 7c.

Removing the model must be done carefully, especially for models that have surfaces with insufficient draft, in which case a flexible mold material is especially desirable. When removing the model from the mold, it is sometimes desirable to direct a stream of compressed air at the parting line. As separation starts, the air will aid in removal. Figure 7d shows the completed mold.

Making the Tile. The mortar mixture consists of 1 part portland cement and 3 parts of concrete sand by volume, with ¼ lb of fine silica or volcanic tuff passing the 325-mesh screen per sack of cement, and just sufficient water to give a barely plastic consistency. Consistency can be judged by squeezing a ball of mortar in the hand; it should be plastic enough to barely stick together without crumbling. If the mix is too wet it will be difficult to obtain the sharp detail necessary on the finished product. If it is too dry, the tile is apt to be porous without sufficient strength. Prepare only sufficient mortar at a time to be used up in ½ or ¾ hr.

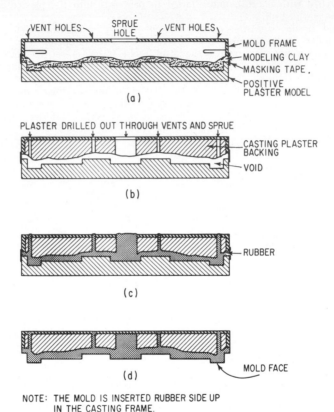

Fig. 7. Cross section through model and mold during four stages of construction: (a) The plaster model has been coated with modeling clay, and the mold frame has been placed on top. Edge is taped with wide masking tape. (b) After the casting plaster for the backing has hardened, the modeling clay is removed and the mold backing is set in place again, leaving a void which is then filled with rubber. Thickness of the void is established by four bolts in the frame that contains the model and the mold frame. (c) The assembly after the rubber has been poured. (d) The finished mold, ready to be inverted and placed in the casting frame.

The mold is mounted in a jig or fixture face (rubber) side up, and the mortar mixture placed in the mold, taking special care to fill the corners and edges. The mortar is thoroughly tamped, either by hand or with a mechanical tamper, then is screeded off and a plywood cover attached to the assembly, permitting the mold and tile to be reversed, placing the tile on the bottom. The mold is immediately removed and the tile, now resting on the plywood cover, is placed in the curing area. These steps are shown in Figs. 8, 9, and 10.

FIG. 8. A four-unit casting frame containing four rubber-faced molds made as shown in Fig. 7. The frame is in the casting machine, ready to receive the concrete mix.

FIG. 9. The tiles have been screeded and smoothed, and the plywood cover is being placed over the tiles, preparatory to inverting the entire assembly to release the tiles.

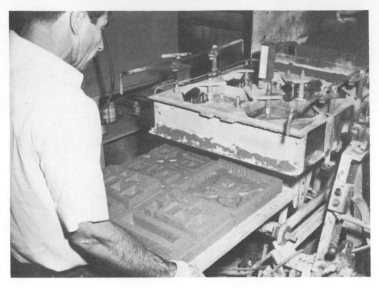

FIG. 10. The mold assembly has been rotated to the inverted position; the tiles have been released onto the plywood; and the tiles are now ready to be placed on the curing shelf.

FIG. 11. Four of the tiles, showing the detail that can be formed with the rubber mold. Note the mold draft on the deep relief.

Tiles in the curing area should be placed on racks or shelves where they can be covered under polyethylene in a moist atmosphere for at least 12 hr. They are then removed from the plywood covers and stored close together on edge, but still under cover in a moist atmosphere for another day, after which they are permitted to dry out slowly. Typical tiles are shown in Fig. 11, and in Fig. 12 a portion of a finished wall can be seen.

MODERN SCULPTURE

In preparation for the construction of a form of concrete sculpture, it is necessary to have a clear drawing and image of what the finished product is to look like

Fig. 12. A portion of a wall that has been faced with Labrado tile.

FIG. 13. "London Forty."

from all angles. The free-form objects that will be used as examples in this chapter are of such size that they can be handled by two or three persons. The "London Forty," shown in Fig. 13, is typical of the modern sculpture that meets these criteria.

The work area on which to place the object should be of ample strength to obviate any bending or warping of the table during fabrication or curing of the object. A pallet on sawhorses is usually a good area for working, remembering to keep the table at a convenient height so the sculpture is near eye level during construction.

Reinforcing. The reinforcing frame must be sufficiently rigid so it will carry the weight of the concrete sculpture without deflection. Preliminary drawings for the reinforcing steel should be made to show the size and shapes. Drawings should clearly indicate the areas of high stress, such as a cantilevered detail where

Fig. 14.   Attaching metal lath to the welded reinforcing steel frame.

added strength is necessary. Extra strength can be added, if needed, after the welded form is complete, by using extra cross members or gussets. The frame shown in Fig. 14 is typical.

Since the steel will rust and expand if exposed to water and air, it is necessary to seal the reinforcing steel completely. This is especially important on forms that will be used in pools and under adverse weather conditions. An epoxy sealer can be used with good results. By sealing the reinforcing steel in this manner, there will be no bond between the concrete or mortar and the steel because of the glaze formed by the sealer; so a bonding agent must be applied after the sealer has dried completely. This is done with good results by using a liquid bonding solution that can be applied easily with a brush. A number of proprietary compounds are available from any building-material supplier.

Application of Wire Mesh. To hold the cement mix in place, a double thickness of galvanized wire mesh is next wrapped around the steel form. Since the object does not rely on the wire mesh for strength, a 1.75-in. size mesh is used because of its flexibility and advantage in shaping the form. If only a single thickness of mesh is desired, a smaller mesh of metal lath should be used. Heavy gloves should be worn, as much hand forming and shaping are necessary. A 16-gage wire is used to secure the mesh to the reinforcing steel at about 1-ft intervals, or closer, thus holding the wire mesh in place while subsequently applying the mortar.

The basic form is now becoming three-dimensional; so it is possible to visualize any changes that need to be made before applying mortar. After the necessary changes and adjustments are made the frame is ready for application of the plaster or concrete.

The Concrete Mix. A mixture of 1 part by volume of portland cement, 2 parts plaster sand, and $\frac{1}{4}$ part of fine silica or volcanic tuff passing the 325-mesh screen will be satisfactory. These materials are mixed dry; then sufficient water to give a good plastic consistency is added (about $\frac{3}{4}$ part by volume) and the batch

FIG. 15. Here the artist is dash-applying a white-cement finish coat. Dashing in this manner assures adequate bond to the brown coat.

is mixed for at least 4 min. In order to accelerate setting and strength development, a portion of aluminous cement can be added in the form of a slurry, the amount being such that the mixture will set in 30 min or so. The amount of aluminous cement required varies with temperature and consistency of the batch, and will have to be determined by trial. About $\frac{1}{2}$ part by volume of aluminous cement is suggested for a starter. The size of the batch should be such that all the mortar can be placed in less than 30 min.

Application of Cement Mixture. When the first coat of plaster is applied it is necessary to force the mix through the wire mesh from both sides, making a solid form. If distance is so great that concrete or cement mixture does not meet and bond, the outside walls can be plastered and the center section can then be poured from an opening that has been provided. This can be done by inserting the nozzle of a plastering machine into the cavity. If no machine is available, the mixture can be pressed into place by hand, taking care to protect the hands with adequate gloves. The mortar should have a thickness of 1 to 2 in. outside the reinforcing steel, and can now be shaped by hand. By this time, the form has reached what is known as the brown-coat stage and is ready for curing to prepare it for a series of finish coats.

Curing. Prompt application of curing is necessary for prevention of shrinkage cracks, especially because of the hardening being hastened by use of the aluminous cement. The sculpture should be covered with polyethylene or similar impervious material and sprayed with water to keep it moist for at least 48 hr. The cover should be tight so as to prevent dry air from reaching the sculpture, but should not come in contact with the form. This can be done by constructing a wood frame around the form and draping the covering over the frame.

Fig. 16. A life-size reinforced-concrete statue of Hebe, cast in a flexible rubber mold.

During the curing period any areas that require more than 2 in. beyond the reinforcing steel can be built up to create detail that requires more time than was available during application of the brown coat.

Finish. The form has now taken on its final shape and is ready for application of color and texture. The artist should consider the rough sculpture from all angles to make sure that he has created what he visualized in the beginning. This condition being satisfied, the finish may now be applied, so that the piece of statuary will have the proper appearance.

Usually white cement is used for the finish mixture because it produces cleaner detail to enhance differences in value or contrast for shadow effects. Usual mix is 1 part white portland cement, 1 part No. 30 silica sand, ¼ part volcanic tuff, by volume, with sufficient water to produce a mushy consistency suitable for dash application with a brush. If color is used, the dry mineral pigment must be thoroughly mixed with the cement before the other ingredients are added. A bonding agent may be added to the mix, following the manufacturer's recommendations.

Application of the finish is accomplished by dashing it onto the form with considerable force, using a brush for this purpose, as shown in Fig. 15. In this way, the finish will be well bonded to the base, making a solid form free of laminations. It is important to dash the finish on and not merely plaster it. The finish coat should be between ⅛ and ¼ in. thick. Additional finish coats can be applied if necessary to create the feeling that the artist is trying to express, so that the completed statuary will have a look or texture that meets any desire for architectural sculptured form. Various textures can be applied to the fresh plaster, or it can be smoothed to a uniform surface.

After the finish has hardened sufficiently so the surface will not be harmed, the object is covered with several layers of cloth, kept continuously damp for at least 7 days. The statue is then permitted to dry out slowly.

When the sculpture has dried out thoroughly following final curing, a sealer can be applied, following the manufacturer's directions. Sealer should be nondiscoloring and should not yellow or peel with age in the environment where the sculpture will be exposed.

CONVENTIONAL STATUARY

A method of casting ornamental objects is to use a rubber mold which is made on a plaster original similar to the method used in making the rubber-tile mold. Intricate shapes can be made by this method, as shown in Fig. 16, requiring considerable artistic ability to produce the original, and dexterity in applying the rubber to the original and subsequently removing it.

Many of the objects, such as the statue shown in Fig. 10, are cast upside down after the reinforcing armature has been placed in the mold. Mixes are quite rich and contain little or no coarse aggregate. The soft rubber mold is peeled off the concrete, to be used again, as soon as the concrete is strong enough to withstand the action. Much handwork is required to finish the rough statue after it is stripped out, touching up the many intricate details, filling voids, and smoothing rough spots. The usual method is to go over the surface with a cement grout, after which the statue is covered and cured for a few days.

Section 12

CRACKING AND SURFACE BLEMISHES

Chapter 44

CRACKING

ROBERT E. PHILLEO

A crack occurs in a material when the tensile stress applied to that material exceeds its tensile strength. Concrete has a low tensile strength; hence it cracks rather easily. Added to its low tensile strength is its tendency to undergo large volume changes. Because of the "gel" structure of its cement paste binder, concrete swells as it becomes wet and shrinks as it dries. While concrete does not have an unusually high coefficient of thermal expansion, it is frequently exposed to extremes of temperature so that it tends to undergo large volume changes from this cause. Whenever a tendency to reduce in size is wholly or partially restrained, tensile stresses result. Such restraint is present in slabs in the form of subgrade friction and is provided in structural members by adjacent members to which they are connected. Furthermore, in large masses of concrete drying or cooling of the surface causes the attempted reduction in size of the surface layers to be restrained by the interior. While concrete is almost never used for structural members in direct tension, whenever it is used in beams or in eccentrically loaded columns (two common uses), bending of the member may place as much as half of the member in tension.

Much of the control and prevention of cracking is a matter of design and therefore beyond the control of those involved in construction. The tension in flexural members, for example, must be carried by steel reinforcement, and that reinforcement must be proportioned in such a manner that the number and size of cracks are not objectionable. In walls and slabs the designer should include grooved joints cast in the concrete or sawed at an early age. The cracks are caused to form in the grooves where their appearance is not objectionable. For this type of crack control it is necessary only that the design be accurately executed during construction. Much of the cracking which occurs in concrete, however, may be attributed to the selection of materials or to construction practices. These are the cracks with which this chapter is concerned. Since Sec. 13 is devoted to the special problems of mass concrete, they are not dealt with here.

1 Cracks Occurring during or Shortly after Construction

As has been mentioned above, concrete swells when it is wetted and shrinks when it dries. Since it begins life as fresh concrete in the thoroughly wet state the early moisture changes are always in the direction of drying. Hence, the early stresses are tensile. Much of the early cracking of concrete results from the early drying.

Plastic-shrinkage Cracks. Most of the problems associated with the shrinkage of concrete are concerned with the shrinkage of the hardened mass after the concrete has set. However, some shrinkage may occur during the first few hours after placing while the concrete is still plastic and before it has attained any significant strength. This shrinkage may be accompanied by cracking which is unsightly and objectionable. It occurs almost entirely on horizontal surfaces exposed to the atmosphere. An example of plastic-shrinkage cracking is shown in Fig. 1. Plastic-shrinkage cracks differ from other early cracks in that they are deeper and form no definite pattern. They are usually at least 2 in. deep and frequently as much as 4 in. in depth. They may extend entirely through a thin slab. Their width may be as great as ⅛ in. and they may extend several feet in length. They do not have the appearance of a clean break as do the cracks that form after the concrete hardens. The latter are sharp-edged and clearly defined, and they frequently break through aggregate particles. Plastic-shrinkage cracks, forming before any bond has been developed between aggregate and mortar, never break through aggregate particles, although sometimes they follow the edges of large aggregate particles or reinforcing bars. Once formed the cracks normally retain their original shape and do not serve as a nucleus for progressive deterioration. They do not usually impair the performance of a slab except that in some

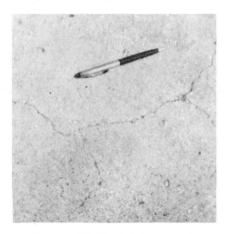

FIG. 1.   Plastic-shrinkage cracks.

building floors, in which cracks have extended entirely through the slab, plastic-shrinkage cracks have been the cause of leakage. The principal objection to the cracks is their poor appearance.

The cause of plastic-shrinkage cracks is the rapid evaporation of water from the surface of the concrete. Immediately after concrete is placed the solid ingredients begin to settle. This process, known as "bleeding," produces a layer of water at the surface since the water moves upward as it is displaced by the settling solids. The process continues until the concrete sets. Under most weather conditions some of the water at the surface evaporates. As long as the rate of bleeding exceeds the rate of evaporation there is a continuous layer of water at the surface as evidenced by the appearance of a "water sheen" on the surface. If evaporation exceeds bleeding, the sheen disappears and the top surface of the slab is placed in tension. Since concrete in the early stages of setting has a negligible tensile strength, cracks form to relieve the tension.

The factors that determine rate of evaporation are the temperature of the concrete and the temperature, relative humidity, and wind velocity of the air adjacent to the concrete. Table 1 shows how these factors affect drying. It may be seen that evaporation increases as the humidity decreases, as the wind velocity increases, as the air temperature decreases, and as the concrete temperature increases. Of particular interest is the fact that rapid evaporation is at least as big a problem in cold weather as in hot weather. Note in group 5 in Table 1 that even when

Table 1. Effect of Variations in Concrete and Air Temperatures, Relative Humidity, and Wind Velocity on Drying Tendency of Air at Jobsite

Case No.	Concrete temp, °F	Air temp, °F	Relative humidity, %	Dew point, °F	Wind velocity, mph	Drying rate, psf/hr
Group 1. Increase in wind velocity						
1	70	70	70	59	0	0.015
2	70	70	70	59	5	0.038
3	70	70	70	59	10	0.062
4	70	70	70	59	15	0.085
5	70	70	70	59	20	0.110
6	70	70	70	59	25	0.135
Group 2. Decrease in relative humidity						
7	70	70	90	67	10	0.020
8	70	70	70	59	10	0.062
9	70	70	50	50	10	0.100
10	70	70	30	37	10	0.135
11	70	70	10	13	10	0.175
Group 3. Increase in concrete and air temperatures						
12	50	50	70	41	10	0.026
13	60	60	70	50	10	0.043
14	70	70	70	59	10	0.062
15	80	80	70	70	10	0.077
16	90	90	70	79	10	0.110
17	100	100	70	88	10	0.180
Group 4. Concrete at 70°F; decrease in air temperature						
18	70	80	70	70	10	0.000
19	70	70	70	59	10	0.062
20	70	50	70	41	10	0.125
21	70	30	70	21	10	0.165
Group 5. Concrete at high temperature, air at 40°F, 100% relative humidity						
22	80	40	100	40	10	0.205
23	70	40	100	40	10	0.130
24	60	40	100	40	10	0.075
Group 6. Concrete at high temperature, air at 40°F, variable wind						
25	70	40	50	23	0	0.035
26	70	40	50	23	10	0.162
27	70	40	50	23	25	0.357
Group 7. Decrease in concrete temperature; air at 70°F						
28	80	70	50	50	10	0.175
29	70	70	50	50	10	0.100
30	60	70	50	50	10	0.045
Group 8. Concrete and air at high temperature, 10% relative humidity, variable wind						
31	90	90	10	26	0	0.070
32	90	90	10	26	10	0.336
33	90	90	10	26	25	0.740

the relative humidity is 100% in cold weather there will be a large amount of evaporation if the concrete is warm. Of all the factors listed in Table 1 only the concrete temperature is subject to control. It may be seen that there is a definite advantage to cool concrete. It should be placed as cool as practical in warm weather and should not be overheated in cold weather. Note in group 7 of Table 1 that if the concrete temperature is reduced from 80 to 60°F, 75% of the evaporation can be eliminated.

Since weather conditions cannot be controlled and since construction personnel have only limited control of the concrete temperature, primary reliance in preventing plastic-shrinkage cracks must be placed in construction techniques. Corrective measures which have been found effective include:

1. Dampening the subgrade and forms
2. Dampening the aggregates if they are dry and absorptive
3. Starting curing as soon as possible after placing
4. Protecting the concrete with temporary coverings or applying a fog spray during any appreciable delay between placing and finishing
5. Erecting windbreaks to reduce the wind velocity over the surface of the concrete
6. Providing sunshades to reduce the temperature at the surface of the concrete

On flatwork not requiring hard steel troweling it is usually possible to adjust construction operations so that finishing is completed and membrane or moist curing begun before the water sheen disappears. On particularly unfavorable days this requires finishing very soon after placing. Where troweling is required, the problem is somewhat different since final troweling and curing must be delayed. Sometimes cracks have occurred prior to final troweling. These cracks might be closed by the troweling. However, a more reliable method is the application of a fog spray or wet burlap to the surface in advance of the final troweling.

Sometimes night construction has been resorted to for preventing plastic-shrinkage cracks. This will be effective only if it produces significantly lower concrete temperatures or if the wind velocity is lower at night. The reduction of air temperature, even with the increase in relative humidity which normally accompanies it, will not reduce cracking.

These control measures are discussed in detail in Chap. 30.

Other Plastic Cracks. In addition to the shrinkage discussed above, cracking of concrete before hardening may result from settlement or other movement of the forms or subgrade, movement of reinforcing steel or embedded items, sagging or slipping of concrete on slopes, and settlement of the concrete near the tops of door and window openings or near beam soffits where beams and slabs are placed monolithically.

For most of these cracks the solution is obvious. The subgrade should be well compacted and form supports should be sufficiently sturdy to prevent movement; reinforcement and embedded items should be secured in place and should be protected after placing from accidental kicks or other movement; concrete placed on steep slopes should be formed. The problem of the settlement of concrete when a single placement involves forms of different depths is one to which special attention should be given. Wherever possible horizontal construction joints should be planned so they pass through the tops of window and door openings. Where these openings are below the tops of lifts, cracks frequently can be seen extending from each top corner to the top of the lift at an angle of approximately 45°. When it is necessary to continue a lift above the openings, the best procedure is to bring the concrete to the top of the openings and then wait 2 hr before placing the remainder of the concrete. This procedure permits a substantial part of the settlement to occur before the concrete where the crack is likely to occur is placed. It is also helpful to place some reinforcing steel at a right angle to the anticipated direction of the crack.

The 2-hr wait is also helpful when beams, walls, or columns are placed monolithically with floor slabs. The use of reinforcing steel perpendicular to anticipated crack directions near corners of openings is also useful to prevent cracks around openings in slabs. Here cracks are produced by shrinkage rather than by settlement.

All cracking resulting from settlement of concrete is aggravated by the use of high-slump concrete. Since settlement results from the downward movement of solids through water, a low water content minimizes settlement.

2 Crazing

Craze cracks are the fine cracks that appear in a hexagonal or octagonal pattern on the surface of concrete. They may be seen to some extent on nearly all surfaces, but they are most prevalent on troweled surfaces. They appear early and remain practically unchanged for the life of the structure. They are especially noticeable when the concrete is drying after the surface has been wet. Figure 2 shows a typical example of crazing. The patterns are usually several inches in diameter but may be smaller.

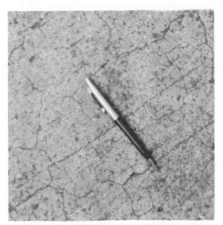

FIG. 2.   Crazing in a sidewalk slab.

Crazing is caused by drying shrinkage and is usually related to finishing and curing procedures. Excessive floating and troweling bring water, cement, and dust from the aggregate to the surface to produce a surface skin which has a higher drying shrinkage than the underlying concrete. Spreading dry cement on a surface that is too wet to trowel and sprinkling water on concrete too dry to trowel both produce a skin likely to suffer from crazing. Use of highly absorptive aggregates batched in the dry state may contribute to crazing. The absorption of water from the cement paste during the early stages of hydration produces drying shrinkage, and therefore tensile stresses, in the paste. This is especially critical at the surface where loss of water through evaporation to the atmosphere is also occurring. In thin slabs on grade, absorption of water by a dry subgrade may produce a similar effect. The use of overwet concrete contributes to all the problems associated with drying shrinkage since shrinkage is almost directly proportional to the amount of mixing water in the concrete.

Even if the concrete is batched, mixed, and finished properly, improper curing can cause crazing. Rapid evaporation of water from the surface during the early stages of hydration is a principal cause of crazing. Although crazing occurs later than plastic-shrinkage cracking, the factors that control evaporation rate enumerated above in the discussion of plastic shrinkage are pertinent here. In addition, the use of salamanders or other unvented heaters during the curing period is a cause of crazing. The high concentration of carbon dioxide in the air carbonates the surface. The chemical process of carbonation, like shrinkage, decreases the volume of the affected material. Another cause of volume decrease is the application of curing water which is much colder than the surface of the concrete to which it is applied.

Crazing can be prevented or minimized by paying attention to the following details:

1. Dampen the subgrade before placing concrete.
2. Batch absorptive aggregates in a moist condition.
3. Delay troweling until surface moisture has disappeared.
4. Do only as much troweling as is necessary to produce a good surface; do not overfinish.
5. Never sprinkle dry cement or water on the surface during finishing operations.
6. Start curing as soon as possible (see paragraph on plastic-shrinkage cracking for curing precautions).
7. Avoid the use of unvented heaters; if they must be used, make sure the surface of the concrete is covered with water or curing compound.
8. Avoid a temperature differential of more than 25°F between the concrete and curing water.

3   Cracks in Hardened Concrete

Concrete structures are, or should be, designed so that objectionable cracking will not occur in service. If the concrete survives the construction and curing period uncracked, it should remain in that condition. Failure to do so reflects improper design, improper selection of materials, improper mix proportions, improper construction methods, or improper maintenance. The immediate cause of cracking is usually load or weather. Design is beyond the scope of this handbook. Cracking associated with materials, mix proportions, construction, and maintenance is discussed below.

Improper Joint Practices. In pavement and in walls the first cracks to be noticed may be those resulting from faulty joint construction. If dummy joints do not extend deep enough or if a portion of the temperature steel in a wall is not cut accurately at the joint, a crack is likely to occur outside the joint and parallel to it. A critical item in highway construction is the accurate alignment of load-transfer dowels. It requires care to keep them aligned during the placing operation, and it is impossible to inspect them after placing. Misalignment will almost certainly cause a transverse crack since the joint cannot possibly function as intended. Weakened plane joints are used only for the purpose of controlling cracking. The crack occurs in the joint where its appearance is not objectionable and where it can be effectively sealed. Anyone interested in minimizing cracking should pay close attention to the joint-construction practice.

Settlement. Another form of structural cracking which appears fairly early is the result of settlement or deflection. Walls may crack near their connections with columns, beams, or slabs when settlement takes place. The structural frame is flexible in the vertical plane; the wall is not. Interior partitions which are supported near the midspan of beams or slabs may crack as the beam or slab deflection increases with time as a result of creep. The ultimate deflection of a beam may be three times its initial deflection. The surest preventive of settlement cracks is adequate foundation design and preparation. If it is felt that some deflection cannot be avoided, placing of the element in which cracking is likely to occur (usually thin walls) should be delayed until well after the heavier parts of the structure have been placed. Interior partitions near midspan of beams should be delayed as long as possible.

Frost Action. Any concrete which is to be subjected to freezing should contain an adequate amount of entrained air. Concrete not protected by entrained air, if frozen repeatedly in the saturated condition, is likely to undergo progressive deterioration. The first stage is the development of fine closely spaced cracks parallel to the edge of the exposed concrete. The cracks soon become filled with a dark deposit consisting largely of calcium carbonate and are commonly called "D-cracks." As the deterioration continues small pieces of concrete between the cracks separate from the body of the concrete and a general raveling of the edges and corners occurs. The deterioration is reduced as the water-cement ratio is reduced, but the only positive ways to prevent the problem is to keep water out of the concrete or to protect the concrete by an adequate air-void system.

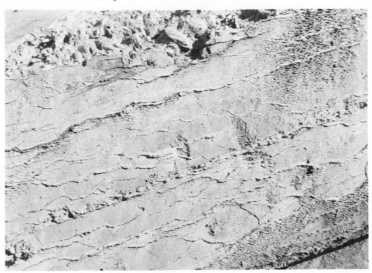

Fig. 3.   D-cracking caused by freezing and thawing.

Fig. 4.   D-cracking in a bridge which has led to complete destruction of the concrete.

In most exposed outdoor locations only the latter procedure is feasible. Figures 3 and 4 show examples of D-cracking. In Fig. 4 both the early stages and final destruction can be seen.

Rusting of Reinforcement. In the presence of moisture and oxygen steel will rust. If the moisture contains salt, the rusting is accelerated. Rusting is accompanied by an increase in volume. Therefore, significant rusting of reinforcing steel embedded in concrete causes cracking or spalling of the concrete. The problem is intensified in concrete exposed to seawater. While some experimenting has been done with galvanized reinforcing rods, the problem is normally considered solvable by keeping water away from the steel. The designer must specify adequate cover and must arrange the reinforcing so that structural cracks which might admit water will not occur. Adequate cover is usually considered to consist of 2 in. of concrete except for seawater exposure, where 3 in. is preferred. The concrete must be placed so that it is well compacted and free from rock pockets or any of the forms of cracking which are associated with materials or construction. A more critical situation exists in connection with heavy reactor-shielding concrete which contains iron or steel aggregate. It is impossible to provide the 2-in. cover. If it is to be used in a moist exposure it should be encased in a layer of conventional concrete or waterproofed.

Galvanic Corrosion. In concrete containing dissimilar metals and chloride ions, a galvanic cell may be set up which produces rapid corrosion of one of the metals.

Fig. 5. Deterioration resulting from alkali-silica reaction.

Specifically, trouble has been encountered in concrete in which calcium chloride has been used as an accelerator and which has embedded in it both steel reinforcing bars and aluminum conduit. The aluminum is involved in an expansive corrosion reaction which cracks or spalls the concrete. The combination of steel, aluminum, and calcium chloride should be avoided. If calcium chloride is used in the concrete, embedded items of aluminum should not be used. Small items of embedded aluminum may be coated so that the aluminum does not come in contact with the concrete. No economical coating for large embedded items such as conduit is available.

Alkali-Aggregate Reaction. Aggregates are normally considered to be chemically inert. Chief reliance is placed on physical tests in selecting aggregates. There are, however, three known types of chemical reactions between certain aggregates and portland cement which produce deterioration of the concrete. All the reactions are expansive. Nonuniform expansion of concrete produces unsightly cracks. While the affected aggregates occur in rather limited geographical areas, the problem is a serious one in the regions where it exists and should never be ignored in the selection of materials for concrete.

Alkali-Silica Reaction. When the term "alkali-aggregate reaction" is used without further qualification, it usually refers to the reaction between certain natural siliceous gravels and the alkalies in portland cement. Such gravels have been encountered principally in the region running from New Mexico to California and in Alabama. Figure 5 shows deterioration of a bridge pier attributable to the alkali-silica reaction. The reaction requires the presence of moisture to continue. While the reaction begins early in the life of a structure, it is usually a few years before evidence of deterioration appears. The chief symptom is "map cracking" on a rather large scale. Once they form, the cracks continue to open until, in severe cases, they may reach a width of ½ in. One of the products of the reaction is a gelatinous substance which sometimes is visible in the cracks. The surface zones of coarse-aggregate particles which participate in the reaction become visibly altered as "reaction rims" form. Deterioration is progressive as long as moisture is present.

Reactive aggregates may be identified by petrographic examination or by two laboratory tests. Details of petrographic detection are given in Ref. 1. The reactive constituents are opal, chalcedony, tridymite, cristobalite, and volcanic glass containing over 55% silica. Rocks containing these constituents include glassy or crypto-crystalline rhyolites, dacites, and andesites and chalcedonic or opaline cherts.

Aggregates may be tested in the laboratory by a quick chemical method, ASTM Designation: C 289, or by measuring expansion of mortar bars containing the aggregate, ASTM Designation: C 227. The latter test provides a more reliable indication of reactivity, but it has the disadvantage of being very time-consuming.

When reactive aggregates are encountered, the most positive means of preventing deterioration is to reject the aggregates. In some cases rejection is not economically feasible. Reactive aggregates of this type may be used if the alkali content of the cement (all alkalies expressed as Na_2O equivalent) does not exceed 0.6% or if a sufficiently active pozzolan is included in the concrete. Because of the ready availability and economy of low-alkali cement, it is used universally when siliceous aggregates of known reactivity are to be used. Pozzolans are sometimes added to provide greater insurance, but they are almost never counted on for the entire job of protection.

Cement-Aggregate Reaction in Sand-Gravel Concrete. In portions of the Midwest, notably in Kansas, Nebraska, western Iowa, and northwestern Missouri, natural coarse aggregates are scarce. As a result much concrete has been made from the so-called "sand-gravel" aggregates. These aggregates have a maximum particle size of ⅜ in. and contain from 80 to 95% material finer than the No. 4 sieve. An adverse expansive reaction has been observed in some of these concretes. While the problems are to some extent undoubtedly associated with volume changes accompanying the abnormally high water content, there is unmistakable evidence of chemical reaction between the aggregates and cement. The reaction is generally thought to be related to the alkali-silica reaction discussed above, but the diagnostic techniques and preventive measures which are so completely effective in dealing with the alkali-silica reaction appear to be inapplicable to the reaction between sand-gravel and cement. The technique which has been most successful has been the "sweetening" of the aggregate with 30% of coarse aggregate. Limestone is the material which has been used most extensively, although other coarse materials have been used. Since most of the sand-gravels have a service record, when a material is proposed for use in new construction, it is usually possible to observe concrete in which it has been used previously. Hence the more reactive combinations of materials can be avoided. Where service record is not available a wetting-drying test described in Ref. 2 may be used to predict the probable reactivity of cement-aggregate combinations.

Alkali-Carbonate Reaction. A more recently discovered reaction involves the alkalies in cement and certain carbonate rocks. The aggregates have been found to occur in Ontario, Virginia, Tennessee, Iowa, and Missouri. The affected rocks generally consist of about equal parts of dolomite and calcite, are fine-textured and contain a highly insoluble residue, a substantial part of which is of the clay particle size. Fortunately, most rocks having these characteristics are considered unsuitable for use as concrete aggregates for reasons having nothing to do with chemical reactivity. The use of low-alkali cement reduces the rate of reaction, but it may not eliminate the problem since there is a two-stage chemical reaction in which the alkalies are regenerated during the second stage. The most reliable diagnostic test is one in which the expansion of rock specimens stored in sodium hydroxide is measured (ASTM Designation: C 586).

REFERENCES

1. Mather, Bryant: Petrographic Identification of Reactive Constituents in Concrete Aggregate, *ASTM Proc.,* vol. 48, p. 1120, 1948.
2. Scholer, C. H.: A Wetting-and-Drying Test for Predicting Cement-Aggregate Reaction, *ASTM Proc.,* vol. 49, p. 942, 1949.
3. Highway Research Board, Symposium on Alkali-Carbonate Rock Reactions, *Highway Res. Board Record* 45, HRB Publication 1167 (1964).

Chapter 45

SURFACE BLEMISHES

ROBERT E. PHILLEO

While cracking of concrete might be attributable to poor design, surface defects are usually the result of deficiencies in construction. They may be traced to poor materials, improper proportioning of mixes, inadequate placing and curing procedures, or poor workmanship. A few blemishes, such as stains, may occur any time the structure is in use. Since the repair of surface defects is difficult, costly, and usually noticeable, every effort should be made to avoid them.

The following are the commonest surface blemishes.

1 Dusting

Dusting, a symptom of surface softness, may occur on either formed or finished surfaces. It is commonest and most objectionable on horizontal finished surfaces. As the surface becomes powdery the dust is removed by foot or wheeled traffic and tracked into other areas. While poor materials and poor finishing practices are important causes of dusting, the major cause is inadequate curing. Hardening is the result of cement hydration. Hydration proceeds only in the presence of moisture. When the surface is permitted to dry out before the surface grains of cement have become well hydrated, there is nothing to bond the cement or fine-aggregate particles near the surface to the rest of the mass, and they are easily removed under traffic. When dusting develops, delayed moist curing may partially correct the situation, but delayed curing is not nearly as effective as an equal amount of curing applied initially.

Aggregates containing an excess of silt or clay may contribute to dusting. The fine particles come to the surface during the finishing operation and produce a surface deficient in cement. Even if the cement becomes well hydrated, there is not enough present in the surface to bond all the silt and clay particles.

Just as insufficient cement in the surface causes dusting, so does too much cement. While concrete as it comes from the mixer almost never contains too much cement, the practice of spreading cement on the surface to dry it for easy finishing produces on the surface more cement particles than can be hydrated in the available space by the available amount of curing water. The unhydrated particles will separate under traffic.

Overly wet concrete produces a weak surface subject to dusting. Even a properly proportioned mix may dust if the surface is finished prematurely or overfinished. Finishing should be delayed until the water sheen has disappeared from the surface and the concrete has attained a slight set. Finishing while water is still on the surface or overmanipulating works excess water and fines to the surface to produce a weak surface with insufficient cement to bond the inert fines.

Lumber used for forms sometimes contributes to dusting, as described in Art. 2. Moisture from the concrete might extract some of the soluble components of new wood, especially if the new lumber was exposed to the weather for several weeks before use. Exposure of an oiled form in a dusty atmosphere may result in a layer of dirt on the form that will cause a dusty concrete surface, if the dirt is not removed before concrete is placed against the form.

During the first 24 hr concrete surfaces are softened if they are exposed to an atmosphere containing significant amounts of carbon dioxide. Carbon dioxide may come from salamanders or other heaters, mixer engines, or power buggies. Heaters should be vented, and proper ventilation should be maintained so that carbon dioxide is removed as it is generated.

For a properly proportioned mix containing good materials properly finished and properly cured, no special surface treatment should be necessary to prevent dusting. Where dusting has developed, the surface may be hardened by surface treatments with a solution of zinc fluosilicate, magnesium fluosilicate, or sodium silicate, as described in Chap. 27.

2 Form-related Defects

Since the surface of formed concrete is a reflection of the original surface of the form, any defects in the forms or form-support systems will be reflected in the concrete surface. The more obvious defects are misaligned or mismatched forms, bulges, knotholes, and broken or split boards. These can be prevented by better form design and workmanship.

Other defects are related to the properties of the form material and the form oil. Concrete may stick to inadequately oiled forms, in which case a portion of the surface concrete may be removed in stripping. During placing concrete should not be permitted to slide down formed surfaces since the sliding concrete might remove oil from the form surface. Improper cleaning of forms between reuses may also promote sticking.

Sap or other constituents of lumber and resin used to bond layers of plywood sometimes stain the concrete or interfere with the set of the cement. Lumber contains organic acids similar to those used as commercial retarders. Although their concentration may be increased by prolonged exposure to sunlight, the quantities present in form boards are usually so small that they are used up in the first use as forms and no problem exists during subsequent use. When boards that interfere with the setting of cement are encountered they may be given a limewater wash prior to their first use to prevent surface softness. Where plywood forms are used, sometimes adjacent plywood panels will produce concrete of different colors. Normally, the color difference disappears as the concrete ages. The problem is attributed to lack of uniformity in the resin. Where it has occurred, repetitions can be prevented through the cooperation of the producer in supplying material in which there is no variation.

3 Air Pockets

Small voids or pits, commonly called "bugholes," usually appear to some extent in vertical or sloping formed surfaces. They are the result of pockets of air or water trapped against the form during placing. Usually their size does not exceed ½ in. in diameter. Figure 1 shows a typical surface. The holes do not lead to deterioration and are not in any way objectionable functionally except in high-velocity water passages where a smooth surface is necessary to prevent cavitation. The chief objection is aesthetic where the concrete is to be observed at close range. The tighter the forms the more pronounced the problem; the bubbles can escape through leaky forms. Unfortunately on surfaces where appearance is especially important, great pains are usually taken to ensure tight forms.

In architectural concrete surfaces the problem may be attacked by preventing the holes or by filling them after the forms are stripped. Since it is usually necessary to perform some sort of cleanup operation to remove construction dirt and stains, the sack-rubbing procedure described at the end of this chapter may be used both to fill the holes and to clean the surface. However, the surface will look better if the holes are minimized.

The most positive means of preventing holes is the use of absorptive form lining. It is used regularly for surfaces of water passages. It is normally considered too expensive for general application and frequently does not produce the texture of surface desired by architects. For most architectural concrete reliance must be placed on mix proportioning and placing technique. The mix should not be over-sanded since it is difficult for bubbles to move upward through oversanded mixes. There is some controversy as to whether the use of air-entrained concrete increases the incidence of voids. However, the problem seems to have been as severe before the introduction of entrained air as it is now. As discussed below, the probability of more serious defects is decreased by the use of air entrainment. The forms should be as smooth as possible and should not contain excessive amounts of form oil. Too much form oil may cause the bubbles to stick to the form.

FIG. 1. Air pockets in a vertical wall surface. (*Courtesy of Kaiser Cement and Gypsum Corp.*)

When placing concrete against vertical forms, it should be deposited in shallow layers, preferably no deeper than 18 in. Each layer should be vibrated somewhat more than is necessary for adequate consolidation. While the concrete is being vibrated a hand spade should be worked up and down adjacent to the form. External form vibrators in the immediate area in which concrete is being placed have been used by some to help force the bubbles upward, and some have advocated light external tapping of the form by a hammer as the concrete is vibrated.

Avoiding holes on a sloping surface where the form slopes inward over the concrete is extremely difficult. All the above techniques, which are intended to promote the upward movement of bubbles, tend to move the bubbles toward the form when the form is above the concrete. The best procedure is to do no more vibration than necessary for consolidation and to employ hand spading while the concrete is being vibrated. Some attempts have been made to build vented forms by leaving small spaces between tongue-and-groove boards which will pass air and water but not mortar.

4 Bolt Holes

Most forming systems require some sort of ties that leave holes in the concrete. If the concrete is to be exposed to water or to public view, the holes should be patched as soon as possible after removal of forms. The holes should be reamed and then filled with dry patching mortar. The cement used in the mortar should be a blend of portland cement and white portland cement proportioned so that the color of the patch after curing and drying matches the color of the surrounding concrete. Usually one-third white cement is satisfactory. The proportions of the mortar should consist of 1 part cement to 3 parts of sand which has been sieved through a No. 16 sieve with enough water so that the mortar will just stick together when molded into a ball by hand. To reduce shrinkage in place, the mortar should be allowed to stand for ½ hr after mixing and then remixed before it is used.

5 Honeycomb and Rock Pockets

Perhaps the commonest and least excusable surface defect is honeycomb. As pictured in Fig. 2, this blemish consists of exposed pockets of coarse aggregate

Fig. 2. Honeycomb on a foundation wall. (*Courtesy of Kaiser Cement and Gypsum Corp.*)

not covered by a surface layer of mortar. Honeycomb may be caused by a harsh unworkable mix which cannot be properly consolidated, by inadequate consolidation, or by leaky forms which allow the mortar to escape. A placing practice which aggravates the situation is the depositing of the concrete too far away from its intended final location and moving it into place with a vibrator.

Honeycomb can be prevented merely by attention to good concrete practice. The concrete should be workable. Air-entrained concrete is more cohesive and more immune to honeycomb than non-air-entrained concrete. The forms should be designed and the sequence of placing planned so that concrete can be placed in its final position and so that there is easy vibrator access to all parts of the section. It may be necessary to use small buckets for a portion of the work if the buckets normally employed cannot fit into tight places. The forms on exposed surfaces should be tight. As the concrete is placed it should be thoroughly vibrated. The spud of the vibrator should be kept moving and not permitted to vibrate in one position for an appreciable time.

6 Sand Streaking

Sand streaking, characterized by irregular vertical lines usually toward the top of a lift, is the result of excessive bleeding. Bleeding can result from too much water in the concrete and from improper sand grading. The most effective way to eliminate or greatly reduce bleeding, however, is the introduction of entrained air. Air-entrained concrete containing well-graded sand and a proper amount of water will not produce sand streaking.

7 Laitance

Laitance is the mixture of bleed water, cement, and fine sand that appears at the top of a lift during the first few hours after placing. All concrete produces some laitance. Since it accompanies bleeding it can be minimized by following the procedures outlined above for minimizing bleeding. Laitance should be removed before another lift is placed on top of it. Wet sandblasting is the established procedure where an appreciable layer must be removed, although high-pressure water jets (3,000 psi or more) are now in common use for mass concrete. The principal objection to laitance is not its appearance but the fact that it provides a weak zone to which additional concrete will not bond well. However, a thick layer of laitance produces an objectionable horizontal band in a vertical surface, and it is an area in which deterioration is likely to start. To prevent a visible layer of laitance, wall forms should be slightly overfilled; this will ensure sound concrete to the top of the form.

8 Popouts

Popouts are blemishes which do not appear during construction. They may start to appear during the first winter following construction and may continue

FIG. 3. Popouts on the surface of a slab. (*Courtesy of Kaiser Cement and Gypsum Corp.*)

to form for several years. A popout is a conical-shaped hole in the surface with a portion of a coarse-aggregate particle exposed at the bottom. Figure 3 shows a surface afflicted with popouts. They occur outdoors on the horizontal and vertical surfaces, although a greater number occur on horizontal surfaces. They are caused by freezing of water in aggregate particles that have an internal pore structure which causes them to expand unduly upon freezing. When these particles are near the surface the expansion forces the thin cover of concrete off and usually ruptures the aggregate particle. In some cases popouts have been attributed to impurities which have found their way into the concrete such as pieces of hard burnt lime, calcined dolomite, iron sulfide, or glass, but freezing of water in unsound coarse-aggregate particles is the predominant cause. Popouts do not harm the concrete, but they are unsightly.

Types of rock that have produced popouts include chert, shale, argillaceous limestone, claystone, mudstone, and siltstone. While chert is a common cause of popouts, many dense cherts are entirely satisfactory as aggregates.

Where concrete is exposed to freezing in a moist state, popouts can be prevented only by avoiding aggregates which cause them. Affected particles can be identified petrographically. However, the simplest and most economical way to evaluate aggregates in this regard is to rely on service records. A sidewalk is a popout-prone structure. If an aggregate proposed for use in a project has ever been used in sidewalks or driveways, an examination of those structures will provide conclusive evidence as to the popout potential of the source. Some gravel deposits contain material that is acceptable except for lightweight particles distributed throughout the deposits. If there is a large difference between the specific gravity of the good material and that which produces popouts, the poor material may be removed by a beneficiation process such as heavy-media separation or hydraulic jigging. There are several commercial beneficiation plants now in operation which produce good aggregate from marginal sources.

9   Efflorescence

Efflorescence is another blemish that usually does not appear immediately in new concrete. It is a whitish crystalline deposit on the surface. It is a more general problem in masonry structures than in concrete structures. As water moves through a structure it picks up salts which it is capable of dissolving. As it reaches the surface it evaporates, leaving the salt deposited on the surface. The commonest manifestation of efflorescence in concrete is that around cracks which allow leaking in hydraulic structures. Water traveling through the crack brings soluble calcium hydroxide to the surface. After the water evaporates, the calcium hydroxide left on the surface reacts with carbon dioxide in the air to form calcium carbonate. This white rocklike material may build up a thickness of several inches. As a corollary problem, the removal of the leached material from the interior of the concrete widens the crack and may increase the flow through the crack.

The surface of concrete may also become partially covered with a thin film much like masonry walls. In addition to calcium carbonate the deposits may consist of the sulfates of sodium, potassium, or calcium. The conditions necessary for development of the film are the presence of salts, the movement of water, and porous concrete. The salts might come from sand or gravel dredged from salt-bearing water, or contaminated in storage, which has not been washed in clean water prior to use. Or they might come from mixing water pumped from a well in sulfate soil. Porous concrete is the result of a high water-cement ratio or poor consolidation. Except for the problem that exists around leaks in hydraulic structures, properly proportioned concrete made of clean materials and well compacted should not be subject to efflorescence.

Deposits of alkali salts can be removed from concrete by washing. Calcium carbonate can be removed by chipping or brushing. Muriatic acid may be used, but care must be taken since it attacks the concrete as well as the deposit. The surface must be thoroughly washed with clean water after the acid treatment.

10   Stains

The principal causes of staining during construction are rust stains from reinforcing rods protruding from a partially completed structure, and impurities in curing water. Iron is one of the frequent contaminants of curing water; hence the stain formed is similar to rust stain. In addition staining may result from certain form oils and from materials inside the concrete. For example, iron pyrites near the surface of concrete hydrate to produce a brown stain. Throughout its life concrete is subject to spillage of liquids that may stain it.

Most stains can be removed. However, the longer they remain the more deeply they penetrate and the more difficult they are to remove. Unless concrete is severely stained during construction, the sack-rubbed finish described at the end of the chapter is generally the only treatment that is necessary. When it is necessary to remove a stain, an effort should be made to identify the material causing the stain. Otherwise a great deal of experimenting must be done to find a method that works. Procedures for removing some of the commoner stains are given below. References 1 and 2 provide additional information.

Iron. The stained area should be saturated with water and then mopped with a solution containing 1 lb of powdered oxalic acid per gal of water. After 3 hr the surface should be scrubbed with stiff brushes as it is flushed with clean water. The surface must be washed until all traces of the acid are gone.

In an alternate method the stain is soaked for $\frac{1}{2}$ hr with a solution containing 1 part of sodium citrate to 6 parts of water. Then, for horizontal surfaces a thin layer of sodium hydrosulfite crystals should be spread on the surface and immediately covered with a stiff paste of whiting and water. For vertical surfaces the paste should be placed on a trowel and the crystals sprinkled on the paste so that as the paste is troweled on the surface the crystals are in contact with the concrete. After 1 hr the paste is scraped off and the surface washed thoroughly. If necessary the process may be repeated.

Aluminum. These stains can be removed by muriatic acid. As with all acid treatments the concrete should be saturated prior to treatment and thoroughly washed following treatment.

Copper or Bronze. A paste consisting of 1 part of powdered ammonium chloride, 4 parts of powdered talc or diatomite, and ammonia water should be applied to the affected areas and allowed to dry before being removed. Three applications may be necessary to remove severe stains. Finally, the area should be thoroughly washed with water.

Linseed Oil. Since linseed oil seals the pores of concrete, it is difficult to remove by solvents. Therefore, an attempt should be made to bleach it rather than remove it. If the oil has not completely penetrated the concrete, the excess oil may be removed by applying hydrated lime, sawdust, talc, or whiting. The residual stain may be bleached by saturating a cloth with hydrogen peroxide and placing it over the stain. A second cloth moistened with ammonia is placed over the first cloth. It will usually require several applications to bleach the stain to a satisfactory color.

Petroleum Oil. Hardened oil or grease should be scraped off. Liquid oil should be soaked up with an absorbent material such as fuller's earth, hydrated lime, portland cement, talc, or whiting. Steam cleaning or scrubbing with a hot 10% caustic soda solution may also be used to remove the surface oil. The residual stain may be removed by applying a paste of benzol and hydrated lime, whiting, or talc. The paste should be allowed to dry thoroughly before being removed.

Asphalt. Petroleum asphalt is best removed by chilling the area with ice until the asphalt becomes brittle and then successively chipping, scraping, and scrubbing with an abrasive powder. Emulsified asphalt can be removed by scrubbing with water containing a scouring powder. It is practically impossible to remove cutback-asphalt stains completely, but their appearance can be greatly improved by several applications of a paste consisting of benzol and diatomaceous earth followed by scrubbing with water and scouring powder.

Ink. Most dark ink stains can be bleached with a sodium hypochlorite solution containing 12% available chlorine. It may be applied by flooding, by saturated cloths, or in paste form. If there is a residual brown stain it can be removed by methods applicable to iron stains, while residual blue stains can be removed by additional bleaching with ammonia water.

Bright-colored ink stains can be bleached by applying a paste made of whiting, a strong solution of sodium perborate and hot water or a paste made of whiting, potassium hypochlorite, potassium chloride, and hot water. The paste should be allowed to dry before it is removed.

Paint. Freshly spilled paint should be soaked up (not wiped) with paper towels or absorbent cloths. No thinners or solvents should be applied since they will merely increase the size of the stain. After the excess paint has been removed the stained area should be alternately scrubbed with water containing scouring powder and rinsed with clear water until the stain is removed.

Dried paint can usually be removed by application of the commercial paint remover applicable to the particular type of paint. The stain should be flooded with the remover for a few minutes, then scrubbed gently and rinsed with water.

Smoke. The stain should be scrubbed, then covered with a paste of trichloroethylene and talc. A nonabsorptive covering should be placed over the paste to prevent evaporation. A final scrubbing with scouring powder and water should remove all traces of the stain.

Rotten Wood. The stain may be bleached by applying white cloths saturated with a sodium hypochlorite solution. The cloths should be renewed as they dry since a long period of moist contact is required.

11 Sack-rubbed Finish

Unavoidable blemishes and discolorations occurring during construction, such as surface air pockets and minor rust stains, may be effectively treated by giving all exposed architectural concrete surfaces a sack-rubbed finish. Such a finish does nothing to improve the concrete functionally, but it improves its appearance. It preferably should be delayed until near the end of the project. Then, if all surfaces are done in a short period of time, there is a better chance of achieving a uniform appearance, and the chance of discoloration caused by subsequent construction operations is eliminated. The area finished in one day should be surrounded by natural breaks in the surface such as corners or grooved joints. The temperature of the air adjacent to the surface should not be less than 50°F for 24 hr prior to and 72 hr following application of the finish. During hot, dry weather the finish should be applied in shaded areas. If this is not possible, the surface to be finished should be covered with damp burlap or subjected to a continuous fog spray for 1 hr before finishing operations are begun.

The sack-rubbed finish consists essentially of thoroughly wetting the surface to prevent absorption of water from the mortar and then coating the surface with mortar. The mortar should consist of 1 part portland cement, 2 parts well-graded sand which passes a No. 30 sieve, and enough water to provide the consistency of thick paint. The mortar will darken the surface unless about one-third of the cement is replaced by white portland cement. The mortar should be applied as soon as the surface of the concrete approaches surface dryness. It should be vigorously and thoroughly rubbed over the area with clean burlap pads so as to fill all voids. While the mortar is still plastic but partially set so that it cannot be easily pulled from the voids, the surface should be sack-rubbed with a dry mix of the same proportions as those given above except that no water is used. The burlap pads used for this operation should be stretched tightly around a board to prevent dishing the mortar in the voids. At the end of the rubbing operation there should be no discernible thickness of mortar on the surface except in the voids. Immediately following the sack-rubbing treatment the surface should be continuously moist-cured for 72 hr. Other surface treatments are discussed in Chap. 27.

REFERENCES

1. Harrison, D.: The Removal of Stains from Concrete, *Research News,* Hydro-Electric Power Commission of Ontario, Toronto, July–September, 1959, pp. 19–24.
2. Kaiser Cement and Gypsum Corp.: Removal of Common Stains from Concrete, *Concrete Topics, Tech. Serv. Dept. Bull.* 21.

Section 13

COOLING AND GROUTING OF CONCRETE IN PLACE

Chapter 46

COOLING OF MASS CONCRETE

ROBERT E. PHILLEO

The hydration of cement, which is the chemical process that gives concrete its strength, is accompanied by the generation of heat. Most structural and pavement concrete sections are thin enough so that this heat is readily dissipated by transfer to the surrounding air. Hence, there is no problem. Therefore, the information in this chapter is applicable only to the specialized field of mass concrete.

It is difficult to give an exact definition of mass concrete. However, it may be said that for structures in which all the concrete is within 5 ft of the nearest boundary—that is, structures consisting entirely of sections no more than 10 ft thick—it is not necessary to employ the procedures outlined in this chapter. This criterion applies to backfilled sections, since normally 10-ft-thick sections will have cooled to the air before the backfill is placed. As the sections increase in size, more special measures are required.

The problem of cracking associated with heat generation is twofold: (1) when a large difference in temperature exists between the interior of a mass and the surface, thermal cracking may result; and (2) even if the concrete attains its maximum temperature without cracking, as it cools to its final stable temperature the decrease in size which accompanies the cooling may result in cracking if the mass is restrained against movement by adjacent concrete or the foundation. The solution is to restrict the maximum temperature attained by the concrete. This may be done by giving attention to any or all of the following measures: mix proportions and maximum aggregate size, selection of cement or other cementitious materials, precooling of the concrete, control of lift thickness and time interval between lifts, and postcooling.

MIX PROPORTIONS AND MAXIMUM AGGREGATE SIZE

The objective in proportioning mixes for mass concrete is to achieve as low a cement content as possible since, for a given type of cement, the heat generated is directly proportional to the amount of cement in the concrete. Close attention should be paid to all the principles set forth in Sec. 3. Particular emphasis should be placed on:

1. Use of the largest practical aggregate size
2. Use of entrained air
3. Placing at the stiffest practical consistency

1 Maximum Aggregate Size

In any concrete the use of the largest practical aggregate size reduces cement content and thereby is economically favorable. The size is normally limited by the spacing of reinforcing bars or the clear space between reinforcing bars and the form. In many mass-concrete sections this criterion would permit an almost unlimited maximum aggregate size. Problems in handling and mixing have led to the adoption of 6 in. as the largest aggregate size normally used in mass concrete in the United States and Canada. When the 6-in. maximum size is used, the coarse aggregate should be stored and batched in at least four sizes. Normal practice is to separate the coarse aggregate into the following sizes: minus ¾ in., ¾ to 1½ in., 1½ to 3 in., and 3 to 6 in. The 6-in. maximum size should be used whenever the distance between bars or between the form and bars exceeds 9 in. When the spacing is between 4½ and 9 in. the maximum aggregate size should be reduced to 3 in., and the coarse aggregate should be separated into at least three sizes.

2 Entrained Air

The principal reason for the use of entrained air in concrete is to improve its durability in a freezing environment. While concrete near the exposed faces of mass-concrete structures should contain entrained air for this purpose, its use should be required in interior mass concrete for quite another reason. It is well known that entrained air improves the workability of concrete. This effect is more pronounced in lean mass concrete than in any other type. The use of entrained air in mass concrete so reduces the mixing-water requirement that it has been possible, without loss of strength, permeability, or other desirable properties, to reduce the cement factor by about 100 lb per cu yd compared with concrete placed before the advent of entrained air. As a result, the temperature rise in the concrete has been reduced about 30%. Mass concrete should never be placed without entrained air. In measuring the air content, it is normal to pass the mass concrete over 1½- or 2-in. sieve and to test only the material passing through the sieve. The air content of that portion of the batch should be between 6 and 7%.

3 Consistency

As in all concrete there is an advantage in placing mass concrete at the stiffest practical consistency since a low water content makes possible a corresponding low cement content. Unfortunately, there is no generally accepted test for the consistency of mass concrete. The most commonly used procedure is to make a slump test on that portion of the batch which will pass a 1½-in. sieve. A slump of 1 to 2 in. in that portion of the batch usually provides adequate placeability of the mass concrete. The slump of the wet-sieved concrete is meaningful, however, only if the coarse aggregate is accurately graded in accordance with accepted grading curves.

By application of the above principles, it has been possible for organizations placing concrete in low- and moderate-height gravity dams to produce concrete regularly having a cement factor of 205 lb per cu yd with rounded aggregates and 260 lb per cu yd with angular aggregates. Arch dams and high gravity dams may require higher cement factors in order to provide adequate strength. Since the concrete is placed at a rather stiff consistency, very close control should be maintained over control of aggregate gradation and batching accuracy. Poor control will result in batches which cannot be placed. In the absence of good control, such a situation can be prevented only by increasing the slump with a resultant increase in both water and cement. Since minimum cement content is the objective, tight control is much preferable to high slump.

SELECTION OF CEMENT

4 Portland Cement

Section 1 contains information on available types of portland cement. It will be noted that Type IV cement is intended for use where a low heat of hydration is required. It has been used on several major dams in the past and still finds occasional use. However, it is not normally carried in stock by cement producers. By far the greatest amount of mass concrete placed today contains Type II portland cement. An optional provision in the specification for portland cement (ASTM Designation: C 150) makes it possible to impose definite limits on the heat of hydration of Type II cement at 7 and 28 days. With the optional provision invoked, Type II cement generates only slightly more heat than Type IV cement and has the advantage of greater availability, lower cost, and higher early strength. The latter advantage facilitates construction by reducing problems in form anchorage and stripping. Type II cement normally produces a 28-day adiabatic-temperature rise (no heat gained by or lost from the concrete) of 14 to 16°F for each cwt of cement in a cubic yard of concrete. Thus, even the leanest mass concretes (205 lb per cu yd) undergo a temperature rise due to cement alone of 30°F or more.

5 Portland Blast-furnace Slag Cement

Type IS (MH) cement (ASTM Designation: C 595) may be used interchangeably with Type II portland cement. It has the same requirements for strength and heat of hydration. Except during periods of cement shortage, it is available only in the vicinity of steel-production centers.

6 Pozzolans

There are several reasons for using pozzolans as a replacement for part of the portland cement in concrete. In mass concrete the principal reasons are economy, a slight decrease in heat generation, improved workability, and reduced permeability to water. Materials used as pozzolans are fly ash, the fine ash which results from burning powdered coal in power plants, and a number of natural materials such as opaline cherts and shales, diatomaceous, earths, tuffs, volcanic ashes, and pumicites. Most of the natural materials require calcining to be effective. Pozzolans by themselves are not cementitious. However, they react with the lime formed during the hydration of portland cement to form further strength-producing hydration products. Effective pozzolans may be used to replace as much as 35% of the portland cement in mass concrete. Pozzolans should comply with the specifications in ASTM Designation: C 618, in which pozzolanic activity is measured by the strength of a lime-pozzolan mortar. The chief difficulties that might be experienced in the use of pozzolans are high mixing-water requirement and high carbon content. The former is common in natural materials and the latter in fly ash. High carbon content greatly increases the required amount of air-entraining agent and increases the difficulty of controlling the air content in concrete. However, at the cost of lower early strength, pozzolans provide more economical mass concrete with reduced early heat-generation potential.

PRECOOLING OF CONCRETE

The maximum temperature attained by mass concrete depends on the initial placing temperatures of the concrete, the adiabatic-temperature-rise potential of the cement, and the heat gained by or lost from the concrete during the period of hydration. One of the most powerful means for controlling the maximum temperature is to restrict the placing temperature. On important mass-concrete projects, this temperature is frequently restricted to 50°F. Means for determining the required placing temperature are discussed later in the chapter. Reduced placing temperature is accomplished by cooling the ingredients prior to batching. Of the four principal ingredients of concrete (cement, water, sand, and coarse aggregate), water and coarse aggregate are the most amenable to precooling. Coarse aggregate occurs in the largest quantity of any of the ingredients, and water has the highest specific heat of all the ingredients. Therefore, these are the two most effective coolants. It is generally not considered practical to cool cement because of the need to keep the cement dry. Many cooling methods involve inundation in water. Such methods are clearly inapplicable to cement. Even dry-cooling methods are seldom used because of the likelihood of condensation. Inundation methods are not applicable to sand because of the difficulty of recovering the sand after cooling and the moisture-control problem imposed on the batching operation. Water is conveniently chilled by conventional refrigeration methods. In addition, if a portion of the mixing water is batched in the form of ice, a large amount of cooling results.

7   Calculation of Required Temperatures and Ingredients

The following heat-balance equation shows the relationship between the weight, specific heat, and initial temperature of each ingredient, amount of ice to be batched, and the final temperature of the concrete. The computed final temperature should be several degrees lower than the specified temperature to compensate for heat gain during batching, mixing, transporting and placing, and for heat resulting from early heat of hydration. The exact amount of this lowering can be determined only experimentally in the field. In the absence of such field data, it may be assumed as 5°F.

$$W_c C_c(t_c - t) + W_s C_s(t_s - t) + W_a C_a(t_a - t) + W_{sm}(t_s - t) + W_{am}(t_a - t)$$
$$+ (W_w - W_i)(t_w - t) - W_i(t - 0.5t_i + 128) = 0$$

where W_c = weight of cement in the batch, lb
   C_c = specific heat of cement
   t_c = initial temperature of cement, °F
   t = final concrete temperature, °F
   W_s = weight of sand in the batch, lb
   C_s = specific heat of sand
   t_s = initial temperature of sand, °F
   W_a = weight of coarse aggregate in the batch, lb
   C_a = specific heat of coarse aggregate
   t_a = initial temperature of coarse aggregate, °F
   W_{sm} = weight of free moisture in sand, lb
   W_{am} = weight of free moisture in coarse aggregate, lb
   W_w = weight of added mixing water (including ice but not including moisture on aggregates), lb
   W_i = weight of ice, lb
   t_w = initial temperature of mixing water, °F
   t_i = initial temperature of ice

In the equation the specific heat of water has been taken as 1.0, the specific heat of ice as 0.5, and the heat of fusion of ice as 144 Btu per lb. The specific heat of the cement and aggregates must be determined for each individual set of materials. Normally the specific heat of cement is between 0.25 and 0.30, while most aggregates have specific heats between 0.20 and 0.25.

Example.   The example shown in the table on page 46–7 illustrates the use of the equation to determine the quantity of ice to be batched for a given set of materials. The desired placing temperature is 50°F. The heat gain after cooling is 5°F. Therefore, the computed final concrete temperature is 45°F.

$$188 \times 0.28(110 - 45) + 700 \times 0.25(80 - 45) + 3{,}300 \times 0.22(40 - 45)$$
$$+ 21(80 - 45) + 33(40 - 45) + (74 - W_i)(35 - 45)$$
$$-W_i(45 - 0.5 \times 25 + 128) = 0$$

$$W_i = 38.1 \text{ lb}$$

Thus, 38 lb of the 74 lb of mixing water must be batched in the form of ice. If, in the solution of the equation, a negative value is obtained for W_i, no ice will

Ingredient	Batch weight, lb	Specific heat	Initial temp, °F
Cement............................	188	0.28	110
Sand..............................	700	0.25	80
Coarse aggregate..................	3,300	0.22	40
Water.............................	74	....	35
Free moisture in sand.............	21	....	80
Free moisture in coarse aggregate..	33	....	40
Ice...............................	?	....	25

be required. The temperature of concrete to which no ice is added will be lower than that specified.

In addition to the discussion in this chapter, additional information on cooling, as well as heating, of concrete will be found in Chap. 20.

8   Methods for Cooling Coarse Aggregate

There have been three basic methods for cooling coarse aggregates: inundation in water, air cooling, and vacuum cooling. There have been several variations of these basic methods. All these methods have one problem in common: regardless of the refrigeration capacity available, the time required to cool a particle of aggregate is a function of the dimensions of the particle and its thermal diffusivity. For small aggregates this condition has little effect on plant design. For 6-in. aggregates, however, the effect is considerable. It requires 20 min for 90% of the total potential cooling to be accomplished when a 6-in. particle of typical aggregate thermal properties is immersed in a liquid at a lower temperature. Thus, a plant must be designed so that the large aggregates will be in contact with the cooling medium for at least 20 or 30 min. If this time is not provided, the particles become only surface-cooled; their centers remain warm. If surface-cooled particles are introduced into the mixer immediately after cooling, they may cause an unrealistically low temperature value to be recorded by a thermometer whose sensing element is placed in mortar. As time passes, heat flows from the center of the large particles into the surrounding mortar, causing the temperature to rise. For this reason, specifications frequently require that the temperature of the concrete be measured 20 min after mixing. Following are brief descriptions of systems that have been used successfully. All these methods can be augmented in arid climates by sprinkling the storage piles. Evaporation cools the piles and reduces the amount of artificial cooling required. Sprinkling, however, is ineffective in humid climates.

Inundation Tanks. Figure 1 is a diagram showing the operation of inundation tanks. Graded aggregate is brought by conveyor from the storage bins to the appropriate tank that is about one-third full of water (step I). As the stone is introduced into the tank more water is added (step II). After the tank is full, additional water is circulated through the tank for a predetermined time (step III), following which the water is drained (step IV). The bottom gate is opened (step V) and the aggregate is loaded on a conveyor for its trip to the mixing plant. On the way it is normally passed over dewatering screens to minimize and stabilize the moisture content. Control of aggregate free moisture is the most difficult problem associated with inundation cooling.

Belt Inundation. This system differs from the tank method in that the aggregates never leave the conveyor belt. The conveyor runs through a trough in which cold water is circulated. After the aggregate leaves the trough it is passed over a dewatering screen and transported to the mixing plant. While this scheme is simple in concept, its execution is hampered by the need for keeping the large-aggregate particles inundated for 20 min. For a high rate of production a very long cooling chamber is required. A typical plant, capable of processing 60 tons per hr of 3- to 6-in. aggregate, requires an inundation trough 300 ft long.

Belt Spraying. In a modification of the above procedure, the aggregate travels a belt through an insulated tunnel under a series of shower heads that spray ice water on the belt. In practice this is a form of inundation cooling since cold water remains on the belt submerging the aggregate. In this procedure it is possible to vary the spacing of the sprays for more efficient cooling. The spacing may be close as the belt enters the tunnel with a gradual increase in spacing toward the exit since it is possible to remove heat more rapidly when there is a large difference in temperature between the aggregate and the water.

Air Cooling. Circulation of refrigerated air through the storage bins in the batch plant is a method which has been used successfully. It is necessary to insulate the bin walls when this practice is adopted, but insulation is a good idea whenever cooled materials are to be handled. Air cooling has the disadvantage

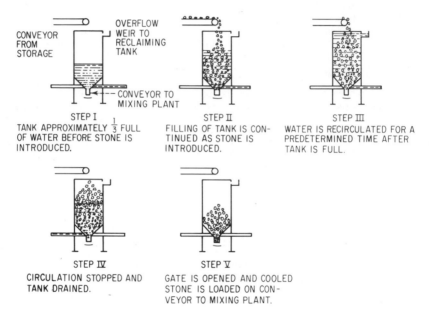

FIG. 1.   Inundation tanks for cooling coarse aggregate.[5]

that heat transfer between the aggregate and air is not so good as the transfer between aggregate and water. It has the advantage that there is no opportunity for heat gain following the cooling, and with air cooling there is less of a problem with moisture control than with inundation methods. It is used where a great deal of cooling is not required or as a supplement to other cooling methods. In the latter application it prevents heat gain after cooling and eliminates the problem of disposing of aggregate remaining in the bin after the completion of a day's concrete placing.

Vacuum Cooling. The vacuum-cooling process has become increasingly popular in recent years. It requires only steam boilers rather than an elaborate refrigeration system. It takes advantage of the fact that the temperature at which water boils is reduced as the atmospheric pressure is reduced. At a pressure of ¼ in. of mercury the boiling point is about 40°F. If aggregate with water on its surface is placed in such an atmosphere, the water will boil. The heat required for the boiling comes from the aggregate itself. Thus, the temperature of the aggregate is reduced. The process continues until the water is all gone or until the temperature of the aggregate is reduced to 40°F, in which case no more heat is available for further boiling. A typical plant consists of three 200 cu yd cooling silos for a 45-min cooling cycle. Aggregate containing 2 to 3% water is delivered to a feed trap at the top of each silo. The aggregate falls over a series of baffles in the silo. These serve as a rock ladder to prevent excessive breakage and a means for preventing packing of the aggregate so that the vacuum will be effective throughout the mass. When the silo is filled, it is sealed and a priming jet is activated for about 5 min. This creates a vacuum such that the pressure in the tank is equal to the vapor pressure of water at the temperature of the aggregate. For a temperature of 78°F the vapor pressure is equal to 1 in. of mercury, or about 97% of a complete vacuum. The vacuum is produced by passing steam through a venturi tube inside a pipe connected to the interior of the silo. Actually four horizontal pipes, connected to an exterior vertical riser, enter the silo at four levels. When the required priming vacuum has been attained the priming jet is turned off and the main steam Evactor jet in the venturi is turned on. The Evactor jet nozzle delivers high-velocity steam to a chamber in the venturi where it meets water vapor coming from the silo. The mixture moves rapidly under pressure, creating a more complete vacuum in the silo. The vacuum is controlled at 0.247 in. of mercury during evaporation. Steam boiler capacity of 10,000 lb per hr is required. At the end of the cycle atmospheric pressure is restored in the silo, and the aggregates pass through a discharge trap at the bottom of the silo for delivery to the batch plant.

The two principal disadvantages to the method are moisture control and aggregate breakage. Moisture content of the processed material can be maintained constant if the input moisture is carefully controlled. The aggregate-breakage problem can be eliminated if the aggregate is finish-screened after processing. However, the screening operation provides opportunity for undesirable heat gain. Plants have been built both with and without screening following processing. At least one plant has been built with the cooling silos erected in pairs over each storage bin in the batching plant. The material flows by gravity from the silo directly into the bin. While the material in one silo is being cooled, the material in the companion silo is being batched.

It is likely that vacuum cooling will increase in use in the future.

9   Cooling of Cement and Sand

As has been mentioned, it is not customary to cool any of the solid materials other than the coarse aggregate. Both cement and fine aggregate, however, have been cooled. None of the above methods is applicable to cement. The vacuum method is applicable to sand and, in fact, is used regularly on those projects where vacuum cooling is used for the coarse aggregate.

Another system which has been used for both cement and sand consists of moving the material by means of a hollow screw conveyor through which chilled water is circulated. The conveyor serves as both a means of transportation and a heat exchanger. It is recommended that the temperature of cement not be lowered to a value below 60°F because of the danger of condensation. A plant in which all solid ingredients of the concrete were precooled is shown schematically in Fig. 2.

10   Determination of Required Concrete Temperature

The purpose of temperature control is to prevent the concrete from undergoing a drop in temperature greater than it can withstand without cracking. While properties of concrete differ with age and constituent materials, on the average concrete can withstand a sudden drop in temperature of about 20°F and a gradual drop of 40 to 60°F. It is the gradual drop with which temperature control is concerned. The concrete will reach a final stable temperature approximately equal to the mean annual air temperature of its environment. The temperature which must be controlled is the peak temperature that will be reached during the period of rapid cement hydration. It is common to design small mass-concrete structures, such as gravity dams not exceeding 50 ft in height, for a temperature drop of 60°F; moderate structures, such as gravity dams between 50 and 150 ft in height, for a temperature drop of 50°F; and larger structures for a temperature drop of 40°F. Whenever possible laboratory tests should be made to determine the susceptibility of the particular combination of materials to temperature cracking. Details of such tests are given in Ref. 1.

The peak temperature depends on the placing temperature, the heat added through cement hydration, and the heat lost or gained externally. The latter includes heat flowing into adjacent concrete or foundation, through forms and escaping from the exposed surface. When the concrete is precooled, the heat transfer at the surface usually involves a gain of heat initially rather than a loss. The external

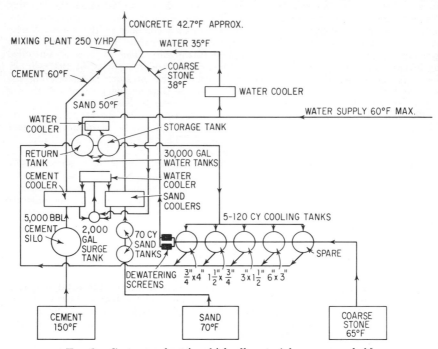

FIG. 2.   Concrete plant in which all materials are precooled.[5]

heat flow may also include the flow of heat into cooling pipes when these are incorporated into the structure.

The peak temperature is therefore controlled by limiting the temperature of the concrete as placed, limiting the lift thickness, controlling the time interval between lifts, removal of heat by cooling coils, or any combination of these methods.

Lift Thickness.   The optimum lift thickness is determined by several economic and technical factors. When precooled concrete is placed, there is an advantage to using as thick a lift as possible since the lift surface is a principal source of heat input. Therefore, the thicker the lift the smaller the average rise in temperature attributable to heat inflow from the surface. The expense of cleaning lift-joint surfaces also provides incentive for thick lifts with the attendant small number of joints to be cleaned. There are, however, disadvantages to the use of thick lifts. When concrete is not precooled, for example, the principal means for reducing the peak temperature is the flow of heat from the concrete through the top surface to the atmosphere. A thin lift is desirable under these conditions for the same reason that a thick lift is desirable when precooled concrete is placed. For uncooled concrete the maximum lift thickness is usually established at 5 ft.

With this thickness the temperature rise at the bottom of the lift is essentially adiabatic for the first 2 days, and the maximum temperature is reached in 3 days, after which slow cooling begins. Another factor which discourages thick lifts in gravity dams is the difficulty of placing concrete under the sloping downstream form. With a thickness greater than 5 ft it is necessary to move the concrete an excessive distance by vibrators. For higher lifts hinged forms are usually required, although they appear to have reached their practical limit at 7½ ft. On arch dams, where the sloping form is not a factor, 10-ft lifts have not been uncommon. Another factor that has a bearing on lift height in a limited number of structures is the spacing of cooling pipes. With a vertical spacing greater than 5 ft the temperature distribution within the concrete is likely to be undesirably nonuniform. Thus, in the commonest mass-concrete structure, the gravity dam, usual practice calls for 5-ft lifts in uncooled concrete or concrete containing cooling pipes and 7½-ft lifts in precooled concrete which is not to be postcooled.

Time Interval between Lifts. When concrete is precooled, there is no reason to require a delay between lifts. Since the concrete in place is initially absorbing heat from the atmosphere, it is advantageous to cover it with cool concrete as soon as possible. The time interval is usually determined by the capacity of the mixing plant in relation to the number of monoliths in which concrete is to be placed. With uncooled concrete, however, it is necessary to enforce a waiting

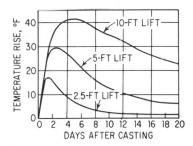

Fig. 3. Variations in temperature with time in the center of uncovered lifts of 10-, 5-, and 2.5-ft depths.[4]

period in order to permit losing an adequate amount of heat to the atmosphere. The commonest requirement is 5 days. This allows for 2 days' cooling after the maximum temperature has been reached. It is also advantageous to avoid an overly long exposure. The concrete may cool too rapidly and crack. It should be covered within 10 days if possible.

Calculation of Concrete-placing Temperature. The temperature of concrete as placed must be such that the net temperature rise resulting from hydration of the cement and exposure to the atmosphere will not produce a peak temperature which exceeds the final stable temperature by more than the allowable temperature drop. Methods for calculation are given in Refs. 2 and 3. Although heat flows in three directions, it is normally adequate in large mass structures built up in relatively thin layers to compute heat flow only in the vertical direction. This procedure greatly simplifies calculation and provides an accurate assessment of the temperature in the central portion of the lift, which normally is the critical region. The calculation may show that no cooling of the concrete is required, although in cool climates, where the final stable temperature is relatively low, some cooling is usually indicated for very massive structures. It is not practical to enforce a specification requiring the routine placing of concrete at temperatures below 45°F. If such low temperatures are indicated by calculation to be necessary, precooling should be augmented by pipe cooling in place.

While it is necessary to perform the calculations for each set of conditions, some typical results may be helpful to visualize the interrelation between lift thickness, time between lifts, cement content, and cement type. Figures 3, 4, and

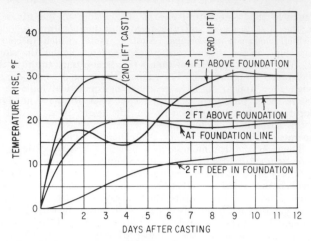

FIG. 4. Variation in temperature with time in concrete cast at a rate of 5 ft every 4 days.[4]

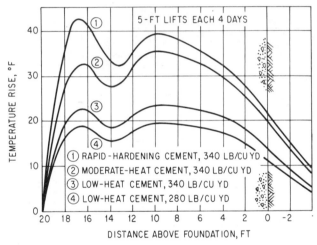

FIG. 5. Temperature distribution at end of fourteenth day showing effect of type and amount of cement in concrete.[4]

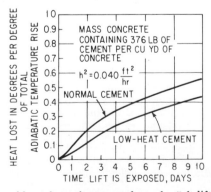

FIG. 6. Loss of heat from the top surface of a 5-ft lift of concrete.[3]

5 are from Ref. 4. Figure 3 demonstrates that thicker lifts attain higher peak temperatures. Figure 4 shows how temperature changes with time at several points in the bottom lift when 5-ft lifts are being added at the rate of one every 4 days. Figure 5 shows, for the same placing schedule, the temperature profile after 14 days for different cements and cement factors. It demonstrates that the peak temperature may be obtained in a given lift when that lift is the top exposed lift or after it is covered, depending on the heat-producing potential of the cement. Figure 6 from Ref. 3 shows the portion of potential adiabatic heat that may be expected to be lost from the surface of a 5-ft lift.

11   Postcooling of Concrete

Another method for the control of concrete temperature is the circulation of cold water through pipes in the concrete. It is used primarily in structures that are to be grouted; but it may also be used where precooling of materials is not capable in itself of producing the required control, in long blocks near the bottom of high gravity dams to make possible construction without longitudinal joints, in high exposed contraction-joint faces made necessary for diversion, and where construction conditions require that uncooled concrete be placed at a rate too fast to take advantage of natural cooling.

Advantages of Pipe Cooling. The chief advantage of pipe cooling is its flexibility. Any part of the structure may be cooled to any desired temperature. If unforeseen ambient conditions produce higher concrete temperatures than those anticipated, it is possible to provide the extra cooling required, whereas with precooling alone no such extra cooling is possible. In structures which are to be grouted it is possible to subcool the concrete below the final stable temperature and to achieve this condition within 2 months. After grouting, as the concrete warms to its final stable temperature, the entire mass is placed under a small compression so that any spurious zones of tensile stress, which might be a starting point of cracks, are eliminated or reduced in severity. During the cooling process proper control of the system makes it possible to maintain a more uniform temperature throughout the mass than is possible when precooling alone is used so that stresses are minimized throughout the cooling period.

Disadvantages of Pipe Cooling. The chief disadvantage is cost. Precooling is a more economical procedure. Therefore, pipe cooling is not normally used in ungrouted structures in which adequate control by precooling is possible. On occasion pipes have been damaged or plugged during embedment or have been clogged during operation by silty water. When pipe cooling is being counted on as the principal control measure, malfunction of this sort produces a serious and insoluble problem. Although proper control makes it possible to minimize stresses, too rapid cooling by pipes may actually cause cracking. Too rapid cooling during the first few days may also retard hydration to the extent that form-anchor failures may occur. During this early stage of hydration, unless a completely impractical pipe spacing is used, the rate of heat evolution from hydration greatly exceeds the rate of removal by pipe cooling. In most cases it is necessary to provide some precooling in addition to postcooling in order to hold the temperature drop to an acceptable value.

Design of a Pipe-cooling System. With the concrete peak temperature, the thermal diffusivity of the concrete, and the desired stable temperature known, the design factors to be determined are:
1. Pipe diameter
2. Pipe spacing
3. Length of each pipe coil
4. Water temperature
5. Rate of flow through the pipe

In selecting the pipe spacing, usually only the horizontal spacing may be varied. The vertical spacing is fixed by the lift thickness since it is impractical to install the pipes anywhere other than on the lift-joint surfaces. However, the incorporation of pipe cooling in the overall design of the structure is a factor in selecting the lift thickness. Where pipe cooling is to be employed, the lift thickness is usually restricted to 5 ft, although recently there has been some use of 7½-ft lifts. Thicker lifts produce the very nonuniformity of temperature which the system is designed to prevent. The selection of water temperature can be an important economic decision. If cool river water is available year-round, it will generally be cheaper to use it instead of refrigerated water provided the required concrete temperature can be obtained within the desired time. The use of river water will require more pipe and a greater pumping capacity but it eliminates the refrigeration plant.

The selection of pipe diameter is almost entirely a problem in hydraulics. The length of coil and volumetric flow rate of water are determined from thermal considerations as explained below. The pipe diameter is selected as that which will most economically pass the required flow of water through the known length of coil. Small diameters minimize the cost of pipe but they increase pumping costs because of high friction losses. Pipes with a 1-in. outside diameter are common for handling the comparatively small flow required for refrigerated water. Larger pipes or closer spacing, or both, are required when river water is used. As will be demonstrated below, pipe diameter has relatively little effect on cooling rate.

The interrelationship of the various design factors is shown in Figs. 7, 8, and 9, taken from Ref. 3. All the figures are based on the assumption that 1-in.-OD tubing will be used and that the vertical pipe spacing will be 5 ft. For any selected concrete diffusion constant, coil length, rate of water flow, and pipe spacing, Fig. 7 shows the fraction of the total initial heat that will remain in the concrete after any given number of days. Since the water is being warmed as it flows through the pipe, the cooling is not uniform along the pipe. Figure 8 shows how the cooling varies along the pipe, and Fig. 9 shows how the water temperature varies along the pipe.

Example. The example solved in Figs. 7, 8, and 9 assumes a concrete diffusivity of 0.6 sq ft per day. The diffusion constant (or diffusivity) is related to conductivity, specific heat, and unit weight as follows:

$$h^2 = \frac{k}{cp}$$

in which h^2 = diffusivity
k = thermal conductivity
c = specific heat
p = weight per unit volume of concrete

In the example

k = 0.825 Btu/(ft)(hr)(°F)
c = 0.22 Btu/(lb)(°F)
p = 150 pcf

then $\qquad h^2 = \dfrac{0.825}{0.22 \times 150} = 0.025$ sq ft/hr, or 0.60 sq ft/day

It is seen in Fig. 7 that if the pipe spacing is 4 ft, the length of the coil 1,600 ft, and the rate of flow 3 gpm, after 30 days the temperature difference between the concrete and the inflow water will be reduced to 0.48 of that existing initially. Figure 8, however, shows that at the outlet end of the pipe the temperature difference will have been reduced only to 0.68 of the original, while Fig. 9 shows that the water temperature will have risen 0.4 of the original temperature difference. If the initial concrete temperature was 80°F and the water inflow temperature 35°F, by the end of 30 days the average concrete temperature will have been reduced to 58°F, but in the vicinity of the outlet the temperature will be 66°F. After 30 days water will be flowing from the pipe at a temperature of 48.5°F. The figures assume that no significant amount of heat is being added to the concrete

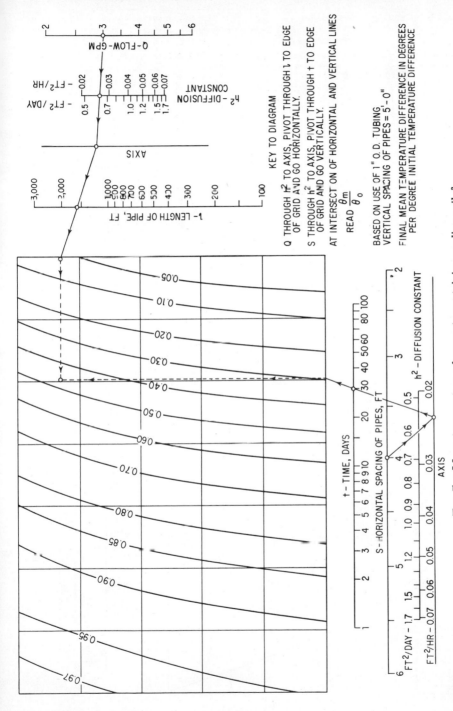

FIG. 7. Mean temperature of concrete containing cooling coils.[2]

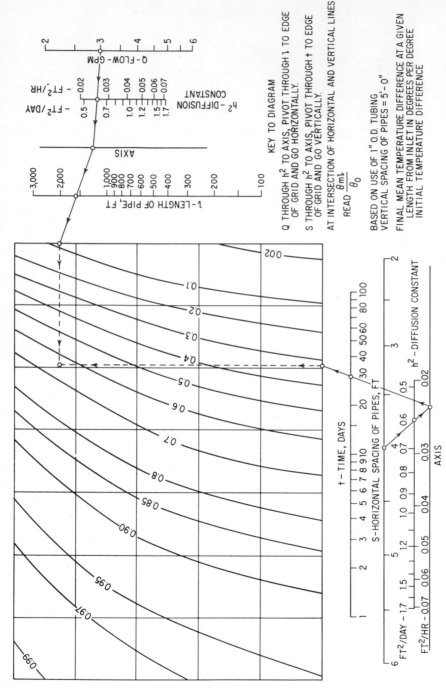

Fig. 8. Mean temperature of concrete at a given length from inlet of cooling coil.[2]

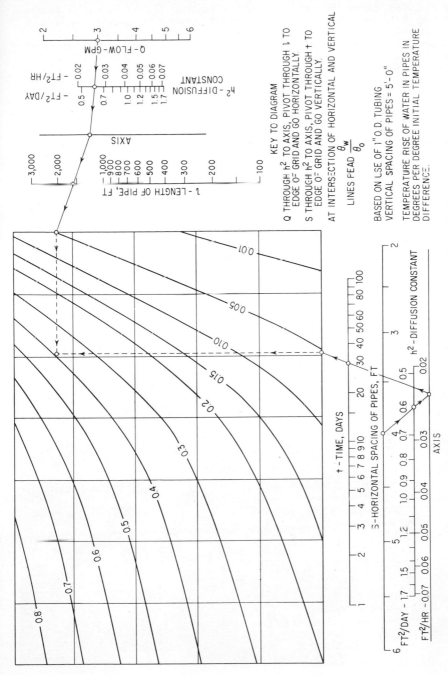

Fig. 9. Temperature rise of water in cooling coils.[2]

by hydration during the period under consideration. If the concrete is gaining heat or if the inflow water temperature is not constant, the curves may be used in a stepwise computation procedure described in Ref. 2.

Correction for Different Pipe Sizes. The effect of varying pipe diameter varies as the logarithm of the ratio of pipe spacing to pipe diameter. The simplest way to introduce pipe diameters other than 1 in. into Figs. 7, 8, and 9 is to compute a fictitious diffusion constant as follows:

$$\frac{h_f{}^2}{h^2} = \frac{\log{(b/a_1)}}{\log{(b/a_2)}}$$

where $h_f{}^2$ = fictitious diffusion constant
$\quad\quad h^2$ = actual diffusion constant
$\quad\quad b$ = pipe spacing
$\quad\quad a_1$ = 1 in.
$\quad\quad a_2$ = actual pipe diameter

If in the example in Fig. 7 the pipe diameter is 2 in.,

$$\frac{h_f{}^2}{h^2} = \frac{\log 48}{\log 24}$$

$$h_f{}^2 = 0.732 \text{ sq ft/day}$$

Then from the figure it follows that the cooling which requires 30 days with a 1-in. pipe may be accomplished in 27 days with a 2-in. pipe. Or it may be shown that in 30 days the concrete temperature on the average is reduced to 0.42 of the original difference between concrete and water temperature rather than to 0.48 of the difference.

Fig. 10. Construction view of Friant Dam, showing cooling coils laid out on lift joints of dam. Forms have not yet been raised on the nearest block. (*Bureau of Reclamation photograph.*)

The example demonstrates the disadvantage of long coils. One technique usually used to alleviate the problem is to provide interconnections between the inlet and outlet headers so that the water may flow in either direction. This procedure not only makes for more uniform cooling, but it also has the effect of flushing the pipe and thereby helps prevent clogging. Since long runs are undesirable from both thermal and hydraulic considerations the lengths of coils are usually limited, if possible, to about 800 ft, although lengths in excess of 1,300 ft have been used. Horizontal spacings have commonly been between 2 and 6 ft. With refrigerated systems the rate of flow seldom exceeds 4 gpm. At higher flows the increased cooling rate is not commensurate with the increased pumping and refrigeration costs. In systems using river water rates of flow as high as 15 gpm have

Fig. 11. Cooling pipe laid out on the rock foundation for a dam, showing the method of anchoring the pipe into the rock. (*Bureau of Reclamation photograph.*)

Fig. 12. Workman is installing wire ties, for holding cooling pipe, into the still-plastic fresh concrete of the lift. (*Bureau of Reclamation photograph.*)

been used. In refrigerated systems if the outflow water is colder than any other available source of water, which is normally the case, the water is merely recirculated through the refrigeration plant in closed circuit. When river water is used the outflow water is normally wasted. Usually, water is pumped through the pipes, although gravity-feed systems have been used.

In the most recent large installation at Glen Canyon Dam, water with a temperature of 45°F was circulated through the pipes during the period from 12 days to 1 month after casting. For the following month 35°F water was circulated to subcool the concrete in preparation for grouting the joints. A concrete temperature of 40°F was achieved in the bottom half of the dam and temperatures in the range of 40 to 50°F in the top half.

Installation and Testing. The pipe system should be laid out on the lift joint as soon as possible after the joint surface is made available so that there will be ample time for testing prior to placing the concrete over the pipes. The pipes should be placed under static pressure to check for leaks. They should also undergo a pumping test at design flow rate to check whether friction losses are greater than calculated. If so, there is a partial stoppage somewhere in the system which should be located and corrected before embedment. After completion of the cooling period it is customary to fill the pipes with grout.

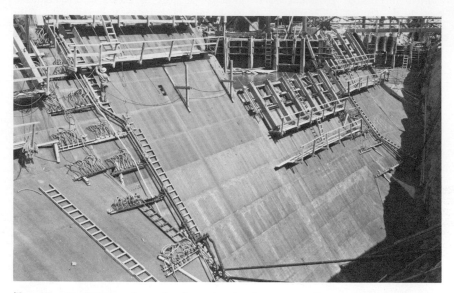

Fig. 13. Downstream face of a dam, showing hose connections to cooling-pipe inlet and outlet manifolds. (*Bureau of Reclamation photograph.*)

Figures 10, 11, 12, and 13 show various details of cooling-pipe installation for a large dam.

INSULATION FOR TEMPERATURE CONTROL

Although the primary purpose of temperature control in mass concrete structures is to limit the drop from maximum temperature to final stable temperature, there are times when short-range effects must be considered. This is true in regions with very cold winters or with rapidly fluctuating temperatures at any time of year and in structures where for reasons such as river diversion it is necessary to leave a high bulkhead face exposed for a long period of time. In such cases, insulation of both vertical and horizontal exposed surfaces can be very helpful. Insulation maintains the concrete temperature at a more uniform level than would otherwise be the case and thereby lessens the likelihood of cracks resulting from rapid surface cooling. It is particularly advantageous with precooled concrete. Both rigid and flexible insulation have been used. The flexible type, which is available in large rolls, is more convenient to handle in the field and is more readily made to conform to the surface of the concrete.

REFERENCES

1. Houk, Ivan E., Jr.; James A., Paxton; and Donald L. Houghton: Prediction of Thermal Stress and Strain Capacity by Tests on Small Beams, *J. ACI*, March, 1970, pp. 253–261.

2. Carlson, R. W.: A Simple Method for the Computation of Temperatures in Concrete Structures, *J. ACI,* November–December, 1937, pp. 89–102.
3. Rawhouser, Clarence: Cracking and Temperature Control of Mass Concrete, *J. ACI,* February, 1945, pp. 305–346.
4. Carlson, R. W.: Temperature and Stresses in Mass Concrete, *J. ACI,* March–April, 1938, pp. 497–515.
5. How Low-temperature Concrete Is Produced, *Eng. News-Record,* Dec. 28, 1950.

Chapter 47

GROUTING OF CONCRETE

ROBERT E. PHILLEO

Grout is normally considered to be a mixture of portland cement and water or a mixture of portland cement, sand, and water to which chemical admixtures may or may not be added. In recent years the term has also been applied to chemical formulations, entirely unrelated to portland cement, which are used for the same purposes as portland-cement grout.

The principal uses of grout are the setting of machine bases, sealing fissures in foundations under hydraulic structures, filling joints in concrete structures, filling cavities behind tunnel linings, and filling the voids between preplaced particles of coarse aggregate in the "prepacked" method of construction. This chapter is concerned only with the grouting of joints and tunnels. Since the use of chemical grouts is confined primarily to foundation grouting or soil stabilization, only cement grouts are discussed here.

GROUTING OF MASS CONCRETE

There are two principal types of joints in mass concrete which must be grouted: (1) vertical radial contraction joints in arch dams and (2) longitudinal vertical joints in high gravity and arch dams. In addition some organizations grout contraction joints in gravity dams, although this is by no means universal practice. Gravity dams are divided into individual monoliths by contraction joints perpendicular to the axis. Each monolith is designed to be stable. Therefore, there is no load transfer across the contraction joints. The only reason for grouting these joints is to prevent leakage. Since it is customary to place water stops in these joints, many authorities consider grouting unnecessary. Some organizations also used keyed joints to increase the path length for water and to increase the probability of the joint's becoming plugged with silt if a leak should occur. When longitudinal joints (joints parallel to the axis) are included in a gravity dam, however, the joints cannot be left open. Shear stresses must be transferred across the joint. The separation of the monolith into two or more individual monoliths by longitudinal joints produces a stress distribution entirely different from that assumed in design and might seriously affect the safety of the structure. Every effort is made to avoid longitudinal joints in gravity dams. As the height of the dam is increased, the length of the lower portion of each monolith is increased and the problem of temperature cracking brought about by the interaction of thermal volume changes in the concrete and foundation restraint increases. Longitudinal joints are introduced to prevent the accidental separation into individual monoliths by cracking and to produce the separation in a controlled location where it can be grouted. In the past longitudinal joints were common in gravity dams. In recent years precooling of materials and postcooling of the concrete have so well controlled the temperature that monoliths over 700 ft high have been built successfully without longitudinal joints.

In arch dams radial contraction joints will close and transmit compression when water pressure is present behind the dam. Although little or no shear stress needs to be transferred, the joints are grouted to ensure a uniform transmission of compressive stress across the joint. Nonuniform stress produces effects not contemplated in design. In very high arch dams longitudinal joints are introduced for the same reason that they are used in high gravity dams.

Grouting is always postponed until the concrete has cooled. Where artificial cooling is employed, the grouting can be done quite early, in some cases within 2 months after placing the concrete. Otherwise, a long delay is necessary. Waiting until the completion of cooling achieves the double objective of opening the joint to its maximum width to facilitate grouting and ensuring that no further contraction will occur following grouting. Such contraction could reopen the joint. Where pipe cooling of the concrete is employed, it is common to cool the concrete to a temperature below its final stable temperature so that, following grouting, the attempt of the concrete to expand on warming will place the concrete in compression and positively prevent reopening of the joint.

1 Design of Layout for Grouting

The objective in planning a grouting program is to achieve complete filling of the joints without damaging the structure. The latter is a definite possibility

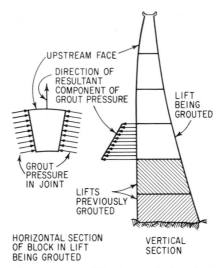

FIG. 1. Forces acting on a monolith of an arch dam during joint grouting.[1]

which must be guarded against. Whereas in foundation grouting it is common to use pressures of several hundred pounds per square inch to fill small remote voids in the underlying rock, such pressure cannot be tolerated within the joints of a structure. Excessive pressure in a joint may force the joint open to an undesirable extent so that it closes parts of adjacent joints or even moves adjacent monoliths out of line. In extreme cases it may cause shear failures along horizontal construction joints and thus increase rather than reduce cracking. Grout pressure should be restricted to about 50 psi. The area of joint to be grouted at one time should be restricted to that area which can be covered with a pressure of 50 psi. As a result of this requirement it has become common practice to grout a height of 50 ft at one time. Means should be provided to ensure that no harm is being done to the structures. These include pressure measurements, overall displacement measurements by theodolite, and joint-opening measurements by dial gages near the top of each grouting lift.

Each area to be grouted at one time should be isolated by metal sealing strips, commonly called grout stops, around the periphery of the joint. The material normally used is 24-oz soft-tempered copper, although annealed No. 20 gage stainless

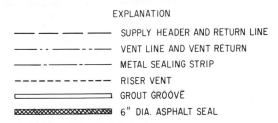

EXPLANATION

— — — — SUPPLY HEADER AND RETURN LINE

———— · · ———— · · — VENT LINE AND VENT RETURN

———— · ———— · —— METAL SEALING STRIP

— — — — — — — — — RISER VENT

⊏━━━━━━━━━━━⊐ GROUT GROOVE

▨▨▨▨▨▨▨▨▨▨ 6" DIA. ASPHALT SEAL

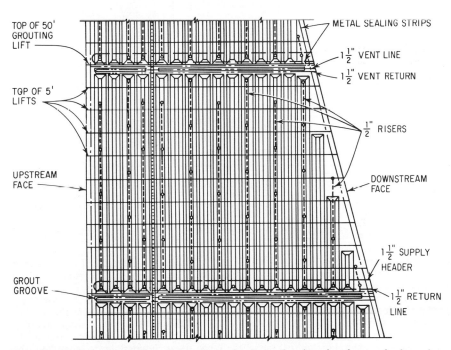

Fig. 2. Radial contraction joint in arch dam showing header and riser-pipe layout.[1]

steel has been used. The strips should include an accordion fold to accommodate changes in joint opening after installation. Connections should be carefully brazed or welded since the grout stops must resist pressure during grouting.

Experience has indicated that it is not advisable to attempt to grout more than 100,000 sq ft of joint area a day. This limitation may influence the grouting procedure. In small arch dams and small gravity dams in which contraction joints are to be grouted, it is desirable to grout all the joints from abutment to abutment in one operation in order to avoid the problem of excessive bending and joint opening. When the size of the dam becomes too large, the normal procedure is to work from each abutment toward the center, although sometimes construction contingencies require some deviation from this plan. Even when it is possible to grout an arch dam from abutment to abutment in one operation, the grout pressure may produce a problem, as illustrated in Fig. 1. The resultant of the pressure on the two sides of the trapezoidal blocks is a force tending to bend the monolith as a cantilever in the upstream direction. This condition is most effectively counteracted by partially filling the reservoir prior to grouting.

Joints may be grouted from a gallery or from the exterior. In large gravity dams it is common to include galleries at 50-ft intervals of height for this purpose. Thinner arch dams are usually grouted from temporary catwalks on the downstream face. Figure 2 shows the piping arrangement for an arch dam and Fig. 3 for a large gravity dam with a gallery, from which contraction joints are to be grouted. The grout is introduced at the bottom of the lift through a supply header, which in turn feeds grout to the riser pipes. Along the risers grout outlets are spaced

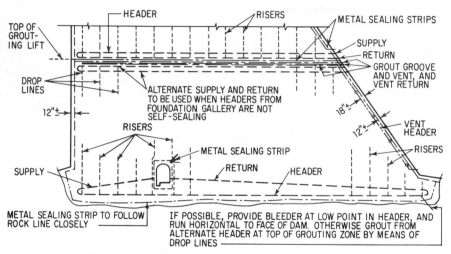

Fig. 3. Installation for grouting a contraction joint in a gravity dam.[2]

uniformly over the face of the joint. The outlets commonly consist of two modified electrical conduit boxes connected by tees to the riser pipes. The supply header makes a complete circuit returning to the gallery or downstream face. During grouting if the supply header should become plugged, grouting can be continued through the return header. At the top of each grouting lift a horizontal groove acts as a vent to collect air, water, and thin grout that must escape as they are displaced by thick grout. The vent is equipped with a pipeline similar to the header at the bottom of the lift. When it is impossible to complete the grouting through the bottom supply pipe, grout may be introduced through the vent pipe. The end of the vent pipe must be equipped with a stopcock and pressure gage so that during the final stages of grouting the joint can be placed under pressure and the pressure may be monitored.

All the piping shown in the figures is embedded during construction. In addition, temporary supply lines must be provided. Except on small structures which can be grouted from abutment to abutment in a single operation, two such temporary systems must be provided. One is for grout, and the other is for water. When the joint adjacent to a joint being grouted is not to be grouted at the same time, it is maintained under water pressure during the grouting operation. This procedure minimizes the pressure difference on the two sides of the block and thus prevents excessive deflection and shear failure. The joint above the one being grouted is also kept filled with water during the grouting operation. The water helps to keep the joint open, minimizes the loss of grout if a leak should exist in the upper grout stop, and provides a means for checking for such leaks during grouting.

2 Installation of Embedded Pipes

The piping is installed just inside the form of the leading monolith prior to placing the concrete. Pipes and outlets must be securely anchored to prevent damage during placing of the concrete. The supply header should be mounted on metal supports. When electrical conduit boxes are used for outlets, the covers are removed and the open boxes attached directly to the forms so that after stripping the covers may be replaced. All plumbing connections should be inspected and tightened, and just before concrete placement the system should be pressure-tested. The metal sealing strips should be carefully inspected and joints or welds tested by the soap-bubble method. During placing of concrete special attention should be given to the pipes and seals to ensure that they are not damaged or displaced. The same care is required during placing of the adjacent following lift since half the grout stop is contained in that lift. Most troubles that occur during grouting are attributable to damage that took place during the embedment.

3 Grouting Operation

There is a considerable amount of preparation to be done the day before the actual grouting. All pipefitting must be in place both for grout and for filling adjacent joints with water. A communications system must be installed linking mixer, pump, vent pipes, and leak inspectors. Dial gages for measuring joint openings and pressure gages for measuring vent pressure must be in place. Both the pipe system and the seal system must be checked. The pipes are checked and cleaned by running water through them until the effluent from the vent pipe appears clean. Then, the vent stopcock is closed, and the joint is placed under pressure. If the pressure cannot be maintained, there is a leak in the system which must be located and caulked. Wetting the joints in advance of grouting also lubricates the surfaces and causes the grout to flow more readily.

4 Composition of Grout

While both neat-cement grout and sanded grouts have been used in grouting joints, the practice now is to use only neat-cement grout. The use of sand increases the minimum width of joint that may be grouted and introduces the problem of keeping the coarser material in suspension prior to pumping. While it is desirable to have joint openings of at least $\frac{1}{16}$ in. at the time of grouting, neat-cement grout has successfully penetrated local areas where the opening was only 0.02 in. The cement should be free from lumps or coarse particles. Some organizations require that 100% of the cement shall pass a No. 100 screen and 98% shall pass a No. 200 screen. Most cement companies can supply a product meeting this specification so that no processing of cement is required on the job. It is important, however, that the cement be properly stored at the site so that contact with moisture does not produce lumps.

Type II portland cement is normally used for joint grouting. Conditions requiring low-heat or oil-well grouting cements usually are not present in joint grouting. The water-cement ratio varies as the grouting operation proceeds. During the course of the operation the water-cement ratio by weight may be reduced from an initial value of 1.5 to a final value 0.5.

5 Mixing and Pumping Equipment

The same type of equipment as is used in foundation grouting is suitable for joint grouting. To grout 100,000 sq ft in an 8-hr shift a mixer and agitator tank with a capacity of 20 cu ft and an air-driven $10 \times 3 \times 10$-in. grout pump are adequate. The mixer size must be large enough in relation to the pump capacity to maintain a steady flow of grout, as an interrupted flow may produce areas of set grout which later impede the flow. The agitator tank must provide sufficient motion to prevent settling of the cement. The pump must be capable of close control of pressure and a continuous uninterrupted flow. A complete set of standby equipment should be available in the event of a breakdown.

6   Pumping Procedure

A few batches of very thin grout (water-cement ratio of 1.5 by weight) should be placed in each joint to lubricate the lower portion of the joint. This is followed by a thicker grout with a water-cement ratio of about 0.7. The grout supply should be rotated among joints so that the level is maintained approximately level throughout the grout zone and the grout has an opportunity to settle in each joint. This can be accomplished by placing a measured amount in each increment. The schedule must be adjusted so that the time lapses are not so great that the pipes become plugged. When the joints are three-quarters filled the water-cement ratio may be reduced to 0.5 or 0.55. The material coming out of the vent pipe should be observed. When grout nearly as thick as that being pumped is flowing from the vent, it may be assumed that the joint is filled. At this time the stopcock at the end of the vent line should be closed. Pumping should be continued until the pressure at the vent reaches the allowable maximum figure or until the opening indicated by the dial gage at the top of the joint reaches the maximum permitted value. Normally for a 50-ft-high grouting lift the increase in joint opening is restricted to 0.02 in. as measured by a dial gage having a least reading of 0.0001 in. During this period leaks in the top grout stop may be detected by drawing off water from the supply header in the lift above that being grouted. If any grout leaks into a joint which has not yet been grouted, water must be circulated through the joint to prevent setting of the grout. During the early stages of the period that the grout is maintained under pressure it usually is necessary to continue intermittent pumping as the pressure drops because of the forcing of water into the concrete. When the pressure remains essentially constant for 30 min, grouting can be considered completed.

Following the grouting, immediate cleanup is important before the remaining grout sets. Not only does the equipment contain grout, but excess grout is always spilled in the galleries or on the face of the structure. If it is not removed promptly, cleanup becomes difficult and unsightly discolorations result. Ordinarily 2 days are required to dismantle one setup and prepare for the next grouting operation.

GROUTING OF TUNNELS

The grouting of tunnel linings differs from the grouting of joints in mass concrete principally in the sizes of the voids to be filled. Whereas the joints in structures have a thickness of the order of $\frac{1}{16}$ in. the space between tunnel lining and the adjacent rock may be several inches. The existence of such a cavity can be a serious problem in a tunnel designed to carry water since it creates an undesirable water passage outside the lining. Because of the volumes to be grouted it is customary to use sanded grouts rather than neat-cement grouts. Sanded grouts are cheaper and have less shrinkage in place. The larger required openings for sanded grouts are not a disadvantage. Where very large openings are involved, the primary grouting with sanded grout may be followed by high-pressure grouting with neat-cement grout to fill the voids left by contraction of the sanded grout. Expansive admixtures, such as aluminum powder, may be added to the sanded grout to minimize the contraction problem.

7 Preparation for Grouting

In long tunnels some means must be found to separate the tunnel into zones for grouting. This can be done by hand-packing a mortar dam at intervals around the periphery of the lining when the forms are stripped. The spacing must be such that the volume between adjacent dams may be grouted in a continuous operation. The period of continuous grouting is usually somewhat longer for tunnel grouting than for joint grouting. By using retarding admixtures in the grout, this period is sometimes as long as 48 hr. Both grout pipes and vent pipes must be in place prior to grouting. Sometimes holes for the pipes are drilled. In other cases they are cast in the concrete. The vent pipes are placed at or near the crown and are extended clear through the void to the overlying rock or soil

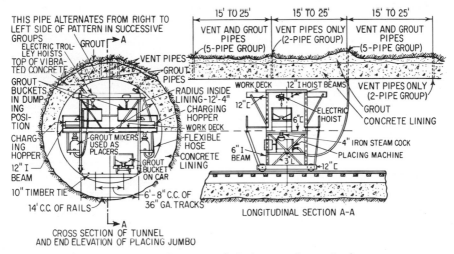

FIG. 4. Jumbo-mounted rig for tunnel grouting.[3]

(Fig. 4). This permits the escape of air but prevents loss of grout until the final stages of grouting. The grout pipes are removed some distance from the crown and extend through the lining and partially through the void. Rows of holes are spaced about 20 ft apart.

8 Grouting Operation

The procedure and equipment for tunnel grouting are less standardized than for joint grouting because conditions are much more variable. Figure 4 illustrates a jumbo-mounted grouting apparatus that was used successfully on a large tunnel project. Grout was mixed outside the tunnel and hauled to the jumbo by rail.

A typical grout mixture contains equal quantities of cement and sand and a water-cement ratio of 0.45 by weight.

It is normal to use a well-graded sand, having a fineness modulus within the range of 1.50 to 2.00, with not more than 5% retained on the No. 16 sieve. For filling large cavities a coarser sand is sometimes used. However, the coarseness of the sand is limited by the difficulty of maintaining it in suspension and of pumping the grout rather than by the size of the cavity.

Pressures of the order of 50 psi are used. Grouting begins at the end of the zone and proceeds in one direction. Initially, connection is made to all the grout pipes at the first location. Pumping is continued to refusal at the specified pressure. If grout starts to flow through the vent pipe, it is plugged. No grout may appear in the vent until grout is pumped at the second location. The operation is continued throughout the zone. Each vent pipe is plugged when it starts to pass grout. If any vent pipe should fail to vent, a grout delivery hose is placed in the vent, and pumping continues to refusal.

REFERENCES

1. Simonds, A. Warren: Arch Dams: Theory; Methods and Details of Joint Grouting, *Proc. ASCE, J. Power Div.*, vol. 82, p. 991–1, 1956.
2. Elston, Judson P.: Grouting Contraction Joints in Dams Controls Cracking and Seepage, *Civil Eng.*, vol. 19, pp. 624–628, 1949.
3. Jacobs, J. D.: Grouting the Tunnels, *Eng. News-Record,* vol. 117, pp. 274–277, Aug. 20, 1936.

Section 14

REPAIR OF CONCRETE

Chapter 48

REPAIR OF CONCRETE

WILLIAM A. CORDON

Concrete is one of the most versatile of basic building materials. It is inexpensive, durable, strong, and will take the shape of the form in which it is placed. A large percentage of concrete members and shapes are manufactured at the site of the finished structure and at the final location of the member in the structure. Poor workmanship and construction procedures cause imperfections that require repair. Subsequent deterioration or damage caused by forces of nature, or normal use, must also be repaired or replaced as part of the maintenance of the structure.[1]

One of the prime reasons small repairs are made is to make the structure more presentable. Repairs will be offensive unless they blend with surrounding untouched surfaces. This requires particular attention to the uniformity of color and texture. In addition, a good serviceable repair must have the following qualifications:

1. It must be thoroughly and permanently bonded to adjacent concrete.
2. It must be sufficiently impermeable to prevent moisture from reaching underlying older concrete.
3. It must, after drying, be free of shrinkage cracks.
4. It must be resistant to freezing where this is a factor in weathering.

On new work the greatest problem is probably with small repairs such as boltholes to fill, an offset to level, a rock pocket to repair, or a rough spot to smooth. These can be treated with a minimum of effort by experienced, well-trained workmen. Too often poor workmanship in making these minor repairs makes the structure unsightly.

Problems of concrete repair in old structures are more serious and generally have to do with deteriorated concrete.[2] At times deteriorated concrete must be carefully isolated and removed, which may result in a major reconstruction project. Some dams, for example, have required complete refacing. The downstream face of Arrowrock Dam in Idaho was refaced with shotcrete, and a new section was placed on the upstream face of Barker Dam in Colorado by prepacked methods.

This chapter includes a discussion of recommended materials and methods for making all types of repairs of concrete.

MATERIALS USED IN CONCRETE REPAIR

Because of the wide variety of concrete repairs which must be made, several different materials may be used.

1 Portland Cement

Repair programs often occur during or following original construction. Common types of portland cement may be used, and although not a specific requirement, many times it is advisable to use the same type of cement as used in the original structure. Type II or Type V portland cement should be used for greater resistance to sulfate attack. Type III high-early-strength cement might be used where there is an urgency for the repair work to have rapid strength gain. Shrinkage-compensating cement and regulated-set cement, which are currently under development, may be used in special situations. These are discussed in Chap. 1. A mixture of portland cement and aluminous (high-alumina) cement sets in a few minutes. Tests of any such mixture should be made under ambient conditions, as the setting time can be extremely short. There are on the market several proprietary quick-setting cements. A typical compound consists of a concentrated polymerized solution containing 60% solids of polyvinyl acetate, wetting agents, plasticizers, and stabilizers. It is used to replace part of the mix water in preparing portland cement grout, mortar, and similar materials. The user is cautioned to investigate any of these materials before using them, especially if the repair is to be in a moist environment. Some proprietary compounds contain calcium sulfate (plaster of paris) and will not be stable in a repair that is exposed to moisture. Because of the very rapid setting and hardening properties of these materials, they should be used in small batches only. Material that has hardened or stiffened should not be retempered to make it plastic again.

Air-entraining agents and other admixtures can be properly used where their use is beneficial.

2 Aggregates

Requirements for mineral aggregates for concrete repair are the same as those required for concrete and concrete mortars used in construction. Desirable properties of mineral aggregates and their use with portland cement are discussed in other sections of this handbook and will not be repeated here.

3 Epoxy Resins

The Highway Research Board Cooperative Report 1[3] describes epoxy resin as follows:

Derived from the Latin word "epi," meaning "on the outside of," and "oxygen," epoxy resins are a family of chemicals containing the epoxide ring. These materials are relatively new and show promise in many different applications to concrete repair.

Commercial epoxy resins are made by reacting, in the presence of a caustic, bisphenol-A and epichlorohydrin, compounds derived from natural gas or coking by-products. The chemistry of the epoxy compounds is complex and beyond the scope of this handbook, but when activated with a curing agent, the epoxy resin polymerizes to form a solid material with special adhesive qualities. The curing or polymerizing process is caused by the use of a catalyst or a reactive hardener. In the reaction, no by-products are formed and no large volume change is experienced.

Epoxy resins have coefficients of expansion that are three to five times greater than that of portland-cement concrete. To prevent shearing failure in the concrete to which they are bonded when temperature changes occur, some epoxy formulations include tar additives that impart stress-releasing qualities to the material. Epoxy resins mixed with the usual quantities of sand used in portland-cement mortar mixes will bring the coefficient of expansion of the epoxy mortar closer to that of concrete mortar, so that the difference in coefficients will not be a serious problem. In some cases thin epoxy overlays have failed because this difference was not compensated for.

Generally, epoxy-resin characteristics include rapid hardening at normal temperatures, a high degree of adhesion to most surfaces, durability and resistance to cracking, and chemical resistance to most acids, alkalies, and solvents. Epoxy resins, being both exothermic and thermosetting, have a relatively short pot life. The heat released during the reaction of the epoxy resin and curing agent acts to increase the reaction; thus the longer the epoxy moisture remains in the batch container, where the heat cannot be released, the faster the curing action. Furthermore, the larger the batch mixture, the greater is the release of heat and the faster the reaction.

Several formulations of epoxy binders are available. Most commonly used materials are either epoxy-resin compounds using coal-tar pitch and amines or those with polysulfide polymers and amines. Mortars made with epoxy-resin binders consist of binder to aggregate ratio 1:2 to 1:5, depending upon the type of mortar required. Mixes containing larger amounts of aggregate are less costly, but they are also much drier and may result in poor adhesion to the underlying concrete. Experience in the field will indicate the best proportions desirable for each particular job.[4] For example, in an open patch where the mortar can be tamped into place a much drier mix (more sand) can be used than in filling a crack where the epoxy must flow into a small confined area.

Curing Agents. The basic types of curing agents used in epoxy resin compounds include the amines, acid anhydrides, and boron trifluorides. The amine curing agents are the most widely used. The resultant mixtures have low viscosity (before curing), are rapid-curing at room temperatures, and are relatively tough and readily mixable. Being nonhygroscopic, they are useful in fresh concrete. The acid anhydrides require high temperatures (120 to 200°C) for proper curing. The acid anhydride-cured epoxies have high resistance to heat and chemicals but are hygroscopic. The boron trifluoride compound is a highly corrosive gas and impractical for use under normal handling conditions.

Flexibilizers. To overcome the inherent brittle qualities of cured epoxy resins, flexibilizing modifiers are used. These modifiers react chemically with the epoxides to form more elastic molecules. Modifiers commonly used include the liquid polysulfides, the polyamides, and the aliphatic amines. The liquid polysulfides are colorless, viscous materials that take part in the chemical reaction. When they are used, a compatible curing agent, such as the tertiary amine, dimethylaminomethyl phenol, is used. The polyamides are viscous brown liquids that also act as a curing agent for the epoxy. The aliphatic amines are low-viscosity hardeners that react readily at room temperatures.

Thickeners and Fillers. In addition to the curing agents and modifiers that react chemically with the epoxy resins, other materials can be added that react mechanically to change the properties of the resulting product.

Thixotropic agents consist of finely divided materials that form a cell-like structure and hold the liquid resin in a relatively immobile position until curing and hardening have occurred. Materials used for this purpose include chopped glass fiber, expanded silica bentonite, and mica platelets.

Inert solids, such as sand, aluminum oxide, calcium carbonate, and other aggregates and metals, are used to fill the mixture and displace, and thus reduce, the amount of resin required. Fillers may be used to alter the strength characteristics of the resin compound.

4 Latex Mortars and Latex Bonding Agents

A number of synthetic-rubber latexes have been developed for use with portland cements. Such latexes are usually produced in the form of water emulsions. Thus, when added to portland-cement mixes, the water in the emulsion reacts with the cement to cause hydration, while the particles of latex influence the strength and characteristics of the resulting mortar. Compressive strengths of latex mortars are decreased, but flexural and tensile strengths increase generally over those of plain portland-cement mortars. Latex mortars also increase initial adhesive qualities on most surfaces.

Among those latex formulations currently available are the styrene-butadiene copolymer particles in a water dispersion, the neoprene elastomer particles in a cationic water dispersion, and the Saran-type particles in a water dispersion. Polyvinyl acetate compounds have been used under several different trade names as bonding agents and as admixtures for concrete mortars. These materials are odorless latex liquids with a viscosity of heavy cream.

5   Polyester Resins

Polyester two-component resinous systems have several characteristics similar to those of epoxy-resin formulations. Polyester resins are rapid-curing at normal temperatures and have high corrosion resistance, high compressive strength, and good bond to properly prepared concrete surfaces. The polyesters differ from most epoxy resins in that they do not adhere well to nonporous surfaces, such as metal or glass, and are not compatible with wet or damp surfaces, such as freshly placed concrete.

The polyester component is somewhat unstable and cannot be stored very long before use. Polyester compounds employ a peroxide as the catalyst.

6   Bituminous Materials

Asphalts and tars have been used to varying degrees for repair work on deteriorated concrete surfaces. Where asphalt cements or liquid asphalts and tars are used, they generally conform to specifications for new construction as adopted by responsible agencies.

Coal-tar-pitch emulsions have been used in a number of concrete-repair programs as protective coatings on repaired decks and other horizontal surfaces. Most coal-pitch emulsions employed in this work meet Interim Federal Specification R-P-00355 B. The material consists of a coal-tar pitch dispersed in water by means of a combination of mineral colloids to form a suspension which, when dried, will not be redispersed by water. The material is usually furnished in a creamy consistency suitable for undiluted application as a heavy film using a squeegee brush or mechanical spray.

Work is being done by several agencies in the development of asphaltic-concrete mixes employing additives to improved impermeability, density, stability, and durability. Asbestos fibers have been used in a number of experimental asphaltic concrete mixes and are reported to permit an increased asphaltic-cement content without loss of stability or danger of bleeding of the mix. This, in turn, provides for increased density and reduced permeability of the asphaltic concrete. The asbestos fibers also appear to contribute some modest reinforcement to the asphaltic-concrete mix which may have structural benefits as well.[3]

Hydrated lime has been used as an additive in hot mixes with particular benefit where marginally acceptable aggregates have been employed. The lime is reported to react chemically with the free silica in the aggregate to form a gelatinous mass that seals cracks and openings in the aggregate and prevents the aggregate from disintegrating. The highly alkaline lime also neutralizes the acid properties of the asphalt and permits better coating of the aggregate.

Rubber and synthetic latex materials have also been used as additives in bituminous-concrete mixes. Other additives employed in asphaltic-concrete mixes include antistripping materials and materials employed as emulsifiers.

Some use has been made in concrete-repair programs of the recently developed, colorless binders which are pigmented and used with aggregates of the same natural color to provide a full-depth-color paving material. These binders are polymeric hydrocarbons, derived from petroleum products. The paving concretes, which consist of binder, plasticizer, pigment, and aggregate, are handled much like asphalt concrete.

7   Surface-penetrating Sealants

To protect newly repaired surface areas from moisture and chemicals, various penetrating sealants are being used.[3] Linseed oil has been used for treating concrete surfaces, generally as boiled linseed oil and petroleum spirits mixture. A number of other penetrating sealants and coatings of a proprietary nature are available, including chlorinated rubber epoxy, silicones, and various protective paints.

A number of state highway departments make use of linseed oil in treating both old and new pavements and bridge decks that are exposed to freezing and thawing and heavy applications of deicing salts. Usual mixture consists of half boiled linseed oil and half mineral spirits by volume applied to old concrete, or new concrete that is at least 2 weeks old. The first coat is applied by either hand spray or power-operated spray at a coverage of 0.025 gal per sq yd. After the first coat has dried, the second is sprayed on at 0.015 gal per sq yd.

PREPARATION OF CONCRETE IMPERFECTIONS FOR REPAIR

Procedures for repair of imperfections of new concrete and reconstruction of disintegrated portions of structures cannot be carelessly performed without detriment to the serviceability of the structure. If not properly performed, the repair will later become loose and drummy, will crack at the edges, will not be watertight, and repair operations will eventually be repeated. Repair of imperfections should be tight, inconspicuous, and of quality and durability comparable with the other portions of the structure.

Most repair procedures involve essentially manual operations. Constant vigilance must be exercised to assure maintenance of the necessary standards of workmanship.

Repairs during construction will develop the best bond and thus have the best chance of being as durable and as permanent as the original work. These repairs should be made immediately after stripping of forms before the concrete is completely hardened. It is recommended that repairs be made within 24 hr after the forms have been removed. Repair of old concrete will differ from the repair of new concrete only in precautions necessary to ensure maximum bond of the repaired work with the original structure.[5]

8 Removal of Damaged or Inferior Concrete

In each case a thorough exploration of imperfections should be made before repairs are started. Effective repair of deteriorated portions of structures cannot

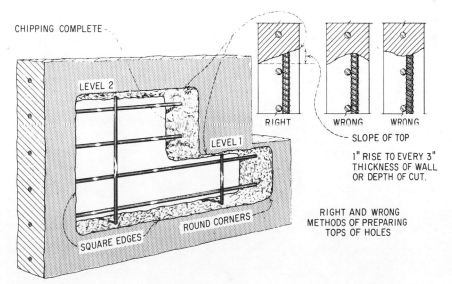

Fig. 1. Proper removal of defective concrete prior to repair. (*U.S. Bureau of Reclamation.*)

be assured unless there is complete removal of all affected concrete. Tapping the surface with a hammer or some such device will locate a drummy, unsound pavement.

All concrete of questionable quality should be removed (Fig. 1). It is far better to remove too much concrete than too little, because affected concrete generally continues to disintegrate. The full nature of imperfections often cannot be determined until defective material has been removed. Air-driven chipping hammers are most satisfactory for this work, although good work can be done by hand methods. The sawtooth bit (Fig. 2) is desirable for cutting and undercutting

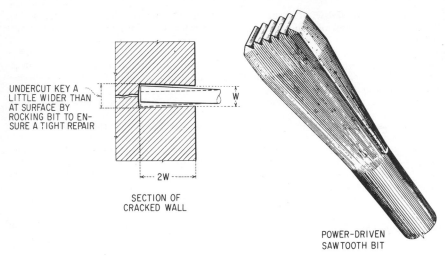

UNDERCUT KEY A LITTLE WIDER THAN AT SURFACE BY ROCKING BIT TO ENSURE A TIGHT REPAIR

W

2W

SECTION OF CRACKED WALL

POWER-DRIVEN SAWTOOTH BIT

FIG. 2. Sawtooth bit is an effective tool to remove deteriorated concrete. (*U.S. Bureau of Reclamation.*)

FIG. 3. Preparation of a concrete slab for repair. (*Master Builders Company.*)

slots in deep holes. A sawed edge is superior in every way to a chipped edge and sawing is generally less costly than chipping, especially if the repair is very extensive. A diamond saw is particularly effective in repair of concrete bridge decks. A saw cut provides a clean vertical face against which patching material can be placed for maximum contact and bond. A slight keying action can also be obtained by slightly tilting the saw blade to undercut the edge of the repair at a slight angle (Fig. 3).

For removal of less than ½ in. of concrete from bridge decks, surface scarification provides good bond. Routing or scarifying tools such as impact hammers and grinding tools have been employed. For greater depth removal is generally accomplished with air hammers. Where minimum vibration and impact of structural concrete are essential the entire area of concrete to be removed can be crisscrossed with a series of saw cuts of desired depth spaced at intervals of 1 in. or less. The saw-cut sections can then be dislodged with very lightweight chipping hammers.

Special care should be taken to prevent damage to surrounding areas during removal of unsound concrete. The bond of remaining sound concrete with reinforcing steel should not be impaired. Precautions should be taken not to permit air hammers and chisels to vibrate on reinforcing steel.

9 Cleaning and Curing

After removal of unsound concrete, patch areas should be thoroughly cleaned, and free from all dust or moisture deposits before the repair material is placed. When repairs are to be made with portland-cement concrete or mortar, surfaces of the trimmed area should be kept continuously wet for several hours, preferably overnight, prior to placing the new concrete. Saturation of the old concrete will help in proper curing of the repaired area, whether large or small. Proper curing is particularly important in repair of old concrete. A reliable means of obtaining saturation of holes in sloping or vertical surfaces is to pack the holes with wet burlap and to keep the burlap wet by occasional sprinkling.

New surfaces to which new concrete is to be bonded can be effectively cleaned by wet sandblasting, followed by washing with an air-water jet and removal of excess water with an air jet. Some organizations prefer water jet for final cleaning instead of air.[3]

In areas where the repair of concrete does not require the removal of old concrete and where the surface is contaminated with grease, oil, or other deposits, strong soap or detergent compounds can be used to flush away the oily deposits, followed by thorough flushing with water. Additional roughness can be given to surfaces by etching. One common form of etching is with commercial hydrochloric acid (20° Baumé scale) brushed or sprinkled on wet concrete at a rate of about 1½ to 1 gal per 100 sq ft. The acid should remain until all chemical action has ceased. Residue is then thoroughly flushed from the concrete to be sure no acid remains. Safety precautions should be stressed to safeguard personnel working with acid.

METHODS OF REPAIRING CONCRETE

The particular method of repair will depend on the size of the job and the type of repair required. Several methods of repair are recommended, each having particular advantages:
1. The dry-pack method
2. Concrete replacement
3. Prepacked aggregate with intruded grout
4. Shotcrete repairs
5. Synthetic patches

10 Dry-pack Method

The dry-pack method can be used in repair of new work where holes have a depth nearly equal to or greater than the smallest surface dimension—for example, cone-bolt holes, she-bolt holes, grout-insert holes, and narrow slots cut for repair of cracks. Dry pack is not used for relatively shallow depressions where lateral restraint cannot be obtained. It is also not recommended for filling back of considerable lengths of exposed reinforcement, or for filling holes which extend entirely through the wall, beam, etc.

After an area to be repaired has been cleaned, cured, and properly prepared, the surface should be thoroughly brushed with a stiff mortar of grout, barely wet enough to wet the surface thoroughly. The dry-pack material should then be immediately packed into the repair area before the bonding grout has dried.

The mix for the bonding grout is equal parts of cement and fine sand mixed to a consistency of thick cream. The bonding coat should not be wet enough or applied heavily enough to make the dry pack more than slightly rubbery.

The dry-pack mix consists of 1 part cement to 2½ parts of sand that will pass a No. 16 sieve. Only enough water should be used to produce a mortar that will stick together on being molded into a ball by slight pressure of the hands. The dry-pack material should become slightly rubbery when the material is solidly packed. Any less water will not make a sound solid pack, and any more water will result in excessive shrinkage and will loosen the patch. Dry-pack material should be placed and packed in layers having a compacted thickness of about ⅜ in. Thicker layers will not be well compacted. The surface of each layer should be scratched to facilitate bonding with the next layer. Subsequent layers may be placed immediately unless appreciable rubberiness develops, in which case work on repairs should be delayed.

Each layer should be solidly compacted over its entire surface by use of hardwood stick and a hammer. The sticks are usually 8 to 12 in. long and not over 1 in. in diameter, and are used on fresh mortar like a caulking tool. Hardwood sticks are used in preference to metal bars, because the latter tend to polish the surface of each layer and thus make the bond less certain and the filling less uniform. Much of the tamping should be directed at a slight angle and toward the sides of the hole to assure maximum compaction in these areas. Holes should not be overfilled, and finishing may usually be completed at once by laying the flat side of a hardwood piece against the fill and striking it several sharp blows. If necessary, a few light strokes with a rag sometimes will improve the appearance. Steel finishing tools should not be used since a dense smooth surface will not blend with the surrounding concrete.

A dry-pack patch is usually darker than the surrounding concrete, unless special precautions are taken to match the colors. In areas where a uniform color is important, the color of the dry-pack can be lightened with white portland cement used in the proper proportion with regular portland cement to produce the desired color.

11 Concrete Replacement

The following method is recommended by the U.S. Bureau of Reclamation.[6]

The concrete-replacement method of repair is used where large holes extend entirely through the concrete section, where holes of unreinforced concrete are more than 1 sq ft in area and 4 in. or more in depth, and where holes in reinforced concrete are more than ½ sq ft in area and deeper than the reinforcement of steel (Fig. 1). When the concrete-replacement method is used, it is necessary to construct forms over the repaired area to bond the surfaces of the original structure (Fig. 4). Additional reinforcement will not be required unless the repair influences the structural properties of the structure.

Forms are important in satisfactory concrete replacement, particularly where concrete must be placed from the side of the structure. Forms for walls are shown in Fig. 4. The following requirements for forms are recommended:

1. Front forms for wall repairs more than 18 in. high should be constructed in horizontal sections so the concrete can be conveniently placed in lifts not more than 12 in. in depth. The back form may be built in one piece. Sections to be set as concreting progresses should be fitted before concrete placement is started.

2. To exert pressure on the largest area of form sheathing, tie bolts should pass through wooden blocks fitted snugly between the walers and the sheathing.

3. For irregularly shaped holes, chimneys may be required at more than one level, and in some cases, such as when beam connections are involved, a chimney may be necessary on both sides of the wall or beam. In all cases the chimney should extend the full width of the hole.

4. Forms should be substantially constructed so that pressure may be applied to the chimney cap at the proper time.

5. Forms must be mortar-tight at all joints and at tie-bolt holes to prevent loss of mortar when pressure is applied to the concrete during final stages of placement. Twisted or stranded caulking cotton, folded canvas strips, or similar material should be used to caulk the forms as they are assembled.

Immediately prior to placing concrete in the repair area, the surface of the old concrete to be covered by new concrete should be coated with a thin layer (about ⅛ in.) of mortar. This mortar should have the same sand and cement content and the same water-cement ratio as the mortar in the replacement concrete. The surface should be damp (but not wet) from the required prewetting, sandblasting, and washing. The mortar can be applied by the most convenient means including an air-suction gun, by brushing, or rubbing into the surface with the hand encased in a rubber glove. Concrete placement should follow immediately.

Concrete for the repair should have the same water-cement ratio as is used for similar new structures. As large a maximum size of aggregate and as low a slump as are consistent with proper placing and thorough vibration should be used to minimize water content and consequent skrinkage. The concrete should contain 3 to 5% entrained air. Where surface color is important, the cement should be carefully selected, or blended with white cement, to obtain the desired

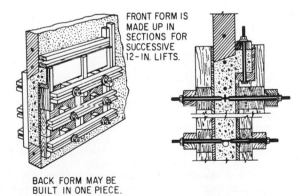

FRONT FORM IS MADE UP IN SECTIONS FOR SUCCESSIVE 12-IN. LIFTS.

BACK FORM MAY BE BUILT IN ONE PIECE.

Fig. 4. Forms used in concrete-replacement method of repair. (*U.S. Bureau of Reclamation.*)

results. To minimize shrinkage, the concrete should be as cool as practicable when placed. Materials should therefore be kept in shaded areas during warm weather. Batching of materials should preferably be by weight. Since batches for this class of work will be small, the uniformity of the materials is important and should receive proper attention. Premixing dry ingredients or commercial presacked dry mix may be used advantageously where repairs are not extensive.

When placing concrete in lifts, placement should not be continuous; a minimum period of 30 min should elapse between lifts. When chimneys are required at more than one level, the lower chimney should be filled and allowed to remain for the 30 min between lifts. When chimneys are required on both faces of a wall or beam, concrete should be placed in one chimney until it flows through, then placed in the other side.

Best repairs are obtained when the lowest practicable slump is used. This is about 3 in. for the first lift in an ordinary large form. Subsequent lifts can be drier, and the top few inches of concrete in the hole and that in the chimney should be placed at almost zero slump. It is usually best to mix enough concrete at the start for the entire hole. Thus the concrete will be ½ hr, 1 hr, or perhaps 1½ hr old when the successive lifts are placed. Such premixed concrete, provided it can be vibrated satisfactorily, will have less settlement, less shrinkage, and greater strength than freshly mixed concrete.

The quality of a repair depends not only on use of low-slump concrete, but also on the thoroughness of the vibration, both during and after depositing the concrete. There is no danger of overvibration. Immersion-type vibrators should be used if accessibility permits. If not, this type of vibrator can be used very effectively on the forms from the outside. Form vibrators can be used to good advantage on forms where the repair area is inaccessible to internal vibration. Immediately after the hole has been completely filled, pressure should be applied and the form vibrated. This operation should be repeated at 30-min intervals until the concrete hardens and no longer responds to vibration. Pressure is applied by wedging or by tightening the bolts extending through the pressure cap (Fig. 4). In filling the top of the form, concrete to a depth of only 2 or 3 in. should be left in the chimney under the pressure cap. A greater depth tends to dissipate the pressure. After the hole has been filled and the pressure cap placed, the concrete should not be vibrated without a simultaneous application of pressure—to do so may produce a film of water at the top of the repair that will prevent bonding.

Addition of aluminum powder (about 2½ g per sack of cement) to concrete causes the latter to expand. Under favorable conditions, this procedure has been successfully used to secure tight, well-bonded repairs in locations where the replacement material had to be introduced from the side. Time should not be allowed for settlement between lifts. When the top lift and the chimney are filled, no pressure need be applied, but the pressure cap should be secured in position so the expanding concrete will be confined and completely fill the hole undergoing repair. There should be no subsequent revibration. Aluminum powder should not be used until tests with job materials and at job temperatures have shown that effective expansion can be obtained and even then only under strict control. When used, the powder should first be blended with 50 parts, by weight, of cement or pozzolan. To secure the required expansion, more of the blend must be used at low temperatures.

Concrete replacement in open-top forms, as used for the reconstruction of the tops of walls, piers, parapets, and curbs, is a comparatively simple operation. Only such materials as will make concrete of proved durability should be used. The water-cement ratio should not exceed 0.45 by weight.[2] For best durability, the maximum size of aggregate should be the largest practicable and the percentage of sand the minimum practicable. No special features are required in the forms, but they should be mortar-tight when vibrated and should give the new concrete a finish similar to the adjacent areas. The slump should be as low as practicable, and the dosage of air-entraining agent increased as necessary to secure the maximum permissible percentage of entrained air, despite the low slump. Top surfaces should be sloped so as to provide rapid drainage. Manipulation in finishing should be held to a minimum, and a wood-float finish is preferable to a steel-trowel finish. Edges and corners should be tooled or chamfered. Use of water to aid in finishing should be prohibited.

Forms for concrete-replacement repairs can usually be removed the day after casting unless form removal would damage the green concrete, in which event stripping should be postponed another day or two. The projections left by the chimneys should be removed as soon as the forms have been removed. These projections should always be removed by working up from the bottom because working down from the top tends to break concrete out of the repair. The rough area resulting from the trimming should be filled and stoned so as to produce a surface comparable with that of the surrounding areas. Plastering of these surfaces should never be permitted.

12 Replacement of Unformed Concrete

The replacement of damaged or deteriorated areas in horizontal slabs, paving, bridge decks, etc., involves no special procedure other than that used in good construction practices for placement of original slabs. Contact edges at the perime-

FIG. 5. Bonding coat about ⅛ in. thick covering exposed concrete and reinforcement. (*Master Builders Company.*)

FIG. 6. Spraying a coat of epoxy resin in bridge-deck repair. Fine aggregate will be applied to produce a wearing seal coat. (*Shell Chemical Company.*)

ter area to be repaired should be clean and perpendicular with the surface (Fig. 3). If the repair is made with epoxy resins or other synthetics or when the repair is made with shotcrete, a featheredge at the contact with the old concrete is sometimes preferred.

As discussed in dry-pack method of repair, repair work can be bonded to old concrete by use of a bond coat of sand and cement in proportions of 1 to 1 (Fig. 5). The sand and cement of the bonding grout should be of the same gradation and type used in the patching mix and should meet specifications required for new construction. Sand and cement are mixed with water to a uniform consistency of whipped cream and scrubbed into the existing concrete face with a stiff-bristled brush or broom. The grout coat should be applied immediately ahead of concrete placing so that it will not dry out or set before the concrete covers it. Synthetic materials may also be employed as bonding coats. Latex emulsions mixed with portland cement are sometimes used as bonding grout coats. Proportioning of the latex cement and water should be made in accordance with the recommendations of the manufacturer of the latex emulsions.

Epoxy-resin compounds or polysulfide polymer–epoxy resin blends (Fig. 6) are also effective as bond coats, brushed or sprayed onto the clean concrete surfaces before placing new concrete patching mixtures. Resins should be placed immediately before placing the concrete, so that they will not harden before the whole patch is completed.

13   Shotcrete

The most popular method of repairing vertical and overhead faces of structural concrete is accomplished by the use of pneumatically placed concrete, or shotcrete, which is described in detail in Chap. 36. Shotcrete consists of a mixture of moistened cement and fine aggregate that is sprayed into the repair area under pressure (Fig. 7).

Two methods are used to add moisture to the mixture.[7]

Dry Mix. In a dry-mix process, mixing water is added to the mixture at the nozzle of the shotcrete equipment. The sand and cement are premixed and conveyed to the nozzle and ejected in a pressurized air stream into which has been injected a fine spray of water. Sand and cement are wetted as they are carried in the air stream to the repair surface. In this method it is necessary for the

Fig. 7. Applying shotcrete. One hose carries water and the other carries compressed air with premixed mortar. Water joins the mortar in the nozzle. (*U.S. Bureau of Reclamation.*)

equipment operator to provide adequate water at the nozzle to hydrate the cement and prevent dusting and blowing of the material. At the same time, it is necessary to maintain a stiff, dense mortar on the surface of the repair area. In applying

shotcrete, the nozzle should be held between 2 and 3 ft from the surface and held in such a position that the stream will strike the face at a right angle.

A bonding coat is not usually required with this type of application, since the initial shotcrete causes the sand to rebound while cement and moisture build up a fine grout coating. A bonding coat is sometimes used, however, to ensure adequate bond in difficult jobs. As soon as the sand is cushioned by the grout coating, the material will build up in a uniform mortar layer.

Where repair areas are deep, shotcrete should be built up in layers and each layer permitted to obtain its initial set before subsequent layers are placed. The maximum thickness of layers will vary depending upon the size of the repair area; they have been placed as thin as ½ in. and as thick as 6 in. In general practice layers do not exceed 3 in.

Wire-mesh reinforcement 2 × 2 in. minimum size should be used to supplement bond of the shotcrete in vertical and overhead faces, if patch thicknesses exceed 2 in. in depth. In large areas wire mesh should be securely fastened to the concrete-masonry area with dowels. Where wire mesh can be fastened to reinforcing steel, it may not be necessary to anchor it to the concrete.

In areas where the repair work is not critical, no additional finishing may be required after the repair area is filled. Where the surface is to be finished to accurate line and elevation, sufficient shotcrete material is applied to permit shaving off the irregular high material with a screed or steel trowel to provide a smooth surface. Finishing should be kept to a minimum to avoid all possible disturbance of the in-place shotcrete material.

A modification of the dry-mix system consists of a compact, lightweight gun, shown in Figs. 8 and 9, and a method of application in which water is added

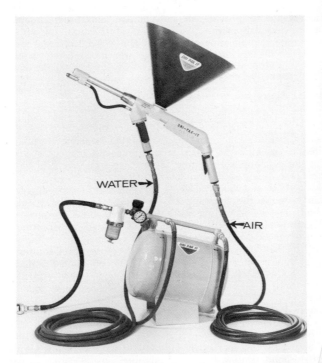

Fig. 8. Portable application gun for dry-mix mortar. (*Dri-pak Limited.*)

while the dry-mix material is simultaneously applied to the area being treated. Water application is automatically controlled by a fixed-output nozzle that ejects the proper amount of water for hydration of the cement. The result is a repair mortar which, because of its low water-cement ratio, is practically nonshrinking.

The mortar consists of 1 part cement to 1½ parts mason's or plaster sand, mixed either by hand or in a mechanical mixer. Sand should be moist to prevent separation. Admixtures can be added to the water in the water tank.

Applied material should be inspected occasionally before it sets up by cutting back from the applied surface. A dark, even color indicates proper application. Light gray streaks indicate improper moistening or excessive drying out. Such material should be removed.

Wet Mix. Water is previously mixed with the sand and cement in the wet-mix process. The plastic mixture is conveyed, under pressure, from the mixing chambers to the nozzle and carried into an air stream to the repair face. The wet-mix process eliminates most dusting and rebound of material. Premixing also permits the use of admixtures and air-entraining agents to provide air entrainment and other desirable properties. The wet-mix process will permit placing of mortars containing up to ¾-in. maximum-size aggregate. Application and use of the wet-mix process are generally the same as those for the dry-mix process. The wet mix must be sufficiently dry so that it will not sag after placement.

Fig. 9. Nozzle of the gun shown in Fig. 8, showing simultaneous application of dry mortar out of top nozzle and water spray from lower nozzle. (*Dri-pak Equipment Limited.*)

14   Prepacked Aggregate with Intruded Mortar

A special technique developed and performed by a private company has been used for massive repairs, particularly underwater repairs of piers, abutments, etc. The process consists of removing the deteriorated concrete, forming the sections to be repaired, prepacking the repair area with coarse aggregate, and finally pressure-grouting the voids between aggregate particles with a cement or a sand-cement mortar.[8] This process is described in detail in Chap. 38.

For thick sections and large masses, the minimum coarse-aggregate size is ⅝ in. For thin sections the minimum coarse aggregate size is ¼ in. Forms used in this method must be securely constructed to withstand pressures of the mortar. Coarse aggregate is placed in the form and consolidated by vibration. The mortar is pumped into the voids between the coarse-aggregate particles, starting at the lowest point so that it can flow upward and outward. Grouting is continued until the grout appears at an outlet located at the uppermost point of the cavity. When the repair area is completely filled, pressure is applied and held for a short time to ensure optimum bond between old concrete and new concrete. Intrusion mortar consists of portland cement, fine sand, which essentially passes the 100-mesh sieve, pozzolanic material of low mixing-water requirement and a plasticizing admixture. Sufficient mixing water is added to produce a mortar which will flow readily to the voids between the aggregate particles.

Concrete formed by this method is reported to have very low shrinkage, because there is point-to-point contact between individual coarse-aggregate particles. High bond strength between old concrete and new concrete is also obtained as a result of mortar application under pressure. Durability is obtained through air entrain-

ment. Approximately 4% entrained air provides resistance to freezing and thawing.

In locations of fast-moving water, where there is a problem of placing and holding cementitious materials or where it is impractical to use conventional forms, the use of grout bags has proved successful. The grout bags, known as "Bagpipe Groutainers,"* are dimensionally engineered to fit specific job requirements. The type of fabric, degree of reinforcement, method of placement, and structural concrete properties are made to rigid specifications. These containers expand when inflated with high-strength "Groutite"* cement mortar to conform to the size and configuration of the cavity to be filled or repaired. Thus the finished product is a cast-in-place structural unit that locks in place and performs its function of supporting a structure, stopping leaks, retaining soils, or preventing water erosion (Fig. 10).

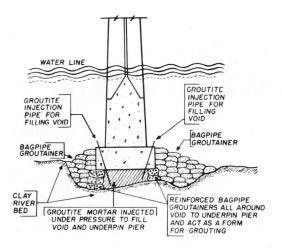

CROSS-SECTION OF BRIDGE PIER

Fig. 10. "Bagpipe Groutainers" being used as forms for pressure grouting for underpinning pier and as riprap. (*Lee Turzillo Contracting Company.*)

This system has been used to reduce the amount of water flow through large voids in dams as well as leakage under dams. Other uses have included underpinning operations to seal larger-sized voids and crevices, to fill underwater scours where confining concrete is extremely difficult, to serve as riprap for protection against scour, and for waterfront and shore protection facilities, including blanket-type revetments for erosion control.

15 Curing

All concrete repairs must be adequately moist-cured in order to be effective. The bond strength of new concrete to old concrete develops much more slowly and for a longer time than do other properties of concrete. The tendency to shrink and loosen is reduced by a long moist-curing period.

Where moist curing is not possible for long periods, the loss of moisture can be reduced by spraying the repaired area with curing compounds.

If high-strength bond is required and moist curing is not practicable, the use of epoxy mortars should be considered.

* "Bagpipe" is a registered trademark, and "Groutainer" and "Groutite" are trademarks of the Lee Turzillo Contracting Co. which developed this patented process.

16 Synthetic Patches

One of the most recent developments in repair of concrete has been the use of synthetic materials in bonding and patching. Most important are the epoxy-resin compounds previously discussed. Epoxy-resin compounds or binders are being used extensively because of their high bonding properties and great strength. The cost of epoxy-resin binders is quite high and necessitates their being used with great care. Because epoxy-resin patches require a short curing time with only a brief interruption of traffic, the overall cost of a patching operation may not exceed that of more conventional methods. Hence, well-bonded epoxy-resin patches appear to offer good service over a number of years and provide for an important need in critical locations (Fig. 11).

In applying epoxy-resin patching mortars, a bonding coat of the epoxy-resin binder is thoroughly brushed into the base of the old concrete before the mortar is placed. The bonding coat is applied immediately before the mortar is placed and should be tacky when the patching material is placed. Epoxy-resin mortars

Fig. 11. Patching with epoxy resin–sand mortar mixture. (*Highway Research Board.*)

are placed and struck off to the elevation of the surrounding material much like portland-cement-concrete mortars. Mortars should be consolidated by tamping or vibrating to increase the durability of the patching mortar.

TYPICAL EXAMPLES OF CONCRETE REPAIR

General considerations and practice for the repair of concrete are presented in this chapter. Specific detailed examples of some of the commonest types of concrete repair are presented for additional clarification.

17 Porous-concrete (Rock Pockets) Dry-pack Method

Unconsolidated concrete placed against a concrete form will later be exposed when the form is stripped. This type of surface blemish is commonly known as a "rock pocket," or porous concrete, and should be repaired as soon as possible after the forms are stripped. The following steps are recommended for the repair of small areas of porous concrete.

1. All porous unconsolidated concrete should be removed from the area. This can be best accomplished with an air-driven chipping hammer. The extent of the porous area will be ascertained as chipping progresses. Concrete should be removed to a minimum depth of 1 in. regardless of the depth of the porous areas.

2. All edges of the area removed should be perpendicular to the surface of the original concrete. This will permit lateral restraint between the patch and the old concrete as the dry packing is applied. A diamond saw can be used to produce a smooth sound edge. Corners within the repair area should be rounded.

3. Old concrete should be well cured and moist when dry packing is applied to permit proper bonding. Saturated burlap or other type of packing packed into the repair area is a satisfactory method of saturating the contact area.

4. The repair area should be cleaned prior to dry packing. All loose material and excess moisture can be removed with an air jet.

5. A satisfactory dry-pack mix is composed of 1 part of cement to 2½ parts of fine sand (passing No. 16 sieve). Where color is important, white portland cement can be substituted for a portion of the regular portland cement to produce matching colors.

6. Immediately prior to packing the repair area, a thin coat of bonding grout should be brushed on all surfaces of the repair area. This grout is composed of 1 part of cement to 1 part fine sand and is mixed with water to the consistency of thick cream.

7. The dry-pack mortar is packed into the repair area in layers about ⅜ in. thick. Compaction can be accomplished with pneumatic hammer or with a hardwood stick and a hammer.

8. The finish of the repaired area should have the appearance and texture of the surrounding concrete. Light-colored repair mortar can be obtained by using sufficient white portland cement to make the mortar color blend with the old concrete after drying. Rich, stiff mortars are darker than regular concrete.

Only wood or stone should be used to finish repairs on formed surfaces. A steel trowel should never be used, because it makes a dark dense surface. Form marks can be "printed" on unformed surfaces by placing a piece of sharply grained form board on it and striking the board with a hammer.

9. Dry-packed concrete must be cured in a manner similar to regular concrete. Moisture can be held in the mortar by application of membrane curing compound, or the wall must be kept moist.

18 Bolt-hole Dry-pack Method*

Filling a bolthole is a minor repair and when attended to as soon as the bolt is removed can be accomplished simply and easily.

1. Preparation of hole—if the concrete is only a few days old the hole will not require moistening. Holes in older concrete will require moistening 24 hr before repair is made.

2. The hole is tamped nearly full using a lean, stiff, light-colored mortar.

3. The surface of the full hole is finished with a piece of wood struck with a hammer. Grout should not be painted or troweled over the surface, since this results in a difference in texture and color.

19 Repair of a Wall Requiring Concrete Replacement†

In some instances inferior concrete, concrete of poor consolidation, or damaged or cracked concrete must be removed from sections too large for dry-pack methods. A typical example of repair is shown in Figs. 2 and 3.

* See also Art. 10.
† See also Art. 11.

1. Remove all inferior loose or damaged concrete with a chipping hammer.

2. All edges of the repair area should be cut perpendicular to the original surface, with the exception of the top surface should slope upward toward the form opening.

3. Concrete in the repair area should be moist-cured for 24 hr before repair is made.

4. Concrete should be throughly cleaned of all loose or foreign material.

5. Place forms over area to be replaced as indicated in Fig. 3 with special emphasis on keeping forms in exact alignment with original concrete.

6. Leave a chimney at the top of the forms through which concrete can be placed and which will allow for plastic shrinkage of the concrete.

7. First place a thin layer of mortar about ⅛ in. thick, having the same water-cement ratio as existing concrete, in the form to act as a cushion to the new concrete and ensure complete contact with existing concrete.

8. Fill the forms with concrete of the same water-cement ratio and the same cement as existing concrete, using any available method of consolidation practical such as form vibration, hammering forms or internal vibration through the access hole. Excess concrete should be placed in the chimney to keep pressure on the concrete in the repair area until initial shrinkage has taken place.

9. When the concrete forms are removed, smooth the green concrete around the chimney area, and align with the original surface of the structure.

10. The surface and form marks of the new concrete can be made to correspond with the structure by aligning a board over the freshly finished area and tapping with a hammer.

11. After 24 hr, the remaining forms can be removed and minor repairs around the edges can be made.

12. Cure the repaired area in the same manner as freshly placed concrete of the original structure.

20   Major Repair of a Deteriorated Bridge Deck

This is an example of repair where an entire section of pavement requires removal because of extensive deterioration.

1. The edges of the repair area are cut with a diamond saw to a depth of 1½ in. (Fig. 3).

Fig. 12. Topping mix being placed in a bridge-deck repair. (*Master Builders Company.*)

2. Deteriorated concrete is chipped and cleaned, exposing but not undercutting the reinforcing.

3. A premixed grout is broomed into the exposed area (Fig. 5), providing a thick bond coat which completely fills small holes and impressions around exposed aggregate and reinforcing.

4. Concrete topping mix is placed with a vibrating screed (Fig. 12) and finished with a wood float.

5. A final broom finish is given the repaired area to correspond with the original bridge deck (Fig. 13).

Fig. 13. Completed repaired area broomed to match original concrete. (*Master Builders Company.*)

21 Repair of Cracks

Concrete being a rigid material, cracks will occur with slight changes in alignment or movement of a structure. Cracks are also produced by dimensional changes that cause excessive tension at the concrete surface. Small shrinkage cracks and crazing are generally not repaired unless they occur in a hydraulic structure and permit leakage. If desired, fine cracks can be sealed with a sealing compound such as those described in Art. 7, and they will require no further repair.

Larger cracks in concrete pavement can be filled with a mastic compound similar to that used in sawed joints. Some attempts have been made to pump epoxy compounds into cracks to reestablish bond. It is important that a crack does not fill with sand or debris; otherwise any subsequent expansion of the pavement causes spalling or further damage to the pavement.

Cracks in structures which carry water require great care and preparation to repair. The following steps are recommended by the U.S. Bureau of Reclamation:[6]

1. Cut a groove along the crack ½ in. wide and 2 to 2½ in. deep (Fig. 14). A sawtooth bit (Fig. 2) works satisfactorily.

2. Rub a thin layer of internal-set-type mastic into the interior surfaces of the groove.

3. Tamp ¼ in. of oakum tightly in the bottom of the slot.

4. Fill in ½ in. of mastic on top of the oakum.

5. Place a ½-in. section of tightly twisted asbestos rope wicking in groove and caulk tightly, using pneumatic tools.

6. Fill the groove with mastic and smooth the surface. Complete repairs are shown in Fig. 14.

A preformed rubber strip provides a rapid method of repairing pavement cracks.[9]

The crack is broken out to a depth of about 1 in. (Fig. 15). The preformed rubber strip is laid along the line of the crack; a wire core enables the strip to take up and retain the irregular contour of the crack. The excavated area on either side of the rubber strip is then filled with an epoxy resin–sand mortar which retains a bond with the strip by virtue of its keyed wedge shape.

The rubber is an extruded wedge-shaped neoprene sponge 1 in. deep and ⅜ in. thick at the base tapering to ½ in. at the top and keyed along the length. It

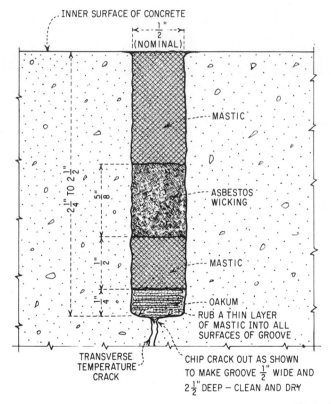

Fig. 14. Repair of a crack in a water-conveying structure. (*U.S. Bureau of Reclamation.*)

is softened and lightened by having holes longitudinally through its length; through these holes are run one or two lengths of malleable wire of suitable gage.

In general, cracks do not influence the structural properties of a concrete structure since reinforced-concrete design methods do not contemplate concrete taking tensile stress. Under compressive stress a crack in concrete merely closes. In the majority of cases, therefore, it is not necessary to restore the bond and tensile strength across a crack. The use of epoxy resins, however, will produce strength across a crack equal to or greater than the surrounding concrete.

For large cracks, concrete can be chipped out along the crack and epoxy-resin mortar tamped into the repair area by dry-tamped methods (Art. 10), using an epoxy mortar consisting of 1 part adhesive to 3 parts of mortar sand by volume. After first cleaning the crack of all loose concrete dirt, and other foreign material, the mortar is troweled or tamped into the crack. Complete filling is assured by thorough tamping and working the mortar into the crack with a knife blade or similar tool.

Epoxy grout can be pumped or poured into a crack without removing adjacent concrete. In this method holes are drilled about 1 in. deep 2 or 3 ft apart in the concrete along the crack. Zerk-type or similar pressure fittings are sealed into these holes by cementing them in place with epoxy resin or by inserting them in a soft metal sleeve which is anchored by tapping the fitting with a hammer. The surface of the crack is then sealed with the resin, leaving small vent holes every 6 in. After the adhesive on the surface has cured, the crack is filled by using a high-pressure hand-operated grease gun to force the adhesive into the crack through the fittings, until it appears in the adjacent vents. Adhesive for this application contains no filler.

Another method, suggested by ACI Committee 403,[4] is to rout out the crack about ½ in. deep and drill ¾-in. diameter holes about ¾ in. deep along the

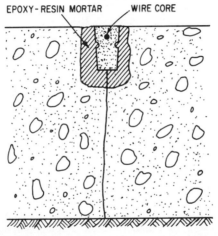

EPOXY-RESIN MORTAR   WIRE CORE

Fig. 15. A method of repairing a crack by using a wire-reinforced rubber strip embedded in epoxy mortar.

crack every 6 to 12 in., the closer spacing being used for narrower cracks. Tire valve stems are cemented in the holes with epoxy adhesive, and the crack is sealed with the resin. After the resin has cured, more resin is injected into the first valve stem until it appears at the next valve, all valves now being capped, and an inert gas (nitrogen) is applied to the first valve using the maximum amount of pressure that will not cause displacement (near 90 psi if possible). Pressure is maintained for as long as 10 min. Pumping epoxy and applying pressure are then continued with successive valves until the entire crack is filled. Sometimes one valve can be pumped and pressurized more than once to fill the crack.

22 Repair of Minor Imperfections

Popouts are caused by inferior aggregate particles that expand either by chemical activity or by freezing of water in the pores of the aggregate. (Porous chert is an example.) The expansion of the aggregate particle fractures the concrete and breaks out a small conical piece of concrete (Fig. 3, Chap. 45). Popouts are unsightly but in many cases do not affect the serviceability of the pavement or structure. Highway pavements have served satisfactorily for 40 years with a severe popout condition. When it is desirable to repair a section of pavement containing popouts the repair can best be accomplished with a thin bonded overlay or seal coat, as described in Art. 20 and in Refs. 10, 11, and 12. Individual popouts are best filled with epoxy mortar, after thorough cleaning of the cavity by removal of loose concrete and dirt.

Small spalls occur frequently on the edges of pavements and slabs, and on the corners cf structures. The primary reason for repairing a small spalled area is to improve the appearance of the structure. Unless the repair is accomplished by experienced workmen, the repaired area might be more unsightly than the original spalled area. Shotcrete and dry-pack patches are sometimes used, but it is extremely difficult to secure adequate bond with the original concrete. Epoxy mortars can be applied to these spalls with reasonable assurance of adequate service life.[13] For any of the methods used, careful attention must be paid to matching the color of the mortar with the surrounding concrete.

REFERENCES

1. ACI Committee 201 Report, Chapter 7, Restoration of Deteriorated Concrete, *J. ACI*, vol. 59, no. 12, p. 1807, December, 1962.
2. Freezing and Thawing of Concrete—Mechanisms and Control, ACI Monograph 3, 1966.
3. Evaluation of Methods of Replacement of Deteriorated Concrete in Structures, National Cooperative Highway Research Program *Rept.* 1, Highway Research Board, National Academy of Sciences, National Research Council, 1964.
4. ACI Committee 403 Report, Guide for Use of Epoxy Compounds with Concrete, *J. ACI*, vol. 59, no. 9, p. 1121, September, 1962.
5. Tuthill, Lewis H.: Conventional Methods of Repairing Concrete, *Proc. ACI*, vol. 57, part 1, pp. 129–138, August, 1960.
6. "Concrete Manual," 7th ed., U.S. Bureau of Reclamation, 1963.
7. ACI Committee 506 Report, Standard Recommended Practice for Shotcreteing, *J. ACI*, vol. 63, no. 2, p. 219, February, 1966.
8. Davis, Raymond E.: Prepakt Method of Concrete Repair, *Proc. ACI*, vol. 57, part 1, pp. 155–172, August, 1960.
9. Report of the Road Research Laboratory, London, England, 1965.
10. Thin Bonded Concrete Patching, *Concrete Paving, Paving Progress,* vol. 259, 1963.
11. Williams, R. I. T.: Thin Bonded Concrete Surfacings Applied to Existing Concrete Road Slabs, *Highway Res. Abstracts,* vol. 30, no. 1, p. 17, January, 1960.
12. Purinton, John B.: A New Look at Thin Bonded Concrete Overlays, *Roads and Streets,* vol. 105, p. 47, February, 1962.
13. Tremper, Bailey: Repair of Damaged Concrete with Epoxy Resins, *Proc. ACI*, vol. 57, part 1, pp. 173–181, August, 1960.

INDEX